COMMON GRAPHS AND MODELS

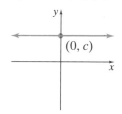

Horizontal Line;
Zero Slope
$y = c$

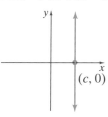

Vertical Line;
Undefined Slope
$x = c$

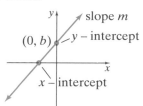

Linear Function;
Positive Slope
$y = mx + b; m > 0$

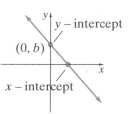

Linear Function;
Negative Slope
$y = mx + b; m < 0$

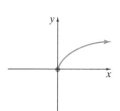

$y = x$

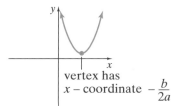

$y = x^2$

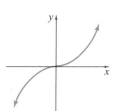

$y = x^3$

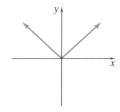

$y = |x|$

$y = \sqrt{x}; x \geq 0$

Quadratic Function
$y = ax^2 + bx + c; a \neq 0$
Parabola opens upward if $a > 0$
Parabola opens downward if $a < 0$

vertex has
x – coordinate $-\dfrac{b}{2a}$

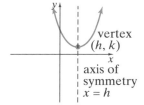

Quadratic Function
$y = a(x - h)^2 + k; a \neq 0$
Parabola opens upward if $a > 0$
Parabola opens downward if $a < 0$

vertex
(h, k)
axis of
symmetry
$x = h$

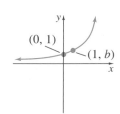

Exponential Function
$y = b^x$ for $b > 1$

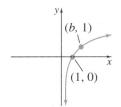

Logarithmic Function
$y = \log_b x$ for $b > 1$

SYSTEMS OF LINEAR EQUATIONS

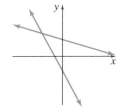

Independent and
consistent; one solution

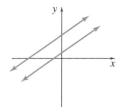

Independent and
inconsistent; no solution

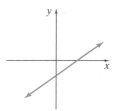

Dependent and consistent;
infinitely many solutions

INTERMEDIATE ALGEBRA
A GRAPHING APPROACH

K. ELAYN MARTIN-GAY
University of New Orleans

MARGARET GREENE
Florida Community College at Jacksonville

Prentice Hall
Upper Saddle River
New Jersey 07458

Library of Congress Cataloging-in-Publication Data

Martin-Gay, K. Elayn
    Intermediate algebra: a graphing approach / K. Elayn Martin-Gay. Margaret Greene.
      p.   cm.
    Includes index.
    ISBN 0-13-281495-1
    1. Algebra.   I. Greene, Margaret.   II. Title..
QA154.2.M385   1997
512.9—dc21

96-53267
CIP

Sponsoring Editor, Ann Marie Jones
Editor-in-Chief, Jerome Grant
Editorial Director, Tim Bozik
Production Editor, Robert C. Walters
Managing Editor, Linda Behrens
Assistant Vice President of Production and Manufacturing, David W. Riccardi
Executive Managing Editor, Kathleen Schiaparelli
Development Editor, Joseph Will
Editor-in-Chief of Development, Ray Mullaney
Director of Marketing, John Tweeddale
Marketing Manager, Jolene Howard
Marketing Assistants, Jennifer Pan, Diane Penha
Creative Director, Paula Maylahn
Art Director, Maureen Eide
Art Manager, Gus Vibal
Interior Design, Geri Davis, The Davis Group, Inc.
Cover Design, Bruce Kenselaar
Photo Editor, Lori Morris-Nantz
Photo Research, Teri Stratford
Manufacturing Buyer, Alan Fischer
Manufacturing Manager, Trudy Pisciotti
Page Formatters, Eric Hulsizer and New England Typographic Services, Inc.
Production Support, Karen Noferi, Richard Foster, Wanda España, Diane Koromhas
Supplements Editor/Editorial Assistant, April Thrower
Cover Photo, Queen Angel Fish, Caribbean, Stuart Westmoreland/Tony Stone Images

Printed in the United States of America

10   9   8   7   6

ISBN 0-13-281495-1

Prentice-Hall International (UK) Limited, *London*
Prentice-Hall of Australia Pty. Limited, *Sydney*
Prentice-Hall Canada Inc., *Toronto*
Prentice-Hall Hispanoamericana, S.A., *Mexico City*
Prentice-Hall of India Private Limited, *New Delhi*
Prentice-Hall of Japan, Inc., *Tokyo*
Pearson Education Asia Pte. Ltd., *Singapore*
Editora Prentice-Hall do Brasil, Ltda., *Rio de Janeiro*

Dedicated to our families

To my husband, Clayton,
and our sons, Eric and Bryan
and all the helpful cousins: Michael, Christopher,
Melissa, Mandy, Matthew, Jessica, and Madison.

To my father, William N. Kramer,
to the loving memory of my mother, Mary M. Kramer
and to my husband, Jack Greene and our sons,
Kevin, Dan, and Matt.

CONTENTS

7

RATIONAL EXPRESSIONS   381

8

RATIONAL EXPONENTS, RADICALS, AND COMPLEX NUMBERS   461

9

QUADRATIC EQUATIONS AND FUNCTIONS   517

PREFACE

ABOUT THIS BOOK

This book was written to provide students with a solid foundation in algebra as well as to help develop their problem solving skills. Specific care has been taken to prepare students to go on to their next course in mathematics, and to help students to succeed in non-mathematical courses which require a grasp of algebraic fundamentals. The basic concepts of graphs and functions are introduced early. These concepts along with problem solving, data interpretation, and geometric concepts are emphasized and integrated throughout the book. This text makes reference to the use of graphing technology to help students better understand important mathematical connections.

In preparing this edition, we considered the comments and suggestions of colleagues throughout the country. The AMATYC Crossroads in Mathematics: Standards for Introductory College Mathematics before Calculus and the NCTM Standards (plus Addenda), together with advances in technology, also influenced the writing of this text.

KEY PEDAGOGICAL FEATURES OF THE SECOND EDITION

Readability and Connections Many reviewers have commented favorably on the readability and clear, organized presentation. We have tried to make the writing style as clear as possible while still retaining the mathematical integrity of the content. When a new topic is presented, an effort has been made to relate the new concepts to those the students may already know. Constant reinforcement and connections within problem-solving strategies, data interpretation, geometry, patterns, graphs, and situations from everyday life can help students gradually master both new and old information.

Problem-Solving Process This is formally introduced in Chapter 2, with a six step process which is integrated throughout the text. The six steps are UNDERSTAND, ASSIGN, ILLUSTRATE, TRANSLATE, COMPLETE, and INTERPRET. The repeated use of these steps in a variety of examples shows its wide applicability. When solving problems, students are encouraged to use technology when applicable during this problem solving process and/or when checking a solution. For instance, with technology, a graph can be generated quickly and accurately in order to enhance the problem solving process. Reinforcing the six steps can increase students' comfort level and confidence in tackling problems.

Applications This book contains a wealth of practical applications in worked-out examples and exercise sets. The applications help to motivate students and strengthen their understanding of mathematics in the real world. They help show connections to a wide range of areas such as biology, environmental issues, consumer applications, allied health, business, entertainment, history, art, literature, finance, sports, music, as well as to related mathematical areas such as geometry. Many involve interesting real life data. Sources for data include newspapers, magazines, government publications, and reference books. Opportunities for obtaining your own real data are also included.

Discover the Concept These explorations, integrated appropriately throughout the text, are often for use with graphing calculators or computer graphing utilities. They promote student involvement and interaction with the text as students are reading. This feature helps student recognize patterns or discover a concept on their own immediately before the concept is formally introduced.

Group Activities Each chapter opens with a photograph and description of a real life situation. At the close of the chapter, students can work cooperatively to apply the algebraic and critical thinking skills they have learned to make decisions and answer the Group Activity which is related to the chapter opening situation. The Group Activity is a multi-part, often hands-on, problem. These situations, designed for student involvement and interaction, allow for a variety of teaching and learning styles.

Answers and tips for instructional strategies for Group Activities are available in the Instructor's Edition. In addition, there are opportunities for group activities within section exercise sets.

Visual Reinforcement of Concepts The text contains numerous graphics, models, and illustrations to visually clarify and reinforce concepts. These include bar charts, line graphs, calculator screens, application illustrations, and geometric figures. The inside back cover of the text includes a quick reference to geometric figures and formulas, and the inside front cover includes a summary of common graphs.

Reminder Reminder boxes contain helpful hints and practical advice on problem solving. Reminders appear in the context of material in the chapter and give students extra help in understanding and working problems. They are highlighted in a box for quick reference.

Technology Note Generally found in the margin, technology notes contain specific suggestions for problem solving with technology. They also contain notes on extra features students might find available on their graphing utilities.

Exercise Sets The exercise sets are graded in difficulty and include computational, conceptual, and applied problems. The first few exercises in each set are carefully keyed to worked examples in the text. A student can gain confidence and then move on to the remaining exercises, which are not keyed to examples. There are ample exercises throughout the book, including end-of-chapter reviews, tests, and cumulative reviews. In addition, each exercise set contains one or more of the following features.

Mental Mathematics These problems are found at the beginning of an exercise set. They are mental warmups that reinforce concepts found in the accompanying section and increase students' confidence before they tackle an exercise set. By relying on their own mental skills, students increase not only their confidence in themselves, but also in their number sense and estimation ability.

Conceptual and Writing Exercises These exercises are identified with the icon ◻. These exercises require students to show an understanding of a concept learned in the corresponding section. This is accomplished by asking students questions that require them to use two or more concepts together. Some require students to stop, think, and explain in their own words the concept(s) used in the exercises they have just completed. Guidelines recommended by the American mathematical Association of Two Year Colleges (AMATYC) and other professional groups recommend incorporating writing in mathematics courses to reinforce concepts.

Data and Graphical Interpretation The ability to interpret data and read and create a variety of types of tables and graphs is developed gradually so students become comfortable with it.

Review Exercises Review Exercises are found at the end of each section after Chapter 1. These problems are keyed to earlier sections and review concepts learned earlier in the text that are needed in the next section or in the next chapter. These exercises show the connections between earlier topics and later material.

A Look Ahead These are examples and problems similar to those found in a "next" algebra course. "A Look Ahead" is presented as a natural extension of the material and contains an example followed by advanced exercises.

Chapter Highlights Found at the end of each chapter, the Chapter Highlights contain key definitions, concepts, and examples to help students understand and retain what they have learned.

Chapter Review and Test The end of each chapter contains a review of topics introduced in the chapter. The review problems are keyed to sections. The chapter test is not keyed to sections.

Cumulative Review Each chapter after the first contains a cumulative review. Each problem contained in the cumulative review is actually an earlier worked example in the text which is referenced in the back of the book along with the answer. Students who need to see a complete worked-out solution with explanation, can do so by turning to the appropriate example in the text.

Functional Use of Color and Design Elements of the text are highlighted with color or design to make it easier for students to read and study.

Videotape and Software Icons At the beginning of each section, videotape and software icons are displayed. These icons help reinforce that these learning aids are available should students wish to use them in reviewing concepts and skills at their own pace. These items have direct correlation to the text.

KEY CONTENT FEATURES

Overview In addition to the traditional topics in intermediate algebra courses, this text contains an early and intuitive introduction to graphs and functions and an emphasis on problem solving. Students are also given the opportunity to see how today's technology can be used to enhance problem solving.

The geometric concepts covered are those most important to a student's understanding of algebra, and we have included many applications and exercises devoted to this topic. Geometric concepts and reading and interpreting graphs and data are integrated throughout. Geometric figures and a review of angles, lines, and special triangles are covered in the appendices. Exercises are a critical part of student learning, and particular care was taken in writing these.

Early and Intuitive Introduction to Graphing As bar and line graphs and tables are gradually introduced in Chapters 1 and 2, an emphasis is slowly placed on the concept of paired data. This leads naturally to the concept of ordered pair and the rectangular system introduced in Chapter 3. Chapter 3 is devoted to graphing and concepts of graphing linear equations such as slope and intercepts. These concepts are reinforced throughout exercise sets in subsequent chapters.

Increased Range of Problem Solving Techniques Access to today's technology lets us expand the range of problems and problem solving techniques. New ways to learn and solve problems include more attention to graphical and numerical representations.

A special emphasis and strong commitment is given to contemporary and practical applications of algebra. Real data was drawn from a variety of sources including magazines, newspapers, government publications, and reference books. Generating and using personal real data is also encouraged.

Data Interpretation Data interpretation via tables and graphs begins in the first section of the book and continues throughout the text. The ability to interpret data from tables, screen displays, and a variety of types of graphs including bar, line, and circle graphs is developed gradually so students become comfortable with it.

Examples Examples are used in one of two ways. Often these are numbered, formal examples, and occasionally an example or application is used to introduce a topic or informally discuss the topic. We have included examples that show *algebraic, numerical,* and/or *graphical* approaches to solving and checking solutions.

Exercises A significant amount of time was spent on the exercise sets. Helping address a wide range of student learning styles and abilities, the exercises include mental math exercises, group activities, conceptual and writing exercises, multi-part exercises, applications, and data analysis from tables and graphs.

ACKNOWLEDGMENTS

First we would like to thank our husbands Clayton and Jack for their constant encouragement. We would also like to thank our children Bryan, Eric, Kevin, Dan, and Matt.

A special thank you to all the following reviewers of this text.

Douglas E. Cameron, *University of Akron*
Celeste Carter, *Richland College*
Elizabeth Chu, *Suffolk Community College*
Linda F. Crabtree, *The Metropolitan Community Colleges*
Brenda Diesslin, *Iowa State University*
Kathy Garrison, *Clayton College and State University*
Sudhir K. Goel, *Valdosta State University*
Kenneth Grace, *Anoka-Ramsey Community College*
Thomas Gruszka, *Western New Mexico University*
Joel K. Haack, *University of Northern Iowa*
Abdi Hajikandi, *State University College at Buffalo*
Diana K. Harke, *State University of New York at Geneseo*
Mickey McClendon, *Blue Mountain Community College*
Iris McMurtry, *Motlow State Community College*
Bonnie Simon, *Naugatuck Valley Community Technical College*
Cora S. West, *Florida Community County at Jacksonville*

James Sellers and Kurt Norlin did an excellent job of providing answers and solutions, and contributing to the overall accuracy of the book. Joe Will offered a variety of extremely helpful suggestions during the development of this text. Emily Keaton's creativity and insights added greatly to this text. We very much appreciated the writers and accuracy checkers of the supplements to accompany this text. Last, but by no means least, a special thanks to the staff at Prentice Hall for their support and assistance: Melissa Acuña, Ann Marie Jones, Robert (Bob) Walters, Eric Hulsizer, April Thrower, Linda Behrens, Jolene Howard, Jennifer Pan, Gary June, Jerome Grant, and Tim Bozik.

K. Elayn Martin-Gay
Margaret (Peg) Greene

ABOUT THE AUTHORS

K. Elayn Martin-Gay has taught Mathematics at the University of New Orleans for over 18 years and has received numerous teaching awards, including the local University Alumni Association's Award for Excellence in Teaching.

Over the years, Elayn has developed videotaped lecture series to help her students understand algebra better. This highly successful video material is the basis for the five book series, *Prealgebra, Beginning Algebra, Intermediate Algebra, Introductory and Intermediate Algebra,* a combined approach, and *Intermediate Algebra, A Graphing Approach.*

Margaret (Peg) Greene has taught Mathematics at Florida Community College at Jacksonville for over 13 years. Peg has also been an instructor for the College Instructor's Training Network, an Ohio State Short Course program, for teaching college instructors how to enhance their teaching by using technology in the mathematics classroom. A recipient of excellence in teaching awards, Peg's experience also includes teaching a telecourse, developing a video series on using a graphing calculator, conducting workshops, numerous publications, and certified completion of courses on cooperative learning through the University of Minnesota.

SUPPLEMENTS FOR THE STUDENT

PRINTED SUPPLEMENTS

Student Solutions Manual (ISBN 0-13-850306-0)

- Detailed step-by-step solutions to odd-numbered text and review exercises
- Solutions to all chapter practice tests
- Solution methods reflect those emphasized in the text.
- Answers checked for accuracy
- Ask your bookstore about ordering.

The New York Times Theme of the Times

- A free newspaper from Prentice Hall and *The New York Times*
- Interesting and current articles on mathematics
- Invites talking and writing about mathematics
- Created new each year

MEDIA SUPPLEMENTS

Videotape Series (Sample Video, ISBN 0-13-887886-2;
Video Series, ISBN 0-13-850281-1)

- Specifically keyed to the textbook by section
- Presentation and step-by-step examples by the textbook co-author, an award-winning teacher

MathPro Tutorial Software
(IBM Network-User, ISBN 0-13-887811-0;
IBM Single-User, ISBN 0-13-887829-3;
Mac, ISBN 0-13-887837-4)

- Text-specific tutorial exercises

- Interactive feedback

- Graded and recorded Practice Problems

- Clear user interface, glossary, and expressions editor for ease of use and flexibility

- Network version available

Graphing Utilities Video (ISBN 0-13-861428-8)

- Introductory keystroke instruction

- Casio and Texas Instruments graphing calculators

SUPPLEMENTS FOR THE INSTRUCTOR

PRINTED SUPPLEMENTS

Instructor's Edition (ISBN 0-13-860404-5)

- Instructor's Answer section contains answers to even-numbered exercises, and answers and pedagogical suggestions for group activities. In combination with the student answer section, the instructor has textbook access to the answers for all the exercises.

Instructor's Resource Manual (ISBN 0-13-850182-5)

- Solutions to even-numbered exercises, review exercises

- Chapter Tests and Final Exams

- Graphics and screen outputs designed for clarity

- Answers checked for accuracy

MEDIA SUPPLEMENTS

TestPro3 Computerized Testing (Sample Disk IBM, ISBN 0-13-887845-5;
Sample Disk Mac, ISBN 0-13-887852-8; IBM, ISBN 0-13-887860-9;
Mac, ISBN 0-13-887878-1)

- Comprehensive text-specific testing

- Generates test questions and worksheets from algorithms keyed to the text learning objectives

- Edit or add your own questions

- Windows-based interface available

- Compatible with Scantron or other possible scanners

Using the Internet and Web Browser

Using the Internet and a Web browser, such as Netscape, can add to your mathematical resources. Below are a list of some of the sites that may be worth your or your students' visit.

- Prentice Hall Math Center   http://www.prenhall.com/mathcenter
- The Mathematical Association   http://www.maa.org
  of America
- The American Mathematical Society   http://www.ams.org
- The National Council of Teachers   http://www.nctm.org
  of Mathematics
- The Census Bureau   http://www.census.gov
- Casio   http://www.casio.com
- Texas Instruments   http://www.ti.com/calc

For your convenience, the Prentice Hall Math Center provides a variety of appropriate resources, including automatic web links. The links connect you to sites, such as some of those listed above, at the touch of a button.

Internet Guide

- This guide provides a brief history of the Internet, discusses the use of the World Wide Web, and describes how to find your way within the Internet and how to reach others on it. Contact your local Prentice Hall representative for the Internet Guide.

HOW TO USE THE TEXT: A GUIDE FOR STUDENTS

Intermediate Algebra: A Graphing Approach, has been designed as one of several tools in a fully integrated learning package to help you develop problem-solving skills. Our goal is to encourage your success and mastery of the mathematical concepts introduced in this text. Take a few moments to see how this text will help you excel.

CHAPTER

5

5.1 SOLVING SYSTEMS OF LINEAR EQUATIONS IN TWO VARIABLES
5.2 SOLVING SYSTEMS OF LINEAR EQUATIONS IN THREE VARIABLES
5.3 SYSTEMS OF LINEAR EQUATIONS AND PROBLEM SOLVING
5.4 SOLVING SYSTEMS OF EQUATIONS BY MATRICES
5.5 SOLVING SYSTEMS OF EQUATIONS BY DETERMINANTS

SYSTEMS OF EQUATIONS

LOCATING LIGHTENING STRIKES

Lightening, most often produced during thunderstorms, is a rapid discharge of high-current electricity into the atmosphere. Around the world, lightening occurs at a rate of approximately 100 flashes per second. Because of lightening's potentially destructive nature, meteorologists track lightening activity by recording and plotting the positions of lightening strikes.

IN THE CHAPTER GROUP ACTIVITY, YOU WILL HAVE THE OPPORTUNITY TO PINPOINT THE LOCATION OF A LIGHTENING STRIKE.

The photo application at the opening of every chapter and **applications throughout** offer real-world scenarios that connect mathematics to your life. In addition, at the end of the chapter, a group activity or discovery-based project further shows the chapter's applicability.

APPLY THE PROBLEM-SOLVING PROCESS

As you study, **make connections**–this text's organization can help you. There are features in this text designed to help you relate material you are learning to previously mastered material. Math topics are tied to real life as often as possible. Key learning objectives are introduced and easily identified in every section.

Save time by having a plan. Follow this **six-step process**, and you will find yourself successfully solving a wide range of problems.

PROBLEM-SOLVING STEPS

1. UNDERSTAND the problem. During this step, don't work with variables, but simply become comfortable with the problem. Some ways of accomplishing this are listed next.
   - Read and reread the problem.
   - Construct a drawing to help visualize the problem.
   - Propose a solution and check. Pay careful attention as to how you check your proposed solution. This will help later when you write an equation to model the problem.

2. ASSIGN a variable to an unknown in the problem. Use this variable to represent any other unknown quantities.

3. ILLUSTRATE the problem. A diagram or chart *using the assigned variables* can often help us visualize the known facts.

4. TRANSLATE the problem into a mathematical model. This is often an equation.

5. COMPLETE the work. This often means to solve the equation.

6. INTERPRET the results. *Check* the proposed solution in the stated problem and *state* your conclusion.

When solving problems, use technology to enhance the problem-solving process and/or when checking your solution.

REMINDER   Note that $f(x)$ is a special symbol in mathematics used to denote a function. The symbol $f(x)$ is read "f of x." It does **not** mean $f \cdot x$ (f times x).

"Reminders" contain practical advice and assist you in understanding as you solve problems.

TECHNOLOGY NOTE

Most graphing utilities have the ability to select and deselect graphs, graph equations simultaneously, and define different graph styles. See your graph-

"Technology Notes" include specific suggestions for problem solving using technology.

VISUALIZE . . . SEE THE CONCEPTS!

Graphing is introduced early and intuitively. An emphasis on data interpretation and opportunities to create your own data via tables and graphs allows you to see how the concept of graphing is gradually developed throughout the text. Knowing how to use and generate data and graphs is a valuable skill in the workplace as well as in other courses.

87. What percent of people ages 18 to 34 have no health insurance?

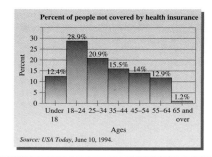

Percent of people not covered by health insurance

Source: USA Today, June 10, 1994.

Real Data is integrated throughout the text, drawn from familiar sources such as magazines and newspapers.

EXAMPLE 6   Graph $y = x^3$ and examine a table of values with x equal to the integers from -3 to 3.

Solution: Define $y_1 = x^3$ and graph the equation in a standard window.

The figure below shows the graph of $y = x^3$ with the trace cursor on $(2, 8)$. The table of values shows ordered pair solutions for $x = -3, -2, -1, 0, 1, 2,$ and 3.

Notice in the table that when x is negative y is also negative. Thus, these points lie in the third quadrant. When x is positive, y is positive; thus, this portion of the graph lies in the first quadrant. When $x = 0$, then $y = 0$, so the graph crosses the axes at the origin.

Many graphics, models and illustrations provide **visual reinforcement** of concepts.

DISCOVER THE CONCEPT

a. Predict how the graph of $y = |x| + 5$ can be obtained from the graph of $y = |x|$. Then check your prediction by graphing both equations using a standard window.

b. Predict how the graph of $y = x^2 - 3$ can be obtained from the graph of $y = x^2$. Use a graphing utility to check your prediction. Next, predict how the graph of $y = x^2 + 5$ can be obtained from $y = x^2$.

c. In general, can you predict how the graph of a basic equation with a constant added can be obtained from the graph of the basic equation? Test your prediction with the basic cubic equation $y = x^3$.

Get involved! **Discover the Concept** on your own using a graphing utility.

CHECK YOUR UNDERSTANDING. EXPAND IT. EXPLORE!

Good exercise sets are essential to the make-up of a solid intermediate algebra textbook. The exercises in this textbook are designed to help you understand skills and concepts as well as challenge and motivate you. Note, too, the Highlights, Test, Review, and Cumulative Review found at the end of each chapter.

MENTAL MATH

Use positive exponents to state each expression.

1. $5x^{-1}y^{-2}$ **2.** $7xy^{-4}$ **3.** $a^2b^{-1}c^{-5}$ **4.** $a^{-4}b^2c^{-6}$ **5.** $\dfrac{y^{-2}}{x^{-4}}$ **6.** $\dfrac{x^{-7}}{z^{-3}}$

Confidence-building **Mental Math** problems are in many sections.

A Look Ahead examples and problems are similar to those found in the *next* algebra course and include more advanced exercises.

Conceptual and Writing Exercises bring together two or more concepts and often require "in your own words" written explanation.

The accompanying graph shows the daily low temperatures for one week in New Orleans, Louisiana.

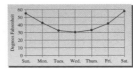

67. Which day of the week shows the greatest decrease in temperature low?

68. Which day of the week shows the greatest increase in temperature low?

69. Which day of the week had the lowest temperature?

70. Use the graph to estimate the low temperature on Thursday.

Notice that the shape of the temperature graph is similar to a parabola (see Section 3.2). In fact, this graph can be modeled by the quadratic function $f(x) = 3x^2 - 18x + 57$, where $f(x)$ is the temperature in degrees Fahrenheit and x is the number of days from Sunday. Use this function to answer Exercises 71 and 72.

71. Use the quadratic function given to approximate the temperature on Thursday. Does your answer agree with the graph above?

72. Use the function given and the quadratic formula to find when the temperature was 35°F. [*Hint:* Let $f(x) = 35$ and solve for x.] Round your answer to one decimal place and interpret your result. Does your answer agree with the graph above?

Review Exercises

Solve each equation. See Sections 7.7 and 8.6.

75. $\sqrt{5x - 2} = 3$ **76.** $\sqrt{y + 2} + 7 = 12$

77. $\dfrac{1}{x} + \dfrac{2}{5} = \dfrac{7}{x}$ **78.** $\dfrac{10}{z} = \dfrac{5}{z} - \dfrac{1}{3}$

Factor. See Section 6.7.

79. $x^4 + x^2 - 20$ **80.** $2y^4 + 11y^2 - 6$

81. $z^4 - 13z^2 + 36$ **82.** $x^4 - 1$

A Look Ahead

EXAMPLE

Solve $x^2 - 3\sqrt{2}x + 2 = 0$.

Solution:

In this equation, $a = 1$, $b = -3\sqrt{2}$, and $c = 2$. By the quadratic formula, we have

$$x = \frac{-b \pm \sqrt{b^2 - 4ac}}{2a}$$

$$= \frac{3\sqrt{2} \pm \sqrt{(-3\sqrt{2})^2 - 4(1)(2)}}{2(1)}$$

$$= \frac{3\sqrt{2} \pm \sqrt{18 - 8}}{2} = \frac{3\sqrt{2} \pm \sqrt{10}}{2}$$

The solution set is $\left\{\dfrac{3\sqrt{2} + \sqrt{10}}{2}, \dfrac{3\sqrt{2} - \sqrt{10}}{2}\right\}$.

Use the quadratic formula to solve each quadratic equation. See the preceding example.

83. $3x^2 - \sqrt{12}x + 1 = 0$

84. $5x^2 - \sqrt{20}x + 1 = 0$

Match each graph with its equation.

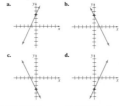

33. $f(x) = 2x + 3$ **34.** $f(x) = 2x - 3$

35. $f(x) = -2x + 3$ **36.** $f(x) = -2x - 3$

Find the slope of each line. See Examples 6 and 7.

37. $x = 1$ **38.** $y = -2$

39. $y = -3$ **40.** $x = 4$

41. $x + 2 = 0$ **42.** $y - 7 = 0$

43. Explain how merely looking at a line can tell us whether its slope is negative, positive, undefined, or zero.

44. Explain why the graph of $y = b$ is a horizontal line.

Find the slope and the y-intercept of each line.

45. $f(x) = -x + 5$ **46.** $f(x) = x + 2$

47. $-6x + 5y = 30$ **48.** $4x - 7y = 28$

Build your confidence with the beginning exercises; the first part of each exercise set is keyed to already worked examples. Then try the remaining exercises.

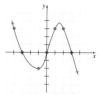

Review Exercises review concepts learned earlier that are needed in the next section or next chapter.

95. List the point(s) of the graph whose coordinates are of the form (, 3).

96. List the *x*-intercept(s) of the graph.

Find the distance between the pairs of points. Then find the midpoint of the line segment joining them. See Examples 3 and 5.

29. $A(1, 2), B(5, 6)$ **30.** $A(-1, -3), B(2, 1)$

31. $A(0, 4), B(3, -2)$ **32.** $A(3, 8), B(-2, 4)$

33. $A(-2, 7), B(-4, 10)$ **34.** $A(7, -5), B(9, -2)$

35. **36.**

37. Find the perimeter of the triangle whose vertices are $A(12, 3)$, $B(5, 3)$, and $C(7, 11)$.

38. Find the perimeter of the triangle whose vertices are $A(2, 8)$, $B(11, -4)$, and $C(7, 4)$.

39. Find the perimeter of the quadrilateral whose vertices are $A(-2, 3)$, $B(4, 8)$, $C(9, 2)$, and $D(3, -3)$.

Graphing utility **screens** are appropriately found throughout the text.

CHAPTER 6   HIGHLIGHTS

DEFINITIONS AND CONCEPTS	EXAMPLES
SECTION 6.1 EXPONENTS AND SCIENTIFIC NOTATION	
Product rule: $a^m \cdot a^n = a^{m+n}$	$x^2 \cdot x^3 = x^5$
Zero exponent: $a^0 = 1, a \neq 0$	$7^0 = 1, (-10)^0 = 1$
Quotient rule: $\dfrac{a^m}{a^n} = a^{m-n}$	$\dfrac{y^{10}}{y^4} = y^{10-4} = y^6$
Negative exponent: $a^{-n} = \dfrac{1}{a^n}$	$3^{-2} = \dfrac{1}{3^2} = \dfrac{1}{9}, \dfrac{x^{-5}}{x^{-7}} = x^{-5-(-7)} = x^2$
A positive number is written in **scientific notation** if it is written as the product of a number a, where $1 \leq a < 10$, and an integer power of 10: $a \times 10^r$.	Numbers written in scientific notation: $568{,}000 = 5.68 \times 10^5$ $0.0002117 = 2.117 \times 10^{-4}$
SECTION 6.2 MORE WORK WITH EXPONENTS AND SCIENTIFIC NOTATION	
Power rules: $(a^m)^n = a^{m \cdot n}$	$(7^8)^2 = 7^{16}$
$(ab)^m = a^m b^m$	$(2y)^3 = 2^3 y^3 = 8y^3$

Chapter Highlights contain key definitions, concepts, *and* examples to help you understand and recall what you have learned.

 GROUP ACTIVITY

LOCATING LIGHTNING STRIKES

There is an opportunity to **explore** an exercise that relates to the chapter-opening photo as a group activity or discovery-based project.

MATERIALS:
• Graphing Utility

Weather-recording stations use a directional antenna to detect and measure the electromagnetic field emitted by a lightning bolt. The

2. A lightning strike is detected by both stations. Station A uses a measured angle to find the slope of the line from the station to the lightning strike as $m = -1.732$. Station B computes a slope of $m = 0.577$ from the angle it measured. Use this information to find the

STUDENT RESOURCES

Seek out these items to match your personal learning style.

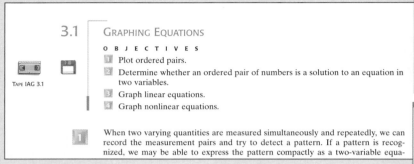

3.1 GRAPHING EQUATIONS

O B J E C T I V E S

1. Plot ordered pairs.
2. Determine whether an ordered pair of numbers is a solution to an equation in two variables.
3. Graph linear equations.
4. Graph nonlinear equations.

TAPE IAG 3.1

1. When two varying quantities are measured simultaneously and repeatedly, we can record the measurement pairs and try to detect a pattern. If a pattern is recognized, we may be able to express the pattern compactly as a two-variable equa-

Text-specific videos hosted by the award-winning teacher and co-author of *Intermediate Algebra: A Graphing Approach,* cover chapter and section objectives as a supplementary review.

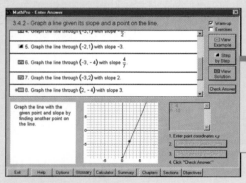

MathPro Tutorial Software, developed around the content of *Intermediate Algebra: A Graphing Approach,* provides interactive warm-up and graded algorithmic practice problems with step-by-step worked solutions.

Graphing Utilities Video provides introductory keystroke instruction for TI and Casio graphing calculators.

ALSO AVAILABLE:

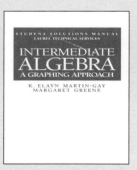

Student Solutions Manual

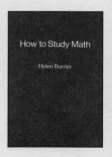

How to Study Math

Helen Burrier

LIFE ON THE INTERNET
MATHEMATICS
A STUDENT'S GUIDE

ANDREW T. STULL
EDWARD WRIGHT

The New York Times

PRENTICE HALL

THEMES OF THE TIMES

The New York Times/Themes of the Times

Newspaper-format supplement—
*ask your professor about
this exciting free supplement!*

REVIEW OF REAL NUMBERS AND ALGEBRAIC EXPRESSIONS

ANALYZING NEWSPAPER CIRCULATION

The number of daily newspapers in business in the United States has declined steadily over the past decade, continuing the trend that started in the mid-1970s. At the turn of the century there were roughly 2300 daily newspapers operating, but by 1995 that number had dropped to only 1548. Average overall daily newspaper circulation also continues to decline, from 60,164,500 in 1992 to 59,811,600 in 1993 to 59,305,400 in 1994. (*Source: Editor & Publisher International Yearbooks.*)

IN THE CHAPTER GROUP ACTIVITY ON PAGE 39, YOU WILL HAVE THE OPPORTUNITY TO ANALYZE THE CIRCULATION OF SEVERAL NEW YORK CITY AREA NEWSPAPERS.

1.1 | REAL NUMBERS

O B J E C T I V E S

TAPE IAG 1.1

1. Identify natural numbers, whole numbers, integers, rational numbers, irrational numbers, and real numbers.
2. Graph numbers on a number line.
3. Write phrases as algebraic expressions.

When solving a problem, it is important to know the kind of number that is appropriate for the solution. For example, if we are asked to determine the maximum number of the parking spaces that is possible for a parking lot being constructed, an answer of $98\frac{1}{10}$ is not appropriate because $\frac{1}{10}$ of a parking space is not realistic. We begin this text with a review of common sets of numbers.

First, recall that a **set** is a collection of objects, called **elements.** When people first counted, they used **natural numbers.** The set of natural numbers is

$$N = \{1, 2, 3, 4, 5, \ldots\}.$$

Notice that the elements of the set are enclosed in braces, and we have assigned the letter N to name the set. The listing of three dots, $\ldots$, in the set is called an **ellipsis** and means that the list continues without end. This means that N has an infinite number of elements (numbers) and this is an **infinite set.**

A set that contains no elements is called the **empty set.** We symbolize the empty set by $\{\ \}$ or $\varnothing$.

> R E M I N D E R   Use $\{\ \}$ to write the empty set. $\{\varnothing\}$ is *not* the empty set because it has one element: $\varnothing$.

With the idea of counting objects in place, the idea came later of counting no object. This resulted in the number 0. If 0 is included with the set of natural numbers, we have the set of **whole numbers.**

$$W = \{0, 1, 2, 3, 4, 5, \ldots\}$$

To write in symbols that 2 is an element of the set $\{0, 1, 2, 3, 4, 5, \ldots\}$, we use the symbol $\in$ (read "is an element of"). Thus,

$$2 \in \{0, 1, 2, 3, 4, 5, \ldots\} \quad \text{or} \quad 2 \in W.$$

To write that 0 is *not* an element of the set of natural numbers N, we write

$$0 \notin \{1, 2, 3, 4, 5, \ldots\} \quad \text{or} \quad 0 \notin N.$$

Whole numbers are not sufficient for describing every situation. For example, to describe the temperature of 3 degrees *below* zero, we use -3, the negative, or opposite, of the natural number 3. The natural numbers, zero, and the negatives of the natural numbers make up the set of **integers.**

$$Z = \{\ldots, -3, -2, -1, 0, 1, 2, 3, \ldots\}$$

2 Let's review these common sets of numbers by graphing them on a number line. To construct a number line, we draw a line and label a point 0 with which we associate the number 0. This point is called the **origin.** Choose a point to the right of 0 and label it 1. The distance from 0 to 1 is called the **unit distance** and can be used to locate more points. The **positive numbers** lie to the right of the origin, and the **negative numbers** lie to the left of the origin. The number 0 is neither positive nor negative.

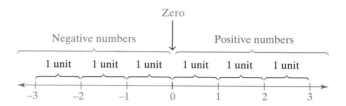

A number is **graphed** on a number line by shading the point on the number line that corresponds to the number. Some common sets of numbers and their graphs include the following:

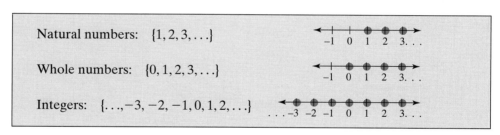

When the elements of a set are listed, as has been done for sets N, W, and Z, we say that sets are written in **roster form.** A set can also be written using **set builder notation,** which describes the numbers of the set but does not list them. In set builder notation, letters are often used to represent numbers. These letters are called **variables.** We describe the next set of numbers using this notation.

When we divide one integer by another (as long as the divisor is not 0), the result is called a rational number. In set builder notation, the **rational numbers** are the elements of the set

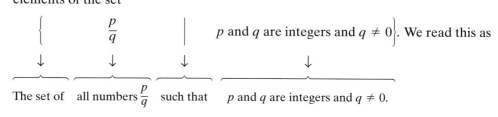

A few examples of rational numbers graphed on a number line are

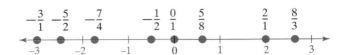

Notice that every integer is also a rational number since each integer can be written as the quotient of itself and 1.

$$2 = \frac{2}{1}, \qquad 0 = \frac{0}{1}, \qquad -3 = \frac{-3}{1}$$

For negative rational numbers such as $-\frac{1}{2}$, the negative sign can be written in the denominator, in the numerator, or in front of the fraction without changing the value of the fraction. Thus,

$$-\frac{1}{2} = \frac{-1}{2} = \frac{1}{-2}$$

> In general, if a and b are real numbers, b fi 0,
>
> $$-\frac{a}{b} = \frac{-a}{b} = \frac{a}{-b}$$

We show why this is so in Section 1.3.

EXAMPLE 1   List the elements in each set.
   a. $\{x \mid x \text{ is a whole number between 1 and 6}\}$
   b. $\{x \mid x \text{ is a natural number greater than 100}\}$

Solution: **a.** $\{2, 3, 4, 5\}$ **b.** $\{101, 102, 103, \ldots\}$

EXAMPLE 2   Determine whether each statement is true or false.
   a. $3 \in \{x \mid x \text{ is a natural number}\}$ **b.** $7 \notin \{1, 2, 3\}$
   c. $5 \notin \{\ldots, -3, -2, -1, 0, 1, 2, 3, \ldots\}$

Solution: **a.** True, since 3 is a natural number and is therefore an element of the set.
   b. True, since 7 is not an element of the set $\{1, 2, 3\}$.
   c. False, since 5 is an element of the set.

We use set builder notation to describe two other common sets of numbers. (The set of rational numbers is repeated here.)

Real numbers: $\{x \mid x$ corresponds to a point on the number line$\}$

Rational numbers: $\left\{\dfrac{p}{q} \;\middle|\; p$ and q are integers and $q \neq 0\right\}$

Irrational numbers: $\{x \mid x$ is a real number and x is not a rational number$\}$

It can be shown that the decimal representation of a rational number either terminates (ends) or repeats in a block of digits.

RATIONAL NUMBERS WHOSE DECIMALS TERMINATE	RATIONAL NUMBERS WHOSE DECIMALS REPEAT
$\dfrac{1}{2} = 0.5$	$\dfrac{2}{3} = 0.66666\ldots = 0.\overline{6}$
$\dfrac{5}{4} = 1.25$	$\dfrac{14}{99} = 0.141414\ldots = 0.\overline{14}$
$\dfrac{11}{8} = 1.375$	$\dfrac{5}{6} = 0.8333\ldots = 0.8\overline{3}$

Notice above that one way to represent a repeating decimal is to place a bar over one block of repeating digits. If we want to write an approximation for a number, we can use the symbol $\approx$, which means "is approximately equal to." Thus,

$$\frac{5}{6} = 0.8\overline{3} \quad \text{and} \quad \frac{5}{6} \approx 0.83$$
$$\uparrow \qquad\qquad\qquad \uparrow$$

  is equal to   is approximately equal to

An irrational number written as a decimal neither terminates nor repeats. When we perform calculations with irrational numbers, we often use rounded decimal approximations. For example, consider the following irrational numbers along with a four-decimal-place approximation of each.

$$\pi \approx 3.1416, \qquad \sqrt{2} \approx 1.4142$$

Earlier we mentioned that every integer is also a rational number. In other words, all the elements of the set of integers are also elements of the set of rational numbers. When this happens, we say that the set of integers, set Z, is a **subset** of the set of rational numbers, set Q. In symbols,

$$Z \subseteq Q$$

  is a subset of

The natural numbers, whole numbers, integers, rational numbers, and irrational numbers are each a subset of the set of real numbers. The relationships among these sets of numbers are shown in the following diagram. Each set is a subset of the sets shown above it.

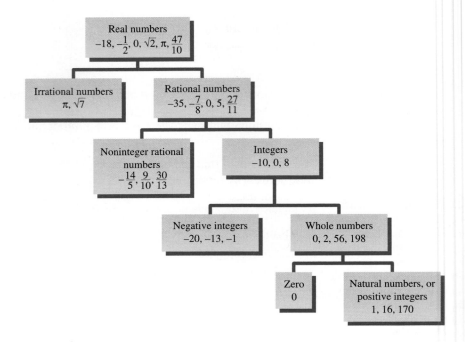

EXAMPLE 3 Determine whether each statement is true or false.
a. Every integer is a real number.

b. Every whole number is a natural number.

c. All real numbers are rational numbers.

d. Some rational numbers are irrational.

e. $N \subseteq Z$.

Solution: **a.** True. Every integer is a real number.

b. False. The whole number 0 is not a natural number.

c. False. The real number $\sqrt{2}$ is not a rational number.

d. False. A number is rational or irrational, but not both.

e. True. Every member of the set of natural numbers N is also a member of the set of integers Z.

For much of this text, we will concentrate on solving problems. Algebraic expressions occur often when solving problems. An **algebraic expression** is formed by numbers and variables connected by the operations of addition, subtraction, multiplication, division, raising to powers, or taking roots. For example,

$$2x + 3, \quad \frac{x + 5}{6} - \frac{z^5}{y^2}, \quad \text{and} \quad \sqrt{y} - 1.6$$

are algebraic expressions or, more simply, expressions.

Solving problems often involves translating a phrase to an algebraic expression. The following list shows key words and phrases and their translations.

ADDITION	SUBTRACTION	MULTIPLICATION	DIVISION
sum	difference of	product	quotient
plus	minus	times	divide
added to	subtracted from	multiply	into
more than	less than	twice	ratio
increased by	decreased by	of	
total	less		

EXAMPLE 4   Translate each phrase to an algebraic expression. Use the variable x to represent each unknown number.

    a. Eight times a number

    b. Three more than eight times a number

    c. The quotient of a number and -7

    d. Negative one and six-tenths subtracted from twice a number

Solution: **a.** $8 \cdot x$ or $8x$ **b.** $8x + 3$ **c.** $x \div -7$ or $\dfrac{x}{-7}$

    d. $2x \underset{\text{subtraction}}{\nearrow} - \underset{\text{negative}}{\nwarrow} (-1.6)$

TECHNOLOGY NOTE

Most calculators have a subtraction key and a separate negative key. If this is so on your calculator, notice the difference in their displays. The subtraction sign is usually a longer horizontal dash and the negative sign is a shorter, raised horizontal dash.

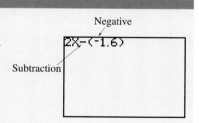

EXERCISE SET 1.1

Tell whether each decimal is terminating or repeating.

1. $0.167\overline{67}$ **2.** 0.125

3. 1.238 **4.** $3.145\overline{45}$

Tell whether each number is rational or irrational.

5. 0.75 **6.** 1.52

7. 2π **8.** $\sqrt{8}$

9. $\dfrac{7}{8}$ **10.** $\dfrac{3}{4}$

List the elements of the set $\{3, 0, \sqrt{7}, \sqrt{36}, \frac{2}{5}, -13.4\}$ that are also elements of the given set.

11. Whole numbers   **12.** Integers

13. Natural numbers   **14.** Rational numbers

15. Irrational numbers   **16.** Real numbers

Tell whether each statement is true or false. See Example 3.

17. Every rational number is a real number.

18. All whole numbers are integers.

19. Every integer is a whole number.

20. All real numbers are irrational numbers.

21. All natural numbers are rational numbers.

22. Every real number is either rational or irrational.

23. In your own words, explain why the graphing utility screen suggests that $\sqrt{25}$ is a rational number and $\sqrt{10}$ is an irrational number.

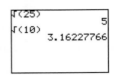

24. In your own words, explain why the graphing utility screen suggests that $\sqrt{\frac{4}{9}}$ is a rational number and $\sqrt{7}$ is an irrational number.

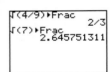

List the elements in each set. See Example 1.

25. $\{x \mid x \text{ is a natural number less than 6}\}$

26. $\{x \mid x \text{ is a natural number greater than 6}\}$

27. $\{x \mid x \text{ is a natural number between 10 and 17}\}$

28. $\{x \mid x \text{ is an odd natural number}\}$

29. $\{x \mid x \text{ is a whole number that is not a natural number}\}$

30. $\{x \mid x \text{ is a natural number less than 1}\}$

31. $\{x \mid x \text{ is an even whole number less than 9}\}$

32. $\{x \mid x \text{ is an odd whole number less than 9}\}$

Graph each set on a number line.

33. $\{0, 2, 4, 6\}$ **34.** $\{-1, -2, -3\}$ **35.** $\left\{\frac{1}{2}, \frac{2}{3}\right\}$

36. $\{1, 3, 5, 7\}$ **37.** $\{-2, -6, -10\}$ **38.** $\left\{\frac{1}{4}, \frac{1}{3}\right\}$

39. In your own words, explain why the empty set is a subset of every set.

40. In your own words, explain why every set is a subset of itself.

Place $\in$ or $\notin$ in the space provided to make each statement true. See Example 2.

41. -11 $\{x \mid x \text{ is an integer}\}$

42. -6 $\{2, 4, 6, \ldots\}$

43. 0 $\{x \mid x \text{ is a positive integer}\}$

44. 12 $\{1, 2, 3, \ldots\}$

45. 12 $\{1, 3, 5, \ldots\}$

46. $\frac{1}{2}$ $\{x \mid x \text{ is an irrational number}\}$

47. 0 $\{1, 2, 3, \ldots\}$

48. 0 $\{x \mid x \text{ is a natural number}\}$

Determine whether each statement is true or false. Use the following sets of numbers:

$$N = \text{set of natural numbers}$$
$$Z = \text{set of integers}$$
$$I = \text{set of irrational numbers}$$
$$Q = \text{set of rational numbers}$$
$$\mathbb{R} = \text{set of real numbers}$$

49. $Z \subseteq \mathbb{R}$ **50.** $\mathbb{R} \subseteq N$

51. $-1 \in Z$ **52.** $\frac{1}{2} \in Q$

53. $0 \in N$ **54.** $Z \subseteq Q$

55. $\sqrt{5} \notin I$ **56.** $\pi \notin \mathbb{R}$

57. $N \subseteq Z$ **58.** $I \subseteq N$

59. $\mathbb{R} \subseteq Q$ **60.** $N \subseteq Q$

61. In your own words, explain why every natural number is also a rational number, but not every rational number is a natural number.

62. In your own words, explain why every irrational number is a real number, but not every real number is an irrational number.

Write each phrase as an algebraic expression. Use the variable x to represent each unknown number. See Example 4.

63. Twice a number

64. Six times a number

65. Five more than twice a number

66. One more than six times a number

67. Ten less than a number

68. A number minus seven

69. The sum of a number and two

70. The difference of twenty-five and a number

71. A number divided by eleven

72. The quotient of twice a number and thirteen

73. Twelve added to three times a number

74. A number subtracted from four

75. Seventeen subtracted from a number

76. Four subtracted from three times a number

77. Twice the sum of a number and three

78. The quotient of four and the sum of a number and one

79. The quotient of five and the difference of four and a number

80. Eight times the difference of a number and nine

81. The following bar graph shows annual revenues of the Walt Disney Company. Use the graph to calculate the annual *increases* in revenue. (Note that dollar amounts are in millions.)

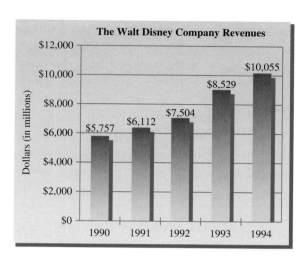

The Walt Disney Company Revenues

YEAR	ANNUAL INCREASE IN REVENUE
1991	
1992	
1993	
1994	

82. GROUP ACTIVITY   Add the numbers in each set. Look for a pattern as you add.

   a. $\{1, -2, 3, -4, 5, -6, 7, -8, 9, -10\}$

   b. $\{-30, 31, -32, 33, -34, 35, -36, 37, -38, 39\}$

   c. $\{1, -3, 5, -7, \ldots, -99\}$

   d. $\{-2, 4, -6, 8, \ldots, 48\}$

1.2 PROPERTIES OF REAL NUMBERS

TAPE IA 1.2

OBJECTIVES

1 Use operation and order symbols to write mathematical sentences.

2 Use interval notation.

3 Identify identity numbers and inverses.

4 Identify and use the commutative, associative, and distributive properties.

1 In Section 1.1, we used the symbol = to mean "is equal to." All the following key words and phrases also imply equality.

EQUALITY

equals	is/was	represents	is the same as
gives	yields	amounts to	is equal to

EXAMPLE 1   Write each sentence using mathematical symbols.

a. The sum of x and 5 is 20.

b. Two times the sum of 3 and y amounts to 4.

c. Subtract 8 from x, and the difference is the same as the product of 2 and x.

d. The quotient of z and 9 is 3 times the difference of z and 5.

Solution: **a.** The sum of x and 5 can be written as "$x + 5$." The word "is" means "is equal to" in this sentence, so we write $x + 5 = 20$.

b. $2(3 + y) = 4$ **c.** $x - 8 = 2x$ **d.** $\dfrac{z}{9} = 3(z - 5)$

If we want to write in symbols that two numbers are not equal, we can use the symbol $\neq$, which means **"is not equal to."** For example,

$$3 \neq 2$$

Graphing two numbers on a number line gives us a way to compare the numbers. For two real numbers a and b, we say a *is less than* b if on the number line a lies to the left of b. Also, if b is to the right of a on the number line, then b *is greater than a*.

The symbol $>$ means **"is greater than."** Since b is greater than a, we can write

$$b > a$$

Also, the symbol $<$ means **"is less than."** Since a is less than b, we can write

$$a < b$$

Notice that if $a < b$, then also $b > a$.

The **trichotomy property** for real numbers assures us that any two real numbers a and b will either be equal, or one will be greater than the other.

TRICHOTOMY PROPERTY

If a and b are real numbers, then exactly one of these statements is true:

$$a = b, \qquad a < b, \qquad \text{or} \qquad a > b$$

EXAMPLE 2   Insert $<$, $>$, or $=$ between each pair of numbers to form a true statement.

    a. -1 -2 **b.** $\dfrac{12}{4}$ 3 **c.** -5 0 **d.** -3.5 -3.05

Solution: **a.** $-1 > -2$ since -1 lies to the right of -2 on the number line.

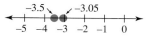

    b. $\dfrac{12}{4} = 3$

    c. $-5 < 0$ since -5 lies to the left of 0 on the number line.

    d. $-3.5 < -3.05$ since -3.5 lies to the left of -3.05 on the number line.

R E M I N D E R   When inserting the $>$ or $<$ symbol, think of each symbol as an arrowhead that "points" toward the smaller number when the statement is true.

In addition to $<$ and $>$, there are the inequality symbols $\leq$ and $\geq$. The symbol

$\leq$ means **"is less than or equal to"**

and the symbol

$\geq$ means **"is greater than or equal to"**

For example, each of the following statements is true.

$$10 \leq 10 \quad \text{since} \quad 10 = 10$$
$$-8 \leq 13 \quad \text{since} \quad -8 < 13$$
$$-5 \geq -5 \quad \text{since} \quad -5 = -5$$
$$-7 \geq -9 \quad \text{since} \quad -7 > -9$$

EXAMPLE 3   Write each sentence using mathematical symbols.

    a. The sum of 5 and y is greater than or equal to 7.

    b. 11 is not equal to z.

    c. 20 is less than the difference of 5 and twice x.

Solution: **a.** $5 + y \geq 7$ **b.** $11 \neq z$ **c.** $20 < 5 - 2x$

We can use inequality symbols to describe sets of real numbers. For example, to describe the set of real numbers greater than or equal to -1, we may write $\{x \mid x \geq -1\}$. To graph this set, we graph all numbers to the right of -1 and place a bracket at -1. This bracket indicates that -1 *is* an element of the set.

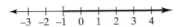

Notice that we have graphed an *interval* on the number line. In **interval notation,** we write $[-1, \infty)$. The infinity symbol ∞ is not a number, but shows that *all* numbers greater than -1 are in the interval $[-1, \infty)$.

   To graph the set $\{x \mid x < 3\}$, we graph all numbers to the left of 3 and place a parenthesis at 3. The parenthesis indicates that 3 is not an element of the set.

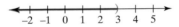

In interval notation, we write $(-\infty, 3)$. The infinity symbol $-\infty$ indicates that *all* numbers less than 3 are in the interval $(-\infty, 3)$. Note that a parenthesis is always used with the infinity symbol in interval notation.

EXAMPLE 4   Graph each set and write it in interval notation.

a. $\{x \mid x \leq 0\}$ **b.** $\{x \mid x > 2\}$

Solution: **a.** The set $\{x \mid x \leq 0\}$ consists of all numbers less than or equal to 0. Place a bracket at 0 to show that it is a member of the set. Shade to the left.

In interval notation, this is $(-\infty, 0]$.

b. The set $\{x \mid x > 2\}$ consists of all numbers greater than 2. Place a parenthesis at 2

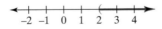

to show that it is not a member of the set. Shade to the right.
In interval notation, this is $(2, \infty)$.

EXAMPLE 5   Graph each set and write it in interval notation.

a. $\{x \mid -5 < x < -1\}$ **b.** $\{x \mid -1.5 \leq x < 4\}$

Solution: **a.** The set $\{x \mid -5 < x < -1\}$ consists of all real numbers between -5 and -1. A parenthesis is placed at both -5 and -1 indicating that neither is in the set. All numbers between -5 and -1 are shaded because they are members of the set.

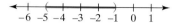

In interval notation, this is $(-5, -1)$.

b. For $\{x \mid -1.5 \le x < 4\}$, a bracket is placed at -1.5, and a parenthesis is placed at 4 since -1.5 is a member of the set and 4 is not. All numbers between -1.5 and 4 are shaded because they are members of the set.

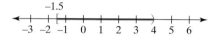

In interval notation, this is $[-1.5, 4)$.

Of all the real numbers, two of them stand out as extraordinary: 0 and 1. The real number 0 is called the **additive identity** since, for every real number a,

$$a + 0 = 0 + a = a$$

For example, $3 + 0 = 0 + 3 = 3$.

The real number 1 is called the **multiplicative identity** since, for every real number a,

$$a \cdot 1 = 1 \cdot a = a$$

For example, $5.2 \cdot 1 = 1 \cdot 5.2 = 5.2$.

The uniqueness of 0 is also demonstrated by the **multiplication property of 0,** which guarantees that, for every real number a,

$$a \cdot 0 = 0 \cdot a = 0$$

For example, $-7 \cdot 0 = 0 \cdot -7 = 0$.

Two numbers whose sum is the additive identity 0 are called **opposites** or **additive inverses** of each other. If a is a real number, then the unique opposite of a is written as $-a$.

$$a + (-a) = (-a) + a = 0$$

On the number line, we picture a real number and its opposite as being the same distance from 0 but on opposite sides of 0.

The opposite of 6 is -6.

The opposite of $-\dfrac{2}{3}$ is $\dfrac{2}{3}$.

-4 and 4 are opposites

We stated earlier that the opposite or additive inverse of a number a is $-a$. This means that the opposite of -4 is $-(-4)$. But as shown above, the opposite of -4 is 4. This means that $-(-4) = 4$, and in general, we have the following property:

DOUBLE NEGATIVE PROPERTY
For every real number $a, -(-a) = a$.

What is the opposite of zero? Recall that the sum of a number and its opposite is 0. This means that opposite of 0 is 0 since

$$0 + 0 = 0$$

EXAMPLE 6   Write the additive inverse, or the opposite, of each number.

a. 8   **b.** $\dfrac{1}{5}$ **c.** -9.6

Solution: **a.** The opposite of 8 is -8.   **b.** The opposite of $\dfrac{1}{5}$ is $-\dfrac{1}{5}$.

c. The opposite of -9.6 is $-(-9.6) = 9.6$

Just as each real number has an opposite, each nonzero real number has a unique reciprocal. Two numbers whose product is 1 are called **reciprocals** or **multiplicative inverses** of each other. If a is a nonzero real number, then its reciprocal is $\dfrac{1}{a}$.

$$a \cdot \frac{1}{a} = \frac{1}{a} \cdot a = 1$$

EXAMPLE 7   Write the multiplicative inverse, or the reciprocal, of each number.

a. 11   **b.** -9 **c.** $\dfrac{7}{4}$

Solution: **a.** The reciprocal of 11 is $\dfrac{1}{11}$.

b. The reciprocal of -9 is $-\dfrac{1}{9}$.

c. The reciprocal of $\dfrac{7}{4}$ is $\dfrac{4}{7}$ because $\dfrac{7}{4} \cdot \dfrac{4}{7} = 1$.

The identities and inverses for addition and multiplication are summarized here.

	ADDITION	**MULTIPLICATION**
IDENTITIES	The additive identity is 0. $a + 0 = 0 + a = a$	The multiplicative identity is 1. $a \cdot 1 = 1 \cdot a = a$
INVERSES	For each number a, there is a unique number $-a$ called the **additive inverse** or **opposite** of a such that $a + (-a) = (-a) + a = 0$	For each nonzero a, there is a unique number $\dfrac{1}{a}$ called the **multiplicative inverse** or **reciprocal** of a such that $a \cdot \dfrac{1}{a} = \dfrac{1}{a} \cdot a = 1$

 All real numbers have certain properties that allow us to write equivalent expressions, that is, expressions that have the same value. These properties will be especially useful in Chapter 2 when we solve equations.

The **commutative properties** state that the order in which two real numbers are added or multiplied does not affect their sum or product.

COMMUTATIVE PROPERTIES

For real numbers a and b,

    Addition: $a + b = b + a$

    Multiplication: $a \cdot b = b \cdot a$

The **associative properties** state that regrouping numbers that are added or multiplied does not affect their sum or product.

ASSOCIATIVE PROPERTIES

For real numbers a, b, and c,

    Addition: $(a + b) + c = a + (b + c)$

    Multiplication: $(a \cdot b) \cdot c = a \cdot (b \cdot c)$

EXAMPLE 8   Use the commutative property of addition to write an expression equivalent to $7x + 5$.

Solution: $7x + 5 = 5 + 7x$

EXAMPLE 9 Use the associative property of multiplication to write an expression equivalent to $4 \cdot (9y)$. Then simplify this equivalent expression.

Solution: $4 \cdot (9y) = (4 \cdot 9)y = 36y$

The **distributive property** states that multiplication *distributes* over addition.

DISTRIBUTIVE PROPERTY

For real numbers a, b, and c,

$$a(b + c) = ab + ac.$$

EXAMPLE 10 Use the distributive property to write an expression equivalent to $3(2x + y)$.

Solution: $3(2x + y) = 3 \cdot 2x + 3 \cdot y$ Apply the distributive property.

$= 6x + 3y$ Apply the associative property of multiplication.

EXAMPLE 11 State the basic property illustrated in each true statement.

  a. $3x + 2y = 2y + 3x$ **b.** $2(5x - 2y) = 2 \cdot 5x - 2 \cdot 2y$

  c. $-5 + 5 = 0$ **d.** $-(-x) = x$

  e. $2x + (3x + 4) = (2x + 3x) + 4$

Solution: **a.** Commutative property of addition   **b.** Distributive property

          c. Additive inverses   **d.** Double negative property

          e. Associative property of addition

In the following example, a calculator screen illustrates some of the properties.

EXAMPLE 12 State the basic property that is being illustrated on each calculator screen:

TECHNOLOGY NOTE

In the third line of Example 12a the multiplicative inverse is indicated by using the x^{-1} key on the calculator. We can refer to the x^{-1} key as the *reciprocal key*, or the *multiplicative inverse* key.

a.
```
11*(1/11)
              1
-9*(-1/9)
              1
5*5⁻¹
              1
```

b.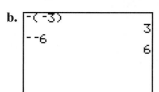

c.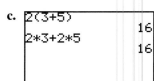

Solution: **a.** Multiplicative inverses   **b.** Double negative property

          c. Distributive property

EXERCISE SET 1.2

Write each sentence using mathematical symbols. See Example 1.

1. The product of 4 and c is 7.

2. The sum of 10 and x is -12.

3. 3 times the sum of x and 1 amounts to 7.

4. 9 times the difference of 4 and m amounts to 1.

5. The quotient of n and 5 is 4 times n.

6. The quotient of 8 and y is 3 more than y.

7. The difference of z and 2 is the same as the product of z and 2.

8. Five added to twice q is the same as 4 more than q.

Insert $<$, $>$, or $=$ in the space provided to form a true statement. See Example 2.

9. 0 -2

10. -5 0

11. $\dfrac{12}{3}$ $\dfrac{8}{2}$

12. $\dfrac{20}{5}$ $\dfrac{20}{4}$

13. -7.9 -7.09

14. -13.07 -13.7

Write each sentence using mathematical symbols. See Example 3.

15. The product of 7 and x is less than or equal to -21.

16. 10 subtracted from x is greater than 0.

17. The sum of -2 and x is not equal to 10.

18. The quotient of y and 3 is less than or equal to y.

19. Twice the difference of x and 6 is greater than the reciprocal of 11.

20. Four times the sum of 5 and x is not equal to the opposite of 15.

Graph each set and write it in interval notation. See Examples 4 and 5.

21. $\{x \mid x < -2\}$

22. $\{x \mid x \geq 3\}$

23. $\{x \mid x \geq 1\}$

24. $\{x \mid x \leq -2.3\}$

25. $\{x \mid x > 4.7\}$

26. $\left\{x \mid x > \dfrac{5}{8}\right\}$

27. $\left\{x \mid x \leq 3\dfrac{1}{2}\right\}$

28. $\{x \mid x < -0.4\}$

29. $\{x \mid -2 < x \leq 5\}$

30. $\{x \mid 0 < x < 3\}$

31. $\{x \mid 1.3 \leq x < 2.3\}$

32. $\left\{x \mid -2\dfrac{1}{3} \leq x \leq -1\dfrac{1}{3}\right\}$

33. $\left\{x \mid -5\dfrac{1}{4} \leq x \leq 0\right\}$

34. $\{x \mid 6.7 < x \leq 9.5\}$

Write the opposite (or additive inverse) of each number. See Example 6.

35. -6.2

36. -7.8

37. $\dfrac{4}{7}$

38. $\dfrac{9}{5}$

39. $-\dfrac{2}{3}$

40. $-\dfrac{14}{3}$

41. 0

42. 10.3

Write the reciprocal (or multiplicative inverse) of each number if one exists. See Example 7.

43. 5

44. 9

45. -8

46. -4

47. $-\dfrac{1}{4}$

48. $\dfrac{1}{9}$

49. 0

50. $\dfrac{0}{6}$

51. $\dfrac{7}{8}$

52. $-\dfrac{23}{5}$

53. Name the only real number that has no reciprocal, and explain why this is so.

54. Name the only real number that is its own opposite, and explain why this is so.

Use a commutative property to write an equivalent expression. See Example 8.

55. $7x + y$

56. $3a + 2b$

57. $z \cdot w$

58. $r \cdot s$

59. $\dfrac{1}{3} \cdot \dfrac{x}{5}$

60. $\dfrac{x}{2} \cdot \dfrac{9}{10}$

61. Is subtraction commutative? Explain why or why not.

62. Is division commutative? Explain why or why not.

Write the property illustrated on each screen. See Example 12.

63.
```
(2.1+5.3)+6.7
           14.1
2.1+(5.3+6.7)
           14.1
```

64.
```
3(5+6)
           33
3*5+3*6
           33
```

65.
```
9*2.75
           24.75
2.75*9
           24.75
```

66.
```
(3*2.1)*4
           25.2
3*(2.1*4)
           25.2
```

Use an associative property to write an equivalent expression. See Example 9.

67. $5 \cdot (7x)$

68. $3 \cdot (10z)$

69. $(x + 1.2) + y$

70. $5q + (2r + s)$

71. $(14z) \cdot y$

72. $(9.2x) \cdot y$

73. Evaluate $12 - (5 - 3)$ and $(12 - 5) - 3$. Use these two expressions and discuss whether subtraction is associative.

74. Evaluate $24 \div (6 \div 3)$ and $(24 \div 6) \div 3$. Use these two expressions and discuss whether division is associative.

Use the distributive property to find the product. See Example 10.

75. $3(x + 5)$

76. $7(y + 2)$

77. $8(2a + b)$

78. $9(c + 7d)$

79. $2(6x + 5y + 2z)$

80. $5(3a + b + 9c)$

Complete the statement to illustrate the given property See Example 11.

81. $3x + 6 =$ _____ Commutative property of addition

82. $8 + 0 =$ _____ Additive identity property

83. $\frac{2}{3} + \left(-\frac{2}{3}\right) =$ _____ Additive inverse property

84. $4(x + 3) =$ _____ Distributive property

85. $7 \cdot 1 =$ _____ Multiplicative identity property

86. $0 \cdot (-5.4) =$ _____ Multiplication property of zero

87. $10(2y) =$ _____ Associative property

88. $9y + (x + 3z) =$ _____ Associative property

89. To demonstrate the distributive property geometrically, represent the area of the larger rectangle in two ways: First as width a times length $b + c$, and second as the sum of the areas of the smaller rectangles.

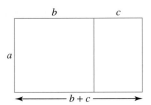

Graph each interval on a number line.

90. $(-\infty, 5]$

91. $(-\infty, -3)$

92. $[8, \infty)$

93. $[6, \infty)$

94. $(-2, 10]$

95. $(3, 8]$

Describe each graph using interval notation.

96.

97.

98.

99.

100.

101.

1.3 | OPERATIONS ON REAL NUMBERS

O B J E C T I V E S

TAPE IAG 1.3

1. Find the absolute value of a number.
2. Add and subtract real numbers.
3. Multiply and divide real numbers.
4. Simplify expressions containing exponents.
5. Find roots of numbers.

In Section 1.2, we used the number line to compare two real numbers. The number line can also be used to visualize distance, which leads to the concept of absolute value.

1

The **absolute value** of a real number a, written $|a|$, is the distance between a and 0 on the number line. Since distance is always positive or zero, $|a|$ is always positive or zero.

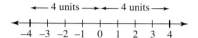

From the number line above, we see that

$$|4| = 4$$

because the distance between 4 and 0 is 4 units. Also,

$$|-4| = 4$$

because the distance between -4 and 0 is 4 units.

An equivalent definition of the absolute value of a real number a is given next.

ABSOLUTE VALUE

The absolute value of a, written as $|a|$, is

$$|a| = \begin{cases} a \text{ if } a \text{ is 0 or a positive number} \\ -a \text{ if } a \text{ is a negative number} \end{cases}$$

EXAMPLE 1   Find each absolute value, and check using your graphing utility.

    a. $|3|$ **b.** $|-5|$ **c.** $-|2|$ **d.** $-|-8|$ **e.** $|0|$

Solution: **a.** $|3| = 3$ since 3 is located 3 units from 0 on the number line.

    b. $|-5| = 5$ since -5 is 5 units from 0 on the number line.

    c. $-|2| = -2$. The negative sign outside the absolute value bars means to take the opposite of the absolute value of 2.

    d. $-|-8| = -8$. Since $|-8|$ is 8, $-|-8| = -8$.

    e. $|0| = 0$ since 0 is located 0 units from 0 on the number line.

The screens below show a check using a graphing utility.

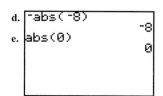

2

The addition of two real numbers may be summarized by the following statements:

ADDING REAL NUMBERS

1. To add two numbers with the same sign, add their absolute values and attach their common sign.

2. To add two numbers with different signs, subtract the smaller absolute value from the larger absolute value and attach the sign of the number with the larger absolute value.

For example, to add $-5 + (-7)$, first add their absolute values.

$$|-5| = 5, |-7| = 7 \quad \text{and} \quad 5 + 7 = 12$$

Next, attach their common negative sign.

$$-5 + (-7) = -12$$

To find $(-4) + 3$, first subtract their absolute values.

$$|-4| = 4, |3| = 3 \quad \text{and} \quad 4 - 3 = 1$$

Next, attach the sign of the number with the larger absolute value.

$$(-4) + 3 = -1$$

EXAMPLE 2   Find each sum.

  a. $-3 + (-11)$ **b.** $3 + (-7)$ **c.** $-10 + 15$

  d. $-8.3 + (-1.9)$ **e.** $-\dfrac{1}{4} + \dfrac{1}{2}$ **f.** $-\dfrac{2}{3} + \dfrac{3}{7}$

Solution: **a.** $-3 + (-11) = -(3 + 11) = -14$ **b.** $3 + (-7) = -4$

  c. $-10 + 15 = 5$ **d.** $-8.3 + (-1.9) = -10.2$

  e. $-\dfrac{1}{4} + \dfrac{1}{2} = -\dfrac{1}{4} + \dfrac{2}{4} = \dfrac{1}{4}$ **f.** $-\dfrac{2}{3} + \dfrac{3}{7} = -\dfrac{14}{21} + \dfrac{9}{21} = -\dfrac{5}{21}$

The screens below show a check using a calculator.

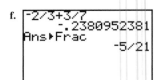

Subtraction of two real numbers may be defined in terms of addition.

> ### SUBTRACTING REAL NUMBERS
>
> If a and b are real numbers, then the difference of a and b, written $a - b$, is defined by
>
> $$a - b = a + (-b)$$

In other words, to subtract a second real number from a first, add the first number and the opposite of the second number.

EXAMPLE 3   Find each difference.

    a. $2 - 8$ **b.** $-8 - (-1)$ **c.** $-11 - 5$ **d.** $10.7 - (-9.8)$

    e. $\dfrac{2}{3} - \dfrac{1}{2}$ **f.** $1 - 0.06$ **g.** Subtract 7 from 4.

Solution:

Add the opposite

a. $2 - 8 = 2 + (-8) = -6$ **b.** $-8 - (-1) = -8 + (1) = -7$

c. $-11 - 5 = -11 + (-5) = -16$ **d.** $10.7 - (-9.8) = 10.7 + 9.8 = 20.5$

e. $\dfrac{2}{3} - \dfrac{1}{2} = \dfrac{2}{3} + \left(-\dfrac{1}{2}\right) = \dfrac{4}{6} + \left(-\dfrac{3}{6}\right) = \dfrac{1}{6}$

f. $1 - 0.06 = 1 + (-0.06) = 0.94$ **g.** $4 - 7 = 4 + (-7) = -3$

To add or subtract three or more real numbers, add or subtract from left to right.

EXAMPLE 4   Simplify each expression.

    a. $11 + 2 - 7$ **b.** $-5 - 4 + 2$

Solution: **a.** $11 + 2 - 7 = 13 - 7 = 6$ **b.** $-5 - 4 + 2 = -9 + 2 = -7$

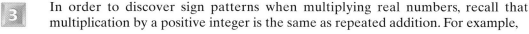

3   In order to discover sign patterns when multiplying real numbers, recall that multiplication by a positive integer is the same as repeated addition. For example,

$$3(2) = 2 + 2 + 2 = 6$$

$$3(-2) = (-2) + (-2) + (-2) = -6$$

Notice here that $3(-2) = -6$. This illustrates that the product of two numbers with different signs is negative.

To discover a sign pattern for the product of two negative numbers, study the following list of products.

$$3(-2) = -6$$
$$2(-2) = -4$$
$$1(-2) = -2$$
$$0(-2) = 0$$

Decrease by 1 each time

Increase by 2 each time

Thus, $-1(-2) = 2.$

The last line above illustrates that the product of two negative numbers is a positive number.

We can summarize sign patterns for multiplying any two real numbers as follows:

SIGN PATTERNS FOR MULTIPLYING TWO REAL NUMBERS

The product of two numbers with the same sign is positive.

The product of two numbers with different signs is negative.

Also recall the multiplication property of zero that states that the product of zero and any real number is zero.

$$0 \cdot a = a \cdot 0 = 0$$

EXAMPLE 5   Find each product.

     a. $(-8)(-1)$ **b.** $(-2)\dfrac{1}{6}$ **c.** $3(-3)$ **d.** $0(11)$ **e.** $\left(\dfrac{1}{5}\right)\left(-\dfrac{10}{11}\right)$

     f. $(7)(1)(-2)(-3)$ **g.** $8(-2)(0)$

Solution: **a.** Since the signs of the two numbers are the same, the product is positive. Thus $(-8)(-1) = +8$, or 8.

     b. Since the signs of the two numbers are different, or unlike, the product is negative. Thus $(-2)\dfrac{1}{6} = -\dfrac{2}{6} = -\dfrac{1}{3}$.

     c. $3(-3) = -9$ **d.** $0(11) = 0$ **e.** $\left(\dfrac{1}{5}\right)\left(-\dfrac{10}{11}\right) = -\dfrac{10}{55} = -\dfrac{2}{11}$

     f. To multiply three or more real numbers, multiply from left to right.

$$(7)(1)(-2)(-3) = 7(-2)(-3)$$
$$= -14(-3)$$
$$= 42$$

     g. By the multiplication property of zero 0, the product is zero.

$$(8)(-2)(0) = 0$$

The screen below shows options for checking parts (b) and (e) where fractions are involved.

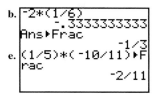

R E M I N D E R   The following sign patterns are helpful for multiplying.
1. An odd number of negative factors gives a negative product.
2. An even number of negative factors gives a positive product.

Recall that $\frac{8}{4} = 2$ because $2 \cdot 4 = 8$. Likewise, $\frac{8}{-4} = -2$ because $(-2)(-4) = 8$. Also, $\frac{-8}{4} = -2$ because $(-2)4 = -8$, and $\frac{-8}{-4} = 2$ because $2(-4) = -8$. From these examples, we can see that the sign patterns for division are the same as for multiplication.

SIGN PATTERNS FOR DIVIDING TWO REAL NUMBERS

The quotient of two numbers with the same sign is positive.
The quotient of two numbers with different signs is negative.

Division of two real numbers may be defined in terms of multiplication.

DIVIDING REAL NUMBERS

Division by a nonzero real number b is the same as multiplication by its reciprocal $\frac{1}{b}$.

$$a \div b = a \cdot \frac{1}{b}$$

In the definition above, notice that b *must* be a nonzero number. We do not define division by 0. For example, $5 \div 0$, or $\frac{5}{0}$, is undefined.

EXAMPLE 6   Find each quotient.

a. $\dfrac{20}{-4}$ b. $\dfrac{-9}{-3}$ c. $-\dfrac{3}{8} \div 3$ d. $\dfrac{-40}{10}$ e. $\dfrac{-1}{10} \div \dfrac{-2}{5}$ f. $\dfrac{8}{0}$

Solution:

a. Since the signs are different or unlike, the quotient is negative and $\dfrac{20}{-4} = -5$.

b. Since the signs are the same, the quotient is positive and $\dfrac{-9}{-3} = 3$.

c. Since division by 3 is the same as multiplication by the reciprocal,

$$-\frac{3}{8} \div 3 = -\frac{3}{8} \cdot \frac{1}{3} = -\frac{3}{24} = -\frac{1}{8}$$

d. $\dfrac{-40}{10} = -4$ **e.** $\dfrac{-1}{10} \div \dfrac{-2}{5} = -\dfrac{1}{10} \cdot -\dfrac{5}{2} = \dfrac{5}{20} = \dfrac{1}{4}$ **f.** $\dfrac{8}{0}$ is undefined.

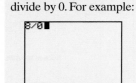
With sign rules for division, we can understand why the positioning of the negative sign in a fraction does not change the value of the fraction, as we originally stated in Section 1.1. For example,

$$\frac{-12}{3} = -4, \quad \frac{12}{-3} = -4, \quad \text{and} \quad -\frac{12}{3} = -4$$

Since all fractions equal -4, we can say that

$$\frac{-12}{3} = \frac{12}{-3} = -\frac{12}{3}$$

In general, the following holds true:

> If a and b are real numbers and $b \neq 0$, then $\dfrac{a}{-b} = \dfrac{-a}{b} = -\dfrac{a}{b}$

4 Recall that when two numbers are multiplied, they are called **factors.** For example, in $3 \cdot 5 = 15$, the 3 and 5 are called factors.

A natural number *exponent* is a shorthand notation for repeated multiplication of the same factor. This repeated factor is called the **base,** and the number of times it is used as a factor is indicated by the **exponent.** For example,

$$\text{base} \quad \overset{\frown \text{exponent}}{4^3} = \underbrace{4 \cdot 4 \cdot 4}_{\text{4 is a factor 3 times}} = 64$$

Also,

$$\text{base} \quad \overset{\frown \text{exponent}}{2^5} = \underbrace{2 \cdot 2 \cdot 2 \cdot 2 \cdot 2}_{\text{2 is a factor 5 times.}} = 32$$

Most graphing utilities use the symbol $\wedge$ for the exponent key. Some graphing utilities also have an x^2 key for squaring, and cubing may be found in a math menu. However, the $\wedge$ key may be used to raise the base to any power, even 2.

EXPONENTS

If a is a real number and n is a natural number, then the **nth power of a**, or **a raised to the nth power,** written as a^n, is the product of n factors, each of which is a.

$$\text{base} \searrow a^n = \overset{\overset{\text{exponent}}{\frown}}{\underbrace{a \cdot a \cdot a \cdot a \cdot \ldots \cdot a}_{a \text{ is a factor } n \text{ times}}}$$

It is not necessary to write 1 when it is an exponent. For example, 3^1 is simply 3.

EXAMPLE 7   Evaluate each expression.

a. 3^2 **b.** $\left(\dfrac{1}{2}\right)^4$ **c.** -5^2 **d.** $(-5)^2$ **e.** -5^3 **f.** $(-5)^3$

Solution: **a.** $3^2 = 3 \cdot 3 = 9$ **b.** $\left(\dfrac{1}{2}\right)^4 = \left(\dfrac{1}{2}\right)\left(\dfrac{1}{2}\right)\left(\dfrac{1}{2}\right)\left(\dfrac{1}{2}\right) = \dfrac{1}{16}$

c. $-5^2 = -(5 \cdot 5) = -25$ **d.** $(-5)^2 = (-5)(-5) = 25$

e. $-5^3 = -(5 \cdot 5 \cdot 5) = -125$ **f.** $(-5)^3 = (-5)(-5)(-5) = -125$

The screens below show a check using a calculator.

```
a. | 3²                    d. | (-5)²
   |              9           |              25
b. | (1/2)^4▶Frac         e. | -5³
   |           1/16          |            -125
c. | -5²                  f. | (-5)³
   |            -25          |            -125
```

> R E M I N D E R   Be very careful when evaluating expressions such as -5^2 and $(-5)^2$.
>
> $$-5^2 = -(5 \cdot 5) = -25 \quad \text{and} \quad (-5)^2 = (-5)(-5) = 25$$
>
> Without parentheses, the base to square is 5, not -5.

5   The opposite of squaring a number is taking the **square root** of a number. For example, since the square of 4, or 4^2, is 16, we say that a square root of 16 is 4. The notation $\sqrt{a}$ is used to denote the **positive**, or **principal**, **square root** of a nonnegative number a. We then have in symbols that

$$\underset{\text{sign}}{\overset{\text{radical}}{\longrightarrow}} \sqrt{16} = 4$$

EXAMPLE 8 Find the following square roots.

    a. $\sqrt{9}$ **b.** $\sqrt{25}$ **c.** $\sqrt{\dfrac{1}{4}}$

Solution: **a.** $\sqrt{9} = 3$ since 3 is positive and $3^2 = 9$

        b. $\sqrt{25} = 5$ since $5^2 = 25$.

        c. $\sqrt{\dfrac{1}{4}} = \dfrac{1}{2}$ since $\left(\dfrac{1}{2}\right)^2 = \dfrac{1}{4}$.

TECHNOLOGY NOTE

Some graphing utilities contain a square root key as well as other root options under a menu such as a math menu.

We can find roots other than square roots. Since 2 cubed, written as 2^3, is 8, we say that the **cube root** of 8 is 2. This is written as

$$\sqrt[3]{8} = 2$$

Also, since $3^4 = 81$ and 3 is positive, the **fourth root** of 81 is

$$\sqrt[4]{81} = 3$$

EXAMPLE 9 Find the following roots.

    a. $\sqrt[3]{27}$ **b.** $\sqrt[5]{32}$ **c.** $\sqrt[4]{16}$

Solution: **a.** $\sqrt[3]{27} = 3$ since $3^3 = 27$.

        b. $\sqrt[5]{32} = 2$ since $2^5 = 32$.

        c. $\sqrt[4]{16} = 2$ since 2 is positive and $2^4 = 16$.

```
a. ³√(27)
                    3
b. 5ˣ√32
                    2
c. 4ˣ√16
                    2
```

Of course, as mentioned in Section 1.1, not all roots are rational numbers. For example, $\sqrt{2} \approx 1.414$ and $\sqrt[3]{105} \approx 4.718$. We study radicals further in Chapter 8.

EXERCISE SET 1.3

Find each absolute value. See Example 1.

1. $-|2|$ **2.** $|8|$ **3.** $|-4|$ **4.** $|-6|$

5. $|0|$ **6.** $|-1|$ **7.** $-|-3|$ **8.** $-|-11|$

9. Explain why $-(-2)$ and $-|-2|$ simplify to different numbers.

10. Explain why $|a|$ is always nonnegative, even though $|a|$ is sometimes equal to $-a$. (See the boxed definition of absolute value.)

Find each sum or difference. See Examples 2 through 4.

11. $-3 + 8$ **12.** $-5 + (-9)$ **13.** $-14 + (-10)$

14. $12 + (-7)$ **15.** $-4.3 - 6.7$ **16.** $-8.2 - (-6.6)$

17. $13 - 17$ **18.** $15 - (-1)$ **19.** $\dfrac{11}{15} - \left(-\dfrac{3}{5}\right)$

20. $\dfrac{7}{10} - \dfrac{4}{5}$ **21.** $19 - 10 - 11$ **22.** $-13 - 4 + 9$

Find each product or quotient. See Examples 5 and 6.

23. $(-5)(12)$ **24.** $6(-3)$ **25.** $(-8)(-10)$

26. $7(0)$ **27.** $\dfrac{-12}{-4}$ **28.** $\dfrac{60}{-6}$

29. $\dfrac{0}{-2}$ **30.** $\dfrac{-2}{0}$ **31.** $(-4)(-2)(-1)$

32. $5(-3)(-2)$ **33.** $\dfrac{-6}{7} \div 2$ **34.** $\dfrac{-9}{13} \div (-3)$

35. $\left(-\dfrac{2}{7}\right)\left(-\dfrac{1}{6}\right)$ **36.** $\dfrac{5}{9}\left(-\dfrac{3}{5}\right)$

Evaluate. See Example 7.

37. -7^2 **38.** $(-7)^2$ **39.** $(-6)^2$

40. -6^2 **41.** $(-2)^3$ **42.** -2^3

43. Explain why -3^2 and $(-3)^2$ simplify to different numbers.

44. Explain why -3^3 and $(-3)^3$ simplify to the same number.

Find the following roots. See Examples 8 and 9.

45. $\sqrt{49}$ **46.** $\sqrt{81}$ **47.** $\sqrt{\dfrac{1}{9}}$ **48.** $\sqrt{\dfrac{1}{25}}$

49. $\sqrt[3]{64}$ **50.** $\sqrt[5]{32}$ **51.** $\sqrt[4]{81}$ **52.** $\sqrt[3]{1}$

Perform the indicated operation.

53. $-4 + 7$ **54.** $-9 + 15$

55. $-9 + (-3)$ **56.** $-17 + (-2)$

57. $-4 - (-19)$ **58.** $-5 - (-17)$

59. $6.3 - 18.5$ **60.** $15.9 - 21.7$

61. $(-4)(-7)(0)$ **62.** $(-9)(0)(-14)$

63. $\left(-\dfrac{2}{3}\right)\left(\dfrac{6}{4}\right)$ **64.** $\left(\dfrac{5}{6}\right)\left(\dfrac{-12}{15}\right)$

65. $-14 - 7$ **66.** $-6 - 31$

67. $-\dfrac{4}{5} - \left(-\dfrac{3}{10}\right)$ **68.** $-\dfrac{5}{2} - \left(-\dfrac{2}{3}\right)$

69. Subtract 14 from 8.   **70.** Subtract 9 from -3.

71. $-\dfrac{34}{2}$ **72.** $\dfrac{48}{-3}$

73. $16 - 8 - 9$ **74.** $-14 - 3 + 6$

75. $\sqrt[3]{8}$ **76.** $\sqrt{36}$

77. $\sqrt{100}$ **78.** $\sqrt[3]{125}$

79. $-\dfrac{1}{6} \div \dfrac{9}{10}$ **80.** $\dfrac{4}{7} \div \left(-\dfrac{1}{8}\right)$

Each circle below represents a whole, or 1. Determine the unknown fractional part of each circle.

81.

82.

83. Most of Mount Kea, a volcano in Hawaii, lies below sea level. If this volcano begins at 5998 meters below sea level and then rises 10,203 meters, find the height of the volcano above sea level.

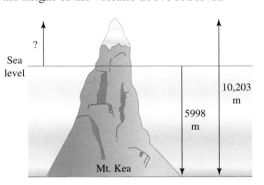

84. The highest point on land on Earth is the top of Mt. Everest, in the Himalayas, at an elevation of 29,028 feet above sea level. The lowest point on land is in the Dead Sea valley, between Israel and Jordan, at 1319 feet below sea level. Find the difference in elevations.

Approximate each square root. Round to four decimal places.

85. $\sqrt{10}$ **86.** $\sqrt{273}$ **87.** $\sqrt{7.9}$ **88.** $\sqrt{19.6}$

Investment firms often advertise their gains and losses in the form of bar graphs such as the one that follows. This graph shows investment risk over time for the S&P 500 Index by showing average annual compound returns for 1 year, 5 years, 15 years, and 25 years. For example, after one year, the annual compound return in percent for an investor is anywhere from a gain of 181.5% to a loss of 64%. Use this graph to answer the questions that follow.

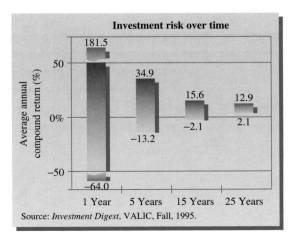

89. A person investing in the S&P 500 Index 15 years ago had at most an average annual gain of what percent?

90. A person investing in the S&P 500 Index 5 years ago lost at most what percent per year on average?

91. Find the difference in the percents of the highest average annual return and the lowest average annual return after 15 years.

92. Find the difference in the percents of the highest average annual return and the lowest average annual return after 25 years.

93. Do you think that the type of investment shown in the figure on page 27 is recommended for short-term investments or long-term investments? Explain your answer.

94. GROUP ACTIVITY. Your group is given $1000 in hypothetical money to purchase shares of two stocks. As a group, agree on the stock(s) to buy and record how many shares of each stock you are able to buy. Assume that there is no commission charged by the stock broker. Keep track of the stock(s) for the next 3 weeks, recording how much they have gained or lost at the end of each week. After 3 weeks, turn in a progress report on the stock(s) and how much money your group has gained or lost.

A fair game is one in which each team or player has the same chance of winning. Suppose that a game consists of three players taking turns spinning a spinner. If the spinner lands on yellow, player 1 gets a point. If the spinner lands on red, player 2 gets a point, and if the spinner lands on blue, player 3 gets a point. After 12 spins, the player with the most points wins.

a.  b. c.  d.

95. Which spinner would provide for a fair game?

96. If you are player 2 and want to win the game, which spinner would you choose?

97. If you are player 1 and want to lose the game, which spinner would you choose?

98. Is it possible for the game to end in a three-way tie? If so, list the possible ending scores.

99. Is it possible for the game to end in a two-way tie? If so, list the possible ending scores.

STOCK NAME	NO. OF SHARES	PURCHASE PRICE	WEEK 1	WEEK 2	WEEK 3
VALUE OF STOCK					

1.4 ORDER OF OPERATIONS AND ALGEBRAIC EXPRESSIONS

TAPE IAG 1.4

O B J E C T I V E S

1. Use the order of operations.
2. Identify and evaluate algebraic expressions.
3. Write word phrases as algebraic expressions.
4. Identify like terms and simplify algebraic expressions.

1 The expression $3 + 2 \cdot 10$ describes the total number of disks shown.

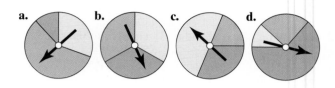

Expressions containing more than one operation are written to follow a particular agreed-upon **order of operations.**

ORDER OF OPERATIONS

Simplify expressions using the following order. If grouping symbols such as parentheses are present, simplify expressions within those first, starting with the innermost set. If fraction bars are present, simplify the numerator and denominator separately.

1. Raise to powers or take roots in order from left to right.

2. Multiply or divide in order from left to right.

3. Add or subtract in order from left to right.

For example, when we write $3 + 2 \cdot 10$, we mean to multiply first and then add.

EXAMPLE 1   Simplify.

a. $3 + 2 \cdot 10$ **b.** $2(1 - 4)^2$ **c.** $\dfrac{|-2|^3 + 1}{-7 - \sqrt{4}}$ **d.** $\dfrac{(6 + 2) - (-4)}{2 - (-3)}$

Solution: **a.** First multiply, then add.

$$3 + 2 \cdot 10 = 3 + 20 = 23$$

b. $2(1 - 4)^2 = 2(-3)^2$ Simplify inside grouping symbols first.

$ = 2(9)$ Write $(-3)^2$ as 9.

$ = 18$ Multiply.

c. Simplify the numerator and the denominator separately, then divide.

$$\frac{|-2|^3 + 1}{-7 - \sqrt{4}} = \frac{2^3 + 1}{-7 - 2}$$ Write $|-2|$ as 2 and $\sqrt{4}$ as 2.

$$= \frac{8 + 1}{-9}$$ Write 2^3 as 8 and simplify the denominator.

$$= \frac{9}{-9} = -1$$ Simplify the numerator; then divide.

d. $\dfrac{(6 + 2) - (-4)}{2 - (-3)} = \dfrac{8 - (-4)}{2 - (-3)}$ Simplify inside grouping symbols first.

$$= \frac{8 + 4}{2 + 3}$$ Write subtraction as equivalent addition.

$$= \frac{12}{5}$$ Add to simplify both numerator and denominator.

TECHNOLOGY NOTE

To evaluate expressions with a calculator, we sometimes need to insert parentheses that may not be shown in the expression. This is especially true when entering a fraction whose numerator or denominator contains more than one term.

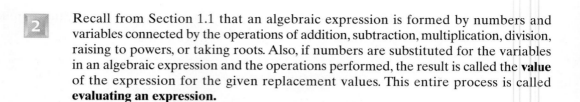

Recall from Section 1.1 that an algebraic expression is formed by numbers and variables connected by the operations of addition, subtraction, multiplication, division, raising to powers, or taking roots. Also, if numbers are substituted for the variables in an algebraic expression and the operations performed, the result is called the **value** of the expression for the given replacement values. This entire process is called **evaluating an expression.**

EXAMPLE 2   Evaluate each algebraic expression when $x = 2$, $y = -1$, and $z = -3$.

a. $|z - y|$ **b.** z^2 **c.** $\dfrac{2x + y}{z}$ **d.** $-x^2 - 4x$

Solution: **a.** $|z - y| = |-3 - (-1)| = |-3 + 1| = |-2| = 2$ **b.** $z^2 = (-3)^2 = 9$

c. $\dfrac{2x + y}{z} = \dfrac{2(2) + (-1)}{-3} = \dfrac{4 + (-1)}{-3} = \dfrac{3}{-3} = -1$

d. $-x^2 - 4x = -2^2 - 4(2) = -4 - 8 = -12$

How can we use a calculator to evaluate the expressions in Example 2? One way to evaluate Example 2d with a calculator is to just enter the numerical expression and evaluate.

Another way to evaluate $-x^2 - 4x$ when $x = 2$ is to store the value 2 in the variable x and then enter and evaluate the algebraic expression.

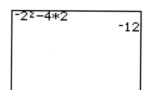

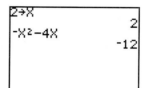

The value of $-x^2 - 4x$ when $x = 2$ is -12.

Example 3 shows how we can use a calculator to evaluate different expressions for the same values of the variables.

EXAMPLE 3   Use a calculator to evaluate each expression when $x = -7.1$ and $y = 2.9$.

a. $x + 3y$ **b.** $2x - 5y$ **c.** $3y^2$

Solution: Store -7.1 in x and 2.9 in y and evaluate the first expression. Since the values for variables remain stored in the respective x and y memory locations, we can simply continue to evaluate the other expressions.

From the previous screen, we have that when $x = -7.1$ and $y = 2.9$,

a. $x + 3y$ evaluates to 1.6

b. $2x - 5y$ evaluates to -28.7

c. $3y^2$ evaluates to 25.23

Check these values

Thus far, we have seen that we can evaluate an expression or perform a calculation mentally, using paper and pencil, or using a calculator. In general, how do we decide which method to use?

The following example shows that all three methods are useful.

EXAMPLE 4   Evaluate the expression $3x^2 + 4$ for the following values of x:

      a. $x = 2$ **b.** $x = 12$ **c.** $x = 3.91$

Solution: **a.** Never underestimate the power and speed of mental calculation. Replace x with 2 and mentally calculate x^2 to be 4, multiply by 3 to get 12, and add 4 to get 16. The value of $3x^2 + 4$ when $x = 2$ is 16.

      b. Replace x with 12 and mentally calculate the square of 12 to be 144. Next, you might use paper and pencil to evaluate $3 \cdot 144$, which equals 432, and then add 4 to get 436. The value of of $3x^2 + 4$ when $x = 12$ is 436.

      c. Since the value of x is a decimal, using a calculator is the most appropriate choice. First, let's mentally approximate the value of the expression. $3.91 \approx 4$, and the value of $3x^2 + 4$ at 4 is 52. The screen at the left shows that the value of $3x^2 + 4$ when $x = 3.91$ is 49.8643. This is close to our approximation of 52 and thus is reasonable.

```
3*3.91²+4
         49.8643
```

EXAMPLE 5   RELATING TEMPERATURES

The algebraic expression $\dfrac{5}{9}(F - 32)$ represents the equivalent temperature in degrees Celsius when x is degrees Fahrenheit. Complete the following table by evaluating this expression at given values of x.

DEGREES FAHRENHEIT	F	-4	10	32
DEGREES CELSIUS	$\dfrac{5}{9}(F - 32)$			

Solution: To complete the table, evaluate $\frac{5}{9}(F - 32)$ at each given replacement value.

When $F = -4$,

$$\frac{5}{9}(F - 32) = \frac{5}{9}(-4 - 32) = \frac{5}{9}(-36) = -20$$

When $F = 10$,

$$\frac{5}{9}(F - 32) = \frac{5}{9}(10 - 32) = \frac{5}{9}(-22) = \frac{-110}{9} = -12\frac{2}{9}$$

When $F = 32$,

$$\frac{5}{9}(F - 32) = \frac{5}{9}(32 - 32) = \frac{5}{9}(0) = 0$$

The completed table is

DEGREES FAHRENHEIT	F	-4	10	32
DEGREES CELSIUS	$\frac{5}{9}(F-32)$	-20	$-12\frac{2}{9}$	0

Thus, $-4°F$ is equivalent to $-20°C$, $10°F$ is equivalent to $-12\frac{2}{9}°C$, and $32°F$ is equivalent to $0°C$.

Algebric expressions occur often during problem solving. For example, suppose that a television commercial for a watch is being filmed on the Golden Gate Bridge. A portion of this commercial consists of dropping a watch from the bridge. In order to determine the best camera angles and also whether the watch will survive the fall, it is important to know the speed of the watch at 1-second intervals. The algebraic expression

$$32t$$

gives the speed of the watch in feet per second for time t.

To find the speed of the watch at 1 second, for example, we replace the variable t with 1 and perform the indicated multiplication.

When $t = 1$ second, $32t = 32 \cdot 1 = 32$ feet per second, the speed of the watch.

When $t = 2$ seconds, $32t = 32 \cdot 2 = 64$ feet per second, the speed of the watch.

When $t = 3$ seconds, $32t = 32 \cdot 3 = 96$ feet per second, the speed of the watch.

EXAMPLE 6   SPEED OF A FALLING OBJECT

Complete the following table to find the number of feet per second at 1-second intervals for the problem above.

TIME $= t$ (SECONDS)	1	2	3	4	5	6
SPEED $= 32t$ (FEET PER SECOND)	32	64	96			

Solution: When $t = 4, 32t = 32 \cdot 4 = 128$.

When $t = 5, 32t = 32 \cdot 5 = 160$.

When $t = 6, 32t = 32 \cdot 6 = 192$.

Our table can now be completed as follows:

```
32*4
         128
32*5
         160
32*6
         192
```

TIME = t (SECONDS)	1	2	3	4	5	6
SPEED = 32t (FEET PER SECOND)	32	64	96	128	160	192

3   As mentioned in section 1.1, an important step in problem solving is writing algebraic expressions from word phrases. Sometimes this involves a direct translation, but often an indicated operation is implied rather than being directly stated.

EXAMPLE 7   Write each as an algebraic expression.

a. A vending machine contains x quarters. Write an expression for the *value in dollars* of the quarters.

b. The number of grams of fat in x slices of bread if each piece of bread contains 2 grams of fat.

c. The cost of x desks if each desk costs $156.

d. Sales tax on a purchase of x dollars if the tax rate is 9%.

Solution: Each of these examples implies finding a product.

a. The value in dollars of the quarters is found by multiplying the value of a quarter (0.25 dollar) by the number of quarters.

In words: Value of quarter   ·   Number of quarters

Translate: 0.25   · x,   or $0.25x$

b. In words: Number of grams of fat in one slice of bread   ·   Number of slices of bread

Translate: 2   · x,   or $2x$

c. In words: Cost of a desk   ·   Number of desks

Translate: 156   · x,   or $156x$

d. In words: Sales tax rate   ·   Purchase price

Translate: 0.09   · x,   or $0.09x$ Here, we wrote 9% as a decimal, 0.09.

Two or more unknown numbers in a problem may sometimes be related. If so, try letting a variable represent one unknown number, and then represent the other unknown number or numbers as expressions containing the same variable.

EXAMPLE 8　Write each as an algebraic expression.

　　a. Two numbers have a sum of 20. If one number is x, represent the other number as an expression in x.

　　b. The older sister is 8 years older than her younger sister. If the age of the younger sister is x, represent the age of the older sister as an expression in x.

　　c. Two angles are complementary if the sum of their measures is 90°. If the measure of one angle is x degrees, represent the measure of the other angle as an expression in x.

　　d. If x is the first of two consecutive integers, represent the second integer as an expression in x.

　　e. A 2-foot section is to be cut from a board whose length is x feet. Express the length of the remaining board as an algebraic expression in x.

Solution:　**a.** If two numbers have a sum of 20 and one number is x, the other number is "the rest of 20."

In words:	Twenty	minus	x
Translate:	20	−	x

b. The older sister's age is

In words:	Eight years	added to	younger sister's age
Translate:	8	+	x

c.

In words:	Ninety	minus	x
Translate:	90	−	x

d. The next consecutive integer is always one more than the previous integer.

In words:	The first integer	plus	one
Translate:	x	+	1

e.

In words:	Total board length	less	2
Translate:	x	−	2

4

Often, an expression may be **simplified** by removing grouping symbols and combining any like terms. The **terms** of an expression are the *addends* of the expression. For example, in the expression $3x^2 + 4x$, the terms are $3x^2$ and $4x$.

EXPRESSION	TERMS
$-2x + y$	$-2x, y$
$3x^2 - \dfrac{y}{5} + 7$	$3x^2, -\dfrac{y}{5}, 7$

Terms with the same variable(s) raised to the same power are called **like terms**. We can add or subtract like terms by using the distributive property. This process is called **combining like terms.**

EXAMPLE 9   Use the distributive property to simplify each expression.

a. $3x - 5x + 4$ **b.** $7yz + yz$ **c.** $4z + 6.1$

Solution: **a.** $3x - 5x + 4 = (3 - 5)x + 4$ Apply the distributive property.
$= -2x + 4$

b. $7zy + yz = (7 + 1)yz = 8yz$

c. $4z + 6.1$ cannot be simplified further since $4z$ and 6.1 are not like terms.

Recall that the distributive property can also be used to multiply. For example,

$$-2(x + 3) = -2(x) + (-2)(3) = -2x - 6$$

The associative and commutative properties are sometimes used to rearrange and group like terms when simplifying expressions.

$$-7x^2 + 5 + 3x^2 - 2 = -7x^2 + 3x^2 + 5 - 2$$
$$= (-7 + 3)x^2 + (5 - 2)$$
$$= -4x^2 + 3$$

EXAMPLE 10   Simplify each expression.

a. $3xy - 2xy + 5 - 7 + xy$ **b.** $7x^2 + 3 - 5(x^2 - 4)$
c. $(2.1x - 5.6) - (-x - 5.3)$

Solution: **a.** $3xy - 2xy + 5 - 7 + xy = 3xy - 2xy + xy + 5 - 7$ Apply the commutative property.
$= (3 - 2 + 1)xy + (5 - 7)$ Apply the distributive property.
$= 2xy - 2$ Simplify.

b. $7x^2 + 3 - 5(x^2 - 4) = 7x^2 + 3 - 5x^2 + 20$ Apply the distributive property.

$$= 2x^2 + 23$$ Simplify.

c. Think of $-(-x - 5.3)$ as $-1(-x - 5.3)$ and use the distributive property.

$$(2.1x - 5.6) - 1(-x - 5.3) = 2.1x - 5.6 + 1x + 5.3$$

$$= 3.1x - 0.3$$ Combine like terms.

EXERCISE SET 1.4

Simplify each expression. Round Exercises 21 and 22 to the nearest thousandth. See Example 1.

1. $3(5 - 7)^4$

2. $7(3 - 8)^2$

3. $-3^2 + 2^3$

4. $-5^2 - 2^4$

5. $\dfrac{3 - (-12)}{-5}$

6. $\dfrac{-4 - (-8)}{-4}$

7. $|3.6 - 7.2| + |3.6 + 7.2|$

8. $|8.6 - 1.9| - |2.1 + 5.3|$

9. $-5^2 - 2[7^2 - 3(4 - 6)]$

10. $1.2[(3 - 6) - (24 - 27)]^2$

11. $\dfrac{(3 - \sqrt{9}) - (-5 - 1.3)}{-3}$

12. $\dfrac{-\sqrt{16} - (6 - 2.4)}{-2}$

13. $\dfrac{|3 - 9| - |-5|}{-3}$

14. $\dfrac{|-14| - |2 - 7|}{-15}$

15. $(-3)^2 + 2^3$

16. $(-15)^2 - 2^4$

17. $\dfrac{3(-2 + 1)}{5} - \dfrac{-7(2 - 4)}{1 - (-2)}$

18. $\dfrac{-1 - 2}{2(-3) + 10} - \dfrac{2(-5)}{-1(8) + 1}$

19. $\dfrac{\left(\dfrac{-3}{10}\right)}{\left(\dfrac{42}{50}\right)}$

20. $\dfrac{\left(\dfrac{-5}{21}\right)}{\left(\dfrac{-6}{42}\right)}$

21. $\dfrac{-1.682 - 17.895}{(-7.102)(-4.691)}$

22. $\dfrac{(-5.161)(3.222)}{7.955 - 19.676}$

Find the value of each expression when $x = -2$, $y = -5$, and $z = 3$. See Examples 2 through 4.

23. $x^2 + z^2$

24. $y^2 - z^2$

25. $-5(-x + 3y)$

26. $-7(-y - 4z)$

27. $\dfrac{3z - y}{2x - z}$

28. $\dfrac{5x - z}{-2y + z}$

Find the value of each expression when $x = 1.4$ and $y = -6.2$. If necessary, round the result to the nearest hundredth.

29. $3x - 2y$

30. $5y^2 + x$

31. $\dfrac{|x - y|}{2y}$

32. $\dfrac{\sqrt{x - y}}{2x}$

33. $-3(x^2 + y^2)$

34. $1.6(3x^2 + 1.8)$

State the expression being evaluated, the values of the variables, and the value of the expression.

35.
```
-2.7→X:0.9→Y
              .9
X²+2Y
            9.09
```

36.
```
5.1→X: -3→Y
              -3
2X-Y²
            1.2
```

37.
```
4→A: -2.5→B
            -2.5
A²+B²
          22.25
```

38.
```
8→B:4.5→H
            4.5
(1/2)BH
            18
```

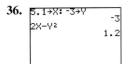 **39.** Consider the expressions $(1/2)x$ and $1/2x$.

   a. Use your calculator to store 2 in x, and then enter and evaluate each expression.

   b. Are the results of part (a) the same or different? Explain why.

40. Consider the expression $12/(x - 2)$ and $12/x - 2$

   a. Use your calculator to store 4 in x, and then enter and evaluate each expression.

   b. Are the results of part (a) the same or different? Explain why.

Solve. See Examples 5 and 6.

41. The algebraic expression $8 + 2y$ represents the perimeter of a rectangle with width 4 and length y.

a. Complete the table that follows by evaluating $8 + 2y$ for given lengths y.

LENGTH	y	5	7	10	100
PERIMETER	$8 + 2y$				

b. Use the results of the table in (a) to answer the following question. As the width of a rectangle remains the same and the length increases, does the perimeter increase or decrease? Explain how you arrived at your answer.

42. The algebraic expression πr^2 represents the area of a circle with radius r.

a. Complete the table below by evaluating πr^2 for given radii r. Round to 2 decimal places.

RADIUS	r	2	3	7	10
AREA	πr^2				

b. As the radius of a circle increases, does its area increase or decrease? Explain your answer.

43. The algebraic expression $\dfrac{100x + 5000}{x}$ represents the cost per bookshelf (in dollars) of producing x bookshelves.

a. Complete the table below.

NUMBER OF BOOKSHELVES	x	10	100	1000
COST PER BOOKSHELF	$\dfrac{100x + 5000}{x}$			

b. As the number of bookshelves increases, does the cost per bookshelf increase or decrease? Why do you think that this is so?

44. If c is temperature in degrees Celsius, the algebraic expression $1.8c + 32$ represents the equivalent temperature in degrees Fahrenheit.

a. Complete the table below.

DEGREES CELSIUS	c	-10	0	50
DEGREES FAHRENHEIT	$1.8c + 32$			

b. As degrees Celsius increase, do equivalent degrees Fahrenheit increase or decrease?

45. The area of a triangle is $\frac{1}{2}$ BH where B is the base length and H is the height of the triangle. Complete the table for the area of a triangle whose base is 8 inches.

BASE (INCHES)	8	8	8	8
HEIGHT (INCHES)	3	6	12	24
AREA = $\frac{1}{2}$BH (SQUARE INCHES)				

46. Complete the table for the area of a triangle whose height is 6 inches.

BASE (INCHES)	8	6.5	2.3	1.9
HEIGHT (INCHES)	6	6	6	6
AREA = $\frac{1}{2}$BH (SQUARE INCHES)				

47. The B747-400 aircraft costs $7098 dollars per hour to operate. The algebraic expression

$$7098t$$

gives the total cost to operate the aircraft for t hours. Find the total cost to operate the B747-400 for 5.2 hours.

48. Flying the SR-71A jet, Capt. Elden W. Joersz, USAF, set a record speed of 2193.16 miles per hours. At this speed, the algebraic expression $2193.16t$ gives the total distance flown in t hours. Find the distance flown by the SR-71A in 1.7 hours.

Write each value as an algebraic expression. See Examples 7 and 8.

49. The value (in dollars) of n nickels.

50. The value (in dollars) of d dimes.

51. The value (in dollars) of n nickels and d dimes.

52. The value (in dollars) of q quarters and d dimes.

53. The cost of x compact disks if each compact disk costs \$6.49.

54. The cost of y books if each book costs \$35.61.

55. Write an expression for the tax on p dollars if the tax rate is 8.25%.

56. Write an expression for the tax on \$45 if the tax rate is r%.

57. Two numbers have a sum of 25. If one number is x, represent the other number as an expression in x.

58. Two numbers have a sum of 112. If one number is x, represent the other number as an expression in x.

59. Two angles are supplementary if the sum of their measures is 180°. If the measure of one angle is x degrees, represent the measure of the other angle as an expression in x.

60. If the measure of an angle is $5x$ degrees, represent the measure of its complement as an expression in x.

61. If x is an odd integer, represent the next odd integer as an expression in x.

62. If $2x$ is an even integer, represent the next even integer as an expression in x.

63. The cost of renting a car from a local agency is \$31 a day and \$0.29 a mile. Write an expression for the total cost for renting a car for one day and driving it x miles.

64. Employees at Word Processors, Inc., are paid an hourly wage of \$9.25 plus \$0.60 for each page typed per hour. Write an expression for the total hourly wage of an employee who types x pages per hour.

Simplify each expression. See Examples 9 and 10.

65. $-9x + 4x + 18 - 10x$ **66.** $5y - 14 + 7y - 20y$

67. $5k - (3k - 10)$ **68.** $-11c - (4 - 2c)$

69. $(3x + 4) - (6x - 1)$ **70.** $(8 - 5y) - (4 + 3y)$

71. $3(x - 2) + x + 15$ **72.** $-4(y + 3) - 7y + 1$

73. $-(n + 5) + (5n - 3)$ **74.** $-(8 - t) + (2t - 6)$

75. $4(6n - 3) - 3(8n + 4)$ **76.** $5(2z - 6) + 10(3 - z)$

77. $3x - 2(x - 5) + x$ **78.** $7n + 3(2n - 6) - 2$

79. $-1.2(5.7x - 3.6) + 8.75x$

80. $5.8(-9.6 - 31.2y) - 18.65$

81. $8.1z + 7.3(z + 5.2) - 6.85$

82. $6.5y - 4.4(1.8x - 3.3) + 10.95$

The following graph is called a broken line graph, or simply a line graph. This particular graph shows past, present, and predicted future population over age 65. Just as with a bar graph, to find the population over 65 for a particular year, read the height of the point for that year. To read the height, follow the point horizontally to the left until you reach the vertical axis.

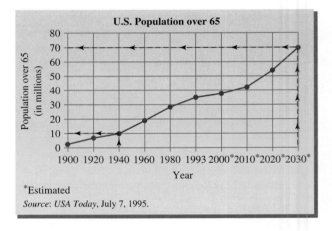

*Estimated
Source: USA Today, July 7, 1995.

83. Estimate the population over 65 in the year 1940.

84. Estimate the predicted population over 65 in the year 2030.

85. Estimate the predicted population over 65 in the year 2010.

86. Estimate the population over 65 in the year 1993.

87. Is the population over 65 increasing as time passes or decreasing? Explain how you arrived at your answer.

88. The percent of Americans over 65 approximately tripled from 1900 to 1993. If this percent in 1900 was 4.1%, estimate the percent of Americans over 65 in the year 1993.

GROUP ACTIVITY

ANALYZING NEWSPAPER CIRCULATION

Use the following data to answer Questions 1 through 6 about newspaper circulation in the New York City metropolitan area.

NEWSPAPER CIRCULATION

NEWSPAPER	1993 DAILY CIRCULATION	1994 DAILY CIRCULATION	DAILY EDITION NEWSSTAND PRICE
Wall Street Journal	1,818,562	1,780,442	75¢
New York Times	1,141,366	1,114,905	60¢
New York Daily News	764,070	753,024	50¢
New York Newsday	747,890	693,556	50¢
New York Post	394,431	405,318	50¢

SOURCES: *1994 Editor & Publisher International Yearbook* and *1995 Editor & Publisher International Yearbook.*

1. Did any newspaper gain circulation in 1994 from 1993? If so, which one(s) and by how much?

2. Construct a bar graph showing the change in circulation from 1993 to 1994 for each newspaper. Which newspaper experienced the largest change in circulation?

3. What was the total circulation of these New York-published newspapers in 1993 and in 1994? Did total circulation increase or decrease from 1993 to 1994? By how much?

4. Discuss factors that may have contributed to the overall change in daily newspaper circulation.

5. The population of New York City is approximately 7,325,000. The population of the New York City metropolitan area is approximately 16,271,000. Find the numbers of newspapers sold per person in 1994 for New York City and for the New York City metropolitan area. Which of these figures do you think is more meaningful? Why? Why might neither of these figures be capable of describing the full circulation situation? (*Sources:* U.S. Bureau of the Census and the United Nations.)

6. Assuming that each copy was sold from a newsstand in New York City metropolitan area, use the daily edition newspaper prices given in the table to find the total amount spent each day on these New York-published newspapers in 1994. Estimate the total amount spent annually on these daily newspapers in 1994. (Assume that daily editions are published only on the five weekdays.)

CHAPTER 1   HIGHLIGHTS

DEFINITIONS AND CONCEPTS	EXAMPLES

SECTION 1.1   REAL NUMBERS

Letters that represent numbers are called **variables**.

Examples of variables are

$$x, a, m, y$$

An **algebraic expression** is formed by numbers and variables connected by the operations of addition, subtraction, multiplication, division, raising to powers, or taking roots.

Examples of algebraic expressions are

$$7y, -3, \frac{x^2 - 9}{-2} + 14x, \sqrt{3} + \sqrt{m}$$

Natural numbers: $\{1, 2, 3, \ldots\}$

Whole numbers: $\{0, 1, 2, 3, \ldots\}$

Integers: $\{\ldots, -3, -2, -1, 0, 1, 2, 3, \ldots\}$

Each listing of three dots above is called an **ellipsis**.

The members of a set are called its **elements**.

Set builder notation describes the elements of a set but does not list them.

Real numbers: $\{x \mid x$ corresponds to a point on a number line$\}$

Rational numbers: $\{\frac{a}{b} \mid a$ and b are integers and $b \neq 0\}$

Irrational numbers: $\{x \mid x$ is a real number and x is not a rational number$\}$

If all the elements of set A are also in set B, we say that set A is a **subset** of set B, and we write $A \subseteq B$.

Given the set $\{-9.6, -5, -\sqrt{2}, 0, \frac{2}{5}, 101\}$, list the elements that belong to the set of

Natural numbers	101
Whole numbers	$0, 101$
Integers	$-5, 0, 101$
Real numbers	$-9.6, -5, -\sqrt{2}, 0, \frac{2}{5}, 101$
Rational numbers	$-9.6, -5, 0, \frac{2}{5}, 101$
Irrational numbers	$-\sqrt{2}$

List the elements in the set $\{x \mid x$ is an integer between -2 and $5\}$.

$$\{-1, 0, 1, 2, 3, 4\}$$

$$\{1, 2, 4\} \subseteq \{1, 2, 3, 4\}.$$

SECTION 1.2   PROPERTIES OF REAL NUMBERS

Symbols:
= is equal to
≠ is not equal to
> is greater than
< is less than
≥ is greater than or equal to
≤ is less than or equal to

$$-5 = -5$$
$$-5 \neq -3$$
$$1.7 > 1.2$$
$$-1.7 < 1.2$$
$$\frac{5}{3} \geq \frac{5}{3}$$
$$-\frac{1}{2} \leq \frac{1}{2}$$

Identity:

$a + 0 = a$ $\quad 0 + a = a$

$a \cdot 1 = a$ $\quad 1 \cdot a = a$

$3 + 0 = 3$ $\qquad 0 + 3 = 3$

$-1.8 \cdot 1 = -1.8$ $\qquad 1 \cdot -1.8 = -1.8$

(continued)

DEFINITIONS AND CONCEPTS	**EXAMPLES**

SECTION 1.2 PROPERTIES OF REAL NUMBERS	

Inverse:

$$a + (-a) = 0 \qquad -a + a = 0$$

$$a \cdot \frac{1}{a} = 1 \qquad \frac{1}{a} \cdot a = 1$$

$$7 + (-7) = 0 \qquad -7 + 7 = 0$$

$$5 \cdot \frac{1}{5} = 1 \qquad \frac{1}{5} \cdot 5 = 1$$

Commutative:

$$a + b = b + a$$

$$a \cdot b = b \cdot a$$

$$x + 7 = 7 + x$$

$$9 \cdot y = y \cdot 9$$

Associative:

$$(a + b) + c = a + (b + c)$$

$$(a \cdot b) \cdot c = a \cdot (b \cdot c)$$

$$(3 + 1) + 10 = 3 + (1 + 10)$$

$$(3 \cdot 1) \cdot 10 = 3 \cdot (1 \cdot 10)$$

Distributive:

$$a(b + c) = ab + ac$$

$$6(x + 5) = 6 \cdot x + 6 \cdot 5$$

$$= 6x + 30$$

Interval notation may be used to write sets of real numbers.

In interval notation, the set $\{x \mid x \geq -2\}$ is written as $[-2, \infty)$.

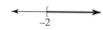

$$\{x \mid x \geq -2\} \quad \text{or} \quad [-2, \infty)$$

SECTION 1.3 OPERATIONS ON REAL NUMBERS	

Absolute value:

$$|a| = \begin{cases} a \text{ if } a \text{ is } 0 \text{ or a positive number} \\ -a \text{ if } a \text{ is a negative number} \end{cases}$$

$$|3| = 3, \ |0| = 0, \ |-7.2| = 7.2$$

```
abs(3)
                    3
abs(0)
                    0
abs(-7.2)
                  7.2
```

Adding real numbers:

1. To add two numbers with different signs, subtract the smaller absolute value from the larger absolute value and attach the sign of the number with the larger absolute value.

2. To add two numbers with the same sign, add their absolute values and attach their common sign.

$$20.8 + (-10.2) = 10.6$$

$$-18 + 6 = -12$$

$$\frac{2}{7} + \frac{1}{7} = \frac{3}{7}$$

$$-5 + (-2.6) = -7.6$$

```
-18+6
                  -12
2/7+1/7▶Frac
                  3/7
-5+(-2.6)
                 -7.6
```

(continued)

DEFINITIONS AND CONCEPTS	**EXAMPLES**

SECTION 1.3 OPERATIONS ON REAL NUMBERS

Subtracting real numbers:

$a - b = a + (-b)$

$18 - 21 = 18 + (-21) = -3$

Multiplying and dividing real numbers:

The product or quotient of two numbers with the same sign is positive.

$(-8)(-4) = 32$ $\qquad \dfrac{-8}{-4} = 2$

$8 \cdot 4 = 32$ $\qquad \dfrac{8}{4} = 2$

The product or quotient of two numbers with different signs is negative.

$-17 \cdot 2 = -34$ $\qquad \dfrac{-14}{2} = -7$

$4(-1.6) = -6.4$ $\qquad \dfrac{22}{-2} = -11$

A natural number **exponent** is a shorthand notation for repeated multiplication of the same factor.

The notation $\sqrt{a}$ is used to denote the **positive,** or **principal, square root** of a nonnegative number a.

$\sqrt{a} = b$ if $b^2 = a$ and b is positive

$3^4 = 3 \cdot 3 \cdot 3 \cdot 3 = 81$

$\sqrt{49} = 7$

```
3^4
             81
√(49)
              7
³√(64)
              4
```

Also,

$\sqrt[3]{a} = b$ if $b^3 = a$

$\sqrt[4]{a} = b$ if $b^4 = a$ and b is positive

$\sqrt[3]{64} = 4$

$\sqrt[4]{16} = 2$

```
4*√16
              2
```

SECTION 1.4 ORDER OF OPERATIONS AND ALGEBRAIC EXPRESSIONS

Order of Operations

Simplify expressions using the order that follows. If grouping symbols such as parentheses are present, simplify expressions within those first, starting with the innermost set. If fraction bars are present, simplify the numerator and denominator separately.

1. Raise to powers or take roots in order from left to right.

2. Multiply or divide in order from left to right.

3. Add or subtract in order from left to right.

Simplify $\dfrac{42 - 2(3^2 - \sqrt{16})}{-8}$.

$$\dfrac{42 - 2(3^2 - \sqrt{16})}{-8} = \dfrac{42 - 2(9 - 4)}{-8}$$

$$= \dfrac{42 - 2(5)}{-8}$$

$$= \dfrac{42 - 10}{-8}$$

$$= \dfrac{32}{-8} = -4$$

To **evaluate** an algebraic expression containing variables, substitute the given numbers for the variables and simplify. The result is called the **value** of the expression.

Evaluate $-2.7x$ if $x = 3$.

$-2.7x = -2.7(3)$

$= -8.1$

```
3→X: -2.7X
            -8.1
```

CHAPTER 1 REVIEW

(1.1) *Write each phrase as an algebraic expression. Use the variable x to represent each unknown number.*

1. Three times a number

2. Thirteen less than a number

3. Twice a number, decreased by one

List the elements in each set.

4. $\{x \mid x$ is an odd integer between -2 and $4\}$

5. $\{x \mid x$ is an even integer between -3 and $7\}$

6. $\{x \mid x$ is a negative whole number$\}$

7. $\{x \mid x$ is a natural number that is not a rational number$\}$

8. $\{x \mid x$ is a whole number greater than $5\}$

9. $\{x \mid x$ is an integer less than $3\}$

Determine whether each statement is true or false if $A = \{6, 10, 12\}$, $B = \{5, 9, 11\}$, $C = \{\ldots, -3, -2, -1, 0, 1, 2, 3, \ldots\}$, $D = \{2, 4, 6, \ldots 16\}$, $E = \{x \mid x$ is a rational number$\}$, $F = \{\ \}$, $G = \{x \mid x$ is an irrational number$\}$, and $H = \{x \mid x$ is a real number$\}$.

10. $10 \in D$ **11.** $B \in 9$ **12.** $\sqrt{169} \notin G$

13. $0 \notin F$ **14.** $\pi \in E$ **15.** $\pi \in H$

16. $\sqrt{4} \in G$ **17.** $-9 \in E$ **18.** $A \subseteq D$

19. $C \nsubseteq B$ **20.** $C \nsubseteq E$ **21.** $F \subseteq H$

22. $B \subseteq B$ **23.** $D \subseteq C$ **24.** $C \subseteq H$

25. $G \subseteq H$ **26.** $\{5\} \in B$ **27.** $\{5\} \subseteq B$

List the elements of the set $\left\{5, -\dfrac{2}{3}, \dfrac{8}{2}, \sqrt{9}, 0.3, \sqrt{7}, 1\dfrac{5}{8},\right.$

$\left. -1, \pi \right\}$ *that are also elements of each given set.*

28. Whole numbers   **29.** Natural numbers

30. Rational numbers   **31.** Irrational numbers

32. Real numbers   **33.** Integers

(1.2) *Write each statement using mathematical symbols.*

34. Twelve is the product of x and negative 4.

35. The sum of n and twice n is negative fifteen.

36. Four times the sum of y and three is -1.

37. The difference of t and five, multiplied by six, is four.

38. Seven subtracted from z is six.

39. Ten less than the product of x and nine is five.

40. The difference of x and five is at least twelve.

41. The opposite of four is less than the product of y and seven.

Graph the set and write it in interval notation.

42. $\{x \mid x < -1.5\}$ **43.** $\{x \mid x \geq 2\}$

44. $\{x \mid -2 < x \leq 5\}$ **45.** $\{x \mid -3.7 < x < -1.2\}$

Name the property illustrated.

46. $(M + 5) + P = M + (5 + P)$

47. $5(3x - 4) = 15x - 20$

48. $(-4) + 4 = 0$

49. $(3 + x) + 7 = 7 + (3 + x)$

50. $(XY)Z = X(YZ)$ **51.** $\left(-\dfrac{3}{5}\right) \cdot \left(-\dfrac{5}{3}\right) = 1$

52. $T \cdot 0 = 0$ **53.** $(ab)c = a(bc)$

Find the additive inverse, or opposite.

54. $-\dfrac{3}{4}$ **55.** 0.6 **56.** 0 **57.** 1

Find the multiplicative inverse, or reciprocal.

58. $-\dfrac{3}{4}$ **59.** 0.6 **60.** 0 **61.** 1

Complete the equation using the given property.

62. $5x - 15z = $ _____ Distributive property

63. $(7 + y) + (3 + x) = $ _____ Commutative property

64. $0 = $ _____ Additive inverse property

65. $1 = $ _____ Multiplicative inverse property

66. $[(3.4)(0.7)]5 = $ _____ Associative property

67. $7 = $ _____ Additive identity property

Write the property illustrated by each screen.

68.
```
3.6+2.5
              6.1
2.5+3.6
              6.1
```

69.
```
9.2(7+1.8)
              80.96
9.2*7+9.2*1.8
              80.96
```

Insert $<$, $>$, or $=$ to make each statement true.

70. -3 ___ $-|-1|$ **71.** 7 ___ $|-7|$

72. -5 ___ $-(-5)$ **73.** $-(-2)$ ___ -2

(1.3) *Simplify*

74. $-7 + 3$

75. $-10 + (-25)$

76. $5(-0.4)$

77. $(-3.1)(-0.1)$

78. $-7 - (-15)$

79. $9 - (-4.3)$

80. $(-6)(-4)(0)(-3)$

81. $(-12)(0)(-1)(-5)$

82. $(-24) \div 0$

83. $0 \div (-45)$

84. $(-36) \div (-9)$

85. $(60) \div (-12)$

86. $\left(-\dfrac{4}{5}\right) - \left(-\dfrac{2}{3}\right)$

87. $\left(\dfrac{5}{4}\right) - \left(-2\dfrac{3}{4}\right)$

88. Determine the unknown fractional part.

(1.4) *Simplify.*

89. $-5 + 7 - 3 - (-10)$

90. $8 - (-3) + (-4) + 6$

91. $3(4 - 5)^4$

92. $6(7 - 10)^2$

93. $\left(-\dfrac{8}{15}\right) \cdot \left(-\dfrac{2}{3}\right)^2$

94. $\left(-\dfrac{3}{4}\right)^2 \cdot \left(-\dfrac{10}{21}\right)$

95. $\dfrac{-\dfrac{6}{15}}{\dfrac{8}{25}}$

96. $\dfrac{\dfrac{4}{9}}{-\dfrac{8}{25}}$

97. $-\dfrac{3}{8} + 3(2) \div 6$

98. $5(-2) - (-3) - \dfrac{1}{6} + \dfrac{2}{3}$

99. $|2^3 - 3^2| - |5 - 7|$

100. $|5^2 - 2^2| + |9 \div (-3)|$

101. $-3[5 - 2(9 - 16)]$

102. $[3^2 - 6(12 - 10)]^2$

103. $\dfrac{(8 - 10)^3 - (-4)^2}{2 + 8(2) \div 4}$

104. $\dfrac{(2 + 4)^2 + (-1)^5}{12 \div 2 \cdot 3 - 3}$

105. $\dfrac{(4 - 9) + 4 - 9}{10 - 12 \div 4 \cdot 8}$

106. $\dfrac{3 - 7 - (7 - 3)}{15 + 30 \div 6 \cdot 2}$

107. $\dfrac{\sqrt{25}}{4 + 3 \cdot 7}$

108. $\dfrac{\sqrt{64}}{24 - 8 \cdot 2}$

Find the value of each expression when $x = 0$, $y = 3$, *and* $z = -2$.

109. $x^2 - y^2 + z^2$

110. $\dfrac{5x + z}{2y}$

111. $\dfrac{-7y - 3z}{-3}$

112. $(x - y + z)^2$

Simplify each expression

113. $-7x + 16 + 4x - 3$

114. $(2y + 3) - (6y + 3)$

115. $5(x + 4) + 4(-2x + 6)$

116. The algebraic expression $2\pi r$ represents the circumference of (distance around) a circle of radius r.

a. Complete the table below by evaluating the expression at given values of r. Round to two decimal places.

RADIUS	r	1	10	100
CIRCUMFERENCE	$2\pi r$			

b. As the radius of a circle increases, does the circumference of the circle increase or decrease?

117. The hummingbird has an average wing speed of 90 beats per second. The expression $90t$ gives the number of wing beats in t seconds. Calculate the number of wing beats in *1 hour* for the hummingbird.

118. In your own words explain to a friend how the following screen from a graphing utility represents an expression being evaluated at a given value.

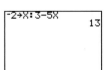

CHAPTER 1 TEST

Determine whether each statement is true or false.

1. $-2.3 > 2.33$

2. $-6^2 = (-6)^2$

3. $-5 - 8 = -(5 - 8)$

4. $(-2)(-3)(0) = \dfrac{(-4)}{0}$

5. All natural numbers are integers.

6. All rational numbers are integers.

Simplify.

7. $5 - 12 \div 3(2)$

8. $|4 - 6|^3 - (1 - 6^2)$

9. $(4 - 9)^3 - |-4 - 6|^2$

10. $[3|4 - 5|^5 - (-9)] \div (-6)$

11. $\dfrac{6(7 - 9)^3 + (-2)}{(-2)(-5)(-5)}$

Evaluate each expression when $q = 4$, $r = -2$, and $t = 1$.

12. $q^2 - r^2$

13. $\dfrac{5t - 3q}{3r - 1}$

14. The algebraic expression $5.75x$ represents the total cost for x adults to attend the theater.

a. Complete the table below.

ADULTS	x	3	7	12	20
TOTAL COST	$5.75x$				

b. As the number of adults increases, does the total cost increase or decrease?

Write each statement using mathematical symbols.

15. Twice the absolute value of the sum of x and five is 30.

16. The square of the difference of six and y, divided by seven, is less than -2.

17. The product of nine and z, divided by the absolute value of -12, is not equal to 10.

18. Three times the quotient of n and five is the opposite of n.

19. Twenty is equal to 6 subtracted from twice x.

20. Negative two is equal to x divided by the sum of x and five.

Name each property illustrated.

21. $6(x - 4) = 6x - 24$

22. $(4 + x) + z = 4 + (x + z)$

23. $(-7) + 7 = 0$

24. $(-18)(0) = 0$

25. Write an expression for the value (in dollars) of n nickels and d dimes.

Simplify each expression.

26. $-4x + 11 - 10x + 13$

27. $8y + 3(2y - 6) + 18$

28. Evaluate the expression $0.2x^3 + 5x^2 - 6.2x + 3$ if $x = -3.1$.

29. The circumference of a circle can be found by multiplying π times the diameter or using the formula, $C = \pi d$. The area of a circle is found by multiplying the radius squared times π or using the formula $A = \pi r^2$. Complete the table below to find the circumference and area, given the following diameters. If necessary, round answers to the nearest hundredth.

DIAMETER	d	2	3.8	10	14.9
RADIUS	r				
CIRCUMFERENCE	πd				
AREA	πr^2				

Graph each set and write it in interval notation.

30. $\{x \mid x < 4.2\}$

31. $\{x \mid -5 \le x \le 0\}$

BASIC CONCEPTS OF ALGEBRA

COMPARISON OF VOLUMES

The volume of a cylinder can be calculated or estimated in different ways. Sometimes we can take advantage of given information about a cylinder to improve the accuracy of volume calculations.

IN THE CHAPTER GROUP ACTIVITY ON PAGE 97, YOU WILL HAVE THE OPPORTUNITY TO INVESTIGATE SEVERAL METHODS OF CALCULATING THE VOLUME OF A CYLINDER.

Mathematics is a tool for solving problems in such diverse fields as transportation, engineering, economics, medicine, business, and biology. We solve problems using mathematics by modeling real-world phenomena with mathematical equations or inequalities. Our ability to solve problems using mathematics, then, depends in part on our ability to solve equations and inequalities. In this chapter, one goal is to solve linear equations in one variable and problems that can be modeled by these equations.

A second goal in this chapter is number sense. Number sense is the ability to determine whether a number is reasonable given a particular situation. This goal is achieved by examining and using equations, tables, and lists to solve problems. Number sense is especially useful when we access the powerful graphing features of a graphing utility beginning in the next chapter.

2.1 | SOLVING LINEAR EQUATIONS ALGEBRAICALLY

O B J E C T I V E S

1. Define linear equations.
2. Solve linear equations using properties of equality.
3. Solve linear equations that can be simplified by combining like terms.
4. Solve linear equations containing fractions.
5. Recognize identities and equations with no solution.
6. Write algebraic expressions that can be simplified.

TAPE IAG 2.1

Suppose you are driving home from the grocery store where you bought $3\frac{1}{2}$ pounds of steak for a special-occasion dinner. On the radio, you hear an ad from another grocery store for a special on steak at $5.60 per pound. You can't remember how much you paid for the steak that you just purchased, but you know you had $25 when you left for the store, and you have $6 left. You wonder if the other store has a better deal.

This situation is a prime example of how we can use known number relationships to reveal unknown measurements. In this case, the unknown measurement is the cost c per pound of steak. Though we don't know the value of c, we do know relationships involving c. We know, for instance, that the total cost of the meat is $3\frac{1}{2} \cdot c$ or $3.5c$, and we know that the difference of $25 and the total cost must be $6.

In words:	$25	minus	total cost of meat	is	$6
Translate:	25	−	3.5c	=	6

This **equation,** like every other equation, is a statement that two expressions are equal. Because we know a relationship between measurement c and other measurements, we can eventually unravel the clues and find the value of c. In this section, we concentrate on solving equations such as this one, called **linear equations** in one variable. Linear equations are also called **first-degree equations** since the exponent on the variable is 1.

LINEAR EQUATIONS IN ONE VARIABLE

$$3x = -15 \qquad 7 - y = 3y \qquad 4n - 9n + 6 = 0 \qquad z = -2$$

LINEAR EQUATION IN ONE VARIABLE

A linear equation in one variable is an equation that can be written in the form

$$ax + b = c$$

where a, b, and c are real numbers and $a \neq 0$.

When a variable in an equation is replaced by a number and the resulting equation is true, that number is called a **solution** of the equation. For example, 1 is a solution of the equation $3x + 4 = 7$, since $3(1) + 4 = 7$ is a true statement. But 2 is not a solution of this equation, since $3(2) + 4 = 7$ is not a true statement. The **solution set** of an equation is the set of solutions of the equation. For example, the solution set of $3x + 4 = 7$ is $\{1\}$.

To **solve an equation** is to find the solution set of an equation. Equations with the same solution set are called **equivalent equations.** For example,

$$3x + 4 = 7, \qquad 3x = 3, \qquad \text{and} \qquad x = 1$$

are equivalent equations because they all have the same solution set $\{1\}$. To solve an equation in x, we start with the given equation and write a series of simpler equivalent equations until we obtain an equation of the form

$$x = \textbf{number}.$$

Two important properties are used to write equivalent equations.

THE ADDITION PROPERTY OF EQUALITY

If a, b, and c, are real numbers, then

$$a = b \quad \text{and} \quad a + c = b + c$$

are equivalent equations.

THE MULTIPLICATION PROPERTY OF EQUALITY

If $c \neq 0$, then

$$a = b \quad \text{and} \quad ac = bc$$

are equivalent equations.

The **addition property of equality** guarantees that the same number may be added to (or subtracted from) both sides of an equation, and the result is an equivalent equation. The **multiplication property of equality** guarantees that both sides of an equation may be multiplied by (or divided by) the same nonzero number, and the result is an equivalent equation.

For example, to solve $2x + 5 = 9$, use the addition and multiplication properties of equality to isolate x, that is, to write an equivalent equation of the form

$x = $ number.

EXAMPLE 1   Solve for x: $2x + 5 = 9$.

Solution: First, use the addition property of equality and subtract 5 from both sides.

$$2x + 5 = 9$$

$$2x + 5 - 5 = 9 - 5 \qquad \text{Subtract 5 from both sides.}$$

$$2x = 4 \qquad \text{Simplify.}$$

To finish solving for x, use the multiplication property of equality and divide both sides by 2.

$$\frac{2x}{2} = \frac{4}{2}$$

$$x = 2 \qquad \text{Simplify.}$$

To check, replace x in the original equation with 2 and see that a true statement results using any method below.

$$2x + 5 = 9 \qquad \text{Original equation.}$$

$2(2) + 5 = 9 \qquad$ Let $x = 2$.

$4 + 5 = 9$

$9 = 9 \qquad$ True.

Let $x = 2$. | Store 2 in x and evaluate the left side of the equation.

| 2*2+5 |
| | 9 |

The left side of the equation evaluates to 9, the value of the right side.

| 2→X:2X+5 |
| | 9 |

The left side of the equation evaluates to 9, the value of the right side.

The solution set is {2}.

EXAMPLE 2   Solve for c: $25 - 3.5c = 6$.

Solution: First, use the addition property of equality and subtract 25 from both sides.

$$25 - 3.5c = 6$$

$$25 - 3.5c - 25 = 6 - 25 \qquad \text{Subtract 25 from both sides.}$$

$$-3.5c = -19 \qquad \text{Simplify.}$$

$$\frac{-3.5c}{-3.5} = \frac{-19}{-3.5} \qquad \text{Divide both sides by } -3.5.$$

$$c = \frac{38}{7} \qquad \text{Simplify.}$$

Using a method shown in Example 1, check to see that the solution set is $\left\{\dfrac{38}{7}\right\}$.

Recall from the beginning of this section that c in the equation given in Example 2 could represent the cost of a pound of steak that you bought. Since $c = \frac{38}{7} \approx 5.43$, you paid approximately \$5.43 per pound which is cheaper than the price of \$5.60 that you heard advertised on the radio.

3   Often, an equation can be simplified by removing any grouping symbols and combining any like terms.

EXAMPLE 3   Solve for x: $-6x - 1 + 5x = 3$.

Solution: First, the left side of this equation can be simplified by combining the like terms $-6x$ and $5x$. Then use the addition property of equality and add 1 to both sides of the equation.

$$-6x - 1 + 5x = 3$$
$$-x - 1 = 3 \qquad \text{Combine like terms.}$$
$$-x - 1 + 1 = 3 + 1 \qquad \text{Add 1 to both sides of the equation.}$$
$$-x = 4 \qquad \text{Simplify.}$$

Notice that this equation is not solved for x since we have $-x$ or $-1x$, not x. To solve for x, divide both sides by -1.

$$\frac{-x}{-1} = \frac{4}{-1} \qquad \text{Divide both sides by } -1.$$
$$x = -4 \qquad \text{Simplify.}$$

Check to see that the solution set is $\{-4\}$.

If an equation contains parentheses, use the distributive property to remove them.

EXAMPLE 4   Solve for x: $2(x - 3) = 5x - 9$.

Solution: First, use the distributive property.

$$\overset{\frown}{2(x - 3)} = 5x - 9$$
$$2x - 6 = 5x - 9 \qquad \text{Apply the distributive property.}$$

Next, get variable terms on the same side of the equation by subtracting $5x$ from both sides.

$$2x - 6 - 5x = 5x - 9 - 5x \qquad \text{Subtract } 5x \text{ from both sides.}$$
$$-3x - 6 = -9 \qquad \text{Simplify.}$$
$$-3x - 6 + 6 = -9 + 6 \qquad \text{Add 6 to both sides.}$$
$$-3x = -3 \qquad \text{Simplify.}$$
$$\frac{-3x}{-3} = \frac{-3}{-3} \qquad \text{Divide both sides by } -3.$$
$$x = 1$$

Let $x = 1$ in the original equation and check that a true statement results.

$$2(x - 3) = 5x - 9 \qquad \text{Original equation}$$

To use a calculator to check, store 1 in x and evaluate the left side and the right side of the equation.

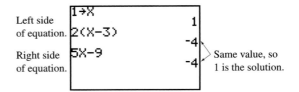

Left side of equation.

Right side of equation.

Same value, so 1 is the solution.

The solution set is {1}.

> **REMINDER**   Check a proposed solution in the original equation.

4   If an equation contains fractions, we first clear the equation of fractions by multiplying both sides of the equation by the *least common denominator* (LCD) of all fractions in the equation.

EXAMPLE 5   Solve for y: $\dfrac{y}{3} - \dfrac{y}{4} = \dfrac{1}{6}$.

Solution: First, clear the equation of fractions by multiplying both sides of the equation by 12, the LCD of denominators 3, 4, and 6.

$$\frac{y}{3} - \frac{y}{4} = \frac{1}{6}$$

$$12\left(\frac{y}{3} - \frac{y}{4}\right) = 12\left(\frac{1}{6}\right) \qquad \text{Multiply both sides by the LCD 12.}$$

$$12\left(\frac{y}{3}\right) - 12\left(\frac{y}{4}\right) = 2 \qquad \text{Apply the distributive property.}$$

$$4y - 3y = 2 \qquad \text{Simplify.}$$

$$y = 2 \qquad \text{Simplify.}$$

```
2→Y:Y/3-Y/4
       .1666666667
Ans▶Frac
             1/6
```

To check using a calculator see the screen to the left. The solution set is {2}.

As a general guideline, the following steps may be used to solve a linear equation in one variable.

To Solve a Linear Equation in One Variable

Step 1. Clear the equation of fractions by multiplying both sides of the equation by the least common denominator (LCD) of all denominators in the equation.

Step 2. Use the distributive property to remove grouping symbols such as parentheses.

Step 3. Combine like terms on each side of the equation.

Step 4. Use the addition property of equality to rewrite the equation as an equivalent equation, with variable terms on one side and numbers on the other side.

Step 5. Use the multiplication property of equality to isolate the variable.

Step 6. Check the proposed solution in the original equation.

EXAMPLE 6   Solve for x: $\dfrac{x + 5}{2} + \dfrac{1}{2} = 2x - \dfrac{x - 3}{8}$.

Solution: First, multiply both sides of the equation by 8, the LCD for 2 and 8.

$$8\left(\frac{x + 5}{2} + \frac{1}{2}\right) = 8\left(2x - \frac{x - 3}{8}\right) \qquad \text{Multiply both sides by 8.}$$

$$4(x + 5) + 4 = 16x - (x - 3) \qquad \text{Apply the distributive property.}$$

$$4x + 20 + 4 = 16x - x + 3 \qquad \begin{array}{l}\text{Use the distributive property}\\ \text{to remove parentheses.}\end{array}$$

$$4x + 24 = 15x + 3 \qquad \text{Combine like terms.}$$

$$-11x + 24 = 3 \qquad \text{Subtract } 15x \text{ from both sides.}$$

$$-11x = -21 \qquad \text{Subtract 24 from both sides.}$$

$$\frac{-11x}{-11} = \frac{-21}{-11} \qquad \text{Divide both sides by } -11.$$

$$x = \frac{21}{11} \qquad \text{Simplify.}$$

```
21/11→X
        1.909090909
(X+5)/2+1/2
        3.954545455
2X-(X-3)/8
        3.954545455
```

To check, verify that replacing x with $\dfrac{21}{11}$ makes the original equation true. If you use a calculator to verify, make sure that parentheses are placed about a numerator or denominator that contains more than one term, as shown to the left.

The solution set is $\left\{\dfrac{21}{11}\right\}$.

5

So far, each linear equation that we have solved has had a single solution. A linear equation in one variable that has exactly one solution is called a **conditional equation.** We will now look at two other types of equations: contradictions and identities.

An equation in one variable that has no solution is called a **contradiction,** and an equation in one variable that has every number (for which the equation is defined) as a solution is called an **identity.** The next examples show how to recognize contradictions and identities.

EXAMPLE 7   Solve for x: $3x + 5 = 3(x + 2)$.

Solution: First, use the distributive property and remove parentheses.

$$3x + 5 = 3(x + 2)$$

$$3x + 5 = 3x + 6 \qquad \text{Apply the distributive property.}$$

$$3x + 5 - 3x = 3x + 6 - 3x \qquad \text{Subtract } 3x \text{ from both sides.}$$

$$5 = 6$$

The equation $5 = 6$ is a false statement no matter what value the variable x might have. Thus, the original equation has no solution. Its solution set is written either as $\{\ \}$ or $\varnothing$. This equation is a contradiction.

EXAMPLE 8   Solve for x: $6x - 4 = 2 + 6(x - 1)$.

Solution: First, use the distributive property and remove parentheses.

$$6x - 4 = 2 + 6(x - 1)$$

$$6x - 4 = 2 + 6x - 6 \qquad \text{Apply the distributive property.}$$

$$6x - 4 = 6x - 4 \qquad \text{Combine like terms.}$$

At this point we might notice that both sides of the equation are the same, so replacing x by any real number gives a true statement. Thus the solution set of this equation is the set of real numbers, and the equation is an identity. Continuing to "solve" $6x - 4 = 6x - 4$, we eventually arrive at the same conclusion.

$$6x - 4 + 4 = 6x - 4 + 4 \qquad \text{Add 4 to both sides.}$$

$$6x = 6x \qquad \text{Simplify.}$$

$$6x - 6x = 6x - 6x \qquad \text{Subtract } 6x \text{ from both sides.}$$

$$0 = 0 \qquad \text{Simplify.}$$

Since $0 = 0$ is a true statement for every value of x, the solution set is the set of all real numbers or $\{x | x \text{ is a real number}\}$, and the equation is an identity.

6 In order to prepare for solving problems, we practice writing algebraic expressions that can be simplified.

EXAMPLE 9 Write the following as algebraic expressions. Then simplify.

    a. The sum of two consecutive integers, if x is the first integer.

    b. The perimeter of the triangle with side lengths x, $5x$, and $6x - 3$.

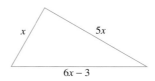

Solution: **a.** Recall that if x is the first integer, then the next consecutive integer is 1 more, or $x + 1$.

In words:	first integer	plus	next consecutive integer
Translate:	x	$+$	$(x + 1)$

Then $x + (x + 1) = x + x + 1$

$\qquad\qquad\qquad\quad = 2x + 1$ Simplify by combining like terms.

    b. The perimeter of a triangle is the sum of the lengths of the sides.

In words:	side	+	side	+	side
Translate:	x	$+$	$5x$	$+$	$(6x - 3)$

Then $x + 5x + (6x - 3) = x + 5x + 6x - 3$

$\qquad\qquad\qquad\qquad\qquad = 12x - 3$ Simplify.

MENTAL MATH

Simplify each expression.

1. $3x + 5x + 6 + 15$ **2.** $8y + 3y + 7 + 11$ **3.** $5n + n + 3 - 10$

4. $m + 2m + 4 - 8$ **5.** $8x - 12x + 5 - 6$ **6.** $4x - 10x + 13 - 16$

EXERCISE SET 2.1

In this exercise set, check the proposed solution of each equation using a method shown in this section.

Solve for the variable. See Examples 1 and 2.

1. $x + 2.8 = 1.9$ **2.** $y - 8.6 = -6.3$

3. $5x - 4 = 26$ **4.** $2y - 3 = 11$

5. $-4.1 - 7z = 3.6$ **6.** $10.3 - 6x = -2.3$

Solve for the variable. See Examples 3 and 4.

7. $5y + 12 = 2y - 3$ **8.** $4x + 14 = 6x + 8$

9. $8x - 5x + 3 = x - 7 + 10$

10. $6 + 3x + x = -x + 2 - 26$

11. $5x + 12 = 2(2x + 7)$ **12.** $2(x + 3) = x + 5$

13. $3(x - 6) = 5x$ **14.** $6x = 4(5 + x)$

15. $-2(5y - 1) - y = -4(y - 3)$

16. $-3(2w - 7) - 10 = 9 - 2(5w + 4)$

17. a. Simplify the expression $4(x + 1) + 1$.

   b. Solve the equation $4(x + 1) + 1 = -7$.

   c. Explain the difference between solving an equation for a variable and simplifying an expression.

18. Explain why the multiplication property of equality does not include multiplying both sides of an equation by 0. (*Hint:* Write down a false statement and then multiply both sides by 0. Is the result true or false? What does this mean?)

Solve for the variable. See Examples 5 and 6.

19. $\dfrac{x}{2} + \dfrac{2}{3} = \dfrac{3}{4}$ **20.** $\dfrac{x}{2} + \dfrac{x}{3} = \dfrac{5}{2}$

21. $\dfrac{3t}{4} - \dfrac{t}{2} = 1$ **22.** $\dfrac{4r}{5} - 7 = \dfrac{r}{10}$

23. $\dfrac{n - 3}{4} + \dfrac{n + 5}{7} = \dfrac{5}{14}$ **24.** $\dfrac{2 + h}{9} + \dfrac{h - 1}{3} = \dfrac{1}{3}$

25. $\dfrac{3x - 1}{9} + x = \dfrac{3x + 1}{3} + 4$

26. $\dfrac{2z + 7}{8} - 2 = z + \dfrac{z - 1}{2}$

Solve the following. See Examples 7 and 8.

27. $4(n + 3) = 2(6 + 2n)$ **28.** $6(4n + 4) = 8(3 + 3n)$

29. $3(x - 1) + 5 = 3x + 2$

30. $5x - (x + 4) = 4 + 4(x - 2)$

31. In your own words, explain why the equation $x + 7 = x + 6$ has no solution while the solution set of the equation $x + 7 = x + 7$ contains all real numbers.

32. In your own words, explain why the equation $x = -x$ has the one solution 0, while the solution set of the equation $x = x$ is all real numbers.

Each screen shows a calculator check of a proposed solution to an equation. Write the equation and the verified solution.

33.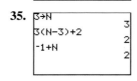

34.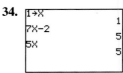

35.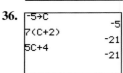

36.

Solve the following.

37. $-5x + 1.5 = -19.5$ **38.** $-3x - 4.7 = 11.8$

39. $x - 10 = -6x + 4$ **40.** $4x - 7 = 2x - 7$

41. $3x - 4 - 5x = x + 4 + x$

42. $13x - 15x + 8 = 4x + 2 - 24$

43. $5(y + 4) = 4(y + 5)$ **44.** $6(y - 4) = 3(y - 8)$

45. $0.6x - 10 = 1.4x - 14$ **46.** $0.3x + 2.4 = 0.1x + 4$

47. $6x - 2(x - 3) = 4(x + 1) + 4$

48. $10x - 2(x + 4) = 8(x - 2) + 6$

49. $\dfrac{3}{8} + \dfrac{b}{3} = \dfrac{5}{12}$ **50.** $\dfrac{a}{2} + \dfrac{7}{4} = 5$

51. $z + 3(2 + 4z) = 6(z + 1) + 5z$

52. $4(m - 6) - m = 8(m - 3) - 5m$

53. $\dfrac{3t + 1}{8} = \dfrac{5 + 2t}{7} + 2$ **54.** $4 - \dfrac{2z + 7}{9} = \dfrac{7 - z}{12}$

55. $\dfrac{m - 4}{3} - \dfrac{3m - 1}{5} = 1$ **56.** $\dfrac{n + 1}{8} - \dfrac{2 - n}{3} = \dfrac{5}{6}$

57. $\dfrac{x}{5} - \dfrac{x}{4} = \dfrac{1}{2}(x - 2)$ **58.** $\dfrac{y}{3} + \dfrac{y}{5} = \dfrac{1}{10}(y + 3)$

59. $5(x - 2) + 2x = 7(x + 4)$

60. $3x + 2(x + 4) = 5(x + 1) + 3$

61. $y + 0.2 = 0.6(y + 3)$

62. $-(w + 0.2) = 0.3(4 - w)$

63. $1.5(4 - x) = 1.3(2 - x)$

64. $2.4(2x + 3) = -0.1(2x + 3)$

65. $\frac{1}{4}(a + 2) = \frac{1}{6}(5 - a)$ **66.** $\frac{1}{3}(8 + 2c) = \frac{1}{5}(3c - 5)$

Write each phrase as an algebraic expression. Then simplify. See Example 9.

67. The perimeter of the square with side length y.

68. The perimeter of the rectangle with length x and width $x - 5$.

69. The sum of three consecutive integers if the first integer is z.

70. The sum of three consecutive odd integers if the first integer is x.

71. The total value (in dollars) of x nickels and $x + 3$ dimes.

72. The total value (in cents) of y quarters and $2y - 1$ nickels.

73. A bayou is a common name in Louisiana for a creek. A piece of land along Bayou Liberty is to be fenced and subdivided as shown so that the rectangles have the same dimensions. Express the total amount of fencing needed as an algebraic expression in x.

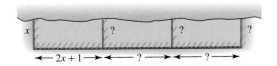

74. Write the total square feet of area of the floor plan shown as an algebraic expression in x.

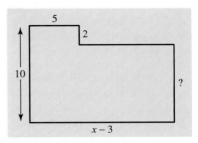

Find the value of K such that the equations are equivalent.

75. $3.2x + 4 = 5.4x - 7$

$$3.2x = 5.4x + K$$

76. $-7.6y - 10 = -1.1y + 12$

$$-7.6y = -1.1y + K$$

77. $\frac{x}{6} + 4 = \frac{x}{3}$ **78.** $\frac{5x}{4} + \frac{1}{2} = \frac{x}{2}$

$x + K = 2x$ $5x + K = 2x$

Solve and check.

79. $2.569x = -12.48534$

80. $-9.112y = -47.537304$

81. $2.86z - 8.1258 = -3.75$

82. $1.25x - 20.175 = -8.15$

83. If $3x - 2 = 16$, find the value of $2x + 1$.

84. If $2(x - 8) + x = 4x - 16$ and $2y + 5(y - 4) = 2y - 20$, which is larger, x or y?

Review Exercises

85. Recall from Section 1.3 that a game is fair if each team or player has an equal chance of winning. Cut or tear a sheet of paper into 10 pieces, numbering the pieces from 1 to 10. Place the pieces into a bag. Draw 2 pieces from the bag, record their sum, and then return them to the bag. If the sum is 10 or less, player 1 gets a point. If their sum is more than 10, player 2 gets a point. Is this a fair game? Try it and see.

Find the value of each expression when $y_1 = -3$, $y_2 = 4$, and $y_3 = 0$. See Section 1.4.

86. $(y_1)^2$ **87.** $y_2 - y_1$ **88.** $y_1 + y_2 + y_3$

89. $\frac{y_3}{3y_1}$ **90.** $\frac{6y_2}{y_1}$ **91.** $y_1 - y_2$

2.2 | An Introduction to Problem Solving

O B J E C T I V E S

1 Apply the steps for problem solving.

Tape IAG 2.2

1 Our main purpose for studying algebra is to solve problems. In previous sections, we have prepared for problem solving by writing phrases as algebraic expressions and sentences as equations. In this section, we now draw upon this experience as we write equations that describe or model a problem. The following problem-solving steps will be used throughout this text and may also be used to solve real-life problems that occur outside the mathematics classroom.

PROBLEM-SOLVING STEPS

1. **UNDERSTAND** the problem. During this step, don't work with variables, but simply become comfortable with the problem. Some ways of accomplishing this are listed next.
   - Read and reread the problem.
   - Construct a drawing to help visualize the problem.
   - Propose a solution and check. Pay careful attention as to how you check your proposed solution. This will help later when you write an equation to model the problem.
2. **ASSIGN** a variable to an unknown in the problem. Use this variable to represent any other unknown quantities.
3. **ILLUSTRATE** the problem. A diagram or chart *using the assigned variables* can often help you visualize the known facts.
4. **TRANSLATE** the problem into a mathematical model.
5. **COMPLETE** the work. This often means to solve the equation.
6. **INTERPRET** the results. *Check* the proposed solution in the stated problem and *state* your conclusion.

In step 4 above, we use the phrase *mathematical model*. A **mathematical model** is a graph, table, list, equation, or inequality that describes a situation.

Let's review these steps by solving a problem involving unknown numbers.

EXAMPLE 1   INTRODUCING THE PROBLEM-SOLVING PROCESS
Find two numbers such that the second number is 3 more than twice the first number and the sum of the two numbers is 72.

Solution: 1. UNDERSTAND the problem. Read and reread the problem and then propose a solution. For example, if the first number is 25, then the second number is 3 more than twice 25, or 53. The sum of 25 and 53 is 78, not the required sum, but we have gained some valuable information about the problem. First, we know that the first number is less than 25 since our guess led to a sum greater than the required sum. Also, we have gained some information as to how to model the problem. Remember: The purpose of guessing a solution is not to guess correctly but to gain confidence and to help understand the problem and how to model it.

2. ASSIGN a variable. Use this variable to represent any other known quantities. If we let

$$\text{the first number} = x, \text{then}$$

$$\text{the second number} = 2x + 3$$

$$\uparrow \qquad \uparrow$$

$$\uparrow \qquad \text{3 more than}$$
$$\text{twice the}$$
$$\text{first number}$$

3. ILLUSTRATE the problem. No illustration is needed here.

4. TRANSLATE the problem into a mathematical model. To do so, we use the fact that the sum of the numbers is 72. First, write this relationship in words and then translate to an equation.

In words:	First number	added to	Second number	is	72
Translate:	x	$+$	$(2x + 3)$	$=$	72

5. COMPLETE the work by solving the equation.

$$x + (2x + 3) = 72$$

$$x + 2x + 3 = 72 \qquad \text{Remove parentheses.}$$

$$3x + 3 = 72 \qquad \text{Combine like terms.}$$

$$3x = 69 \qquad \text{Subtract 3 from both sides.}$$

$$x = 23 \qquad \text{Divide both sides by 3.}$$

6. INTERPRET. Here, we *check* our work and *state* the solution. Recall that if the first number $x = 23$, then the second number $2x + 3 = 2 \cdot 23 + 3 = 49$.

Check: Is the second number 3 more than twice the first number? Yes, since 3 more than twice 23 is $46 + 3$, or 49. Also, their sum, $23 + 49 = 72$, is the required sum.

State: The two numbers are 23 and 49.   ▬▬▬▬

Much of today's rates and statistics are given as percents. Interest rates, tax rates, nutrition labeling, and percent of households in a given category are just a few examples. Before we practice solving problems containing percents, let's take a moment and review the meaning of percent and how to find a percent of a number.

The word *percent* means *per hundred*, and the symbol % is used to denote percent. This means that 23% is 23 per hundred, or $\frac{23}{100}$. Also,

$$41\% = \frac{41}{100} = 0.41$$

To find a percent of a number, multiply. For example, to find 16% of 25, find the product of 16% (written as a decimal) and 25.

EXAMPLE 2 Find 16% of 25.

Solution:
$$16\% \cdot 25 = 0.16 \cdot 25$$
$$= 4$$

Thus, 16% of 25 is 4.

```
.16*25
                    4
```

Next, we solve a problem containing percent.

EXAMPLE 3 FINDING A PERCENT DECREASE

Suppose that Service Merchandise just announced an 8% decrease in the price of their Compaq Presario computers. If one particular computer model sells for $2162 after the decrease, find the original price of this computer.

Solution:
1. UNDERSTAND. Read and reread the problem. Recall that a percent decrease means a percent of the original price. Let's guess that the original price of the computer is $2500. The amount of decrease is then 8% of $2500, or (0.08)($2500) = $200. This means that the new price of the computer is the original price minus the decrease, or $2500 − $200 = $2300. Our guess is incorrect, but we now have an idea of how to model this problem.
2. ASSIGN a variable. Let x = the original price of the computer.
3. ILLUSTRATE the problem. No illustration is needed.
4. TRANSLATE.

In words:	Original price of computer	minus	8% of original price	is	new price
Translate:	x	−	$0.08x$	=	2162

5. COMPLETE. Solve the equation.

$$x - 0.08x = 2162$$
$$0.92x = 2162 \qquad \text{Combine like terms.}$$
$$x = \frac{2162}{0.92} = 2350 \qquad \text{Divide both sides by 0.92.}$$

6. INTERPRET. *Check:* If the original price of the computer is $2350, the new price is

$$\$2350 - (0.08)(\$2350) = \$2350 - \$188$$
$$= \$2162, \text{ the given new price.}$$

State: The original price of the computer is $2350.

EXAMPLE 4   **SOLVING A PROBLEM CONTAINING A GEOMETRIC FIGURE**
A pennant in the shape of an isosceles triangle is to be constructed for the Slidell
High School Athletic Club and sold as a fund-raiser. The company manufacturing
the pennant charges according to perimeter, and the athletic club has determined
that a perimeter of 149 centimeters should make a nice profit. If each equal side of
the triangle is 12 centimeters more than twice the length of the third side, find the
lengths of the sides of the triangular pennant.

Solution: **1. UNDERSTAND.** Read and reread the problem. Recall that the perimeter of a
triangle is the distance around. Let's guess that the third side of the triangular
pennant is 20 centimeters. This means that each equal side is twice 20 centimeters,
increased by 12 centimeters, or $2(20) + 12 = 52$ centimeters.

This gives a perimeter of $20 + 52 + 52 = 124$ centimeters. Our guess is incorrect, but we have a better understanding of how to model this problem.

2. ASSIGN a variable. Let

the third side of the triangle $= x$; then

the first side $=$ twice the third side increased by 12

$$= 2 \quad x \quad + \quad 12, \text{ or } 2x + 12$$

and the second side $= 2x + 12$

3. ILLUSTRATE.

4. TRANSLATE.

In words: | First side | + | Second side | + | Third side | = | 149 |

Translate: $(2x + 12)$ $+$ $(2x + 12)$ $+$ x $=$ 149

5. COMPLETE. Here, we solve the equation.

$$(2x + 12) + (2x + 12) + x = 149$$

$$2x + 12 + 2x + 12 + x = 149 \qquad \text{Remove parentheses.}$$

$$5x + 24 = 149 \qquad \text{Combine like terms.}$$

$$5x = 125 \qquad \text{Subtract 24 from both sides.}$$

$$x = 25 \qquad \text{Divide both sides by 5.}$$

6. INTERPRET. If the third side is 25 centimeters, then the first side is 2(25) + 12 = 62 centimeters and the second side is 62 centimeters also.

Check: The first and second sides are each twice 25 centimeters increased by 12 centimeters, or 62 centimeters. Also, the perimeter is 25 + 62 + 62 = 149 centimeters, the required perimeter.

State: The dimensions of the triangular pennant are 62 centimeters, 62 centimeters, and 25 centimeters.

EXERCISE SET 2.2

Solve. See Example 1.

1. Four times the difference of a number and 2 is the same as 6 times the number increased by 2. Find the number.

2. Twice the sum of a number and 3 is the same as 1 subtracted from the number. Find the number.

3. One number is 5 times another number. If the sum of the two numbers is 270, find the numbers.

4. One number is 6 less than another number. If the sum of the two numbers is 150, find the numbers.

Solve. See Example 2.

5. Find 30% of 260.

6. Find 70% of 180.

7. Find 12% of 16.

8. Find 22% of 12.

9. The United States consists of 1943 million acres of land. Approximately 21% of this land is federally owned. Find the number of acres that are federally owned. (*Source: USA Today,* July 24, 1995.)

10. The state of Nevada contains the most federally owned land in the United States. If 85% of the state's 71 million acres of land is federally owned, find the number of federally owned acres. (*Source: USA Today,* July 4, 1995.)

Nevada

11. At this writing, 47% of homes in the United States contain computers. If Charlotte, North Carolina, contains 110,000 homes, how many of these homes would you expect to have computers? (*Source: Telecommunication Research* survey.)

12. At this writing, 12% of homes in the United States contain on-line services. If Abilene, Texas, contains 40,000 homes, how many of these homes would you expect to have on-line services? (*Source: Telecommunication Research* survey.)

The following graph is called a circle graph or a pie chart. The circle represents a whole, or 100%. This particular graph shows the kinds of loans that customers get from credit unions.

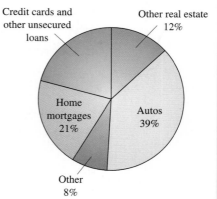

Credit Union Loans

Credit cards and other unsecured loans

Other real estate 12%

Home mortgages 21%

Autos 39%

Other 8%

Source: National Credit Union Administration.

13. What percent of credit union loans are for credit cards and other unsecured loans?

14. What type of loan makes up most credit union loans?

15. If the University of New Orleans Credit Union processed 300 loans last year, how many of these might we expect to be automobile loans?

16. If Homestead's Credit Union processed 537 loans last year, how many of these do you expect to be for either home mortgages or other real estate? (Round to the nearest whole number.)

Solve. See Examples 3 and 4.

17. The B767-300 aircraft has twice as many seats as the B737-200 aircraft. If their total number of seats is 336, find the number of seats for each aircraft.

18. The governor of Delaware makes $17,000 more per year than the governor of Connecticut. If the total of these salaries is $173,000, find the salary of each governor.

19. A new fax machine was recently purchased for an office in Hopedale for $464.40 including tax. If the tax rate in Hopedale is 8%, find the price of the fax machine before taxes.

20. A premedical student at a local university was complaining that she had just paid $86.11 for her human anatomy book, including tax. Find the price of the book before taxes if the tax rate at this university is 9%.

21. Two frames are needed with the same outside perimeter; one frame in the shape of a square and one in the shape of an equilateral triangle. Each side of the triangle is 6 centimeters longer than each side of the square. Find the dimensions of each frame.

22. The length of a rectangular sign is 2 feet less than three times its width. Find the dimensions if the perimeter is 28 feet.

23. In a blueprint of a rectangular room, the length is to be 2 centimeters greater than twice the width. Find the dimensions if the perimeter is to be 40 centimeters.

24. A plant food solution contains 5 cups of water for every 1 cup of concentrate. If the solution contains 78 cups of these two ingredients, find the number of cups of concentrate in the solution.

25. Manufacturers claim that a CD-ROM disk will last 20 years. Recently, statements made by the U.S. National Archives and Records Administration suggest that 20 years decreased by 75% is a more realistic lifespan because the aluminum substratum on which the data are recorded can be affected by oxidation. Find the lifespan of a CD-ROM according to the U.S. National Archives and Records Administration.

CD-ROM disk

26. In one year, 2.7% of India's forest was lost to deforestation. This percent represents 10,000 square kilometers of forest. Find the total square kilometers of forest in India before this decrease. (Round to the nearest whole square kilometer.)

27. The external tank of a NASA Space Shuttle contains the propellants used for the first 8.5 minutes after launch. Its height is 5 times the sum of its width and 1. If the sum of the height and width is 55.4 meters, find the dimensions of this tank.

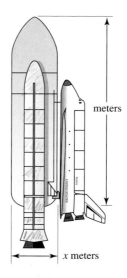

meters

x meters

28. The blue whale is the largest of whales. Its average weight is 3 times the difference of the average weight of a humpback whale and 5 tons. If the total of the average weights is 117 tons, find the average weight of each type of whale.

29. Recall that the sum of the angle measures of a triangle is 180°. Find the measures of the angles of a triangle if the measure of one angle is twice the measure of a second angle and the third angle measures 3 times the second angle decreased by 12.

30. One angle is twice its complement increased by 30°. Find the measures of the two complementary angles. (Two angles are complimentary if the sum of their measures is 90°.)

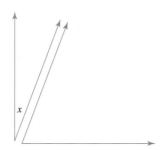

31. It is estimated that American taxpayers spent 1.5 billion hours preparing their 1995 tax forms. About 44% of this time is spent on the 1040 form. Estimate the time that American taxpayers spent preparing the 1040 form. (*Source: USA Today,* July 25, 1995.)

32. It takes about 2.9 hours for an American taxpayer to learn how to use the 1040 form. This time represents 25% of the average time spent on the form. Find the average time an American taxpayer spends on the 1040 form. (*Source: USA Today,* July 25, 1995.)

33. The zip codes of three Nevada locations–Fallon, Fernley, and Gardnerville Ranchos–are three consecutive even integers. If twice the first integer added to the third is 268,222, find each zip code.

34. During a recent year, the average SAT scores in math for the states of Alabama, Louisiana, and Michigan were 3 consecutive integers. If the sum of the first integer, second integer, and three times the third integer is 2637, find each average score.

35. Determine whether there are three consecutive integers such that their sum is three times the second integer.

36. Determine whether there are two consecutive odd integers such that 7 times the first exceeds 5 times the second by 54.

37. The sum of the angles of a triangle is 180°. Find the angles of a triangle whose two base angles are equal and whose third angle is 10° less than three times a base angle.

38. Find an angle such that its supplement is equal to twice its complement increased by 50°. (Two angles are supplementary if the sum of their measures is 180°.)

39. According to the Labor Department, the number of telephone operators will decrease to 46,000 by the year 2005. This represents a decrease of 23.3% from the number of telephone operators in 1995. Find the number of telephone operators in 1995. Round to the nearest whole number.

40. The number of deaths by tornadoes from the 1940s to the 1980s has decreased by 70.86%. If the number of deaths in the 1980s was 521, find the number of deaths in the 1940s. Round to the nearest whole number. (*Source:* National Weather Service.)

41. In your own words, explain why you think that the need for telephone operators is decreasing. See Exercise 39.

42. In your own words, explain why you think that the number of deaths by tornadoes has decreased so much since the 1940s. See Exercise 40.

43. According to the U.S. Census Bureau, the population of Chicago decreased by 22.2% from the time of the Democratic National Convention in 1968 to the time of the Democratic National Convention in 1995. Find the population of Chicago in 1968 if the 1995 population was reported to be 2.8 million people. Round to the nearest tenth of a million.

44. The United States Army National Guard personnel increased 10.2% from 1980 to 1993. Find the number of Army National Guard personnel in 1980 if the number in 1993 was 410 thousand. Round to the nearest thousand. (*Source*: National Guard Bureau.)

To break even in a manufacturing business, income or revenue R must equal the cost of production C. Use this information for Exercises 45 through 50.

45. The cost C to produce x number of skateboards is $C = 100 + 20x$. The skateboards are sold wholesale for $24 each, so revenue R is given by $R = 24x$. Find how many skateboards the manufacturer needs to produce and sell to break even. (*Hint:* Set the cost expression equal to the revenue expression and solve for x.)

46. The revenue R from selling x number of computer boards is given by $R = 60x$, and the cost C of making them is given by $C = 50x + 5000$. Find how many boards must be made and sold to break even. Find how much money is needed to make the break-even number of boards.

47. The cost C of producing x number of paperback books is given by $C = 4.50x + 2400$. Income R from these books is given by $R = 7.50x$. Find how many books should be produced and sold to break even.

48. Find the break-even quantity for a company that makes x number of computer monitors at a cost C given by $C = 875 + 70x$ and receives revenue R given by $R = 105x$.

49. In your own words, explain what happens if a company makes and sells fewer products than the number required to break even.

50. In your own words, explain what happens if more products than the break-even number are made and sold.

Exercises 51 through 54 do not contain enough information for a single solution. Supply one additional piece of information so that each exercise has a single solution. Find the solution.

51. Two numbers have a sum of 10. Find the numbers.

52. Two numbers have a sum of 18. Find the numbers.

53. Two angles are supplementary. Find the measures of the angles.

54. Two angles are complementary. Find the measures of the angles.

Review Exercises

Find the value of each expression for the given values of the variables. See Section 1.4.

55. $2a + b - c$; $a = 5$, $b = -1$, and $c = 3$

56. $-3a + 2c - b$; $a = -2$; $b = 6$, and $c = -7$

57. $4ab - 3bc$; $a = -5$, $b = -8$, and $c = 2$

58. $ab + 6bc$; $a = 0$, $b = -1$ and $c = 9$

59. $n^2 - m^2$; $n = -3$ and $m = -8$

60. $2n^2 + 3m^2$; $n = -2$ and $m = 7$

Use a calculator to find the value of each expression for the given values of the variables. See Section 1.4.

61. $P + Prt$; $P = 3000$, $r = 0.0325$, $t = 2$

62. $\frac{1}{3}lwh$; $l = 37.8$, $w = 5.6$, $h = 7.9$

63. $\frac{1}{3}Bh$; $B = 53.04$, $h = 6.89$

64. $\left(1 + \frac{r}{n}\right)^{nt}$; $r = 0.06$, $n = 3$, $t = 1$

2.3 | A NUMERICAL APPROACH: MODELING WITH TABLES

O B J E C T I V E S

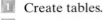 Create tables.

  Use tables to solve problems.

TAPE IAG 2.3

In this section, we introduce a numerical approach to problem solving. This approach may be used instead of or in addition to an algebraic approach. A numerical approach is invaluable for developing estimation skills and number sense.

Recall that in Chapter 1, we introduced methods for using a calculator to evaluate an algebraic expression at given values of the variable. Often we organized our results in tables such as the one below.

TIME (SECONDS)	x	2	3	4	5	6
SPEED (FEET PER SECOND)	$32x$	64	96	128	160	192

Notice for this particular table that the minimum x-value is 2, and the x-values increase by 1. Another method for completing a table such as this is the table feature of your calculator.

When generating a **table** on a calculator, we need to give the calculator specific instructions as to a minimum x-value to *start* with and the *increment* or change in the x-values. (For many calculators, the Greek letter delta, Δ, is used in the table setup menu. Here, Δ indicates the change in x.) In this text, we will use the term *table start* to indicate the first x-value to appear in the table and the term *table increment* to indicate the increment or change in x-values.

TECHNOLOGY NOTE

On some graphing utilities, table increment is designated by ΔTbl or Pitch.

To display the table above on your calculator, go to the Y= editor and enter the expression $32x$. Next, go to the table setup menu. There, let table start = 2 and table increment = 1. Then the displayed table should look like the table below.

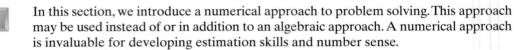

Next, we generate tables to help us solve problems.

EXAMPLE 1 FINDING COMMISSION

Bette Meish is considering a job as a salesperson for a company that manufactures plumbing equipment. She is offered monthly gross pay of $1500 plus 5% commission on sales. An expression that models her gross pay is

In words: $1500 | plus | 5% of sales

Translate: 1500 + 0.05x

When using the table feature, it is necessary to use x as the variable. We enter the expression to be evaluated in y_1 and then indicate where the table is to start, and the increment, or change in x, under the table setup menu. When viewing the table, we can scroll up or down in the x column and down in the y_1 column.

To help her decide about the job, she considers her monthly pay for sales amounts of $1000 to $9000 in $1000 increments.

SALES X	1000	2000	3000	4000	5000	6000	7000	8000	9000
GROSS PAY 1500 + 0.05x									

a. Complete the table.

b. The company informs her that the average sales per month for its sales force is $6000. Find her annual gross pay if she maintains this average.

c. Look for a pattern in the table. Complete the following sentence. For each additional $1000 in sales, Bette's gross pay is increased by _____.

d. What could Bette expect her annual gross pay to be if she averaged $5000 in sales per month?

e. If Bette decides that she needs to earn a gross pay of $1900 per month how much must she average in sales per month?

Solution:

a. To complete the table, go to the Y= editor and enter $y_1 = 1500 + 0.05x$. Notice that the sales values in the given table start at 1000 and the change or increment is 1000. In the table setup menu, let table start = 1000 and table increment = 1000.

X	Y1
1000	1550
2000	1600
3000	1650
4000	1700
5000	1750
6000	1800
7000	1850

Y1■1500+.05X

X	Y1
4000	1700
5000	1750
6000	1800
7000	1850
8000	1900
9000	1950
10000	2000

X=9000

From the screens above we can complete the table as follows:

SALES X	1000	2000	3000	4000	5000	6000	7000	8000	9000
GROSS PAY 1500 + 0.05x	1550	1600	1650	1700	1750	1800	1850	1900	1950

b. To find her annual gross pay if she maintains $6000 in sales per month, find 6000 under x, and the corresponding y_1 entry is 1800. Therefore, if she averages $6000 worth of sales per month, she will average $1800 gross pay per month. To find her annual gross pay, find 12($1800) = $21,600.

c. Observing the table entries, we see that for each additional $1000 sales, Bette's gross pay is increased by $50.

d. To find her annual gross pay if she averages $5000 in sales monthly, scroll to the 5000 entry in the x column. Read the corresponding y_1-value to find a monthly gross pay of $1750. The annual gross pay is 12(1750) = $21,000.

e. Find the 1900 entry in the y_1 (gross pay) column and read the corresponding x-value which is sales. She must average $8000 in sales per month to earn a gross pay of $1900 per month.

EXAMPLE 2   CALCULATING COSTS

Tom Sabo, a licensed electrician, charges $45 per house visit and $30 per hour.

a. Write an equation for Tom's total charge given the number of hours x on the job.

b. Use a table to find Tom's total charge for a job that could take from 1 to 6 hours.

Solution: **a.** To model this problem, recall that we are given that x is number of hours.

In words:	Total charge	is	$45	plus	$30 per hour
Translate:	Total charge	=	45	+	30x

b. To produce a table to model what Tom charges, go to the Y= editor and enter $y_1 = 45 + 30x$. Notice that the x-values (hours) we are interested in start at 1. In the table setup menu, set table start at 1 and table increment at 1.

X	Y1
1	75
2	105
3	135
4	165
5	195
6	225
7	255

Y1▉45+30X

From the table, we read that Tom charges $75 for 1 hour, $105 for 2 hours, $135 for 3 hours, and so on, up to $225 for 6 hours.

There are numerous advantages to producing a table rather than evaluating the algebraic expression for each value of x, but one of the most advantageous situations occurs when you want to change the increment slightly. Let's say that Tom decides to calculate his fee every 15 minutes, or in 15-minute increments. In this case, let's start the table at 0 and have the table increment be 0.25, since every 15 minutes is a quarter of an hour.

X	Y1
0	45
.25	52.5
.5	60
.75	67.5
1	75
1.25	82.5
1.5	90

Y1▉45+30X

From the table we see that it would cost us $45 just for having Tom show up. We can look down the table (scrolling if necessary) to find other charges. For example, we see that the charge is $82.50 to hire him for 1 hour and 15 minutes. This is a reasonable number since we found earlier that the charge is $75 for 1 hour and $105 for 2 hours.

Recall that a *mathematical model* is a graph, table, list, equation, or inequality that describes a situation. We call the equations that model Examples 1 and 2 **linear models** because they can be written in the form $y = mx + b$.

	Equation	$y = mx + b$
Example 1	$y = 1500 + 0.05x$	$y = 0.05x + 1500$
Example 2	$y = 45 + 30x$	$y = 30x + 45$

For linear models, as the x-values increase, the y-values or expressions in x always increase, always decrease, or always stay the same. Check the tables in Examples 1 and 2 to see that this is true. We study linear models and the equation $y = mx + b$ further in Chapter 3.

It is possible to enter more than one expression in the Y= editor.

EXAMPLE 3   **FINDING SALES TAX AND TOTAL COST**
Rod and Karen Pasch are shopping for holiday gifts. On their list are gifts with prices of $39.95, $45, and $62.75 shown in the table below. If the tax rate is 7%, the tax and the total cost is shown below. Let

$$x = \text{the price on the price tag. Then}$$

$$0.07x = \text{amount of tax and}$$

$$x + 0.07x = \text{total cost.}$$

Use a table to find the amount of tax and the total cost of each gift.

PRICE ON TAG	TAX	TOTAL COST
$39.95		
$45.00		
$62.75		

Solution: Since we are looking for tax and total cost, define $y_1 = 0.07x$, the amount of tax, and $y_2 = x + 0.07x$, the total cost of the gift. Notice that there is no constant change in the three price tag values. To keep from scrolling over a large span of numbers in a table, we introduce the **ask** feature. This feature allows us to enter x-values one at a time and see the corresponding y-values. Access the ask feature of the table and enter the tag prices in the x column. The corresponding y_1 entry represents the tax and the y_2 entry represents the total price.

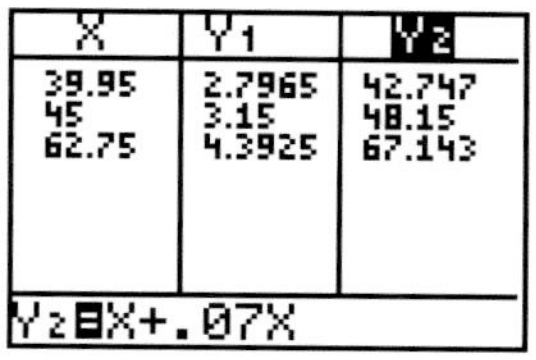

X	Y1	Y2
39.95	2.7965	42.747
45	3.15	48.15
62.75	4.3925	67.143

Y₂■X+.07X

The completed table is given next. Dollar amounts have been rounded to hundredths.

PRICE ON TAG	TAX (y_1)	TOTAL COST (y_2)
$39.95	$2.80	$42.75
$45.00	$3.15	$48.15
$62.75	$4.39	$67.14

Thus far, we have studied linear models only. Another common model is a quadratic model. A quadratic model has the form $y = ax^2 + bx + c, a \neq 0$. For quadratic models, as x-values increase, y-values increase and then decrease, or decrease and then increase. We study quadratic models further in Chapters 3, 6, and 9.

EXAMPLE 4   **HEIGHT OF A ROCKET**
A small rocket is fired straight up from ground level on a campus parking lot with a velocity of 80 feet per second. Neglecting air resistance, the rocket's height y feet at time x seconds is given by the equation

$$y = -16x^2 + 80x.$$

If you are standing at a window that is 64 feet high,

a. In how many seconds will you see the rocket pass by the window?

b. When will you see the rocket pass by the window again?

c. From firing, how many seconds does it take the rocket to hit the ground?

d. Approximate to the nearest tenth of a second when the rocket reaches its maximum height. Approximate the maximum height to the nearest foot.

Solution: Enter $y_1 = -16x^2 + 80x$ in the Y= editor. Using the table feature, let table start = 0 and table increment = 1.

a. To find the number of seconds at which the rocket passes the window, recall that the window height is 64 feet and look for a y_1 value of 64. From the table we can see that it will first pass the window 1 second after it is shot off.

b. Scrolling down the y_1 column in the table we see that at 4 seconds the rocket again passes the window on the way down.

c. We see that at 5 seconds the height is again 0, indicating that the rocket has hit the ground.

d. From the table, the rocket appears to reach its maximum height somewhere between 2 and 3 seconds. To get a better approximation, start the table at 2 and set the table increment to be 0.1 to find one-tenth of a second intervals of time.

X	Y₁
2	96
2.1	97.44
2.2	98.56
2.3	99.36
2.4	99.84
2.5	100
2.6	99.84

$Y_1=100$

The maximum height, 100 feet, occurs 2.5 seconds after the rocket is launched.

Notice that the table above displays a relation between two values. For example, the x-value 2 is paired with the y-value 96 and so on. We study relations and paired data further in Chapter 3.

MENTAL MATH

For each table below, the y_1 column is generated by evaluating an expression in x for corresponding values in the x column. Match each table with the expression in x that generates the y_1 column.

a. $5x$ **b.** $3x + 1$ **c.** x^2

d. $x^2 + 1$ **e.** $-x + 5$ **f.** $-2x$

1.

x	y_1
1	−2
2	−4
3	−6

2.

x	y_1
1	5
2	10
3	15

3.

x	y_1
1	4
2	7
3	10

4.

x	y_1
1	4
2	3
3	2

5.

x	y_1
1	2
2	5
3	10

6.

x	y_1
1	1
2	4
3	9

EXERCISE SET 2.3

Use the table feature to complete each table for the given expression. If necessary, round results to two decimal places. See Examples 1 through 4.

1.

SIDE LENGTH OF A CUBE	x	7	8	9	10	11
VOLUME	x^3					

a. From the table, give the volume of a cube whose side measures 9 centimeters. Use correct units.

b. From the table, give the volume of a cube whose side measures 11 feet. Use correct units.

c. If a cube has volume 343 cubic inches, what is the length of its side?

2.

RADIUS OF CYLINDER	x	2	3	4	5	6
VOLUME (IF HEIGHT IS 3 UNITS)	$3\pi x^2$					

a. Give the volume of a cylinder whose height is 3 inches and whose radius is 5 inches. Use correct units.

b. Give the volume of a cylinder whose height is 3 meters and whose radius is 4 meters. Use correct units.

c. If a cylinder with height 3 kilometers has a volume of 150.8 cubic kilometers, approximate the radius of the cylinder.

3.

HOURS WORKED	x	39	39.5	40	40.5
GROSS PAY (DOLLARS)	$25+7x$				

a. Scroll down the table on your calculator to find how many hours worked gives a gross pay of $333.

b. Scroll up the table to find how many hours worked gives a gross pay of $270.

c. If gross pay is $284, how many hours were worked?

4.

MINUTES ON PHONE	x	7	7.5	8	8.5
TOTAL CHARGE (DOLLARS)	$1.30+0.15x$				

a. Scroll on your calculator to find how long a person with $3.25 can talk on a pay phone.

b. If a person wants to spend no more than $4 for a single phone call, how long can that person stay on the phone?

Solve. If necessary, round amounts to two decimal places.

5. Emily Keaton has a job typing term papers for students on her computer. She charges $5.25 per hour for a rough draft and $7 per hour for a final bound manuscript. Her price increases in 15-minute intervals with a 1-hour minimum.

a. Let x = number of hours and write an algebraic representation for y_1, the cost of preparing a rough draft.

b. Let x = number of hours and write an algebraic representation for y_2, the cost of preparing a final manuscript.

c. Using a table, complete the following chart for the costs of jobs ranging from 1 to 3 hours.

HOURS	1	1.25	1.5	1.75	2	2.25	2.5	2.75	3
ROUGH DRAFT									
MANUSCRIPT									

d. If Emily can type at a rate of 2 pages per 15 minutes (0.25 of an hour), how much does she charge for a 10-page rough draft?

e. If a job that consists of typing a bound manuscript takes Emily 10.75 hours, how much does she charge?

6. The cost of tuition at the local community college is $35.80 per credit hour for in-state students and $134.20 per credit hour for out-of-state students.

a. Write an algebraic representation, y_1, for the cost of in-state tuition dependent on x, the number of credit hours.

b. Write an algebraic representation, y_2, for the cost of out-of-state tuition dependent on x, the number of credit hours.

c. Complete the following table given the number of credit hours.

CREDIT HOURS	2	3	4	5	6
IN-STATE COST (DOLLARS)					
OUT-OF-STATE COST (DOLLARS)					

d. What is the total tuition due if Mandy, an in-state student, is taking a 3-hour math course, a 4-hour science course, a 2-hour study skills course, a 3-hour history course, and a 3-hour Spanish course. The science course has an additional $15 lab fee.

Use the table and the ask feature of your calculator to complete each table for the given expression. If necessary, round to two decimal places. See example 3.

7.

RADIUS OF CIRCLE	x	2	5	21	94.2
CIRCUMFERENCE	$2\pi x$				
AREA	πx^2				

a. Find the circumference and the area of a circle whose radius is 30 yards. Use correct units.

b. Find the circumference and the area of a circle whose radius is 78.5 millimeters. Use correct units.

8.

RADIUS OF SPHERE	x	3	7.1	43	50
VOLUME	$\frac{4}{3}\pi x^3$				
SURFACE AREA	$4\pi x^2$				

a. Find the volume and surface area of a sphere whose radius is 1.7 centimeters. Use correct units.

b. Find the volume and surface area of a sphere whose radius is 25 feet. Use correct units.

Solve.

9. Dan Tweedale is considering a job as an assistant manager of a local department store. He is offered a monthly gross pay of $1100 plus 0.05% commission on sales. He would like to know his gross pay based on the sales amounts shown in the table.

a. Write an algebraic expression for the gross pay, y_1, if x represents the amount of sales.

b. Complete the following table:

SALES (THOUSAND OF DOLLARS)	100	200	300	400	500	600	700
GROSS PAY (DOLLARS)							

c. Find his annual salary if he averages $500,000 in sales per month.

10. Sally Fogelburg is considering a part-time job selling encyclopedias. She is offered a monthly gross pay of $400 plus 20% commission on sales. She would like to know her gross pay based on the sales amounts shown in the table.

a. Write an algebraic expression for the gross pay, y_1, if x represents the amount of sales.

b. Complete the following table:

SALES (DOLLARS)	1000	1500	2000	2500	3000	3500	4000
GROSS PAY (DOLLARS)							

c. Find her gross annual salary if she averages $2500 in sales per month.

11. Kinsley Water Service charges a basic water fee of $12.96 per month plus $1.10 per thousand gallons of water used.

a. Write an algebraic representation for the total cost, y_1, in terms of the number of thousands of gallons of water used, x.

b. Make a table for the cost each month of 0 to 6 thousand gallons of water used.

GALLONS OF WATER USED (THOUSANDS)	0	1	2	3	4	5	6
MONTHLY COST (DOLLARS)							

12. Jewell Water Service charges a basic water fee of $6.00 per month plus $1.72 per thousand gallons of water used.

a. Write an algebraic representation for the total cost, y_1, in terms of the number of thousands of gallons of water used, x.

b. Make a table for the cost each month of 0 to 6 thousand gallons of water.

GALLONS OF WATER USED (THOUSANDS)	0	1	2	3	4	5	6
MONTHLY COST (DOLLARS)							

13. A recipe for cookies calls for the following amounts of ingredients: $\frac{3}{4}$ cup granulated sugar, $\frac{1}{4}$ cup brown sugar, $\frac{1}{2}$ cup butter, 3 cups flour, $\frac{1}{8}$ teaspoon soda. Nancy wants to make a recipe and a half. Find the amount of each ingredient if the recipe is to be 1.5 times greater. Write the new amounts as fractions or mixed numbers.

INGREDIENT	Gran. Sugar	Brown Sugar	Butter	Flour	Soda
RECIPE	$\frac{3}{4}$ cup	$\frac{1}{4}$ cup	$\frac{1}{2}$ cup	3 cups	$\frac{1}{8}$ teaspoon
NEW AMOUNT					

14. A recipe for rice calls for the following ingredients in the following measures: 4 cups water, $2\frac{1}{2}$ cups rice, $\frac{1}{4}$ teaspoon salt, $\frac{3}{8}$ teaspoon butter. The recipe is for serving 3 people. Calculate the amounts needed for 5 people. Write the amounts as fractions or mixed numbers.

INGREDIENT	Water	Rice	Salt	Butter
3 PERSONS	4 cups	$2\frac{1}{2}$ cups	$\frac{1}{4}$ teaspoon	$\frac{3}{8}$ teaspoon
5 PERSONS				

15. Kelsey is considering two job offers. The first job pays a starting salary of $14,500, with raises of $1000 every year. The second job pays a starting salary of $20,000 with raises of $575 each year.

a. Make a chart showing both salaries at the end of 5, 10, 15, and 20 years.

b. Which job do you think pays her the most money if she plans to stay with the company until retirement in 20 years?

16. Rory Baruch considers two different salary plans that his company offers. Plan A starts at $25,000, with raises of $650 per year. Plan B starts at $22,000, with raises of $1000 per year.

a. If Rory plans to stay with the company less than 10 years, make a chart showing his salary for both plans for 0 to 9 years.

b. Which plan looks more advantageous for Rory?

17. The Jackson Electric Authority charges a basic fee of $15.00 per month plus $0.01924 per kilowatt-hour of electricity used.

a. Write an algebraic representation for the total cost, y_1, in terms of the number of kilowatt-hours used, x.

b. Complete the table for the cost each month of 0 to 2000 kilowatt-hours.

KILOWATT-HOURS	0	500	1000	1500	2000
MONTHLY COST (DOLLARS)					

18. The First Coast Electric charges a basic fee of $10.00 per month plus $0.01850 per kilowatt-hour of electricity used.

a. Write an algebraic representation for the total cost, y_1, in terms of the number of kilowatt-hours used, x.

b. Complete the table for the cost each month of 0 to 2000 kilowatt-hours.

KILOWATT-HOURS	0	500	1000	1500	2000
MONTHLY COST (DOLLARS)					

19. Judy Martinez has been shopping for school clothes and has picked out several outfits. She knows that the following Wednesday there will be a Super Wednesday sale and everything in the store will be discounted 35%.

a. Write an algebraic representation for the sale price, y_1, in terms of the price, x, on the price tag.

b. Complete the table and find the cost of each item on sale, given the price on the price tag. Round dollar amounts to the nearest cent.

ITEM	Blouse	Skirt	Shorts	Shoes	Purse	Earrings	Backpack
PRICE TAG	$29.95	$35.95	$19.25	$39.95	$17.95	$9.95	$25.75
SALE PRICE							

c. If she purchases all the items when on sale, what will her total bill be before taxes?

d. What is the total after taxes if a 7% tax is added to the bill?

e. How much did she save before taxes by waiting for the sale?

20. Saumil has been shopping for tools and has picked out several at Sears. Next month, Sears is having a tool sale and all tools in the store will be discounted 30%.

a. Write an algebraic representation for the sale price, y_1, in terms of the price, x, on the price tag.

b. Complete the table and find the cost of each item on sale, given the price on the price tag.

ITEM	Hammer	Drill	Sander	Glue gun	Screwdriver set	Socket wrench set
PRICE TAG	$9.95	$29.95	$49.75	$19.95	$27.95	$79.95
SALE PRICE						

c. If he purchases all the items when on sale, what will his total bill be before taxes?

d. What is the total after taxes if a 7% tax is added to the bill?

e. How much did he save before taxes by waiting for the sale?

21. GROUP ACTIVITY: The owner of Descartes Pizzaria is in the process of changing his pizza prices. He is considering two methods for determining the new price for a pizza. One method is to charge 5¢ per square inch of area of pizza. Another method is to charge 25¢ per inch of circumference. Fill in the table below and compare the costs of pizzas for the given *diameters* using the two methods. Which price do you think the owner should choose for each size of pizza? Why?

DIAMETER (INCHES)	AREA	CIRCUMFERENCE	COST (AREA)	COST (CIRCUMFERENCE)
6				
12				
18				
24				
30				
36				

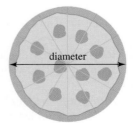

diameter

22. A firecracker rocket is fired at ground level with an initial velocity of 100 feet per second. Neglecting air resistance the height of the rocket at time x seconds is given by the equation $y = -16x^2 + 100x$ where y is measured in feet.

a. Find the height of the rocket for the first five seconds in one second intervals.

SECONDS	1	2	3	4	5
HEIGHT (FEET)					

b. Explain why the rocket heights increase and then decrease.

c. If it doesn't explode, determine to the nearest tenth of a second when the rocket hits the ground.

23. A rocket with an initial velocity of 85 feet per second is launched from the top of a mountain 1500 feet above sea level. Neglecting air resistance, the height of the rocket above sea level at the time x seconds is given by the equation $y = -16x^2 + 85x + 1500$ where y is measured in feet.

a. Find its height for the first 5 seconds after it is launched in one second intervals.

b. Find the rocket's maximum height to the nearest foot.

24. Dan Newman is going to the county fair. The advertisement in the paper states that there are two different ticket options. The first option is to pay $4.00 for an admission ticket and $0.35 for each ride. The second option is to pay $1.50 for an admission ticket and $0.75 for each ride.

a. For 5 rides, determine the price of each option.

b. After how many rides is option 1 the better buy?

25. Anne Holloway rents a car for $25 per day plus $0.08 per mile. Michelle Goods rents a similar car from a different rental agency for $35 per day plus $0.05 per mile.

a. Find both rental charges if they each drive 300 miles.

b. For what distance is Anne's rental agreement cheaper than Michelle's?

26. Matthew Aires repairs computers and charges $35 per house visit plus $25 per hour.

a. Write an expression for Matthew's total charge given the number of hours, x, on the job.

b. Complete the following table for the total charges given the number of hours on the job.

HOURS ON JOB	1.5	2	2.5	3
COST (DOLLARS)				

27. William Kramer, a lawyer, charges an initial consultation fee of $90 plus $75 per hour after this consultation.

a. Write an expression for William's total charges given the number of hours, x, on the job.

b. Complete the table below for the total charges given the number of hours *after* the initial consultation.

NO. OF HOURS	2	4	6	8
TOTAL FEE (DOLLARS)				

28. An object is *dropped* from the top of the Nations Bank Tower in Atlanta, Georgia. Neglecting air resistance, the height of the object at time x seconds is given by

$$y = -16x^2 + 1050$$

where y is measured in feet. Use a table to determine the following.

a. What is the object's maximum height? How did you arrive at your answer?

b. To the nearest tenth of a second, determine when the object hits the ground.

29. The path of some leaping animals can be described by a quadratic model. A particular leap of a frog can be modeled by the equation

$$y = -0.14x^2 + 1.4x$$

where x is the number of feet horizontally from a starting point and y is the vertical height of the frog in feet. Use a table to determine the following.

Leap of frog

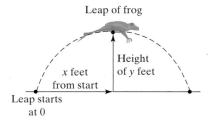

a. Determine to the nearest tenth of a foot the maximum height reached by the frog.

b. What is the frog's horizontal distance from the starting point when the maximum height is reached?

c. How far is the frog from the starting point when it lands? (By the way, the longest frog leap on record is 21 feet $5\frac{3}{4}$ inches.)

30. The high temperature for a particular week in Jackson, Mississippi can be described by the quadratic model

$$y = 1.05x^2 - 8.38x + 89.43$$

where x is the day of the week (Sunday, $x = 1$; Monday, $x = 2$, and so on) and y is the temperature in degrees Fahrenheit.

a. Find the day of the week with the lowest temperature.

b. What is the lowest temperature for the week rounded to the nearest tenth?

c. Do you think that this equation models the temperature for much longer than a week, say 2 weeks? Why or why not?

 31. The sunrise times in Valdivia, Chile for the months of April through October can be described by the quadratic model

$$y = -0.14x^2 + 1.09x + 5.11$$

where $x = 1$ is the first day of April, $x = 2$ is the first day of May, $x = 3$ is the first day of June, and so on and y is the time of sunrise in hours A.M.

South
America

Valdivia

a. Approximate the time of sunrise on May 1. Give the time in hundredths of an hour and also in hours and minutes.

b. For what integer value of x is the sunrise the latest? What calendar date does this correspond to?

c. Give this latest sunrise time in hours and minutes.

d. Approximate the sunrise time in hours and minutes for September 15.

e. Do you think that this equation models the sunrise times for all the months of the year? Why or why not?

Review Exercises

Solve each equation. See Section 2.1.

32. $x + 4.8 = 9.6$

33. $y - 6.8 = 2.3$

34. $2(n + 6) - 5n = -3(n - 4)$

35. $-8x + 2(x + 3) = -6(x - 1)$

36. $\dfrac{y}{7} + 2 = \dfrac{y + 1}{14}$

37. $\dfrac{z}{3} - \dfrac{1}{6} = \dfrac{z}{12}$

2.4   FORMULAS AND PROBLEM SOLVING

O B J E C T I V E S

1   Solve a formula for a specified variable.

2   Use formulas to solve problems.

TAPE IAG 2.4

1   Solving problems that we encounter in the real world sometimes requires us to express relationships among measured quantities. A **formula** is an equation that describes a known relationship among measured phenomena, such as time, area, and gravity. Some examples of formulas are

FORMULA	MEANING
$I = PRT$	Interest = principal $\cdot$ rate $\cdot$ time
$A = lw$	Area of a rectangle = length $\cdot$ width
$d = rt$	Distance = rate $\cdot$ time
$C = 2\pi r$	Circumference of a circle = $2 \cdot \pi \cdot$ radius
$V = lwh$	Volume of a rectangular solid = length $\cdot$ width $\cdot$ height

Other formulas are listed inside the back cover of this text. Notice that the formula for the volume of a rectangular solid, $V = lwh$, is solved for V, since V is by itself on one side of the equation with no, V's on the other side of the equation. Suppose that the volume of a rectangular solid is known as well as its width and its length, and we wish to find its height. One way to find its height is by solving the formula $V = lwh$ for h.

EXAMPLE 1   Solve $V = lwh$ for h.

Solution: To solve $V = lwh$ for h, isolate h on one side of the equation. To do so, divide both sides of the equation by lw.

$$V = lwh$$

$$\frac{V}{lw} = \frac{lwh}{lw} \qquad \text{Divide both sides by } lw.$$

$$\frac{V}{lw} = h \qquad \text{Simplify.}$$

Thus, to find the height of a rectangular solid, divide the volume by the product of its length and its width.

The following steps may be used to solve formulas, and equations in general, for a specified variable.

TO SOLVE EQUATIONS FOR A SPECIFIED VARIABLE

Step 1.   Clear the equation of fractions by multiplying each side of the equation by the least common denominator.

Step 2.   Use the distributive property to remove grouping symbols such as parentheses.

Step 3.   Combine like terms on each side of the equation.

Step 4.   Use the addition property of equality to rewrite the equation as an equivalent equation with terms containing the specified variable on one side and all other terms on the other side.

Step 5.   Use the distributive property and the multiplication property of equality to isolate the specified variable.

EXAMPLE 2   Solve $3y - 2x = 7$ for y.

Solution: This is a linear equation in two variables. Often an equation such as this is solved for y in order to reveal some properties about the graph of this equation, which we will learn more about in Chapter 3. Since there are no fractions, grouping symbols, or uncombined terms, we begin with step 4. Isolate the term containing the specified variable y by adding $2x$ to both sides of the equation.

$$3y - 2x = 7$$

$$3y - 2x + 2x = 7 + 2x \qquad \text{Add } 2x \text{ to both sides.}$$

$$3y = 7 + 2x$$

To solve for y, divide both sides by 3.

$$\frac{3y}{3} = \frac{7 + 2x}{3}$$ Divide both sides by 3.

$$y = \frac{2x + 7}{3} \quad \text{or} \quad y = \frac{2}{3}x + \frac{7}{3}$$

EXAMPLE 3   Solve $A = \frac{1}{2}(B + b)h$ for b.

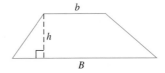

Solution: Since this formula for finding the area of a trapezoid contains a fraction, we begin by multiplying both sides of the equation by the LCD 2.

$$A = \frac{1}{2}(B + b)h$$

$$2 \cdot A = 2 \cdot \frac{1}{2}(B + b)h$$ Multiply both sides by 2.

$$2A = (B + b)h$$ Simplify.

Next, use the distributive property and remove parentheses.

$$2A = (B + b)h$$

$$2A = Bh + bh$$ Apply the distributive property.

$$2A - Bh = bh$$ Isolate the term containing b by subtracting Bh from both sides.

$$\frac{2A - Bh}{h} = \frac{bh}{h}$$ Divide both sides by h.

$$\frac{2A - Bh}{h} = b \text{ or } b = \frac{2A - Bh}{h}$$

> R E M I N D E R   Remember that we may isolate the specified variable on either side of the equation.

2   In this section, we also solve problems that can be modeled by known formulas. We use the same problem-solving steps that were introduced in Section 2.2. These steps have been slightly revised to include tables and formulas.

PROBLEM-SOLVING STEPS

1. UNDERSTAND the problem. During this step, don't work with variables (except for known formulas), but simply become comfortable with the problem. Some ways of accomplishing this are listed below.
   - Read and reread the problem.
   - Construct a drawing.
   - Look up an unknown formula.
   - Propose a solution and check. Pay careful attention as to how to check your proposed solution. This will help later when you write an equation to model the problem.
   - A table may help you to recognize any patterns.

2. ASSIGN a variable to an unknown in the problem. Use this variable to represent any other unknown quantities.

3. ILLUSTRATE the problem. A diagram or chart using the assigned variables can often help you visualize the known facts.

4. TRANSLATE the problem into a mathematical model. This is often an equation.

5. COMPLETE the work. This often means to solve the equation.

6. INTERPRET the results: *Check* the proposed solution in the stated problem and *state* your conclusion.

EXAMPLE 4   The formula $C = \dfrac{5}{9}(F - 32)$ converts degrees Fahrenheit to degrees Celsius. Use this formula and the table feature of your calculator to complete the given table. If necessary, round values to the nearest hundredth.

FAHRENHEIT		−4	10	32	70	100
CELSIUS						

Solution: Let x = degrees Fahrenheit. Then enter $y_1 = \dfrac{5}{9}(x - 32)$ to find the corresponding degrees Celsius. Notice that there is no constant increment change in the Fahrenheit values in the given table. To keep from scrolling over a large span of numbers in a table (not a good use of time), we activate the ask feature in table setup.

Enter each x-value from the table. The corresponding y_1-values are shown to the left. The completed table is

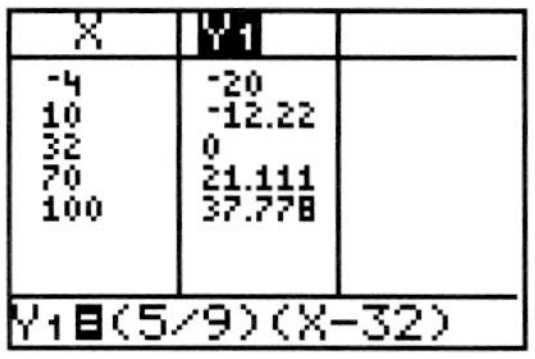

X	Y1	
-4	-20	
10	-12.22	
32	0	
70	21.111	
100	37.778	

$Y_1 \boxminus (5/9)(X-32)$

FAHRENHEIT	x	−4	10	32	70	100
CELSIUS	$\frac{5}{9}(x - 32)$	−20	−12.22	0	21.11	37.78

Formulas are very useful in problem solving. For example, the compound interest formula

$$A = P\left(1 + \frac{r}{n}\right)^{nt}$$

is used by banks to compute the amount A in an account that pays compound interest. The variable P represents the principal, or amount invested in the account, r is the annual rate of interest, t is the time in years, and n is the number of times compounded per year.

EXAMPLE 5   **CALCULATING INTEREST**
Marial Callier just received an inheritance of $10,000 and plans to place all the money in a savings account that pays 5% compounded quarterly to help her son go to college in 3 years. How much money will be in the account in 3 years?

Solution: 1. UNDERSTAND. Read and reread the problem. The appropriate formula needed to solve this problem is the compound interest formula

$$A = P\left(1 + \frac{r}{n}\right)^{nt}$$

Make sure that you understand the meaning of all the variables in this formula.

2. ASSIGN. Since this is a direct application of the compound interest formula, the variables and their meaning are already assigned. To review, they are

A = amount in the account after t years
P = principal or amount invested
t = time in years
r = rate of interest
n = number of times compounded per year

3. ILLUSTRATE. No illustration is needed.

4. TRANSLATE. Use the compound interest formula and let P = $10,000, r = 5% = 0.05, t = 3 years, and n = 4 since the account is compounded quarterly, or 4 times a year.

Formula: $$A = P\left(1 + \frac{r}{n}\right)^{nt}$$

Substitute: $$A = 10,000\left(1 + \frac{0.05}{4}\right)^{4\cdot3}$$

5. COMPLETE. Here, we can store the known values in the calculator and evaluate the algebraic expression, or we can use our calculator to evaluate the numerical expression.

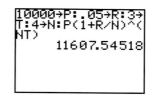

```
10000→P:.05→R:3→
T:4→N:P(1+R/N)^(
NT)
            11607.54518
```

6. INTERPRET. To *check* here, repeat your calculations to make sure that no error was made. Notice that $11,607.55 is a reasonable amount to have in the account after 3 years. *State:* In 3 years, the account contains $11,607.55.

DISCOVER THE CONCEPT

Finding the Maximum Area. While this discovery can be done by an individual, we suggest working in groups of 2 or 3 if feasible. Each group/person begins with 12 toothpicks.

Suppose that each toothpick represents a 1-foot piece of fencing. Use all the toothpicks (without bending or breaking) to form a rectangular fence. First, form a rectangular fence of width 1 foot, next, a rectangular fence of width 2 feet, then 3 feet and so on. For each rectangular fence constructed, record the width, length, area, and perimeter in a table such as the one below.

WIDTH	LENGTH	AREA	PERIMETER
1			
2			
3			
4			
5			
6			

a. Find the dimensions of the rectangle with the largest area.

b. Compare the areas of the rectangles with their perimeters. What patterns do you notice?

From the discovery above, we see that the rectangle with largest area has dimensions 3 feet by 3 feet and an area of 9 square feet. Notice that although a change in dimensions results in a change in area, the perimeter remains constant.

EXAMPLE 6

FINDING MAXIMUM AREA

Sylvia Daschle has purchased 60 feet of fencing in 1-foot sections. She wants to enclose the largest possible rectangular garden using a portion of a side of her house as a side of the rectangle. Use a table to find the dimensions that will give Sylvia the largest area.

Fencing

House

Solution: **1.** UNDERSTAND. Read and reread the problem and then try some dimensions in particular. For example, if the width is 1 foot, the length is $60 - 2(1)$ or 58 feet. The area of the rectangle is $w \cdot l$ or $1(58) = 58$ square feet. Next, let the width be 2 feet and continue until a pattern is found. A table can be useful when looking for a pattern.

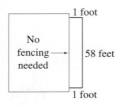

WIDTH	LENGTH	AREA
1	$60 - 2(1)$ or 58	$1 \cdot 58 = 58$
2	$60 - 2(2)$ or 56	$2 \cdot 56 = 112$
3	$60 - 2(3)$ or 54	$3 \cdot 54 = 162$

2. **ASSIGN.** Let x = the width of the garden. From the table, we can see that
$60 - 2x$ = the length of the garden

3. **ILLUSTRATE.** No other illustration is needed.

4. **TRANSLATE.**

In words: | Area | = | width | · | length | or

Translate: Area = x · $(60 - 2x)$

5. **COMPLETE.** Since Sylvia has fencing in 1-foot sections, we know that the width and the length are natural numbers. To solve, we can use a table. We can enter the expression for length in y_1 and the area in y_2.

$y_1 = 60 - 2x$ length

$y_2 = x(60 - 2x)$ area

Set table start to be 1 and table increment to be 1. Scroll down the y_2 or area column to find the largest area. Notice that equation $y_2 = x(60 - 2x)$ is a quadratic model so for increasing values of x, the y_2 values will increase and then decrease, or decrease and then increase.

X	Y₁	Y₂
1	58	58
2	56	112
3	54	162
4	52	208
5	50	250
6	48	288
7	46	322

Y₁ = 60 − 2X

X	Y₁	Y₂
10	40	400
11	38	418
12	36	432
13	34	442
14	32	448
15	30	450
16	28	448

Y₂ = 450

The largest area is 450 square feet.

6. **INTERPRET.** *Check* the solution and *state*. From the table, we see that the dimensions that gives the largest area are 15 feet by 30 feet. The largest area is 450 square feet.

MENTAL MATH

Solve each equation for the specified variable. See Examples 1 and 2.

1. $2x + y = 5$; for y
2. $7x - y = 3$; for y
3. $a - 5b = 8$; for a
4. $7r + s = 10$; for s
5. $5j + k - h = 6$; for k
6. $w - 4y + z = 0$; for z

EXERCISE SET 2.4

Solve each equation for the specified variable. See
Examples 1 through 3.

1. $D = rt$; for t **2.** $W = gh$; for g

3. $I = PRT$; for R **4.** $V = LWH$; for L

5. $9x - 4y = 16$; for y **6.** $2x + 3y = 17$; for y

7. $P = 2L + 2W$; for W **8.** $A = 3M - 2N$; for N

9. $J = AC - 3$; for A **10.** $y = mx + b$; for x

11. $W = gh - 3gt^2$; for g **12.** $A = Prt + P$; for P

13. $T = C(2 + AB)$; for B **14.** $A = 5H(b + B)$; for b

15. $C = 2\pi r$; for r

16. $S = 2\pi r^2 + 2\pi rh$; for h

17. $E = I(r + R)$; for r **18.** $A = P(1 + rt)$; for t

19. $s = \dfrac{n}{2}(a + L)$; for L **20.** $\dfrac{3}{4}(b - 2c) = a$; for b

21. $\dfrac{1}{u} - \dfrac{1}{v} = \dfrac{1}{w}$; for w **22.** $\dfrac{1}{r_1} + \dfrac{1}{r_2} = \dfrac{1}{R}$; for R

23. $N = 3st^4 - 5sv$; for v **24.** $L = a + (n - 1)d$; for d

25. $S = \dfrac{a}{1 - r}$; for r **26.** $m = \dfrac{y_2 - y_1}{x_2 - x_1}$; for y_1

27. $S = 2LW + 2LH + 2WH$; for H

28. $T = 3vs - 4ws + 5vw$; for v

In this exercise set, round all dollar amounts to two
decimal places.

Solve. See Examples 4 through 6.

29. Complete the table and find the balance A if \$3500
is invested at an annual percentage rate of 3% for
10 years and compounded n times a year.

n	1	2	4	12	365
A					

30. Complete the table and find the balance A if \$5000
is invested at an annual percentage rate of 6% for
15 years and compounded n times a year.

n	1	2	4	12	365
A					

31. If you are investing money in a savings account
paying a rate of r, which account should you

choose—one compounded 4 times a year or one
compounded 12 times a year? Explain your choice.

32. To borrow money at a rate of r, which bank should
you choose—one compounding 4 times a year or
one compounding 12 times a year? Explain your
choice.

33. A principal of \$6000 is invested in an account
paying an annual percentage rate of 4%. Find the
amount in the account after 5 years if the account is
compounded

a. semiannually   **b.** quarterly   **c.** monthly

34. A principal of \$25,000 is invested in an account
paying an annual percentage rate of 5%. Find the
amount in the account after 2 years if the account is
compounded

a. semiannually   **b.** quarterly   **c.** monthly

35. The day's high temperature in Phoenix, Arizona, was
recorded as 104°F. Write 104°F as degrees Celsius.

36. The day's low temperature in Nome, Alaska, was
recorded as −15°C. Write −15°C as degrees
Fahrenheit.

37. Helen Schmitz wants to enclose a rectangular
garden. She has 56 feet of fencing in 1 foot lengths.
She plans to use one side of her house as one side of
the fence so that she only needs to fence three sides.

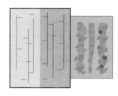

a. Find the dimensions that will give Helen the
largest area.

b. What is the largest area?

38. Kevin Elliott has 48 feet of fencing that comes in 1
foot lengths. He wants to use the entire amount of
fencing to construct a rectangular pen for his new
puppy. He needs to fence all four sides of the
rectangular pen.

a. Find the dimensions of the rectangular pen that
will give him the largest area.

b. What is the largest area?

39. Omaha, Nebraska, is about 90 miles from Lincoln, Nebraska. Irania must go to the law library in Lincoln to get a document for the law firm she works for. Find how long it takes her to drive **round-trip** if she averages 50 mph.

40. It took the Selby family $5\frac{1}{2}$ hours round-trip to drive from their house to their beach house 154 miles away. Find their average speed.

41. A package of floor tiles contains 24 one-foot-square tiles. Find how many packages should be bought to cover a square ballroom floor whose side measures 64 feet.

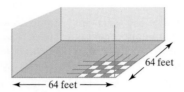

42. One-foot-square ceiling tiles are sold in packages of 50. Find how many packages must be bought for a rectangular ceiling 18 feet by 12 feet.

43. The deepest hole drilled into the ocean floor is beneath the Pacific Ocean and is called Hole 504B. It is located off the coast of Ecuador. Scientists are drilling it to learn more about Earth's history. Currently, the hole is in the shape of a cylinder whose volume is approximately 3800 cubic feet and whose length is 1.3 miles. Find the radius of the hole to the nearest hundredth of a foot. (*Hint:* Make sure the same units of measurement are used.)

44. The deepest hole drilled into Earth is called the Kola Superdeep Borehole. It is approximately 8 miles deep and is located near a small Russian town in the Arctic Circle. If it takes 7.5 hours to remove the drill from the bottom of the hole, find the rate that the drill can be retrieved in feet per second. Round to the nearest tenth. (*Hint:* Write 8 miles as feet, 7.5 hours as seconds, then use the formula $d = rt$.)

45. The formula for the measure of an angle y (in degrees) of a regular polygon with x sides is given by

$$y = \frac{(x - 2)180}{x}$$

(Recall that a regular polygon has sides the same length and angles with the same measure.)

Complete the table and find the measure of an angle of a polygon with x sides.

REGULAR POLYGON					
NUMBER OF SIDES (x)	3	5	8	10	12
MEASURE OF AN ANGLE (DEGREES)					

46. The formula for finding distance traveled at the rate of 47 miles per hour for x hours is given by

$$y = 47x$$

where y is measured in miles. Complete the table and find the distances traveled for the given times.

HOURS (x)	2	3.5	5.2	8.7	12.1
DISTANCE TRAVELED (MILES)					

47. On April 1, 1985, *Sports Illustrated* published an April Fool's story by writer George Plimpton. He wrote that the New York Mets had discovered a man who could throw a 168-miles-per-hour fastball. If the distance from the pitcher's mound to the plate is 60.5 feet, how long would it take for a ball thrown at that rate to travel that distance? (*Hint:* Write the rate 168 miles per hour in feet per second.

$$
\begin{aligned}
168 \text{ miles per hour} &= \frac{168 \text{ miles}}{1 \text{ hour}} \\
&= \frac{\_\_\_ \text{ feet}}{\_\_\_ \text{ seconds}} \\
&= \frac{\_\_\_ \text{ feet}}{1 \text{ second}} \\
&= \_\_\_ \text{ feet per second.}
\end{aligned}
$$

Then use the formula $d = r \cdot t$.)

48. In 1945, Arthur C. Clarke, a scientist and science-fiction writer, calculated that an artificial satellite placed at a height of 22,248 miles directly above the equator would orbit the globe at the same speed with which Earth was rotating. This belt along the equator is known as the Clarke belt. Use the formula for circumference of a circle and find the "length" of the Clarke belt. (*Hint:* Recall that the radius of Earth is approximately 4000 miles. Round to the nearest whole mile.)

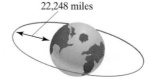

22,248 miles

49. An orbit such as Clarke's belt in Exercise 48 is called a geostationary orbit. In your own words, why do you think that communications satellites are placed in geostationary orbits?

50. How much do you think it costs each American to build a space shuttle? Write down your estimate. The Space Shuttle *Endeavour* was completed in 1992 and cost approximately $1.7 billion. If the population of the United States in 1992 was 250 million, find the cost per person to build the *Endeavour*. How close was your estimate?

51. The formula for the volume of a rectangular box is $V = \text{length} \cdot \text{width} \cdot \text{height}$. A box with no lid is formed from a 9 inch by 15 inch sheet of cardboard by cutting a 2 inch square out of each side and folding up the sides to form a box with no lid.

  a. What are the dimensions of the box?

  b. What is the volume of the box?

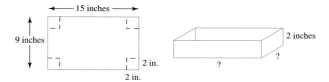

52. Find the volume of a rectangular box with no lid that is made from a 8.5 inch by 11 inch sheet of cardboard by cutting a 1.5 inch square out of each corner and folding up the resulting sides.

  a. Find the dimensions of the box with no lid.

  b. Find the volume of the box.

53. Find *how much interest* $10,000 earns in 2 years in a certificate of deposit paying 8.5% interest compounded quarterly.

54. Bryan, Eric, Mandy, and Melissa would like to go to Disneyland in 3 years. Their total cost should be $4500. If each invests $1000 in a savings account paying 5.5% interest compounded semiannually, will they have enough in 3 years?

55. A gallon of latex paint can cover 500 square feet. Find how many gallon containers of paint should be bought to paint two coats on all four walls of a rectangular room whose dimensions are 14 feet by 16 feet (assume 8-foot ceilings).

56. A gallon of enamel paint can cover 300 square feet. Find how many gallon containers of paint should be bought to paint three coats on a wall measuring 21 feet by 8 feet.

57. A portion of the external tank of the Space Shuttle *Endeavour* is a liquid hydrogen tank. If the ends of the tank are hemispheres, find the volume of the tank. To do so, answer parts (a) through (c).

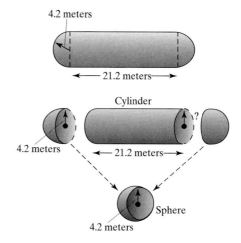

  a. Find the volume of the cylinder shown. Round to 2 decimal places.

  b. Find the volume of the sphere shown. Round to 2 decimal places.

  c. Add the results of parts (a) and (b) This sum is the approximate volume of the tank.

58. The space probe Pioneer 10 traveled 619 million miles from Mars to Jupiter in 21 months. Find the average speed of the probe in miles per hour. [*Hint:* Convert 21 months to hours (use 1 month = 30 days) and then use the formula $d = rt$.]

59. Find how long it takes Mark to drive 135 miles on I-10 if he merges onto I-10 at 10 A.M. and drives nonstop with his cruise control set at 60 mph.

60. If the area of a triangular kite is 18 square feet and its base is 4 feet, find the height of the kite.

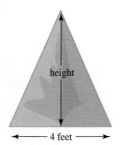

61. Solar System distances are so great that units other than miles or kilometers are often used. For example, the astronomical unit (AU) is the average distance between Earth and the Sun, or 92,900,000 miles. Use this information to convert each planet's distance in miles from the Sun to astronomical units. Round to three decimal places.

	MILES FROM THE SUN	AU FROM THE SUN
Mercury	36 million	
Venus	67.2 million	
Earth	92.9 million	
Mars	141.5 million	
Jupiter	483.3 million	
Saturn	886.1 million	
Uranus	1,783 million	
Neptune	2,793 million	
Pluto	3,670 million	

62. The Space Shuttle has a cargo bay that is in the shape of a cylinder whose length is 18.3 meters and whose diameter is 4.6 meters. Find its volume rounded to the nearest tenth of a cubic meter.

A Look Ahead

The measure of the chance, or likelihood, of an event occurring is its **probability***. A formula basic to the study of probability is the formula for the probability of an event when all the outcomes are equally likely. This formula is*

$$\text{Probability of an event} = \frac{\text{number of ways that the event can occur}}{\text{number of possible outcomes}}$$

For example, to find the probability that a single spin on the spinner will result in red, notice first that the spinner is divided into 8 equal parts, so there are 8 possible equally likely outcomes. Next, notice that there is only one sector of the spinner colored red, so the number of ways that the spinner will land on red is 1. Then this probability denoted by P(red) is

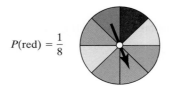

$$P(\text{red}) = \frac{1}{8}$$

Find each probability in simplest form.

63. $P(\text{green})$

64. $P(\text{yellow})$

65. $P(\text{black})$

66. $P(\text{blue})$

67. $P(\text{green or blue})$

68. $P(\text{black or yellow})$

69. $P(\text{red, green, or black})$

70. $P(\text{yellow, blue, or black})$

71. $P(\text{white})$

72. $P(\text{red, yellow, green, blue, or black})$

73. From the previous probability formula, what do you think is always the probability of an event that is impossible to occur?

74. What do you think is always the probability of an event that is sure to occur?

Review Exercises

Determine which numbers in the set $\{-3, -2, -1, 0, 1, 2, 3\}$ are solutions of each inequality. See Section 1.2.

75. $x < 0$

76. $x > 1$

77. $x + 5 \leq 6$

78. $x - 3 \geq -7$

2.5 | INTERPRETING DATA AND READING BAR, LINE, AND CIRCLE GRAPHS

TAPE IAG 2.5

O B J E C T I V E S

1. Find the mean, median, and mode of a set of data.
2. Interpret circle graphs.
3. Interpret bar graphs.
4. Interpret line graphs.

Many real-world situations involve problem solving based on collecting, graphing, and analyzing data. Computers can quickly organize and analyze the data whether they are collected by means of a survey, a collection device such as the Calculator Based Laboratory that can be linked to the TI calculators, or simply by using numerical methods. Once we collect data we need to organize and interpret the results.

To organize and analyze data, it is sometimes desirable to be able to describe the set of data, or a set of numbers, by a single "middle" number. Three such **measures of central tendency** are the *mean,* the *median,* and the *mode.*

The most common measure of central tendency is the mean (sometimes called the arithmetic mean or the average). The **mean** of a set of data items, denoted by $\bar{x}$, is the sum of the items divided by the number of items.

EXAMPLE 1   **ANALYZING A PSYCHOLOGY EXPERIMENT**
Seven students in a psychology class conducted an experiment on mazes. Each student was given a pencil and asked to successfully complete the same maze. The time results are below.

STUDENT	Ann	Thanh	Carlos	Jesse	Melinda	Ramzi	Dayni
TIME (SECONDS)	13.2	11.8	10.7	16.2	15.9	13.8	18.5

a. What was the shortest time and the longest time for the maze to be completed?

b. Find the mean of the times.

c. How many students took more than the mean time? How many students took less than the mean time?

Solution: **a.** Carlos completed the maze in 10.7 seconds, the shortest time. Dayni completed the maze in 18.5 seconds, the longest time.

b. To find the mean, $\bar{x}$, find the sum of the data items and divide by 7, the number of items.

$$\bar{x} = \frac{13.2 + 11.8 + 10.7 + 16.2 + 15.9 + 13.8 + 18.5}{7} = \frac{100.1}{7} = 14.3 \text{ seconds}$$

c. Three students, Jesse, Melinda, and Dayni, had times longer than the mean time. Four students, Ann, Thanh, Carlos, and Ramzi, had times shorter than the mean time.

Two other measures of central tendency are the median and the mode.

The **median** of an ordered set of numbers is the middle number. If the number of items is even, the median is the mean of the two middle numbers. The **mode** of a set of numbers is the number that occurs most often. It is possible for a data set to have no mode or more than one mode.

EXAMPLE 2   ANALYZING TEMPERATURE

Find the median and the mode of the following set of numbers. These numbers were high temperatures for fourteen consecutive days in a city in Montana.

$$76, 80, 85, 86, 89, 87, 82, 77, 76, 79, 82, 89, 89, 92$$

Solution   First, write the numbers in order.

$$76, 76, 77, 79, 80, 82, 82, 85, 86, 87, 89, 89, 89, 92$$

two   mode
middle numbers

Since there are an even number of items, the median is the mean of the two middle numbers.

$$\text{median} = \frac{82 + 85}{2} = 83.5$$

The mode is 89, since 89 occurs most often.

If a data set contains a large number of items, calculating the mean, median, and mode by hand can be tedious. In this section we introduce **lists** to help us to calculate these measures of central tendency. A list is different from a table in many ways. Whereas a *table* may give x and y_1 values that are related by some equation, it is possible to enter a *list* of single values only.

EXAMPLE 3   FINDING THE MEAN AND THE MEDIAN OF A SET OF DATA

Find the mean and the median of the following set of temperatures for a given week.

Mon	Tues	Wed	Thurs	Fri	Sat	Sun
89	82	76	80	68	70	74

Solution: To find the mean by hand, find the sum of the temperatures and divide by 7.

$$\frac{89 + 82 + 76 + 80 + 68 + 70 + 74}{7} = \frac{539}{7} = 77, \text{ or } 77°F.$$

To find the median by hand, arrange the data in numerical order and find the middle entry.

$$68 \quad 70 \quad 74 \quad \mathbf{76} \quad 80 \quad 82 \quad 89$$

Since there is an odd number of entries, the median is the middle entry, or fourth entry, 76, which represents 76°F.

To find the mean and the median using a graphing utility, access the list in the statistics feature of your calculator and enter the given temperatures in the first list L_1. Next, access a list math menu and select the mean and median entries.

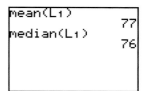

We see that the mean is 77°F and the median is 76°F.

EXAMPLE 4   **PRESIDENTS' AGE PROBLEM**
Here is a list of the first 42 presidents of the United States and their ages when they took office. Use a graphing utility to find the mean, median, and mode of the ages.

1.	George Washington	57	22.	Grover Cleveland	47
2.	John Adams	61	23.	Benjamin Harrison	55
3.	Thomas Jefferson	57	24.	Grover Cleveland	55
4.	James Madison	58	25.	William McKinley	54
5.	James Monroe	58	26.	Theodore Roosevelt	42
6.	John Quincy Adams	57	27.	William Taft	51
7.	Andrew Jackson	61	28.	Woodrow Wilson	56
8.	Martin Van Buren	54	29.	Warren Harding	55
9.	William Harrison	68	30.	Calvin Coolidge	51
10.	John Tyler	51	31.	Herbert Hoover	54
11.	James Polk	49	32.	Franklin Roosevelt	51
12.	Zachary Taylor	64	33.	Harry Truman	60
13.	Millard Fillmore	50	34.	Dwight Eisenhower	62
14.	Franklin Pierce	48	35.	John Kennedy	43
15.	James Buchanan	65	36.	Lyndon Johnson	55
16.	Abraham Lincoln	52	37.	Richard Nixon	56
17.	Andrew Johnson	56	38.	Gerald Ford	61
18.	Ulysses S. Grant	46	39.	Jimmy Carter	52
19.	Rutherford Hayes	54	40.	Ronald Reagan	69
20.	James Garfield	49	41.	George Bush	64
21.	Chester Arthur	50	42.	William Clinton	46

Solution: Enter the data in a list, and find the mean, median, and mode of the ages.

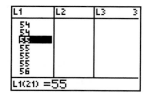

We can see that the mean is about 54.9 years and the median is 55 years. To find the mode, use a sort feature to *arrange the ages in ascending order* and scroll to count the number of multiple entries.

The modes are 51, 54, and 55 years.

2   Graphs can be used to visualize data. The following graph is called a circle graph or pie chart. The circle is divided into sectors that represent different parts of the whole. The circle represents the whole, in this case 100%.

EXAMPLE 5   INTERPRETING DATA FROM A CIRCLE GRAPH
United States astronaut Shannon Lucid spent more hours in space than any other American at the time of her return in September 1996. The circle graph below represents the percent of time spent in space by Lucid, Norm Thagard, Skylab 4 crew, Skylab 3 crew, and Skylab 2 crew. If the total number of hours recorded by the five components is 11,398 hours, approximate the number of hours that each spent in space. (*Source*: Lyndon B. Johnson Space Center.)

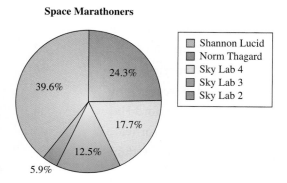

Space Marathoners

Solution: To approximate the number of hours that each spent in space, we find percents of the total hours.

Lucid: 39.6% of 11,398 or (0.396)(11,398) ≈ 4514 hours

Thagard: 24.3% of 11,398 or (0.243)(11,398) ≈ 2770 hours

Skylab 4 crew: 17.7% of 11,398 or (0.177)(11,398) ≈ 2017 hours

Skylab 3 crew: 12.5% of 11,398 or (0.125)(11,398) ≈ 1425 hours

Skylab 2 crew: 5.9% of 11,398 or (0.059)(11,398) ≈ 672 hours

Notice that the percents have a sum of 100% and the hours found, although they are rounded, have a sum of 11,398.

 Another common type of graph is the bar graph. A bar graph consists of a series of bars arranged vertically or horizontally as shown next.

EXAMPLE 6   **INTERPRETING DATA FROM A BAR GRAPH**
The bar graph below shows the monthly average temperatures for Albany, New York. (*Source:* The World Almanac, 1996.)

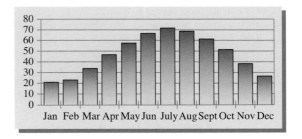

a. Approximate the highest monthly average temperature.
b. Approximate the lowest monthly average temperature.
c. Which month shows the greatest increase in temperature?
d. Which month shows the greatest decrease in temperature?
e. Discuss any patterns noticed in this graph.

Solution:

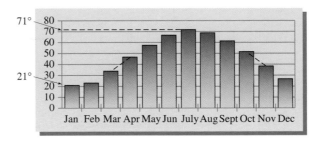

a. The highest monthly average temperature is approximately 71°F in July.
b. The lowest monthly average temperature is approximately 21°F in January.

c. The month of April shows the greatest increase in monthly temperature. (To see this, use a straightedge and see that the temperature is increasing the fastest when the straightedge is the steepest).

d. The month of November shows the greatest decrease.

e. The graph clearly shows the temperature climbing from January until July and dropping from July to January.

A third type of graph is a broken line graph, or simply a line graph.

EXAMPLE 7   INTERPRETING DATA FROM A LINE GRAPH

The broken line graph shows the incidence of cancer among Americans from 1975 to 1995. (*Source*: Cancer Facts and Figures, American Cancer Society, 1995.)

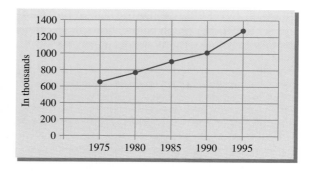

a. Approximate the number of people with cancer in 1980.

b. Approximate the number of people with cancer in 1995.

c. Which year shows the greatest increase in the incidence of cancer?

d. In what year shown did the cancer rate rise above 1,000,000?

e. Discuss any patterns that you notice.

Solution:

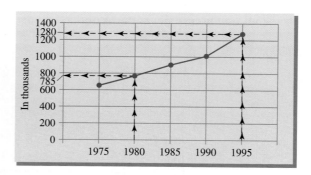

a. The incidence of cancer in 1980 was approximately 785,000.

b. The incidence of cancer in 1995 was approximately 1,280,000.

c. The greatest increase appears to be in the year 1995.

d. The cancer rate rose above 1,000,000 in the year 1990.

e. This graph clearly shows that the incidence of cancer is increasing.   ▬▬▬▬▬

An equation that models the graph in Example 7 is the linear model

$$y = 28.58x + 644.6$$

where x is the number of years past 1975 and y is the incidence of cancer in thousands.
To approximate the cancer incidence in 1995 using the model, let $x = 20$ (number of years past 1975).

$$y = 28.58(20) + 644.6$$

$$y = 1216.2$$

How close is this value to your approximation?

EXERCISE SET 2.5

The following sets of data come from the U.S. Department of Commerce, Bureau of Labor Statistics, 1995, and show the average weekly earnings for an employee in the industries indicated. Find the mean weekly earnings for each of the following industries. See Examples 1 through 4.

1. Mining

YEAR	1970	1980	1985	1990	1992	1993	1994
WEEKLY EARNING (DOLLARS)	164	397	520	604	638	647	666

2. Manufacturing

YEAR	1970	1980	1985	1990	1992	1993	1994
WEEKLY EARNING (DOLLARS)	133	289	386	442	470	486	507

3. Construction

YEAR	1970	1980	1985	1990	1992	1993	1994
WEEKLY EARNING (DOLLARS)	195	368	464	526	538	551	570

4. Retail Trade

YEAR	1970	1980	1985	1990	1992	1993	1994
WEEKLY EARNING (DOLLARS)	82	147	175	195	205	210	215

According to the U.S. Bureau of the Census, Statistical Abstract of the United States, 1994, the unemployment percents for the industries indicated are as given below.

Find the median unemployment percent for each industry for the given years. See Examples 1 through 4.

5. Agriculture

YEAR	1975	1980	1985	1990	1992	1993
UNEMPLOYMENT	10.4%	11%	3.2%	9.7%	12.3%	11.6%

6. Mining

YEAR	1975	1980	1985	1990	1992	1993
UNEMPLOYMENT	4.1%	6.4%	9.5%	4.8%	7.9%	7.3%

7. Construction

YEAR	1975	1980	1985	1990	1992
UNEMPLOYMENT	18%	14.1%	13.1%	11.1%	16.7%

8. Manufacturing

YEAR	1975	1980	1985	1990	1992
UNEMPLOYMENT	10.9%	8.5%	7.7%	5.8%	7.8%

9. Given the following grades for a particular history exam, find the mode of the set of data:

92, 75, 85, 91, 83, 85, 92, 90, 85, 87, 82, 76, 93, 81

10. Given the following temperatures recorded for a given week, find the mode of the set of data.

90, 86, 85, 82, 85, 86, 85

Financial planners say that "empty nesters" should save the money that they previously spent on their children in a mutual fund portfolio. A possible mutual fund portfolio for someone 45 to 50 years old is exhibited in the following circle graph. Use this graph for Exercises 11 and 12. See Example 5.

Mutual Fund Portfolio

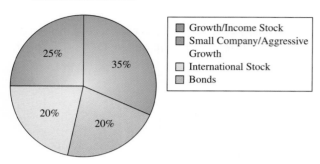

11. If an "empty nester" has $15,000 to invest, how much should be invested in each category?

12. If an "empty nester" has $40,000 to invest, how much should be invested in each category?

13. **GROUP ACTIVITY:** Form groups of 9 to 10 people. Measure each person's height. Record the heights in L_1 and find the mean, median, and mode of the data.

14. **GROUP ACTIVITY:** Form groups of 9 to 10 people. Measure each person's length of foot in centimeters. Record the lengths in L_1 and find the mean, median, and mode of the data.

The ten tallest buildings in the United States are listed below. Use this table for Exercises 15-18.

BUILDING	HEIGHT (FEET)
Sears Tower, Chicago, IL	1454
One World Trade Center (1972), New York, NY	1368
One World Trade Center (1973), New York, NY	1362
Empire State, New York, NY	1250
Amoco, Chicago, IL	1136
John Hancock Center, Chicago, IL	1127
First Interstate World Center, Los Angelos, CA	1107
Chrysler, New York, NY	1046
Nations Bank Tower, Atlanta, GA	1023
Texas Commerce Tower, Houston, TX	1002

15. Find the mean height for the five tallest buildings.

16. Find the median height for the five tallest buildings.

17. Find the median height for the ten tallest buildings.

18. Find the mean height for the ten tallest buildings.

During an experiment, the following times (in seconds) were recorded: 7.8, 6.9, 7.5, 4.7, 6.9, 7.0.

19. Find the mean. Round to the nearest tenth.

20. Find the median.   21. Find the mode.

In a mathematics class, the following test scores were recorded for a student: 86, 95, 91, 74, 77, 85.

22. Find the mean. Round to the nearest hundredth.

23. Find the median.   24. Find the mode.

The following pulse rates were recorded for a group of fifteen students: 78, 80, 66, 68, 71, 64, 82, 71, 70, 65, 70, 75, 77, 86, 72.

25. Find the mean.

26. Find the median.   27. Find the mode.

28. How many rates were higher than the mean?

29. How many rates were lower than the mean?

30. Have each student in your algebra class take his or her pulse rate. Record the data and find the mean, the median, and the mode.

Find the missing numbers in each set of numbers. (These numbers are not necessarily in numerical order.)

31. __, __, 16, 18, __
    The mode is 21.
    The median is 20.

32. __, __, __, __, 40
    The mode is 35.
    The median is 37.
    The mean is 38.

The following bar graph shows the professions that walk the farthest, in average miles walked per year. Use this graph for Exercise 33 through 38. See Example 6. (Source: Dr. Scholl's and the American Podiatry Association)

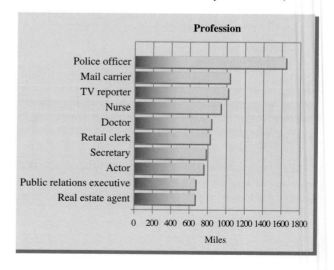

33. Which profession walks the farthest?

34. Which profession does the least amount of walking?

35. Estimate the average miles walked per year by a nurse.

36. Estimate the average miles walked per year by a retail clerk.

37. Estimate the number of miles you walk in a week. Multiply this number by 52 to find the average number of miles you walk in a year. How does your distance compare with the distances in the bar graph?

38. Estimate how much farther a police officer walks in a year than a mail carrier.

The following graph shows the political party identification of the adult population for the years shown. Use this graph for Exercises 39 through 46. See Example 7.

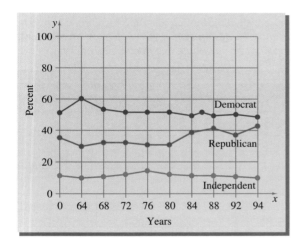

39. Which political party shows the greatest percent of the adult population for all the years shown?

40. Which political party shows the least percent of the adult population for all the years shown?

41. What year shows the greatest difference in percent of Democrats and percent of Republicans?

42. What year shows the least difference in percent of Democrats and percent of Republicans?

43. Estimate the percent of adult population identified as Democrat, Republican, and Independent in the year 1992.

44. Estimate the percent of adult population identified as Democrat, Republican, and Independent in the year 1964.

45. Discuss any patterns you notice in the graph.

46. Do you think the percents for any given year have a sum of 100%? Why or why not?

According to USA Today, parents spend an average of $363 per child on back-to-school items for a school year, and 58% of this cost will be spent on clothes. The following circle graph shows the distribution method used to pay for the clothes. Use this graph for Exercises 47 through 50. If necessary, round amounts to the nearest hundredth.

School Clothes Budget

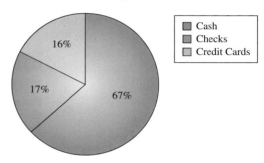

47. What was the average amount of money spent on purchasing clothes?

48. What was the average amount paid in cash for clothes?

49. What was the average amount paid by credit card for clothes?

50. What was the average amount paid by checks for clothes?

The data below represents the 2015 projected population in millions of the world's 15 largest cities. Use this data to answer Exercises 51 through 54. (Source: The World Almanac, 1996.)

28.7	17.6	20.8	18.8	23.4	27.4
14.3	19.4	17.6	13.1	21.2	12.4
10.6	17.0	11.6			

51. Find the mean of the 5 most populous cities.

52. Find the median of the 5 most populous cities.

53. Find the median of the 15 most populous cities.

54. Find the mean of the 15 most populous cities.

55. A survey was conducted to determine which age category knows their history best. When asked to name the state where the final battle of the Revolutionary War was fought, these percents of each age group correctly named Virginia. Make a bar graph using the data listed below. (*Source*: USA

Today and Opinion Research for Colonial Williamsburg Foundation.)

AGE	18-24	25-34	35-44	45-54	55-64	65+
PERCENT	44	38	43	43	51	43

56. Many business and government computers are not set up to handle the year 2000. Their computers will falsely read 00 as 1900. A survey was conducted to see how aware companies are of the "century problem." The data listed below is from a USA Today and Ofsten Corporation survey of senior executives. Construct a circle graph using the data.

Conversion program under way	28%
Planning conversion	34%
Unaware of the problem	13%
Not a problem for our company	21%
Don't know	4%

57. GROUP ACTIVITY Decide on a topic and survey your algebra class. Some suggested topics are favorite brand of soft drink given 5 choices, favorite leisure activity, number of hours spent each week watching television, or the best time of day to study. Tabulate your data and use it to construct a bar graph or a circle graph. Make sure that your graph is labeled properly.

58. The annual starting salaries for 5 levels of employees at a local factory are $15,000, $22,000, $30,000, $60,000, and $110,000. During a salary dispute, the employees claim that a typical employee makes approximately $30,000 a year (the median). The owners of the factory claim that a typical employee makes almost $50,000 a year (the mean). Which number do you think is a better representation of the situation? Why?

59. Suppose that a set of data contains only numbers between 21.6 and 46.8. Can the mean of this data set be 20.2?

60. In a set of data, suppose that the largest number (not the mode) is replaced with a larger number. How does this affect the mean, median, and mode?

61. In a set of data, suppose that the smallest number (not the mode) is replaced with a smaller number. How does this affect the mean, median, and mode?

62. Suppose that a set of data consists of positive integers. Is it possible for the mean not to be an integer? Is it possible for the median not to be an integer? Is it possible for the mode not to be an integer?

Review Exercises

Complete each table. Calculate the difference between adjacent x-values and the difference between adjacent expression, or y_1, values. See Section 2.3.

63.

x	2	4	6	8	10
$2x+1$					

64.

x	1	3	5	7	9
$-x-2$					

65.

x	-7	-6	-5	-4	-3
$-5x+2$					

66.

x	-3	-2	-1	0	1
$1.5x-3$					

GROUP ACTIVITY

COMPARISON OF VOLUMES

MATERIALS
- Two sheets of 8.5″ × 11″ paper
- tape
- ruler

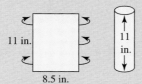

FIGURE 1   Cylinder 1

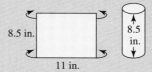

FIGURE 2   Cylinder 2

Make a different cylinder from each sheet of paper, as shown in Figures 1 and 2. Be careful not to overlap the paper when joining the ends with tape. In this activity you will investigate the relationship between the surface areas and volumes of these two cylinders. Recall that the circumference of a circle is given by the formula $C = 2\pi r$ (where r is the radius of the circle), area of a circle by the formula $A = \pi r^2$, and volume of a cylinder by the formula $V = $ Area of base × height $ = \pi r^2 h$.

1. Measure the height and radius of each cylinder and record these values in Table 1. Use these measurements to calculate the volume to the nearest tenth of a cubic inch. Record the volume of each cylinder in the table.

TABLE 1

	MEASURED HEIGHT (INCHES)	MEASURED RADIUS (INCHES)	VOLUME (CUBIC INCHES)
Cylinder 1			
Cylinder 2			

2. Compare the volumes of the two cylinders. Based on your calculations in Question 1, which cylinder has the larger volume? The volume of the smaller cylinder is what percent of the volume of the larger cylinder?

3. What is the circumference of each cylinder? Discuss how you know these values without measuring.

4. In Question 1, you estimated each cylinder's radius in order to approximate its volume. Now *calculate* each radius; use the formula for the circumference of a circle and the known value of the circumference, and solve for r. For each cylinder, store the calculated radius value in R on your calculator. Use this stored value of the radius to recalculate the volume of each cylinder. Record the heights, the calculated radii rounded to seven decimal places, and the recalculated volumes rounded to the nearest tenth of a cubic inch in Table 2.

TABLE 2

	HEIGHT (INCHES)	CALCULATED RADIUS (INCHES)	RECALCULATED VOLUME (CUBIC INCHES)
Cylinder 1			
Cylinder 2			

5. For each cylinder, compare the volumes recorded in Table 1 and Table 2. Is there any difference in results between the two methods of finding the value of *r*? If so, which do you think is more accurate? Why? Discuss, and record your observations.

6. The total surface area of a cylinder includes the area of the side plus the area of both circular bases. The lateral surface area of a cylinder includes just the area of the side. Using the value of *r* that you believe to be more accurate, find the lateral surface area and the total surface area of each cylinder and record these

values rounded to the nearest tenth of a square inch in Table 3.

TABLE 3

	LATERAL SURFACE AREA (SQUARE INCHES)	TOTAL SURFACE AREA (SQUARE INCHES)
Cylinder 1		
Cylinder 2		

7. What do you observe about the lateral surface areas of the two cylinders? Does the cylinder with the larger volume also have the larger total surface area? Do these observations surprise you? Why or why not?

CHAPTER 2   HIGHLIGHTS

DEFINITIONS AND CONCEPTS	EXAMPLES
SECTION 2.1 SOLVING LINEAR EQUATIONS ALGEBRAICALLY	

An **equation** is a statement that two expressions are equal.

Equations:

$$5 = 5 \qquad 7x + 2 = -14 \qquad 3(x-1)^2 = 9x^2 - 6$$

A **linear equation in one variable** is an equation that can be written in the form $ax + b = c$, where a, b, and c are real numbers and a is not 0.

Linear equations:

$$7x + 2 = -14 \qquad x = -3$$
$$5(2y - 7) = -2(8y - 1)$$

A **solution** of an equation is a value for the variable that makes the equation a true statement.

Check to see that -1 is a solution of $3(x - 1) = 4x - 2$.

$$3(-1 - 1) = 4(-1) - 2$$
$$3(-2) = -4 - 2$$
$$-6 = -6 \qquad \text{True.}$$

Thus, -1 is a solution.

Equivalent equations have the same solution set.

$x - 12 = 14$ and $x = 26$ are equivalent equations.

The **addition property of equality** guarantees that the same number may be added to (or subtracted from) both sides of an equation, and the result is an equivalent equation.

Solve for x: $-3x - 2 = 10$.

$$-3x - 2 + 2 = 10 + 2 \quad \text{Add 2 to both sides.}$$
$$-3x = 12$$

The **multiplication property of equality** guarantees that both sides of an equation may be multiplied by (or divided by) the same nonzero number, and the result is an equivalent equation.

$$\frac{-3x}{-3} = \frac{12}{-3} \qquad \text{Divide both sides by } -3$$
$$x = -4$$

(continued)

DEFINITIONS AND CONCEPTS	EXAMPLES

SECTION 2.1   SOLVING LINEAR EQUATIONS ALGEBRAICALLY

To solve linear equations in one variable:

Solve for x:

$$x - \frac{x-2}{6} = \frac{x-7}{3} + \frac{2}{3}$$

1. Clear the equation of fractions.

1. $6\left(x - \frac{x-2}{6}\right) = 6\left(\frac{x-7}{3} + \frac{2}{3}\right)$ Multiply both sides by 6.

$$6x - (x-2) = 2(x-7) + 2(2)$$

2. Remove grouping symbols such as parentheses.

2. $6x - x + 2 = 2x - 14 + 4$ Remove grouping symbols.

3. Simplify by combining like terms.

3. $5x + 2 = 2x - 10$

4. Write variable terms on one side and numbers on the other side using the addition property of equality.

4. $5x + 2 - 2 = 2x - 10 - 2$ Subtract 2.

$$5x = 2x - 12$$

$$5x - 2x = 2x - 12 - 2x$$ Subtract $2x$.

$$3x = -12$$

5. Isolate the variable using the multiplication property of equality.

5. $\dfrac{3x}{3} = \dfrac{-12}{3}$ Divide by 3.

$$x = -4$$

6. Check the proposed solution in the original equation.

6. $-4 - \dfrac{-4-2}{6} = \dfrac{-4-7}{3} + \dfrac{2}{3}$ Replace x with -4 in the original equation

$$-4 - \frac{-6}{6} = \frac{-11}{3} + \frac{2}{3}$$

$$-4 - (-1) = \frac{-9}{3}$$

$$-3 = -3$$ True.

SECTION 2.2   AN INTRODUCTION TO PROBLEM SOLVING

The state of Colorado is in the shape of a rectangle whose length is approximately 1.3 times its width. If the perimeter of Colorado is 2070 kilometers, find its dimensions.

1. UNDERSTAND the problem.

1. Read and reread the problem. Guess a solution and check your guess.

2. ASSIGN a variable.

2. Let x = width of Colorado in kilometers. Then $1.3x$ = length of Colorado in kilometers.

3. ILLUSTRATE the problem.

3.

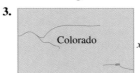

1.3x

(continued)

DEFINITIONS AND CONCEPTS	**EXAMPLES**

SECTION 2.2 AN INTRODUCTION TO PROBLEM SOLVING

4. TRANSLATE the problem.

4. In words: $\boxed{\text{twice the length}} + \boxed{\text{twice the width}} = \boxed{\text{perimeter}}$

Translate: $\quad 2(1.3x) \quad + \quad 2x \quad = \quad 2070$

5. COMPLETE by solving.

5. $2.6x + 2x = 2070$

$4.6x = 2070$

$x = 450$

6. INTERPRET the results.

6. If $x = 450$ kilometers, then $1.3x = 1.3(450) = 585$ kilometers. *Check:* The perimeter of a rectangle whose width is 450 kilometers and length is 585 kilometers is $2(450) + 2(585) = 2070$ kilometers, the required perimeter. *State:* The dimensions of Colorado are approximately 450 kilometers by 585 kilometers.

SECTION 2.3 A NUMERICAL APPROACH: MODELING WITH TABLES

To generate a table on a calculator, a **table start** x-value and a **table increment**, or change in x-values, must be given.

Use your calculator to complete the given table.

HOURS WORKED	x	8	10	12	14	16
GROSS PAY	8+6.25x					

Go to the Y= editor and enter $y_1 = 8 + 6.25x$. In the table setup menu, let table start $= 8$ and table increment $= 2$ since the change in x-values in the table is 2. The table generated is shown below.

x	y_1
8	58
10	70.5
12	83
14	95.5
16	108

SECTION 2.4 FORMULAS AND PROBLEM SOLVING

An equation that describes a known relationship among quantities is called a **formula.**

Formulas:

$$A = \pi r^2 \text{ (area of a circle)}$$

$$I = P \cdot R \cdot T \text{ (interest = principal} \cdot \text{rate} \cdot \text{time)}$$

To solve a formula for a specified variable, use the steps for solving an equation. Treat the specified variable as the only variable of the equation.

Solve $A = 2HW + 2LW + 2LH$ for H.

$A - 2LW = 2HW + 2LH \qquad$ Subtract $2LW$.

$A - 2LW = H(2W + 2L) \qquad$ Factor out H.

$\dfrac{A - 2LW}{2W + 2L} = \dfrac{H(2W + 2L)}{2W + 2L} \qquad$ Divide by $2W + 2L$.

$\dfrac{A - 2LW}{2W + 2L} = H \qquad$ Simplify.

(continued)

DEFINITIONS AND CONCEPTS	EXAMPLES

SECTION 2.5   INTERPRETING DATA AND READING BAR, LINE, AND CIRCLE GRAPHS

Find the mean, the median, and the mode of the following list of numbers.

$$23, 23, 31, 35, 46, 63$$

The **mean** of a set of data items, denoted by $\bar{x}$, is the sum of the data items divided by the number of items.

$$\bar{x} = \frac{23 + 23 + 31 + 35 + 46 + 63}{6} \approx 36.8$$

The **median** of an ordered set of numbers is the middle number. If the number of items is even, the median is the mean of the two middle numbers.

$$\text{median} = \frac{31 + 35}{2} = 33$$

The **mode** of a set of numbers is the number that occurs most often.

$$\text{mode} = 23$$

CHAPTER 2 REVIEW

(2.1)   *Solve each equation. Check the solution.*

1. $4(x - 5) = 2x - 14$ **2.** $x + 7 = -2(x + 8)$

3. $3(2y - 1) = -8(6 + y)$

4. $-(z + 12) = 5(2z - 1)$

5. $n - (8 + 4n) = 2(3n - 4)$

6. $4(9v + 2) = 6(1 + 6v) - 10$

7. $0.3(x - 2) = 1.2$ **8.** $1.5 = 0.2(c - 0.3)$

9. $-4(2 - 3h) = 2(3h - 4) + 6h$

10. $6(m - 1) + 3(2 - m) = 0$

11. $6 - 3(2g + 4) - 4g = 5(1 - 2g)$

12. $20 - 5(p + 1) + 3p = -(2p - 15)$

13. $\dfrac{x}{3} - 4 = x - 2$ **14.** $\dfrac{9}{4}y = \dfrac{2}{3}y$

15. $\dfrac{3n}{8} - 1 = 3 + \dfrac{n}{6}$ **16.** $\dfrac{z}{6} + 1 = \dfrac{z}{2} + 2$

17. $\dfrac{y}{4} - \dfrac{y}{2} = -8$ **18.** $\dfrac{2x}{3} - \dfrac{8}{3} = x$

19. $\dfrac{b - 2}{3} = \dfrac{b + 2}{5}$ **20.** $\dfrac{2t - 1}{3} = \dfrac{3t + 2}{15}$

21. $\dfrac{2(t + 1)}{3} = \dfrac{2(t - 1)}{3}$ **22.** $\dfrac{3a - 3}{6} = \dfrac{4a + 1}{15} + 2$

23. $\dfrac{x - 2}{5} + \dfrac{x + 2}{2} = \dfrac{x + 4}{3}$

24. $\dfrac{2z - 3}{4} - \dfrac{4 - z}{2} = \dfrac{z + 1}{3}$

(2.2)   *Solve.*

25. Twice the difference of a number and 3 is the same as 1 added to three times the number. Find the number.

26. One number is 5 more than another number. If the sum of the numbers is 285, find the numbers.

27. Find 40% of 130.   **28.** Find 1.5% of 8.

29. In a recent year, the average annual earnings for a worker with a college degree were $37,000. This represents an 85% increase over the average annual earnings for a high school graduate that year. Find the average annual earnings for a high school graduate that year. (*Source:* U.S. Bureau of the Census.)

30. Find four consecutive integers such that twice the first subtracted from the sum of the other three integers is sixteen.

31. Determine whether there are two consecutive odd integers such that 5 times the first exceeds 3 times the second by 54.

32. The length of a rectangular playing field is 5 meters less than twice its width. If 230 meters of fencing go around the field, find the dimensions of the field.

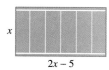

33. A car rental company charges $19.95 per day for a compact car plus 12 cents per mile for every mile over 100 miles driven per day. If Mr. Woo's bill for 2 days is $46.86, find how many miles he drove.

34. The cost C of producing x number of scientific calculators is given by $C = 4.50x + 3000$, and the revenue R from selling them is given by $R = 16.50x$. Find the number of calculators that must be sold to break even. (To break even, revenue = cost.)

35. An entrepreneur can sell her musically vibrating plants for $40 each, while her cost C to produce x number of plants is given by $C = 20x + 100$. Find the number she needs to sell to break-even. Find her revenue if she sells exactly that number of plants.

(2.3) *Use your calculator and the table feature to complete each table. If necessary, round any amounts to 2 decimal places.*

36.

RADIUS OF CONE	x	1	1.5	2	2.5	3
VOLUME (IF HEIGHT IS **10** INCHES)	$\frac{10}{3}\pi x^2$					

37.

WIDTH OF RECTANGLE	x	2	4.68		9.5	12.68
PERIMETER (IF LENGTH IS **15** UNITS)	$30+2x$					

38. Gina Saltalamacchia works as a hostess at a restaurant and makes $5.75 per hour. Complete the table and find her gross pay for the hours given.

HOURS	5	10	15	20	25	30
GROSS PAY (DOLLARS)						

39. Coast Waterworks, Inc. charges $8.00 for the first 4 thousand gallons of water used and then 1.10 per thousand gallons used over 4 thousand gallons. Cross Gates Water Company charges $12.00 for the first 5 thousand gallons and $1.50 per thousand gallons used over 5 thousand gallons. Complete the table and find the total charges for the gallons used.

GALLONS (IN THOUSANDS USED)	5	6	7	8
COAST CHARGE (DOLLARS)				
CROSS GATES CHARGE (DOLLARS)				

40. A rock is thrown upward from a bridge. Neglecting air resistance, the height of the rock at time x seconds is given by

$$y = -16x^2 + 40x + 500$$

where y is measured in feet.

a. Complete the table to find the height of the rock at the given times.

SECONDS	0	1	2	3	4	5
HEIGHT IN FEET						

b. Find the maximum height of the rock to the nearest foot.

c. Find to the nearest tenth of a second when the rock strikes the ground.

(2.4) *Solve each equation for the specified variable.*

41. $V = LWH$; W

42. $C = 2\pi r$; r

43. $5x - 4y = -12$; y

44. $5x - 4y = -12$; x

45. $y - y_1 = m(x - x_1)$; m

46. $y - y_1 = m(x - x_1)$; x

47. $E = I(R + r)$; r

48. $S = vt + gt^2$; g

49. $T = gr + gvt$; g

50. $I = Prt + P$; P

51. $A = \frac{h}{2}(B + b)$; B

52. $V = \frac{1}{3}\pi r^2 h$; h

53. $\frac{1}{a} + \frac{1}{b} = \frac{1}{c}$; b

54. $\frac{2}{x} - \frac{3}{y} = \frac{1}{z}$; y

55. $\frac{x - y}{5} + \frac{y}{4} = \frac{2x}{3}$; y

56. $\frac{b + c}{d} - \frac{b}{c} = \frac{5}{c}$; d

Solve.

57. A principal of $3000 is invested in an account paying an annual percentage rate of 3%. Find the amount in the account after 7 years if the interest is compounded

a. semiannually   **b.** weekly
(Approximate to the nearest cent.)

58. Angie has a photograph in which the length is 2 inches longer than the width. If she increases each dimension by 4 inches, the area is increased by 88 square inches. Find the original dimensions.

59. The formula for converting Celsius temperatures to Fahrenheit temperatures is $F = \dfrac{9C + 160}{5}$.

a. Complete the following table to find Fahrenheit temperatures given the Celsius temperatures.

CELSIUS	-40	-15	10	60
FAHRENHEIT				

b. If the boiling point of water is 100 degrees Celsius, what is this in degrees Fahrenheit?

c. If the freezing point of water is 0 degrees Celsius, what is this in degrees Fahrenheit?

60. The formula for finding the distance traveled at the rate of 55 miles per hour for x hours is given by $y = 55x$, where y is measured in miles.

a. Find the distance traveled in 15-minute increments for 0 to 1.5 hours.

HOURS	0	.25	.50	.75	1	1.25	1.5
MILES							

b. If the distance traveled was 192.5 miles, how many hours did it take to travel this distance at 55 mph?

61. One-square-foot floor tiles come 24 to a package. Find how many packages are needed to cover a rectangular floor 18 feet by 21 feet.

62. Determine which container holds more ice cream, an 8 inch × 5 inch × 3 inch box or a cylinder with radius 3 inches and height 6 inches.

63. Angie left Los Angeles at 11 A.M. and drove nonstop to San Diego, 130 miles away. If she arrived at 1:15 P.M., find her average speed, rounded to the nearest mile per hour.

The volume of a cone can be found using the formula $V = \frac{1}{3}\pi x^2 h$. *Find the volumes of the cones with the following heights and radii.*

64.

HEIGHT h	6	6	6	6	6
RADIUS x	1	1.5	2	2.25	3
VOLUME					

65.

HEIGHT h	10	10	10	10	10
RADIUS x	1.5	2.1	2.75	3	3.5
VOLUME					

(2.5) *For each of the following sets of numbers, find the mean, the median, and the mode. If necessary, round the mean to one decimal place.*

66. 21, 28, 16, 42, 38   **67.** 42, 35, 36, 40, 50

68. 7.6, 8.2, 8.2, 9.6, 5.7, 9.1   **69.** 4.9, 7.1, 6.8, 6.8, 5.3, 4.9

70. 0.2, 0.3, 0.5, 0.6, 0.6, 0.9, 0.2, 0.7, 1.1

71. 0.6, 0.6, 0.8, 0.4, 0.5, 0.3, 0.7, 0.8, 0.1

72. 231, 543, 601, 293, 588, 109, 334, 268

73. 451, 356, 478, 776, 892, 500, 467, 780

The circle graph below shows percents of financial aid for college students from federal, state, and institutional sources. Use this graph to answer Exercises 74 and 75. (Source: College Board Trends in Student Aid 1986-1996.)

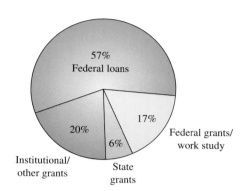

74. What type of financial aid do most students receive?

75. If the total financial aid from the categories shown is $50.3 billion for 1995-96 find the amount of money spent in each category. Round to the nearest hundredth of a billion.

The bar graph below compares the size of a company (number of employees) with the percent of those employees receiving health benefits. Use this graph for Exercises 76 through 79. (Source: Lewin Group for American Hospital Association.)

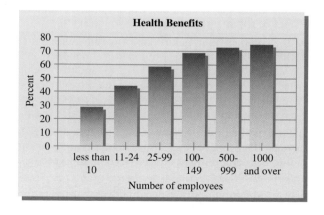

76. Estimate the percent of employees receiving health benefits for companies with 11-24 employees.

77. Which size company has the lowest percent of employees receiving health benefits? Estimate the percent.

78. Which size company has the highest percent of employees receiving health benefits? Estimate the percent.

79. Discuss a trend shown by this bar graph.

CHAPTER 2 TEST

Solve each equation. Check the solution.

1. $8x + 14 = 5x + 44$

2. $3(x + 2) = 11 - 2(2 - x)$

3. $3(y - 4) + y = 2(6 + 2y)$

4. $7n - 6 + n = 2(4n - 3)$

5. $\dfrac{z}{2} + \dfrac{z}{3} = 10$

6. $\dfrac{7w}{4} + 5 = \dfrac{3w}{10} + 1$

Solve each equation for the specified variable.

7. $3x - 4y = 8$; y

8. $4(2n - 3m) - 3(5n - 7m) = 0$; n

9. $S = gt^2 + gvt$; g

10. $F = \dfrac{9}{5}C + 32$; C

Solve.

11. Find 12% of 80.

12. In 1995, Latter & Blum sold approximately $939.25 million in residential property in southeast Louisiana. This represents a 121% increase over Prudential La. Properties. Find the 1995 residential sales for Prudential La. Properties. (*Source: Times/Picayune Newspaper*, March 10, 1996.)

13. A circular dog pen has a circumference of 78.5 feet. Estimate how many hunting dogs could be safely kept in the pen if each dog needs at least 60 square feet of room.

14. The company that makes Photoray sunglasses figures that the cost C to make x number of sunglasses weekly is given by $C = 3910 + 2.8x$, and the weekly revenue R is given by $R = 7.4x$. Find the number of sunglasses that must be made and sold to break even. (Revenue must equal cost in order to break even.)

15. Find the amount of money in an account after 10 years if a principal of $2500 is invested at 3.5% interest compounded quarterly. (Round to the nearest cent.)

16. Ann-Margaret Tober is deciding whether to accept a sales position at Campo Electronics. She is offered a gross monthly pay of $1500 plus 5% commission on her sales.

    a. Complete the gross monthly pay table for the given amounts of sales.

SALES (DOLLARS)	8,000	9,000	10,000	11,000	12,000
GROSS MONTHLY PAY (DOLLARS)					

b. If she has sales of $11,000 per month, what is her gross annual pay?

c. If Ann-Margaret decides she needs a gross monthly pay of $2200, how much must she sell every month?

17. Fencing is to be installed around part of the parking lot of a chemical manufacturing company to form a rectangular storage area. The fence is to be formed from 80 feet of fencing. If a side of the factory is used together with the fencing, what is the largest area that can be enclosed given that the dimensions of the rectangle are to be natural numbers?

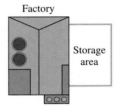

18. The fireworks display for a small community is launched from a platform that is 20 feet from ground level. A particular rocket is launched with a velocity of 80 feet per second and explodes 0.5 second after it reaches its maximum height. The height of the rocket at time x seconds is given by

$$y = -16x^2 + 80x + 20$$

where y is measured in feet from the ground.

a. When is the height of the rocket 116 feet?

b. What is the rocket's maximum height?

c. When does the rocket reach its maximum height?

d. When does the rocket explode?

The following chart shows the number of books a graduate student checked out of the library for a period of 12 weeks while working on a term paper.

WEEK	1	2	3	4	5	6	7	8	9	10	11	12
NUMBER OF BOOKS	12	0	6	5	2	10	3	1	5	7	1	5

19. Find the mean number of books checked out per week.

20. Find the median of the number of books checked out per week.

21. Find the mode of the number of books checked out per week.

The graph to the right shows U.S. consumption of fruits and vegetables per person. Use this graph for Exercises 22 through 25.

22. Approximate the pounds of fresh fruit consumed per person in 1970.

23. Approximate the pounds of fresh vegetables consumed per person in 1993.

24. Estimate the year in which there was the greatest difference in consumption of fresh fruits and vegetables.

25. Discuss any trends shown in the graph.

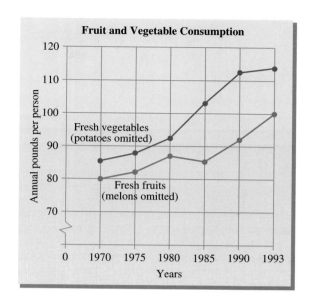

Fruit and Vegetable Consumption

CHAPTER 2   CUMULATIVE REVIEW

1. List the elements in each set.

  a. $\{x \mid x$ is a whole number between 1 and 6$\}$

  b. $\{x \mid x$ is a natural number greater than 100$\}$

2. Determine whether each statement is true or false.

  a. Every integer is a real number.

  b. Every whole number is a natural number.

  c. All real numbers are rational numbers.

  d. Some rational numbers are irrational.

  e. $N \subseteq Z$ (*N* equals natural numbers, *Z* equals integers)

3. Insert $<, >,$ or $=$ between each pair of numbers to form a true statement.

  a. $-1 \quad -2$ **b.** $\dfrac{12}{4} \quad 3$

  c. $-5 \quad 0$ **d.** $-3.5 \quad -3.05$

4. Find the following roots

  a. $\sqrt[3]{27}$ **b.** $\sqrt[5]{32}$ **c.** $\sqrt[4]{16}$

5. Write the additive inverse, or the opposite, of each number.

  a. 8 **b.** $\dfrac{1}{5}$ **c.** -9.6

6. Write the multiplicative inverse, or the reciprocal, of each number.

  a. 11 **b.** -9 **c.** $\dfrac{7}{4}$

7. Use the distributive property to write an expression equivalent to $3(2x + y)$.

8. State the basic property that is being illustrated in each true statement.

  a. $3x + 2y = 2y + 3x$

  b. $2(5x - 2y) = 2 \cdot 5x - 2 \cdot 2y$

  c. $-5 + 5 = 0$

  d. $-(-x) = x$

  e. $2x + (3x + 4) = (2x + 3x) + 4$

9. Find each sum.

a. $-3 + (-11)$ **b.** $3 + (-7)$ **c.** $-10 + 15$

d. $-8.3 + (-1.9)$ **e.** $-\dfrac{1}{4} + \dfrac{1}{2}$ **f.** $-\dfrac{2}{3} + \dfrac{3}{7}$

10. Find each product.

a. $(-8)(-1)$ **b.** $(-2)\dfrac{1}{6}$ **c.** $3(-3)$

d. $0(11)$ **e.** $\left(\dfrac{1}{5}\right)\left(-\dfrac{10}{11}\right)$ **f.** $(7)(1)(-2)(-3)$

g. $8(-2)(0)$

11. Evaluate each algebraic expression when $x = 2$, $y = -1$, and $z = -3$.

a. $|z - y|$ **b.** z^2 **c.** $\dfrac{2x + y}{z}$ **d.** $-x^2 - 4x$

12. The algebraic expression

$$32t$$

gives the speed of a falling object in feet per second for time t. Complete the following table to find the number of feet per second at 1-second intervals.

TIME $= t$ (SECONDS)	1	2	3	4	5	6
SPEED $= 32t$ (FEET PER SECOND)	32	64	96			

13. Use the distributive property to simplify each expression.

a. $3x - 5x + 4$ **b.** $7yz + yz$ **c.** $4z + 6.1$

14. Solve for x: $2x + 5 = 9$.

15. Tom Sabo, a licensed electrician, charges $45 per house visit and $30 per hour.

a. Write an equation for Tom's total charge given the number of hours x on the job.

b. Use a table to find Tom's total charge for a job that could take from 1 to 6 hours.

16. Rod and Karen Pasch are shopping for holiday gifts. On their list are gifts with prices of $39.95, $45, and $62.75 shown in the table below. If the tax rate is 7%, the tax and the total cost is shown below. Let

$$x = \text{the price on the price tag. Then}$$
$$0.07x = \text{amount of tax and}$$
$$\underset{\underset{\text{price}}{\uparrow}}{x} + \underset{\underset{\text{tax}}{\uparrow}}{0.07x} = \text{total cost.}$$

Use a table to find the amount of tax and the total cost of each gift.

PRICE ON TAG	TAX	TOTAL COST
$39.95		
$45.00		
$62.75		

17. Find 16% of 25.

18. Solve $V = lwh$ for h.

19. The formula $C = \dfrac{5}{9}(F - 32)$ converts degrees Fahrenheit to degrees Celsius. Use this formula and the table feature of your calculator to complete the given table. If necessary, round values to the nearest hundredth.

FAHRENHEIT	-4	10	32	70	100
CELSIUS					

20. Find the median and the mode of the following set of numbers. These numbers were high temperatures for fourteen consecutive days in a city in Montana.

$$76, 80, 85, 86, 89, 87, 82, 77, 76, 79, 82, 89, 89, 92$$

INTRODUCTION TO GRAPHS AND FUNCTIONS

MODELING JAPANESE AUTOMOBILE IMPORTS

Annual Japanese automobile imports into the United States topped 1 million cars in 1976 and grew to an all-time high of 2.6 million cars per year in 1986. Since 1986, annual imports have steadily decreased, and a linear function adequately describes this decline.

IN THE CHAPTER GROUP ACTIVITY ON PAGE 189, YOU WILL HAVE THE OPPORTUNITY TO FIND THE LINEAR FUNCTION THAT FITS THIS PATTERN OF DECLINE IN JAPANESE IMPORTS.

W hat makes a graphing utility uniquely different from a scientific calculator? One difference is its ability to graph equations quickly. This not only helps us to explore patterns in graphs but also provides an additional way to solve and verify solutions.

Reading graphs is similar to reading maps. Geographic locations can be described by a gridwork of lines called latitudes and longitudes, as shown on the map below. For example, the location of Houston, Texas, is latitude 30° north and longitude 95° west.

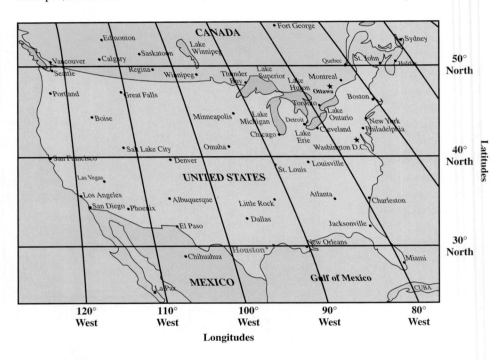

Similarly, as we see in this chapter, we locate a point on a graph by stating its *two coordinates*.

3.1 | INTRODUCTION TO GRAPHING AND GRAPHING UTILITIES

TAPE IAG 3.1

O B J E C T I V E S

1. Plot ordered pairs.
2. Introduce the window setting capability of a graphing utility.
3. Find the distance between two points.
4. Find the midpoint of a line segment.

When two varying quantities are measured simultaneously and repeatedly, we can record the measurement pairs and try to detect a pattern. If a pattern is found, we may be able to express it compactly with a two-variable equation. For example, suppose that two quantities, products sold x and monthly salary y, are measured simultaneously and repeatedly for a salesperson and recorded as in the following table.

PRODUCTS SOLD ($)	x	0	100	200	300	400
MONTHLY SALARY ($)	y	1500	1510	1520	1530	1540

The x and y pairs suggest a pattern. For example, the 1510 entry is $1500 + 0.1 \cdot 100$ and $1520 = 1500 + 0.1 \cdot 200$, and so on. We can represent the monthly salary in the form of an equation. Recall that

x = the amount of products sold, and

y = monthly salary

Then the monthly salary and the amount of sales are related by the equation

$$y = 1500 + 0.1x$$

This is an example of a **linear equation in two variables.** We discuss such equations in the next section. Let's first review the rectangular or Cartesian coordinate system.

The **rectangular** or **Cartesian coordinate system** consists of two number lines that intersect at right angles at their 0 coordinates. We position these axes on paper such that one number line is horizontal and the other number line is then vertical. The horizontal number line is called the **x-axis** (or the axis of the **abscissa**), and the vertical number line is called the **y-axis** (or the axis of the **ordinate**). The point of intersection of these axes is called the **origin.**

Notice that the axes divide the plane into four regions, called **quadrants,** which are numbered as shown.

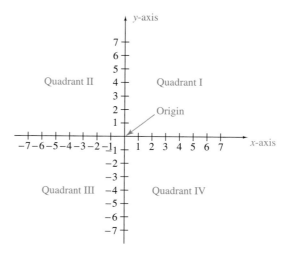

Each point in the plane can be located, or **plotted,** by describing its position in terms of distances along each axis from the origin. An **ordered pair,** represented by the notation (x, y), records these distances.

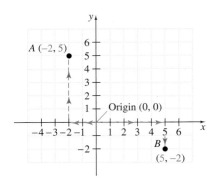

For example, the location of point A is described as 2 units to the left of the origin along the x-axis and 5 units upward parallel to the y-axis. Thus, we identify point A with the ordered pair $(-2, 5)$. Notice that the order of these numbers is critical. The x-value -2 is called the **x-coordinate** and is associated with the x-axis. The y-value 5 is called the **y-coordinate** and is associated with the y-axis. Compare the location of point A with the location of point B, which corresponds to the ordered pair $(5, -2)$. The x-coordinate 5 indicates that we move 5 units to the right of the origin along the x-axis. The y-coordinate -2 indicates that we move 2 units down parallel to the y-axis. Point A is in a different position than point B. Two ordered pairs are considered equal and correspond to the same point if and only if their x-coordinates are equal and their y-coordinates are equal.

Keep in mind that **each ordered pair corresponds to exactly one point in the real plane and that each point in the plane corresponds to exactly one ordered pair.** Thus, we may refer to the ordered pair (x, y) as the point (x, y).

EXAMPLE 1   Plot each ordered pair on a rectangular coordinate system and name the quadrant in which the point is located.

a. $(2, -1)$ **b.** $(0, 5)$ **c.** $(-3, 5)$ **d.** $(-2, 0)$ **e.** $\left(-\dfrac{1}{2}, -4\right)$ **f.** $(1.5, 1.5)$

Solution: The five points are graphed as shown:

a. $(2, -1)$ lies in quadrant IV.

b. $(0, 5)$ is not in any quadrant.

c. $(-3, 5)$ lies in quadrant II.

d. $(-2, 0)$ is not in any quadrant.

e. $\left(-\dfrac{1}{2}, -4\right)$ is in quadrant III.

f. $(1.5, 1.5)$ is in quadrant I.

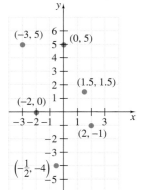

Notice that the y-coordinate of any point on the x-axis is 0. For example, the point with coordinates $(-2, 0)$ lies on the x-axis. Also, the x-coordinate of any point on the y-axis is 0. For example, the point with coordinates $(0, 5)$ lies on the y-axis. A point on an axis is called a **quadrantel** point.

2 The rectangular coordinate system extends infinitely in all directions with the number lines that form it. A graphing utility can be used to display a portion of the system. The portion being viewed is called a **viewing window** or simply a **window.**
Below, the rectangular coordinate system is shown with a viewing window indicated.

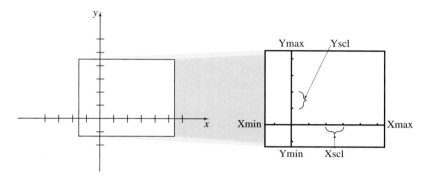

Yscl indicates the number of units per tick mark on the y-axis.
Xscl indicates the number of units per tick mark on the x-axis.

The dimensions of a window are selected by entering the minimum value, maximum value, and scale for both x and y under the window setting.
Several viewing windows can be accessed automatically on most graphing utilities. The screen below illustrates viewing the rectangular coordinate system from -10 to 10 on both the x-axis and the y-axis, with each tick mark representing 1 unit. This particular window setting is called the **standard window** on some graphing utilities.

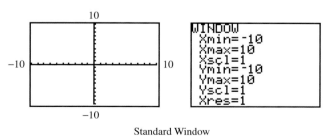

Standard Window

To refer to the window dimensions throughout this text, we will use the following notation. $[-10, 10, 1]$ by $[-10, 10, 1]$ refers to $[\text{Xmin}, \text{Xmax}, \text{Xscl}]$ by $[\text{Ymin}, \text{Ymax}, \text{Yscl}]$.

If we select the same standard window setting but change the Xmin to −20, the Xmax to 20, and the Xscl to 5, observe to the left how the window has been changed. Next, select the standard window setting but change the Ymin to −50, the Ymax to 50, and the Yscl to 10 and observe the changes in the window to the right.

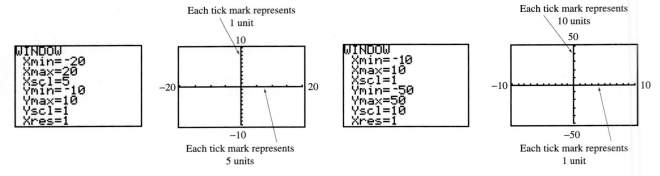

Many graphing utilities have the ability to show placement (via coordinates) and movement about the viewing window by means of a **cursor.**

DISCOVER THE CONCEPT

Clear any equations from the Y= editor and access the **standard window.** The screen below shows the standard window with the cursor located at the ordered pair (1.9148936, 3.5483871).

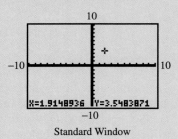

Standard Window

Use the arrow keys and move the cursor around the rectangular coordinate system. Using the arrow keys, move to some points in quadrant I. What do you observe about the signs of the *x*- and *y*-coordinates in this quadrant? Repeat this process in quadrants II, III, and IV. What observations can you make about the signs of the coordinates in each quadrant?

From the discovery above, we see that the signs of the coordinates of an ordered pair depends on the quadrant in which it is located.

Quadrant I: $(+, +)$ Quadrant III: $(-, -)$

Quadrant II: $(-, +)$ Quadrant IV: $(+, -)$

x-axis: $(x, 0)$ *y*-axis: $(0, y)$

An **integer window** is another window that you may be able to automatically access on your graphing utility. In an integer window the *x*-coordinates are integers as you move the cursor around the coordinate system.

TECHNOLOGY NOTE

The settings for an integer window vary among graphing utilities. Look in your graphing utility manual to find the integer setting for your particular graphing utility.

Access an integer window on your graphing utility and observe how the x- and y-coordinates are different from the x- and y- coordinates in a standard window. The screen below shows an integer window with the ordered pair $(10, 7)$ indicated.

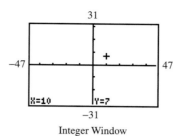

Integer Window

Another common window on many graphing utilities is a **decimal window.** Access this window and then move the cursor around the screen. Notice that the x-coordinates are incrementing or changing by 0.1 as you move from one point to another. The screen below shows an example of a decimal window on a graphing utility and its window settings.

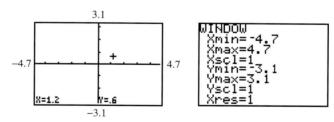

Decimal Window

TECHNOLOGY NOTE

For many graphing utilities, both the standard window and the decimal window are centered at the origin. However, your graphing utility may allow you to center an integer window anywhere in the rectangular coordinate system. The center may depend on where you last left a cursor before selecting the integer setting. Check your graphing utility manual to see what window options are available to you.

Experiment on your own by changing the window settings and predicting the resulting windows.

REMINDER   Remember that each point in the rectangular coordinate system corresponds to a unique ordered pair (x, y), and each ordered pair corresponds to a unique point.

EXAMPLE 2   Display a window of the rectangular coordinate system with a setting of $[-20, 40, 5]$ by $[-30, 30, 10]$.

Solution: The window setting and the resulting window are shown next.

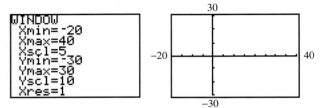

3   The distance between two points in a rectangular coordinate system can be found by using the Pythagorean theorem. First, we need to consider the distance between two points on a horizontal line. To find the distance between these points, we subtract the x-values and take the absolute value of this difference. To find the distance between two points on a vertical line, we subtract the y-values and take the absolute value of this difference. The figure below illustrates the process.

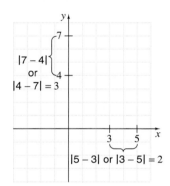

To find the distance between the points $A(5, 7)$ and $B(3, 4)$, plot the points and complete the right triangle that is shown in the figure below.

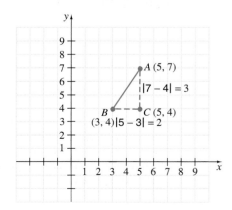

The length of $\overline{AC}$ is 3 units and the length of $\overline{BC}$ is 2 units. Using the Pythagorean theorem,

$$(AB)^2 = (AC)^2 + (BC)^2$$

or

$$(AB)^2 = 3^2 + 2^2$$

or

$$(AB)^2 = 9 + 4$$
$$AB = \sqrt{13}$$

The distance between points $A(5, 7)$ and $B(3, 4)$ is exactly $\sqrt{13}$ units or approximately 3.61 units.

In general, to find the distance d between two points (x_1, y_1) and (x_2, y_2) as shown to the left, notice that the length of leg a is $|x_2 - x_1|$ and that the length of leg b is $|y_2 - y_1|$.

Thus, the Pythagorean theorem tells us that

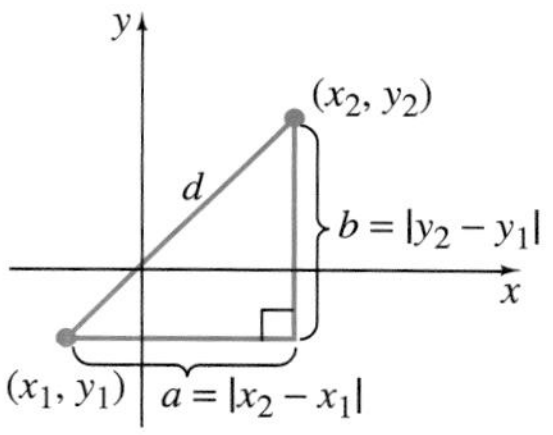

$$d^2 = a^2 + b^2$$

or

$$d^2 = (x_2 - x_1)^2 + (y_2 - y_1)^2$$

No absolute value bars are needed since we are squaring the distance.

or

$$d = \sqrt{(x_2 - x_1)^2 + (y_2 - y_1)^2}$$

This formula gives us the distance between any two points on the plane.

DISTANCE FORMULA

The distance d between two points (x_1, y_1) and (x_2, y_2) is given by

$$d = \sqrt{(x_2 - x_1)^2 + (y_2 - y_1)^2}$$

EXAMPLE 3   Find the distance between the two points $A(3, 10)$ and $B(-2, 4)$.

Solution: Using the distance formula $d = \sqrt{(x_2 - x_1)^2 + (y_2 - y_1)^2}$ with $x_1 = 3, x_2 = -2, y_1 = 10$, and $y_2 = 4$, we have

$$d = \sqrt{(-2 - 3)^2 + (4 - 10)^2}$$
$$= \sqrt{(-5)^2 + (-6)^2}$$
$$= \sqrt{25 + 36}$$
$$= \sqrt{61} \approx 7.81$$

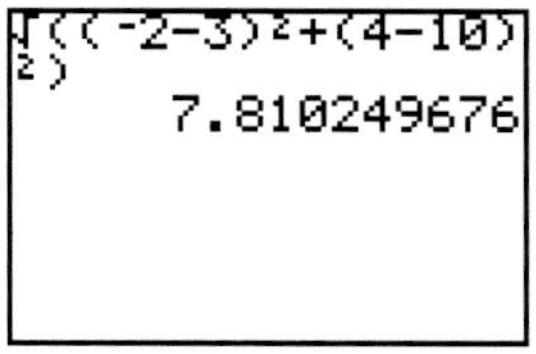

Therefore, the distance between points $A(3, 10)$ and $B(-2, 4)$ is exactly $\sqrt{61}$, or approximately 7.81 units.

To check using your graphing utility, evaluate $\sqrt{(-2 - 3)^2 + (4 - 10)^2}$. Correct placement of parentheses is vital! The screen to the left indicates that you need parentheses around the entire expression under the radical.

```
√((-2-3)²+(4-10)
²)
     7.810249676
```

We will study radicals and roots further in Chapter 8.

In Example 4 we use the distance formula to identify types of triangles. Recall that triangles can be classified according to the number of sides that are equal. If no sides are equal, a triangle is said to be **scalene.** If two sides are equal, it is said to be **isosceles.** If all three sides are equal, it is said to be **equilateral.**

EXAMPLE 4   Given the points $A(2, -1)$, $B(4, -4)$, and $C(5, -3)$, determine whether the triangle ($\triangle$) ABC is scalene, isosceles, or equilateral. Sketch the graph.

Solution:

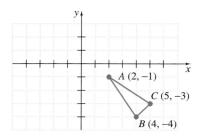

The sketch above shows $\triangle ABC$ formed by joining the points. Below are the calculations used to find the length of each side of the triangle.

$$AB = \sqrt{(4 - 2)^2 + (-4 - (-1))^2} = \sqrt{4 + 9} = \sqrt{13} \approx 3.61$$

$$BC = \sqrt{(5 - 4)^2 + (-3 - (-4))^2} = \sqrt{1 + 1} = \sqrt{2} \approx 1.41$$

$$AC = \sqrt{(5 - 2)^2 + (-3 - (-1))^2} = \sqrt{9 + 4} = \sqrt{13} \approx 3.61$$

Since sides AB and AC have equal lengths, the triangle is isosceles.

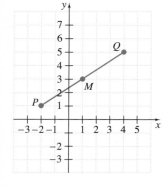

The **midpoint** of a line segment is the *point* located exactly halfway between the two end points of the line segment. On the graph to the left, the point M is the midpoint of line segment PQ. Thus, the distance between M and P equals the distance between M and Q.

The x-coordinate of M is at half the distance between the x-coordinates of P and Q, and the y-coordinate of M is at half the distance between the y-coordinates of P and Q. That is, the x-coordinate of M is the average of the x-coordinates of P and Q; the y-coordinate of M is the average of the y-coordinates of P and Q.

4

MIDPOINT FORMULA

The midpoint of the line segment whose endpoints are (x_1, y_1) and (x_2, y_2) is the point with coordinates

$$\left(\frac{x_1 + x_2}{2}, \frac{y_1 + y_2}{2}\right)$$

EXAMPLE 5   Find the midpoint of the line segment that joins points $P(-3, 3)$ and $Q(1, 0)$.

Solution: Use the midpoint formula. It makes no difference which point we call (x_1, y_1), or which point we call (x_2, y_2). Let $(x_1, y_1) = (-3, 3)$ and $(x_2, y_2) = (1, 0)$.

$$\text{Midpoint} = \left(\frac{x_1 + x_2}{2}, \frac{y_1 + y_2}{2}\right)$$

$$= \left(\frac{-3 + 1}{2}, \frac{3 + 0}{2}\right)$$

$$= \left(\frac{-2}{2}, \frac{3}{2}\right)$$

$$= \left(-1, \frac{3}{2}\right)$$

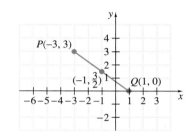

The midpoint of the segment is $\left(-1, \frac{3}{2}\right)$.

MENTAL MATH

Determine the coordinates of each point on the graph.

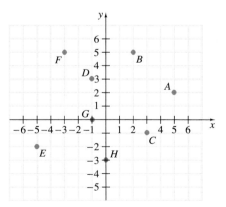

1. Point A **2.** Point B **3.** Point C **4.** Point D

5. Point E **6.** Point F **7.** Point G **8.** Point H

EXERCISE SET 3.1

Plot each point and name the quadrant or axis in which the point lies. See Example 1.

1. $(3, 2)$ **2.** $(2, -1)$

3. $(-5, 3)$ **4.** $(-3, -1)$

5. $\left(5\frac{1}{2}, -4\right)$ **6.** $\left(-2, 6\frac{1}{3}\right)$

7. $(0, 3.5)$ **8.** $(-2, 4)$

9. $(-2, -4)$ **10.** $(-4.2, 0)$

Given that x is a positive number and that y is a positive number, determine the quadrant or axis in which each point lies.

☐ 11. $(x, -y)$ **☐ 12.** $(-x, y)$

☐ 13. $(x, 0)$ **☐ 14.** $(0, -y)$

☐ 15. $(-x, -y)$ **☐ 16.** $(0, 0)$

Determine the coordinates of the point shown and name the quadrant in which the point lies.

17. 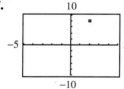   **18.**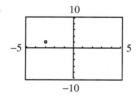

19. 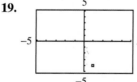   **20.**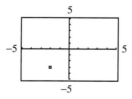

Determine a window setting so that the three given points will lie in that window. Report your answer in the [Xmin, Xmax, Xscl] by [Ymin, Ymax, Yscl] format. See Example 2.

21. $(-5, 2), (3, 8), (10, 15)$ **22.** $(-10, 12), (0, 9), (6, -15)$

23. $(-25, 0), (5, 7), (20, 100)$ **24.** $(0, 50), (25, 75), (50, 150)$

Match the following screens with the window settings listed below.

25. 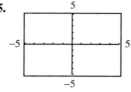   **26.**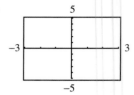

27. 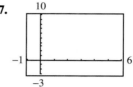   **28.**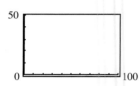

(a)    **(b)**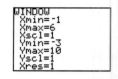

(c)   **(d)**

Find the distance between the pairs of points. Then find the midpoint of the line segment joining them. See Examples 3 and 5.

29. $A(1, 2), B(5, 6)$ **30.** $A(-1, -3), B(2, 1)$

31. $A(0, 4), B(3, -2)$ **32.** $A(3, 8), B(-2, 4)$

33. $A(-2, 7), B(-4, 10)$ **34.** $A(7, -5), B(9, -2)$

35. 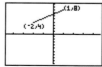   **36.**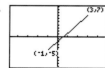

37. Find the perimeter of the triangle whose vertices are $A(12, 3)$, $B(5, 3)$, and $C(7, 11)$.

38. Find the perimeter of the triangle whose vertices are $A(2, 8)$, $B(11, -4)$, and $C(7, 4)$.

39. Find the perimeter of the quadrilateral whose vertices are $A(-2, 3)$, $B(4, 8)$, $C(9, 2)$, and $D(3, -3)$.

40. Find the perimeter of the quadrilateral whose vertices are $A(1, 6)$, $B(7, 11)$, $C(12, 5)$, and $D(6, 0)$.

Using the distance formula, classify the triangle formed by joining the following points as scalene, isosceles, or equilateral. See Example 4.

41. $A(-6, 15)$, $B(0, 3)$, $C(10, 9)$

42. $A(-3, 18)$, $B(3, 6)$, $C(13, 12)$

43. $A(-4, 1)$, $B(2, 6)$, $C(10, -4)$

44. $A(7, 8)$, $B(3, 5)$, $C(11, 0)$

45. $A(5, 11)$, $B(7, 5)$, $C(13, 7)$

46. $A(2, 8)$, $B(4, 2)$, $C(10, 4)$

47. Show that the triangle formed by joining the points $A(7, 10)$, $B(8, 3)$, and $C(5, 4)$ is a right triangle. Use the Pythagorean theorem.

48. Show that the triangle formed by joining the points $A(3, 6)$, $B(4, -1)$, and $C(1, 0)$ is a right triangle. Use the Pythagorean theorem.

Review Exercises

Write each statement as an equation in two variables. See Section 2.2.

49. The y-value is 5 more than three times the x-value.

50. The y-value is -3 decreased by twice the x-value.

51. The y-value is 2 more than the square of the x-value.

52. The y-value is 5 decreased by the square of the x-value.

Solve the following equations. See Section 2.1.

53. $3(x - 2) + 5x = 6x - 16$

54. $5 + 7(x + 1) = 12 + 10x$

55. $3x + \dfrac{2}{5} = \dfrac{1}{10}$

56. $\dfrac{1}{6} + 2x = \dfrac{2}{3}$

A Look Ahead

We can plot particular ordered pairs on some graphing utilities by using the statistics features, such as a statistical plotting feature.

EXAMPLE

Plot the ordered pair $(-10, -20)$ and $(15, 30)$ using an integer window.

Solution:

Enter the ordered pairs using a list feature and activate a statistical plotting feature. The result using an integer window is shown.

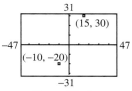

Integer Window

Plot the ordered pairs on a graphing utility.

57. $(-10, 20)$ and $(10, 20)$ using an integer window.

58. $(0, -30)$, and $(30, 0)$ using an integer window.

59. $(-2, 1)$, and $(4, -1)$ using a decimal window.

60. $(-3.2, 3)$, and $(0.6, 1)$ using a decimal window.

3.2 | GRAPHING EQUATIONS

TAPE IAG 3.2

O B J E C T I V E S

1 Determine whether an ordered pair of numbers is a solution to an equation in two variables.

2 Identify and graph linear equations.

3 Identify and graph quadratic equations.

4 Identify and graph cubic equations.

5 Identify and graph absolute value equations.

Solutions of equations in two variables consist of two numbers that can be written as ordered pairs of numbers. For example, we say that the ordered pair $(100, 1510)$ is a solution of the equation $y = 1500 + 0.1x$ because, when x is replaced with 100 and y is replaced with 1510, a true statement results.

$$y = 1500 + 0.1x$$

$$1510 = 1500 + 0.1(100)$$

$$1510 = 1500 + 10$$

$$1510 = 1510 \qquad \text{True}$$

EXAMPLE 1   Determine whether $(0, -12)$, $(1, 9)$, and $(2, -6)$ are solutions of the equation $3x - y = 12$.

Solution: To check each ordered pair, replace x with the x-coordinate and y with the y-coordinate and see whether a true statement results.

Let $x = 0$, and $y = -12$.	Let $x = 1$ and $y = 9$.	Let $x = 2$ and $y = -6$.
$3x - y = 12$	$3x - y = 12$	$3x - y = 12$
$3(0) - (-12) = 12$	$3(1) - (9) = 12$	$3(2) - (-6) = 12$
$0 + 12 = 12$	$3 - 9 = 12$	$6 + 6 = 12$
$12 = 12$ True	$-6 = 12$ False	$12 = 12$ True

Therefore $(1, 9)$ is not a solution, but both $(2, -6)$ and $(0, -12)$ are solutions.

 In fact, the equation $3x - y = 12$ has an infinite number of ordered pair solutions. Since it is impossible to list all solutions, we visualize them by graphing them.

A few more ordered pairs that satisfy $3x - y = 12$ are $(4, 0)$, $(3, -3)$, $(5, 3)$, and $(1, -9)$. These ordered pair solutions along with the ordered pair solutions from Example 1 are listed in a table and are plotted on the following graph. The graph of

x	y
4	0
1	-9
0	-12
2	-6
5	3
3	-3

[Graph showing the line $3x - y = 12$ with plotted points $(5, 3)$, $(4, 0)$, $(3, -3)$, $(2, -6)$, $(1, -9)$, and $(0, -12)$.]

$3x - y = 12$ is the single line containing these points. Every ordered pair solution of the equation corresponds to a point on this line, and every point on this line corresponds to an ordered pair solution.

The equation $3x - y = 12$ is called a linear equation in two variables, and **the graph of every linear equation in two variables is a line.**

LINEAR EQUATION IN TWO VARIABLES

A linear equation in two variables is an equation that can be written in the form

$$Ax + By = C$$

where A, B, and C are real numbers, and A and B are not both 0.

A linear equation written in the form $Ax + By = C$ is said to be written in **standard form.**

EXAMPLES OF LINEAR EQUATIONS IN STANDARD FORM

$$3x - y = 12$$

$$-2.1x + 5.6y = 0$$

As we mentioned in Section 3.1, the equation $y = 1500 + 0.1x$ is also a linear equation in two variables. This means that the equation $y = 1500 + 0.1x$ can be written in standard form $Ax + By = C$ and that its graph is a line. Its solutions in ordered pair form are shown next along with a portion of its graph. Recall that x is products sold and y is monthly salary. Since we assume that the smallest amount of product sold is none, or 0, then x must be greater than or equal to 0. Therefore, only the part of the graph that lies in quadrant I is shown. Notice that the graph gives a visual picture of the correspondence between products sold and salary.

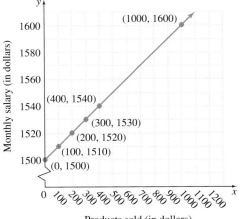

(The $\lessgtr$ symbol on the y-axis means that some values are missing along the axis.)

EXAMPLE 2 FINDING SALARY AND PRODUCTS SOLD

Use the graph of $y = 1500 + 0.1x$ to answer the following questions:

a. If the salesperson has $800 of products sold for a particular month, what is the salary for that month?

b. If the salesperson wants to make more than $1600 per month, what must be the total amount of products sold?

Solution: a. Since x is products sold, find 800 along the x-axis and move vertically up until you reach a point on the line. From this point on the line, move horizontally to the left until you reach the y-axis. The value on the y-axis is 1580, which means that if $800 worth of products is sold the salary for the month is $1580.

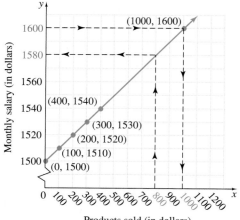

b. Since y is monthly salary, find 1600 along the y-axis and move horizontally to the right until you reach a point on the line. Either read the corresponding x-value from the labeled ordered pair or move vertically downward until you reach the x-axis. The corresponding x-value is 1000. This means that $1000 worth of products sold gives a salary of $1600 for the month. For the salary to be greater than $1600, products sold must be greater than $1000.

EXAMPLE 3 Sketch the graph of $2x + y = 7$. Then use a graphing utility to check the graph.

Solution: Find three ordered pair solutions, and plot the ordered pairs. The line through the plotted points is the graph. Let x be 0, let y be 0, and then let x be 2 to find your three ordered pair solutions.

Let $x = 0$.	Let $y = 0$.	Let $x = 2$.
$2x + y = 7$	$2x + y = 7$	$2x + y = 7$
$2 \cdot 0 + y = 7$	$2x + 0 = 7$	$2 \cdot 2 + y = 7$
$y = 7$	$\dfrac{2x}{2} = \dfrac{7}{2}$	$4 + y = 7$
	$x = 3.5$	$y = 3$

The ordered pair solutions found are $(0, 7)$, $(3.5, 0)$ and $(2, 3)$. They are listed in table form and the graph of $2x + y = 7$ is shown next.

x	0	2	3.5
y	7	3	0

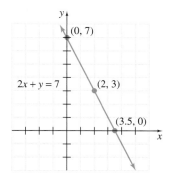

Next we graph $2x + y = 7$ using a graphing utility. To do so, first solve the equation for y in order to enter it into the Y= editor.

$$2x + y = 7$$

$$2x + y - 2x = -2x + 7 \qquad \text{Subtract } 2x \text{ from both sides.}$$

$$y = -2x + 7$$

Define $y_1 = -2x + 7$ and select the integer window. The screen below illustrates the graph of the linear equation. Access a trace feature to locate the ordered pair solutions in the table above. Trace to verify that the point $(0, 7)$ is a solution for the equation as well as $(2, 3)$.

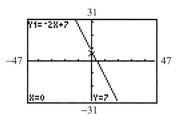

EXAMPLE 4   Graph $5x - 3y = 10$ in an integer window centered at the origin. Find the y-coordinate that makes the ordered pair $(8, ?)$ a solution for the given equation.

Solution: To enter the equation in the Y= editor, solve for y.

$$5x - 3y = 10$$

$$5x - 3y - 5x = -5x + 10 \qquad \text{Subtract 5x from both sides.}$$

$$-3y = -5x + 10$$

$$y = \frac{-5x + 10}{-3} \qquad \text{Divide both sides by } -3.$$

Define $y_1 = \dfrac{-5x + 10}{-3}$ and graph in an integer window centered at $(0, 0)$. Trace to locate the point with x-coordinate 8 and find the corresponding y-value to be 10.

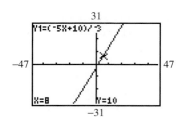

The ordered pair $(8, 10)$ is a solution of the given solution. To check, let $x = 8$ and $y = 10$ in the equation $5x - 3y = 10$ and see that a true statement results. ▬▬▬▬

In Example 3, notice that the graph crosses the x-axis at the point $(3.5, 0)$. This point is called the **x-intercept point,** and 3.5 is called the **x-intercept.** This graph also crosses the y-axis at the point $(0, 7)$. This point is called the **y-intercept point,** and 7 is called the **y-intercept.**

3 Not all equations in two variables are linear equations. Recall that the **degree of an equation** is the highest degree of any of its terms. Therefore, equations in the form $Ax + By = C$ are first-degree equations. Notice that all first-degree equations in two variables are linear equations, and their graph is a line. We begin looking at the graphs of nonlinear equations by first examining the graphs of equations of the form $y = ax^2 + bx + c$. Equations in this form are second-degree or **quadratic equations.** An equation such as $y = x^3$ is a third-degree or **cubic equation.** Once we know the shape of the graph of these common equations, we can learn to predict the general shape of a curve before graphing it. This will help us to recognize whether a graph we produce on a graphing utility is reasonable.

EXAMPLE 5 Make a table of values for $y = x^2$ for $x = -3, -2, -1, 0, 1, 2$, and 3. Use the table to sketch the graph of $y = x^2$. Then use a graphing utility to check the table of values and the shape of the graph.

Solution: Notice that the equation is not linear, and its graph is therefore not a line. We begin by finding ordered pair solutions for $x = -3, -2, -1, 0, 1, 2$, and 3.

	x	y
If $x = -3$, then $y = (-3)^2$, or 9.	-3	9
If $x = -2$, then $y = (-2)^2$, or 4.	-2	4
If $x = -1$, then $y = (-1)^2$, or 1.	-1	1
If $x = 0$, then $y = 0^2$, or 0.	0	0
If $x = 1$, then $y = 1^2$, or 1.	1	1
If $x = 2$, then $y = 2^2$, or 4.	2	4
If $x = 3$, then $y = 3^2$, or 9.	3	9

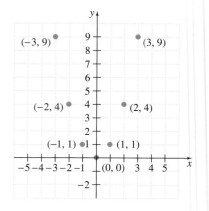

Study the table a moment and look for patterns. Notice that the ordered pair solution $(0, 0)$ contains the smallest y-value because any *other* x-value squared will give a positive result. This means that the point $(0, 0)$ will be the lowest point on the graph. Also notice that all other y-values correspond to two opposite x-values. For example, $3^2 = 9$ and also $(-3)^2 = 9$.

This means that the graph will be a mirror image of itself across the y-axis. Connect the plotted points with a smooth curve to sketch its graph as shown.

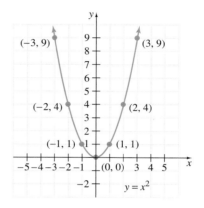

To check, graph $y_1 = x^2$ in an integer window. Compare the graph and table shown with the calculations done by hand.

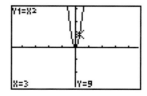

This curve is given a special name called a **parabola.** It can be shown that the graph of a quadratic equation of the form $y = ax^2 + bx + c$ is a parabola. We will study more about parabolas in later chapters.

The graphing utility finds and plots various ordered pair solutions in the same manner that we find and plot ordered pair solutions. The advantage of a graphing utility is that it can perform this process much faster and with greater accuracy than we can.

DISCOVER THE CONCEPT

a. Compare the graphs of $y = x^2$ and $y = -x^2$ by graphing each in a standard window.

b. How can the graph of $y = x^2$ be transformed to look like the graph of $y = -x^2$?

In the above discovery, we see that if the coefficient of x^2 is positive, the parabola opens upward, and if the coefficient is negative, the parabola opens downward. Compare this fact with the table of values below. We defined $y_1 = x^2$ and $y_2 = -x^2$.

For $y = -x^2$, the y-value is always negative or 0 and the parabola opens downward.

X	Y1	Y2
-3	9	-9
-2	4	-4
-1	1	-1
0	0	0
1	1	-1
2	4	-4
3	9	-9

Y1◻X2

For $y = x^2$, the y-value is always positive or 0 and the parabola opens upward.

We say the graph of $y = x^2$ is reflected across the x-axis to obtain the graph of $y = -x^2$.

4 Another nonlinear equation is a third-degree or cubic equation.

EXAMPLE 6 Graph $y = x^3$ and examine a table of values with x equal to the integers from -3 to 3.

Solution: Define $y_1 = x^3$ and graph the equation in a standard window.

The figure below shows the graph of $y = x^3$ with the trace cursor on $(2, 8)$. The table of values shows ordered pair solutions for $x = -3, -2, -1, 0, 1, 2,$ and 3.

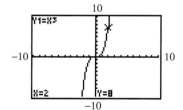

X	Y1
-3	-27
-2	-8
-1	-1
0	0
1	1
2	8
3	27

Y1◻X3

Notice in the table that when x is negative y is also negative. Thus, these points lie in the third quadrant. When x is positive, y is positive; thus, this portion of the graph lies in the first quadrant. When $x = 0$, then $y = 0$, so the graph crosses the axes at the origin.

DISCOVER THE CONCEPT

How do you think the graph of $y = x^3$ is affected when we place a negative sign in front of the x^3? Test your prediction by graphing $y = x^3$ and $y = -x^3$ separately in a standard window. State in words how you can obtain the graph of $y = -x^3$ from the graph of $y = x^3$.

5

We next graph the basic absolute value equation, $y = |x|$. Begin by finding and plotting ordered pair solutions and analyze any patterns.

EXAMPLE 7   Make a table of values for $y = |x|$. Let x be the integers from -3 to 3. Use the table to graph the equation $y = |x|$. Use a graphing utility to check the table of values and the shape of the graph.

Solution: This is not a linear equation, so its graph is not a line. We list x-values and substitute to find corresponding y-values. Then we plot the ordered pair solutions.

		x	y
If $x = -3$, then $y = |-3|$, or 3.		-3	3
If $x = -2$, then $y = |-2|$, or 2.		-2	2
If $x = -1$, then $y = |-1|$, or 1.		-1	1
If $x = 0$, then $y = |0|$, or 0.		0	0
If $x = 1$, then $y = |1|$, or 1.		1	1
If $x = 2$, then $y = |2|$, or 2.		2	2
If $x = 3$, then $y = |3|$, or 3.		3	3

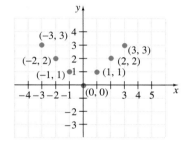

Study the table of values and the plotted points and notice any patterns.

From the plotted ordered pairs, we see that the graph of this absolute value equation is V-shaped. The completed graph is shown next. To check, we show the graph of $y = |x|$ using a standard window along side.

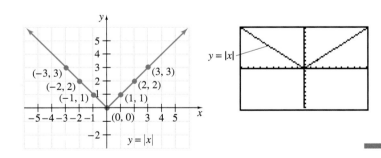

Notice that the basic absolute value graph is V-shaped as compared to the U-shape of a parabola.

We now examine how a graph is affected when a constant is added.

EXAMPLE 8   Graph the equation $y = |x| - 3$ and examine a table of values for x equal to the integers from -3 to 3. Compare the graph and the table to the graph and table for $y = |x|$.

Solution: Graph both $y_1 = |x|$ and $y_2 = |x| - 3$ using a standard window. Notice that compared with the equation $y = |x|$, the equation $y = |x| - 3$ decreases the y-values by 3 units for the same x-values. This means that the graph of $y = |x| - 3$ is the same as the graph of $y = |x|$ lowered by 3 units.

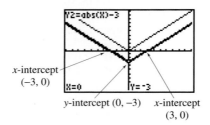

x-intercept
$(-3, 0)$

y-intercept $(0, -3)$ x-intercept $(3, 0)$

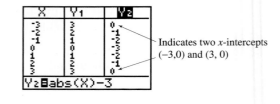

Indicates two x-intercepts $(-3,0)$ and $(3, 0)$

This graph of $y = |x| - 3$ shows us that it is possible to have more than one x- or y- intercept. Notice that this graph has one y-intercept point $(0, -3)$ and two x-intercept points $(-3, 0)$ and $(3, 0)$.

DISCOVER THE CONCEPT

a. Predict how the graph of $y = |x| + 5$ can be obtained from the graph of $y = |x|$. Then check your prediction by graphing both equations using a standard window.

b. Predict how the graph of $y = x^2 - 3$ can be obtained from the graph of $y = x^2$. Use a graphing utility to check your prediction. Next, predict how the graph of $y = x^2 + 5$ can be obtained from $y = x^2$.

c. In general, can you predict how the graph of a basic equation with a constant added can be obtained from the graph of the basic equation? Test your prediction with the basic cubic equation $y = x^3$.

If we add a constant K to a basic graph, it is said to cause a **vertical translation** of the graph. That is, if the constant K is positive, the basic graph is moved or translated K units upward. Similarly, if the constant K is negative, the basic graph is moved, or translated, $|K|$ units downward.

To summarize, the graphs of the basic equations we have discussed so far are shown below.

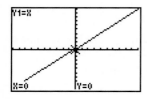

Basic linear Equation
$y = x$

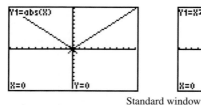

Absolute Value Equation
$y = |x|$

Standard window

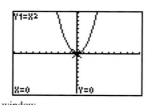

Quadratic Equation (Parabola)
$y = x^2$

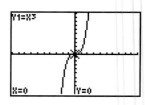

Cubic Equation
$y = x^3$

EXERCISE SET 3.2

Determine whether each ordered pair is a solution of the given equation. See Example 1.

1. $y = 3x - 5$; $(0, 5), (-1, -8)$

2. $y = -2x + 7$; $(1, 5), (-2, 3)$

3. $-6x + 5y = -6$; $(1, 0), \left(2, \dfrac{6}{5}\right)$

4. $5x - 3y = 9$; $(0, 3), \left(\dfrac{12}{5}, -1\right)$

5. $y = 2x^2$; $(1, 2), (3, 18)$

6. $y = 2|x|$; $(-1, 2), (0, 2)$

7. $y = x^3$; $(2, 8), (3, 9)$

8. $y = x^4$; $(-1, 1), (2, 16)$

9. $y = \sqrt{x} + 2$; $(1, 3), (4, 4)$

10. $y = \sqrt[3]{x} - 4$; $(1, -3), (8, 6)$

Determine whether each equation is linear or nonlinear and tell the basic shape of the graph (line, parabola, cubic, V-shaped). See Examples 2 through 8.

11. $x + y = 3$

12. $y - x = 8$

13. $y = 4x$

14. $y = 6x$

15. $y = 4x - 2$

16. $y = 6x - 5$

17. $y = |x| + 3$

18. $y = |x| + 2$

19. $2x - y = 5$

20. $4x - y = 7$

21. $y = 2x^2$

22. $y = 3x^2$

23. $y = x^2 - 3$

24. $y = x^2 + 3$

25. $y = -2x$

26. $y = -3x$

27. $y = -2x + 3$

28. $y = -3x + 2$

29. $y = |x + 2|$

30. $y = |x - 1|$

31. $y = x^3$

32. $y = x^3 - 2$

33. $y = -|x|$

34. $y = -x^2$

35. $y = -1.3x + 5.6$

36. $y = 7.6x - 1.4$

Rewrite each equation as an equivalent equation that can be entered in the Y= editor of your graphing utility. See Example 3.

37. $2x + y = 10$

38. $-6x + y = 2$

39. $-7x - 3y = 4$

40. $5x - 11y = -1.2$

41. The graph of $y = 2x^2 + 1.2x - 5.6$ has two x-intercepts and one y-intercept. Use your graphing utility and graph this equation using a decimal window. Find the coordinates of the intercepts.

42. The graph of $y = x^2 + 7x - 6$ has two ordered pair solutions whose y-value is 12. Use your graphing

utility and graph this equation using an integer window. Find the coordinates of these points.

43. If you trace along the graph of $y = 1.5x - 6$ using an integer window, which of the following coordinates would not be displayed?

   a. $(0, -6)$ **b.** $(1, -4.5)$

   c. $(4, 0)$ **d.** $(1.2, -4.2)$

44. If you trace along the graph of $y = x + 2.3$ using an integer window, which of the following coordinates would not be displayed?

   a. $(1.7, 4)$ **b.** $(0, 2.3)$

   c. $(2, 4.3)$ **d.** $(-3, -0.7)$

Match each equation with its graph without using a graphing utility. Then check your answers by graphing each equation using a standard window.

45. $y = x^2 - 4$

46. $y = x + 5$

47. $y = x^3 + x^2 - 4$

48. $y = |x| + 3$

a.    **b.**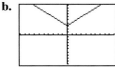

c. 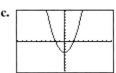   **d.**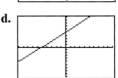

Match each equation with its graph without using a graphing utility. Then check your answers by graphing each equation using a standard window.

49. $y = x - 5$

50. $y = -x^2 + 4$

51. $y = -x^3 + x^2 + 2x - 3$ **52.** $y = |x + 3|$

a. 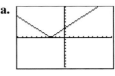   **b.**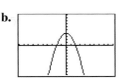

c.    **d.**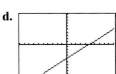

Match each graph with its equation without using a graphing utility. Then graph using a standard window to check your answers.

53.

54.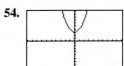

55.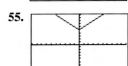

56.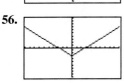

a. $y = x^2 + 3$ **b.** $y = x^2 - 3$

c. $y = |x| + 5$ **d.** $y = |x| - 3$

Sketch the graphs of the three equations by hand on the same coordinate plane. Then use your graphing utility to check your results. Describe the patterns you see in these graphs.

57. $y = x$

   $y = x + 2$

   $y = x - 3$

58. $y = x^2$

   $y = x^2 + 1$

   $y = x^2 - 3$

59. $y = -x^2$

   $y = -x^2 + 1$

   $y = -x^2 - 2$

60. $y = -|x|$

   $y = -|x| + 2$

   $y = -|x| - 4$

Graph each equation by plotting ordered pair solutions by hand. If the equation is not linear, suggested x-values have been given for generating ordered pair solutions. Check the graph using a graphing utility. See Examples 2 through 8.

61. $x + y = 3$

62. $y - x = 8$

63. $y = 4x$

64. $y = 6x$

65. $y = 4x - 2$

66. $y = 6x - 5$

67. $y = |x| + 3$

   Let $x = -3, -2, -1, 0, 1, 2, 3$.

68. $y = |x| + 2$

   Let $x = -3, -2, -1, 0, 1, 2, 3$.

69. $2x - y = 5$

70. $4x - y = 7$

71. $y = 2x^2$

   Let $x = -3, -2, -1, 0, 1, 2, 3$.

72. $y = 3x^2$

   Let $x = -3, -2, -1, 0, 1, 2, 3$.

73. $y = x^2 - 3$

   Let $x = -3, -2, -1, 0, 1, 2, 3$.

74. $y = x^2 + 3$

   Let $x = -3, -2, -1, 0, 1, 2, 3$.

75. $y = -2x$ **76.** $y = -3x$

77. $y = -2x + 3$ **78.** $y = -3x + 2$

79. $y = |x + 2|$

   Let $x = -4, -3, -2, -1, 0, 1$.

80. $y = |x - 1|$

   Let $x = -1, 0, 1, 2, 3, 4$.

81. $y = x^3$

   Let $x = -3, -2, -1, 0, 1, 2$.

82. $y = x^3 - 2$

   Let $x = -3, -2, -1, 0, 1, 2$.

83. $y = -|x|$

   Let $x = -3, -2, -1, 0, 1, 2, 3$.

84. $y = -x^2$

   Let $x = -3, -2, -1, 0, 1, 2, 3$.

85. Given the following table of values for ordered pairs satisfying an absolute value equation, plot the points and join them to complete the graph. Then write an equation of the graph.

86. Given the following table of values for ordered pairs satisfying a second degree equation, plot the points and join them to complete the graph. Then write an equation of the graph.

87. The perimeter y of a rectangle whose width is a constant 3 inches and whose length is x inches is given by the equation

$$y = 2x + 6$$

a. Draw a graph of this equation.

b. Read from the graph the perimeter y of a rectangle whose length x is 4 inches.

88. The distance y traveled in a train moving at a constant speed of 50 miles per hour is given by the equation

$$y = 50x$$

a. Draw a graph of this equation.

b. Read from the graph the distance y traveled after 6 hours.

*For income tax purposes, Rob Calcutta, owner of Copy Services, uses a method called **straight-line depreciation** to show the loss in value of a copy machine he recently purchased. Rob assumes that he can use the machine for 7 years. The following graph shows the value of the machine over the years. Use this graph to answer the following questions.*

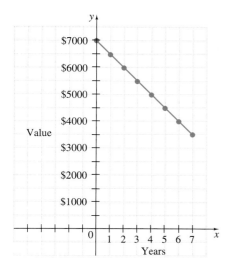

89. What was the purchase price of the copy machine?

90. What is the depreciated value of the machine in 7 years?

91. What loss in value occurred during the first year?

92. What loss in value occurred during the second year?

93. Why do you think that this method of depreciating is called straight-line depreciation?

94. Why is the line tilted downward?

Review Exercises

Use the graph for Exercises 95–98. See Section 3.1

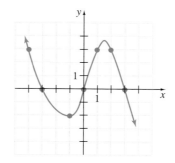

95. List the point(s) of the graph whose coordinates are of the form $(\ , 3)$.

96. List the x-intercept(s) of the graph.

97. List the y-intercept(s) of the graph.

98. List the point(s) of the graph whose coordinates are of the form $(-1,\ )$.

A Look Ahead

As mentioned in A Look Ahead for Section 3.1, many graphing utilities allow us to plot ordered pairs by using the statistics features, such as a statistical plotting feature. Plot the ordered pairs on a graphing utility. Tell whether they could belong to the graph of a quadratic equation or an absolute value equation.

99. $(-1, -6), (0, -3), (1, -2), (2, -3), (3, -6)$

100. $(-1, 4), (0, 3), (1, 2), (2, 3), (3, 4)$

3.3 | INTRODUCTION TO FUNCTIONS

O B J E C T I V E S

1. Define relation, domain, and range.
2. Identify functions.
3. Use the vertical line test for functions.
4. Find the domain and range of functions.
5. Use function notation.

TAPE IAG 3.3

1

Equations in two variables, such as $y = 2x + 1$, describe relations between x-values and y-values. For example, if $x = 1$, then $y = 3$. In other words the equation pairs an x-value with a particular y-value that makes the equation a true statement. We represent this pairing as the ordered pair, $(1, 3)$.

Other ordered pairs described by the relation $y = 2x + 1$ are $(2, 5), (3, 7), (4, 9)$, $(5, 11), (6, 13)$, and $(7, 15)$.

Another example of a relation is

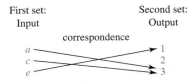

First set: Input   Second set: Output

correspondence

This relation shows the pairing $(a, 3), (c, 3)$, and $(e, 1)$. Notice that no element in the first set is assigned to the element 2 of the second set.

A **relation** is a pairing of elements from a first set (set of inputs) with elements from a second set (set of outputs). The **domain of a relation** is the set of all first elements in the pairing. The **range of a relation** is the set of all second elements in the pairing.

> **R E M I N D E R**   A relation is simply a pairing, and it can be a pairing of the elements of any two sets, not necessarily sets of numbers.

Example 1 illustrates different ways that relations may be described.

EXAMPLE 1   Determine the domain and range of each relation.

a. $\{(2, 3), (2, 4), (0, -1), (3, -1)\}$

b.

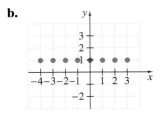

c.

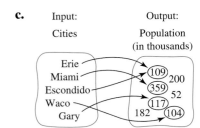

Input: Cities   Output: Population (in thousands)

Solution: **a.** The domain is the set of all first coordinates of the ordered pairs, {2, 0, 3}. The range is the set of all second coordinates, {3, 4, −1}.

b. Ordered pairs are not listed here, but are given in graph form. The relation is {(−4, 1), (−3, 1), (−2, 1), (−1, 1), (0, 1), (1, 1), (2, 1), (3, 1)}. The domain is {−4, −3, −2, −1, 0, 1, 2, 3}.

The range is {1}.

c. The domain is the first set, {Erie, Escondido, Gary, Miami, Waco}.

The range is the numbers in the second set that correspond to elements in the first set, {104, 109, 117, 359}.

2   Now we consider a special kind of relation called a function.

> A **function** is a relation in which each element of the first set corresponds to *exactly one* element of the second set.

> REMINDER   A function is a special type of relation, so all functions are relations, but not all relations are functions.

EXAMPLE 2   Which of the following relations are also functions?

a. {(−2, 5), (2, 7), (−3, 5), (9, 9)}   **b.**

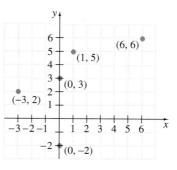

c.

INPUT:	CORRESPONDENCE:	OUTPUT
People in a certain city	Each person's age	Nonnegative integers

Solution: **a.** Although the ordered pairs (−2, 5) and (−3, 5) have the same *y*-value, each *x*-value is assigned to only one *y*-value, so this set of ordered pairs is a function.

b. The *x*-value 0 is assigned to two *y*-values, −2 and 3, in this graph, so this relation is not a function.

c. This relation is a function because, although two different people may have the same age, each person has only one age. Thus, each element in the first set is assigned to only one element in the second set.

Recall that we will call an equation such as $y = 2x + 1$ a relation since this equation defines a set of ordered pair solutions.

EXAMPLE 3   Is the relation $y = 2x + 1$ also a function?

Solution: The relation $y = 2x + 1$ is a function if each x-value corresponds to just one y-value. For each x-value substituted in the equation $y = 2x + 1$, the multiplication and addition performed each give a single result, so only one y-value will be associated with each x-value. Thus, $y = 2x + 1$ is a function.

EXAMPLE 4   Is the relation $x = y^2$ also a function?

Solution: In $x = y^2$, if $y = 3$, then $x = 9$. Also, if $y = -3$, then $x = 9$. In other words, the x-value 9 corresponds to two y-values, 3 and -3. Thus, $x = y^2$ is not a function.

As we have seen so far, not all relations are functions. Consider the graphs of $y = 2x + 1$ and $x = y^2$. On the graph of $y = 2x + 1$, notice that each x-value corresponds to only one y-value. Recall from Example 3 that $y = 2x + 1$ is a function.

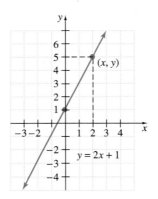

On the graph of $x = y^2$, the x-value 9, for example, corresponds to two y-values, 3 and -3, as shown by the vertical line on the hand-drawn sketch below. Recall from Example 4 that $x = y^2$ is not a function.

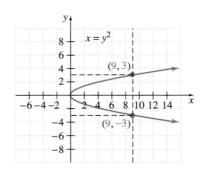

Graphs can be used to help determine whether a relation is also a function by the following vertical line test.

VERTICAL LINE TEST

If no vertical line can be drawn so that it intersects a graph more than once, the graph is the graph of a function.

EXAMPLE 5   Which of the following graphs are graphs of functions?

a. 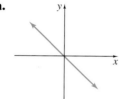   b. 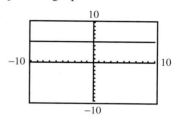   c.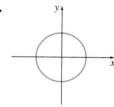

d. 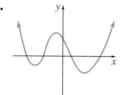   e. 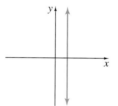   f.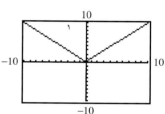

Solution: **a.** This graph is the graph of a function since no vertical line will intersect this graph more than once.

b. This graph is also the graph of a function.

c. This graph is not the graph of a function. Note that vertical lines can be drawn that intersect the graph in two points.

d. This graph is the graph of a function.

e. This graph is not the graph of a function. A vertical line can be drawn that intersects the graph at every point.

f. This graph is the graph of a function.

Recall that the graph of a linear equation in two variables is a line, and a line that is not vertical will pass the vertical line test. Thus, **all linear equations are functions except those whose graphs are vertical lines.**

Next, we practice finding the domain and range of a relation and deciding whether a relation is a function from its graph.

EXAMPLE 6   **FINDING THE DOMAIN AND RANGE**

Find the domain and range of each relation. Determine whether the relation is also a function.

a.
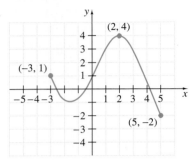

b.
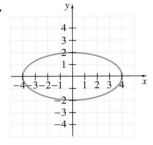

c.
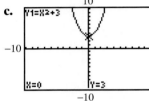

d.
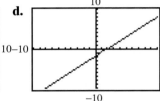

e.
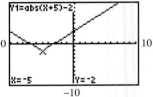

Solution: **a.** By the vertical line test, this is the graph of a function. From the graph, x takes on values from -3 to 5 inclusive; thus, the domain of the function is $[-3, 5]$. The range of the function is $[-2, 4]$ since y takes on the values -2 to 4 inclusive.

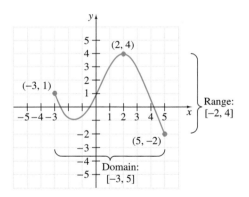

b. By the vertical line test, this is not the graph of a function. The domain is $[-4, 4]$. The range is $[-2, 2]$.

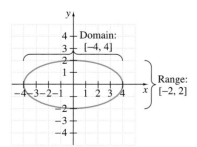

c. This graph of $y = x^2 + 3$ passes the vertical line test so $y = x^2 + 3$ is a function. The domain is all real numbers or $(-\infty, \infty)$ using interval notation. The range is $[3, \infty)$ since the y values begin with 3 and increase to infinity.

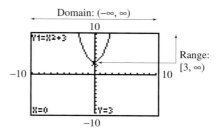

d. All lines except vertical lines are functions. The domain and the range of each are all real numbers.

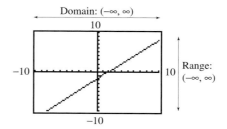

e. By the vertical line test, $y = |x + 5| - 2$ is a function. The domain is $(-\infty, \infty)$. The range is $[-2, \infty)$.

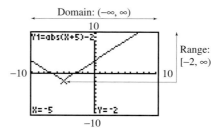

5 The graph of $y = 4x + 3$, for example, is a nonvertical linear equation, so it describes a function. Many times letters such as f, g, and h are used to name functions. To denote that y is a function of x in the equation $y = 4x + 3$, we write $y = f(x)$. Then $y = 4x + 3$ can be written as

$$f(x) = 4x + 3$$

The symbol $f(x)$ means *function of x* and is read "f of x." This notation is called **function notation**.

The notation $f(1)$ (read "f of 1") means to replace x with 1 and find the resulting y or function value. Since

$$f(x) = 4x + 3$$

then

$$f(1) = 4(1) + 3 = 7$$

This means that when $x = 1$ then y or $f(x) = 7$. Here the input is 1 and the output is $f(1)$ or 7. Thus, the process of finding function values is a pairing of an input value and an output value. In this case, the ordered pair $(1, 7)$ is an ordered pair solution of the equation $y = 4x + 3$.

Now find $f(2)$, $f(0)$, and $f(-1)$.

$$f(x) = 4x + 3$$

$$f(2) = 4(2) + 3 = 11 \qquad \text{and} \qquad (2, 11) \text{ is an ordered pair solution.}$$

$$f(0) = 4(0) + 3 = 3 \qquad \text{and} \qquad (0, 3) \text{ is an ordered pair solution.}$$

$$f(-1) = 4(-1) + 3 = -1 \qquad \text{and} \qquad (-1, -1) \text{ is an ordered pair solution.}$$

There are many ways of evaluating function values using a graphing utility. One way is to use the store feature. For example, to find $f(2)$ when $f(x) = 4x + 3$, store the number 2, enter the expression, and calculate. To find $f(0)$ and $f(-1)$ for the same function and save time, have your graphing utility copy (replay) the last entry and then edit it.

Replay, then replace 2 with 0 and enter.
Replay, then replace 0 with −1 and enter.

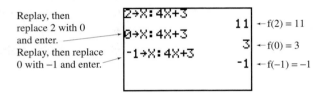

We can also use the graph to evaluate a function at a given value. The graph of $y_1 = 4x + 3$ in an integer window is shown below. Move the cursor to the point with x-coordinate 2 to find $f(2)$. Continue in this manner to find $f(0)$ and $f(-1)$.

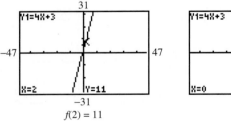

$f(2) = 11$ $\qquad\qquad$ $f(0) = 3$ $\qquad\qquad$ $f(-1) = -1$

A third method for finding function values is by using a table.

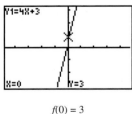

$f(-1) = -1 \rightarrow$
$f(0) = 3 \rightarrow$
$f(2) = 11 \rightarrow$

The method you use to find function values depends on the particular situation. For instance, in Example 7, we evaluate several different functions for various values of x, so we calculate by hand and use the store feature to check. However, in Example 8 we evaluate the same function, but for different values of x, so we choose to look at the graph of the function. These are just a few methods that can be used to evaluate functions at given values.

EXAMPLE 7   If $f(x) = 7x^2 - 3x + 1$, $g(x) = 3x - 2$, and $h(x) = x^2$, find the following.

a. $f(1)$ **b.** $g(3)$ **c.** $h(-2)$

Solution: **a.** Substitute 1 for x in $f(x)$ and simplify.

$$f(x) = 7x^2 - 3x + 1$$

$$f(1) = 7(1)^2 - 3(1) + 1 = 5$$

b. Substitute 3 for x in $g(x)$.

$$g(x) = 3x - 2$$

$$g(3) = 3(3) - 2 = 7$$

c. Substitute -2 for x in $h(x)$.

$$h(x) = x^2$$

$$h(-2) = (-2)^2 = 4$$

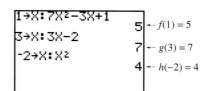

The screen to the right shows a check of each result using a graphing utility.

EXAMPLE 8   If $f(x) = 0.5x - 25$, find

a. $f(-10)$ **b.** $f(18)$ **c.** $f(11)$ **d.** $f(0)$

Solution: **a.** We choose to use a graphical method to evaluate. Define $y_1 = 0.5x - 25$ and graph using an integer window. Move the cursor to find $x = -10$. In the screen below we see that -10 is paired with $y = -30$; therefore $f(-10) = -30$.

b. Move the cursor to $x = 18$ and see that it is paired with -16 and thus $f(18) = -16$.

c. Move the cursor to $x = 11$ and see that $y = -19.5$, so $f(11) = -19.5$.

d. Move the cursor to $x = 0$ and see that $y = -25$, so $f(0) = -25$.

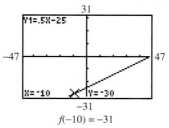

$f(-10) = -31$

Many formulas that are familiar to you describe functions. For example, we have used the formula for finding the area of a circle, $A = \pi r^2$. The area of the circle is actually a function of the length of the radius. Using this function notation, we write

$$A(r) = \pi r^2 \qquad .$$

$A(r)$ can be read as the area with respect to r. To find the area of the circle whose radius is 3 cm, we write

$$A(r) = \pi r^2$$

$$A(3) = \pi(3)^2 = 9\pi \text{ square centimeters}$$

An approximation to two decimal places is

$$9\pi \text{ sq cm} \approx 28.27 \text{ sq cm}$$

EXAMPLE 9   If the area of a circle with respect to its radius can be described by the formula $A(r) = \pi r^2$, find the following:

        a. $A(0.5)$ **b.** $A(3.1)$ **c.** $A(3.5)$

Solution: We choose to evaluate the function $A(r) = \pi r^2$ by graphing $y_1 = \pi x^2$ in a decimal window. Here, $y_1 =$ area and $x =$ radius. The screens below illustrate entering the function in the Y= editor, and the various ordered pair solutions found by using the trace feature. We round each function value to two decimal places.

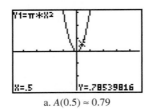

a. $A(0.5) \approx 0.79$

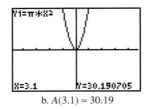

b. $A(3.1) \approx 30.19$

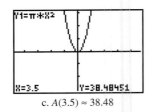

c. $A(3.5) \approx 38.48$

When a store manager is setting the retail price of an article, the price may be dependent on the wholesale price of the article. When this happens, we say that the retail price is a function of the wholesale price.

EXAMPLE 10   FINDING RETAIL PRICES

Sabrina manages the college bookstore and purchases the books from several wholesale companies. She finds that she needs to mark up the wholesale cost by 25%.

a. Write the retail price as a function of the wholesale cost.

b. Find the retail price of the following books given the wholesale cost.

WHOLESALE COST	$15.00	$22.75	$38.50	$53.00
RETAIL PRICE				

Solution: **a.** Let $w =$ the wholesale cost of a book. To denote that the retail price is a function of wholesale cost, we define the function $R(w)$.

In words:	Retail price (function of wholesale cost)	is	wholesale cost	plus	25% of wholesale cost
Translate:	$R(w)$	$=$	w	$+$	$0.25w$

b. Here we evaluate $R(w) = w + 0.25w$ for the given values of w.

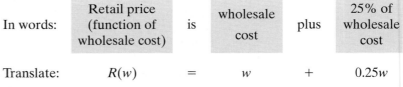

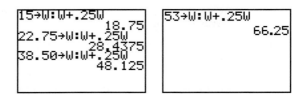

Completing the table, we have

WHOLESALE COST	W	$15.00	$22.75	$38.50	$53.00
RETAIL PRICE	R (w)	$18.75	$28.44	$48.13	$66.25

EXAMPLE 11

The following graph shows the research and development expenditures by the Pharmaceutical Manufacturers Association as a function of time.

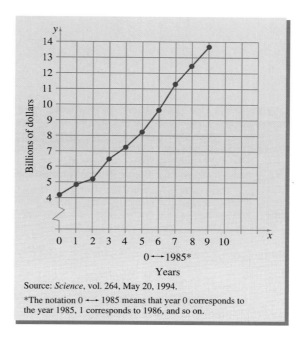

Source: *Science*, vol. 264, May 20, 1994.

*The notation 0 ←→ 1985 means that year 0 corresponds to the year 1985, 1 corresponds to 1986, and so on.

a. Approximate the money spent on research and development in 1992.

b. In 1958, research and development expenditures were $200 million. Find the increase in expenditures from 1958 to 1994.

Solution:

a. In 1992, approximately $11.3 billion was spent.

b. In 1994, approximately $13.8 billion, or $13,800 million, was spent. The increase in spending from 1958 to 1994 is $13,800 − $200 = $13,600 million.

Notice that the graph in Example 11 is the graph of a function since each year there is only one total amount of money spent by the Pharmaceutical Manufacturers Association on research and development. Also notice that the graph resembles the graph of a line. Often, businesses depend on equations that "closely fit" data-defined functions like this one in order to model the data and predict future trends. For example, by a method called least squares, the linear function $f(x) = 1.087x + 3.44$ approximates or models the data shown. Its graph and the actual data function are shown next. Later in the text, we will use a graphing utility to model real data such as shown in this example.

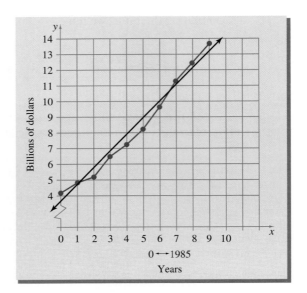

To predict the amount of money spent in the year 2000, we use $f(x) = 1.087x + 3.44$ and find $f(15)$. (Notice that year 0 on the graph corresponds to the year 1985, so year 15 corresponds to the year 2000.)

$$f(x) = 1.087x + 3.44$$

$$f(15) = 1.087(15) + 3.44$$

$$= 19.745$$

We predict that, in the year 2000, $19,745 billion will be spent on research and development by the Pharmaceutical Manufacturers Association.

EXERCISE SET 3.3

Find the domain and the range of each relation. Also determine whether the relation is a function. See Examples 1 and 2.

1. $\{(-1, 7), (0, 6), (-2, 2), (5, 6)\}$

2. $\{(4, 9), (-4, 9), (2, 3), (10, -5)\}$

3. $\{(-2, 4), (6, 4), (-2, -3), (-7, -8)\}$

4. $\{(6, 6), (5, 6), (5, -2), (7, 6)\}$

5. $\{(1, 1), (1, 2), (1, 3), (1, 4)\}$

6. $\{(1, 1), (2, 1), (3, 1), (4, 1)\}$

7. $\left\{\left(\frac{3}{2}, \frac{1}{2}\right), \left(1\frac{1}{2}, -7\right), \left(0, \frac{4}{5}\right)\right\}$

8. $\{(\pi, 0), (0, \pi), (-2, 4), (4, -2)\}$

9. $\{(-3, -3), (0, 0), (3, 3)\}$

10. $\left\{\left(\frac{1}{2}, \frac{1}{4}\right), \left(0, \frac{7}{8}\right), (0.5, \pi)\right\}$

11.

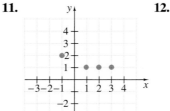

12.

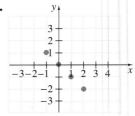

13.

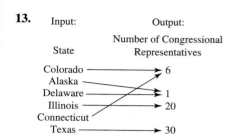

Input: Output:

                Number of Congressional
State   Representatives

Colorado ——————→ 6
Alaska
Delaware ——————→ 1
Illinois ——————→ 20
Connecticut
Texas ——————————→ 30

14.

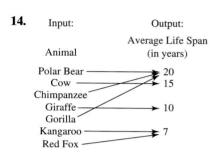

Input: Output:

                Average Life Span
Animal   (in years)

Polar Bear ——————→ 20
Cow ——————————→ 15
Chimpanzee
Giraffe ——————→ 10
Gorilla
Kangaroo ——————→ 7
Red Fox

Use the vertical line test to determine whether each graph is the graph of a function. See Example 5.

15.

16.

17.

18.

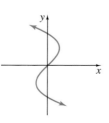

19.

20.

21.

22.

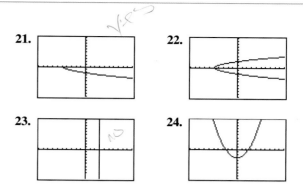

23.

24.

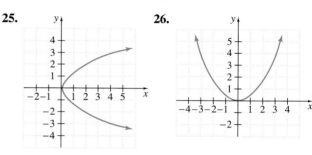

Find the domain and the range of each relation. Use the vertical line test to determine whether each graph is the graph of a function. See Example 6.

25.

26.

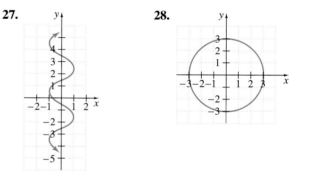

27.

28.

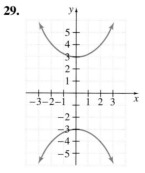

29.

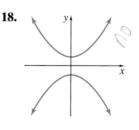

30.

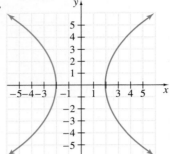

31.

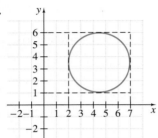

32.

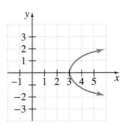

33.

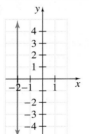

34.

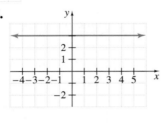

35.

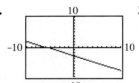

36.

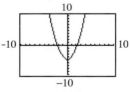

🗂 **37.** In your own words define **(a)** function; **(b)** domain; **(c)** range.

🗂 **38.** Explain the vertical line test and how it is used.

Decide whether each is a function. See Examples 3 and 4.

39. $y = x + 1$

40. $y = x - 1$

41. $x = 2y^2$

42. $y = x^2$

43. $y - x = 7$

44. $2x - 3y = 9$

45. $y = \dfrac{1}{x}$

46. $y = \dfrac{1}{x - 3}$

47. $y = 5x - 12$

48. $y = \dfrac{1}{2}x + 4$

49. $x = y^2$

50. $x = |y|$

If $f(x) = 3x + 3, g(x) = 4x^2 - 6x + 3,$ and $h(x) = 5x^3 - 7,$ find the following. See Examples 7 through 9.

51. $f(4)$

52. $f(-1)$

53. $h(-3)$

54. $h(0)$

55. $g(2)$

56. $g(1)$

57. $g(0)$

58. $h(-2)$

Given the following functions, find the indicated values. See Examples 7 through 9.

59. $f(x) = \dfrac{x}{2}$;

    a. $f(0)$ **b.** $f(2)$ **c.** $f(-2)$

60. $g(x) = -\dfrac{1}{3}x$;

    a. $g(0)$ **b.** $g(-1)$ **c.** $g(3)$

61. $f(x) = -5$;

    a. $f(2)$ **b.** $f(0)$ **c.** $f(606)$

62. $h(x) = 7$;

    a. $h(7)$ **b.** $h(542)$ **c.** $h\left(-\dfrac{3}{4}\right)$

63. $f(x) = 1.3x^2 - 2.6x + 5.1$;

    a. $f(2)$ **b.** $f(-2)$ **c.** $f(3.1)$

64. $g(x) = 2.7x^2 + 6.8x - 10.2$;

    a. $g(1)$ **b.** $g(-5)$ **c.** $g(7.2)$

65. Given the following table of values for $f(x) = |2x - 5| + 8$, find the indicated values.

    a. $f(-5)$ **b.** $f(10)$ **c.** $f(15)$

X	Y1
-5	23
0	13
5	13
10	23
15	33
20	43
25	53

Y1◻abs(2X-5)+8

66. Given the following table of values for $f(x) = -2x^3 + x$ find the indicated values.

    a. $f(-4)$ **b.** $f(2)$ **c.** $f(6)$

X	Y1
-4	124
-2	14
0	0
2	-14
4	-124
6	-426
8	-1016

Y1◻-2X³+X

67. Given the graph of the function $f(x) = x^2 + x + 1$, find the value of $f(2)$.

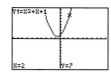

68. Given the graph of the function $f(x) = x^3 + x^2 + 1$, find $f(-2)$.

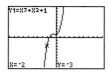

Solve. See Example 10.

69. Julie needs to hire a plumber and realizes that the cost of hiring a plumber, C, in dollars is a function of the time spent on the job, t, in hours. If the plumber charges $20 plus $35 per hour, we can write this in function notation as $C(t) = 20 + 35t$. Complete the table.

TIME IN HOURS	t	1	2	3	4	5
TOTAL COST	$C(t)$					

70. The cost of hiring a secretary to type a term paper is $10 plus $5.25 per hour. Therefore, the cost, C, in dollars is a function of time, t, in hours, or $C(t) = 10 + 5.25t$. Complete the table.

TIME IN HOURS	t	0.5	1	1.5	2.5	3
TOTAL COST	$C(t)$					

71. The height h of a firecracker being shot into the air from ground level is a function of time t and can be represented using the following formula from physics: $h = -16t^2 + vt$ where v is initial velocity. If the velocity is 80 feet per second, the height in terms of time is $h(t) = -16t^2 + 80t$. Find the following:

a. $h(0.5)$ **b.** $h(1)$ **c.** $h(1.5)$
d. $h(2)$ **e.** $h(2.5)$ **f.** $h(5)$

72. The height h of a rocket launched from a 200-foot high building with velocity of 120 feet per second can be expressed as a function of time t using the formula $h = -16t^2 + vt + h_0$, where h_0 stands for the initial height. If $h(t) = -16t^2 + 120t + 200$, find the following:

a. $h(0.2)$ **b.** $h(0.6)$ **c.** $h(2.25)$
d. $h(3)$ **e.** $h(4)$

Use the graph in Example 11 to answer the following.

73. a. Use the graph to approximate the money spent on research and development in 1988.

b. Recall that the function $f(x) = 1.087x + 3.44$ approximates the graph of Example 11. Use this equation to approximate the money spent on research and development in 1988. [*Hint:* Find $f(3)$.]

74. a. Use the graph to approximate the money spent on research and development in 1993.

b. Use the function $f(x) = 1.087x + 3.44$ to approximate the money spent on research and development in 1993. [*Hint:* Find $f(8)$.]

75. Use the function $f(x) = 1.087x + 3.44$ to predict the money spent on research and development in 2005.

76. Use the function $f(x) = 1.087x + 3.44$ to predict the money spent on research and development in 2010.

77. Since $y = x + 7$ describes a function, rewrite the equation using function notation.

78. In your own words, explain how to find the domain of a function given its graph.

The function $V(x) = x^3$ may be used to find the volume of a cube if we are given the length x of a side.

79. Find the volume of a cube whose side is 14 inches.

80. Find the volume of a game die whose side is 1.7 centimeters.

Forensic scientists use the following functions to find the height of a woman in centimeters if they are given the height of her femur bone f or her tibia bone t in centimeters.

$$H(f) = 2.59f + 47.24$$
$$H(t) = 2.72t + 61.28$$

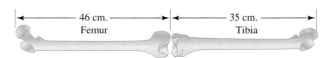

81. Find the height of a woman whose femur measures 46 centimeters.

82. Find the height of a woman whose tibia measures 35 centimeters.

The dosage D in milligrams of Ivermectin, a heartworm preventive, for a dog who weighs x pounds is given by

$$D(x) = \frac{136}{25}x$$

83. Find the proper dosage for a dog that weighs 30 pounds.

84. Find the proper dosage for a dog that weighs 50 pounds.

Review Exercises

Complete the given table and use the table to graph the linear equation.

85. $x - y = -5$

x	0		1
y		0	

86. $2x + 3y = 10$

x	0		
y		0	2

87. $7x + 4y = 8$

x	0		
y		0	-1

88. $5y - x = -15$

x	0		-2
y		0	

89. $y = 6x$

x	0		-1
y		0	

90. $y = -2x$

x	0		-2
y		0	

91. Is it possible to find the perimeter of the following geometric figure? If so, find the perimeter.

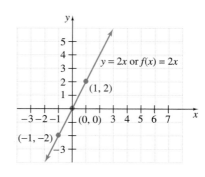

45 meters

40 meters

3.4 | GRAPHING LINEAR FUNCTIONS

TAPE IAG 3.4

O B J E C T I V E S

1. Graph linear functions.
2. Sketch the graph of linear functions by finding their intercepts.
3. Graph vertical and horizontal lines.
4. Use a graph to solve problems.

1

In this section we identify and graph linear functions. By the vertical line test, we know that all linear equations are functions except those whose graphs are vertical lines. For example, we know from Section 3.2 that $y = 2x$ is a linear equation in two variables. Its graph is shown next.

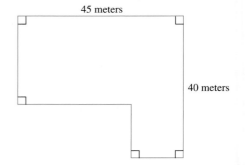

x	y = 2x
1	2
0	0
-1	-2

$y = 2x$ or $f(x) = 2x$

(1, 2)

(0, 0)

(-1, -2)

Because this graph passes the vertical line test, we know that $y = 2x$ is a function. If we want to emphasize that this equation describes a function we may write $y = 2x$ as $f(x) = 2x$.

A graphing utility is a very versatile tool for exploring the graphs of functions.

DISCOVER THE CONCEPT

a. On your graphing utility, graph both $f(x) = 2x$ as $y_1 = 2x$, and $g(x) = 2x + 10$ as $y_2 = 2x + 10$ using an integer window.

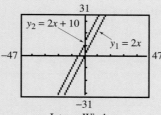

Integer Window

b. Trace on the graph of $y_1 = 2x$ to find the point whose ordered pair has x-coordinate 3 and find the corresponding y-coordinate. Now, press the down arrow key and find the corresponding y-coordinate on the graph of $y_2 = 2x + 10$. Compare the two y-coordinates for the same x-coordinate.

c. Trace to another point on the graph of y_1 and repeat this process. Compare the two lines, and see if you can state how we could sketch the graph of $g(x) = 2x + 10$ from the graph of $f(x) = 2x$.

d. Predict how the graph of $h(x) = 2x - 15$ could be drawn. Graph $y_3 = 2x - 15$ and see if your prediction is correct.

In the above discovery, we noticed that the y-values for the graph of $g(x)$ or $y = 2x + 10$ are obtained by adding 10 to the y-value of each corresponding point of the graph of $f(x)$ or $y = 2x$. The graph of $g(x) = 2x + 10$ is the same as the graph of $f(x) = 2x$ shifted upward 10 units.

x	−1	0	1
f(x) = 2x	−2	0	2
g(x) = 2x + 10	8	10	12

add 10

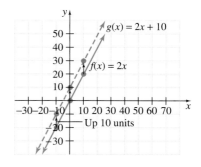

Also, the graph of $h(x)$ or $y = 2x - 15$ is obtained by subtracting 15 from the y-value of each corresponding point of the graph of $f(x)$ or $y = 2x$. Thus, the graph of $h(x) = 2x - 15$ is the same as the graph of $f(x) = 2x$ shifted downward 15 units.

The functions $f(x) = 2x$, $g(x) = 2x + 10$, and $h(x) = 2x - 15$ are called linear functions—"linear" because each graph is a line, and "function" because each graph passes the vertical line test. In general, a **linear function** is a function that can be written in the form $f(x) = mx + b$. For example, $g(x) = 2x + 10$ is in this form, with $m = 2$ and $b = 10$.

EXAMPLE 1   Graph both linear functions $f(x) = 0.5x$ and $g(x) = 0.5x + 8$ using the same integer window.

a. Use the graphs to complete the following table of solution pairs.

x	−6	0	5	13	18
f(x) = 0.5x					
g(x) = 0.5x + 8					

b. Complete the following sentence. The graph of $g(x) = 0.5x + 8$ can be obtained from the graph of $f(x) = 0.5x$ by _____ .

Solution: **a.** Define $y_1 = 0.5x$ and $y_2 = 0.5x + 8$ and graph in the integer window. The screens below show the values of $f(x)$ and $g(x)$ when $x = 4$. In other words, $f(4) = 2$ and $g(4) = 10$.

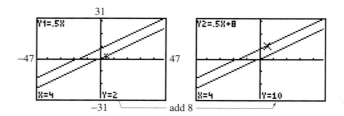

Trace along each graph to complete the table as follows:

x	−6	0	5	13	18
f(x) = 0.5x	−3	0	2.5	6.5	9
g(x) = 0.5x + 8	5	8	10.5	14.5	17

b. The graph of $g(x) = 0.5x + 8$ can be obtained from the graph of $f(x) = 0.5x$ by shifting it upward 8 units.

EXAMPLE 2   Graph both linear functions $f(x) = -x$ and $g(x) = -x - 16$ using an integer window. Complete the following statement: the graph of $g(x) = -x - 16$ can be obtained from the graph of $f(x) = -x$ by_____ .

Solution: Graph $y_1 = -x$ and $y_2 = -x - 16$ using an integer window as shown below. We can complete the statement as follows. The graph of $g(x) = -x - 16$ can be obtained from the graph of $f(x)$ by shifting it downward 16 units. To further illustrate this fact, see the table below and compare the y_1 and y_2 values for the same x-value.

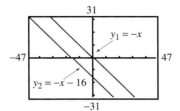

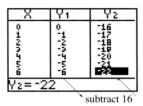

subtract 16

In general, for any function $f(x)$, the graph of $y = f(x) + K$ is the same as the graph of $y = f(x)$ shifted $|K|$ units upward if K is positive and downward if K is negative.

2   The graph of $y = 2x + 10$ is shown in both screens below. Notice that this graph crosses both the x-axis and the y-axis. Recall that a point where a graph crosses the x-axis is called the **x-intercept point,** and a point where a graph crosses the y-axis is called the **y-intercept point.**

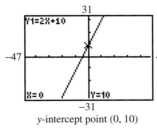

y-intercept point $(0, 10)$

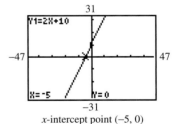

x-intercept point $(-5, 0)$

By tracing along the graph, we can see that the y-intercept point of the graph of $y = 2x + 10$ is $(0, 10)$, or we can say simply that the y-intercept is 10. Also the x-intercept point of the graph is $(-5. 0)$ or the x-intercept is -5.

One way to find the y-intercept of the graph of an equation is to let $x = 0$, since a point on the y-axis has an x-coordinate of 0. To find the x-intercept, let $y = 0$, since a point on the x-axis has a y-coordinate of 0.

FINDING X- AND Y-INTERCEPTS

To find an x-intercept, let $y = 0$ or $f(x) = 0$ and solve for x.
To find a y-intercept, let $x = 0$ and solve for y.

Intercept points are usually easy to find and plot since one coordinate is 0. In the next example we sketch a linear function by plotting the x- and y-intercepts.

EXAMPLE 3   Graph $x - 3y = 6$ by plotting intercept points. Check using a graphing utility.

Solution: Let $y = 0$ to find the x-intercept and $x = 0$ to find the y-intercept.

$$\text{If } y = 0 \quad \text{then} \quad \text{If } x = 0 \text{ then}$$

$$x - 3(0) = 6 \qquad 0 - 3y = 6$$

$$x - 0 = 6 \qquad -3y = 6$$

$$x = 6 \qquad y = -2$$

The x-intercept is 6, and the y-intercept is -2. We find a third ordered pair solution to check our work. If we let $y = -1$, then $x = 3$. Plot the points $(6, 0)$, $(0, -2)$, and $(3, -1)$. The graph of $x - 3y = 6$ is the line drawn through these points, as shown.

x	y
6	0
0	-2
3	-1

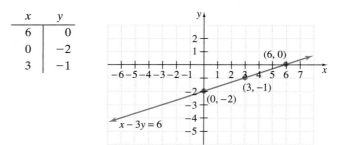

To check using a graphing utility, solve for y to enter the equation in the Y= editor.

$$x - 3y = 6$$

$$-3y = -x + 6$$

$$y = \frac{-x + 6}{-3}$$

Define $y_1 = \dfrac{-x + 6}{-3}$ and graph in an integer window as shown below.

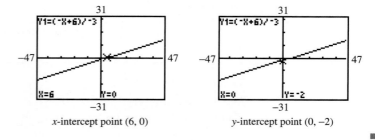

x-intercept point (6, 0)   y-intercept point (0, −2)

> REMINDER   Recall that when generating a graph using a graphing utility
> we first solve for the variable y. When a function is given in the $f(x)$ notation, it is
> already solved for y and can be entered in the Y= editor directly.

DISCOVER THE CONCEPT

a. Use an integer setting and graph each linear function.

$$y_1 = x + 15 \qquad y_2 = x \qquad y_3 = x - 11$$

b. Decide how the y-intercept of the graph of the equation compares with the equation.

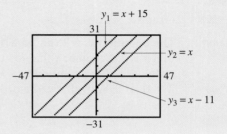

From the discovery above, we found the following:

EQUATION	Y-INTERCEPT POINT	Y-INTERCEPT
$y = x + 15$	$(0, 15)$	15
$y = x$	$(0, 0)$	0
$y = x - 11$	$(0, -11)$	-11

In general,

$y = mx + b$	$(0, b)$	b

Notice that the y-intercept of the graph of the equation of the form $y = mx + b$ is
b each time. This is because we find the y-intercept of the graph of an equation by
letting $x = 0$. Thus,

$$y = mx + b$$
$$y = m \cdot 0 + b \qquad \text{Let } x = 0.$$
$$y = b$$

The intercept point is $(0, b)$ and the intercept is b.

EXAMPLE 4   Find the y-intercept of the graph of each equation.

a. $f(x) = \dfrac{1}{2}x + \dfrac{3}{7}$ **b.** $y = -2.5x - 3.2$

Solution: Both equations are of the form $y = mx + b$, so to find the y-intercept, we identify b.

a. The y-intercept of $f(x) = \dfrac{1}{2}x + \dfrac{3}{7}$ is $\dfrac{3}{7}$. The y-intercept point is $\left(0, \dfrac{3}{7}\right)$.

b. The y-intercept of $y = -2.5x - 3.2$ is -3.2. The y-intercept point is $(0, -3.2)$.

The equation $x = c$, where c is a real number constant, is a linear equation in two variables because it can be written in the form $Ax + By = C$; that is, $x + 0y = c$. The graph of this equation is a vertical line as shown in the next example. Since a vertical line does not pass the vertical line test, the equation $x = c$ does *not* describe a function.

EXAMPLE 5 Graph $x = 2$.

Solution: The equation $x = 2$ can be written as $x + 0y = 2$. For any y-value chosen, notice that x is 2. No other value for x satisfies $x + 0y = 2$. Any ordered pair whose x-coordinate is 2 is a solution to $x + 0y = 2$ because 2 added to 0 times any value of y is $2 + 0$, or 2. We will use the ordered pairs $(2, 3)$, $(2, 0)$, and $(2, -3)$ to graph $x = 2$.

TECHNOLOGY NOTE
Since the graph of $x = c$ is a vertical line and is not a function, we cannot enter this equation in the Y= editor. However, most graphing utilities have a draw feature that allows you to draw the vertical line, but you cannot use trace on it since it is not a function. See if your graphing utility has this feature.

x	y
2	3
2	0
2	-3

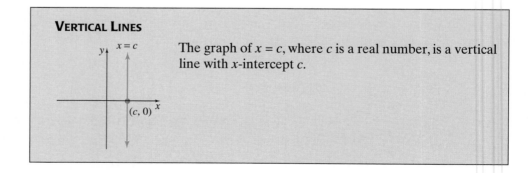

The graph is a vertical line with x-intercept 2. Notice that this graph is not the graph of a function, and it has no y-intercept because x is never 0.

VERTICAL LINES

The graph of $x = c$, where c is a real number, is a vertical line with x-intercept c.

Does the equation $y = c$ describe a function? Yes, because the graph is a horizontal line, as shown next.

EXAMPLE 6   Graph $y = -3$.

Solution: The equation $y = -3$ can be written as $0x + y = -3$. For any x-value chosen, y is -3. If we choose 4, 1, and -2 as x-values, the ordered pair solutions are $(4, -3)$, $(1, -3)$, and $(-2, -3)$. Use these ordered pairs to graph $y = -3$. The graph is a horizontal line with y-intercept -3 and no x-intercept. Recall that we may write $y = -3$ as $f(x) = -3$. Since $y = -3$ is a function, we can also graph this equation using a graphing utility. Trace along the line to find ordered pair solutions and observe that all solution pairs have y-coordinate -3.

x	y
4	−3
1	−3
−2	−3

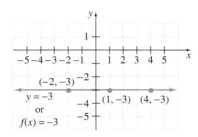

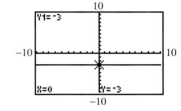

HORIZONTAL LINES

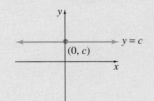

The graph of $y = c$ or $f(x) = c$, where c is a real number, is a horizontal line with y-intercept c.

  We can use the graph of a linear equation to solve problems.

EXAMPLE 7   **COST OF RENTING A CAR**
The cost of renting a car for a day is given by the linear function $C(x) = 35 + 0.15x$ where $C(x)$ represents the cost in dollars and x is the number of miles driven. Use the graph of the function to complete the table below in dollars, and find the cost $C(x)$ for the given number of miles.

No. of Miles	x	150	200	325	500
Cost	$C(x) = 35 + 0.15x$				

Solution: Define $y_1 = 35 + 0.15x$ and graph in a $[0, 600, 100]$ by $[0, 200, 50]$ window. We choose $0 \le x \le 600$ since the values in the table are in this interval. We choose $0 \le y \le 200$ as we are estimating our cost to be less than $200. The graph below illustrates how to obtain the cost for 150 miles.

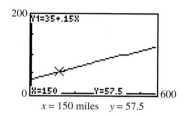

$x = 150$ miles $y = 57.5$

Finding the remaining values from the graph, we can complete the table.

No. OF MILES	x	150	200	325	500
COST	$C(x)=35+0.15x$	57.5	65	83.75	110

Notice from the graph that the cost $C(x)$ increases as the number of miles, x, increases.

EXERCISE SET 3.4

The graph of $f(x) = 5x$ in a standard window follows. Use this graph to match each linear function with its graph. See Examples 1 and 2.

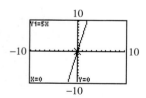

a. 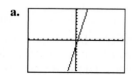 **b.**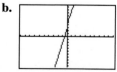

c. 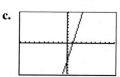 **d.**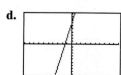

1. $f(x) = 5x - 6$ **2.** $f(x) = 5x - 2$
3. $f(x) = 5x + 7$ **4.** $f(x) = 5x + 3$

Sketch the graph of each linear function by finding x- and y-intercepts. See Examples 3 and 4.

5. $x - y = 3$ **6.** $x - y = -4$
7. $x = 5y$ **8.** $2x = y$
9. $-x + 2y = 6$ **10.** $x - 2y = -8$
11. $2x - 4y = 8$ **12.** $2x + 3y = 6$

13. In your own words, explain how to find x- and y-intercepts using an equation.

14. In your own words, explain how to find x- and y-intercepts from a graph using a graphing utility.

Graph each linear equation. See Examples 5 and 6.

15. $x = -1$ **16.** $y = 5$
17. $y = 0$ **18.** $x = 0$
19. $x - 3 = 0$ **20.** $y + 4 = 0$

Explain how the graph of y_2 and y_3 can be obtained from the graph of y_1.

21. $y_1 = 4x$
$y_2 = 4x + 3$
$y_3 = 4x - 5$

22. $y_1 = -3x$
$y_2 = -3x - 2$
$y_3 = -3x + 4$

23. If $f(x) = -2x$, write the equation of the line whose graph results from shifting the graph of $f(x)$ 4 units upward.

24. If $f(x) = 0.5x$, write the equation of the line whose graph results from shifting the graph of $f(x)$ 2.5 units downward.

Match each equation with its graph.

a.

b.

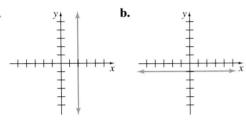

c.

d.

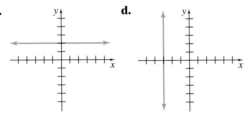

25. $y = 2$

26. $x = -3$

27. $x - 2 = 0$

28. $y + 1 = 0$

Sketch the graph of each equation by finding x- and y-intercepts. Then use a graphing utility to check.

29. $x + 2y = 8$

30. $x - 3y = 3$

31. $f(x) = \frac{3}{4}x + 2$

32. $f(x) = \frac{4}{3}x + 2$

33. $x = -3$

34. $f(x) = 3$

35. $3x + 5y = 7$

36. $3x - 2y = 5$

37. $f(x) = x$

38. $f(x) = -x$

39. $x + 8y = 8$

40. $x - 3y = 9$

41. $5 = 6x - y$

42. $4 = x - 3y$

43. $-x + 10y = 11$

44. $-x + 9 = -y$

45. $y = 1$

46. $x = 1$

47. $f(x) = \frac{1}{2}x$

48. $f(x) = -2x$

49. $x + 3 = 0$

50. $y - 6 = 0$

51. $f(x) = 4x - \frac{1}{3}$

52. $f(x) = -3x + \frac{3}{4}$

53. $2x + 3y = 6$

54. $4x + y = 5$

55. Given the graph of $f(x)$, which of the following statements are true?

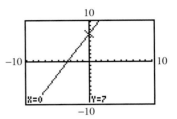

a. $f(7) = 0$

b. $f(0) = 7$

c. The x-intercept is 7.   **d.** The y-intercept is 7.

56. Given the graph of $g(x)$, which of the following statements are true?

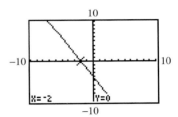

a. $g(-2) = 0$

b. $g(0) = -2$

c. The x-intercept is -2.

d. The y-intercept is -2.

57. Broyhill Furniture found that it takes 2 hours to manufacture each table for one of its special dining room sets. Each chair takes 3 hours to manufacture. A total of 1500 hours is available to produce tables and chairs of this style. The linear equation that models this situation is $2x + 3y = 1500$, where x represents the number of tables produced and y the number of chairs produced.

a. Complete the ordered pair solution $(0, \ )$ of this equation. Describe the manufacturing situation that this solution corresponds to.

b. Complete the ordered pair solution $(\ , 0)$ for this equation. Describe the manufacturing situation that this solution corresponds to.

c. If 50 tables are produced, find the greatest number of chairs the company can make.

58. While manufacturing two different camera models, Kodak found that the basic model costs $55 to produce, whereas the deluxe model costs $75. The weekly budget for these two models is limited to $33,000 in production costs. The linear equation that models this situation is $55x + 75y = 33,000$, where x represents the number of basic models and y the number of deluxe models.

a. Complete the ordered pair solution $(0,\ )$ of this equation. Describe the manufacturing situation that this solution corresponds to.

b. Complete the ordered pair solution $(\ ,0)$ of this equation. Describe the manufacturing situation that this solution corresponds to.

c. If 350 deluxe models are produced, find the greatest number of basic models that can be made in one week.

59. The cost of renting a car for a day is given by the linear function $C(x) = 0.2x + 24$, where $C(x)$ is in dollars and x is the number of miles driven.

a. Find the cost of driving the car 200 miles.

b. Graph $C(x) = 0.2x + 24$.

c. How can you tell from the graph of $C(x)$ that as the number of miles driven increases the total cost increases also?

60. The cost of renting a piece of machinery is given by the linear function $C(x) = 4x + 10$, where $C(x)$ is in dollars and x is given in hours.

a. Find the cost of renting the piece of machinery for 8 hours.

b. Graph $C(x) = 4x + 10$.

c. How can you tell from the graph of $C(x)$ that as the number of hours increases the total cost increases also?

61. The yearly cost of tuition and required fees for attending a two-year college full time can be estimated by the linear function $f(t) = 91.7t + 747.8$, where t is the number of years after 1990 and $f(t)$ is the total cost.

a. Use this function to approximate the yearly cost of attending a two-year college in the year 2010. [Hint: Find $f(20)$.]

b. Use the given function to predict the year in which the yearly cost of tuition and required fees will exceed $2000. [*Hint:* Graph $f(t)$ and trace to see when $f(t)$ or y exceeds 2000.]

c. Use this function to approximate the yearly cost of attending a two-year college in the present year. If you attend a two-year college, is this amount greater than or less than the amount that is currently charged by the college that you attend?

62. The yearly cost of tuition and required fees for attending a four-year college can be estimated by the linear function $f(t) = 201.9t + 2002.2$, where t is the number of years after 1990 and $f(t)$ is the total cost in dollars.

a. Use this function to approximate the yearly cost of attending a four-year college in the year 2010. [*Hint:* Find $f(20)$.]

b. Use the given function to predict the year in which the yearly cost of tuition and required fees will exceed $4000. [*Hint:* Graph $f(t)$ and trace to see when $f(t)$ or y exceeds 4000.]

c. Use this function to approximate the yearly cost of attending a four-year college in the present year. If you attend a four-year college, is this amount greater than or less than the amount that is currently charged by the college that you attend?

Source for Exercises 61 and 62: U.S. Bureau of the Census, *Statistical Abstract of the United States: 1995:* (115th edition.) Washington, DC, 1995.

Review Exercises

Simplify.

63. $\dfrac{-6 - 3}{2 - 8}$

64. $\dfrac{4 - 5}{-1 - 0}$

65. $\dfrac{-8 - (-2)}{-3 - (-2)}$

66. $\dfrac{12 - 3}{10 - 9}$

67. $\dfrac{0 - 6}{5 - 0}$

68. $\dfrac{2 - 2}{3 - 5}$

Graph each linear function using an integer window. Use the graph to find ordered pair solutions of the form $(-3,\ )$, $(5,\ )$, and $(8,\ )$. If necessary, round answers to the nearest hundredth.

69. $f(x) = 0.5x$

70. $f(x) = \dfrac{1}{3}x$

71. $g(x) = 0.5x - 4$

72. $g(x) = \dfrac{1}{3}x - 2$

3.5 | THE SLOPE OF A LINE

O B J E C T I V E S

1. Find the slope of a line given two points on the line.
2. Find the slope of a line given the equation of the line.
3. Compare the slopes of parallel and perpendicular lines.
4. Find the slopes of horizontal and vertical lines.

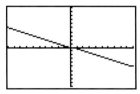

TAPE IAG 3.5

You may have noticed by now that different lines tilt differently as shown.

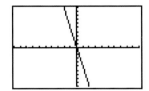 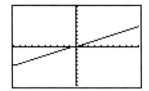 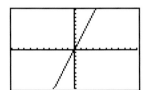

In many fields, it is very important to be able to measure and compare the tilt, or steepness, of lines. For example, if the x-axes, shown above, represent years and the y-axes represent thousands of dollars, most business owners would prefer the graph of their profit to look like the first quadrant of the graph at the far right. In mathematics, the steepness, or tilt, of the line is also known as its **slope.** We measure the slope of a line as a ratio of *vertical change* to *horizontal change*. Slope is usually designated by the letter m. Often we use the Greek symbol **delta,** Δ, to represent change or increment.

Suppose that we want to measure the slope of the following line.

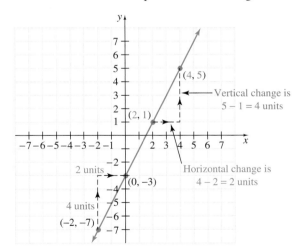

The vertical change between both pairs of points on the line is 4 units per horizontal change of 2 units. Then

$$\text{Slope } m = \frac{\text{change in } y \text{ (vertical change)}}{\text{change of } x \text{ (horizontal change)}} = \frac{\Delta y}{\Delta x} = \frac{4}{2} = 2$$

Notice that slope is a rate of change between points. A slope of 2, or $\frac{2}{1}$, means that the rate of change between points on the line is 2 units vertically per horizontal change of 1 unit.

In general, consider the following line, which passes through the points (x_1, y_1) and (x_2, y_2). The vertical change, or *rise,* between these points is the difference in the y-coordinates: $y_2 - y_1$. The horizontal change, or *run,* between the points is the difference of the x-coordinates: $x_2 - x_1$. The slope of the line is the ratio of these differences.

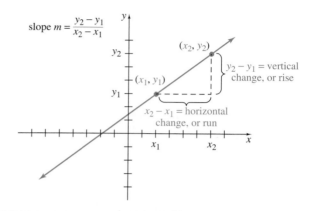

$$\text{slope } m = \frac{y_2 - y_1}{x_2 - x_1}$$

SLOPE OF A LINE

Given a line passing through points (x_1, y_1) and (x_2, y_2), the slope m of the line is

$$m = \frac{\text{rise}}{\text{run}} = \frac{\Delta y}{\Delta x} = \frac{y_2 - y_1}{x_2 - x_1}, \qquad \text{as long as } x_2 \neq x_1$$

EXAMPLE 1   Find the slope of the line containing the points $(0, 3)$ and $(2, 5)$. Graph the line.

Solution: Use the slope formula. It does not matter which point we call (x_1, y_1) and which we call (x_2, y_2). Let $(x_1, y_1) = (0, 3)$ and $(x_2, y_2) = (2, 5)$.

$$m = \frac{y_2 - y_1}{x_2 - x_1}$$

$$= \frac{5 - 3}{2 - 0} = \frac{2}{2} = 1$$

Notice in this example that the slope is positive and that the graph of the line containing $(0, 3)$ and $(2, 5)$ slants upward as we go from left to right.

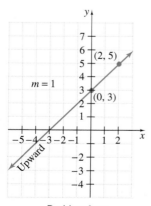

Positive slope

> R E M I N D E R   When we are trying to find the slope of a line through two given points, it makes no difference which given point is called (x_1, y_1) and which is called (x_2, y_2). Once an x-coordinate is called x_1, however, make sure that its corresponding y-coordinate is called y_1.

EXAMPLE 2   Find the slope of the line containing the points $(5, -7)$ and $(-3, 6)$. Graph the line.

Solution: Use the slope formula. Let $(x_1\ y_1) = (5, -7)$ and $(x_2, y_2) = (-3, 6)$.

$$m = \frac{y_2 - y_1}{x_2 - x_1}$$

$$= \frac{6 - (-7)}{-3 - 5} = \frac{13}{-8} = -\frac{13}{8}$$

Notice in this example that the slope is negative and that the graph of the line through $(5, -7)$ and $(-3, 6)$ slants downward as we go from left to right.

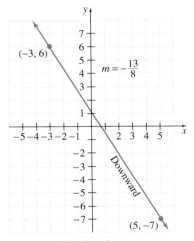

Negative slope

To give meaning to the fact that the slope of a line is constant, see the following.

Linear Equation

$$y = 2x + 3$$

	x	y	
+1 (	-2	-1	) +2
+1 (	-1	1	) +2
+1 (	0	3	) +2
+1 (	1	5	) +2
	2	7	

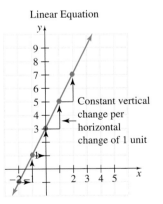

For linear equations, there is a constant vertical change per constant horizontal change. This means that the slope is the same between any two points of the graph.

2   The slope of a line indicates the direction of the line, that is, whether it slants upward or downward from left to right, as well as the amount of tilt to the line. In the following discovery we investigate how the equation itself can indicate direction and amount of tilt.

DISCOVER THE CONCEPT

a. Graph $y_1 = x$, $y_2 = 3x$, and $y_3 = 0.5x$ using a standard window. Examine the graphs. What observation can you make with regard to the slant of the lines? What observation can you make about each line with respect to the amount of tilt and the coefficient of x in the equation?

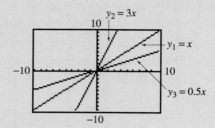

b. Can you change the equations so that the graphs slant downward instead of slanting upward from left to right?

c. Graph each equation separately. Predict the direction of slant and the amount of tilt of the line before seeing its graph on a graphing utility.

   i. $y = x + 3$　　　**ii.** $y = 2x + 3$　　　**iii.** $y = -2x + 3$　　　**iv.** $y = 0.5x + 3$

In part (a) of the discovery, we see that each line moves upward from left to right. Also, notice that the greater the positive coefficient of x is, the steeper the tilt of the line. From part (b), we discover that if the coefficient of x is a negative number, the line moves downward from left to right.

    As we have seen, the slope of a line is defined by two points on the line. Thus, if we know an equation of a line, we can find its slope.

EXAMPLE 3　　Find the slope of the line whose equation is $f(x) = \dfrac{2}{3}x + 4$.

Solution:　　Two points are needed on the line defined by $f(x) = \dfrac{2}{3}x + 4$ or $y = \dfrac{2}{3}x + 4$

to find its slope. We will use intercepts as our two points.

    If $x = 0$, then　　　　If $y = 0$, then

$$y = \frac{2}{3} \cdot 0 + 4 \qquad\qquad 0 = \frac{2}{3}x + 4$$

$$y = 4 \qquad\qquad\qquad -4 = \frac{2}{3}x \qquad\qquad \text{Subtract 4.}$$

$$\frac{3}{2}(-4) = \frac{3}{2} \cdot \frac{2}{3}x \qquad\qquad \text{Multiply by } \frac{3}{2}.$$

$$-6 = x$$

Use the points $(0, 4)$ and $(-6, 0)$ to find the slope. Let (x_1, y_1) be $(0, 4)$ and (x_2, y_2) be $(-6, 0)$. Then

$$m = \frac{y_2 - y_1}{x_2 - x_1} = \frac{0 - 4}{-6 - 0} = \frac{-4}{-6} = \frac{2}{3}$$

Analyzing the results of Example 3, you may notice a striking pattern: The slope of $y = \frac{2}{3}x + 4$ is $\frac{2}{3}$, the same as the coefficient of x.

Also, the y-intercept is 4, the same as the constant term, as expected. When a linear equation is written in the form $f(x) = mx + b$ or $y = mx + b$, m is the slope of the line and b is its y-intercept. The form $y = mx + b$ is appropriately called the **slope-intercept form.**

SLOPE–INTERCEPT FORM

When a linear equation in two variables is written in slope-intercept form,

$$y = mx + b$$

then m is the slope of the line and b is the y-intercept of the line.

EXAMPLE 4   Given the equation $3x - 4y = 4$,

a. Find the slope and the y-intercept of the line.

b. Use the slope to determine whether the line slants upward or downward from left to right.

c. Use a graphing utility to check the results.

Solution: **a.** Write the equation in slope-intercept form by solving for y.

$$3x - 4y = 4$$

$$-4y = -3x + 4 \qquad \text{Subtract } 3x \text{ from both sides.}$$

$$\frac{-4y}{-4} = \frac{-3x}{-4} + \frac{4}{-4} \qquad \text{Divide both sides by } -4.$$

$$y = \frac{3}{4}x - 1 \qquad \text{Simplify.}$$

The coefficient of x, $\frac{3}{4}$, is the slope, and the constant term -1 is the y-intercept.

b. Since the slope is positive, the line slants upward from left to right.

c. We graph the equation $y_1 = (-3x + 4)/-4$ in a decimal window. The graph below indicates that the line does indeed slant upward from left to right and the y-intercept point is $(0, -1)$.

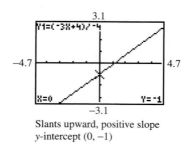

Slants upward, positive slope
y-intercept $(0, -1)$

3 Slopes of lines can help us determine whether lines are parallel or perpendicular. Recall that parallel lines are distinct lines with the same tilt that do not meet, and perpendicular lines are lines that intersect to form right angles.

TECHNOLOGY NOTE

Although the graphs of the equations shown below to the left are perpendicular lines, they do not appear to be so when graphed using a standard viewing window.

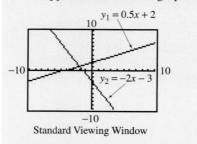

Standard Viewing Window

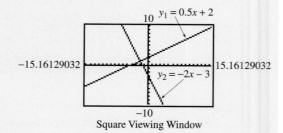

Square Viewing Window

This is because a graphing calculator screen is rectangular, so the tick marks along the x-axis are farther apart than the tick marks along the y-axis. The graphs of the same lines at the right do appear to be perpendicular because the viewing window has been selected so that there is equal spacing between tick marks on both axes. We say that the lines to the right are graphed using a **square viewing window**. Some graphing utilities can automatically provide a square setting. (This technology note shows the necessity of being able to algebraically determine whether lines are perpendicular. A graphing utility can then be used to partially verify the result.)

DISCOVER THE CONCEPT

Graph each pair of equations in a square window, such as an integer window. Do the lines appear to be parallel, perpendicular, or neither? Find the relationship between the slopes of the lines and whether they appear parallel, perpendicular, or neither.

a. $y = 3x - 5$ and $y = 3x + 7$

b. $y = 2x + 3$ and $y = 4x - 5$

c. $y = 0.5x - 9$ and $y = 2x - 3$

d. $y = -5x - 6$ and $y = -5x - 2$

e. $y = -\dfrac{1}{3}x - 7$ and $y = 3x + 5$

In the above discovery, we see that parallel lines have the same tilt and therefore have the same slope.

PARALLEL LINES

Nonvertical parallel lines have the same slope.

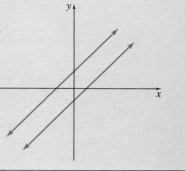

How do the slopes of perpendicular lines compare? The product of the slopes of two perpendicular lines is -1.

PERPENDICULAR LINES

If the product of the slopes of two lines is -1, then the lines are perpendicular. (Two nonvertical lines are perpendicular if the slope of one is the negative reciprocal of the slope of the other.)

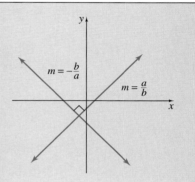

EXAMPLE 5   Given the following pairs of equations, determine whether their graphs are parallel lines, perpendicular lines, or neither. Then use a graphing utility and a square setting to visualize your results.

a. $3x + 7y = 21$ **b.** $-x + 3y = 2$ **c.** $2x - 3y = 12$

    $6x + 14y = 7$ $2x + 6y = 5$ $6x + 4y = 16$

Solution: Find the slope of each line by solving each equation for y.

a. $3x + 7y = 21$ $6x + 14y = 7$

    $7y = -3x + 21$ $14y = -6x + 7$

    $\dfrac{7y}{7} = \dfrac{-3x}{7} + \dfrac{21}{7}$ $\dfrac{14y}{14} = \dfrac{-6x}{14} + \dfrac{7}{14}$

    $y = -\dfrac{3}{7}x + 3$ $y = -\dfrac{3}{7}x + \dfrac{1}{2}$

The slope of both lines is $-\dfrac{3}{7}$. The y-intercept of one line is 3, whereas the y-intercept of the other line is $\dfrac{1}{2}$. Since these lines have the same slope and different y-intercepts, the lines are parallel.

b. Solve each equation for y.

$$-x + 3y = 2 \qquad \text{and} \qquad 2x + 6y = 5$$

$$y = \dfrac{1}{3}x + \dfrac{2}{3} \qquad\qquad\qquad y = -\dfrac{1}{3}x + \dfrac{5}{6}$$

$$\text{slope, } m = \dfrac{1}{3} \qquad\qquad\qquad \text{slope, } m = -\dfrac{1}{3}$$

The slopes are not equal nor are they negative reciprocals of each other, so they are simply intersecting lines.

c. Solve each equation for y.

$$2x - 3y = 12 \qquad \text{and} \qquad 6x + 4y = 16$$

$$y = \dfrac{2}{3}x - 4 \qquad\qquad\qquad y = -\dfrac{3}{2}x + 4$$

$$\text{slope, } m = \dfrac{2}{3} \qquad\qquad\qquad \text{slope, } m = -\dfrac{3}{2}$$

The slopes of the lines are $\dfrac{2}{3}$ and $-\dfrac{3}{2}$, which are negative reciprocals of each other, so the lines are perpendicular.

To visualize the results of parts (a), (b) and (c), see the graphs below using square settings.

$$y_1 = -\dfrac{3}{7}x + 3 \qquad\qquad y_1 = \dfrac{1}{3}x + \dfrac{2}{3} \qquad\qquad y_1 = \dfrac{2}{3}x - 4$$

$$y_2 = -\dfrac{3}{7}x + \dfrac{1}{2} \qquad\qquad y_2 = -\dfrac{1}{3}x + \dfrac{5}{6} \qquad\qquad y_2 = -\dfrac{3}{2}x + 4$$

Parallel lines   Neither parallel   Perpendicular lines
                             nor perpendicular

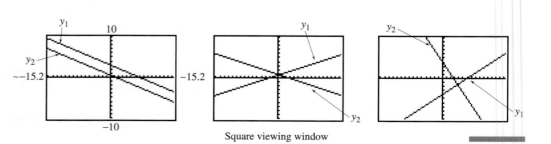

Square viewing window

4   Next we investigate the slopes of vertical and horizontal lines.

EXAMPLE 6   Find the slope of the line $x = -5$.

Solution: Recall that the graph of $x = -5$ is a vertical line with x-intercept -5.

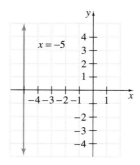

To find the slope, find two ordered pair solutions of $x = -5$. Solutions of $x = -5$ must have an x-value of -5.
    Let $(x_1, y_1) = (-5, 0)$ and $(x_2, y_2) = (-5, 4)$. Then

$$m = \frac{\Delta y}{\Delta x} = \frac{y_2 - y_1}{x_2 - x_1} = \frac{4 - 0}{-5 - (-5)} = \frac{4}{0}$$

Since $\frac{4}{0}$ is undefined, we say that the slope of the vertical line $x = -5$ is undefined.
Since all vertical lines are parallel, we can say that all **vertical lines have undefined slope.**

EXAMPLE 7   Graph the line $y = 2$ and find the slope.

Solution: The graph of $y = 2$ is shown. Trace to find two ordered pair solutions. We select $(0, 2)$ and $(1, 2)$.

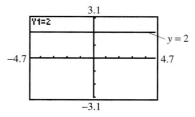

Use these points to find the slope.

$$m = \frac{\Delta y}{\Delta x} = \frac{y_2 - y_1}{x_2 - x_1} = \frac{2 - 2}{1 - 0} = \frac{0}{1} = 0$$

The slope of the line $y = 2$ is 0. Since all horizontal lines are parallel, we can say that all **horizontal lines have a slope of 0.**

REMINDER   Slope of 0 and undefined slope are not the same. Vertical lines have undefined slope or no slope, whereas horizontal lines have a slope of 0.

The following four graphs summarize the overall appearance of lines with positive, negative, zero, or undefined slopes.

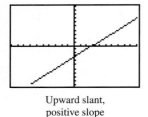

Upward slant,
positive slope

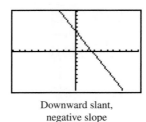

Downward slant,
negative slope

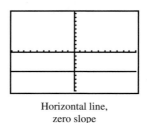

Horizontal line,
zero slope

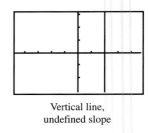

Vertical line,
undefined slope

Finally, we examine further the relationship between the steepness of a line and its slope. The graphs of $y = \frac{1}{2}x + 1$ and $y = 5x + 1$ are shown below. Recall that the graph of $y = \frac{1}{2}x + 1$ has a slope of $\frac{1}{2}$ and that the graph of $y = 5x + 1$ has a slope of 5.

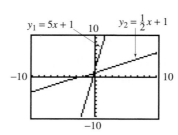

Notice that the line with the slope of 5 is steeper than the line with the slope of $\frac{1}{2}$. This is true in general for positive slopes.

For a line with positive slope m, as m increases, the line becomes steeper.

MENTAL MATH

Decide whether a line with the given slope moves upward, downward, horizontally, or vertically from left to right.

1. $m = \dfrac{7}{6}$

2. $m = -3$

3. $m = 0$

4. m is undefined.

EXERCISE SET 3.5

Give the slope of the line that goes through the given points and tell if the line joining these points slants upward or downward from left to right or neither. See Examples 1, 2, 6, and 7.

1. $(3, 2), (8, 11)$

2. $(1, 6), (7, 11)$

3. $(3, 1), (1, 8)$

4. $(2, 9), (6, 4)$

5. $(-2, 8), (4, 3)$

6. $(3, 7), (-2, 11)$

7. $(-2, -6), (4, -4)$

8. $(-3, -4), (-1, 6)$

9. $(-3, -1), (-12, 11)$

10. $(3, -1), (-6, 5)$

11. $(-2, 5), (3, 5)$

12. $(4, 2), (4, 0)$

A linear function is graphed below. Using the trace feature, we found two points on the line. Give the slope of the line.

13.

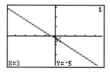

A linear function is graphed below. Using the trace feature we found two points on the line. Give the slope of the line.

14.

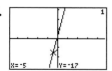

Examine the table of values for a linear function. Select two ordered pairs and find the slope of the line joining these points.

15.

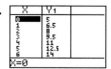

16.

17.

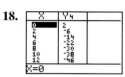

18.

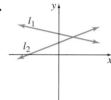

Two lines are graphed on each set of axes. Decide whether l_1 or l_2 has the greater slope.

19.

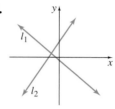

20.

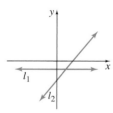

21.

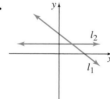

22.

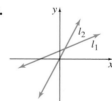

23.

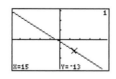

24.

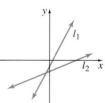

Find the slope and the y-intercept of each line. See Examples 3 and 4.

25. $f(x) = 5x - 2$

26. $f(x) = -2x + 6$

27. $2x + y = 7$

28. $-5x + y = 10$

29. $2x - 3y = 10$

30. $-3x - 4y = 6$

31. $f(x) = \dfrac{1}{2}x$

32. $f(x) = -\dfrac{1}{4}x$

Match each graph with its equation.

a.

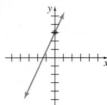

b.

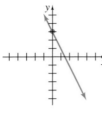

c.

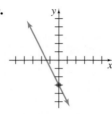

d.

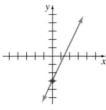

33. $f(x) = 2x + 3$ **34.** $f(x) = 2x - 3$

35. $f(x) = -2x + 3$ **36.** $f(x) = -2x - 3$

Find the slope of each line. See Examples 6 and 7.

37. $x = 1$ **38.** $y = -2$

39. $y = -3$ **40.** $x = 4$

41. $x + 2 = 0$ **42.** $y - 7 = 0$

43. Explain how merely looking at a line can tell us whether its slope is negative, positive, undefined, or zero.

44. Explain why the graph of $y = b$ is a horizontal line.

Find the slope and the y-intercept of each line.

45. $f(x) = -x + 5$ **46.** $f(x) = x + 2$

47. $-6x + 5y = 30$ **48.** $4x - 7y = 28$

49. $3x + 9 = y$ **50.** $2y - 7 = x$

51. $y = 4$ **52.** $x = 7$

53. $f(x) = 7x$ **54.** $f(x) = \dfrac{1}{7}x$

55. $6 + y = 0$ **56.** $x - 7 = 0$

57. $2 - x = 3$ **58.** $2y + 4 = -7$

Determine whether the lines are parallel, perpendicular, or neither. See Example 5.

59. $f(x) = -3x + 6$ **60.** $f(x) = 5x - 6$

    $g(x) = 3x + 5$ $g(x) = 5x + 2$

61. $-4x + 2y = 5$ **62.** $2x - y = -10$

    $2x - y = 7$ $2x + 4y = 2$

63. $-2x + 3y = 1$ **64.** $x + 4y = 7$

    $3x + 2y = 12$ $2x - 5y = 0$

65. Explain whether two lines, both with positive slopes, can be perpendicular.

66. Explain why it is reasonable that nonvertical parallel lines have the same slope.

Determine the slope of each line.

67.

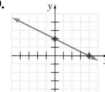

68.

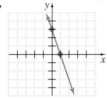

69.

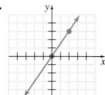

70.

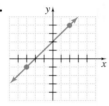

Find each slope.

71. Construction laws state that when constructing access ramps for the handicapped, every vertical rise of 1 foot must have at least a horizontal run of 12 feet. Find the slope or grade of the following ramp. Express the grade as a percent.

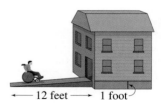

12 feet ⟶ 1 foot

72. Find the pitch, or slope, of the roof shown.

8 feet

←12 feet→

73. Upon takeoff, a Delta Airlines jet climbs to 3 miles as it passes over 25 miles of land below it. Find the slope of its climb.

3 miles

25 miles

74. Driving down Ball Mountain in Wyoming, Bob Dean finds that he descends 1600 feet in elevation by the time he is 2.5 miles (horizontally) away from the high point on the mountain road. Find the slope of his descent (1 mile = 5280 feet).

75. GROUP ACTIVITY   Locate an access ramp for the handicapped on your campus. Measure the horizontal run and the vertical run to find the slope of the ramp. Express the slope as a percent and see that it is $8\frac{1}{3}\%$ or less (See Exercise 71).

76. Find the grade, or slope, of the road shown. Express the grade as a percent.

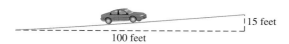

15 feet

100 feet

77. Find the slope of a line parallel to the line $f(x) = -\frac{7}{2}x - 6.$

78. Find the slope of a line parallel to the line $f(x) = x$.

79. Find the slope of a line perpendicular to the line $f(x) = -\frac{7}{2}x - 6.$

80. Find the slope of a line perpendicular to the line $f(x) = x$.

81. Find the slope of a line parallel to the line $5x - 2y = 6.$

82. Find the slope of a line parallel to the line $-3x + 4y = 10.$

83. Find the slope of a line perpendicular to the line $5x - 2y = 6.$

84. The following graph shows the altitude of a seagull in flight over a time period of 30 seconds.

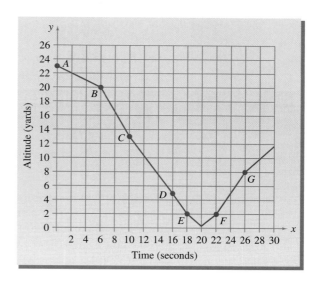

a. Find the coordinates of point B.

b. Find the coordinates of point C.

c. Find the rate of change of altitude between points B and C. (Recall that the rate of change between points is the slope between points. This rate of change will be in yards per second.)

85. Find the rate of change of altitude (in yards per second) between points F and G.

86. Support the result of Exercise 61 by graphing the pair of equations on a graphing utility.

87. Support the result of Exercise 62 by graphing the pair of equations on a graphing utility.

88. a. Graph $y = \frac{1}{2}x + 1$, $y = x + 1$, and $y = 2x + 1$ on a single screen. Notice the change in slope for each graph.

b. Graph $y = -\frac{1}{2}x + 1$, $y = -x + 1$, and $y = -2x + 1$ on a single screen. Notice the change in slope for each graph.

c. Determine whether the following statement is true or false for slopes m of given lines. As $|m|$ becomes greater, the line become steeper.

Review Exercises

Simplify and solve for y.

89. $y - 2 = 5(x + 6)$

90. $y - (-1) = 2(x - 0)$

91. $y - 0 = -3[x - (-10)]$

92. $y - 9 = -8[x - (-4)]$

3.6 | EQUATIONS OF LINES

O B J E C T I V E S

TAPE IAG 3.6

1. Use the slope-intercept form to find an equation of a line.
2. Graph a line given its slope and y-intercept.
3. Use the point-slope form to find an equation of a line.
4. Write equations of vertical and horizontal lines.
5. Find equations of parallel and perpendicular lines.

In the previous section, we learned that the slope-intercept form of a linear equation is $y = mx + b$. When an equation is written in this form, the slope of the line is the same as the coefficient m of x. Also, the y-intercept of the line is the same as the constant term b. For example, the slope of the line defined by $y = 2x + 3$ is 2, and its y-intercept is 3.

We may also use the slope-intercept form to write an equation of a line given its slope and y-intercept.

EXAMPLE 1   Write an equation of the line with y-intercept -3 and slope of $\dfrac{1}{4}$.

Solution: We are given the slope and the y-intercept. Let $m = \dfrac{1}{4}$ and $b = -3$, and write the equation in slope-intercept form, $y = mx + b$.

$$y = mx + b$$

$$y = \frac{1}{4}x + (-3) \qquad \text{Let } m = \frac{1}{4} \text{ and } b = -3.$$

$$y = \frac{1}{4}x - 3 \qquad \text{Simplify.}$$

2   Given the slope and y-intercept of a line, we may graph the line as well as write its equation. Let's graph the line from Example 1. We are given that it has slope $\frac{1}{4}$ and that its y-intercept is -3. First plot the y-intercept point $(0, -3)$. To find another point on the line, recall that slope is $\frac{\text{rise}}{\text{run}} = \frac{1}{4}$. Another point may then be plotted by starting at $(0, -3)$, rising 1 unit up, and then running 4 units to the right. We are now at the point $(4, -2)$. The graph of $y = \frac{1}{4}x - 3$, which follows, is the line through points $(0, -3)$ and $(4, -2)$.

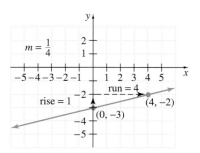

EXAMPLE 2   Graph the line through $(-1, 5)$ with slope -2.

Solution: To graph the line, we need two points. One point is $(-1, 5)$, and we will use the slope -2, which can be written as $\frac{-2}{1}$, to find another point.

$$m = \frac{\text{rise}}{\text{run}} = \frac{-2}{1}$$

To find another point, start at $(-1, 5)$ and move vertically 2 units down, since the numerator of the slope is -2; then move horizontally 1 unit to the right, since the denominator of the slope is 1. We arrive at the point $(0, 3)$. The line through $(-1, 5)$ and $(0, 3)$ will have the required slope of -2.

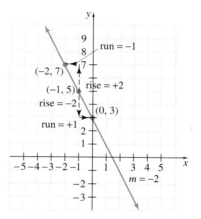

In example 2, the slope -2 can also be written as $\frac{2}{-1}$, so to find another point we could start at $(-1, 5)$ and move 2 units up and then 1 unit left. We would stop at the point $(-2, 7)$. The line through $(-1, 5)$ and $(-2, 7)$ will have the required slope and will be the same line as shown previously through $(-1, 5)$ and $(0, 3)$.

3   When the slope of a line and a point on the line are known, an equation of the line can also be found. To do this, use the slope formula to write the slope of the line that passes through points (x, y) and (x_1, y_1). We have

$$m = \frac{y - y_1}{x - x_1}$$

Multiply both sides of this equation by $x - x_1$ to obtain

$$y - y_1 = m(x - x_1)$$

This form is called the **point-slope form** of an equation of a line.

POINT-SLOPE FORM OF AN EQUATION OF A LINE

The point-slope form of an equation of a line is $y - y_1 = m(x - x_1)$, where m is the slope of the line and (x_1, y_1) is a point on the line.

EXAMPLE 3   Find an equation of the line with slope -3 containing the point $(1, -5)$. Write the equation in standard form: $Ax + By = C$. Then use a graphing utility to check your result.

Solution: Because we know the slope and a point of the line, we use the point-slope form with $m = -3$ and $(x_1, y_1) = (1, -5)$.

$$y - y_1 = m(x - x_1) \qquad \text{Point-slope form.}$$

$$y - (-5) = -3(x - 1) \qquad \text{Let } m = -3 \text{ and } (x_1, y_1) = (1, -5).$$

$$y + 5 = -3x + 3 \qquad \text{Apply the distributive property.}$$

$$3x + y = -2 \qquad \text{Write in standard form.}$$

In standard form, the equation is $3x + y = -2$.

Solving the equation for y we have $y = -3x - 2$. Define $y_1 = -3x - 2$ and graph in an integer window. From the equation we see that $m = -3$, so the slope is -3. Next, check on the graph to see that it contains the point $(1, -5)$ as in the screen below to the left.

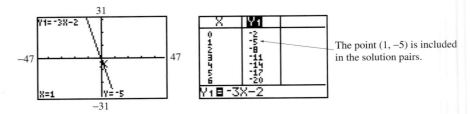

The point $(1, -5)$ is included in the solution pairs.

A second way to see that $(1, -5)$ is an ordered pair solution is to examine a table of values as shown above, to the right.

> **REMINDER**   If we multiply both sides of the equation $3x + y = -2$ by -1, it becomes $-3x - y = 2$. Both equations are in standard form, and their graphs are the same line.

EXAMPLE 4   Find an equation of the line through points $(4, 0)$ and $(-4, -5)$. Write the equation using function notation. Then use a graphing utiliy to check your result.

Solution: First, find the slope of the line.

$$m = \frac{-5 - 0}{-4 - 4} = \frac{-5}{-8} = \frac{5}{8}$$

Next, make use of the point-slope form. Replace (x_1, y_1) by either $(4, 0)$ or $(-4, -5)$ in the point-slope equation. We will choose the point $(4, 0)$. The line through $(4, 0)$ with slope $\frac{5}{8}$ is

$$y - y_1 = m(x - x_1)$$ Point-slope form.

$$y - 0 = \frac{5}{8}(x - 4)$$ Let $m = \frac{5}{8}$ and $(x_1, y_1) = (4, 0)$.

$$8y = 5(x - 4)$$ Multiply both sides by 8.

$$8y = 5x - 20$$ Apply the distributive property.

To write the equation using function notation, we solve for y.

$$y = \frac{5}{8}x - \frac{20}{8}$$ Divide both sides by 8.

$$f(x) = \frac{5}{8}x - \frac{5}{2}$$ Write using function notation.

To check, graph $y_1 = \frac{5}{8}x - \frac{5}{2}$ in an integer window. Trace to find $(4, 0)$ and $(-4, -5)$ on the graph of the line. We can also examine a table of ordered pair solutions to check.

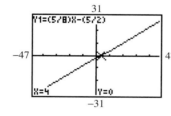 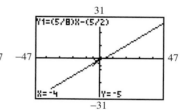 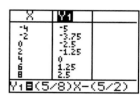

 4 Special types of linear equations are those whose graphs are vertical or horizontal lines.

EXAMPLE 5   Find an equation of the horizontal line containing the point $(2, 3)$.

Solution: Recall that a horizontal line has an equation of the form $y = b$. Since the line contains the point $(2, 3)$, the equation is $y = 3$. Below is a graph of $y = 3$ with the point $(2, 3)$ indicated.

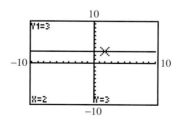

EXAMPLE 6   Find the equation of the line with no slope that contains the point $(2, 3)$.

Solution: Since the line has no slope, the line must be vertical. A vertical line has an equation of the form $x = c$, and since the line contains the point $(2, 3)$, the equation is $x = 2$.

5   Next, we find equations of parallel and perpendicular lines.

EXAMPLE 7   Find an equation of the line containing the point $(4, 4)$ and parallel to the line $2x + 3y = -6$. Write the equation in standard form.

Solution: Because the line we want to find is *parallel* to the line $2x + 3y = -6$, the two lines must have equal slopes. Find the slope of $2x + 3y = -6$ by writing it in the form $y = mx + b$.

TECHNOLOGY NOTE

Recall that a vertical line does not represent the graph of a function and cannot be graphed in the Y = editor. Vertical lines may be drawn using the draw menu. Below is an example of using a draw vertical feature.

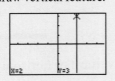

$$2x + 3y = -6$$

$$3y = -2x - 6 \qquad \text{Subtract } 2x \text{ from both sides.}$$

$$y = \frac{-2x}{3} - \frac{6}{3} \qquad \text{Divide by 3.}$$

$$y = -\frac{2}{3}x - 2 \qquad \text{Write in slope-intercept form.}$$

The slope of this line is $-\frac{2}{3}$. Thus a line parallel to this line will also have a slope of $-\frac{2}{3}$. The equation we are asked to find describes a line containing the point $(4, 4)$ with a slope of $-\frac{2}{3}$. We use the point-slope form.

$$y - y_1 = m(x - x_1)$$

$$y - 4 = -\frac{2}{3}(x - 4) \qquad \text{Let } m = -\frac{2}{3}, x_1 = 4, \text{ and } y_1 = 4.$$

$$3(y - 4) = -2(x - 4) \qquad \text{Multiply both sides by 3.}$$

$$3y - 12 = -2x + 8 \qquad \text{Apply the distributive property.}$$

$$2x + 3y = 20 \qquad \text{Write in standard form.}$$

EXAMPLE 8   Write a function that describes the line containing the point $(4, 4)$ and is perpendicular to the line $2x + 3y = -6$.

Solution: Recall that the slope of the line $2x + 3y = -6$ is $-\frac{2}{3}$. A line perpendicular to this line will have a slope that is the negative reciprocal of $-\frac{2}{3}$, or $\frac{3}{2}$. Using the point-slope equation, we have

$$y - y_1 = m(x - x_1)$$

$$y - 4 = \frac{3}{2}(x - 4) \qquad \text{Let } x_1 = 4, y_1 = 4, \text{ and } m = \frac{3}{2}.$$

$$2(y - 4) = 3(x - 4) \qquad \text{Multiply both sides by 2.}$$

$$2y - 8 = 3x - 12 \qquad \text{Apply the distributive property.}$$

$$2y = 3x - 4 \qquad \text{Add 8 to both sides.}$$

$$y = \frac{3}{2}x - 2 \qquad \text{Divide both sides by 2.}$$

$$f(x) = \frac{3}{2}x - 2 \qquad \text{Write using function notation.}$$

EXAMPLE 9   **A LINEAR MODEL FOR SALES PRODUCTION**
Southern Star Realty is an established real estate company that has enjoyed constant linear growth in sales since 1988. In 1990 the company sold 200 houses, and in 1995 the company sold 275 houses. Use these figures to predict the number of houses that this company will sell in the year 2000.

Solution: **1.** UNDERSTAND. Read and reread this problem.

2. ASSIGN. Let

$x = $ the number of years after 1988

$y = $ the number of houses sold in year x

The information provided then gives the ordered pairs $(2, 200)$ and $(7, 275)$.

3. ILLUSTRATE. To illustrate the sales of Southern Star Realty, we graph the linear equation that passes through the points $(2, 200)$ and $(7, 275)$.

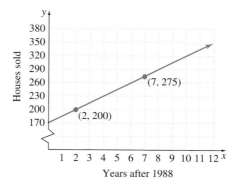

4. TRANSLATE. Write a linear equation that passes through the points $(2, 200)$ and $(7, 275)$. To do so, first find the slope of the line.

$$m = \frac{275 - 200}{7 - 2} = \frac{75}{5} = 15$$

Then, using the point-slope form to write the equation, we have

$$y - y_1 = m(x - x_1)$$

$$y - 200 = 15(x - 2) \qquad \text{Let } m = 15 \text{ and } (x_1, y_1) = (2, 200).$$

$$y - 200 = 15x - 30 \qquad \text{Multiply.}$$

$$y = 15x + 170 \qquad \text{Add 200 to both sides.}$$

$$f(x) = 15x + 170 \qquad \text{Write using function notation.}$$

5. **COMPLETE.** To predict the number of houses that the company will sell in the year 2000, we find $f(12)$, since $2000 - 1988 = 12$.

$$f(12) = 15(12) + 170$$

$$= 350$$

6. **INTERPRET.** *Check:* Graph $y_1 = 15x + 170$ in the window shown below. Use the trace feature to find that the line passes through the given ordered pairs $(2, 200)$ and $(7, 275)$, as well as $(12, 350)$. *State:* In the 12th year, or the year 2000, the company may be expected to sell 350 houses.

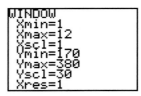
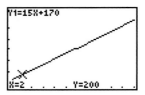

FORMS OF LINEAR EQUATIONS

$Ax + By = C$	**Standard form** of a linear equation A and B are not both 0.
$y = mx + b$	**Slope-intercept form** of a linear equation The slope is m and the y-intercept is b.
$y - y_1 = m(x - x_1)$	**Point-slope form** of a linear equation The slope is m, and (x_1, y_1) is a point on the line.
$y = c$	**Horizontal line** The slope is 0, and the y-intercept is c.
$x = c$	**Vertical line** The slope is undefined, and the x-intercept is c.

PARALLEL AND PERPENDICULAR LINES

Nonvertical parallel lines have the same slope.
The product of the slopes of two nonvertical perpendicular lines is -1.

MENTAL MATH

State the slope and the y-intercept of each line with the given equation.

1. $y = -4x + 12$

2. $y = \frac{2}{3}x - \frac{7}{2}$

3. $y = 5x$

4. $y = -x$

5. $y = \frac{1}{2}x + 6$

6. $y = -\frac{2}{3}x + 5$

Decide whether the lines are parallel, perpendicular, or neither.

7. $y = 12x + 6$

$y = 12x - 2$

8. $y = -5x + 8$

$y = -5x - 8$

9. $y = -9x + 3$

$y = \frac{3}{2}x - 7$

10. $y = 2x - 12$

$y = \frac{1}{2}x - 6$

EXERCISE SET 3.6

Use the slope-intercept form of the linear equation to write an equation of the line with the given slope and y-intercept. See Example 1.

1. Slope -1; y-intercept 1

2. Slope $\frac{1}{2}$; y-intercept -6

3. Slope 2; y-intercept $\frac{3}{4}$

4. Slope -3; y-intercept $-\frac{1}{5}$

5. Slope $\frac{2}{7}$; y-intercept 0

6. Slope $-\frac{4}{5}$; y-intercept 0

Graph the line passing through the given point with the given slope. See Example 2.

7. Through $(1, 3)$ with slope $\frac{3}{2}$

8. Through $(-2, -4)$ with slope $\frac{2}{5}$

9. Through $(0, 0)$ with slope 5

10. Through $(-5, 2)$ with slope 2

11. Through $(0, 7)$ with slope -1

12. Through $(3, 0)$ with slope -3

Find an equation of the line with the given slope and containing the given point. Write the equation in standard form. See Example 3.

13. Slope 3; through $(1, 2)$

14. Slope 4; through $(5, 1)$

15. Slope -2; through $(1, -3)$

16. Slope -4; through $(2, -4)$

17. Slope $\frac{1}{2}$; through $(-6, 2)$

18. Slope $\frac{2}{3}$; through $(-9, 4)$

19. Slope $-\frac{9}{10}$; through $(-3, 0)$

20. Slope $-\frac{1}{5}$; through $(4, -6)$

Find an equation of each line graphed. Write the equation in standard form.

21.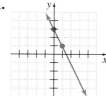

22.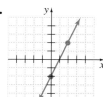

23.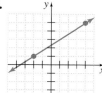

24.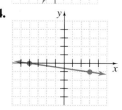

Find an equation of the line passing through the given points. Use function notation to write the equation. See Example 4.

25. $(2, 0), (4, 6)$

26. $(3, 0), (7, 8)$

27. $(-2, 5), (-6, 13)$

28. $(7, -4), (2, 6)$

29. $(-2, -4), (-4, -3)$

30. $(-9, -2), (-3, 10)$

31. $(-3, -8), (-6, -9)$ **32.** $(8, -3), (4, -8)$

33. Describe how to check whether the graph of $2x - 4y = 7$ passes through the points $(1.4, -1.05)$ and $(0, -1.75)$. Then follow your directions and check these points.

Use the graph of the following function $f(x)$ to find each value.

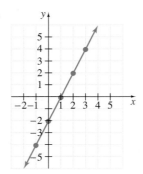

34. $f(1)$ **35.** $f(0)$ **36.** $f(-1)$ **37.** $f(2)$

38. Find x such that $f(x) = 4$.

39. Find x such that $f(x) = -6$

Write an equation of each line. See Examples 5 and 6.

40. Vertical; through $(2, 6)$

41. Slope 0; through $(-2, -4)$

42. Horizontal; through $(-3, 1)$

43. Vertical; through $(4, 7)$

44. Undefined slope; through $(0, 5)$

45. Horizontal; through $(0, 5)$

46. Answer the following true or false. A vertical line is always perpendicular to a horizontal line.

Find an equation of each line. Write the equation using function notation. See Examples 7 and 8.

47. Through $(3, 8)$; parallel to $f(x) = 4x - 2$

48. Through $(1, 5)$; parallel to $f(x) = 3x - 4$

49. Through $(2, -5)$; perpendicular to $3y = x - 6$

50. Through $(-4, 8)$; perpendicular to $2x - 3y = 1$

51. Through $(-2, -3)$; parallel to $3x + 2y = 5$

52. Through $(-2, -3)$; perpendicular to $3x + 2y = 5$

Find an equation of each line. Write the equation in standard form unless indicated otherwise.

53. Slope 2; through $(-2, 3)$

54. Slope 3; through $(-4, 2)$

55. Through $(1, 6)$ and $(5, 2)$; use function notation

56. Through $(2, 9)$ and $(8, 6)$

57. With slope $-\dfrac{1}{2}$; y-intercept 11

58. With slope -4; y-intercept $\dfrac{2}{9}$; use function notation

59. Through $(-7, -4)$ and $(0, -6)$

60. Through $(2, -8)$ and $(-4, -3)$

61. Slope $-\dfrac{4}{3}$; through $(-5, 0)$

62. Slope $-\dfrac{3}{5}$; through $(4, -1)$

63. Vertical line; through $(-2, -10)$

64. Horizontal line; through $(1, 0)$

65. Through $(6, -2)$; parallel to the line $2x + 4y = 9$

66. Through $(8, -3)$; parallel to the line $6x + 2y = 5$

67. Slope 0; through $(-9, 12)$

68. Undefined slope; through $(10, -8)$

69. Through $(6, 1)$; parallel to the line $8x - y = 9$

70. Through $(3, 5)$; perpendicular to the line $2x - y = 8$

71. Through $(5, -6)$; perpendicular to $y = 9$

72. Through $(-3, -5)$; parallel to $y = 9$

73. Through $(2, -8)$ and $(-6, -5)$; use function notation

74. Through $(-4, -2)$ and $(-6, 5)$; use function notation

Solve. See Example 9.

75. A rock is dropped from the top of a 400-foot building. After 1 second, the rock is traveling 32 feet per second. After 3 seconds, the rock is traveling 96 feet per second. Let $R(x)$ be the rate of descent and x be the number of seconds since the rock was dropped.

  a. Write a linear function that relates time x to rate $R(x)$. [*Hint:* Use the ordered pairs $(1, 32)$ and $(3, 96)$ and use function notation to write an equation.]

  b. Use this function to determine the rate of the rock 4 seconds after it was dropped.

76. The Whammo Company has learned that, by pricing a newly released Frisbee at $6, sales will reach 2000 per day. Raising the price to $8 will cause the sales to fall to 1500 per day. Assume that the ratio of change in daily sales to change in price is constant, and let x be the price of the Frisbee and S be number of sales.

a. Find the linear function $S(x)$ that models the price–sales relationship for this Frisbee. [*Hint:* The line must pass through $(6, 2000)$ and $(8, 1500)$.]

b. Predict the daily sales of this Frisbee if the price is set at $7.50.

77. In 1990, the median price of an existing home in the United States was $97,500. In 1994, the median price of an existing home was $109,800. Let $P(x)$ be the median price of an existing home in the year x, where $x = 0$ represents 1990. (Source: National Association of REALTORS®)

a. Find a linear function $P(x)$ that models the median existing home price in terms of the year x. (See the hint for Exercise 75.)

b. Predict the median existing home price for 1999.

78. The number of births (in thousands) in the United States in 1994 was 3797. The number of births (in thousands) in the United States in 1990 was 4158. Let $B(x)$ be the number of births (in thousands) in the year x, where $x = 0$ represents 1990. (Source: National Center for Health Statistics)

a. Find a linear function $B(x)$ that models the number of births (in thousands) in terms of the year x. (See the hint for Exercise 76.)

b. Predict the number of births in the United States for the year 2000.

79. Del Monte Fruit Company recently released a new applesauce. By the end of its first year, profit on this product amounted to $30,000. The anticipated profit for the end of the fourth year is $66,000. The ratio of change in profit to change in time is constant. Let x be years and P be profit.

a. Write a linear function $P(x)$ that expresses profit as a function of time.

b. Use this function to predict the company's profit at the end of the seventh year.

c. Predict when the profit should reach $126,000.

80. The value of a computer bought in 1990 depreciates, or goes down, as time passes. Two years after the computer was bought, it was worth $2600; 5 years after it was bought, it was worth $2,000.

a. Assume that the relationship between number of years past 1990 and value of computer is linear, and write an equation describing this relationship. [Use ordered pairs of the form (years past 1990, value of computer).]

b. Use this equation to estimate the value of the computer in the year 2000.

81. The Pool Fun Company has learned that, by pricing a newly released Fun Noodle at $3, sales will reach 10,000 Fun Noodles per day during the summer. Raising the price to $5 will cause sales to fall to 8000 Fun Noodles per day.

a. Assume that the relationship between sales price and number of Fun Noodles sold is linear, and write an equation describing this relationship.

b. Predict the daily sales of Fun Noodles if the price is $3.50.

82. The value of a building bought in 1980 appreciates, or increases, as time passes. Seven years after the building was bought, it was worth $165,000; 12 years after it was bought, it was worth $180,000.

a. Assume that the relationship between number of years past 1980 and value of the building is linear, and write an equation describing this relationship. [Use ordered pairs of the form (years past 1980, value of building).]

b. Use this equation to estimate the value of the building in the year 2000.

Use a graphing utility and the trace feature to verify the results of each exercise.

83. Exercise 55; graph the function and verify that it passes through $(1, 6)$ and $(5, 2)$.

84. Exercise 56; graph the equation and verify that it passes through $(2, 9)$ and $(8, 6)$.

85. Exercise 61; graph the equation. See that it has a negative slope and passes through $(-5, 0)$.

86. Exercise 62; graph the equation. See that it has a negative slope and passes through $(4, -1)$.

Review Exercises

Solve. See Section 2.1.

87. $2x - 7 = 21$

88. $-3x + 1 = 0$

89. $5(x - 2) = 3(x - 1)$

90. $-2(x + 1) = -x + 10$

91. $\dfrac{x}{2} + \dfrac{1}{4} = \dfrac{1}{8}$

92. $\dfrac{x}{5} - \dfrac{3}{10} = \dfrac{x}{2} - 1$

3.7 | SOLVING LINEAR EQUATIONS GRAPHICALLY

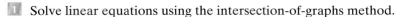

O B J E C T I V E S

1 Solve linear equations using the intersection-of-graphs method.

2 Solve problems using graphing methods.

3 Solve an identity or a contradiction.

4 Solve linear equations using the *x*-intercept method.

TAPE IAG 3.7

In Chapter 2 we solved linear equations algebraically and checked the solution numerically with a calculator. Recall that a solution of an equation is a value for the variable that makes the equation a true statement. In this section we solve equations algebraically and graphically. Then we use various methods to check the solution. We need to stress the importance of the check. Technology allows us to reduce the time spent on solving and aids us in verifying solutions. Consequently, by using technology we can increase the accuracy of our work.

1 There are two methods we use to graphically obtain a solution of an equation. The first method we refer to as the **intersection-of-graphs method.** For this method, we graph

$$y_1 = \text{left side of equation and}$$

$$y_2 = \text{right side of equation.}$$

Recall that when a solution is substituted for a variable in an equation, the left side of the equation is equal to the right side of the equation. This means that graphically a solution occurs when $y_1 = y_2$ or where the graphs of y_1 and y_2 intersect.

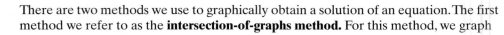

DISCOVER THE CONCEPT

Consider the equation $2x - 5 = 27$ and graph $y_1 = 2x - 5$ and $y_2 = 27$ in an integer window.

a. Use the trace feature to estimate the point of intersection of the two graphs.

b. Solve the equation algebraically and compare the *x*-coordinate of the point of intersection found in part (a) to the algebraic solution of the equation.

c. Locate the intersect feature on your graphing utility, sometimes found on the calculate menu. Use this feature to find the point of intersection of the two graphs.

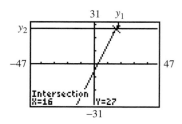

The above discovery indicates that the solution of the equation $2x - 5 = 27$ is 16, the *x*-coordinate of the point of intersection, as shown above.

To check this solution, we may replace x with 16 in the original equation.

$$2x - 5 = 27$$
$$2 \cdot 16 - 5 = 27 \qquad \text{Let } x = 16.$$
$$32 - 5 = 27$$
$$27 = 27 \qquad \text{True.}$$

Recall that 27 is the y-coordinate of the point of intersection. Why? Because it is the value of both the left side and the right side of the equation when x is replaced with the solution.

The steps below may be used to solve an equation by the intersection of graphs method.

INTERSECTION-OF-GRAPHS METHOD FOR SOLVING AN EQUATION

Step 1. Graph $y_1 =$ left side of the equation and
$y_2 =$ right side of the equation.

Step 2. Find the point(s) of intersection of the two graphs.

Step 3. The x-coordinate of a point of intersection is a solution to the equation.

Step 4. The y-coordinate of the point of intersection is the value of both the left side and the right side of the original equation when x is replaced with the solution.

EXAMPLE 1   Solve the equation $5(x - 2) + 15 = 20$.

Algebraic Solution:

$$5(x - 2) + 15 = 20$$
$$5x - 10 + 15 = 20 \qquad \text{Use the distributive property.}$$
$$5x + 5 = 20 \qquad \text{Combine like terms.}$$
$$5x = 15 \qquad \text{Subtract 5 from both sides.}$$
$$x = 3 \qquad \text{Divide both sides by 5.}$$

To check, we replace x with 3 and see that a true statement results.

$$5(x - 2) + 15 = 20$$
$$5(3 - 2) + 15 = 20$$
$$5(1) + 15 = 20$$
$$20 = 20 \qquad \text{True}$$

The solution set is {3}.

Graphical Solution:

Graph $y_1 = 5(x - 2) + 15$ (left side of equation)
$y_2 = 20$ (right side of equation)

Since the graph of the equation $y = 20$ is a horizontal line with y-intercept 20, use the window $[-25, 25, 5]$ by $[-25, 25, 5]$

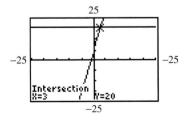

The x-coordinate of the point of intersection is 3. The solution set is {3}.

We can also use a table feature to check the solution. If $y_1 = 5(x - 2) + 15$ and $y_2 = 20$, scroll to $x = 3$ and see that y_1 and y_2 are both 20.

When solving equations graphically, we often have no indication where the intersection lies before looking at the graphs of the two equations. The zoom feature of your graphing utility may be used to quickly look for an appropriate window. If the graphs are present but the intersection point is missing, you can assess the situation and decide which component to adjust on the window setting. You will quickly get used to this assessment process as you solve more equations.

EXAMPLE 2   Graphically solve the equation $2(x - 30) + 6(x - 10) - 70 = 35 - x$.

Solution: Graph $y_1 = 2(x - 30) + 6(x - 10) - 70$ and

$$y_2 = 35 - x$$

in an integer window.

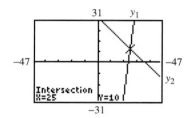

The point of intersection $(25, 10)$ indicates that the solution is 25 and the solution set is $\{25\}$.

To check numerically with a calculator, see the screen below.

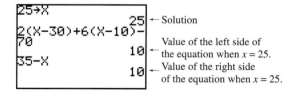

In the above example, recall that the intersection point $(25, 10)$ indicates that if $x = 25$, then the value of the expression on the left side of the equation $2(x - 30) + 6(x - 10) - 70$ is 10, and the value of the expression on the right side of the equation, $35 - x$, is also 10.

> R E M I N D E R   The integer window allows the cursor to move along a graph with integer x-values only. Using this window does *not* mean that the calculator will give only integer values when calculating a point of intersection.

R E M I N D E R   In general, when using the intersection method, the x-coordinate of the point of intersection is a solution of the equation, and the y- coordinate of the point of intersection is the value of each side of the original equation when the variable is replaced with the solution.

EXAMPLE 3   Solve the equation using the intersection-of-graphs method.

$$5.1x + 3.78 = x + 4.7$$

Solution: Define $y_1 = 5.1x + 3.78$ and $y_2 = x + 4.7$ and graph in a standard window. The screen below shows the intersection point.

TECHNOLOGY NOTE

The point of intersection shown suggests that either the exact solution of the equation is a fraction with decimal equivalence greater than or equal to 8 decimal places, or the solution is an irrational number. To see if it is a fraction, you may want to see if your graphing utility can convert to fractions.

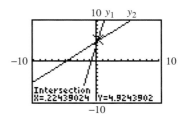

Notice the coordinates of the intersection point. When this happens, we will approximate the solution. For this example, the solution rounded to four decimal places is 0.2244.

To check when the solution is an approximation, note that the value of the left side of the equation and the value of the right side may differ slightly because of rounding. In the screen below we see a check of the equation for $x \approx 0.2244$.

```
.2244→X
              .2244      ← Approximate solution
5.1X+3.78
            4.92444      ← Value of left
X+4.7                       side of equation
             4.9244      ← Value of right
                            side of equation
```

The solution is approximately 0.2244.

2   Next, we use the intersection method to solve a problem.

EXAMPLE 4

PURCHASING A REFRIGERATOR

The McDonalds are purchasing a new refrigerator. They find two models that fit their needs. Model 1 sells for $575 and costs $0.07 per hour to run. Model 2 is the energy efficient model that sells for $825, but only costs $0.04 per hour to run. If x represents number of hours, then the costs to purchase and run the refrigerators are modeled by the equations

$$C_1(x) = 575 + 0.07x \quad \text{Cost to run model 1.}$$

$$C_2(x) = 825 + 0.04x \quad \text{Cost to run model 2.}$$

$0.07
per hour
Model 1

$0.04
per hour
Model 2

a. If the McDonalds buy Model 2, how many hours must it run before they save money?

b. How many days must it run before they save money?

Solution:

Graph $y_1 = 575 + 0.07x$ and $y_2 = 825 + 0.04x$. Here, the ordered pairs represent (x, y).
        ↑   ↑
     hours cost

a. Find the point of intersection of the graphs.

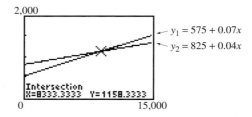

The point of intersection is $(8333.\overline{3}, 1158.\overline{3})$. This means that in approximately 8,333.33 hours, both machines cost a total of approximately $1158.33 to buy and run. Before that point of intersection, the graph of y_1 is below the graph of y_2. This means that the cost of Model 1 is less than the cost of Model 2. Likewise, after the point of intersection, the graph of y_2 is below the graph of y_1. This means that the cost of Model 2 is less than the cost of Model 1. Thus, in approximately 8,333.33 hours, the McDonalds start saving money.

b. To find the approximate number of days, find

$$\frac{8,333.33}{24} \approx 347.22 \text{ days.}$$

3 Recall in Section 2.1 that we solved two special types of equations: contradictions and identities. An equation in one variable that has no solution is called a contradiction and an equation in one variable that has every number (for which the equation is defined) as a solution is called an identity. We now look at these two special types of equations graphically.

EXAMPLE 5 Solve $4(x + 6) - 2(x - 3) = 2x + 30$.

Algebraic Solution:

$$4(x + 6) - 2(x - 3) = 2x + 30$$

$$4x + 24 - 2x + 6 = 2x + 30 \qquad \text{Multiply.}$$

$$2x + 30 = 2x + 30 \qquad \text{Simplify.}$$

Since both sides are the same, we see that replacing x with any real number will result in a true statement.

Graphical Solution:

Graph $y_1 = 4(x + 6) - 2(x - 3)$ and

$$y_2 = 2x + 30.$$

The graphs shown below appear to be the same since there appears to be a single line only. When we trace along y_1, we have the same ordered pairs as when we trace along y_2. (If 2 points of y_1 are the same as 2 points of y_2, we know the lines are identical because 2 points uniquely determine a line.)

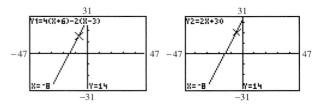

Both solutions show that the equation is an identity and the solution set is the set of real numbers or $\{x | x$ is a real number$\}$.

EXAMPLE 6   Solve $3x - 8 = 5(x - 1) - 2(x + 6)$.

Algebraic Solution:

$$3x - 8 = 5(x - 1) - 2(x + 6)$$

$$3x - 8 = 5x - 5 - 2x - 12 \quad \text{Multiply.}$$

$$3x - 8 = 3x - 17 \qquad\qquad \text{Simplify.}$$

$$3x - 8 - 3x = 3x - 17 - 3x \quad \text{Subtract } 3x.$$

$$-8 = -17$$

This equation is a false statement no matter what value the variable x might have.

Graphical Solution:

Graph $y_1 = 3x - 8$ and

$$y_2 = 5(x - 1) - 2(x + 6)$$

The graph below shows that the lines appear to be parallel and, therefore, will never intersect.

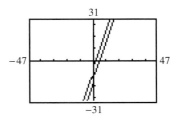

To be sure, check the slope of each graph.

$$y_1 = 3x - 8 \qquad y_2 = 5(x - 1) - 2(x + 6)$$

$$y_2 = 3x - 17$$

The lines have the same slope, 3, and their graphs are distinct, so they are indeed parallel.

The equation has no solution and is a contradiction. The solution set is $\{\ \}$ or $\varnothing$.

We now look at another method for solving equations graphically, called the **x-intercept method.** It is actually a form of the intersection-of-graphs method, but for this method we write an equivalent equation with one side of the equation 0. Since $y = 0$ is the x-axis, we look for points where the graph intersects the x-axis, or x-intercept points. Recall that an x-intercept point is of the form $(x, 0)$. Thus, the solutions that lie on the x-axis are called the **zeros** of the equation since this is where $y = 0$. They are also referred to as **roots** of the equation. The built-in feature on graphing utilities to find these solutions is referred to as root or zero on different graphing utilities.

EXAMPLE 7   Solve the following equation using the x-intercept method.

$$-3.1(x + 1) + 8.3 = -x + 12.4$$

Solution: We begin by writing the equation so that one side equals 0 by adding x to both sides and subtracting 12.4 from both sides.

$$-3.1(x + 1) + 8.3 = -x + 12.4$$

$$-3.1(x + 1) + 8.3 + x - 12.4 = 0$$

Define $y_1 = -3.1(x + 1) + 8.3 + x - 12.4$ and graph in a standard window. Choose the root, or zero, feature to solve. You will be prompted to indicate a left bound—an x-value less than the x-intercept and a right bound—an x-value greater than the x-intercept. The guess portion of the prompting asks you to move the cursor close to the x-intercept point.

TECHNOLOGY NOTE

The real solutions of an equation in the form $f(x) = 0$ occur at the x-intercepts of the graph $y = f(x)$. Because of this, most graphing utilities have a root or zero feature to find the x-intercepts.

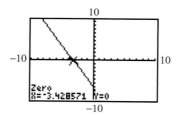

The x-intercept point is approximately $(-3.428571, 0)$, which means that the solution to the equation is approximately -3.428571.

To check, replace x with -3.428571 and see that the left side of the equation is approximately the right side. Rounded to two decimal places, $x \approx -3.43$ or the solution is approximately -3.43.

MENTAL MATH

Mentally solve the following equations.

1. $4x = 24$

2. $6x = -12$

3. $2x + 10 = 20$

4. $5x + 25 = 30$

5. $-3x = 0$

6. $-2x = -14$

EXERCISE SET 3.7

Solve each equation algebraically and graphically. See Examples 1 through 3, and 7.

1. $5x + 2 = 3x + 6$ **2.** $2x + 9 = 3x + 7$

3. $9 - x = 2x + 12$ **4.** $3 - 4x = 2 - 3x$

5. $8 - (2x - 1) = 13$ **6.** $3(2 - x) + 4 = 2x + 3$

For each given screen, write an equation in x and its solution.

7.

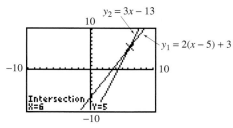

8.

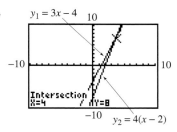

9.

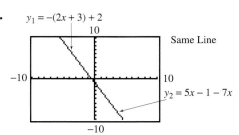

10.

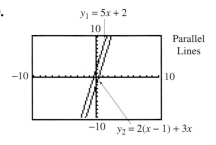

Solve each equation algebraically and graphically. See Examples 5 and 6.

11. $7(x - 6) = 5(x + 2) + 2x$

12. $6x - 9 = 3(x - 3)$

13. $3x - (6x + 2) = -(3x + 2)$

14. $8x - (2x + 3) = 6(x + 5)$

15. $5(x + 1) - 3(x - 7) = 2(x + 4) - 3$

16. $(5x + 8) - 2(x + 3) = (7x - 4) - (4x + 6)$

17. $3(x + 2) - 6(x - 5) = 36 - 3x$

18. $2(x + 3) - 7 = 4 + 2x$

Solve each equation. If necessary, round solutions to two decimal places.

19. $(x + 2.1) - (0.5x + 3) = 12$

20. $2(x + 1.3) - 4(5 - x) = 15$

21. $5(a - 12) + 2(a + 15) = a - 9$

22. $8(b + 2) - 4(b - 5) = b + 12$

23. $8(p - 4) - 5(2p + 3) = 3.5(2p - 5)$

24. $3(x + 2) - 5(x - 7) = x + 3$

25. $5(x - 2) + 2x = 7(x + 4)$

26. $3x + 2(x + 4) = 5(x + 1) + 3$

27. $y + 0.2 = 0.6(y + 3)$

28. $-(w + 0.2) = 0.3(4 - w)$

29. $2y + 5(y - 4) = 4y - 2(y - 10)$

30. $9c - 3(6 - 5c) = c - 2(3c + 9)$

31. $2(x - 8) + x = 3(x - 6) + 2$

32. $4(x + 5) = 3(x - 4) + x$

33. $\dfrac{5x - 1}{6} - 3x = \dfrac{1}{3} + \dfrac{4x + 3}{9}$

34. $\dfrac{2r - 5}{3} - \dfrac{r}{5} = 4 - \dfrac{r + 8}{10}$

35. $-2(b - 4) - (3b - 1) = 5b + 3$

36. $4(t - 3) - 3(t - 2) = 2t + 8$

37. $1.5(4 - x) = 1.3(2 - x)$

38. $2.4(2x + 3) = -0.1(2x + 3)$

39. $\dfrac{1}{4}(a + 2) = \dfrac{1}{6}(5 - a)$ **40.** $\dfrac{1}{3}(8 + 2c) = \dfrac{1}{5}(3c - 5)$

Solve. See Example 4.

41. Acme Mortgage Company wants to hire a student computer consultant. A first consultant charges an initial fee of $30 plus $20 per hour. A second consultant charges a flat fee of $25 per hour. If x represents number of hours, then the costs of the consultants are modeled by

$$C_1(x) = 30 + 20x \quad \text{First consultant's total cost.}$$
$$C_2(x) = 25x \quad \text{Second consultant's total cost.}$$

a. Find which consultant's cost is lower if the job takes 2 hours.

b. Find which consultant's cost is lower if the job takes 8 hours.

c. When is the cost of hiring each consultant the same?

42. Rod Pasch needs to hire a graphic artist. A first graphic artist charges $50 plus $35 per hour. A second one charges $50 per hour. If x represents number of hours, then the costs of the graphic artists can be modeled by

$$G_1(x) = 50 + 35x \quad \text{First graphic artist's total cost}$$
$$G_2(x) = 50x \quad \text{Second graphic artist's total cost.}$$

a. Which graphic artist costs less if the job takes 3 hours?

b. Which graphic artist costs less if the job takes 5 hours?

c. When is the cost of hiring each artist the same?

43. One car rental agency charges $25 a day plus $0.30 a mile. A second car rental agency charges $28 a day plus $0.25 a mile. If x represents the number of miles driven, then the costs of the car rental agencies for a 1-day rental can be modeled by

$$R_1(x) = 25 + 0.30x \quad \text{First agency's cost}$$
$$R_2(x) = 28 + 0.25x \quad \text{Second agency's cost}$$

a. Which agency's cost is lower if 50 miles are driven.

b. Which agency's cost is lower if 100 miles are driven.

c. When is the cost of using each agency the same?

44. Copycat Printing charges $18 plus $0.03 per page for making copies. Duplicate Inc. charges $0.05 per page copied. If x represents the number of pages copied, then the charges for the companies can be modeled by

$$C_1(x) = 18 + 0.03x \quad \text{Copycat}$$
$$C_2(x) = 0.05x \quad \text{Duplicate, Inc.}$$

a. For 500 copies, which company charges less?

b. For 1000 copies, which company charges less?

c. When is the cost of using each company the same?

The given screen shows the graphs of y_1 and y_2 and their intersection. Use this screen to answer the questions below.

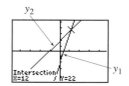

45. Complete the ordered pair for the graphs of both y_1 and y_2: (12,).

46. If x is less than 12, is y_1 less than, greater than, or equal to y_2?

47. If x is greater than 12, is y_1 less than, greater than, or equal to y_2?

48. True or false? If x is 45, $y_1 < y_2$.

Solve each equation graphically and check by a method of your choice. Round solutions to the nearest hundredth.

49. $1.75x - 2.5 = 0$

50. $3.1x + 5.6 = 0$

51. $2.5x + 3 = 7.8x - 5$

52. $4.8x - 2.3 = 6.8x + 2.7$

53. $3x + \sqrt{5} = 7x - \sqrt{2}$

54. $0.9x + \sqrt{3} = 2.5x - \sqrt{5}$

55. $2\pi x - 5.6 = 7(x - \pi)$

56. $-\pi x + 1.2 = 0.3(x - 5)$

57. If the intersection-of-graphs method leads to parallel lines, explain what this means in terms of the solution of the original equation.

58. If the intersection-of-graphs method leads to the same line, explain what this means in terms of the solution of the original equation.

Review Exercises

Graph each inequality on a number line and then write the inequality using interval notation. See Section 1.2.

59. $x < -4$

60. $x > -1$

61. $x \geq 2$

62. $x \leq 0$

63. $-3 < x < 5$

64. $1 \leq x \leq 7$

65. $x \leq 3.5$

66. $x < 0.5$

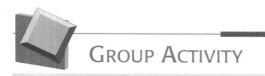

GROUP ACTIVITY

MODELING JAPANESE AUTOMOBILE IMPORTS

OPTIONAL, MATERIALS:

- Graphing utility with least squares curve-fitting capability

The following table shows the number (in millions) of Japanese automobiles imported into the United States during the years 1986-1994. Use the table along with your answers to the questions below to find a linear function $f(x)$ that represents the data.

JAPANESE AUTOMOBILE IMPORTS (IN MILLIONS)

YEAR	1986	1987	1988	1989	1990	1991	1992	1993	1994
x	6	7	8	9	10	11	12	13	14
IMPORTS, Y	2.6	2.4	2.1	2.1	1.9	1.8	1.6	1.5	1.5

Source: U.S. Department of Commerce.

1. Plot the data given in the table as ordered pairs.

2. Use a straightedge to draw on your graph what appears to be the line that "best fits" the data you plotted.

3. Estimate the coordinates of two points that fall on your best-fitting line. Use these points to find a linear function $f(x)$ for the line.

4. Find the value of $f(19)$ and interpret its meaning in context.

5. Compare your group's linear function with other groups' functions. Are they different?

   If so, explain why.

6. (Optional) Enter the data from the table as ordered pairs into lists L_1 and L_2 on your graphing utility. Use the linear regression feature to compute the line that "best fits" the graph of the ordered pairs. Compare this function with the one you found in Question 3. How are they alike or different? Find $f(19)$ using the model you found with the graphing utility and compare it with the value of $f(19)$ you found in Question 4. (For further help, see A Look Ahead at the end of Exercise Sets 3.1 and 3.2.)

CHAPTER 3 HIGHLIGHTS

DEFINITIONS AND CONCEPTS	EXAMPLES

SECTION 3.1 INTRODUCTION TO GRAPHING AND GRAPHING UTILITIES

The **rectangular coordinate system,** or **Cartesian coordinate system,** consists of a vertical and a horizontal number line on a plane intersecting at their 0 coordinates. The vertical number line is called the **y-axis,** and the horizontal number line is called the **x-axis.** The point of intersection of the axes is called the **origin.** The axes divide the plane into four regions called **quadrants.**

To **plot** or **graph** an ordered pair means to find its corresponding point on a rectangular coordinate system.

To plot or graph the ordered pair $(-2, 5)$, start at the origin. Move 2 units to the left along the x-axis, then 5 units upward parallel to the y-axis.

A graphing utility can be used to display a portion of the rectangular coordinate system. The portion being viewed is called a **viewing window** or simply **window.**

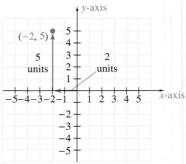

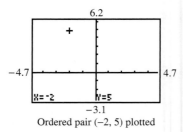

Ordered pair $(-2, 5)$ plotted

SECTION 3.2 GRAPHING EQUATIONS

A ordered pair is a **solution** of an equation in two variables if replacing the variables by the corresponding coordinates results in a true statement.

Determine whether $(-2, 3)$ is a solution of
$$3x + 2y = 0$$

$$3(-2) + 2(3) = 0$$

$$-6 + 6 = 0$$

$$0 = 0 \qquad \text{True.}$$

$(-2, 3)$ is a solution.

(continued)

DEFINITIONS AND CONCEPTS	**EXAMPLES**

SECTION 3.2   GRAPHING EQUATIONS

A **linear equation in two variables** is an equation that can be written in the form $Ax + By = C$, where A, B, and C are real numbers and A and B are not both 0. The form $Ax + By = C$ is called **standard form.**

Linear Equations in Two Variables
$$y = -2x + 5, \; x = 7$$
$$y - 3 = 0, \; 6x - 4y = 10$$
$6x - 4y = 10$ is in standard form.

The graph of a linear equation in two variables is a line. To graph a linear equation in two variables, find three ordered pair solutions. Plot the solution points, and draw the line through the points.

Graph $3x + y = -6$.

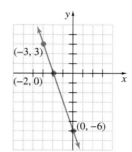

x	y
0	-6
-2	0
-3	3

To graph an equation that is not linear, find a sufficient number of ordered pair solutions so that a pattern may be discovered.

Graph $y = x^3 + 2$.

x	y
-2	-6
-1	1
0	2
1	3
2	10

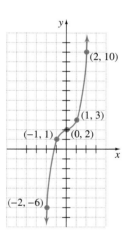

To graph an equation using a graphing utility, solve the equation for y and enter it into the Y= editor.

Graph $y = x^3 + 2$.

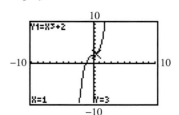

(continued)

DEFINITIONS AND CONCEPTS	EXAMPLES

SECTION 3.3   INTRODUCTION TO FUNCTIONS

A **relation** is a correspondence between a first set (set of inputs) and a second set (set of outputs) that assigns to each element of the first set an element of the second set. The **domain** of the relation is the first set, and the **range** of the relation is the set of elements that correspond to some element of the first set.

Relation

Input: Output:
Words   Number of Vowels

cat ⟶ 1
dog ⟶ 3
too ⟶ 2
give ⟶

Domain: {cat, dog, too, give}
Range: {1, 2}

A **function** is a relation in which each element of the first set corresponds to exactly one element of the second set.

The previous relation is a function. Each word contains one exact number of vowels.

Vertical Line Test:

If no vertical line can be drawn so that it intersects a graph more than once, the graph is the graph of a function.

Find the domain and the range of the relation. Also determine whether the relation is a function.

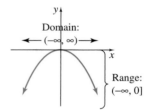

Domain: $(-\infty, \infty)$

Range: $(-\infty, 0]$

By the vertical line test, this graph is the graph of a function.

The symbol $f(x)$ means **function of x** and is called **function notation.**

If $f(x) = 2x^2 - 5$, find $f(-3)$.

$$f(-3) = 2(-3)^2 - 5 = 2(9) - 5 = 13$$

SECTION 3.4   GRAPHING LINEAR FUNCTIONS

A **linear function** is a function that can be written in the form $f(x) = mx + b$.

Linear Functions:

$$f(x) = -3, g(x) = 5x, h(x) = -\frac{1}{3}x - 7$$

To graph a linear function, find three ordered pair solutions. Graph the solutions and draw a line through the plotted points.

Graph $f(x) = -2x$.

x	y or $f(x)$
-1	2
0	0
2	-4

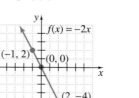

(continued)

DEFINITIONS AND CONCEPTS	**EXAMPLES**

SECTION 3.4   GRAPHING LINEAR FUNCTIONS

For any function $f(x)$, the graph of $y = f(x) + K$ is the same as the graph of $y = f(x)$ shifted $|K|$ units up if K is positive and $|K|$ units down if K is negative.

Graph $g(x) = -2x + 3$.

This is the same as the graph of $f(x) = -2x$ shifted 3 units up.

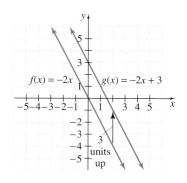

The x-coordinate of a point where a graph crosses the x-axis is called an **x-intercept.** The y-coordinate of a point where a graph crosses the y-axis is called a **y-intercept.**

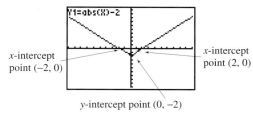

The x-intercepts of the graph are -2 and 2. The y-intercept is -2.

To find an x-intercept, let $y = 0$, or $f(x) = 0$, and solve for x.

To find a y-intercept, let $x = 0$ and solve for y.

Graph $5x - y = -5$ by finding intercepts.

If $x = 0$, then   If $y = 0$, then

$$5x - y = -5 \qquad\qquad 5x - y = -5$$
$$5 \cdot 0 - y = -5 \qquad\qquad 5x - 0 = -5$$
$$-y = -5 \qquad\qquad 5x = -5$$
$$y = 5 \qquad\qquad x = -1$$

Ordered pairs are $(0, 5)$ and $(-1, 0)$.

To find an x-intercept graphically, trace on the graph to find point(s) where $y = 0$. To find a y-intercept, trace on the graph to find point(s) where $x = 0$.

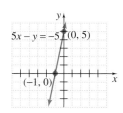

(continued)

DEFINITIONS AND CONCEPTS	EXAMPLES

SECTION 3.4 GRAPHING LINEAR FUNCTIONS

The graph of $x = c$ is a vertical line with x-intercept c.

The graph of $y = c$ is a horizontal line with y-intercept c.

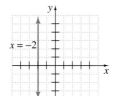 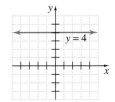

SECTION 3.5 THE SLOPE OF A LINE

The **slope** m of the line through (x_1, y_1) and (x_2, y_2) is given by

$$m = \frac{\Delta y}{\Delta x} = \frac{y_2 - y_1}{x_2 - x_1} \text{ as long as } x_2 \neq x_1$$

Find the slope of the line through $(-1, 7)$ and $(-2, -3)$.

$$m = \frac{y_2 - y_1}{x_2 - x_1} = \frac{-3 - 7}{-2 - (-1)} = \frac{-10}{-1} = 10$$

The **slope-intercept form** of a linear equation is $y = mx + b$, where m is the slope of the line and b is the y-intercept.

Find the slope and the y-intercept of $-3x + 2y = -8$.

$$2y = 3x - 8$$

$$\frac{2y}{2} = \frac{3x}{2} - \frac{8}{2}$$

$$y = \frac{3}{2}x - 4$$

The slope of the line is $\frac{3}{2}$, and the y-intercept is -4.

Nonvertical parallel lines have the same slope.

If the product of the slopes of two lines is -1, then the lines are perpendicular.

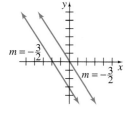 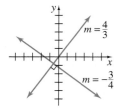

The slope of a horizontal line is 0.

The slope of a vertical line is undefined.

The slope of $y = -2$ is 0.

The slope of $x = 5$ is undefined.

SECTION 3.6 EQUATIONS OF LINES

We can use the slope-intercept form to write an equation of a line given its slope and y-intercept.

Write an equation of the line with y-intercept -1 and slope $\frac{2}{3}$.

$$y = mx + b$$

$$y = \frac{2}{3}x - 1$$

(continued)

DEFINITIONS AND CONCEPTS	EXAMPLES

SECTION 3.6   EQUATIONS OF LINES

The **point slope form** of the equation of a line is $y - y_1 = m(x - x_1)$, where m is the slope of the line and (y_1, y_1) is a point on the line.

Find an equation of the line with slope 2 containing the point $(1, -4)$. Write the equation in standard form: $Ax + By = C$.

$$y - y_1 = m(x - x_1)$$
$$y - (-4) = 2(x - 1)$$
$$y + 4 = 2x - 2$$
$$-2x + y = -6 \qquad \text{Standard form.}$$

SECTION 3.7   SOLVING LINEAR EQUATIONS GRAPHICALLY

To solve an equation graphically by the **intersection-of-graphs method:**

- Graph the left side as y_1.
- Graph the right side as y_2.
- Find any points of intersection, or where $y_1 = y_2$.
- The x-coordinate of an intersection point is a solution.
- The y-coordinate of an intersection point is the value of each side of the original equation when the variable is replaced with the solution.

Solve $5(x - 2) + 1 = 2(x - 1) + 2$

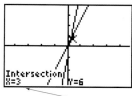

Solution $x = 3$

To solve an equation using the **x-intercept method:**

- Write the equation so that one side is 0.
- For the equation $y_1 = 0$, graph y_1.
- The x-intercepts of the graph are solutions of $y_1 = 0$.

Solve $5(x - 2) + 1 = 2(x - 1) + 2$
$$5(x - 2) + 1 - 2(x - 1) - 2 = 0$$

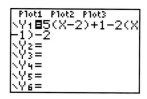
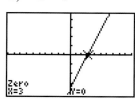

CHAPTER 3 REVIEW

(3.1)   *Plot the points and name the quadrant in which each point lies.*

1. $A(2, -1), B(-2, 1), C(0, 3), D(-3, -5)$

2. $A(-3, 4), B(4, -3), C(-2, 0), D(-4, 1)$

Find the distance between each pair of points. Approximate 7 and 8 to the nearest hundredth.

3. $(-6, 3)$ and $(8, 4)$ **4.** $(3, 5)$ and $(8, 9)$

5. $(-4, -6)$ and $(-1, 5)$ **6.** $(-1, 5)$ and $(2, -3)$

7. $(7.4, -8.6)$ and $(-1.2, 5.6)$

8. $(2.3, 1.8)$ and $(10.7, -9.2)$

Find the midpoint of the line segment whose endpoints are given.

9. $(2, 6)$ and $(-12, 4)$　　**10.** $(-3, 8)$ and $(11, 24)$

11. $(-6, -5)$ and $(-9, 7)$　　**12.** $(4, -6)$ and $(-15, 2)$

13. $\left(0, -\frac{3}{8}\right)$ and $\left(\frac{1}{10}, 0\right)$　　**14.** $\left(\frac{3}{4}, -\frac{1}{7}\right)$ and $\left(-\frac{1}{4}, -\frac{3}{7}\right)$

(3.2) *Determine whether each ordered pair is a solution to the given equation.*

15. $7x - 8y = 56; (0, 56), (8, 0)$

16. $-2x + 5y = 10; (-5, 0), (1, 1)$

17. $x = 13; (13, 5), (13, 13)$　**18.** $y = 2; (7, 2), (2, 7)$

Complete the ordered pairs so that each is a solution of the given equation.

19. $-2 + y = 6x; (7, \ )$　　**20.** $y = 3x + 5; (\ , -8)$

Determine whether each equation is linear or not.

21. $3x - y = 4$　　　　　　**22.** $x - 3y = 2$

23. $y = |x| + 4$　　　　　**24.** $y = x^2 + 4$

Graph each equation by plotting ordered pair solutions. In some exercises, suggested x-values have been given. Then check the graph using a graphing utility.

25. $3x - y = 4$　　　　　**26.** $x - 3y = 2$

27. $y = |x| + 4$　Let $x = -3, -2, -1, 0, 1, 2, 3$.

28. $y = x^2 + 4$　Let $x = -3, -2, -1, 0, 1, 2, 3$.

29. $y = -\frac{1}{2}x + 2$　　　　　**30.** $y = -x + 5$

Match each graph with its equation.

31.
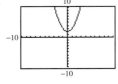

32.
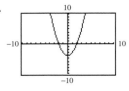

33.
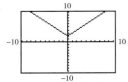

34.
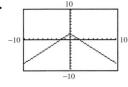

a. $y = x^2 - 4$　　　　**b.** $y = -|x| + 2$

c. $y = |x| + 2$　　　　**d.** $y = x^2 + 2$

(3.3) *Find the domain and range of each relation. Also determine whether the relation is a function.*

35. $\left\{\left(-\frac{1}{2}, \frac{3}{4}\right), (6, 0.75), (0, -12), (25, 25)\right\}$

36. $\left\{\left(\frac{3}{4}, -\frac{1}{2}\right), (0.75, 6), (-12, 0), (25, 25)\right\}$

37.

38.

39.

40.

41.

42.

If $f(x) = x - 5$, $g(x) = -3x$, and $h(x) = 2x^2 - 6x + 1$, find the following:

43. $f(2)$

44. $g(0)$

45. $g(-6)$

46. $h(-1)$

47. $h(1)$

48. $f(5)$

The function $J(x) = 2.54x$ may be used to calculate the weight J of an object on Jupiter given its weight x on Earth.

49. If a person weighs 150 pounds on Earth, find the equivalent weight on Jupiter.

50. A 2000-pound probe on Earth weighs how many pounds on Jupiter?

(3.4) *Graph each linear function.*

51. $f(x) = -\dfrac{1}{3}x$

52. $g(x) = 4x - 1$

The graph of $f(x) = 3x$ is sketched next. Use this graph to match each linear function with its graph.

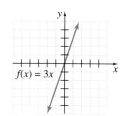

A.

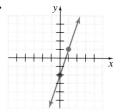

B.

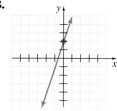

C.

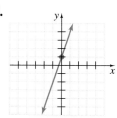

D.

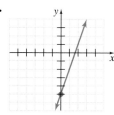

53. $f(x) = 3x + 1$

54. $f(x) = 3x - 2$

55. $f(x) = 3x + 2$

56. $f(x) = 3x - 5$

Graph each linear equation by finding intercepts if possible. Then use a graphing utility to check.

57. $4x + 5y = 20$

58. $3x - 2y = -9$

59. $4x - y = 3$

60. $2x + 6y = 9$

61. $y = 5$

62. $x = -2$

63. The cost C in dollars of renting a minivan for a day is given by the linear function $C(x) = 0.3x + 42$, where x is number of miles driven.

    a. Find the cost of renting the minivan for a day and driving it 150 miles.

    b. Graph $C(x) = 0.3x + 42$.

    c. How can you tell from the graph of $C(x)$ that as the number of miles driven increases, the total cost increases.

(3.5) *Find the slope of the line through each pair of points.*

64. $(2, 8)$ and $(6, -4)$

65. $(-3, 9)$ and $(5, 13)$

66. $(-7, -4)$ and $(-3, 6)$

67. $(7, -2)$ and $(-5, 7)$

Find the slope and y-intercept of each line.

68. $6x - 15y = 20$

69. $4x + 14y = 21$

Find the slope of each line.

70. $y - 3 = 0$

71. $x = -5$

Two lines are graphed on each set of axes. Decide whether l_1 or l_2 has the greater slope.

72.

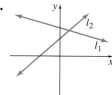

73.

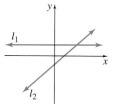

74.

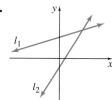

75.

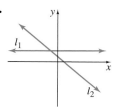

Decide whether the lines are parallel, perpendicular, or neither.

76. $f(x) = -2x + 6$

    $g(x) = 2x - 1$

77. $-x + 3y = 2$

    $6x - 18y = 3$

(3.6)   *Find an equation of the line satisfying the conditions given.*

78. 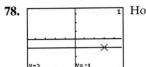 Horizontal, through $(3, -1)$

79. 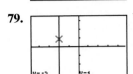 Vertical, through $(-2, 1)$

80. Parallel to the line $x = 6$; through $(-4, -3)$

81. Slope 0; through $(2, 5)$

Find the standard-form equation of each line satisfying the conditions given.

82. Through $(-3, 5)$; slope 3

83. Slope 2; through $(5, -2)$

84. Through $(-6, -1)$ and $(-4, -2)$

85. Through $(-5, 3)$ and $(-4, -8)$

86. Through $(-2, 3)$; perpendicular to $x = 4$

87. Through $(-2, -5)$; parallel to $y = 8$

Find the equation of each line satisfying the given conditions. Write each equation using function notation.

88. Slope $-\dfrac{2}{3}$; y-intercept 4

89. Slope -1; y-intercept -2

90. Through $(2, -6)$; parallel to $6x + 3y = 5$

91. Through $(-4, -2)$; parallel to $3x + 2y = 8$

92. Through $(-6, -1)$; perpendicular to $4x + 3y = 5$

93. Through $(-4, 5)$; perpendicular to $2x - 3y = 6$

(3.7)   *Solve each equation algebraically and graphically.*

94. $4(x - 6) + 3 = 27$ **95.** $15(x + 2) - 6 = 18$

96. $5x + 15 = 3(x + 2) + 2(x - 3)$

97. $2x - 5 + 3(x - 4) = 5(x + 2) - 27$

98. $14 - 2(x + 3) = 3(x - 9) + 18$

99. $16 + 2(5 - x) = 19 - 3(x + 2)$

Solve each equation graphically. Round solutions to the nearest hundredth.

100. $0.4(x - 6) = \pi x + \sqrt{3}$

101. $1.7x + \sqrt{7} = -0.4x - \sqrt{6}$

CHAPTER 3 TEST

1. Plot the points, and name the quadrant in which each is located: $A(6, -2)$, $B(4, 0)$, $C(-1, 6)$.

2. Complete the ordered pair solution $(-6, \ )$ of the equation $2y - 3x = 12$.

Graph each line. Then check using a graphing utility.

3. $2x - 3y = -6$ **4.** $4x + 6y = 7$

5. $f(x) = \dfrac{2}{3}x$ **6.** $y = -3$

7. Find the slope of the line that passes through $(5, -8)$ and $(-7, 10)$.

8. Find the slope and the y-intercept of the line $3x + 12y = 8$.

Graph each function by plotting ordered pair solutions. Suggested x-values have been given for ordered pair solutions. Then check using a graphing utility.

9. $f(x) = (x - 1)^2$

Let $x = -2, -1, 0, 1, 2, 3, 4$

10. $g(x) = |x| + 2$

Let $x = -3, -2, -1, 0, 1, 2, 3$

11. Find the distance between the points $(-6, 3)$ and $(-8, -7)$.

12. Find the midpoint of the line segment whose end points are $\left(-\dfrac{2}{3}, -\dfrac{1}{5}\right)$ and $\left(-\dfrac{1}{3}, \dfrac{4}{5}\right)$.

Find an equation of each line satisfying the conditions given. Write Exercises 13–17 in standard form. Write Exercises 18-20 using function notation.

13. Horizontal; through $(2, -8)$

14. Vertical; through $(-4, -3)$

15. Perpendicular to $x = 5$; through $(3, -2)$

16. Through $(4, -1)$; slope -3

17. Through $(0, -2)$; slope 5

18. Through $(4, -2)$ and $(6, -3)$

19. Through $(-1, 2)$; perpendicular to $3x - y = 4$

20. Parallel to $2y + x = 3$; through $(3, -2)$

21. Line L_1 has the equation $2x - 5y = 8$. Line L_2 passes through the points $(1, 4)$ and $(-1, -1)$. Determine whether these lines are parallel lines, perpendicular lines, or neither.

Match each graph with its equation.

22.
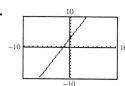

23.
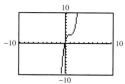

24.
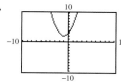

25.
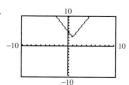

a. $y = 2|x - 1| + 3$ **b.** $y = x^2 + 2x + 3$

c. $y = 2(x - 1)^3 + 3$ **d.** $y = 2x + 3$

Find the domain and range of each relation. Also determine whether the relation is a function.

26.
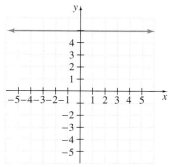

27.
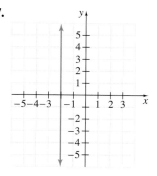

28.
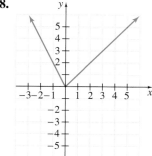

29.
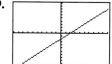

30. The average first year earnings for high school graduates age 18 and older are given by the linear function

$$f(x) = 708x + 13{,}570$$

where x is the number of years since 1985 that a person graduated.

a. Find the average earnings in 1990 for high school graduates.

b. Predict the average earnings for high school graduates in the year 2000.

c. Graph $f(x)$ and use the graph to predict the first year that the average earnings for high school graduates will be greater than $30,000.

31. Solve $15x + 26 = -2(x + 1) - 1$ algebraically and graphically.

32. Solve $-3x - \sqrt{5} = \pi(x - 1)$ graphically. Round the solution to the nearest hundredth.

CHAPTER 3   CUMULATIVE REVIEW

1. Simplify each expression.

   a. $11 + 2 - 7$ **b.** $-5 - 4 + 2$

2. Graph the set and write it in interval notation.

   a. $\{x \mid x \le 0\}$ **b.** $\{x \mid x > 2\}$

3. Evaluate the expression $3x^2 + 4$ for the following values of x:

   a. $x = 2$ **b.** $x = 12$ **c.** $x = 3.91$

4. Solve for x: $-6x - 1 + 5x = 3$.

5. Find two numbers such that the second number is 3 more than twice the first number and the sum of the two numbers is 72.

6. Solve $3y - 2x = 7$ for y.

7. Plot each ordered pair on a rectangular coordinate system and name the quadrant in which the point is located.

   a. $(2, -1)$ **b.** $(0, 5)$ **c.** $(-3, 5)$

   d. $(-2, 0)$ **e.** $\left(-\dfrac{1}{2}, -4\right)$ **f.** $(1.5, 1.5)$

8. Find the distance between $A(3, 10)$ and $B(-2, 4)$.

9. Graph the equation $y = |x|$.

10. Determine the domain and range of each relation.

   a. $\{(2, 3), (2, 4), (0, -1), (3, -1)\}$

   b.

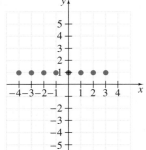

   c.

11. Which of the following graphs are graphs of functions?

   a.

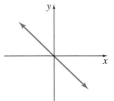

   b.

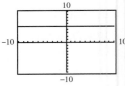

   c.

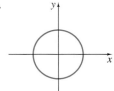

   d.

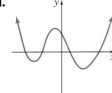

   e.

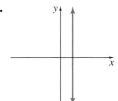

   f.

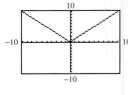

12. Graph $x - 3y = 6$ by plotting intercept points.

13. Graph $y = -3$.

14. Find the slope of the line containing the points $(5, -7)$ and $(-3, 6)$.

15. Given the equation $3x - 4y = 4$.

   a. Find the slope and the y-intercept of the line.

   b. Use the slope to determine whether the line slants upward or downward from left to right.

   c. Use a graphing utility to check the results.

16. Find an equation of the line with slope -3 containing the point $(1, -5)$. Write the equation in standard form: $Ax + By = C$.

17. Find the equation of the horizontal line containing the point $(2, 3)$.

18. Solve the equation $5(x - 2) + 15 = 20$ algebraically and graphically.

19. Solve the equation $3x - 8 = 5(x - 1) - 2(x + 6)$ algebraically and graphically.

20. Solve the equation $5.1x + 3.78 = x + 4.7$ graphically. Round the solution to two decimal places.

INEQUALITIES AND ABSOLUTE VALUE

ANALYZING MUNICIPAL BUDGETS

Nearly all cities, towns, and villages operate with an annual budget. Budget items might include expenses for fire and police protection as well as for street maintenance and parks. No matter how big or small the budget, city officials need to know if municipal spending is over or under budget.

IN THE CHAPTER GROUP ACTIVITY ON PAGE 239, YOU WILL HAVE THE OPPORTUNITY TO ANALYZE A MUNICIPAL BUDGET
AND MAKE BUDGETARY RECOMMENDATIONS.

4.1 | SOLVING LINEAR INEQUALITIES ALGEBRAICALLY AND GRAPHICALLY

O B J E C T I V E S

1. Define a linear inequality in one variable.
2. Use interval notation.
3. Solve linear inequalities using the addition property of inequality.
4. Solve linear inequalities using the multiplication property of inequality.
5. Solve problems that can be modeled by linear inequalities.

TAPE IAG 4.1

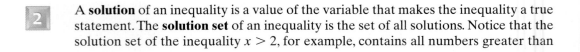

Relationships among measureable quantities are not always described by equations. For example, suppose a salesperson earns a monthly gross pay of $600 plus a commission of 20% of sales. Find the minimum amount of sales needed for the salesperson to receive a total income of *at least* $1500 per month. Here, the phrase "at least" implies that an income of $1500 *or more* is acceptable. In symbols, we can write

$$\text{income} \geq 1500$$

This is an example of an inequality, and we will solve this problem in Example 7.

A **linear inequality** is similar to a linear equation except that the equality symbol is replaced with an inequality symbol, such as $<$, $>$, $\leq$, or $\geq$.

LINEAR INEQUALITIES IN ONE VARIABLE

$$3x + 5 \geq 4 \qquad 2y < 0 \qquad 4n \geq n - 3 \qquad 3(x - 4) < 5x \qquad \frac{x}{3} \leq 5$$

LINEAR INEQUALITY IN ONE VARIABLE

A linear inequality in one variable is an inequality that can be written in the form

$$ax + b < c$$

where $a, b,$ and c are real numbers and $a \neq 0$.

In this section, when we make definitions, state properties, or list steps about an inequality containing the symbol $<$, we mean that the definition, property, or steps also apply to inequalities containing the symbols $>$, $\leq$, and $\geq$.

A **solution** of an inequality is a value of the variable that makes the inequality a true statement. The **solution set** of an inequality is the set of all solutions. Notice that the solution set of the inequality $x > 2$, for example, contains all numbers greater than

2. Its graph is an interval on the number line since an infinite number of values satisfy the inequality. The graph of $\{x\,|\,x > 2\}$ looks like the following.

2

In interval notation, this is $(2, \infty)$. The following table shows three equivalent ways to describe an interval: in set notation, as a graph, and in interval notation. For further review, see Section 1.2.

SET NOTATION	GRAPH	INTERVAL NOTATION
$\{x\,|\,x < a\}$	a	$(-\infty, a)$
$\{x\,|\,x > a\}$	a	(a, ∞)
$\{x\,|\,x \le a\}$	a	$(-\infty, a]$
$\{x\,|\,x \ge a\}$	a	$[a, \infty)$
$\{x\,|\,a < x < b\}$	$a \quad b$	(a, b)
$\{x\,|\,a \le x \le b\}$	$a \quad b$	$[a, b]$
$\{x\,|\,a < x \le b\}$	$a \quad b$	$(a, b]$
$\{x\,|\,a \le x < b\}$	$a \quad b$	$[a, b)$

> R E M I N D E R   Notice that a parenthesis is always used to enclose ∞ and $-\infty$.

EXAMPLE 1   Graph the solution set of each inequality on a number line and then write it in interval notation.

a. $\{x\,|\,x \ge 2\}$ **b.** $\{x\,|\,x < -1\}$ **c.** $\{x\,|\,0.5 < x \le 3\}$

Solution: **a.** $-2 \;-1 \;\; 0 \;\; 1 \;\; 2 \;\; 3 \;\; 4$ $[2, \infty]$

b. $-3 \;-2 \;-1 \;\; 0 \;\; 1 \;\; 2 \;\; 3$ $(-\infty, -1)$

c. $-1 \quad 0\;0.5\;1 \;\; 2 \;\; 3$ $(0.5, 3]$

3   Interval notation can be used to write solutions of linear inequalities. To solve a linear inequality, we use a process similar to the one used to solve a linear equation. We use properties of inequalities to write equivalent inequalities until the variable is isolated.

ADDITION PROPERTY OF INEQUALITY

If a, b, and c are real numbers, then

$$a < b \quad \text{and} \quad a + c < b + c$$

are equivalent inequalities.

In other words, we may add the same real number to both sides of an inequality and the resulting inequality will have the same solution set. This property also allows us to subtract the same real number from both sides.

EXAMPLE 2 Solve for x: $3x + 4 \geq 2x - 6$. Graph the solution set and write it in interval notation.

Solution:

$$3x + 4 \geq 2x - 6$$

$3x + 4 - 2x \geq 2x - 6 - 2x$ Subtract $2x$ from both sides.

$x + 4 \geq -6$ Combine like terms.

$x + 4 - 4 \geq -6 - 4$ Subtract 4 from both sides.

$x \geq -10$ Simplify.

The solution set is $\{x | x \geq -10\}$, which in interval notation is $[-10, \infty)$. The graph of the solution set is

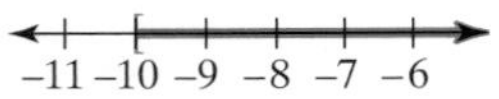

$[-10, \infty)$

From the example above, we find that *every* real number greater than or equal to -10 is a solution of the linear inequality $3x + 4 \geq 2x - 6$. For example $-10, 11, 38,$ 502.7, and 1,000,000 are a few examples of solutions. To see this, replace x in $3x + 4 \geq 2x - 6$ with each of the numbers and see that the result is a true inequality.

DISCOVER THE CONCEPT

Let's use a graphing utility to check the solution of $3x + 4 \geq 2x - 6$ from Example 2. Let $y_1 = 3x + 4$ and $y_2 = 2x - 6$ and we can now think of the inequality as $y_1 \geq y_2$.

a. Graph y_1 and y_2 using the window $[-15, 20, 5]$ by $[-40, 30, 10]$.

b. Find the point of intersection of the graphs. What does this point represent?

c. Determine where the graph of y_1 is above the graph of y_2.

d. Now find the x-values for which $y_1 \geq y_2$.

e. Write the solution set of the inequality $y_1 \geq y_2$.

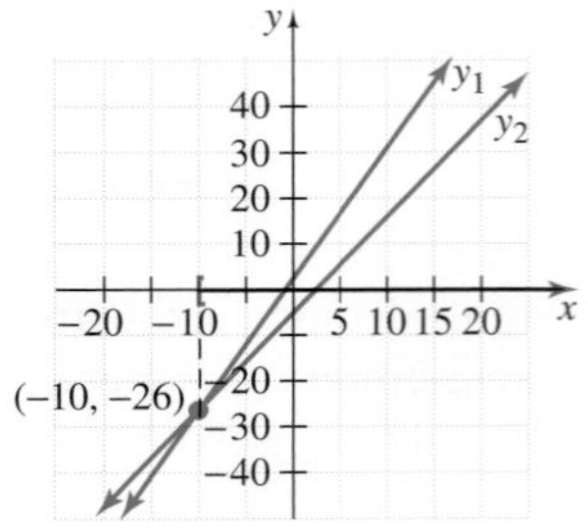

The graphs of y_1 and y_2 are shown to the left. From the above discovery, we find that

b. The point of intersection is $(-10, -26)$. This means that the solution of the equation $y_1 = y_2$ is -10.

c. The graph of y_1 is above the graph of y_2 to the right of the point of intersection, $(-10, -26)$.

d. The x-values for which $y_1 \geq y_2$ are -10 and those x-values to the right of -10 on the x-axis.

e. The solution set is $[-10, \infty)$.

Next, we introduce and use the multiplication property of inequality to solve linear inequalities. To understand this property, let's start with the true statement $-3 < 7$ and multiply both sides by 2.

$$-3 < 7$$
$$-3(2) < 7(2) \qquad \text{Multiply by 2.}$$
$$-6 < 14 \qquad \text{True.}$$

The statement remains true.

Notice what happens if both sides of $-3 < 7$ are multiplied by -2.

$$-3 < 7$$
$$-3(-2) < 7(-2) \qquad \text{Multiply by } -2.$$
$$6 < -14 \qquad \text{False.}$$

The inequality $6 < -14$ is a false statement. However, **if the direction of the inequality sign is reversed,** the result is

$$6 > -14, \text{a true statement.}$$

These examples suggest the following property.

MULTIPLICATION PROPERTY OF INEQUALITY

If $a, b,$ and c are real numbers and c is **positive,** then $a < b$ and $ac < bc$ are equivalent inequalities.

If $a, b,$ and c are real numbers and c is **negative,** then $a < b$ and $ac > bc$ are equivalent inequalities.

In other words, we may multiply both sides of an inequality by the same positive real number and the result is an equivalent inequality.

We may also multiply both sides of an inequality by the same **negative number** and **reverse the direction of the inequality symbol,** and the result is an equivalent inequality. The multiplication property holds for division also, since division is defined in terms of multiplication.

R E M I N D E R   Whenever both sides of an inequality are multiplied or divided by a negative number, the direction of the inequality symbol *must be* reversed to form an equivalent inequality.

EXAMPLE 3   Solve algebraically for x.

  a. $\dfrac{1}{4}x \leq \dfrac{3}{8}$ **b.** $-2.3x < 6.9$

Solution: **a.** $\dfrac{1}{4}x \leq \dfrac{3}{8}$

  $4 \cdot \dfrac{1}{4}x \leq 4 \cdot \dfrac{3}{8}$ Multiply both sides by 4.

  $x \leq \dfrac{3}{2}$ Simplify.

The solution set is $\left\{ x \mid x \leq \dfrac{3}{2} \right\}$, which in interval notation is $\left(-\infty, \dfrac{3}{2} \right]$. The graph of the solution set is

$\left(-\infty, \dfrac{3}{2} \right]$

  b. $-2.3x < 6.9$

  $\dfrac{-2.3x}{-2.3} > \dfrac{6.9}{-2.3}$ Divide both sides by -2.3 and reverse the inequality symbol.

  $x > -3$ Simplify.

The solution set is $\{x \mid x > -3\}$, which is $(-3, \infty)$ in interval notation. The graph of the solution set is

$(-3, \infty)$

To solve linear inequalities algebraically, we follow steps similar to those for solving linear equations.

TO SOLVE A LINEAR INEQUALITY IN ONE VARIABLE ALGEBRAICALLY

Step 1.   Clear the equation of fractions by multiplying both sides of the inequality by the least common denominator (LCD) of all fractions in the inequality.

Step 2.   Use the distributive property to remove grouping symbols such as parentheses.

Step 3.   Combine like terms on each side of the inequality.

Step 4.   Use the addition property of inequality to rewrite the inequality as an equivalent inequality with variable terms on one side and numbers on the other side.

Step 5.   Use the multiplication property of inequality to isolate the variable.

EXAMPLE 4   Solve algebraically for x: $-(x - 3) + 2 < 3(2x - 5) + x$. Then use a graphical approach to check the solution.

Solution:

$$-(x - 3) + 2 < 3(2x - 5) + x$$

$-x + 3 + 2 < 6x - 15 + x$	Apply the distributive property.
$5 - x < 7x - 15$	Combine like terms.
$5 - x + x < 7x - 15 + x$	Add x to both sides.
$5 < 8x - 15$	Combine like terms.
$5 + 15 < 8x - 15 + 15$	Add 15 to both sides.
$20 < 8x$	Combine like terms.
$\dfrac{20}{8} < \dfrac{8x}{8}$	Divide both sides by 8.
$\dfrac{5}{2} < x \quad \text{or} \quad x > \dfrac{5}{2}$	Simplify.

To check, graph $y_1 = -(x - 3) + 2$ and $y_2 = 3(2x - 5) + x$ and find x values such that $y_1 < y_2$.

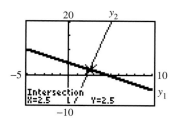

The point of intersection is $(2.5, 2.5)$. The graph of y_1 is below the graph of y_2 for all x-values greater than 2.5, or for the interval $\left(\dfrac{5}{2}, \infty\right)$.

EXAMPLE 5   Solve algebraically for x: $\dfrac{2}{5}(x - 6) \geq x - 1$.

Solution:

$$\dfrac{2}{5}(x - 6) \geq x - 1$$

$5\left[\dfrac{2}{5}(x - 6)\right] \geq 5(x - 1)$	Multiply both sides by 5 to eliminate fractions.
$2x - 12 \geq 5x - 5$	Apply the distributive property.
$-3x - 12 \geq -5$	Subtract $5x$ from both sides.
$-3x \geq 7$	Add 12 to both sides.

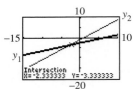

The intersection of y_1 and y_2

is the point $\left(\dfrac{-7}{3}, \dfrac{-10}{3}\right)$.

$y_1 \geq y_2$ for all x values in the

interval $\left(-\infty, \dfrac{-7}{3}\right]$ since this

is where y_1 is equal to or above y_2.

$$\frac{-3x}{-3} \leq \frac{7}{-3} \qquad \text{Divide both sides by } -3 \text{ and reverse}$$
$$\text{the inequality symbol.}$$

$$x \leq -\frac{7}{3} \qquad \text{Simplify.}$$

The solution written in interval notation is $\left(-\infty, -\dfrac{7}{3}\right]$. A graphing utility check is shown to the left.

Recall that when the graph of the left side and the right side of an equation do not intersect, the equation has no solution. Let's see what a similar situation means for inequalities.

DISCOVER THE CONCEPT

a. Use a graphing utility to solve $2(x + 3) > 2x + 1$. To do so, graph $y_1 = 2(x + 3)$ and $y_2 = 2x + 1$ and find the x-values for which the graph of y_1 is above the graph of y_2.

b. Next, solve $2(x + 3) < 2x + 1$. Notice that the graphs of y_1 and y_2 found in part (a) can be used. Find the x-values for which the graph of y_1 is below the graph of y_2.

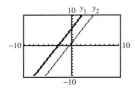

The graphs of parallel lines y_1 and y_2 from the discovery above are shown at the left. Since the graph of y_1 is *always* above the graph of y_2, all x-values satisfy the inequality $y_1 > y_2$ or $2(x + 3) > 2x + 1$. The solution set is all real numbers.

Also, notice that the graph of y_2 is *never* above the graph of y_1. This means that no x-values satisfy $y_1 < y_2$ or $2(x + 3) < 2x + 1$. The solution set is the empty set. These results can also be obtained algebraically as we see in Example 6.

EXAMPLE 6   Solve algebraically for x:

a. $2(x + 3) > 2x + 1$.   **b.** $2(x + 3) < 2x + 1$

Solution: **a.** $2(x + 3) > 2x + 1$

$2x + 6 > 2x + 1$ Distribute on the left side.

$2x + 6 - 2x > 2x + 1 - 2x$ Subtract $2x$ from both sides.

$6 > 1$ Simplify.

Since $6 > 1$ is a true statement for all values of x, this inequality and the original inequality are true for all numbers. The solution set is $\{x \mid x \text{ is a real number}\}$, or $(-\infty, \infty)$ in interval notation.

b. Solving $2(x + 3) < 2x + 1$ in a similar fashion leads us to the statement $6 < 1$. This statement, as well as the original inequality, is false for all values of x. This means that the solution set is the empty set, or $\{\ \ \}$.

Application problems containing words such as "at least," "at most," "between," "no more than," and "no less than" usually indicate that an inequality is to be solved instead of an equation. In solving applications involving linear inequalities, we use the same procedure as when we solved applications involving linear equations. We solve the inequality produced in Example 7 graphically in order to gain further insight about appropriate windows.

EXAMPLE 7   MONTHLY GROSS PAY

A salesperson earns a monthly gross pay of $600 plus 20% of sales. Find the amount of sales needed to receive a total income of at least $1500 per month.

Solution: **1.** UNDERSTAND. Read and reread the problem. Let's calculate a salesperson's gross pay for a few sales figures. If the salesperson has sales of $2000 in a month, for example, the gross pay is $600 plus 20% of $2000, or

$$\text{gross pay} = \$600 + 0.20(\$2000) = \$1000$$

Also, for sales of $5000,

$$\text{gross pay} = \$600 + 0.20(\$5000) = \$1600$$

2. ASSIGN. Since the unknown is the amount of sales,

let x = amount of sales

3. ILLUSTRATE. No illustration needed.

4. TRANSLATE. As stated in the beginning of this section, we want the income to be greater than or equal to $1500. To write an inequality, notice that the salesperson's income consists of a base plus a commission (20% of sales).

In words:	base	+	commission (20% of sales)	$\geq$	1500
Translate:	600	+	0.20x	$\geq$	1500

5. COMPLETE. Solve the inequality for x.

First, let's decide on a window.

In part 1. UNDERSTAND, we discovered that for sales of $2000, gross pay is $1000, and for sales of $5000, gross pay is $1600. This means that a sales amount between $2000 and $5,000 will give a gross pay of $1500. Although many windows are acceptable, for this example, we graph $y_1 = 600 + 0.20x$ and $y_2 = 1500$ on a $[2000, 5000, 1000]$ by $[1000, 2000, 100]$ window.

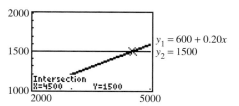

From the graph we see that $y_1 \geq y_2$ for x-values greater than or equal to 4500, or for x-values in the interval $[4500, \infty)$.

6. INTERPRET. *Check:* The income for sales of $4500 is

$$600 + 0.20(4500), \text{ or } 1500$$

Thus, if sales are greater than or equal to $4500, income is greater than or equal to $1500.

State: The amount of sales needed for the salesperson to earn at least $1500 per month is at least $4500.

EXERCISE SET 4.1

Graph the solution set of each inequality on a number line, and write the solution set in interval notation. See Example 1.

1. $\{x \mid x < -3\}$
2. $\{x \mid x \geq -7\}$

3. $\{x \mid x \geq 0.3\}$
4. $\{x \mid x < -0.2\}$

5. $\{x \mid 5 < x\}$
6. $\{x \mid -7 \geq x\}$

7. $\{x \mid -2 < x < 5\}$
8. $\{x \mid -5 \leq x \leq -1\}$

9. $\{x \mid 5 > x > -1\}$
10. $\{x \mid -3 \geq x \geq -7\}$

11. When graphing the solution set of an inequality, explain how you know whether to use a parenthesis or a bracket.

12. Explain what is wrong with the interval notation $(-6, -\infty)$.

Use the given screen to solve each inequality. Write the solution in interval notation.

13. $y_1 < y_2$

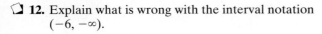

14. $y_1 \geq 0$

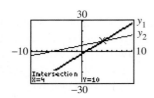

15. $y_1 \geq y_2$

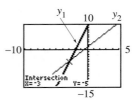

16. $y_1 > y_2$

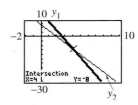

Use the graph of parallel lines y_1 and y_2 to solve each inequality. Write the solution in interval notation.

17. $y_1 > y_2$
18. $y_1 < y_2$

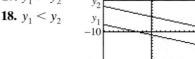

Solve each inequality for the variable. Write the solution set in interval notation and graph the solution set on a number line. See Examples 2 and 3.

19. $5x + 3 > 2 + 4x$
20. $7x - 1 \geq 6x - 1$

21. $8x - 7 \leq 7x - 5$
22. $12x + 14 < 11x - 2$

23. $5x < -23.5$
24. $-4x > -11.2$

25. $-3x \geq \dfrac{1}{2}$
26. $7x \leq -\dfrac{7}{3}$

Solve each inequality for the variable. Write the solution set in interval notation and graph the solution set on a number line. See Examples 4 through 6.

27. $15 + 2x \geq 4x - 7$ **28.** $20 + x < 6x$

29. $\dfrac{3x}{4} \geq 2$ **30.** $\dfrac{5}{6}x \geq -8$

31. $3(x - 5) < 2(2x - 1)$ **32.** $5(x + 4) \leq 4(2x + 3)$

33. $\dfrac{1}{2} + \dfrac{2}{3} \geq \dfrac{x}{6}$ **34.** $\dfrac{3}{4} - \dfrac{2}{3} > \dfrac{x}{6}$

35. $4(x - 1) \geq 4x - 8$ **36.** $3x + 1 < 3(x - 2)$

37. $7x < 7(x - 2)$ **38.** $8(x + 3) \leq 7(x + 5) + x$

39. Explain how solving a linear inequality is similar to solving a linear equation.

40. Explain how solving a linear inequality is different from solving a linear equation.

Solve each inequality. Write the solution set in interval notation.

41. $7.3x > 43.8$ **42.** $9.1x < 45.5$

43. $-4x \leq \dfrac{2}{5}$ **44.** $-6x \geq \dfrac{3}{4}$

45. $-2x + 7 \geq 9$ **46.** $8 - 5x \leq 23$

47. $4(2x + 1) > 4$ **48.** $6(2 - x) \geq 12$

49. $\dfrac{x + 7}{5} > 1$ **50.** $\dfrac{2x - 4}{3} \leq 2$

51. $\dfrac{-5x + 11}{2} \leq 7$ **52.** $\dfrac{4x - 8}{7} < 0$

53. $8x - 16.4 \leq 10x + 2.8$ **54.** $18x - 25.6 < 10x + 60.8$

55. $2(x - 3) > 70$ **56.** $3(5x + 6) \geq -12$

57. $-5x + 4 \leq -4(x - 1)$ **58.** $-6x + 2 < -3(x + 4)$

59. $\dfrac{1}{4}(x - 7) \geq x + 2$ **60.** $\dfrac{3}{5}(x + 1) \leq x + 1$

61. $\dfrac{2}{3}(x + 2) < \dfrac{1}{5}(2x + 7)$ **62.** $\dfrac{1}{6}(3x + 10) > \dfrac{5}{12}(x - 1)$

63. $4(x - 6) + 2x - 4 \geq 3(x - 7) + 10x$

64. $7(2x + 3) + 4x \leq 7 + 5(3x - 4)$

65. $\dfrac{5x + 1}{7} - \dfrac{2x - 6}{4} \geq -4$

66. $\dfrac{1 - 2x}{3} + \dfrac{3x + 7}{7} > 1$

Solve. See Example 7.

67. Shureka has scores of $72, 67, 82,$ and 79 on her algebra tests. Use an inequality to find the possible scores she can make on the final exam to pass the course with an average of 60 or higher, given that the final exam counts as two tests.

68. In a Winter Olympics speed-skating event, Hans scored times of 3.52, 4.04, and 3.87 minutes on his first three trials. Use an inequality to find the possible times he can score on his last trial so that his average time is under 4.0 minutes.

69. A small plane's maximum takeoff weight is 2000 pounds. Six passengers weigh an average of 160 pounds each. Use an inequality to find the possible weights of luggage and cargo the plane can carry.

70. A clerk must use the elevator to move boxes of paper. The elevator's weight limit is 1500 pounds. If each box of paper weighs 66 pounds and the clerk weighs 147 pounds, use an inequality to find the possible numbers of boxes she can move on the elevator at one time.

71. To mail an envelope first class, the U.S. Postal Service charges 32 cents for the first ounce and 23 cents per ounce for each additional ounce. Use an inequality to find the possible weights that can be mailed for $4.00.

72. A shopping mall parking garage charges $1 for the first half hour and 60 cents for each additional half hour or a portion of a half hour. Use an inequality to find how long you can park if you have only $4.00 in cash.

73. Northeast Telephone Company offers two billing plans for local calls. Plan 1 charges $25 per month for unlimited calls, and plan 2 charges $13 per month plus 6 cents per call. Use an inequality to find the number of monthly calls for which plan 1 is more economical than plan 2.

74. A car rental company offers two subcompact rental plans. Plan A charges $32 per day for unlimited mileage, and plan B charges $24 per day plus 15 cents per mile. Use an inequality to find the number of daily miles for which plan A is more economical than plan B.

75. At room temperature, glass used in windows actually has some properties of a liquid. It has a very slow, viscous flow. (Viscosity is the property of a fluid that resists internal flow. For example, lemonade flows more easily than fudge syrup. Fudge syrup has a higher viscosity than lemonade.)

Glass does not become a true liquid until temperatures are greater than or equal to 500°C. Find the Fahrenheit temperatures for which glass is a liquid. [Use the formula $F = \frac{9}{5}C + 32$).]

76. Stibnite is a silvery white mineral with a metallic luster. It is one of the few minerals that melts easily in the match flame or at temperatures of approximately 977°F or greater. Find the Celsius temperatures for which stibnite melts. [Use the formula $C = \frac{5}{9}(F - 32)$.]

77. Although beginning salaries vary greatly according to your field of study, the equation $s = 651.2t + 27,821$ can be used to approximate and to predict average beginning salaries for candidates for bachelor's degrees. The variable s is the starting salary and t is the number of years after 1989.

a. Approximate when beginning salaries for candidates will be greater than 30,000.

b. Determine the year you plan to graduate from college. Use this year to find the corresponding value of t and approximate your beginning salary.

The average consumption per year per person of whole milk w can be approximated by the equation

$$w = -3.26t + 87.79$$

where t is the number of years after 1990. The average consumption of skim milk s per person per year can be approximated by the equation

$$s = 1.25t + 22.75$$

where t is the number of years after 1990. The consumption of whole milk is shown on the graph in blue and the consumption of skim milk is shown on the graph in red. Use this information for Exercises 78–86.

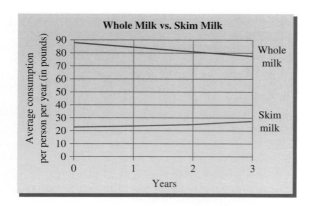

78. From the graph shown, is the consumption of whole milk increasing or decreasing over time? Explain how you arrived at your answer.

79. From the graph shown, is the consumption of skim milk increasing or decreasing over time? Explain how you arrived at your answer.

80. Predict the consumption of whole milk in the year 2000. (*Hint:* First use the year 2000 and find the value of t.)

81. Predict the consumption of skim milk in the year 2000. (*Hint:* First use the year 2000 and find the value of t.)

82. Determine when the consumption of whole milk will be less than 50 pounds per person per year.

83. Determine when the consumption of skim milk will be greater than 50 pounds per person per year.

84. For 1990 to 1993, the consumption of whole milk was greater than the consumption of skim milk. Explain how this can be determined from the graph shown.

85. How will the two lines in the graph appear when the consumption of whole milk is the same as the consumption of skim milk?

86. The consumption of whole milk will be the same as the consumption of skim milk when $w = s$. Find when this will occur, by substituting the given equivalent expression for w and the given equivalent expression for s and solving for t. Round the value of t to the nearest whole and estimate the year when this will occur.

Review Exercises

List or describe the integers that make both inequalities true.

87. $x < 5$ and $x > 1$ **88.** $x \geq 0$ and $x \leq 7$
89. $x \geq -2$ and $x \geq 2$ **90.** $x < 6$ and $x < -5$

Graph each set on a number line and write it in interval notation. See Section 1.2.

91. $\{x \mid 0 \leq x \leq 5\}$ **92.** $\{x \mid -7 < x \leq 1\}$

93. $\left\{x \mid -\frac{1}{2} < x < \frac{3}{2}\right\}$ **94.** $\{x \mid -2.5 \leq x < 5.3\}$

4.2 | COMPOUND INEQUALITIES

O B J E C T I V E

1 Solve compound inequalities.

1 Two inequalities joined by the words **and** or **or** are called **compound inequalities**.

COMPOUND INEQUALITIES

$$x + 3 < 8 \quad \text{and} \quad x > 2$$

$$\frac{2x}{3} \geq 5 \quad \text{or} \quad -x + 10 < 7$$

The solution set of a compound inequality formed by the word **and** is the intersection of the solution sets of the two inequalities. The **intersection** of two sets, denoted by $\cap$, is the set of elements common to both sets. For example, given the sets

$$A = \{2, 4, 6, 8\}$$

$$B = \{3, 4, 5, 6\}, \text{then}$$

$$A \cap B = \{4, 6\}.$$

This is true for compound inequalities formed by the word *and*. A value of x is a solution of a compound inequality formed by the word *and* if it is a solution of *both* inequalities.

For example, the solution set of the compound inequality $x \leq 5$ and $x \geq 3$ contains all values of x that make the inequality $x \leq 5$ a true statement *and* the inequality $x \geq 3$ a true statement. The first graph shown next is the graph of $x \leq 5$, the second graph is the graph of $x \geq 3$, and the third graph shows the intersection of the two graphs. It is the graph of $x \leq 5$ *and* $x \geq 3$.

$\{x | x \leq 5\}$ $(\infty, 5]$

$\{x | x \geq 3\}$ $[3, \infty)$

$\{x | x \leq 5 \text{ and } x \geq 3\}$ $[3, 5]$

In interval notation, the set $\{x | x \leq 5 \text{ and } x \geq 3\}$ is written as $[3, 5]$.

EXAMPLE 1   Solve for x: $x - 7 < 2$ and $2x + 1 < 9$.

Solution: First solve each inequality separately.

$$x - 7 < 2 \quad \text{and} \quad 2x + 1 < 9$$

$$x < 9 \quad \text{and} \quad 2x < 8$$

$$x < 9 \quad \text{and} \quad x < 4$$

Graph the two intervals on two number lines and find their intersection.

$\{x \mid x < 9\}$ $(-\infty, 9)$

$\{x \mid x < 4\}$ $(-\infty, 4)$

                                        $(-\infty, 4)$

$\{x \mid x < 9 \text{ and } x < 4\}$
$= \{x \mid x < 4\}$

The solution set written as an interval is $(-\infty, 4)$.

Compound inequalities containing the word *and* can be written in a more compact form. The compound inequality $2 \le x$ and $x \le 6$ can be written as

$$2 \le x \le 6$$

Recall from Sections 1.2 and 4.1 that the graph of $2 \le x \le 6$ is all numbers between 2 and 6, including 2 and 6.

The set $\{x \mid 2 \le x \le 6\}$ written in interval notation is $[2, 6]$.

To solve a compound inequality like $2 < 4 - x < 7$, we isolate x in the "middle side." Since a compound inequality is really two inequalities in one statement, we must perform the same operation to all three "sides" of the inequality.

EXAMPLE 2   Solve $2 < 4 - x < 7$.

Solution: To isolate x, first subtract 4 from all three sides.

$$2 < 4 - x < 7$$

$2 - 4 < 4 - x - 4 < 7 - 4$ Subtract 4 from all three sides.

$-2 < -x < 3$ Simplify.

$\dfrac{-2}{-1} > \dfrac{-x}{-1} > \dfrac{3}{-1}$ Divide all three sides by -1 and reverse the inequality symbols.

$2 > x > -3$

This is equivalent to $-3 < x < 2$, and its graph is shown. The solution set in interval notation is $(-3, 2)$.

DISCOVER THE CONCEPT

For the compound inequality $2 < 4 - x < 7$, in Example 2, graph $y_1 = 2$, $y_2 = 4 - x$, and $y_3 = 7$. With this notation, we can think of our inequality as $y_1 < y_2 < y_3$.

a. Find the point of intersection of y_1 and y_2.

b. Find the point of intersection of y_2 and y_3.

c. Determine where the graph of y_2 is between the graphs of y_1 and y_3.

d. Find the x-values for which $y_1 < y_2 < y_3$.

e. Write the solution set of $y_1 < y_2 < y_3$.

In the discovery above, we find the points of intersection as shown.

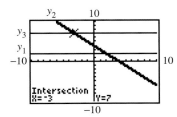
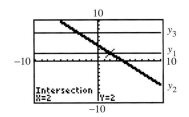

The solution of $y_1 < y_2 < y_3$, or $2 < 4 - x < 7$ contains the x-values where the graph of $y_2 = 4 - x$ is between $y_1 = 2$ and $y_3 = 7$. These x-values in interval notation are $(-3, 2)$. (Parentheses are used because of the inequality symbols $<$.)

EXAMPLE 3   Solve algebraically for x: $-1 \le \dfrac{2x}{3} + 5 \le 2$. Then use a graphical approach to check the solution.

Solution: First, clear the inequality of fractions by multiplying all three sides by the LCD of 3.

$$-1 \le \frac{2x}{3} + 5 \le 2$$

$$3(-1) \le 3\left(\frac{2x}{3} + 5\right) \le 3(2) \qquad \text{Multiply by the LCD of 3.}$$

$$-3 \le 2x + 15 \le 6 \qquad \text{Apply the distributive property and multiply.}$$

$$-3 - 15 \le 2x + 15 - 15 \le 6 - 15 \qquad \text{Subtract 15 from all three sides.}$$

$$-18 \le 2x \le -9 \qquad \text{Simplify.}$$

$$\frac{-18}{2} \le \frac{2x}{2} \le \frac{-9}{2} \qquad \text{Divide all three sides by 2.}$$

$$-9 \le x \le -\frac{9}{2} \qquad \text{Simplify.}$$

To check, graph $y_1 = -1$, $y_2 = \dfrac{2x}{3} + 5$, and $y_3 = 2$ in a $[-15, 5, 3]$ by $[-5, 5, 1]$ window and solve $y_1 \leq y_2 \leq y_3$.

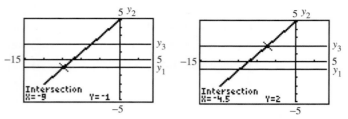

The solution of $y_1 \leq y_2 \leq y_3$ consists of the x-values where the graph of y_2 is between or equal to the graphs of y_1 and y_3. The solution set in interval notation is $[-9, -4.5]$ or $\left[-9, -\dfrac{9}{2}\right]$.

The solution set of a compound inequality formed by the word **or** is the union of the solution sets of the two inequalities. The **union** of two sets, denoted by $\cup$, is the set of elements that belong to either of the sets.

For example, given the sets

$$A = \{2, 4, 6, 8\}$$

$$B = \{3, 4, 5, 6\}, \text{ then}$$

$$A \cup B = \{2, 3, 4, 5, 6, 8\}$$

This is true for compound inequalities formed by the word *or.* A value of x is a solution of a compound inequality formed by the word *or* if it is a solution of *either* inequality.

For example, the solution set of the compound inequality $x \leq 1$ or $x \geq 3$ contains all numbers that make the inequality $x \leq 1$ a true statement *or* the inequality $x \geq 3$ a true statement.

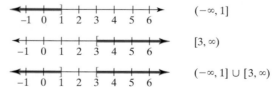

In interval notation, the set $\{x \mid x \leq 1 \text{ or } x \geq 3\}$ is written as $(-\infty, 1] \cup [3, \infty)$.

EXAMPLE 4   Solve $5x - 3 \leq 10$ or $x + 1 \geq 5$.

Solution: Solve each inequality separately.

$$5x - 3 \leq 10 \quad \text{or} \quad x + 1 \geq 5$$

$$5x \leq 13 \quad \text{or} \quad x \geq 4$$

$$x \leq \frac{13}{5} \quad \text{or} \quad x \geq 4$$

Graph each interval on a number line. Then find their union.

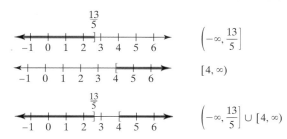

The solution set in interval notation is $\left(-\infty, \dfrac{13}{5}\right] \cup [4, \infty)$.

EXERCISE SET 4.2

If $A = \{x \mid x$ is an even integer$\}$, $B = \{x \mid x$ is an odd integer$\}$, $C = \{2, 3, 4, 5\}$, and $D = \{4, 5, 6, 7\}$, list the elements of each set.

1. $C \cup D$ **2.** $C \cap D$ **3.** $A \cap D$

4. $A \cup D$ **5.** $A \cup B$ **6.** $A \cap B$

7. $B \cap D$ **8.** $B \cup D$

Solve each compound inequality. Graph the solution set. See Example 1.

9. $x < 5$ and $x > -2$

10. $x \leq 7$ and $x \leq 1$

11. $x + 1 \geq 7$ and $3x - 1 \geq 5$

12. $-2x < -8$ and $x - 5 < 5$

13. $4x + 2 \leq -10$ and $2x \leq 0$

14. $x + 4 > 0$ and $4x > 0$

Solve each compound inequality algebraically and check graphically. See Examples 2 and 3.

15. $5 < x - 6 < 11$ **16.** $-2 \leq x + 3 \leq 0$

17. $-2 \leq 3x - 5 \leq 7$ **18.** $1 < 4 + 2x < 7$

19. $1 \leq \dfrac{2}{3}x + 3 \leq 4$ **20.** $-2 < \dfrac{1}{2}x - 5 < 1$

21. $-5 \leq \dfrac{x + 1}{4} \leq -2$ **22.** $-4 \leq \dfrac{2x + 5}{3} \leq 1$

Solve each compound inequality. Graph the solution set. See Example 4.

23. $x < -1$ or $x > 0$ **24.** $x \leq 1$ or $x \leq -3$

25. $-2x \leq -4$ or $5x - 20 \geq 5$

26. $x + 4 < 0$ or $6x > -12$

27. $3(x - 1) < 12$ or $x + 7 > 10$

28. $5(x - 1) \geq -5$ or $5 - x \leq 11$

29. Explain how solving an *and*-compound inequality is similar to finding the intersection of two sets.

30. Explain how solving an *or*-compound inequality is similar to finding the union of two sets.

Solve each compound inequality. Graph the solution set.

31. $x < 2$ and $x > -1$ **32.** $x < 5$ and $x < 1$

33. $x < 2$ or $x > -1$ **34.** $x < 5$ or $x < 1$

35. $x \geq -5$ and $x \geq -1$ **36.** $x \leq 0$ or $x \geq -3$

37. $x \geq -5$ or $x \geq -1$ **38.** $x \leq 0$ and $x \geq -3$

39. $0 \leq 2x - 3 \leq 9$ **40.** $3 < 5x + 1 < 11$

41. $\dfrac{1}{2} < x - \dfrac{3}{4} < 2$ **42.** $\dfrac{2}{3} < x + \dfrac{1}{2} < 4$

43. $x + 3 \geq 3$ and $x + 3 \leq 2$ **44.** $2x - 1 \geq 3$ and $-x > 2$

45. $3x \geq 5$ or $-x - 6 < 1$ **46.** $\dfrac{3}{8}x + 1 \leq 0$ or $-2x < -4$

47. $-x + 5 > 6$ and $1 + 2x \leq -5$

48. $5x \leq 0$ and $-x + 5 < 8$

49. $3x + 2 \leq 5$ or $7x > 29$

50. $-x < 7$ or $3x + 1 < -20$

51. $5 - x > 7$ and $2x + 3 \geq 13$

52. $-2x < -6$ or $1 - x > -2$

Solve each compound inequality using the graphing utility screens.

53. a. $y_1 < y_2 < y_3$ **b.** $y_2 < y_1$ or $y_2 > y_3$

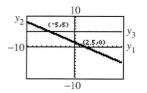

54. a. $y_1 < y_2 < y_3$ **b.** $y_2 < y_1$ or $y_2 > y_3$

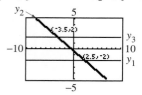

55. a. $y_1 \leq y_2 \leq y_3$ **b.** $y_2 \leq y_1$ or $y_2 \geq y_3$

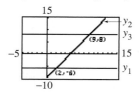

56. a. $y_1 \leq y_2 \leq y_3$ **b.** $y_2 \leq y_1$ or $y_2 \geq y_3$

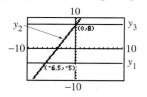

Solve each compound inequality algebraically and then check graphically.

57. $-\dfrac{1}{2} \leq \dfrac{4x - 1}{6} < \dfrac{5}{6}$ **58.** $-\dfrac{1}{2} \leq \dfrac{3x - 1}{10} < \dfrac{1}{2}$

59. $\dfrac{1}{15} < \dfrac{8 - 3x}{15} < \dfrac{4}{5}$ **60.** $-\dfrac{1}{4} < \dfrac{6 - x}{12} < -\dfrac{1}{6}$

61. $0.3 < 0.2x - 0.9 < 1.5$ **62.** $-0.7 \leq 0.4x + 0.8 < 0.5$

The formula for converting Fahrenheit temperatures to Celsius temperatures is $C = \frac{5}{9}(F - 32)$. Use this formula for Exercises 63 and 64.

63. During a recent year, the temperatures in Chicago ranged from $-29°C$ to $35°C$. Use a compound inequality to convert these temperatures to Fahrenheit temperatures.

64. In Oslo, the temperature ranges from $-10°$ to $18°$ Celsius. Use a compound inequality to convert these temperatures to the Fahrenheit scale.

Solve.

65. Christian D'Angelo has scores of 68, 65, 75, and 78 on his algebra tests. Use a compound inequality to find the scores he can make on his final exam to receive a C in the course. The final exam counts as two tests, and a C is received if the final course average is from 70 to 79.

66. Wendy Wood has scores of 80, 90, 82, and 75 on her chemistry tests. Use a compound inequality to find the range of scores she can make on her final exam to receive a B in the course. The final exam counts as two tests, and a B is received if the final course average is from 80 to 89.

Use the graph shown to answer Exercises 67 and 68.

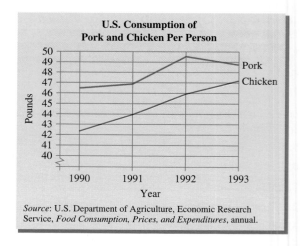

Source: U.S. Department of Agriculture, Economic Research Service, *Food Consumption, Prices, and Expenditures*, annual.

67. For what years was the consumption of pork greater than 45 pounds per person *and* the consumption of chicken greater than 45 pounds per person?

68. For what years was the consumption of pork less than 47 pounds per person *or* the consumption of chicken greater than 47 pounds per person?

Review Exercises

Evaluate the following. See Section 1.4.

69. $|-7| - |19|$ **70.** $|-7 - 19|$

71. $-(-6) - |-10|$ **72.** $|-4| - (-4) + |-20|$

Find by inspection all values for x that make each equation true.

73. $|x| = 7$

74. $|x| = 5$

75. $|x| = 0$

76. $|x| = -2$

A Look Ahead

EXAMPLE

Solve: $x - 6 < 3x < 2x + 5$

Solution

Notice that this inequality contains a variable on the left, on the right, and in the middle. To solve, rewrite the inequality using the word **and**.

$$x - 6 < 3x \quad \text{and} \quad 3x < 2x + 5$$
$$-6 < 2x \quad \text{and} \quad x < 5$$
$$-3 < x$$
$$x > -3 \quad \text{and} \quad x < 5$$

$x > -3$

$x < 5$

$-3 < x < 5,$

or $(-3, 5)$

Solve each compound inequality for x. See the previous example.

77. $2x - 3 < 3x + 1 < 4x - 5$

78. $x + 3 < 2x + 1 < 4x + 6$

79. $-3(x - 2) \le 3 - 2x \le 10 - 3x$

80. $7x - 1 \le 7 + 5x \le 3(1 + 2x)$

81. $5x - 8 < 2(2 + x) < -2(1 + 2x)$

82. $1 + 2x < 3(2 + x) < 1 + 4x$

4.3 | **ABSOLUTE VALUE EQUATIONS**

O B J E C T I V E

1 Solve absolute value equations.

TAPE IAG 4.3

1 In Chapter 1, we defined the absolute value of a number as its distance from 0 on a number line.

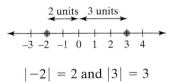

$$|-2| = 2 \text{ and } |3| = 3$$

In this section, we concentrate on solving equations containing the absolute value of a variable or a variable expression. Examples of absolute value equations are

$$|x| = 3 \qquad -5 = |2y + 7| \qquad |z - 6.7| = |3z + 1.2|$$

Absolute value equations and inequalities (see Section 4.4) are extremely useful in data analysis, especially for calculating acceptable measurement error, and errors that result from the way numbers are sometimes represented in computers.

Let us now consider the absolute value equation $|x| = 3$. The solution set of this equation will contain all numbers whose distance from 0 is 3 units. Two numbers are 3 units away from 0 on the number line: 3 and -3.

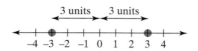

Thus, the solution set of the equation $|x| = 3$ is $\{3, -3\}$. This suggests the following absolute value property:

SOLVING EQUATIONS OF THE FORM $|x| = a$

If a is a positive number, then $|x| = a$ is equivalent to $x = a$ or $x = -a$.

EXAMPLE 1 Solve $|x| = 2$.

Solution Since 2 is positive, $|x| = 2$ is equivalent to $x = 2$ or $x = -2$.

To check, let $x = 2$ and then $x = -2$ in the original equation.

$|x| = 2$	Original equation.	$|x| = 2$	Original equation.
$|2| = 2$	Let $x = 2$.	$|-2| = 2$	Let $x = -2$.
$2 = 2$	True.	$2 = 2$	True.

The solution set is $\{2, -2\}$.

To visualize the solution of Example 1, we solve $|x| = 2$ by the intersection-of-graphs method. Graph $y_1 = |x|$ and $y_2 = 2$.

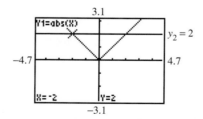 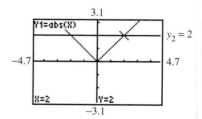

The graphs of $y_1 = |x|$ and $y_2 = 2$ intersect at $(-2, 2)$ and $(2, 2)$. The solutions are thus, the x-values, -2 and 2.

If the expression inside the absolute value bars is more complicated than a single variable x, we can still apply the absolute value property.

EXAMPLE 2 Solve $|5x + 3| = 7$ algebraically. Then use a graphical approach to check.

Solution: For the equation $|5x + 3| = 7$ to be true, the expression $|5x + 3|$ must equal 7 or -7.

$$5x + 3 = 7 \quad \text{or} \quad 5x + 3 = -7$$

Solve these two equations for x.

$$5x + 3 = 7 \quad \text{or} \quad 5x + 3 = -7$$
$$5x = 4 \quad \text{or} \quad 5x = -10$$
$$x = \frac{4}{5} \quad \text{or} \quad x = -2$$

To check, graph $y_1 = |5x + 3|$ and $y_2 = 7$ in a standard window as shown.

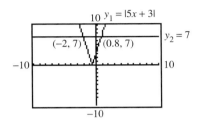

The intersections of the two graphs are the points $(-2, 7)$ and $(0.8, 7)$. Therefore, the solutions to the equation $|5x + 3| = 7$ are $x = 0.8$, or $\frac{4}{5}$ and $x = -2$.

Both solutions check, and the solution set is $\left\{-2, \frac{4}{5}\right\}$.

EXAMPLE 3   Solve $\left|\dfrac{x}{2} - 1\right| = 11$.

Solution: $\left|\dfrac{x}{2} - 1\right| = 11$ is equivalent to

$$\frac{x}{2} - 1 = 11 \qquad \text{or} \qquad \frac{x}{2} - 1 = -11$$

$$2\left(\frac{x}{2} - 1\right) = 2(11) \quad \text{or} \quad 2\left(\frac{x}{2} - 1\right) = 2(-11) \qquad \text{Clear fractions.}$$

$$x - 2 = 22 \qquad \text{or} \qquad x - 2 = -22 \qquad \text{Apply the distributive}$$
$$x = 24 \qquad \text{or} \qquad x = -20 \qquad \qquad \text{property.}$$

Try using mental math to check these solutions. Mentally replace x with 24 in the original equation and see that a true statement results. Verify the solution -20 mentally, also.

The solution set is $\{24, -20\}$.

To apply the absolute value property, first make sure that the absolute value expression is isolated.

R E M I N D E R   If the equation has a single absolute value expression containing variables, isolate the absolute value expression first.

EXAMPLE 4   Solve $|2x| + 5 = 7$.

Solution: We want the absolute value expression alone on one side of the equation, so begin by subtracting 5 from both sides. Then apply the absolute value property.

$$|2x| + 5 = 7$$

$$|2x| = 2 \qquad \text{Subtract 5 from both sides.}$$

$$2x = 2 \quad \text{or} \quad 2x = -2$$

$$x = 1 \quad \text{or} \quad x = -1$$

The solution set is $\{-1, 1\}$.

EXAMPLE 5   Solve $|y| = 0$.

Solution: We are looking for all numbers whose distance from 0 is zero units. The only number is 0. The solution set is $\{0\}$

The next example illustrates a special case for absolute value equations. This special case occurs when an isolated absolute value is equal to a negative number.

EXAMPLE 6   Solve $2|x| + 25 = 23$ algebraically and graphically.

Algebraic Solution:

First, isolate the absolute value.

$$2|x| + 25 = 23$$

$$2|x| = -2 \qquad \text{Subtract 25 from both sides.}$$

$$|x| = -1 \qquad \text{Divide both sides by 2.}$$

The absolute value of a number is never negative, so this equation has no solution.

Graphical Solution:

Graph $y_1 = 2|x| + 25$ and $y_2 = 23$ in a $[-47, 47, 10]$ by $[-30, 50, 10]$ window.

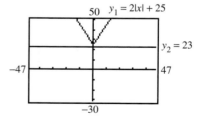

Since there is no point of intersection, there is no solution.

The solution set is $\{\ \}$ or $\varnothing$.

Given two absolute value expressions, we might ask, when are the absolute values of the two expressions equal? To see the answer, notice that

$$|2| = |2|, \quad |-2| = |-2|, \quad |-2| = |2|, \quad \text{and} \quad |2| = |-2|.$$

$$\underset{\text{same}}{\uparrow \quad \uparrow} \qquad \underset{\text{same}}{\uparrow \quad \uparrow} \qquad \underset{\text{opposites}}{\uparrow \quad \uparrow} \qquad \underset{\text{opposites}}{\uparrow \quad \uparrow}$$

Two absolute value expressions are equal when the expressions inside the absolute value bars are equal to or are opposites of each other.

EXAMPLE 7   Solve $|3x + 2| = |5x - 8|$ algebraically and check graphically.

Solution: This equation is true if the expressions inside the absolute value bars are equal to or are opposites of each other.

$$3x + 2 = 5x - 8 \quad \text{or} \quad 3x + 2 = -(5x - 8)$$

Next, solve each equation.

$$3x + 2 = 5x - 8 \quad \text{or} \quad 3x + 2 = -5x + 8$$
$$-2x + 2 = -8 \quad \text{or} \quad 8x + 2 = 8$$
$$-2x = -10 \quad \text{or} \quad 8x = 6$$
$$x = 5 \quad \text{or} \quad x = \frac{3}{4}$$

Using the x-intercept method to check, rewrite the equation as

$$|3x + 2| - |5x - 8| = 0$$

and graph $y_1 = |3x + 2| - |5x - 8|$. The x-intercepts are $x = 0.75$ and $x = 5$.

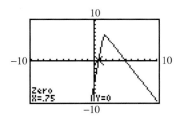   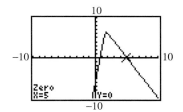

Both solutions check and the solution set is $\left\{\dfrac{3}{4}, 5\right\}$.

The following box summarizes the methods shown for solving absolute value equations algebraically.

SOLVING ABSOLUTE VALUE EQUATIONS ALGEBRAICALLY

$|x| = a$
- If a is positive, then solve $x = a$ or $x = -a$.
- If a is 0, solve $x = 0$.
- If a is negative, the equation $|x| = a$ has no solution.

$|x| = |y|$ Solve $x = y$ or $x = -y$.

To solve absolute value equations graphically, use either the intersection-of-graphs method or the x-intercept method.

MENTAL MATH

Simplify each expression.

1. $|-7|$ **2.** $|-8|$ **3.** $-|5|$

4. $-|10|$ **5.** $-|-6|$ **6.** $-|-3|$

7. $|-3| + |-2| + |-7|$ **8.** $|-1| + |-6| + |-8|$

EXERCISE SET 4.3

Solve each absolute value equation. See Examples 1 through 6.

1. $|x| = 7$ **2.** $|y| = 15$

3. $|3x| = 12.6$ **4.** $|6n| = 12.6$

5. $|2x - 5| = 9$ **6.** $|6 + 2n| = 4$

7. $\left|\dfrac{x}{2} - 3\right| = 1$ **8.** $\left|\dfrac{n}{3} + 2\right| = 4$

9. $|z| + 4 = 9$ **10.** $|x| + 1 = 3$

11. $|3x| + 5 = 14$ **12.** $|2x| - 6 = 4$

13. $|2x| = 0$ **14.** $|7z| = 0$

15. $|4n + 1| + 10 = 4$ **16.** $|3z - 2| + 8 = 1$

17. $|5x - 1| = 0$ **18.** $|3y + 2| = 0$

19. Write an absolute value equation representing all numbers x whose distance from 0 is 5 units.

20. Write an absolute value equation representing all numbers x whose distance from 0 is 2 units.

Solve. See Example 7.

21. $|5x - 7| = |3x + 11|$ **22.** $|9y + 1| = |6y + 4|$

23. $|z + 8| = |z - 3|$ **24.** $|2x - 5| = |2x + 5|$

25. Describe how to solve an absolute value equation such as $|3x| = 4$ by the intersection-of-graphs method.

26. Describe how algebraically solving an absolute value equation such as $|2x - 1| = 3$ is similar to solving an absolute value equation such as $|2x - 1| = |x - 5|$.

Use the given graphing utility screens to solve the equations shown. Write the solution set of the equation.

27. $|2x - 3| = 5$, where $y_1 = |2x - 3|$ and $y_2 = 5$

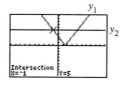 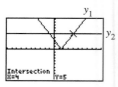

28. $|3x - 4| = 14$, where $y_1 = |3x - 4|$ and $y_2 = 14$

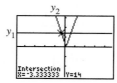 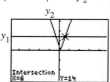

29. $|x - 4| = |1 - x|$, where $y_1 = |x - 4|$ and $y_2 = |1 - x|$

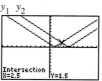

30. $|x + 2| = |3 - x|$, where $y_1 = |x + 2|$ and $y_2 = |3 - x|$

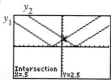

Solve each absolute value equation.

31. $|z| = -2$

32. $|y| = -9$

33. $|7 - 3x| = 7$

34. $|4m + 5| = 5$

35. $|6x| - 1 = 11$

36. $|7z| + 1 = 22$

37. $|4p| = -8$

38. $|5m| = -10$

39. $|x - 3| + 3 = 7$

40. $|x + 4| - 4 = 1$

41. $\left|\dfrac{z}{4} + 5\right| = -7$

42. $\left|\dfrac{c}{5} - 1\right| = -2$

43. $|9v - 3| = -8$

44. $|1 - 3b| = -7$

45. $|8n + 1| = 0$

46. $|5x - 2| = 0$

47. $|1 + 6c| - 7 = -3$

48. $|2 + 3m| - 9 = -7$

49. $|5x + 1| = 11$

50. $|8 - 6c| = 1$

51. $|4x - 2| = |-10|$

52. $|3x + 5| = |-4|$

53. $|5x + 1| = |4x - 7|$

54. $|3 + 6n| = |4n + 11|$

55. $|6 + 2x| = -|-7|$

56. $|4 - 5y| = -|-3|$

57. $|2x - 6| = |10 - 2x|$

58. $|4n + 5| = |4n + 3|$

59. $\left|\dfrac{2x - 5}{3}\right| = 7$

60. $\left|\dfrac{1 + 3n}{4}\right| = 4$

61. $2 + |5n| = 17$

62. $8 + |4m| = 24$

63. $\left|\dfrac{2x - 1}{3}\right| = |-5|$

64. $\left|\dfrac{5x + 2}{2}\right| = |-6|$

65. $|2y - 3| = |9 - 4y|$

66. $|5z - 1| = |7 - z|$

67. $\left|\dfrac{3n + 2}{8}\right| = |-1|$

68. $\left|\dfrac{2r - 6}{5}\right| = |-2|$

69. $|x + 4| = |7 - x|$

70. $|8 - y| = |y + 2|$

71. $\left|\dfrac{8c - 7}{3}\right| = -|-5|$

72. $\left|\dfrac{5d + 1}{6}\right| = -|-9|$

Use a graphical approach to approximate the solutions of each equation. Round the solutions to two decimal places.

73. $|2.3x - 1.5| = 5$

74. $|-7.6x + 2.6| = 1.9$

75. $3.6 - |4.1x - 2.6| = |x - 1.4|$

76. $-1.2 + |5x + 12.1| = -|x + 7.3| + 10$

77. Explain why some absolute value equations have two solutions.

78. Explain why some absolute value equations have one solution.

Review Exercises

The following circle graph shows the sources of Walt Disney Company's income. Use this graph to answer the questions below. See Section 2.5.

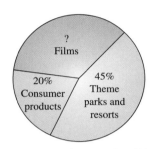

Source: USA Today, August 25, 1994.

79. What percent of Disney's income comes from films?

80. A circle contains 360°. Find the number of degrees found in the 20% sector.

81. If Disney's income one year is $9.75 billion, find the income for the year from theme parks and resorts.

List five integer solutions of each inequality.

82. $|x| \leq 3$

83. $|x| \geq -2$

84. $|y| > -10$

85. $|y| < 0$

4.4 | ABSOLUTE VALUE INEQUALITIES

TAPE IAG 4.4

O B J E C T I V E S

1. Solve absolute value inequalities of the form $|x| < a$.
2. Solve absolute value inequalities of the form $|x| > a$.

1 The solution set of an absolute value inequality such as $|x| < 2$ contains all numbers whose distance from 0 is less than 2 units, as shown next.

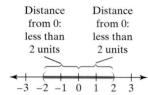

Distance from 0: less than 2 units   Distance from 0: less than 2 units

The solution set is $\{x \mid -2 < x < 2\}$, or $(-2, 2)$ in interval notation.

EXAMPLE 1   Solve $|x| < 7$.

Solution: The solution set of this inequality contains all numbers whose distance from 0 is less than 7. Thus, all numbers between 7 and -7 are in the solution set.

The solution set is $(-7, 7)$.

In general, we have the following property:

SOLVING ABSOLUTE VALUE INEQUALITIES OF THE FORM $|x| < a$

If a is a positive number, then $|x| < a$ is equivalent to $-a < x < a$.

This property also holds true for the inequality symbol $\leq$.

DISCOVER THE CONCEPT

a. Solve the equation $|x| = 7$ by graphing $y_1 = |x|$, $y_2 = 7$, and finding the point(s) of intersection.

b. Move the trace cursor along y_1 from the left-hand point of intersection of the two graphs to the right-hand point of intersection. Observe the y-values of the points as you move the cursor.

c. Find the x-values for which $|x| < 7$ (or $y_1 < y_2$).

d. Write the solution set of $|x| < 7$ (or $y_1 < y_2$).

In the discovery above, we see that the solutions of $|x| = 7$ are 7 and -7, the x-values of the points of intersection of y_1 and y_2. The solutions of $|x| < 7$ consist of all x-values for which $y_1 < y_2$ or for which the graph of y_1 is below the graph of y_2. These x-values are between -7 and 7, or $(-7, 7)$.

Also notice that the solutions of $|x| > 7$ consist of all x-values for which $y_1 > y_2$ or for which the graph of y_1 is above the graph of y_2. These x-values are less than -7 or greater than 7, or $(-\infty, -7) \cup (7, \infty)$.

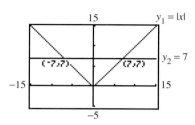

The solution set of the inequality $|x| < 7$ consists of the x-values where $y_1 < y_2$, or in interval notation $(-7, 7)$.

The solution set of the inequality $|x| > 7$ consists of the x-values where $y_1 > y_2$, or in interval notation $(-\infty, -7) \cup (7, \infty)$.

EXAMPLE 2   Solve algebraically and graphically for m: $|m - 6| < 2$.

Algebraic Solution:

From the preceding property, we see that

$$|m - 6| < 2 \text{ is equivalent to } -2 < m - 6 < 2$$

Solve this compound inequality for m by adding 6 to all three sides.

$$-2 < m - 6 < 2$$

$$-2 + 6 < m - 6 + 6 < 2 + 6 \quad \text{Add 6 to all three sides.}$$

$$4 < m < 8 \quad \text{Simplify.}$$

The solution set is $(4, 8)$.

Graphical Solution:

In the inequality $|m - 6| < 2$, replace m with x and proceed as usual. Graph $y_1 = |x - 6|$, $y_2 = 2$, and find x-values for which $y_1 < y_2$.

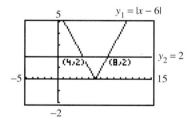

y_1 is below y_2 for x-values between 4 and 8. Thus, the solution set of $|x - 6| = 2$ in interval notation is $(4, 8)$.

> REMINDER   When solving an absolute value inequality algebraically, first isolate the absolute value expression on one side of the inequality.

EXAMPLE 3   Solve algebraically for x: $|5x + 1| + 1 \le 10$.

Solution: First, isolate the absolute value expression by subtracting 1 from both sides.

$$|5x + 1| + 1 \le 10$$

$$|5x + 1| \le 10 - 1 \quad \text{Subtract 1 from both sides.}$$

$$|5x + 1| \le 9 \quad \text{Simplify.}$$

Since 9 is positive, we apply the absolute value property for $|x| \leq a$.

$$-9 \leq 5x + 1 \leq 9$$

$$-9 - 1 \leq 5x + 1 - 1 \leq 9 - 1 \qquad \text{Subtract 1 from all three sides.}$$

$$-10 \leq 5x \leq 8 \qquad \text{Simplify.}$$

$$-2 \leq x \leq \frac{8}{5} \qquad \text{Divide all three sides by 5.}$$

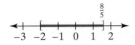

The solution set is $\left[-2, \dfrac{8}{5}\right]$, and the number line graph is shown at the left.

EXAMPLE 4 Solve for x: $\left|2x - \dfrac{1}{10}\right| < -13$.

Solution: The absolute value of a number is always nonnegative and can never be less than -13. Thus, this absolute value inequality has no solution. The solution set is $\{\ \}$, or $\varnothing$.

2 Let us now solve an absolute value inequality of the form $|x| > a$, such as $|x| \geq 3$. The solution set contains all numbers whose distance from 0 is 3 or more units. Thus the graph of the solution set contains 3 and all points to the right of 3 on the number line or -3 and all points to the left of -3 on the number line.

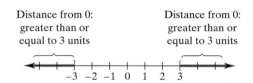

This solution set is written as $\{x \mid x \leq -3 \text{ or } x \geq 3\}$. In interval notation, the solution set is $(-\infty, -3] \cup [3, \infty)$, since "or" means "union." In general, we have the following.

SOLVING ABSOLUTE VALUE INEQUALITIES OF THE FORM $|x| > a$

If a is a positive number, then $|x| > a$ is equivalent to $x < -a$ or $x > a$.

This property also holds true for the inequality symbol $\geq$.

EXAMPLE 5   Solve algebraically and graphically for x: $|x - 3| \geq 7$.

Algebraic Solution:

Since 7 is positive,

$$|x - 3| \geq 7 \text{ is equivalent to}$$

$$x - 3 \leq -7 \text{ or } x - 3 \geq 7$$

Next, solve the compound inequality.

$$x - 3 \leq -7 \qquad \text{or} \qquad x - 3 \geq 7$$

$$x - 3 + 3 \leq -7 + 3 \quad \text{or} \quad x - 3 + 3 \geq 7 + 3$$

$$x \leq -4 \qquad \text{or} \qquad x \geq 10$$

The solution set is $(-\infty, -4] \cup [10, \infty)$, and its graph is shown.

Graphical Solution:

Graph $y_1 = |x - 3|$ and $y_2 = 7$. Find the x-values for which $y_1 \geq y_2$.

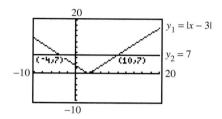

y_1 intersects or is above y_2 for x-values less than or equal to -4 and also x-values greater than or equal to 10, or $(-\infty, -4] \cup [10, \infty)$.

Examples 6 and 8 illustrate special cases of absolute value inequalities. These special cases occur when an isolated absolute value inequality is equal to a negative number or 0.

EXAMPLE 6   Solve $|2x + 9| + 5 > 3$ algebraically and check graphically.

Solution: First isolate the absolute value expression by subtracting 5 from both sides.

$$|2x + 9| + 5 > 3$$

$$|2x + 9| + 5 - 5 > 3 - 5 \qquad \text{Subtract 5 from both sides.}$$

$$|2x + 9| > -2 \qquad \text{Simplify.}$$

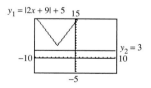

The absolute value of any number is always nonnegative and thus is always greater than -2. This inequality and the original inequality are true for all values of x. The solution set is $\{x \mid x \text{ is a real number}\}$ or $(-\infty, \infty)$. To check graphically, see the screen to the left. Notice that $y_1 > y_2$ or $|2x + 9| + 5 > 3$ for all real numbers.

EXAMPLE 7   Use a graphical approach to solve $\left|\dfrac{x}{3} - 1\right| - 2 \geq 0$.

Solution: Graph $y_1 = \left|\dfrac{x}{3} - 1\right| - 2$ and use the graph of y_1 to solve $y_1 \geq 0$.

Find x-values where the graph of y_1 is above the x-axis.

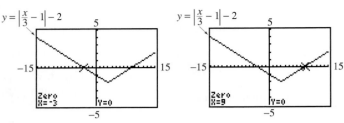

The graph of $y = \left|\dfrac{x}{3} - 1\right| - 2$ is above or on the x-axis for x-values less than or equal to -3 or greater than or equal to 9.

The solution set is $(-\infty, -3] \cup [9, \infty)$.

EXAMPLE 8   Solve for x: $\left|\dfrac{2(x + 1)}{3}\right| \le 0$.

Solution: Recall that "$\le$" means "less than or equal to." The absolute value of any expression will never be less than 0, but it may be equal to 0. Thus, to solve $\left|\dfrac{2(x + 1)}{3}\right| \le 0$ we solve $\left|\dfrac{2(x + 1)}{3}\right| = 0$.

$$\frac{2(x + 1)}{3} = 0$$

$$3\left[\frac{2(x + 1)}{3}\right] = 3(0) \qquad \text{Clear the equation of fractions.}$$

$$2x + 2 = 0 \qquad \text{Apply the distributive property.}$$

$$2x = -2 \qquad \text{Subtract 2 from both sides.}$$

$$x = -1 \qquad \text{Divide both sides by 2.}$$

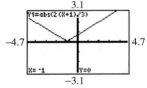

The graph touches the x-axis at $x = -1$ and is never below the x-axis. Therefore, the only point that satisfies the inequality has x-value -1.

The solution set is $\{-1\}$. See the screen to the left to check this solution set graphically.

The following box summarizes the types of absolute value equations and inequalities.

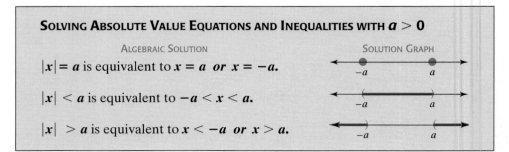

EXERCISE SET 4.4

Solve each inequality. Then graph the solution set on a number line. See Examples 1 through 4.

1. $|x| \le 4$

2. $|x| < 6$

3. $|x - 3| < 2$

4. $|y| \le 5$

5. $|x + 3| < 2$

6. $|x + 4| < 6$

7. $|2x + 7| \le 13$

8. $|5x - 3| \le 18$

9. $|x| + 7 \le 12$

10. $|x| + 6 \le 7$

11. $|3x - 1| < -5$

12. $|8x - 3| < -2$

13. $|x - 6| - 7 \le -1$

14. $|z + 2| - 7 < -3$

Solve each inequality. Graph the solution set on a number line. See Examples 5 through 7.

15. $|x| > 3$

16. $|y| \ge 4$

17. $|x + 10| \ge 14$

18. $|x - 9| \ge 2$

19. $|x| + 2 > 6$

20. $|x| - 1 > 3$

21. $|5x| > -4$

22. $|4x - 11| > -1$

23. $|6x - 8| + 3 > 7$

24. $|10 + 3x| + 1 > 2$

Solve each inequality. Graph the solution set. See Example 8.

25. $|x| \le 0$

26. $|x| \ge 0$

27. $|8x + 3| > 0$

28. $|5x - 6| < 0$

29. Write an absolute value inequality representing all numbers x whose distance from 0 is less than 7 units.

30. Write an absolute value inequality representing all numbers x whose distance from 0 is greater than 4 units.

31. Write $-5 \le x \le 5$ as an equivalent inequality containing an absolute value.

32. Write $x > 1$ or $x < -1$ as an equivalent inequality containing an absolute value.

Solve each inequality.

33. $|x| \le 2$

34. $|z| < 6$

35. $|y| > 1$

36. $|x| \ge 10$

37. $|x - 3| < 8$

38. $|-3 + x| \le 10$

39. $|6x - 8| > 4$

40. $|1 + 0.3x| \ge 0.1$

41. $5 + |x| \le 2$

42. $8 + |x| < 1$

43. $|x| > -4$

44. $|x| \le -7$

45. $|2x - 7| \le 11$

46. $|5x + 2| < 8$

47. $|x + 5| + 2 \ge 8$

48. $|-1 + x| - 6 > 2$

49. $|x| > 0$

50. $|x| < 0$

51. $9 + |x| > 7$

52. $5 + |x| \ge 4$

53. $6 + |4x - 1| \le 9$

54. $-3 + |5x - 2| \le 4$

55. $|2(3 + x)| > 6$

56. $|5(x - 3)| \ge 10$

57. $\left|\dfrac{5(x + 2)}{3}\right| < 7$

58. $\left|\dfrac{6(3 + x)}{5}\right| \le 4$

59. $-15 + |2x - 7| \le -6$ **60.** $-9 + |3 + 4x| < -4$

61. $\left|2x + \dfrac{3}{4}\right| - 7 \le -2$ **62.** $\left|\dfrac{3}{5} + 4x\right| - 6 < -1$

Use the given graphing utility screen to solve each equation or inequality.

63.

a. $|x - 3| - 2 = 6$

b. $|x - 3| - 2 < 6$

c. $|x - 3| - 2 \ge 6$

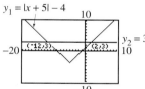

64.

a. $|x + 5| - 4 = 3$

b. $|x + 5| - 4 \le 3$

c. $|x + 5| - 4 > 3$

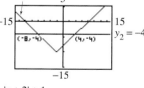

65.

a. $|x + 2| - 10 = -4$

b. $|x + 2| - 10 \le -4$

c. $|x + 2| - 10 > -4$

66.

a. $|x + 2| + 1 = -5$

b. $|x + 2| + 1 < -5$

c. $|x + 2| + 1 > -5$

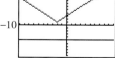

Solve each equation or inequality for x.

67. $|2x - 3| < 7$

68. $|2x - 3| > 7$

69. $|2x - 3| = 7$

70. $|5 - 6x| = 29$

71. $|x - 5| \ge 12$

72. $|x + 4| \ge 20$

73. $|9 + 4x| = 0$

74. $|9 + 4x| \geq 0$

75. $|2x + 1| + 4 < 7$

76. $8 + |5x - 3| \geq 11$

77. $|3x - 5| + 4 = 5$

78. $|8x| = -5$

79. $|x + 11| = -1$

80. $|4x - 4| = -3$

81. $\left|\dfrac{2x - 1}{3}\right| = 6$

82. $\left|\dfrac{6 - x}{4}\right| = 5$

83. $\left|\dfrac{3x - 5}{6}\right| > 5$

84. $\left|\dfrac{4x - 7}{5}\right| < 2$

The expression $|x_T - x|$ is defined to be the absolute error in x, where x_T is the true value of a quantity and x is the measured value or value as stored in a computer.

85. If the true value of a quantity is 3.5 meters and the absolute error must be less than 0.05 meter, find the acceptable measured values.

86. If the true value of a quantity is 0.2 and the approximate value stored in a computer is $\frac{51}{256}$, find the absolute error.

87. GROUP ACTIVITY   Bring to class a few manufactured items that are cylinder-shaped such as cans, lids, or washers. Measure the diameter of an item. Suppose that your measurement is the true diameter of the cylinder and that when manufacturing the item, the diameter may vary from your measured amount by no more than 1%. Write an absolute value inequality stating the result.

Review Exercises

Recall the formula

$$\text{Probability of an event} = \frac{\substack{\text{number of ways that} \\ \text{the event can occur}}}{\substack{\text{number of possible} \\ \text{outcomes}}}$$

Find the probability of rolling each number on a single toss of a die. (Recall that a die is a cube with each of its six sides containing 1, 2, 3, 4, 5, and 6 black dots, respectively.) See Section 2.4.

88. $P(\text{rolling a 2})$

89. $P(\text{rolling a 5})$

90 $P(\text{rolling a 7})$

91. $P(\text{rolling a 0})$

92. $P(\text{rolling a 1 or 3})$

93. $P(\text{rolling a 1, 2, 3, 4, 5, or 6})$

Consider the equation $3x - 4y = 12$. For each value of x or y given, find the corresponding value of the other variable that makes the statement true. See Section 2.4.

94. If $x = 2$, find y.

95. If $y = -1$, find x.

96. If $y = -3$, find x.

97. If $x = 4$, find y.

4.5   GRAPHING LINEAR INEQUALITIES IN TWO VARIABLES

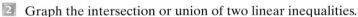

1 Graph linear inequalities.

2 Graph the intersection or union of two linear inequalities.

TAPE IAG 4.5

1   Recall that the graph of a linear equation in two variables is the graph of all ordered pairs that satisfy the equation, and we determined that the graph is a line. Here we graph **linear inequalities** in two variables; that is, we graph all the ordered pairs that satisfy the inequality.

If the equals sign in a linear equation in two variables is replaced with an inequality symbol, the result is a linear inequality in two variables.

EXAMPLES OF LINEAR INEQUALITIES IN TWO VARIABLES

$$3x + 5y \geq 6 \qquad 2x - 4y < -3$$

$$4x > 2 \qquad y \leq 5$$

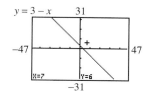

$y = 3 - x$

The graph of $y = 3 - x$ and a point $(7, 6)$ above the line.

DISCOVER THE CONCEPT

a. Graph $x + y = 3$ by graphing $y_1 = 3 - x$ in an integer window.

b. Access the cursor that results from using the arrow key pad (not the trace cursor). Move the cursor to points above the line and notice the sum of the x- and y-coordinates.

c. Move the cursor to points below the line and notice the sum of the x- and y-coordinates.

d. Describe the set of points that satisfies $x + y < 3$ and the set of points that satisfies $x + y > 3$.

We call the line defined by $x + y = 3$ a boundary line. This boundary line contains all ordered pairs the sum of whose coordinates is 3. This line separates the plane into two **half-planes.** All points *above* the boundary line $x + y = 3$ have coordinates that satisfy the inequality $x + y > 3$, and all points *below* the line have coordinates that satisfy the inequality $x + y < 3$.

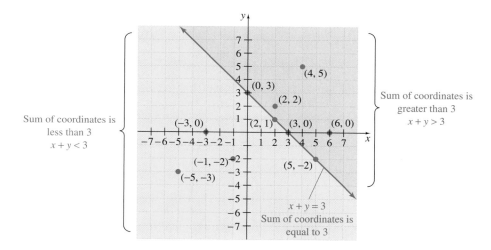

The graph, or **solution region,** of $x + y < 3$, for example, is the half-plane below the boundary line and is shown shaded. The boundary line is shown dashed since it is not a part of the solution region. These ordered pairs on this line satisfy $x + y = 3$ and not $x + y < 3$.

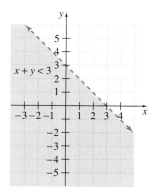

The following steps may be used to graph linear inequalities in two variables.

TO GRAPH A LINEAR INEQUALITY IN TWO VARIABLES

Step 1. Graph the boundary line found by replacing the inequality sign with an equal sign. If the inequality sign is $<$ or $>$, graph a dashed line, indicating that points on the line are not solutions of the inequality. If the inequality sign is $\leq$ or $\geq$, graph a solid line indicating that points on the line are solutions of the inequality.

Step 2. Choose a **test point not on the boundary line** and substitute the coordinates of this test point into the **original inequality.**

Step 3. If a true statement is obtained in *step 2*, shade the half-plane that contains the test point. If a false statement is obtained, shade the half-plane that does not contain the test point.

EXAMPLE 1 Graph $3x \geq y$.

Solution: First, graph the boundary line $3x = y$. Graph a solid boundary line because the in-equality symbol is $\geq$. Test a point not on the boundary line to determine which half-plane contains points that satisfy the inequality. We choose $(0, 1)$ as our test point.

$$3x \geq y$$
$$3(0) \geq 1 \qquad \text{Let } x = 0 \text{ and } y = 1.$$
$$0 \geq 1 \qquad \text{False.}$$

This point does not satisfy the inequality, so the correct half-plane is on the oppo-site side of the boundary line from $(0, 1)$. The graph of $3x \geq y$ is the boundary line together with the shaded region shown next.

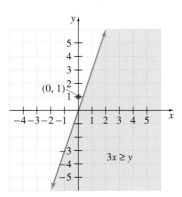

If an inequality is written in the form $y < mx + b$ or $y > mx + b$, the inequality symbol tells us which half-plane to shade.

For $y < mx + b$, shade below the line, and for

$y > mx + b$, shade above the line.

EXAMPLE 2   Graph $2x - y < 6$.

Solution: First, the boundary line for this inequality is the graph of $2x - y = 6$. Graph a dashed boundary line because the inequality symbol is $<$. Next, solve the inequality for y.

$$2x - y < 6$$

$$-y < -2x + 6 \qquad \text{Subtract } 2x \text{ from both sides.}$$

$$\frac{-y}{-1} > \frac{-2x}{-1} + \frac{6}{-1} \qquad \text{Divide both sides by } -1.$$

$$y > 2x - 6 \qquad \text{Simplify.}$$

Because the inequality symbol is $>$, we shade above the line. The half-plane graph of the inequality is shown next.

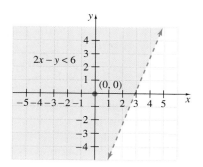

Every point in the shaded half-plane satisfies the original inequality. Notice that the inequality $2x - y < 6$ does not describe a function since its graph does not pass the vertical line test.

In general, linear inequalities of the form $Ax + By \leq C$, when A and B are not both 0, do not describe functions.

To graph Example 2, $2x - y < 6$, using your graphing utility, recall that $2x - y < 6$ is equivalent to $y > 2x - 6$. Graph the related equation $y_1 = 2x - 6$ and shade above the line. The graph below is shown on a standard window.

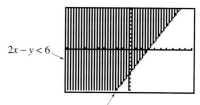

On most graphing utilities, the boundary line will always appear solid. Thus, simply note that the boundary line should be dashed.

2 The intersection and the union of linear inequalities can also be graphed, as shown in the next two examples.

EXAMPLE 3   Graph the intersection of $x \geq 1$ and $y \geq 2x - 1$.

Solution: Graph each inequality. The intersection of the two graphs is all points common to both regions, as shown by the heaviest shading in the third graph.

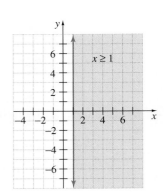 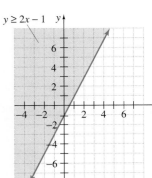 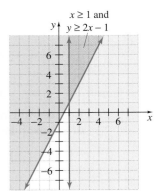

EXAMPLE 4   Graph the union of $x + \dfrac{1}{2}y \geq -4$ or $y \leq -2$.

Solution: Graph each inequality. The union of the two inequalities is both shaded regions, including the solid boundary lines shown in the third graph.

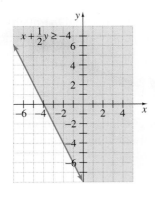 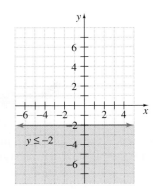 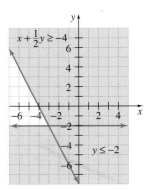

EXERCISE SET 4.5

Graph each inequality. See Examples 1 and 2.

1. $x < 2$

2. $x > -3$

3. $x - y \geq 7$

4. $3x + y \leq 1$

5. $3x + y > 6$

6. $2x + y > 2$

7. $y \leq -2x$

8. $y \leq 3x$

9. $2x + 4y \geq 8$

10. $2x + 6y \leq 12$ **11.** $5x + 3y > -15$ **12.** $2x + 5y < -20$

13. Explain when a dashed boundary line should be used in the graph of an inequality.

14. Explain why, after the boundary line is sketched, we test a point on either side of this boundary in the original inequality.

Graph each union or intersection. See Examples 3 and 4.

15. The intersection of $x \geq 3$ and $y \leq -2$

16. The union of $x \geq 3$ or $y \leq -2$

17. The union of $x \leq -2$ or $y \geq 4$

18. The intersection of $x \leq -2$ and $y \geq 4$

19. The intersection of $x - y < 3$ and $x > 4$

20. The intersection of $2x > y$ and $y > x + 2$

21. The union of $x + y \leq 3$ or $x - y \geq 5$

22. The union of $x - y \leq 3$ or $x + y > -1$

Graph each inequality.

23. $y \geq -2$ **24.** $y \leq 4$ **25.** $x - 6y < 12$

26. $x - 4y < 8$ **27.** $x > 5$ **28.** $y \geq -2$

29. $-2x + y \leq 4$ **30.** $-3x + y \leq 9$ **31.** $x - 3y < 0$

32. $x + 2y > 0$ **33.** $3x - 2y \leq 12$ **34.** $2x - 3y \leq 9$

35. The union of $x - y \geq 2$ or $y < 5$

36. The union of $x - y < 3$ or $x > 4$

37. The intersection of $x + y \leq 1$ and $y \leq -1$

38. The intersection of $y \geq x$ and $2x - 4y \geq 6$

39. The union of $2x + y > 4$ or $x \geq 1$

40. The union of $3x + y < 9$ or $y \leq 2$

41. The intersection of $x \geq -2$ and $x \leq 1$

42. The intersection of $x \geq -4$ and $x \leq 3$

43. The union of $x + y \leq 0$ or $3x - 6y \geq 12$

44. The intersection of $x + y \leq 0$ and $3x - 6y \geq 12$

45. The intersection of $2x - y > 3$ and $x \geq 0$

46. The union of $2x - y > 3$ or $x \geq 0$

Match each graph with its inequality.

A

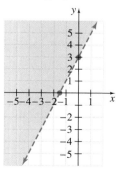

B

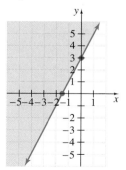

C

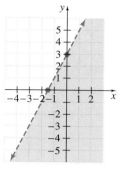

D

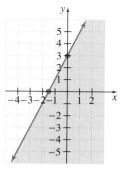

47. $y \leq 2x + 3$ **48.** $y < 2x + 3$

49. $y > 2x + 3$ **50.** $y \geq 2x + 3$

Write the inequality whose graph is given.

51.

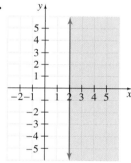

52.

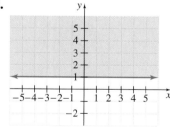

53.

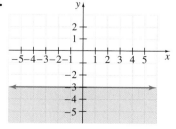

54.

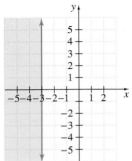

55.

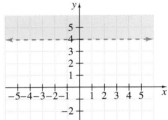

56.

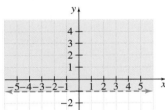

57.

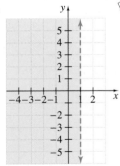

58.

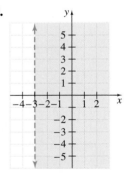

Solve.

59. Rheem Abo-Zahrah decides that she will study at most 20 hours every week and that she must work at least 10 hours every week. Let x represent the hours studying and y represent the hours working. Write two inequalities that model this situation and graph their intersection.

60. The movie and TV critic for the *New York Times* spends between 2 and 6 hours daily reviewing movies and fewer than 5 hours reviewing TV shows. Let x represent the hours watching movies and y represent the time spent watching TV. Write two inequalities that model this situation and graph their intersection.

61. Chris-Craft manufactures boats out of Fiberglas and wood. Fiberglas hulls require 2 hours work, whereas wood hulls require 4 hours work. Employees work at most 40 hours a week. The following inequalities model these restrictions, where x represents the number of Fiberglas hulls produced and y represents the number of wood hulls produced.

$$\begin{cases} x \geq 0 \\ y \geq 0 \\ 2x + 4y \leq 40 \end{cases}$$

Graph the intersection of these inequalities.

Review Exercises

Evaluate each expression. See Section 1.3.

62. 2^3　　　**63.** 3^2　　　**64.** -5^2　　　**65.** $(-5)^2$

66. $(-2)^4$　　**67.** -2^4　　**68.** $\left(\dfrac{3}{5}\right)^3$　　**69.** $\left(\dfrac{2}{7}\right)^2$

Find the domain and the range of each relation. Determine whether the relation is also a function. See Section 3.3.

70.

71.

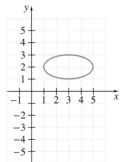

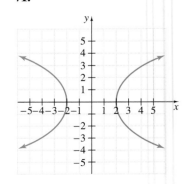

GROUP ACTIVITY

ANALYZING MUNICIPAL BUDGETS

MATERIALS
• Calculator

Suppose that each year your town creates a municipal budget. The next year's annual municipal budget is submitted for approval by the town's citizens at the annual town meeting. This year's budget was printed in the town newspaper earlier in the year.

THE TOWN CRIER
Annual Budget Set at Town Meeting
ANYTOWN, USA (MG)–This year's annual budget is as follows:

	Amount Budgeted
BOARD OF HEATH	
Immunization Programs	$15,000
Inspections	$50,000
FIRE DEPARTMENT	
Equipment	$450,000
Salaries	$275,000
LIBRARIES	
Book/Periodical Purchases	$90,000
Equipment	$30,000
Salaries	$120,000
PARKS AND RECREATION	
Maintenance	$70,000
Playground Equipment	$50,000
Salaries	$140,000
Summer Programs	$80,000
POLICE DEPARTMENT	
Equipment	$300,000
Salaries	$400,000
PUBLIC WORKS	
Recycling	$50,000
Sewage	$100,000
Snow Removal & Road Salt	$200,000
Street Maintenance	$250,000
Water Treatment	$100,000
TOTAL	**$2,770,000**

You have joined a group of citizens who are concerned about your town's budgeting and spending processes. Your group plans to analyze this year's budget along with what was actually spent by the town this year. You hope to present your findings at the annual town meeting and make some budgetary recommendations for next year's budget. The municipal budget contains many different areas of spending. To help focus your group's analysis, you have decided to research spending habits only for categories in which the actual expenses differ from the budgeted amount by more than 12% of the budgeted amount.

1. For each category in the budget, write a specific absolute value inequality that describes the condition that must be met before your group will research spending habits for that category. In each case, let the variable x represent the actual expense for a budget category.

2. For each category in the budget, write an equivalent compound inequality for the condition described in Question 1. Again, let the variable x represent the actual expense for a budget category.

3. Below is a listing of the actual expenditures made this year for each budget category. Use the inequalities from either Question 1 or 2 to fill in the given table at the end of this section. (The first category has been filled in.) From the table, decide which categories must be researched.

	ACTUAL EXPENDITURES			ACTUAL EXPENDITURES
Board of Health			Parks and Recreation *(continued)*	
Immunization Programs	$14,800		Salaries	$118,000
Inspections	$41,900		Summer Programs	$96,200
Fire Department			Police Department	
Equipment	$375,000		Equipment	$328,000
Salaries	$268,500		Salaries	$405,000
Libraries			Public Works	
Book/Periodical Purchases	$107,300		Recycling	$48,100
Equipment	$29,000		Sewage	$92,500
Salaries	$118,400		Snow Removal & Road Salt	$268,300
Parks and Recreation			Street Maintenance	$284,000
Maintenance	$82,500		Water Treatment	$94,100
Playground Equipment	$45,000		TOTAL	$2,816,600

4. Can you think of possible reasons why spending in the categories that must be researched might be over or under budget?
5. Based on this year's municipal budget and actual expenses, what recommendations would you make for next year's budget? Explain your reasoning.
6. (Optional) Research the annual budget used by your town or your college. Conduct a similar analysis of the budget with respect to actual expenses. What can you conclude?

	BUDGET VALUE	MINIMUM	ACTUAL VALUE	MAXIMUM	WITHIN BUDGET	OVER/ UNDER
IMMUNIZATION PROGRAMS	15,000	13,200	14,800	16,800	yes	under 200

CHAPTER 4 HIGHLIGHTS

DEFINITIONS AND CONCEPTS	EXAMPLES
SECTION 4.1 SOLVING LINEAR INEQUALITIES AND PROBLEM SOLVING	
A **linear inequality in one variable** is an inequality that can be written in the form $ax + b < c$, where a, b, and c are real numbers and $a \neq 0$. (The inequality symbols $\leq$, $>$, and $\geq$ also apply here.)	Linear inequalities: $$5x - 2 \leq -7 \qquad 3y > 1 \qquad \frac{z}{7} < -9(z - 3)$$
The **addition property of inequality** guarantees that the same number may be added to (or subtracted from) both sides of an inequality, and the resulting inequality will have the same solution set.	$$x - 9 \leq -16$$ $$x - 9 + 9 \leq -16 + 9 \qquad \text{Add 9.}$$ $$x \leq -7$$
	(continued)

DEFINITIONS AND CONCEPTS	**EXAMPLES**

SECTION 4.1   SOLVING LINEAR INEQUALITIES AND PROBLEM SOLVING

The **multiplication property of inequality** guarantees that both sides of an inequality may be multiplied by (or divided by) the same **positive** number, and the resulting inequality will have the same solution set. We may also multiply (or divide) both sides of an inequality by the same **negative** number and **reverse the direction of the inequality symbol,** and the result is an inequality with the same solution set.	$6x < -66$ $\dfrac{6x}{6} < \dfrac{-66}{6}$ Divide by 6. Do not reverse direction of inequality symbol. $x < -11$ $-6x < -66$ $\dfrac{-6x}{-6} > \dfrac{-66}{-6}$ Divide by -6. Reverse direction of inequality symbol. $x > 11$

To solve a linear inequality in one variable:

Solve for x:

$$\frac{3}{7}(x - 4) \geq x + 2$$

1. Clear the equation of fractions.	**1.** $7\left[\dfrac{3}{7}(x - 4)\right] \geq 7(x + 2)$ Multiply by 7. $3(x - 4) \geq 7(x + 2)$
2. Remove grouping symbols such as parentheses.	**2.** $3x - 12 \geq 7x + 14$ Apply the distributive property.
3. Simplify by combining like terms.	
4. Write variable terms on one side and numbers on the other side using the addition property of inequality.	**4.** $-4x - 12 \geq 14$ Subtract $7x$. $-4x \geq 26$ Add 12.
5. Isolate the variable using the multiplication property of inequality.	**5.** $\dfrac{-4x}{-4} \leq \dfrac{26}{-4}$ Divide by -4. Reverse direction of inequality symbol. $x \leq -\dfrac{13}{2}$

SECTION 4.2   COMPOUND INEQUALITIES

Two inequalities joined by the words *and* or *or* are called **compound inequalities.**	Compound inequalities: $x - 7 \leq 4$ and $x \geq -21$ $2x + 7 > x - 3$ or $5x + 2 > -3$
The solution set of a compound inequality formed by the word **and** is the **intersection** $\cap$ of the solution sets of the two inequalities.	Solve for x: $x < 5$ and $x < 3$ $\{x \mid x < 5\}$ ⟵————————→ 5 $(-\infty, 5)$ $\{x \mid x < 3\}$ ⟵————————→ 3 $(-\infty, 3)$ $\{x \mid x < 3$ and $x < 5\}$ ⟵————————→ 3 $(-\infty, 3)$

(continued)

DEFINITIONS AND CONCEPTS	**EXAMPLES**

SECTION 4.2   COMPOUND INEQUALITIES

Solve for x: $x - 2 \geq -3$ or $2x \leq -4$

$\qquad\qquad\qquad\quad x \geq -1$ or $x \leq -2$

The solution set of a compound inequality formed by the word **or** is the **union** $\cup$ of the solution sets of the two inequalities.

$\{x \mid x \geq -1\}$

$(-1, \infty)$

$\{x \mid x \leq -2\}$

$(-\infty, -2)$

$\{x \mid x \geq -2 \text{ or } x \geq -1\}$

$(-\infty, 3)$

To solve a compound inequality $y_1 < y_2 < y_3$ graphically,

1. Graph separately each of the three parts y_1, y_2, and y_3 respectively, in an appropriate window.

2. Observe where the graph of y_2 is between the graphs of y_1 and y_3.

3. Find the x-coordinates of the points of intersection and determine the appropriate interval of the solution.

Solve for x: $-13 < 3x - 4 \leq 8$

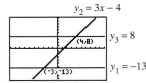

$y_2 = 3x - 4$

$(4, 8)$

$y_3 = 8$

$y_1 = -13$

$(-3, -13)$

The solution in interval notation is $(-3, 4]$.

SECTION 4.3   ABSOLUTE VALUE EQUATIONS

If a is a positive number, then $|x| = a$ is equivalent to $x = a$ or $x = -a$.

Solve for y:

$|5y - 1| - 7 = 4$

$\qquad |5y - 1| = 11$ Add 7.

$\qquad 5y - 1 = 11$ or $5y - 1 = -11$

$\qquad 5y = 12$ or $5y = -10$ Add 1.

$\qquad y = \dfrac{12}{5}$ or $y = -2$ Divide by 5.

The solution set is $\left\{-2, \frac{12}{5}\right\}$.

If a is negative, then $|x| = a$ has no solution.

Solve for x:

$$\left|\frac{x}{2} - 7\right| = -1$$

The solution set is $\{\ \}$ or $\varnothing$.

If an absolute value equation is of the form $|x| = |y|$, solve $x = y$ or $x = -y$.

Solve for x:

$\qquad\qquad |x - 7| = |2x + 1|$

$x - 7 = 2x + 1$ or $x - 7 = -(2x + 1)$

$\qquad x = 2x + 8$ $\qquad\qquad x - 7 = -2x - 1$

$\qquad -x = 8$ $\qquad\qquad\qquad x = -2x + 6$

$\qquad x = -8$ or $\qquad 3x = 6$

$\qquad\qquad\qquad\qquad\qquad\qquad x = 2$

The solution set is $\{-8, 2\}$.

(continued)

DEFINITIONS AND CONCEPTS	**EXAMPLES**

SECTION 4.4   ABSOLUTE VALUE INEQUALITIES

If a is a positive number, then $\lvert x \rvert < a$ is equivalent to $-a < x < a$.	Solve for y: $$\lvert y - 5 \rvert \leq 3$$ $$-3 \leq y - 5 \leq 3$$ $$-3 + 5 \leq y - 5 + 5 \leq 3 + 5 \qquad \text{Add 5.}$$ $$2 \leq y \leq 8$$ The solution set is $[2, 8]$.
If a is a positive number, then $\lvert x \rvert > a$ is equivalent to $x < -a$ or $x > a$.	Solve for x: $$\left\lvert \frac{x}{2} - 3 \right\rvert > 7$$ $$\frac{x}{2} - 3 < -7 \quad \text{or} \quad \frac{x}{2} - 3 > 7$$ $$x - 6 < -14 \quad \text{or} \quad x - 6 > 14 \qquad \begin{array}{l}\text{Multiply}\\\text{by 2.}\end{array}$$ $$x < -8 \quad \text{or} \quad x > 20 \qquad \text{Add 6.}$$ The solution set is $(-\infty, -8) \cup (20, \infty)$

SECTION 4.5   GRAPHING LINEAR INEQUALITIES IN TWO VARIABLES

If the equal sign in a linear equation in two variables is replaced with an equality symbol, the result is a **linear inequality in two variables.**	Linear inequalities in two variables: $$x \leq -5y \qquad y \geq 2$$ $$3x - 2y > 7 \qquad x < -5$$
To graph a linear inequality:	Graph $2x - 4y > 4$.
1. Graph the boundary line by graphing the related equation. Draw the line solid if the inequality symbol is $\leq$ or $\geq$. Draw the line dashed if the inequality symbol is $<$ or $>$.	**1.** Graph $2x - 4y = 4$. Draw a dashed line because the inequality symbol is $>$.
2. Choose a test point not on the line. Substitute its coordinates into the original inequality.	**2.** Check the test point $(0, 0)$ in the inequality $2x - 4y > 4$. $$2 \cdot 0 - 4 \cdot 0 > 4 \qquad \text{Let } x = 0 \text{ and } y = 0.$$ $$0 > 4 \qquad \text{False.}$$
3. If the resulting inequality is true, shade the **half-plane** that contains the test point. If the inequality is not true, shade the half-plane that does not contain the test point.	**3.** The inequality is false, so we shade the half-plane that does not contain $(0, 0)$. 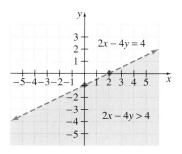

CHAPTER 4 REVIEW

(4.1) *Solve each linear inequality.*

1. $3(x - 5) > -(x + 3)$ **2.** $-2(x + 7) \geq 3(x + 2)$

3. $4x - (5 + 2x) < 3x - 1$

4. $3(x - 8) < 7x + 2(5 - x)$

5. $24 \geq 6x - 2(3x - 5) + 2x$

6. $48 - x \geq 5(2x + 4) - 2x$

7. $\dfrac{x}{3} + \dfrac{1}{2} > \dfrac{2}{3}$ **8.** $x + \dfrac{3}{4} < \dfrac{-x}{2} + \dfrac{9}{4}$

9. $\dfrac{x - 5}{2} \leq \dfrac{3}{8}(2x + 6)$ **10.** $\dfrac{3(x - 2)}{5} > \dfrac{-5(x - 2)}{3}$

Solve.

11. George Boros can pay his housekeeper $15 per week to do his laundry, or he can have the laundromat do it at a cost of 50 cents per pound for the first 10 pounds and 40 cents for each additional pound. Use an inequality to find the weight at which it is more economical to use the housekeeper than the laundromat.

12. Ceramic firing temperatures usually range from 500° to 1000° Fahrenheit. Use a compound inequality to convert this range to the Celsius scale. Round to the nearest degree.

13. In the Olympic gymnastics competition, Nana must average a score of 9.65 to win the silver medal. Seven of the eight judges have reported scores of 9.5, 9.7, 9.9, 9.7, 9.7, 9.6, and 9.5. Use an inequality to find the minimum score that the last judge can give so that Nana wins the silver medal.

14. Carol would like to pay cash for a car when she graduates from college and estimates that she can afford a car that costs between $4000 and $8000. She has saved $500 so far and plans to earn the rest of the money by working the next two summers. If Carol plans to save the same amount each summer, use a compound inequality to find the range of money she must save each summer to buy the car.

(4.2) *Solve each inequality.*

15. $1 \leq 4x - 7 \leq 3$ **16.** $-2 \leq 8 + 5x < -1$

17. $-3 < 4(2x - 1) < 12$ **18.** $-6 < x - (3 - 4x) < -3$

19. $\dfrac{1}{6} < \dfrac{4x - 3}{3} \leq \dfrac{4}{5}$ **20.** $0 \leq \dfrac{2(3x + 4)}{5} \leq 3$

21. $x \leq 2$ and $x > -5$ **22.** $x \leq 2$ or $x > -5$

23. $3x - 5 > 6$ or $-x < -5$

24. $-2x \leq 6$ and $-2x + 3 < -7$

(4.3) *Solve each absolute value equation.*

25. $|x - 7| = 9$ **26.** $|8 - x| = 3$

27. $|2x + 9| = 9$ **28.** $|-3x + 4| = 7$

29. $|3x - 2| + 6 = 10$ **30.** $5 + |6x + 1| = 5$

31. $-5 = |4x - 3|$ **32.** $|5 - 6x| + 8 = 3$

33. $|7x| - 26 = -5$ **34.** $-8 = |x - 3| - 10$

35. $\left|\dfrac{3x - 7}{4}\right| = 2$ **36.** $\left|\dfrac{9 - 2x}{5}\right| = -3$

37. $|6x + 1| = |15 + 4x|$ **38.** $|x - 3| = |7 + 2x|$

(4.4) *Solve each absolute value inequality. Graph the solution set on a number line and write in interval notation.*

39. $|5x - 1| < 9$ **40.** $|6 + 4x| \geq 10$

41. $|3x| - 8 > 1$ **42.** $9 + |5x| < 24$

43. $|6x - 5| \leq -1$ **44.** $|6x - 5| \geq -1$

45. $\left|3x + \dfrac{2}{5}\right| \geq 4$ **46.** $\left|\dfrac{4x - 3}{5}\right| < 1$

47. $\left|\dfrac{x}{3} + 6\right| - 8 > -5$ **48.** $\left|\dfrac{4(x - 1)}{7}\right| + 10 < 2$

(4.5) *Graph each linear inequality.*

49. $3x + y > 4$ **50.** $\dfrac{1}{2}x - y < 2$ **51.** $5x - 2y \leq 9$

52. $3y \geq x$ **53.** $y < 1$ **54.** $x > -2$

55. Graph the union of $y > 2x + 3$ or $x \leq -3$.

56. Graph the intersection of $2x < 3y + 8$ and $y \geq -2$.

CHAPTER 4 TEST

Use the given screen to solve each inequality. Write the solution in interval notation.

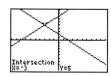

1. $y_1 < y_2$

2. $y_1 > y_2$

Solve each equation or inequality.

3. $3(2x - 7) - 4x > -(x + 6)$ **4.** $8 - \dfrac{x}{2} \le 7$

5. $-3 < 2(x - 3) \le 4$

6. $|3x + 1| > 5$

7. $x \ge 5$ and $x \ge 4$

8. $x \ge 5$ or $x \ge 4$

9. $-x > 1$ and $3x + 3 \ge x - 3$

10. $|6x - 5| = 1$

11. $|8 - 2t| = -6$

12. $|3x - 2| = |6x + 11|$

13. $\left|\dfrac{2x - 6}{5}\right| = 4$ **14.** $\left|\dfrac{7x - 1}{2}\right| \le 3$

Graph each inequality.

15. $x \le -4$ **16.** $y > -2$ **17.** $2x - y > 5$

18. The intersection of $2x + 4y < 6$ and $y \le -4$

Solve.

19. In the United States, the annual consumption of cigarettes is declining. The consumption c in billions of cigarettes per year since the year 1985 can be approximated by the formula.

$$c = -14.25t + 598.69$$

where t is the number of years after 1985. Use this formula to predict the years that the consumption of cigarettes will be less than 200 billion per year.

CHAPTER 4   CUMULATIVE REVIEW

1. Determine whether each statement is true or false.

   a. $3 \in \{x \,|\, x \text{ is a natural number}\}$

   b. $7 \notin \{1, 2, 3\}$ **c.** $5 \notin \{\dots, -3, -2, -1, 0, 1, 2, 3, \dots\}$

2. Graph each set and write it in interval notation.

   a. $\{x \,|\, -5 < x < -1\}$ **b.** $\{x \,|\, -1.5 \le x < 4\}$

3. Find each difference.

   a. $2 - 8$

   b. $-8 - (-1)$

   c. $-11 - 5$

   d. $10.7 - (-9.8)$

   e. $\dfrac{2}{3} - \dfrac{1}{2}$

   f. $1 - 0.06$

   g. Subtract 7 from 4.

4. Find each quotient.

   a. $\dfrac{20}{-4}$

   b. $\dfrac{-9}{-3}$

   c. $-\dfrac{3}{8} \div 3$

   d. $\dfrac{-40}{10}$

   e. $\dfrac{-1}{10} \div \dfrac{-2}{5}$

   f. $\dfrac{8}{0}$

5. Find the following square roots.

   a. $\sqrt{9}$

   b. $\sqrt{25}$

   c. $\sqrt{\dfrac{1}{4}}$

6. Simplify.

   a. $3 + 2 \cdot 10$

   b. $2(1 - 4)^2$

   c. $\dfrac{|-2|^3 + 1}{-7 - \sqrt{4}}$

   d. $\dfrac{(6 + 2) - (-4)}{2 - (-3)}$

7. Write each as an algebraic expression.

   a. A vending machine contains x quarters. Write an expression for the value of the quarters in dollars.

   b. The number of grams of fat in x slices of bread if each slice of bread contains 2 grams of fat.

c. The cost of x desks if each desk costs $156.

d. Sales tax on a purchase of x dollars if the tax rate is 9%.

8. Solve for x: $6x - 4 = 2 + 6(x - 1)$

9. Suppose that Service Merchandise just announced an 8% decrease in the price of their Compaq Presario computers. If one particular computer model sells for $2162 after the decrease, find the original price of this computer.

10. Solve $A = \dfrac{1}{2}(B + b)h$ for b.

11. Make a table of values for $y = x^2$ for $x = -3, -2, -1, 0, 1, 2, 3$. Use the table to sketch the graph of $y = x^2$. Then use a graphing utility to check.

12. Graph the equation $y = |x| - 3$.

13. Find the slope of the line containing the points $(0, 3)$ and $(2, 5)$.

14. Solve algebraically for x.

a. $\dfrac{1}{4}x \le \dfrac{3}{8}$ **b.** $-2.3x < 6.9$

15. Solve algebraically for x: (a) $2(x + 3) > 2x + 1$ (b) $2(x + 3) < 2x + 1$.

16. A salesperson earns a monthly gross pay of $600 plus 20% of sales. Find the minimum amount of sales needed to receive a total income of at least $1500 per month.

17. Solve for x: $x - 7 < 2$ and $2x + 1 < 9$.

18. Solve for x: $-1 \le \dfrac{2x}{3} + 5 \le 2$.

19. Solve $5x - 3 \le 10$ or $x + 1 \ge 5$.

20. Solve $|x| = 2$.

21. Solve algebraically and graphically for m: $|m - 6| < 2$.

22. Solve $|2x + 9| + 5 > 3$ algebraically and check graphically.

SYSTEMS OF EQUATIONS

LOCATING LIGHTNING STRIKES

Lightning, most often produced during thunderstorms, is a rapid discharge of high-current electricity into the atmosphere. Around the world, lightning occurs at a rate of approximately 100 flashes per second. Because of lightning's potentially destructive nature, meteorologists track lightning activity by recording and plotting the positions of lightning strikes.

IN THE CHAPTER GROUP ACTIVITY ON PAGE 295, YOU WILL HAVE THE OPPORTUNITY TO PINPOINT THE LOCATION OF A LIGHTNING STRIKE.

Ｉn this chapter, two or more equations in two or more variables are solved simultaneously. Such a collection of equations is called a **system of equations.**
Systems of equations are good mathematical models for many real-world
problems because these problems may involve several related patterns. ▬▬▬▬▬

5.1　SOLVING SYSTEMS OF LINEAR EQUATIONS IN TWO VARIABLES

O B J E C T I V E S

1 Determine whether an ordered pair is a solution of a system of equations.
2 Solve a system by graphing.
3 Solve a system by substitution.
4 Solve a system by elimination.

TAPE IAG 5.1

1 An important problem that often occurs in the fields of business and economics
concerns the concepts of revenue and cost. For example, suppose that a small
manufacturing company begins to manufacture and sell compact disc storage
units. The revenue of a company is the company's income from selling these units,
and the cost is the amount of money that a company spends to manufacture these
units. The following coordinate system shows the graph of revenue and cost for
the storage units.

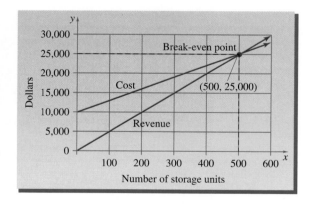

These lines intersect at the point (500, 25,000). This means that when 500 storage units are manufactured and sold, both cost and revenue are $25,000. In busi-
ness, this point of intersection is called the *break-even point*. Notice that for
x-values (units sold) less than 500, the cost graph is above the revenue graph,
meaning that cost of manufacturing is greater than revenue, and so the company is
losing money. For *x*-values (units sold) greater than 500, the revenue graph is
above the cost graph, meaning that revenue is greater than cost, and so the company is making money.

Recall from Chapter 3 that each line is a graph of some linear equation in two variables. Both equations together form a **system of equations.** The common point of intersection is called the **solution of the system.** Some examples of systems of linear equations in two variables are

SYSTEMS OF LINEAR EQUATIONS IN TWO VARIABLES

$$\begin{cases} x - 2y = -7 \\ 3x + y = 0 \end{cases} \qquad \begin{cases} x = 5 \\ x + \dfrac{y}{2} = 9 \end{cases} \qquad \begin{cases} x - 3 = 2y + 6 \\ y = 1 \end{cases}$$

Recall that a solution of an equation in two variables is an ordered pair (x, y) that makes the equation true. A *solution of a system* of two equations in two variables is an ordered pair (x, y) that makes both equations true.

EXAMPLE 1   Determine whether $(4, -3)$ is a solution of the following system of equations.

$$\begin{cases} 2x - y = 11 \\ x + 3y = -5 \end{cases}$$

Solution: Replace x with 4 and y with -3 in each equation.

$$\begin{array}{ll}
2x - y = 11 & x + 3y = -5 \\
2 \cdot 4 - (-3) = 11 & 4 + 3(-3) = -5 \\
8 + 3 = 11 & 4 + (-9) = -5 \\
11 = 11 \quad \text{True.} & -5 = -5 \quad \text{True.}
\end{array}$$

Since *both* equations are true, the ordered pair $(4, -3)$ is a solution of the system of equations.

In the above example, if both equations are not true when a proposed ordered pair solution is substituted into the equations of the system, then the ordered pair is not a solution of the system.

2
Example 1 above shows how to determine that an ordered pair is a solution of a system of equations, but how do we find such a solution? Actually, there are various methods to find the solution. We will investigate several in this chapter: graphing, substitution, elimination, matrices, and determinants.

To solve by graphing, we graph each equation in an appropriate window and find the coordinates of any points of intersection.

EXAMPLE 2   Solve the system by graphing.

$$\begin{cases} x + y = 2 \\ 3x - y = -2 \end{cases}$$

Solution: Since the graph of a linear equation in two variables is a line, graphing two such equations yields two lines in a plane. To use a graphing utility, solve each equation for y.

$$\begin{cases} y = -x + 2 & \text{First equation.} \\ y = 3x + 2 & \text{Second equation.} \end{cases}$$

Graph $y_1 = -x + 2$ and $y_2 = 3x + 2$ and find the point of intersection.

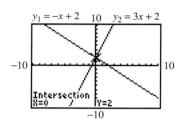

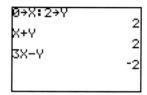

Verify the ordered pair solution $(0, 2)$ by replacing x with 0 and y with 2 in both original equations and seeing that true statements result each time. The screen on the right above shows that the ordered pair $(0, 2)$ does satisfy both equations. We conclude therefore that $(0, 2)$ is the solution of the system. A system that has at least one solution, such as this one, is said to be **consistent.**

DISCOVER THE CONCEPT

Use your graphing utility to solve the system
$$\begin{cases} x - 2y = 4 \\ x \quad\;\; = 2y \end{cases}$$

In the discovery above, we see that solving each equation for y produces the following:

$x - 2y = 4$	First equation.	$x = 2y$	Second equation.
$-2y = -x + 4$	Subtract x from both sides.	$\dfrac{1}{2}x = y$	Divide both sides by 2.
$y = \dfrac{1}{2}x - 2$	Divide both sides by -2.	$y = \dfrac{1}{2}x$	

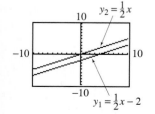

Notice that each equation is in the form $y = mx + b$. From this form, we see that both lines have the same slope, $\frac{1}{2}$, but different y-intercepts, so they are parallel as shown to the left. Therefore, the system has no solution since the equations have no common solution (there are no intersection points). A system that has no solution is said to be **inconsistent.**

DISCOVER THE CONCEPT

Use your graphing utility to solve the system

$$\begin{cases} 2x + 4y = 10 \\ x + 2y = 5 \end{cases}$$

In the discovery above, we see that solving each equation for y produces the following:

$$2x + 4y = 10 \quad \text{First equation.} \qquad x + 2y = 5 \qquad \qquad \text{Second equation.}$$

$$y = -\frac{1}{2}x + \frac{5}{2} \qquad\qquad\qquad\qquad y = -\frac{1}{2}x + \frac{5}{2}$$

Notice that both lines have the same slope, $-\frac{1}{2}$, and the same y-intercept, $\frac{5}{2}$. This means that the graph of each equation is the same line.

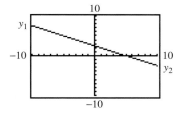

To confirm this, notice that the entries for y_1 and y_2 are the same in the table shown. The equations have identical solutions and any ordered pair solution of one equation satisfies the other equation also. Thus, these equations are said to be **dependent equations.** The solution set of the system is $\{(x, y) \mid x + 2y = 5\}$ or, equivalently, $\{(x, y) \mid 2x + 4y = 10\}$ since the equations describe identical ordered pairs. Written this way, the solution set is read "the set of all ordered pairs (x, y), such that $2x + 4y = 10$." There are therefore an infinite number of solutions to this system.

We can summarize the information discovered above as follows:

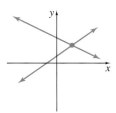

One solution:
consistent system;
independent equations

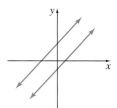

No solution:
inconsistent system;
independent equations

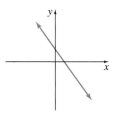

Infinite number
of solutions:
consistent system;
dependent equations

3

We next review two algebraic methods for solving a system of equations: the substitution method and the elimination method.

We introduce the first method, the **substitution method,** below.

EXAMPLE 3   Use the substitution method to solve the system

$$\begin{cases} y = x + 5 & \text{First equation.} \\ 3x = 2y - 9 & \text{Second equation.} \end{cases}$$

Solution: Remember that we are looking for an ordered pair, if there is one, that satisfies both equations. Satisfying the first equation, $y = x + 5$, means that y must be $x + 5$. **Substituting** the expression $x + 5$ for y in the second equation yields an equation in one variable, which we can solve for x.

$$3x = 2y - 9 \qquad \text{Second equation.}$$

$$3x = 2(x + 5) - 9 \qquad \text{Replace } y \text{ with } x + 5 \text{ in the second equation.}$$

$$3x = 2x + 10 - 9$$

$$x = 1$$

The x-coordinate of the solution of the system is 1. The y-coordinate is the y-value corresponding to the x-value 1. Choose either equation of the system and solve for y when x is 1.

$$y = x + 5 \qquad \text{First equation.}$$

$$= 1 + 5 \qquad \text{Let } x = 1.$$

$$= 6$$

The y-coordinate is 6, so the solution of the system is $(1, 6)$. This means that, when both equations are graphed, the one point of intersection occurs at the point with coordinates $(1, 6)$. We can check this solution by substituting 1 for x and 6 for y in both equations of the system, as shown below.

TO SOLVE A SYSTEM OF TWO EQUATIONS BY THE SUBSTITUTION METHOD

Step 1. Solve one of the equations for one of its variables.

Step 2. Substitute the expression for the variable found in *step 1* into the other equation.

Step 3. Find the value of one variable by solving the equation from *step 2*.

Step 4. Find the value of the other variable by substituting the value found in *step 3* in any equation of the system.

Step 5. Check the ordered pair solution in **both** of the original equations.

EXAMPLE 4   Use the substitution method to solve the system

$$\begin{cases} 2x + 4y = -6 & \text{First equation.} \\ x - 2y = -5 & \text{Second equation.} \end{cases}$$

Solution: We begin by solving the second equation for x since the coefficient 1 on the x-term keeps us from introducing tedious fractions. The equation $x - 2y = -5$ solved for x is $x = 2y - 5$. Substitute $2y - 5$ for x in the first equation.

$$2x + 4y = -6 \qquad \text{First equation.}$$

$$2(2y - 5) + 4y = -6 \qquad \text{Substitute } 2y - 5 \text{ for } x.$$

$$4y - 10 + 4y = -6$$

$$8y = 4$$

$$y = \frac{4}{8} = \frac{1}{2} \qquad \text{Solve for } y.$$

The y-coordinate of the solution is $\frac{1}{2}$. To find the x-coordinate, replace y with $\frac{1}{2}$ in the equation $x = 2y - 5$.

$$x = 2y - 5$$

$$x = 2\left(\frac{1}{2}\right) - 5 = 1 - 5 = -4$$

The solution set is $\left\{\left(-4, \frac{1}{2}\right)\right\}$. To check, see that $\left(-4, \frac{1}{2}\right)$ satisfies both equations of the system.

4

The **elimination method,** or **addition method,** is a second algebraic technique for solving systems of equations. For this method, we rely on a version of the addition property of equality, which states that "equals added to equals are equal." In symbols,

$$\text{if } A = B \text{ and } C = D \text{ then } A + C = B + D.$$

EXAMPLE 5

Use the elimination method to solve the system

$$\begin{cases} x - 5y = -12 & \text{First equation.} \\ -x + y = 4 & \text{Second equation.} \end{cases}$$

Solution:

Since the left side of each equation is equal to the right side, we add equal quantities by adding the left sides of the equations and the right sides of the equations. This sum gives us an equation in one variable, y, which we can solve for y.

$$\begin{array}{ll} x - 5y = -12 & \text{First equation.} \\ \underline{-x + y = 4} & \text{Second equation.} \\ -4y = -8 & \text{Add.} \\ y = 2 & \text{Solve for } y. \end{array}$$

The y-coordinate of the solution is 2. To find the corresponding x-coordinate, replace y with 2 in either original equation of the system.

$$\begin{array}{ll} -x + y = 4 & \text{Second equation.} \\ -x + 2 = 4 & \text{Let } y = 2. \\ -x = 2 \\ x = -2 \end{array}$$

The solution set is $\{(-2, 2)\}$. Check to see that $(-2, 2)$ satisfies both equations of the system.

TO SOLVE A SYSTEM OF TWO LINEAR EQUATIONS BY THE ELIMINATION METHOD

Step 1. Rewrite each equation in standard form, $Ax + By = C$.

Step 2. If necessary, multiply one or both equations by some nonzero number so that the coefficient of one variable in one equation is the opposite of its coefficient in the other equation.

Step 3. Add the equations.

Step 4. Find the value of one variable by solving the equation from *step 3*.

Step 5. Find the value of the second variable by substituting the value found in *step 4* into either of the original equations.

Step 6. Check the proposed ordered pair solution in both of the original equations.

EXAMPLE 6   Use the elimination method to solve the system

$$\begin{cases} 3x + \dfrac{y}{2} = 2 \\ 6x + y = 5 \end{cases}$$

Solution: If we add the two equations, the sum will still be an equation in two variables. Notice that if we multiply both sides of the first equation by -2, the coefficients of x in the two equations will be opposites. Then

$$\begin{cases} -2\left(3x + \dfrac{y}{2}\right) = -2(2) \\ 6x + y = 5 \end{cases} \quad \text{simplifies to} \quad \begin{cases} -6x - y = -4 \\ 6x + y = 5 \end{cases}$$

Next, add the left sides and add the right sides.

$$\begin{aligned} -6x - y &= -4 \\ \underline{6x + y} &= \underline{5} \\ 0 &= 1 \qquad \text{False.} \end{aligned}$$

The resulting equation, $0 = 1$, is false for all values of y or x. Thus, the system has no solution. The solution set is $\{\ \}$ or $\varnothing$.

This system is inconsistent, and the graphs of the equations are parallel lines.

EXAMPLE 7   Use the elimination method to solve the system

$$\begin{cases} 3x - 2y = 10 \\ 4x - 3y = 15 \end{cases}$$

Solution: To eliminate y when the equations are added, multiply both sides of the first equation by 3 and both sides of the second equation by -2. Then

$$\begin{cases} 3(3x - 2y) = 3(10) \\ -2(4x - 3y) = -2(15) \end{cases} \quad \text{simplifies to} \quad \begin{cases} 9x - 6y = 30 \\ -8x + 6y = -30 \end{cases}$$

Next, add the left sides and add the right sides.

$$\begin{aligned} 9x - 6y &= 30 \\ \underline{-8x + 6y} &= \underline{-30} \\ x &= 0 \end{aligned}$$

To find y, let $x = 0$ in either equation of the system.

$$\begin{aligned} 3x - 2y &= 10 \\ 3(0) - 2y &= 10 \qquad \text{Let } x = 0. \\ -2y &= 10 \\ y &= -5 \end{aligned}$$

The solution set is $\{(0, -5)\}$. Check to see that $(0, -5)$ satisfies both equations.

EXAMPLE 8   Use the elimination method to solve

$$\begin{cases} -5x - 3y = 9 \\ 10x + 6y = -18 \end{cases}$$

Solution: To eliminate x when the equations are added, multiply both sides of the first equation by 2. Then

$$\begin{cases} 2(-5x - 3y) = 2(9) \\ \quad 10x + 6y = -18 \end{cases} \quad \text{simplifies to} \quad \begin{cases} -10x - 6y = 18 \\ \quad 10x + 6y = -18 \end{cases}$$

Next, add the equations.

$$\begin{array}{r} -10x - 6y = 18 \\ 10x + 6y = -18 \\ \hline 0 = 0 \end{array}$$

The resulting equation, $0 = 0$, is true for all possible values of y or x. Notice in the original system that if both sides of the first equation are multiplied by -2 the result is the second equation. This means that the two equations are equivalent and that they have the same solution set. Thus, the equations of this system are dependent, and the solution set of the system is

$$\{(x, y) \,|\, -5x - 3y = 9\} \text{ or equivalently } \{(x, y) \,|\, 10x + 6y = -18\}$$

To algebraically solve a system of equations that has coefficients that are fractions, we first clear the equation of fractions and then solve by the substitution or elimination method.

EXAMPLE 9   Use the elimination method to solve the system

$$\begin{cases} -\dfrac{x}{6} + \dfrac{y}{2} = \dfrac{1}{2} \\ \dfrac{x}{3} - \dfrac{y}{6} = -\dfrac{3}{4} \end{cases}$$

Solution: First, multiply each equation by its least common denominator in order to write this system as an equivalent system without fractions. We multiply the first equation by 6 and the second equation by 12. (Just as for equations, equivalent systems are systems that have the same solution.)

$$\begin{cases} 6\left(-\dfrac{x}{6} + \dfrac{y}{2}\right) = 6\left(\dfrac{1}{2}\right) \\ 12\left(\dfrac{x}{3} - \dfrac{y}{6}\right) = 12\left(-\dfrac{3}{4}\right) \end{cases}$$

simplifies to

$$\begin{cases} -x + 3y = 3 & \text{First equation.} \\ 4x - 2y = -9 & \text{Second equation.} \end{cases}$$

To eliminate x when the equations are added, multiply both sides of the first equation by 4.

$$\begin{cases} 4(-x + 3y) = 4(3) \\ 4x - 2y = -9 \end{cases} \quad \text{simplifies to} \quad \begin{cases} -4x + 12y = 12 \\ 4x - 2y = -9 \end{cases}$$

Next, add the left sides and the right sides.

$$\begin{array}{r} -4x + 12y = 12 \\ 4x - 2y = -9 \\ \hline 10y = 3 \\ y = \dfrac{3}{10} \end{array}$$

The y-coordinate is $\dfrac{3}{10}$. To find the x-coordinate, replace y with $\dfrac{3}{10}$ in the equation $x = 3y - 3$. Then

$$x = 3\left(\dfrac{3}{10}\right) - 3 = \dfrac{9}{10} - 3 = \dfrac{9}{10} - \dfrac{30}{10} = -\dfrac{21}{10}.$$

The solution set is $\left\{\left(-\dfrac{21}{10}, \dfrac{3}{10}\right)\right\}$. Check to see that $\left(-\dfrac{21}{10}, \dfrac{3}{10}\right)$ satisfies both equations.

EXAMPLE 10 FINDING THE BREAK-EVEN POINT
A small manufacturing company manufactures and sells compact disc storage units. The revenue equation for these units is

$$y = 50x$$

where x is the number of units sold and y is the revenue, or income, in dollars for selling x units. The cost equation for these units is

$$y = 30x + 10,000$$

where x is the number of units manufactured and y is the total cost in dollars for manufacturing x units. Use these equations to find the number of units to be sold for the company to break even.

Solution: The break-even point is found by solving the system

$$\begin{cases} y = 50x & \text{First equation.} \\ y = 30x + 10{,}000 & \text{Second equation.} \end{cases}$$

To solve the system, graph $y_1 = 50x$ and $y_2 = 30x + 10{,}000$ and find the point of intersection, the break-even point.

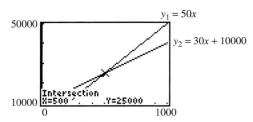

The ordered pair solution is $(500, 25{,}000)$. This means that the business must sell 500 compact disc storage units to break even. A hand-drawn graph of the equations in this system can be found at the beginning of this section.

EXERCISE SET 5.1

A system of equations and the graph of each equation of the system is given below. Find the solution of the system and verify that it is the solution. See Example 1.

1. $\begin{cases} 2x + 5y = 8 \\ 6x + y = 10 \end{cases}$

2. $\begin{cases} x + y = 1 \\ x - 2y = 4 \end{cases}$

3. $\begin{cases} x - 4y = -5 \\ -3x - 8y = 0 \end{cases}$

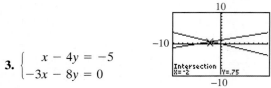

4. $\begin{cases} 2x - y = 8 \\ x - 3y = 11 \end{cases}$

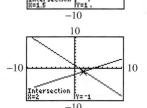

Solve each system by graphing. See Example 2.

5. $\begin{cases} x + y = 1 \\ x - 2y = 4 \end{cases}$

6. $\begin{cases} 2x - y = 8 \\ x + 3y = 11 \end{cases}$

7. $\begin{cases} 2y - 4 = 0 \\ x + 2y = 5 \end{cases}$

8. $\begin{cases} 4x - y = 6 \\ x - y = 0 \end{cases}$

9. $\begin{cases} 3x - y = 4 \\ 6x - 2y = 4 \end{cases}$

10. $\begin{cases} -x + 3y = 6 \\ 3x - 9y = 9 \end{cases}$

❑ 11. Can a system consisting of two linear equations have exactly two solutions? Explain why or why not.

❑ 12. Suppose the graph of the equations in a system of two equations in two variables consists of a circle and a line. Discuss the possible number of solutions for this system.

Solve each system of equations by the substitution method. See Examples 3 and 4.

13. $\begin{cases} x + y = 10 \\ \quad y = 4x \end{cases}$

14. $\begin{cases} 5x + 2y = -17 \\ \quad x = 3y \end{cases}$

15. $\begin{cases} 4x - y = 9 \\ 2x + 3y = -27 \end{cases}$

16. $\begin{cases} 3x - y = 6 \\ -4x + 2y = -8 \end{cases}$

17. $\begin{cases} \dfrac{x}{3} + y = \dfrac{4}{3} \\ -x + 2y = 11 \end{cases}$

18. $\begin{cases} \dfrac{x}{8} - \dfrac{y}{2} = 1 \\ \dfrac{x}{3} - y = 2 \end{cases}$

Solve each system of equations by the elimination method. See Examples 5–9.

19. $\begin{cases} 2x - 4y = 0 \\ \quad x + 2y = 5 \end{cases}$

20. $\begin{cases} 2x - 3y = 0 \\ 2x + 6y = 3 \end{cases}$

21. $\begin{cases} 5x + 2y = 1 \\ \quad x - 3y = 7 \end{cases}$

22. $\begin{cases} 6x - y = -5 \\ 4x - 2y = 6 \end{cases}$

23. $\begin{cases} 3x - 5y = 11 \\ 2x - 6y = 2 \end{cases}$

24. $\begin{cases} 6x - 3y = -3 \\ 4x + 5y = -9 \end{cases}$

25. $\begin{cases} x - 2y = 4 \\ 2x - 4y = 4 \end{cases}$

26. $\begin{cases} -x + 3y = 6 \\ 3x - 9y = 9 \end{cases}$

27. $\begin{cases} 3x + y = 1 \\ 2y = 2 - 6x \end{cases}$

28. $\begin{cases} y = 2x - 5 \\ 8x - 4y = 20 \end{cases}$

🔲 29. Write a system of two linear equations in x and y that has the ordered pair solution (2, 5).

🔲 30. Which method would you use to solve the system

$$\begin{cases} 5x - 2y = 6 \\ 2x + 3y = 5 \end{cases}$$

Explain your choice.

Solve each system of equations.

31. $\begin{cases} 2x + 5y = 8 \\ 6x + y = 10 \end{cases}$

32. $\begin{cases} x - 4y = -5 \\ -3x - 8y = 0 \end{cases}$

33. $\begin{cases} x + y = 1 \\ x - 2y = 4 \end{cases}$

34. $\begin{cases} 2x - y = 8 \\ x + 3y = 11 \end{cases}$

35. $\begin{cases} \dfrac{1}{3}x + y = \dfrac{4}{3} \\ -\dfrac{1}{4}x - \dfrac{1}{2}y = -\dfrac{1}{4} \end{cases}$

36. $\begin{cases} \dfrac{3}{4}x - \dfrac{1}{2}y = -\dfrac{1}{2} \\ x + y = -\dfrac{3}{2} \end{cases}$

37. $\begin{cases} 2x + 6y = 8 \\ 3x + 9y = 12 \end{cases}$

38. $\begin{cases} x = 3y - 1 \\ 2x - 6y = -2 \end{cases}$

39. $\begin{cases} 4x + 2y = 5 \\ 2x + y = -1 \end{cases}$

40. $\begin{cases} 3x + 6y = 15 \\ 2x + 4y = 3 \end{cases}$

41. $\begin{cases} 10y - 2x = 1 \\ 5y = 4 - 6x \end{cases}$

42. $\begin{cases} 3x + 4y = 0 \\ 7x = 3y \end{cases}$

43. $\begin{cases} \dfrac{3}{4}x + \dfrac{5}{2}y = 11 \\ \dfrac{1}{16}x - \dfrac{3}{4}y = -1 \end{cases}$

44. $\begin{cases} \dfrac{2}{3}x + \dfrac{1}{4}y = -\dfrac{3}{2} \\ \dfrac{1}{2}x - \dfrac{1}{4}y = -2 \end{cases}$

45. $\begin{cases} x = 3y + 2 \\ 5x - 15y = 10 \end{cases}$

46. $\begin{cases} y = \dfrac{1}{7}x + 3 \\ x - 7y = -21 \end{cases}$

47. $\begin{cases} 2x - y = -1 \\ \quad y = -2x \end{cases}$

48. $\begin{cases} x = \dfrac{1}{5}y \\ x - y = -4 \end{cases}$

49. $\begin{cases} 2x = 6 \\ y = 5 - x \end{cases}$

50. $\begin{cases} x = 3y + 4 \\ -y = 5 \end{cases}$

51. $\begin{cases} \dfrac{x + 5}{2} = \dfrac{6 - 4y}{3} \\ \dfrac{3x}{5} = \dfrac{21 - 7y}{10} \end{cases}$

52. $\begin{cases} \dfrac{y}{5} = \dfrac{8 - x}{2} \\ x = \dfrac{2y - 8}{3} \end{cases}$

53. $\begin{cases} 4x - 7y = 7 \\ 12x - 21y = 24 \end{cases}$

54. $\begin{cases} 2x - 5y = 12 \\ -4x + 10y = 20 \end{cases}$

55. $\begin{cases} 0.7x - 0.2y = -1.6 \\ 0.2x - y = -1.4 \end{cases}$

56. $\begin{cases} -0.7x + 0.6y = 1.3 \\ 0.5x - 0.3y = -0.8 \end{cases}$

Use a graphing utility to solve each system of equations. Approximate the solutions to two decimal places.

57. $y = -1.65x + 3.65$
$y = 4.56x - 9.44$

58. $y = 7.61x + 3.48$
$y = -1.26x - 6.43$

59. $2.33x - 4.72y = 10.61$
$5.86x + 6.22y = -8.89$

60. $-7.89x - 5.68y = 3.26$
$-3.65x + 4.98y = 11.77$

The concept of supply and demand is used often in business. In general, as the unit price of a commodity increases, the demand for that commodity decreases. Also, as a commodity's unit price increases, the manufacturer normally increases the supply. The point where supply is equal to demand is called the equilibrium point. The following graph shows the graph of a de-

mand equation and a supply equation for ties. The x-axis represents number of ties in thousands, and the y-axis represents the cost of a tie. Use this graph for Exercises 61–64.

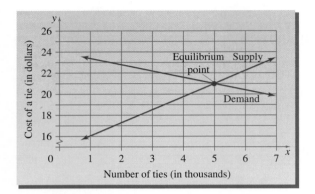

61. Find the number of ties and the price per tie when supply equals demand.

62. When x is between 3 and 4, is supply greater than demand or is demand greater than supply?

63. When x is greater than 7, is supply greater than demand or is demand greater than supply?

64. For what x-values are the y-values corresponding to the supply equation greater than the y-values corresponding to the demand equation?

The revenue equation for a certain brand of toothpaste is $y = 2.5x$, where x is the number of tubes of toothpaste sold and y is the total income for selling x tubes. The cost equation is $y = 0.9x + 3000$, where x is the number of tubes of toothpaste manufactured and y is the cost of producing x tubes. The following set of axes shows the graph of the cost and revenue equations. Use this graph for Exercises 65–70. See Example 10.

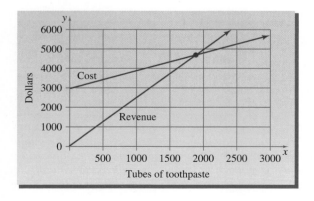

65. Find the coordinates of the point of intersection by solving the system

$$\begin{cases} y = 2.5x \\ y = 0.9x + 3000 \end{cases}$$

66. Explain the meaning of the ordered pair point of intersection.

67. If the company sells 2000 tubes of toothpaste, does the company make money or lose money?

68. If the company sells 1000 tubes of toothpaste, does the company make money or lose money?

69. For what x-values will the company make a profit? (*Hint:* For what x-values is the revenue graph "higher" than the cost graph?)

70. For what x-values will the company lose money? (*Hint:* For what x-values is the revenue graph "lower" than the cost graph?)

71. The amount of U.S. federal government income y (in billions of dollars) for the fiscal year x, from 1991 through 1994 ($x = 0$ represents 1990), can be modeled by the linear equation $y = 67x + 971$. The amount of U.S. federal government expenditures (in billions of dollars) for the same period can be modeled by the linear equation $y = 44x + 1285$. If these patterns in income and expenditures continue into the future, will income ever equal expenditures? If so, in what year? (*Source*: U.S. Department of the Treasury)

72. The number of milk cows y (in thousands) on farms in the United States for the year x, from 1970 through 1995 ($x = 0$ represents 1980), can be modeled by the linear equation $107x + y = 11{,}096$. The number of sheep (in thousands) on farms in the United States for the same period can be modeled by the linear equation $y = -399x + 15{,}149$. In which year were there the same number of milk cows as sheep? (*Source*: National Agricultural Statistics Service)

Review Exercises

Determine whether the given replacement values make each equation true or false. See Section 2.1.

73. $3x - 4y + 2z = 5$; $x = 1$, $y = 2$, and $z = 5$

74. $x + 2y - z = 7$; $x = 2$, $y = -3$, and $z = 3$

75. $-x - 5y + 3z = 15$; $x = 0$, $y = -1$, and $z = 5$

76. $-4x + y - 8z = 4; x = 1, y = 0,$ and $z = -1$

Add the equations. See Section 5.1.

77. $3x + 2y - 5z = 10$ **78.** $x + 4y - 5z = 20$
$\quad\;\; -3x + 4y + \;\; z = 15$ $\qquad\;\; 2x - 4y - 2z = -17$

79. $10x + 5y + 6z = 14$ **80.** $-9x - 8y - z = 31$
$\quad\;\; -9x + 5y - 6z = -12$ $\qquad\;\; 9x + 4y - z = 12$

A Look Ahead

EXAMPLE   Solve the system $\begin{cases} -\dfrac{4}{x} - \dfrac{4}{y} = -11 \\ \dfrac{1}{x} + \dfrac{1}{y} = 1 \end{cases}$

Solution: First, make the following substitution.

Let $a = \dfrac{1}{x}$ and $b = \dfrac{1}{y}$ in both equations. Then

$$\begin{cases} -4\left(\dfrac{1}{x}\right) - 4\left(\dfrac{1}{y}\right) = -11 \\ \dfrac{1}{x} + \dfrac{1}{y} = 1 \end{cases}$$

is equivalent to $\begin{cases} -4a - 4b = -11 \\ a + b = 1 \end{cases}$

We solve by the elimination method. Multiplying both sides of the second equation by 4 and adding the left sides and the right sides of the equations,

$$\begin{cases} -4a - 4b = -11 \\ 4(a + b) = 4(1) \end{cases}$$

simplifies to $\begin{cases} -4a - 4b = -11 \\ 4a + 4b = 4 \end{cases}$

$$0 = -7 \quad \text{False.}$$

The equation $0 = -7$ is false for all values of a and hence for all values of $\dfrac{1}{x}$ and all values of x. This system has no solution.

Solve each system. See the preceding example.

81. $\begin{cases} \dfrac{1}{x} + y = 12 \\ \dfrac{3}{x} - y = 4 \end{cases}$ **82.** $\begin{cases} x + \dfrac{2}{y} = 7 \\ 3x + \dfrac{3}{y} = 6 \end{cases}$

83. $\begin{cases} \dfrac{1}{x} + \dfrac{1}{y} = 5 \\ \dfrac{1}{x} - \dfrac{1}{y} = 1 \end{cases}$ **84.** $\begin{cases} \dfrac{2}{x} + \dfrac{3}{y} = 5 \\ \dfrac{5}{x} - \dfrac{3}{y} = 2 \end{cases}$

85. $\begin{cases} \dfrac{2}{x} + \dfrac{3}{y} = -1 \\ \dfrac{3}{x} - \dfrac{2}{y} = 18 \end{cases}$ **86.** $\begin{cases} \dfrac{3}{x} - \dfrac{2}{y} = -18 \\ \dfrac{2}{x} + \dfrac{3}{y} = 1 \end{cases}$

87. $\begin{cases} \dfrac{2}{x} - \dfrac{4}{y} = 5 \\ \dfrac{1}{x} - \dfrac{2}{y} = \dfrac{3}{2} \end{cases}$ **88.** $\begin{cases} \dfrac{5}{x} + \dfrac{7}{y} = 1 \\ -\dfrac{10}{x} - \dfrac{14}{y} = 0 \end{cases}$

5.2   SOLVING SYSTEMS OF LINEAR EQUATIONS IN THREE VARIABLES

TAPE IAG 5.2

O B J E C T I V E S

1 Recognize a linear equation in three variables.

2 Solve a system of three linear equations in three variables.

1 In this section, the algebraic methods of solving systems of two linear equations in two variables are extended to systems of three linear equations in three variables. We call the equation $3x - y + z = -15$, for example, a **linear equation in three variables** since there are three variables and each variable is raised only to the power 1. A solution of this equation is an **ordered triple (x, y, z)** that makes the equation a true statement. For example, the ordered triple $(2, 0, -21)$ is a solution of $3x - y + z = -15$ since replacing x with 2, y with 0, and z with -21 yields the

true statement $3(2) - 0 + (-21) = -15$. The graph of this equation is a plane in three-dimensional space, just as the graph of a linear equation in two variables is a line in two-dimensional space.

Although we will not discuss the techniques for graphing equations in three variables, visualizing the possible patterns of intersecting planes gives us insight into the possible patterns of solutions of a system of three three-variable linear equations. There are four possible patterns.

1. Three planes have a single point in common. This point represents the single solution of the system. The system is **consistent.**

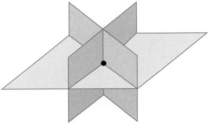

2. Three planes intersect at no point common to all three. This system has no solution. A few ways that this can occur are shown. This system is **inconsistent.**

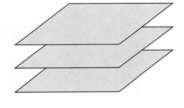
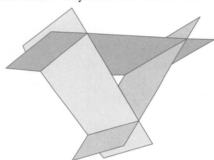

3. Three planes intersect at all the points of a single line. The system has infinitely many solutions. This system is **consistent.**

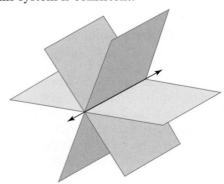

4. Three planes coincide at all points on the plane. The system is consistent, and the equations are **dependent.**

2   Using the elimination method to solve a system in three variables, we eliminate a variable and obtain a system in two variables.

EXAMPLE 1   Solve the system:

$$\begin{cases} 3x - y + z = -15 & \text{Equation (1)} \\ x + 2y - z = 1 & \text{Equation (2)} \\ 2x + 3y - 2z = 0 & \text{Equation (3)} \end{cases}$$

Solution: Add equations (1) and (2) to eliminate z.

$$\begin{array}{r} 3x - y + z = -15 \\ x + 2y - z = 1 \\ \hline 4x + y = -14 \end{array} \quad \text{Equation (4)}$$

Next, add two *other* equations and *eliminate z again*. To do so, multiply both sides of equation (1) by 2 and add this resulting equation to equation (3). Then

$$\begin{cases} 2(3x - y + z) = 2(-15) \\ 2x + 3y - 2z = 0 \end{cases} \quad \begin{matrix} \text{simplifies} \\ \text{to} \end{matrix} \quad \begin{cases} 6x - 2y + 2z = -30 \\ 2x + 3y - 2z = 0 \\ \hline 8x + y = -30 \end{cases} \text{Equation (5)}$$

Now solve equations (4) and (5) for x and y. To solve by elimination, multiply both sides of equation (4) by -1 and add this resulting equation to equation (5). Then

$$\begin{cases} -1(4x + y) = -1(-14) \\ 8x + y = -30 \end{cases} \quad \begin{matrix} \text{simplifies} \\ \text{to} \end{matrix} \quad \begin{cases} -4x - y = 14 \\ 8x + y = -30 \\ \hline 4x = -16 \quad \text{Add the equations.} \\ x = -4 \quad \text{Solve for } x. \end{cases}$$

Replace x with -4 in equation (4) or (5).

$$\begin{array}{rl} 4x + y = -14 & \text{Equation (4)} \\ 4(-4) + y = -14 & \text{Let } x = -4. \\ y = 2 & \text{Solve for } y. \end{array}$$

Finally, replace x with -4 and y with 2 in equation (1), (2), or (3).

$$x + 2y - z = 1 \qquad \text{Equation (2)}$$
$$-4 + 2(2) - z = 1 \qquad \text{Let } x = -4 \text{ and } y = 2.$$
$$-4 + 4 - z = 1$$
$$-z = 1$$
$$z = -1$$

The solution is $(-4, 2, -1)$. To check, let $x = -4$, $y = 2$, and $z = -1$ in all three original equations of the system. We can perform this check by paper and pencil or by calculator. Both methods are shown below.

EQUATION (1)	EQUATION (2)	EQUATION (3)
$3x - y + z = -15$	$x + 2y - z = 1$	$2x + 3y - 2z = 0$
$3(-4) - 2 + (-1) = -15$	$-4 + 2(2) - (-1) = 1$	$2(-4) + 3(2) - 2(-1) = 0$
$-12 - 2 - 1 = -15$	$-4 + 4 + 1 = 1$	$-8 + 6 + 2 = 0$
$-15 = -15$	$1 = 1$	$0 = 0$
True.	True.	True.

Checking by calculator,

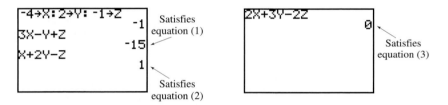

Satisfies equation (1)

Satisfies equation (2)

Satisfies equation (3)

All three statements are true, so the solution set is $\{(-4, 2, -1)\}$.

EXAMPLE 2   Solve the system:

$$\begin{cases} 2x - 4y + 8z = 2 & (1) \\ -x - 3y + z = 11 & (2) \\ x - 2y + 4z = 0 & (3) \end{cases}$$

Solution: Add equations (2) and (3) to eliminate x, and the new equation is

$$-5y + 5z = 11 \quad (4)$$

To eliminate x again, multiply both sides of equation (2) by 2, and add the resulting equation to equation (1). Then

$$\begin{cases} 2x - 4y + 8z = 2 \\ 2(-x - 3y + z) = 2(11) \end{cases} \quad \begin{matrix} \text{simplifies} \\ \text{to} \end{matrix} \quad \begin{cases} 2x - 4y + 8z = 2 \\ -2x - 6y + 2z = 22 \end{cases}$$
$$\overline{ -10y + 10z = 24 \quad (5)}$$

Next, solve for y and z using equations (4) and (5). Multiply both sides of equation (4) by -2, and add the resulting equation to equation (5).

$$\begin{cases} -2(-5y + 5z) = -2(11) \\ -10y + 10z = 24 \end{cases} \quad \begin{matrix} \text{simplifies} \\ \text{to} \end{matrix} \quad \begin{cases} 10y - 10z = -22 \\ -10y + 10z = 24 \end{cases}$$

$$\overline{ 0 = 2} \quad \text{False.}$$

Since the statement is false, this system is inconsistent and has no solution. The solution set is the empty set { } or $\varnothing$.

The elimination method is summarized next.

TO SOLVE A SYSTEM OF THREE LINEAR EQUATIONS BY THE ELIMINATION METHOD

Step 1.   Write each equation in standard form: $Ax + By + Cz = D$.

Step 2.   Choose a pair of equations and use the equations to eliminate a variable.

Step 3.   Choose any other pair of equations and eliminate the **same variable** as in *step 2*.

Step 4.   Two equations in two variables should be obtained from *step 2* and *step 3*. Use methods from Section 5.1 to solve this system for both variables.

Step 5.   To solve for the third variable, substitute the values of the variables found in *step 4* into any of the original equations containing the third variable.

Step 6.   Check the proposed ordered triple solution in all original equations of the system.

EXAMPLE 3   Solve the system:

$$\begin{cases} 2x + 4y = 1 & (1) \\ 4x - 4z = -1 & (2) \\ y - 4z = -3 & (3) \end{cases}$$

Solution: Notice that equation (2) has no term containing the variable y. Let us eliminate y using equations (1) and (3). Multiply both sides of equation (3) by -4, and add the resulting equation to equation (1). Then

$$\begin{cases} 2x + 4y = 1 \\ -4(\ y - 4z) = -4(-3) \end{cases} \quad \begin{matrix} \text{simplifies} \\ \text{to} \end{matrix} \quad \begin{array}{r} 2x + 4y = 1 \\ - 4y + 16z = 12 \\ \hline 2x + 16z = 13 \quad (4) \end{array}$$

Next, solve for z using equations (4) and (2). Multiply both sides of equation (4) by -2 and add the resulting equation to equation (2).

$$\begin{cases} -2(2x + 16z) = -2(13) \\ 4x - 4z = -1 \end{cases} \quad \begin{array}{c} \text{simplifies} \\ \text{to} \end{array} \quad \begin{cases} -4x - 32z = -26 \\ 4x - 4z = -1 \end{cases}$$

$$\begin{aligned} -36z &= -27 \\ z &= \frac{3}{4} \end{aligned}$$

Replace z with $\dfrac{3}{4}$ in equation (3) and solve for y.

$$y - 4\left(\frac{3}{4}\right) = -3 \qquad \text{Let } z = \frac{3}{4} \text{ in equation (3).}$$

$$y - 3 = -3$$

$$y = 0$$

Replace y with 0 in equation (1) and solve for x.

$$2x + 4(0) = 1$$

$$2x = 1$$

$$x = \frac{1}{2}$$

The solution set is $\left\{\left(\dfrac{1}{2}, 0, \dfrac{3}{4}\right)\right\}$. Check to see that this solution satisfies all three equations of the system.

EXAMPLE 4   Solve the system:

$$\begin{cases} x - 5y - 2z = 6 & (1) \\ -2x + 10y + 4z = -12 & (2) \\ \dfrac{1}{2}x - \dfrac{5}{2}y - z = 3 & (3) \end{cases}$$

Solution: Multiply both sides of equation (3) by 2 to eliminate fractions, and multiply both sides of equation (2) by $-\dfrac{1}{2}$ so that the coefficient of x is 1. The resulting system is then

$$\begin{cases} x - 5y - 2z = 6 & (1) \\ x - 5y - 2z = 6 & \text{Multiply (2) by } -\frac{1}{2}. \\ x - 5y - 2z = 6 & \text{Multiply (3) by 2.} \end{cases}$$

All three equations are identical, and therefore equations (1), (2), and (3) are all equivalent. There are infinitely many solutions of this system. The equations are dependent. The solution set can be written as $\{(x, y, z) \mid x - 5y - 2z = 6\}$.

EXERCISE SET 5.2

Solve each system. See Examples 1 and 3.

1. $\begin{cases} x + y = 3 \\ 2y = 10 \\ 3x + 2y - 3z = 1 \end{cases}$ **2.** $\begin{cases} 5x = 5 \\ 2x + y = 4 \\ 3x + y - 4z = -15 \end{cases}$

3. $\begin{cases} 2x + 2y + z = 1 \\ -x + y + 2z = 3 \\ x + 2y + 4z = 0 \end{cases}$ **4.** $\begin{cases} 2x - 3y + z = 5 \\ x + y + z = 0 \\ 4x + 2y + 4z = 4 \end{cases}$

Solve each system. See Examples 2 and 4.

5. $\begin{cases} x - 2y + z = -5 \\ -3x + 6y - 3z = 15 \\ 2x - 4y + 2z = -10 \end{cases}$

6. $\begin{cases} 3x + y - 2z = 2 \\ -6x - 2y + 4z = -2 \\ 9x + 3y - 6z = 6 \end{cases}$

7. $\begin{cases} 4x - y + 2z = 5 \\ 2y + z = 4 \\ 4x + y + 3z = 10 \end{cases}$ **8.** $\begin{cases} 5y - 7z = 14 \\ 2x + y + 4z = 10 \\ 2x + 6y - 3z = 30 \end{cases}$

🔲 **9.** Write a linear equation in three variables that has $(-1, 2, -4)$ as a solution. (There are many possibilities.)

🔲 **10.** Write a system of three linear equations in three variables that has $(2, 1, 5)$ as a solution. (There are many possibilities.)

Solve each system.

11. $\begin{cases} x + 5z = 0 \\ 5x + y = 0 \\ y - 3z = 0 \end{cases}$ **12.** $\begin{cases} x - 5y = 0 \\ x - z = 0 \\ -x + 5z = 0 \end{cases}$

13. $\begin{cases} 6x - 5z = 17 \\ 5x - y + 3z = -1 \\ 2x + y = -41 \end{cases}$ **14.** $\begin{cases} x + 2y = 6 \\ 7x + 3y + z = -33 \\ x - z = 16 \end{cases}$

15. $\begin{cases} x + y + z = 8 \\ 2x - y - z = 10 \\ x - 2y - 3z = 22 \end{cases}$ **16.** $\begin{cases} 5x + y + 3z = 1 \\ x - y + 3z = -7 \\ -x + y = 1 \end{cases}$

17. $\begin{cases} x + 2y - z = 5 \\ 6x + y + z = 7 \\ 2x + 4y - 2z = 5 \end{cases}$ **18.** $\begin{cases} 4x - y + 3z = 10 \\ x + y - z = 5 \\ 8x - 2y + 6z = 10 \end{cases}$

19. $\begin{cases} 2x - 3y + z = 2 \\ x - 5y + 5z = 3 \\ 3x + y - 3z = 5 \end{cases}$ **20.** $\begin{cases} 4x + y - z = 8 \\ x - y + 2z = 3 \\ 3x - y + z = 6 \end{cases}$

21. $\begin{cases} -2x - 4y + 6z = -8 \\ x + 2y - 3z = 4 \\ 4x + 8y - 12z = 16 \end{cases}$

22. $\begin{cases} -6x + 12y + 3z = -6 \\ 2x - 4y - z = 2 \\ -x + 2y + \dfrac{z}{2} = -1 \end{cases}$

23. $\begin{cases} 2x + 2y - 3z = 1 \\ y + 2z = -14 \\ 3x - 2y = -1 \end{cases}$ **24.** $\begin{cases} 7x + 4y = 10 \\ x - 4y + 2z = 6 \\ y - 2z = -1 \end{cases}$

25. $\begin{cases} \dfrac{3}{4}x - \dfrac{1}{3}y + \dfrac{1}{2}z = 9 \\ \dfrac{1}{6}x + \dfrac{1}{3}y - \dfrac{1}{2}z = 2 \\ \dfrac{1}{2}x - y + \dfrac{1}{2}z = 2 \end{cases}$ **26.** $\begin{cases} \dfrac{1}{3}x - \dfrac{1}{4}y + z = -9 \\ \dfrac{1}{2}x - \dfrac{1}{3}y - \dfrac{1}{4}z = -6 \\ x - \dfrac{1}{2}y - z = -8 \end{cases}$

🔲 **27.** The fraction $\frac{1}{24}$ can be written as the following sum:

$$\frac{1}{24} = \frac{x}{8} + \frac{y}{4} + \frac{z}{3}$$

where the numbers x, y, and z are solutions of

$$\begin{cases} x + y + z = 1 \\ 2x - y + z = 0 \\ -x + 2y + 2z = -1 \end{cases}$$

Solve the system and see that the sum of the fractions is $\frac{1}{24}$.

🔲 **28.** The fraction $\frac{1}{18}$ can be written as the following sum:

$$\frac{1}{18} = \frac{x}{2} + \frac{y}{3} + \frac{z}{9}$$

where the numbers x, y, and z are solutions of

$$\begin{cases} x + 3y + z = -3 \\ -x + y + 2z = -14 \\ 3x + 2y - z = 12 \end{cases}$$

Solve the system and see that the sum of the fractions is $\frac{1}{18}$.

Review Exercises

Solve. See Section 2.2.

29. The sum of two numbers is 45 and one number is twice the other. Find the numbers.

30. The difference between two numbers is 5. Twice the smaller number added to five times the larger number is 53. Find the numbers.

Solve. See Section 2.1.

31. $2(x - 1) - 3x = x - 12$

32. $7(2x - 1) + 4 = 11(3x - 2)$

33. $-y - 5(y + 5) = 3y - 10$

34. $z - 3(z + 7) = 6(2z + 1)$

A Look Ahead

Solve each system.

35. $\begin{cases} x + y \quad\;\; - w = 0 \\ y + 2z + w = 3 \\ x \quad\;\; - z \quad\;\; = 1 \\ 2x - y \quad\;\; - w = -1 \end{cases}$

36. $\begin{cases} 5x + 4y \quad\;\; = 29 \\ y + z - w = -2 \\ 5x \quad\;\; + z \quad\;\; = 23 \\ y - z + w = 4 \end{cases}$

37. $\begin{cases} x + y + z + w = 5 \\ 2x + y + z + w = 6 \\ x + y + z \quad\;\; = 2 \\ x + y \quad\;\; = 0 \end{cases}$

38. $\begin{cases} 2x \quad\;\; - z \quad\;\; = -1 \\ y + z + w = 9 \\ y \quad\;\; - 2w = -6 \\ x + y \quad\;\; = 3 \end{cases}$

TAPE IAG 5.3

5.3 SYSTEMS OF LINEAR EQUATIONS AND PROBLEM SOLVING

O B J E C T I V E

1 Solve problems that can be modeled by a system of linear equations.

Thus far, we have solved problems by writing one-variable equations and solving for the variable. Some of these problems can be solved, perhaps more easily, by writing a system of equations, as illustrated in this section. In order to review our problem-solving steps, we begin with a problem about numbers.

EXAMPLE 1 **FINDING UNKNOWN NUMBERS**
A first number is 4 less than a second number. Four times the first number is 6 more than twice the second. Find the numbers.

Solution: **1.** UNDERSTAND. Read and reread the problem and guess a solution. If one number is 10 and this is 4 less than a second number, the second number is 14. Four times the first number is 4(10), or 40. This is not equal to 6 more than twice the second number, which is 2(14) + 6 or 34. Although we guessed incorrectly, we now have a better understanding of the problem.

2. ASSIGN. Since we are looking for two numbers, we will let

x = first number

y = second number

4. TRANSLATE. Since we have assigned two variables to this problem, we will translate the given facts into two equations.

In words:	the first number	is	4 less than the second number
Translate:	x	$=$	$y - 4$

Next, we translate the second statement into an equation.

In words:	four times the first number	is	6 more than twice the second number

Translate: $\qquad 4x \qquad = \qquad 2y + 6$

5. COMPLETE. Here we solve the system

$$\begin{cases} x = y - 4 \\ 4x = 2y + 6 \end{cases}$$

Thus far we know three methods for solving systems of two equations in two unknowns. They are graphing, elimination, and substitution. Since the first equation expresses x in terms of y, we use substitution.

Substitute $y - 4$ for x in the second equation and solve for y.

$$4x = 2y + 6$$

$$4(y - 4) = 2y + 6 \qquad \text{Let } x = y - 4.$$
$$4y - 16 = 2y + 6$$
$$2y = 22$$
$$y = 11$$

Now replace y with 11 in the equation $x = y - 4$ and solve for x. Then $x = y - 4$ becomes $x = 11 - 4 = 7$. The solution of the system is (7, 11).

6. INTERPRET. Since the solution of the system is (7, 11), then the first number we are looking for is 7 and the second number is 11. *Checking,* notice that 7 *is* 4 less than 11, and 4 times 7 *is* 6 more than twice 11. The proposed numbers, 7 and 11, are correct.

State: The numbers are 7 and 11.

EXAMPLE 2   **FINDING THE SPEED OF A CAR**

Two cars leave Indianapolis, one traveling east and the other west. After 3 hours they are 297 miles apart. If one car is traveling 5 mph faster than the other, what is the speed of each?

Solution: **1. UNDERSTAND.** Read and reread the problem. Let's guess a solution and use the formula $d = r \cdot t$ to check. Suppose that one car is traveling at a rate of 55 miles per hour. This means that the other car is traveling at a rate of 50 miles per hour since we are told that one car is traveling 5 mph faster than the other. To find the distance apart after 3 hours, we will first find the distance traveled by each car. One car's distance is rate · time = 55(3) = 165 miles. The other car's distance is rate · time = 50(3) = 150 miles. Since one car is traveling east and the other west, their distance apart is the sum of their distances, or 165 miles + 150 miles = 315 miles. This distance apart is not the required distance of 297 miles.

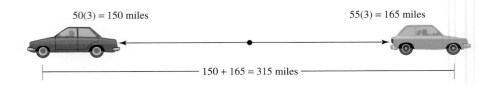

50(3) = 150 miles   55(3) = 165 miles

150 + 165 = 315 miles

Now that we have a better understanding of the problem, let's model it with a system of equations.

2. ASSIGN.   Let

x = speed of one car

y = speed of the other car

3. ILLUSTRATE.   We summarize the information on the following chart. Both cars have traveled 3 hours. Since distance = rate · time, their distances are $3x$ and $3y$ miles, respectively.

	RATE ·	**TIME** =	**DISTANCE**
ONE CAR	x	3	$3x$
OTHER CAR	y	3	$3y$

4. TRANSLATE.   We translate into two equations.

In words:	one car's distance	added to	the other car's distance	is	297
Translate:	$3x$	$+$	$3y$	$=$	297

In words:	one car	is	5 mph faster than the other
Translate:	x	$=$	$y + 5$

5. COMPLETE.   Here we solve the system.

$$\begin{cases} 3x + 3y = 297 \\ x \quad\quad = y + 5 \end{cases}$$

Although other methods may be used to solve the system, let's review the inter-section-of-graphs method. To use a graphing utility, solve each equation for y.

$y = -x + 99$ First equation.
$y = x - 5$ Second equation.

Graph $y_1 = -x + 99$ and $y_2 = x - 5$ and find the coordinates of the point of inter-section.

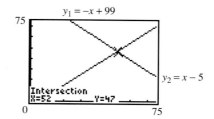

6. INTERPRET.   The solution (52, 47) means that the cars are traveling at 52 mph and 47 mph, respectively. To *check,* notice that one car is traveling 5 mph faster than the other. Also, if one car travels 52 mph for 3 hours, the distance is 3(52) = 156 miles. The other car traveling for 3 hours at 47 mph travels a distance of 3(47) = 141 miles. The sum of the distances 156 + 141 is 297 miles, the required distance.

State: The cars are traveling at 52 mph and 47 mph.

EXAMPLE 3

MIXING ALCOHOL SOLUTIONS
Lynn Pike, a pharmacist, needs 70 liters of a 50% alcohol solution. She has available a 30% alcohol solution and an 80% alcohol solution. How many liters of each solution should she mix to obtain 70 liters of a 50% alcohol solution?

Solution:
1. UNDERSTAND.   Read and reread the problem. Next, guess the solution. Suppose that we need 20 liters of the 30% solution. Then we need 70 − 20 = 50 liters of the 80% solution. To see if this gives us 70 liters of a 50% alcohol solution, let's find the amount of pure alcohol in each solution.

number of liters	×	alcohol strength	=	amount of pure alcohol
20 liters	×	0.30	=	6 liters
50 liters	×	0.80	=	40 liters
70 liters	×	0.50	=	35 liters

Since 6 liters + 40 liters = 46 liters and not 35 liters, our guess is incorrect, but we have gained some insight as to how to model and check this problem.

2. ASSIGN.   Let

x = amount of 30% solution, in liters

y = amount of 80% solution, in liters

3. ILLUSTRATE.   Use a table to organize the given data.

	NUMBER OF LITERS	ALCOHOL STRENGTH	AMOUNT OF PURE ALCOHOL
30% SOLUTION	x	30%	$0.30x$
80% SOLUTION	y	80%	$0.80y$
50% SOLUTION NEEDED	70	50%	$(0.50)(70)$

4. TRANSLATE. We translate into two equations.

In words: | amount of 30% solution | + | amount of 80% solution | = 70

Translate: x + y = 70

In words: | amount of pure alcohol in 30% solution | + | amount of pure alcohol in 80% solution | = | amount of pure alcohol in 50% solution |

Translate: $0.30x$ + $0.80y$ = $(0.50)(70)$

5. COMPLETE. Here we solve the system

$$\begin{cases} x + y = 70 \\ 0.30x + 0.80y = (0.50)(70) \end{cases}$$

To solve this system, use the elimination method. Multiply both sides of the first equation by -3 and both sides of the second equation by 10. Then

$$\begin{cases} -3(x + y) = -3(70) \\ 10(0.30x + 0.80y) = 10(0.50)(70) \end{cases}$$ simplifies to $\begin{cases} -3x - 3y = -210 \\ 3x + 8y = 350 \end{cases}$

$$5y = 140$$
$$y = 28$$

Replace y with 28 in the equation $x + y = 70$ and find that $x + 28 = 70$, or $x = 42$.

The solution of the system is (42, 28).

6. INTERPRET. To *check*, recall how we checked our guess. *State:* The pharmacist needs to mix 42 liters of 30% solution and 28 liters of 80% solution to obtain 70 liters of 50% solution.   ▬▬▬▬▬

   Recall that businesses are often computing cost and revenue functions or equations in order to predict needed sales, to determine whether prices need to be adjusted, and also to see whether the company is making money or losing money. Recall also that the value at which revenue equals cost is called the break-even point. When revenue is less than cost, the company is losing money, and when revenue is greater than cost, the company is making money.

EXAMPLE 4   **FINDING THE BREAK-EVEN POINT**

A manufacturing company recently purchased $3000 worth of new equipment in order to offer new personalized stationery to its customers. The cost of producing a package of personalized stationery is $3.00, and it is sold for $5.50. Find the number of packages that must be sold to break even.

Solution: **1. UNDERSTAND.** Read and reread the problem. Notice that the total production cost to the company includes a one-time cost of $3000 for the equipment and then $3.00 per package produced. The revenue is $5.50 per package sold.

2. ASSIGN.

Let x = number of packages of personalized stationery

Let $C(x)$ = total cost for producing x packages of stationery

Let $R(x)$ = total revenue for selling x packages of stationery

4. TRANSLATE.   The revenue equation is

	revenue for selling x packages of stationery	=	price per package	⋅	number of packages
In words:					

Translate: $R(x)$ =   5.5   ⋅ x

The cost equation is

	cost for producing x packages of stationery	=	cost per package	⋅	number of packages	+	cost for equipment
In words:							

Translate: $C(x)$ =   3   ⋅ x +   3000

Since the break-even point is when $R(x) = C(x)$, we solve the equation

$5.5x = 3x + 3000$

5. COMPLETE.

$5.5x = 3x + 3000$

$2.5x = 3000$ Subtract $3x$ from both sides.

$x = 1200$ Divide both sides by 2.5.

6. INTERPRET.   *Check:* To see whether the break-even point occurs when 1200 packages are produced and sold, see if revenue equals cost when $x = 1200$. When $x = 1200$, $R(x) = 5.5x = 5.5(1200) = 6600$ and $C(x) = 3x + 3000 = 3(1200) + 3000 = 6600$. Since $R(x) = C(x) = 6600$, the break-even point is (1200, 6600).

State: The company must sell 1200 packages of stationery in order to break even. The graph of this system follows.

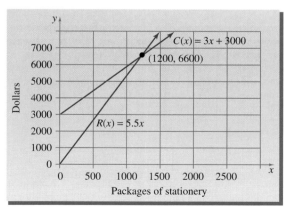

In order to introduce problem solving by writing a system of three linear equations in three unknowns, we solve a problem about triangles.

EXAMPLE 5

FINDING ANGLE MEASURE

The measure of the largest angle of a triangle is 80° more than the measure of the smallest angle, and the measure of the remaining angle is 10° more than the measure of the smallest angle. Find the measure of each angle.

Solution:

1. **UNDERSTAND.** Read and reread the problem. Recall that the sum of the measures of the angles of a triangle is 180°. Then guess a solution. If the smallest angle measures 20°, the measure of the largest angle is 80° more, or 20° + 80° = 100°. The measure of the remaining angle is 10° more than the measure of the smallest angle, or 20° + 10° = 30°. The sum of these three angles is 20° + 100° + 30° = 150°, not the required 180°. We now know that the measure of the smallest angle is greater than 20°.

2. **ASSIGN.** Let

x = degree measure of the smallest angle
y = degree measure of the largest angle
z = degree measure of the remaining angle

3. **ILLUSTRATE.**

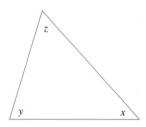

4. **TRANSLATE.** We translate into three equations.

In words:	the sum of the angles	=	180
Translate:	$x + y + z$	=	180

In words:	the largest angle	is	80 more than the smallest angle
Translate:	y	=	$x + 80$

In words:	the remaining angle	is	10 more than the smallest angle
Translate:	z	=	$x + 10$

5. COMPLETE.   We solve the system

$$\begin{cases} x + y + z = 180 \\ y = x + 80 \\ z = x + 10 \end{cases}$$

Since y and z are both expressed in terms of x, we will solve using the substitution method.

Substitute $y = x + 80$ and $z = x + 10$ in the first equation. Then

becomes

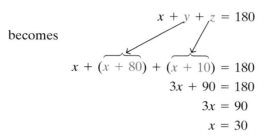

$$x + y + z = 180$$

$$x + (x + 80) + (x + 10) = 180$$
$$3x + 90 = 180$$
$$3x = 90$$
$$x = 30$$

Then $y = x + 80 = 30 + 80 = 110$, and $z = x + 10 = 30 + 10 = 40$. The angles measure 30°, 40°, and 110°.

6. INTERPRET.   To *check,* notice that $30° + 40° + 110° = 180°$. Also, the measure of the largest angle, 110°, is 80° more than the measure of the smallest angle, 30°. The measure of the remaining angle, 40°, is 10° more than the measure of the smallest angle, 30°.

State: The angles measure 30°, 40°, and 110°.

EXERCISE SET 5.3

Solve. See Examples 1–3.

1. One number is two more than a second number. Twice the first is 4 less than 3 times the second. Find the numbers.

2. Three times one number minus a second is 8, and the sum of the numbers is 12. Find the numbers.

3. A Delta 727 traveled 560 mph with the wind and 480 mph against the wind. Find the speed of the plane in still air and the speed of the wind.

4. Terry Watkins can row about 10.6 kilometers in 1 hour downstream and 6.8 kilometers upstream in 1 hour. Find how fast he can row in still water, and find the speed of the current.

5. Find how many quarts of 4% butterfat milk and 1% butterfat milk should be mixed to yield 60 quarts of 2% butterfat milk.

6. A pharmacist needs 500 milliliters of a 20% phenobarbital solution but has only 5% and 25% phenobarbital solutions available. Find how many milliliters of each he should mix to get the desired solution.

7. Karen Karlin bought some large frames for $15 each and some small frames for $8 each at a closeout sale. If she bought 22 frames for $239, find how many of each type she bought.

8. Hilton University Drama Club sold 311 tickets for a play. Student tickets cost 50 cents each; nonstudent tickets cost $1.50. If total receipts were $385.50, find how many tickets of each type were sold.

9. One number is two less than a second number. Twice the first is 4 more than 3 times the second. Find the numbers.

10. Twice one number plus a second number is 42, and

the one number minus the second number is −6. Find the numbers.

11. An office supply store in San Diego sells seven tablets and 4 pens for $6.40. Also, two tablets and 19 pens cost $5.40. Find the price of each.

12. A Candy Barrel shop manager mixes M&M's worth $2.00 per pound with trail mix worth $1.50 per pound. Find how many pounds of each she should use to get 50 pounds of a party mix worth $1.80 per pound.

13. An airplane takes 3 hours to travel a distance of 2160 miles with the wind. The return trip takes 4 hours against the wind. Find the speed of the plane in still air and the speed of the wind.

14. Two cyclists start at the same point and travel in opposite directions. One travels 4 mph faster than the other. In 4 hours they are 112 miles apart. Find how fast each is traveling.

15. The perimeter of a quadrilateral (four-sided polygon) is 29 inches. The longest side is twice as long as the shortest side. The other two sides are equally long and are 2 inches longer than the shortest side. Find the length of all four sides.

16. The perimeter of a triangle is 93 centimeters. If two sides are equally long and the third side is 9 centimeters longer than the others, find the lengths of the three sides.

17. The sum of three numbers is 40. One number is five more than a second and twice the third. Find the numbers.

18. The sum of the digits of a three-digit number is 15. The tens-place digit is twice the hundreds-place digit, and the ones-place digit is 1 less than the hundreds-place digit. Find the three-digit number.

19. Jack Reinholt, a car salesman, has a choice of two pay arrangements: a weekly salary of $200 plus 5% commission on sales, or a straight 15% commission. Find the amount of sales for which Jack's earnings are the same regardless of the pay arrangement.

20. Hertz car rental agency charges $25 daily plus 10 cents per mile. Budget charges $20 daily plus 25 cents per mile. Find the daily mileage for which the Budget charge for the day is twice that of the Hertz charge for the day.

21. Carroll Blakemore, a drafting student, bought 3 templates and a pencil one day for $6.45. Another day he bought 2 pads of paper and 4 pencils for

$7.50. If the price of a pad of paper is three times the price of a pencil, find the price of each type of item.

Given the cost function $C(x)$ and the revenue function $R(x)$, find the number of units x that must be sold to break even. See Example 4.

22. $C(x) = 30x + 10{,}000$
$R(x) = 46x$

23. $C(x) = 12x + 15{,}000$
$R(x) = 32x$

24. $C(x) = 1.2x + 1500$
$R(x) = 1.7x$

25. $C(x) = 0.8x + 900$
$R(x) = 2x$

26. $C(x) = 75x + 160{,}000$
$R(x) = 200x$

27. $C(x) = 105x + 70{,}000$
$R(x) = 245x$

28. The planning department of Abstract Office Supplies has been asked to determine whether the company should introduce a new computer desk next year. The department estimates that $6000 of new equipment will need to be purchased and that the cost of manufacturing each desk will be $200. The department also estimates that the revenue from each desk will be $450.

   a. Determine the revenue function $R(x)$ from the sale of x desks.

   b. Determine the cost function $C(x)$ for manufacturing x desks.

   c. Find the break-even point.

29. Baskets, Inc., is planning to introduce a new woven basket. The company estimates that $500 worth of new equipment will be needed to manufacture this new type of basket and that it will cost $15 per basket to manufacture. The company also estimates that the revenue from each basket will be $31.

   a. Determine the revenue function $R(x)$ from the sale of x baskets.

   b. Determine the cost function $C(x)$ for manufacturing x baskets.

   c. Find the break-even point.

30. Line l and line m are parallel lines cut by transversal t. Find the values of x and y.

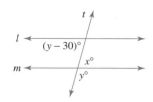

31. Find the values of x and y in the following isosceles triangle.

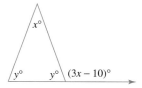

32. Office Station rents a printer for $35 plus $10 per hour. Stationery Plus rents the same printer for $25 per hour.

a. After how many hours would the charge for renting each printer be the same?

b. What would the charge be?

33. Distinctive Car Rental rents a particular car for $29 a day plus $0.35 a mile. A competing agency rents the same model car for $43 a day plus $0.21 a mile.

a. Assuming a one-day rental, after how many miles would the charge for renting the car be the same?

b. What would the charge be?

Solve. See Example 5.

34. Rabbits in a lab are to be kept on a strict daily diet to include 30 grams of protein, 16 grams of fat, and 24 grams of carbohydrates. The scientist has only three food mixes available with the following grams of nutrients per unit.

	PROTEIN	FAT	CARBOHYDRATE
MIX A	4	6	3
MIX B	6	1	2
MIX C	4	1	12

Find how many units of each mix are needed daily to meet one rabbit's dietary needs.

35. Gerry Gundersen mixes different solutions with concentrations of 25%, 40%, and 50% to get 200 liters of a 32% solution. If he uses twice as much of the 25% solution as of the 40% solution, find how many liters of each kind he uses.

36. Find the values of a, b, and c such that the equation $y = ax^2 + bx + c$ has ordered pair solutions $(1, 6)$, $(-1, -2)$, and $(0, -1)$. To do so, substitute each ordered pair solution into the equation. Each time, the result is an equation in three unknowns: a, b, and c. Then solve the resulting system of three linear equations in three unknowns, a, b, and c.

37. Find the values of a, b, and c such that the equation $y = ax^2 + bx + c$ has ordered pair solutions $(1, 2)$, $(2, 3)$ and $(-1, 6)$. (See Exercise 36.)

38. Find the values of x, y, and z in the following triangle.

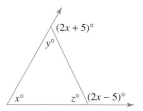

39. The sum of the measures of the angles of a quadrilateral is $360°$. Find the value of x, y, and z in the following quadrilateral.

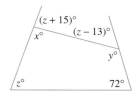

40. Data (x, y) for the total number y (in thousands) of college-bound students who took the ACT assessment in the year x are $(-1, 855)$, $(1, 796)$, $(4, 892)$, where $x = 0$ represents 1990. Find the values of a, b, and c such that the equation $y = ax^2 + bx + c$ models these data. According to your model, how many students will take the ACT assessment in 1999? (*Source*: The American College Testing Program)

41. Monthly normal rainfall data (x, y) for Portland, Oregon, are $(4, 2.47)$, $(7, 0.6)$, $(8, 1.1)$, where x represents time in months (with $x = 1$ representing January) and y represents rainfall in inches. Find the values of a, b, and c rounded to 2 decimal places such that the equation $y = ax^2 + bx + c$ models these data. According to your model, how much rain should Portland expect during September? (*Source*: National Climatic Data Center)

Review Exercises

Multiply both sides of equation (1) by 2, and add the resulting equation to equation (2). See Section 5.2.

42. $3x - y + z = 2$ (1)
$-x + 2y + 3z = 6$ (2)

43. $2x + y + 3z = 7$ (1)
$-4x + y + 2z = 4$ (2)

Multiply both sides of equation (1) by -3, and add the resulting equation to equation (2). See Section 5.2.

44. $x + 2y - z = 0$ (1)
$3x + y - z = 2$ (2)

45. $2x - 3y + 2z = 5$ (1)
$x - 9y + z = -1$ (2)

Given the spinner below, find the probability of the spinner landing on the indicated color in one spin. See Section 2.4.

46. $P(\text{red})$

47. $P(\text{green})$

48. $P(\text{white})$

49. $P(\text{red or blue})$

| 5.4 | SOLVING SYSTEMS OF EQUATIONS BY MATRICES |

TAPE IAG 5.4

O B J E C T I V E S

1. Use matrices to solve a system of two equations.
2. Use matrices to solve a system of three equations.
3. Use a calculator to solve a system of equations by matrices.

By now, you have seen that the solution of a system of equations depends on the coefficients of the equations in the system and not on the variables. In this section, we introduce solving a system of equations by a **matrix.**

1 A matrix (plural: **matrices**) is a rectangular array of numbers. The following are examples of matrices.

$$\begin{bmatrix} 1 & 0 \\ 0 & 1 \end{bmatrix}$$

2 × 2 matrix
2 rows, 2 columns

$$\begin{bmatrix} 2 & 1 & 3 & -1 \\ 0 & -1 & 4 & 5 \\ -6 & 2 & 1 & 0 \end{bmatrix}$$

3 × 4 matrix
3 rows, 4 columns

$$\begin{bmatrix} a & b & c \\ d & e & f \end{bmatrix}$$

2 × 3 matrix
2 rows, 3 columns

To see the relationship between systems of equations and matrices, consider this system of equations, written in standard form.

$$\begin{cases} 2x - 3y = 6 \\ x + y = 0 \end{cases}$$

A corresponding matrix associated with this system is

$$\begin{bmatrix} 2 & -3 & \vdots & 6 \\ 1 & 1 & \vdots & 0 \end{bmatrix}$$

The coefficients of each variable are placed to the left of a vertical dashed line. The constants are placed to the right. This 2 × 3 matrix is called the **augmented matrix**

of the system. Observe that the rows of this augmented matrix correspond to the equations in the system. The first equation corresponds to the first row; the second equation corresponds to the second row. Each number in the matrix is called an **element.**

The method of solving systems by matrices is to write the augmented matrix as an equivalent matrix from which we easily identify the solution. Two matrices are equivalent if they represent systems that have the same solution set. The following **row operations** can be performed on matrices, and the result is an equivalent matrix.

ELEMENTARY ROW OPERATIONS

1. Any two rows in a matrix may be interchanged.

2. The elements of any row may be multiplied (or divided) by the same nonzero number.

3. The elements of any row may be multiplied (or divided) by a nonzero number and added to its corresponding elements in any other row.

EXAMPLE 1 Solve the system using matrices.

$$\begin{cases} x + 3y = 5 \\ 2x - y = -4 \end{cases}$$

Solution: The augmented matrix is $\begin{bmatrix} 1 & 3 & | & 5 \\ 2 & -1 & | & -4 \end{bmatrix}$. Use elementary row operations to write an equivalent matrix that has 1's along the main diagonal and 0's below each 1 in the main diagonal. The main diagonal of a matrix is the left-to-right diagonal starting with row 1, column 1. For the matrix given, the element in the first row, first column is already 1, as desired. Next we write an equivalent matrix with a 0 below the 1. To do this, multiply row 1 by -2 and add to row 2. *We will change only row 2.*

$$\begin{bmatrix} 1 & 3 & | & 5 \\ -2(1)+2 & -2(3)+(-1) & | & -2(5)+(-4) \end{bmatrix} \text{ simplifies to } \begin{bmatrix} 1 & 3 & | & 5 \\ 0 & -7 & | & -14 \end{bmatrix}$$

row 1, row 2, row 1, row 2 elements; row 1, row 2 elements

Now continue down the main diagonal and change the -7 to a 1 by use of an elementary row operation. Divide row 2 by -7. Then

$$\begin{bmatrix} 1 & 3 & | & 5 \\ \frac{0}{-7} & \frac{-7}{-7} & | & \frac{-14}{-7} \end{bmatrix} \text{ simplifies to } \begin{bmatrix} 1 & 3 & | & 5 \\ 0 & 1 & | & 2 \end{bmatrix}$$

This last matrix corresponds to the system

$$\begin{cases} 1x + 3y = 5 \\ 0x + 1y = 2 \end{cases} \text{ or } \begin{cases} x + 3y = 5 \\ y = 2 \end{cases}$$

To find x, let $y = 2$ in the first equation, $x + 3y = 5$.

$$x + 3y = 5 \qquad \text{First equation.}$$
$$x + 3(2) = 5 \qquad \text{Let } y = 2.$$
$$x = -1$$

The solution set is $\{(-1, 2)\}$. Check to see that this ordered pair satisfies both equations.

---

2 Solving a system of three equations in three variables using matrices means writing the corresponding matrix and finding an equivalent matrix that has 1's along the main diagonal and 0's below the 1's.

EXAMPLE 2 Solve the system using matrices.

$$\begin{cases} x + 2y + z = 2 \\ -2x - y + 2z = 5 \\ x + 3y - 2z = -8 \end{cases}$$

Solution: The corresponding matrix is $\begin{bmatrix} 1 & 2 & 1 & | & 2 \\ -2 & -1 & 2 & | & 5 \\ 1 & 3 & -2 & | & -8 \end{bmatrix}$. Our goal is to write an equivalent matrix with 1's on the main diagonal and 0's below the 1's. The element in row 1, column 1 is already 1. Next we must get 0's for each element in the rest of column 1. To do this, first we multiply the elements of row 1 by 2 and add the new elements to row 2. Also, we multiply the elements of row 1 by -1 and add the new elements to the elements of row 3. We *do not change row 1*. Then

$$\begin{bmatrix} 1 & 2 & 1 & | & 2 \\ 2(1)-2 & 2(2)-1 & 2(1)+2 & | & 2(2)+5 \\ -1(1)+1 & -1(2)+3 & -1(1)-2 & | & -1(2)-8 \end{bmatrix} \text{ simplifies to } \begin{bmatrix} 1 & 2 & 1 & | & 2 \\ 0 & 3 & 4 & | & 9 \\ 0 & 1 & -3 & | & -10 \end{bmatrix}$$

We continue down the diagonal and use elementary row operations to get 1 where the element 3 is now. To do this, interchange rows 2 and 3.

$$\begin{bmatrix} 1 & 2 & 1 & | & 2 \\ 0 & 3 & 4 & | & 9 \\ 0 & 1 & -3 & | & -10 \end{bmatrix} \text{ is equivalent to } \begin{bmatrix} 1 & 2 & 1 & | & 2 \\ 0 & 1 & -3 & | & -10 \\ 0 & 3 & 4 & | & 9 \end{bmatrix}$$

Next, we want the new row 3, column 2 element to be 0. We multiply the elements of row 2 by -3 and add the result to the elements of row 3.

$$\begin{bmatrix} 1 & 2 & 1 & | & 2 \\ 0 & 1 & -3 & | & -10 \\ -3(0)+0 & -3(1)+3 & -3(-3)+4 & | & -3(-10)+9 \end{bmatrix} \text{ simplifies to } \begin{bmatrix} 1 & 2 & 1 & | & 2 \\ 0 & 1 & -3 & | & -10 \\ 0 & 0 & 13 & | & 39 \end{bmatrix}$$

Finally, we divide the elements of row 3 by 13 so that the final main diagonal element is 1.

$$\begin{bmatrix} 1 & 2 & 1 & \vdots & 2 \\ 0 & 1 & -3 & \vdots & -10 \\ \dfrac{0}{13} & \dfrac{0}{13} & \dfrac{13}{13} & \vdots & \dfrac{39}{13} \end{bmatrix} \quad \text{simplifies to} \quad \begin{bmatrix} 1 & 2 & 1 & \vdots & 2 \\ 0 & 1 & -3 & \vdots & -10 \\ 0 & 0 & 1 & \vdots & 3 \end{bmatrix}$$

This matrix corresponds to the system

$$\begin{cases} x + 2y + z = 2 \\ \quad\;\; y - 3z = -10 \\ \qquad\quad z = 3 \end{cases}$$

We identify the z-coordinate of the solution as 3. Next, we replace z with 3 in the second equation and solve for y.

$$\begin{aligned} y - 3z &= -10 & &\text{Second equation.} \\ y - 3(3) &= -10 & &\text{Let } z = 3. \\ y &= -1 \end{aligned}$$

To find x, we let $z = 3$ and $y = -1$ in the first equation.

$$\begin{aligned} x + 2y + z &= 2 & &\text{First equation.} \\ x + 2(-1) + 3 &= 2 & &\text{Let } z = 3 \text{ and } y = -1. \\ x &= 1 \end{aligned}$$

The solution set is $\{(1, -1, 3)\}$. Check to see that it satisfies the original system.

EXAMPLE 3   Solve the system using matrices.

$$\begin{cases} 2x - y = 3 \\ 4x - 2y = 5 \end{cases}$$

Solution: The corresponding augmented matrix is $\begin{bmatrix} 2 & -1 & \vdots & 3 \\ 4 & -2 & \vdots & 5 \end{bmatrix}$. To get 1 in the row 1, column 1 position, divide the elements of row 1 by 2.

$$\begin{bmatrix} \dfrac{2}{2} & -\dfrac{1}{2} & \vdots & \dfrac{3}{2} \\ 4 & -2 & \vdots & 5 \end{bmatrix} \quad \text{simplifies to} \quad \begin{bmatrix} 1 & -\dfrac{1}{2} & \vdots & \dfrac{3}{2} \\ 4 & -2 & \vdots & 5 \end{bmatrix}$$

To get 0 under the 1, multiply the elements of row 1 by -4 and add the new elements to the elements of row 2.

$$\begin{bmatrix} 1 & -\dfrac{1}{2} & \vdots & \dfrac{3}{2} \\ -4(1) + 4 & -4\left(-\dfrac{1}{2}\right) - 2 & \vdots & -4\left(\dfrac{3}{2}\right) + 5 \end{bmatrix} \quad \text{simplifies to} \quad \begin{bmatrix} 1 & -\dfrac{1}{2} & \vdots & \dfrac{3}{2} \\ 0 & 0 & \vdots & -1 \end{bmatrix}$$

The corresponding system is $\begin{cases} x - \dfrac{1}{2}y = \dfrac{3}{2}. \\ 0 = -1 \end{cases}$

The equation $0 = -1$ is false for all y or x values; hence the system is inconsistent and has no solution.

3 Thus far, we have solved systems of equations by matrices by writing the augmented matrix of the system as an equivalent matrix with 1's along the main diagonal and 0's below the 1's. Another way to solve a system of equations by matrices is to use the elementary row operations and write the augmented matrix of the system as an equivalent matrix with 1's along the main diagonal and 0's *above* and below these 1's. This form of a matrix is given a special name: the **reduced row echelon form.**

Although we will not practice writing matrices in reduced row echelon form in this text, we can easily see an advantage of this form.

Below is a system of equations, its corresponding augmented matrix and an equivalent matrix in reduced row echelon form found by performing elementary row operations on the augmented matrix.

SYSTEM OF EQUATIONS	AUGMENTED MATRIX OF THE SYSTEM	MATRIX IN REDUCED ROW ECHELON FORM
$\begin{aligned} 3x + y + 2z &= 5 \\ x - y + z &= 0 \\ -2x + 2y - z &= -1 \end{aligned}$	$\begin{bmatrix} 3 & 1 & 2 & \vert & 5 \\ 1 & -1 & 1 & \vert & 0 \\ -2 & 2 & -1 & \vert & -1 \end{bmatrix}$	$\begin{bmatrix} 1 & 0 & 0 & \vert & 2 \\ 0 & 1 & 0 & \vert & 1 \\ 0 & 0 & 1 & \vert & -1 \end{bmatrix}$

The corresponding system equivalent to the original system is

$$\begin{cases} 1x + 0y + 0z = 2 \\ 0x + 1y + 0z = 1 \\ 0x + 0y + 1z = -1 \end{cases} \quad \text{or} \quad \begin{cases} x \qquad\quad = 2 \\ \quad y \qquad = 1 \\ \qquad z = -1 \end{cases}$$

Notice that the solution of the system is $(2, 1, -1)$, which is found by reading the last column of the associated matrix written in reduced row echelon form.

Although using elementary row operations to write an augmented matrix in equivalent reduced row echelon form may be tedious for us, a calculator can do this quickly and accurately.

EXAMPLE 4 Solve the system of equations using matrices and a calculator.

$$\begin{cases} x + 3y = 5 \\ 2x - y = -4 \end{cases}$$

Solution: This is the same system of equations solved in Example 1. Recall that the augmented matrix associated with this system is

$$\begin{bmatrix} 1 & 3 & \vert & 5 \\ 2 & -1 & \vert & -4 \end{bmatrix}$$

Using the matrix edit feature, put the corresponding entries in matrix A. Then use a feature of your calculator to find the equivalent reduced row echelon form.

Augmented matrix in
reduced row echelon form

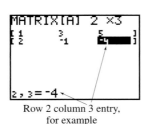

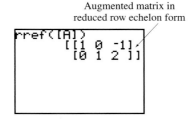

Row 2 column 3 entry,
for example

The augmented matrix in reduced row echelon form is $\begin{bmatrix} 1 & 0 & \vdots & -1 \\ 0 & 1 & \vdots & 2 \end{bmatrix}$, which corresponds to the system

$$\begin{cases} 1x + 0y = -1 \\ 0x + 1y = 2 \end{cases} \quad \text{or} \quad \begin{cases} x \quad\quad = -1 \\ \quad y = 2 \end{cases}$$

The solution set of the system is $\{(-1, 2)\}$, as confirmed in Example 1.

EXAMPLE 5   Solve the system using matrices and a calculator.

$$\begin{cases} x + 2y + z = 2 \\ -2x - y + 2z = 5 \\ x + 3y - 2z = -8 \end{cases}$$

Solution: This is the same system of equations solved in Example 2.

Enter the augmented matrix in matrix A and then use your calculator to find the reduced row echelon form of matrix A.

Dimensions of augmented
matrix

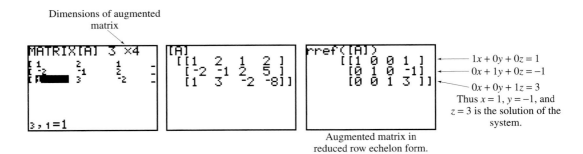

Augmented matrix in
reduced row echelon form.

$1x + 0y + 0z = 1$
$0x + 1y + 0z = -1$
$0x + 0y + 1z = 3$
Thus $x = 1$, $y = -1$, and
$z = 3$ is the solution of the
system.

This confirms that the solution set of the system of equations is $\{(1, -1, 3)\}$.

EXERCISE SET 5.4

Solve each system of linear equations using matrices. See Example 1.

1. $\begin{cases} x + y = 1 \\ x - 2y = 4 \end{cases}$

2. $\begin{cases} 2x - y = 8 \\ x + 3y = 11 \end{cases}$

3. $\begin{cases} x + 3y = 2 \\ x + 2y = 0 \end{cases}$

4. $\begin{cases} 4x - y = 5 \\ 3x - 3 = 0 \end{cases}$

Solve each system of linear equations using matrices. See Example 2.

5. $\begin{cases} x + y = 3 \\ 2y = 10 \\ 3x + 2y - 4z = 12 \end{cases}$

6. $\begin{cases} 5x = 5 \\ 2x + y = 4 \\ 3x + y - 5z = -15 \end{cases}$

7. $\begin{cases} 2y - z = -7 \\ x + 4y + z = -4 \\ 5x - y + 2z = 13 \end{cases}$

8. $\begin{cases} 4y + 3z = -2 \\ 5x - 4y = 1 \\ -5x + 4y + z = -3 \end{cases}$

Solve each system of linear equations using matrices. See Example 3.

9. $\begin{cases} x - 2y = 4 \\ 2x - 4y = 4 \end{cases}$

10. $\begin{cases} -x + 3y = 6 \\ 3x - 9y = 9 \end{cases}$

11. $\begin{cases} 3x - 3y = 9 \\ 2x - 2y = 6 \end{cases}$

12. $\begin{cases} 9x - 3y = 6 \\ -18x + 6y = -12 \end{cases}$

Solve each system of linear equations using matrices. See Examples 1 through 5.

13. $\begin{cases} x - 4 = 0 \\ x + y = 1 \end{cases}$

14. $\begin{cases} 3y = 6 \\ x + y = 7 \end{cases}$

15. $\begin{cases} x + y + z = 2 \\ 2x - z = 5 \\ 3y + z = 2 \end{cases}$

16. $\begin{cases} x + 2y + z = 5 \\ x - y - z = 3 \\ y + z = 2 \end{cases}$

17. $\begin{cases} 5x - 2y = 27 \\ -3x + 5y = 18 \end{cases}$

18. $\begin{cases} 4x - y = 9 \\ 2x + 3y = -27 \end{cases}$

19. $\begin{cases} 4x - 7y = 7 \\ 12x - 21y = 24 \end{cases}$

20. $\begin{cases} 2x - 5y = 12 \\ -4x + 10y = 20 \end{cases}$

21. $\begin{cases} 4x - y + 2z = 5 \\ 2y + z = 4 \\ 4x + y + 3z = 10 \end{cases}$

22. $\begin{cases} 5y - 7z = 14 \\ 2x + y + 4z = 10 \\ 2x + 6y - 3z = 30 \end{cases}$

23. $\begin{cases} 4x + y + z = 3 \\ -x + y - 2z = -11 \\ x + 2y + 2z = -1 \end{cases}$

24. $\begin{cases} x + y + z = 9 \\ 3x - y + z = -1 \\ -2x + 2y - 3z = -2 \end{cases}$

Solve using matrices.

25. An office supply store in San Diego sells 7 tablets and 4 pens for $6.40. You can also buy 2 tablets and 19 pens for $5.40. Find the price of each. Compare your result with the result of Exercise 11, Section 5.3.

26. A Candy Barrel shop manager mixes M&M's worth $2.00 per pound with trail mix worth $1.50 per pound. Find how many pounds of each she should use to get 50 pounds of a party mix worth $1.80 per pound. Compare your result with the result of Exercise 12, Section 5.3.

27. Karen Karlin bought some large frames for $15 each and some small frames for $8 each at a close-out sale. If she bought 22 frames for $239, find how many of each type she bought. Compare your result with the result of Exercise 7, Section 5.3.

28. Hilton University Drama Club sold 311 tickets for a play. Student tickets cost 50¢ each; nonstudent tickets cost $1.50 each. If the total receipts were $385.50, find how many tickets of each type were sold. Compare your result with the result of Exercise 8, Section 5.3.

29. Carra is investing $4800 in three different investments. The first investment pays 6.2% simple interest, the second pays 7.5% simple interest, and the third pays 8% simple interest. She invests $200 more in the one paying 6.2% than in the one paying 7.5% interest. The total amount of annual interest earned is $352.80. Find how much she invests at each rate.

30. Lisa invests $3800 in three different investments paying 7.2%, 6.8%, and 5.9%, simple interest respectively. The amount invested at 5.9% is equal to the sum of the amounts invested at the 7.2% and 6.8% rates. The total annual return on the investments is $244.10. Find how much is invested at each rate.

Review Exercise

Determine whether each graph is the graph of a function. See Section 3.3.

31.

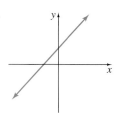

32.

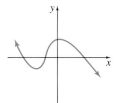

33.

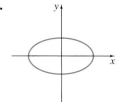

34.

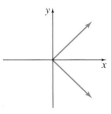

Evaluate. See Section 1.3.

35. $(-1)(-5) - (6)(3)$ **36.** $(2)(-8) - (-4)(1)$

37. $(4)(-10) - (2)(-2)$ **38.** $(-7)(3) - (-2)(-6)$

39. $(-3)(-3) - (-1)(-9)$ **40.** $(5)(6) - (10)(10)$

5.5 SOLVING SYSTEMS OF EQUATIONS BY DETERMINANTS

O B J E C T I V E S

TAPE IAG 5.5

1. Define and evaluate a 2×2 determinant.
2. Use Cramer's rule to solve a system of two linear equations in two variables.
3. Define and evaluate a 3×3 determinant.
4. Use Cramer's rule to solve a linear system of three equations in three variables.
5. Use a calculator to find determinants.

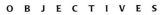

 We have solved systems of two linear equations in two variables in four different ways: graphically, by substitution, by elimination, and by matrices. Now we analyze another method called **Cramer's rule.**

Recall that a matrix is a rectangular array of numbers. If a matrix has the same number of rows and columns, it is called a **square matrix.** Examples of square matrices are

$$\begin{bmatrix} 1 & 6 \\ 5 & 2 \end{bmatrix} \qquad \begin{bmatrix} 2 & 4 & 1 \\ 0 & 5 & 2 \\ 3 & 6 & 9 \end{bmatrix}$$

A **determinant** is a real number associated with a square matrix. The determinant of a square matrix is denoted by placing vertical bars about the array of numbers.

Thus,

The determinant of the square matrix $\begin{bmatrix} 1 & 6 \\ 5 & 2 \end{bmatrix}$ is denoted by $\begin{vmatrix} 1 & 6 \\ 5 & 2 \end{vmatrix}$.

The determinant of the square matrix $\begin{bmatrix} 2 & 4 & 1 \\ 0 & 5 & 2 \\ 3 & 6 & 9 \end{bmatrix}$ is denoted by $\begin{vmatrix} 2 & 4 & 1 \\ 0 & 5 & 2 \\ 3 & 6 & 9 \end{vmatrix}$.

We define the determinant of a 2×2 matrix first.

DETERMINANT OF A 2×2 MATRIX

$$\begin{vmatrix} a & b \\ c & d \end{vmatrix} = ad - bc$$

EXAMPLE 1 Evaluate.

a. $\begin{vmatrix} -1 & 2 \\ 3 & -4 \end{vmatrix}$ 　　　　　　**b.** $\begin{vmatrix} 2 & 0 \\ 7 & -5 \end{vmatrix}$

Solution: First, identify the values of a, b, c, and d.

a. Here $a = -1$, $b = 2$, $c = 3$, and $d = -4$.

$$\begin{vmatrix} -1 & 2 \\ 3 & -4 \end{vmatrix} = ad - bc = (-1)(-4) - (2)(3) = -2$$

b. In this example, $a = 2$, $b = 0$, $c = 7$, and $d = -5$.

$$\begin{vmatrix} 2 & 0 \\ 7 & -5 \end{vmatrix} = ad - bc = 2(-5) - (0)(7) = -10$$

2　To develop Cramer's rule, we solve by elimination the system $\begin{cases} ax + by = h \\ cx + dy = k \end{cases}$.
First, eliminate y by multiplying both sides of the first equation by d and both sides of the second equation by $-b$ so that the coefficients of y are opposites. The result is the following system:

$$\begin{cases} d(ax + by) = d \cdot h \\ -b(cx + dy) = -b \cdot k \end{cases} \quad \text{simplifies to} \quad \begin{cases} adx + bdy = hd \\ -bcx - bdy = -kb \end{cases}$$

We now add the two equations and solve for x.

$$
\begin{aligned}
adx + bdy &= hd \\
-bcx - bdy &= -kb \\
\hline
adx - bcx &= hd - kb \qquad \text{Add the equations.} \\
(ad - bc)x &= hd - kb \\
x &= \frac{hd - kb}{ad - bc} \qquad \text{Solve for } x.
\end{aligned}
$$

When we replace x with $\dfrac{hd - kb}{ad - bc}$ in the equation $ax + by = h$ and solve for y, we find that $y = \dfrac{ak - ch}{ad - bc}$.

Notice that the numerator of the value of x is the determinant

$$\begin{vmatrix} h & b \\ k & d \end{vmatrix} = hd - kb$$

Also, the numerator of the value of y is the determinant

$$\begin{vmatrix} a & h \\ c & k \end{vmatrix} = ak - hc$$

Finally, the denominator of the values of x and y is the same and is the determinant

$$\begin{vmatrix} a & b \\ c & d \end{vmatrix} = ad - bc$$

This means that the values of x and y can be written in determinant notation:

$$x = \dfrac{\begin{vmatrix} h & b \\ k & d \end{vmatrix}}{\begin{vmatrix} a & b \\ c & d \end{vmatrix}} \quad \text{and} \quad y = \dfrac{\begin{vmatrix} a & h \\ c & k \end{vmatrix}}{\begin{vmatrix} a & b \\ c & d \end{vmatrix}}$$

For convenience, we label the determinants D, D_x, and D_y.

x-coefficients

y-coefficients

$$\begin{vmatrix} a & b \\ c & d \end{vmatrix} = D \qquad \begin{vmatrix} h & b \\ k & d \end{vmatrix} = D_x \qquad \begin{vmatrix} a & h \\ c & k \end{vmatrix} = D_y$$

x-column replaced by constants

y-column replaced by constants

These determinant formulas for the coordinates of the solution of a system are known as **Cramer's rule.**

Cramer's Rule for Two Linear Equations in Two Variables

The solution of the system $\begin{cases} ax + by = h \\ cx + dy = k \end{cases}$ is given by

$$x = \dfrac{\begin{vmatrix} h & b \\ k & d \end{vmatrix}}{\begin{vmatrix} a & b \\ c & d \end{vmatrix}} = \dfrac{D_x}{D}; \qquad y = \dfrac{\begin{vmatrix} a & h \\ c & k \end{vmatrix}}{\begin{vmatrix} a & b \\ c & d \end{vmatrix}} = \dfrac{D_y}{D}$$

as long as $D = ad - bc$ is not 0.

When $D = 0$, the system is either inconsistent or the equations are dependent. When this happens, use another method to see which is the case.

EXAMPLE 2   Use Cramer's rule to solve each system.

a. $\begin{cases} 3x + 4y = -7 \\ x - 2y = -9 \end{cases}$ **b.** $\begin{cases} 5x + y = 5 \\ -7x - 2y = -7 \end{cases}$

Solution: **a.** Find D, D_x, and D_y.

$$\begin{array}{ccc} a & b & h \\ \downarrow & \downarrow & \downarrow \end{array}$$
$$\begin{cases} 3x + 4y = -7 \\ x - 2y = -9 \end{cases}$$
$$\begin{array}{ccc} \uparrow & \uparrow & \uparrow \\ c & d & k \end{array}$$

$$D = \begin{vmatrix} a & b \\ c & d \end{vmatrix} = \begin{vmatrix} 3 & 4 \\ 1 & -2 \end{vmatrix} = 3(-2) - 4(1) = -10$$

$$D_x = \begin{vmatrix} h & b \\ k & d \end{vmatrix} = \begin{vmatrix} -7 & 4 \\ -9 & -2 \end{vmatrix} = (-7)(-2) - 4(-9) = 50$$

$$D_y = \begin{vmatrix} a & h \\ c & k \end{vmatrix} = \begin{vmatrix} 3 & -7 \\ 1 & -9 \end{vmatrix} = 3(-9) - (-7)(1) = -20$$

Then $x = \dfrac{D_x}{D} = \dfrac{50}{-10} = -5$ and $y = \dfrac{D_y}{D} = \dfrac{-20}{-10} = 2$. The solution set is $\{(-5, 2)\}$.

As always, check the solution in both original equations.

b. $\begin{cases} 5x + y = 5 \\ -7x - 2y = -7 \end{cases}$ Find D, D_x, and D_y.

$$D = \begin{vmatrix} 5 & 1 \\ -7 & -2 \end{vmatrix} = 5(-2) - (-7)(1) = -3$$

$$D_x = \begin{vmatrix} 5 & 1 \\ -7 & -2 \end{vmatrix} = 5(-2) - (-7)(1) = -3$$

$$D_y = \begin{vmatrix} 5 & 5 \\ -7 & -7 \end{vmatrix} = 5(-7) - 5(-7) = 0$$

$$x = \frac{D_x}{D} = \frac{-3}{-3} = 1, \qquad y = \frac{D_y}{D} = \frac{0}{-3} = 0$$

The solution set is $\{(1, 0)\}$.

3   Three-by-three determinants can be used to solve systems of three equations in three variables. The determinant of a 3×3 matrix, however, is considerably more complex than the 2×2 case.

DETERMINANT OF A 3×3 MATRIX

$$\begin{vmatrix} a_1 & b_1 & c_1 \\ a_2 & b_2 & c_2 \\ a_3 & b_3 & c_3 \end{vmatrix} = a_1 \cdot \begin{vmatrix} b_2 & c_2 \\ b_3 & c_3 \end{vmatrix} - a_2 \cdot \begin{vmatrix} b_1 & c_1 \\ b_3 & c_3 \end{vmatrix} + a_3 \cdot \begin{vmatrix} b_1 & c_1 \\ b_2 & c_2 \end{vmatrix}$$

The determinant of a 3×3 matrix, then, is related to the determinants of three 2×2 matrices. Each determinant of these 2×2 matrices is called a **minor,** and every element of a 3×3 matrix has a minor associated with it. For example, the minor of c_2 is the determinant of the 2×2 matrix found by deleting the row and column containing c_2.

$$\begin{matrix} a_1 & b_1 & c_1 \\ a_2 & b_2 & c_2 \\ a_3 & b_3 & c_3 \end{matrix}$$

The minor of c_2 is

$$\begin{vmatrix} a_1 & b_1 \\ a_3 & b_3 \end{vmatrix}$$

Also, the minor of element a_1 is the determinant of the 2×2 matrix found by deleting the row or column containing a_1.

$$\begin{matrix} a_1 & b_1 & c_1 \\ a_2 & b_2 & c_2 \\ a_3 & b_3 & c_3 \end{matrix}$$

The minor of a_1 is

$$\begin{vmatrix} b_2 & c_2 \\ b_3 & c_3 \end{vmatrix}$$

So the determinant of a 3×3 matrix can be written as

$$a_1(\text{minor of } a_1) - a_2(\text{minor of } a_2) + a_3(\text{minor of } a_3)$$

Finding the determinant by using minors of elements in the first column is called **expanding** by the minors of the first column. *The value of a determinant can be found by expanding by the minors of any row or column.* The following **array of signs** is helpful in determining whether to add or subtract the product of an element and its minor.

$$\begin{matrix} + & - & + \\ - & + & - \\ + & - & + \end{matrix}$$

If an element is in a position marked $+$, we add. If marked $-$, we subtract.

EXAMPLE 3 Evaluate by expanding by the minors of the given row or column.

$$\begin{vmatrix} 0 & 5 & 1 \\ 1 & 3 & -1 \\ -2 & 2 & 4 \end{vmatrix}$$

a. First column   **b.** Second row

Solution: **a.** The elements of the first column are 0, 1, and -2. The first column of the array of signs is $+, -, +$.

$$\begin{vmatrix} 0 & 5 & 1 \\ 1 & 3 & -1 \\ -2 & 2 & 4 \end{vmatrix} = 0 \cdot \begin{vmatrix} 3 & -1 \\ 2 & 4 \end{vmatrix} - 1 \cdot \begin{vmatrix} 5 & 1 \\ 2 & 4 \end{vmatrix} + (-2) \cdot \begin{vmatrix} 5 & 1 \\ 3 & -1 \end{vmatrix}$$

$$= 0(12 + 2) - 1(20 - 2) + (-2)(-5 - 3)$$

$$= 0 - 18 + 16 = -2$$

b. The elements of the second row are 1, 3, and -1. This time, the signs begin with $-$ and again alternate.

$$\begin{vmatrix} 0 & 5 & 1 \\ 1 & 3 & -1 \\ -2 & 2 & 4 \end{vmatrix} = -1 \cdot \begin{vmatrix} 5 & 1 \\ 2 & 4 \end{vmatrix} + 3 \cdot \begin{vmatrix} 0 & 1 \\ -2 & 4 \end{vmatrix} - (-1) \cdot \begin{vmatrix} 0 & 5 \\ -2 & 2 \end{vmatrix}$$

$$= -1(20 - 2) + 3(0 - (-2)) - (-1)(0 - (-10))$$

$$= -18 + 6 + 10 = -2$$

Notice that the determinant of the 3×3 matrix is the same regardless of the row or column you select to expand by.

4 A system of three equations in three variables may be solved with Cramer's rule also. Using the elimination process to solve a system with unknown constants as coefficients leads to the following:

CRAMER'S RULE FOR THREE EQUATIONS IN THREE VARIABLES

The solution of the system $\begin{cases} a_1x + b_1y + c_1z = k_1 \\ a_2x + b_2y + c_2z = k_2 \\ a_3x + b_3y + c_3z = k_3 \end{cases}$ is given by

$$x = \frac{D_x}{D}, \quad y = \frac{D_y}{D}, \quad \text{and} \quad z = \frac{D_z}{D},$$

where

$$D = \begin{vmatrix} a_1 & b_1 & c_1 \\ a_2 & b_2 & c_2 \\ a_3 & b_3 & c_3 \end{vmatrix} \qquad D_x = \begin{vmatrix} k_1 & b_1 & c_1 \\ k_2 & b_2 & c_2 \\ k_3 & b_3 & c_3 \end{vmatrix}$$

$$D_y = \begin{vmatrix} a_1 & k_1 & c_1 \\ a_2 & k_2 & c_2 \\ a_3 & k_3 & c_3 \end{vmatrix} \qquad D_z = \begin{vmatrix} a_1 & b_1 & k_1 \\ a_2 & b_2 & k_2 \\ a_3 & b_3 & k_3 \end{vmatrix}$$

as long as D is not 0.

EXAMPLE 4   Use Cramer's rule to solve the system

$$\begin{cases} x - 2y + z = 4 \\ 3x + y - 2z = 3 \\ 5x + 5y + 3z = -8 \end{cases}$$

Solution: First, find D, D_x, D_y, and D_z. Beginning with D, we expand by the minors of the first column.

$$D = \begin{vmatrix} 1 & -2 & 1 \\ 3 & 1 & -2 \\ 5 & 5 & 3 \end{vmatrix} = 1 \cdot \begin{vmatrix} 1 & -2 \\ 5 & 3 \end{vmatrix} - 3 \cdot \begin{vmatrix} -2 & 1 \\ 5 & 3 \end{vmatrix} + 5 \cdot \begin{vmatrix} -2 & 1 \\ 1 & -2 \end{vmatrix}$$

$$= 1(3 - (-10)) - 3(-6 - 5) + 5(4 - 1)$$

$$= 13 + 33 + 15 = 61$$

$$D_x = \begin{vmatrix} 4 & -2 & 1 \\ 3 & 1 & -2 \\ -8 & 5 & 3 \end{vmatrix} = 4 \cdot \begin{vmatrix} 1 & -2 \\ 5 & 3 \end{vmatrix} - 3 \cdot \begin{vmatrix} -2 & 1 \\ 5 & 3 \end{vmatrix} + (-8) \cdot \begin{vmatrix} -2 & 1 \\ 1 & -2 \end{vmatrix}$$

$$= 4(3 - (-10)) - 3(-6 - 5) + (-8)(4 - 1)$$

$$= 52 + 33 - 24 = 61$$

$$D_y = \begin{vmatrix} 1 & 4 & 1 \\ 3 & 3 & -2 \\ 5 & -8 & 3 \end{vmatrix} = 1 \cdot \begin{vmatrix} 3 & -2 \\ -8 & 3 \end{vmatrix} - 3 \cdot \begin{vmatrix} 4 & 1 \\ -8 & 3 \end{vmatrix} + 5 \cdot \begin{vmatrix} 4 & 1 \\ 3 & -2 \end{vmatrix}$$

$$= 1(9 - 16) - 3(12 + 8) + 5(-8 - 3)$$

$$= -7 - 60 - 55 = -122$$

$$D_z = \begin{vmatrix} 1 & -2 & 4 \\ 3 & 1 & 3 \\ 5 & 5 & -8 \end{vmatrix} = 1 \cdot \begin{vmatrix} 1 & 3 \\ 5 & -8 \end{vmatrix} - 3 \cdot \begin{vmatrix} -2 & 4 \\ 5 & -8 \end{vmatrix} + 5 \cdot \begin{vmatrix} -2 & 4 \\ 1 & 3 \end{vmatrix}$$

$$= 1(-8 - 15) - 3(16 - 20) + 5(-6 - 4)$$

$$= -23 + 12 - 50 = -61$$

From these determinants, we calculate the solution:

$$x = \frac{D_x}{D} = \frac{61}{61} = 1, \quad y = \frac{D_y}{D} = \frac{-122}{61} = -2, \quad z = \frac{D_z}{D} = \frac{-61}{61} = -1$$

The solution set of the system is $\{(1, -2, -1)\}$. Check this solution by verifying that it satisfies each equation of the system.

A calculator may be used to find determinants. In Example 1(a), we found that $\begin{vmatrix} -1 & 2 \\ 3 & -4 \end{vmatrix} = -2.$

To check with a calculator, enter the elements of the determinant $\begin{vmatrix} -1 & 2 \\ 3 & -4 \end{vmatrix}$ in matrix A and then find the determinant of matrix A.

```
[A]
        [[-1  2 ]
         [3  -4]]
det([A])
             -2
```

Since calculators may be used to find determinants, and therefore, may also be used to solve systems of equations by Cramer's rule.

EXAMPLE 5 Solve the system using Cramer's rule and a calculator.

$$\begin{cases} 3x + 4y = -7 \\ x - 2y = -9 \end{cases}$$

Solution: This is the same system as Example 2(a). Recall that

$$D = \begin{vmatrix} 3 & 4 \\ 1 & -2 \end{vmatrix}, \quad D_x = \begin{vmatrix} -7 & 4 \\ -9 & -2 \end{vmatrix}, \quad D_y = \begin{vmatrix} 3 & -7 \\ 1 & -9 \end{vmatrix}$$

To evaluate D, D_x, and D_y, enter the elements of D in matrix A, D_x in matrix B, and D_y in matrix C and evaluate the determinant of each matrix. Then

$$x = \frac{D_x}{D} = \frac{\det [B]}{\det [A]} = -5 \quad \text{and} \quad y = \frac{D_y}{D} = \frac{\det [C]}{\det [A]} = 2$$

```
det([B])/det([A]
)
             -5
det([C])/det([A]
)
              2
```

Thus, the solution set is $\{(-5, 2)\}$.

EXERCISE SET 5.5

Evaluate. See Example 1.

1. $\begin{vmatrix} 3 & 5 \\ -1 & 7 \end{vmatrix}$

2. $\begin{vmatrix} -5 & 1 \\ 0 & -4 \end{vmatrix}$

3. $\begin{vmatrix} 9 & -2 \\ 4 & -3 \end{vmatrix}$

4. $\begin{vmatrix} 4 & 0 \\ 9 & 8 \end{vmatrix}$

5. $\begin{vmatrix} -2 & 9 \\ 4 & -18 \end{vmatrix}$

6. $\begin{vmatrix} -40 & 8 \\ 70 & -14 \end{vmatrix}$

Use Cramer's rule, if possible, to solve each system of linear equations. See Examples 2 and 5.

7. $\begin{cases} 2y - 4 = 0 \\ x + 2y = 5 \end{cases}$

8. $\begin{cases} 4x - y = 5 \\ 3x - 3 = 0 \end{cases}$

9. $\begin{cases} 3x + y = 1 \\ 2y = 2 - 6x \end{cases}$

10. $\begin{cases} y = 2x - 5 \\ 8x - 4y = 20 \end{cases}$

11. $\begin{cases} 5x - 2y = 27 \\ -3x + 5y = 18 \end{cases}$

12. $\begin{cases} 4x - y = 9 \\ 2x + 3y = -27 \end{cases}$

Evaluate. See Example 3.

13. $\begin{vmatrix} 2 & 1 & 0 \\ 0 & 5 & -3 \\ 4 & 0 & 2 \end{vmatrix}$

14. $\begin{vmatrix} -6 & 4 & 2 \\ 1 & 0 & 5 \\ 0 & 3 & 1 \end{vmatrix}$

15. $\begin{vmatrix} 4 & -6 & 0 \\ -2 & 3 & 0 \\ 4 & -6 & 1 \end{vmatrix}$

16. $\begin{vmatrix} 5 & 2 & 1 \\ 3 & -6 & 0 \\ -2 & 8 & 0 \end{vmatrix}$

17. $\begin{vmatrix} 3 & 6 & -3 \\ -1 & -2 & 3 \\ 4 & -1 & 6 \end{vmatrix}$

18. $\begin{vmatrix} 2 & -2 & 1 \\ 4 & 1 & 3 \\ 3 & 1 & 2 \end{vmatrix}$

Use Cramer's rule, if possible, to solve each system of linear equations. See Examples 4 and 5.

19. $\begin{cases} 3x \quad\quad + z = -1 \\ -x - 3y + z = 7 \\ \quad\quad 3y + z = 5 \end{cases}$

20. $\begin{cases} \quad\quad 4y - 3z = -2 \\ 8x - 4y \quad\quad = 4 \\ -8x + 4y + z = -2 \end{cases}$

21. $\begin{cases} x + y + z = 8 \\ 2x - y - z = 10 \\ x - 2y + 3z = 22 \end{cases}$

22. $\begin{cases} 5x + y + 3z = 1 \\ x - y - 3z = -7 \\ -x + y \quad\quad = 1 \end{cases}$

Evaluate.

23. $\begin{vmatrix} 10 & -1 \\ -4 & 2 \end{vmatrix}$

24. $\begin{vmatrix} -6 & 2 \\ 5 & -1 \end{vmatrix}$

25. $\begin{vmatrix} 1 & 0 & 4 \\ 1 & -1 & 2 \\ 3 & 2 & 1 \end{vmatrix}$

26. $\begin{vmatrix} 0 & 1 & 2 \\ 3 & -1 & 2 \\ 3 & 2 & -2 \end{vmatrix}$

27. $\begin{vmatrix} \frac{3}{4} & \frac{5}{2} \\ -\frac{1}{6} & \frac{7}{3} \end{vmatrix}$

28. $\begin{vmatrix} \frac{5}{7} & \frac{1}{3} \\ \frac{6}{7} & \frac{2}{3} \end{vmatrix}$

29. $\begin{vmatrix} 4 & -2 & 2 \\ 6 & -1 & 3 \\ 2 & 1 & 1 \end{vmatrix}$

30. $\begin{vmatrix} 1 & 5 & 0 \\ 7 & 9 & -4 \\ 3 & 2 & -2 \end{vmatrix}$

31. $\begin{vmatrix} -2 & 5 & 4 \\ 5 & -1 & 3 \\ 4 & 1 & 2 \end{vmatrix}$

32. $\begin{vmatrix} 5 & -2 & 4 \\ -1 & 5 & 3 \\ 1 & 4 & 2 \end{vmatrix}$

33. If all the elements in a single row of a determinant are zero, to what does the determinant evaluate? Explain your answer.

34. If all the elements in a single column of a determinant are 0, to what does the determinant evaluate? Explain your answer.

Find the value of x such that each is a true statement.

35. $\begin{vmatrix} 1 & x \\ 2 & 7 \end{vmatrix} = -3$

36. $\begin{vmatrix} 6 & 1 \\ -2 & x \end{vmatrix} = 26$

Use Cramer's rule, if possible, to solve each system of linear equations.

37. $\begin{cases} 2x - 5y = 4 \\ x + 2y = -7 \end{cases}$

38. $\begin{cases} 3x - y = 2 \\ -5x + 2y = 0 \end{cases}$

39. $\begin{cases} 4x + 2y = 5 \\ 2x + y = -1 \end{cases}$

40. $\begin{cases} 3x + 6y = 15 \\ 2x + 4y = 3 \end{cases}$

41. $\begin{cases} 2x + 2y + z = 1 \\ -x + y + 2z = 3 \\ x + 2y + 4z = 0 \end{cases}$

42. $\begin{cases} 2x + 3y + z = 5 \\ x + y + z = 0 \\ 4x + 2y + 4z = 4 \end{cases}$

43. $\begin{cases} \dfrac{2}{3}x - \dfrac{3}{4}y = -1 \\ -\dfrac{1}{6}x + \dfrac{3}{4}y = \dfrac{5}{2} \end{cases}$

44. $\begin{cases} \dfrac{1}{2}x - \dfrac{1}{3}y = -3 \\ \dfrac{1}{8}x + \dfrac{1}{6}y = 0 \end{cases}$

45. $\begin{cases} 0.7x - 0.2y = -1.6 \\ 0.2x - y = -1.4 \end{cases}$

46. $\begin{cases} -0.7x + 0.6y = 1.3 \\ 0.5x - 0.3y = -0.8 \end{cases}$

47. $\begin{cases} -2x + 4y - 2z = 6 \\ x - 2y + z = -3 \\ 3x - 6y + 3z = -9 \end{cases}$

48. $\begin{cases} -x - y + 3z = 2 \\ 4x + 4y - 12z = -8 \\ -3x - 3y + 9z = 6 \end{cases}$

49. $\begin{cases} x - 2y + z = -5 \\ 3y + 2z = 4 \\ 3x - y = -2 \end{cases}$

50. $\begin{cases} 4x + 5y = 10 \\ 3y + 2z = -6 \\ x + y + z = 3 \end{cases}$

Review Exercises

Simplify each expression. See Section 1.4.

51. $5x - 6 + x - 12$

52. $4y + 3 - 15y - 1$

53. $2(3x - 6) + 3(x - 1)$

54. $-3(2y - 7) - 1(11 + 12y)$

Graph each function. See Section 3.4.

55. $f(x) = 5x - 6$

56. $g(x) = -x + 1$

57. $h(x) = 3$

58. $f(x) = -3$

A Look Ahead

EXAMPLE Evaluate the determinant.

$$\begin{vmatrix} 2 & 0 & -1 & 3 \\ 0 & 5 & -2 & -1 \\ 3 & 1 & 0 & 1 \\ 4 & 2 & -2 & 0 \end{vmatrix}$$

Solution:

To evaluate a 4×4 determinant, select any row or column and expand by the minors. The array of signs for a 4×4 determinant is the same as for a 3×3 determinant except expanded. We expand using the fourth row.

$$\begin{vmatrix} 2 & 0 & -1 & 3 \\ 0 & 5 & -2 & -1 \\ 3 & 1 & 0 & 1 \\ \rightarrow 4 & 2 & -2 & 0 \end{vmatrix}$$

$$= -4 \cdot \begin{vmatrix} 0 & -1 & 3 \\ 5 & -2 & -1 \\ 1 & 0 & 1 \end{vmatrix} + 2 \cdot \begin{vmatrix} 2 & -1 & 3 \\ 0 & -2 & -1 \\ 3 & 0 & 1 \end{vmatrix}$$

$$- (-2) \cdot \begin{vmatrix} 2 & 0 & 3 \\ 0 & 5 & -1 \\ 3 & 1 & 1 \end{vmatrix} + 0 \cdot \begin{vmatrix} 2 & 0 & -1 \\ 0 & 5 & -2 \\ 3 & 1 & 0 \end{vmatrix}$$

Now find the value of each 3×3 determinant. The value of the 4×4 determinant is

$$-4(12) + 2(17) + 2(-33) + 0 = -80$$

Find the value of each determinant. See the preceding example.

59. $\begin{vmatrix} 5 & 0 & 0 & 0 \\ 0 & 4 & 2 & -1 \\ 1 & 3 & -2 & 0 \\ 0 & -3 & 1 & 2 \end{vmatrix}$

60. $\begin{vmatrix} 1 & 7 & 0 & -1 \\ 1 & 3 & -2 & 0 \\ 1 & 0 & -1 & 2 \\ 0 & -6 & 2 & 4 \end{vmatrix}$

61. $\begin{vmatrix} 4 & 0 & 2 & 5 \\ 0 & 3 & -1 & 1 \\ 0 & 0 & 2 & 0 \\ 0 & 0 & 0 & 1 \end{vmatrix}$

62. $\begin{vmatrix} 2 & 0 & -1 & 4 \\ 6 & 0 & 4 & 1 \\ 2 & 4 & 3 & -1 \\ 4 & 0 & 5 & -4 \end{vmatrix}$

GROUP ACTIVITY

LOCATING LIGHTNING STRIKES

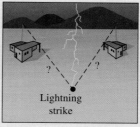

MATERIALS:
- Graphing Utility

Weather-recording stations use a directional antenna to detect and measure the electromagnetic field emitted by a lightning bolt. The antenna can determine the angle between a fixed point and the position of the lightning strike (see figure) but cannot determine the distance to the lightning strike. However, the angle measured by the antenna can be used to find the slope of the line connecting the positions of the weather station and the lightning strike. From there, the equation of the line connecting the positions of the weather station and the lightning strike may be found. If two such lines may be found—that is, if another weather station's antenna detects the same lightning flash—the coordinates of the lightning strike's position may be pinpointed.

1. A weather-recording station A is located at the coordinates $(35, 28)$, and a second station B is at $(52, 12)$. Plot the positions of the two weather-recording stations.

2. A lightning strike is detected by both stations. Station A uses a measured angle to find the slope of the line from the station to the lightning strike as $m = -1.732$. Station B computes a slope of $m = 0.577$ from the angle it measured. Use this information to find the equations of the lines connecting each station to the position of the lightning strike.

3. Have each group member solve the resulting system of equations in one of the following four ways:
   (a) Using a graphing utility to graph the lines and an intersect feature to find the coordinates of their point of intersection
   (b) Using either the method of substitution or of elimination (whichever you prefer)
   (c) Using matrices
   (d) Using Cramer's rule

Compare your results. What are the coordinates of the lightning strike?

CHAPTER 5 HIGHLIGHTS

DEFINITIONS AND CONCEPTS	**EXAMPLES**

| **SECTION 5.1 SOLVING SYSTEMS OF LINEAR EQUATIONS IN TWO VARIABLES** ||

A **system of linear equations** consists of two or more linear equations.

Systems of Linear Equations:

$$\begin{cases} x - 3y = 6 \\ y = \dfrac{1}{2}x \end{cases} \qquad \begin{cases} x + 2y - z = 1 \\ 3x - y + 4z = 0 \\ 5y + z = 6 \end{cases}$$

A **solution** of a system of two equations in two variables is an ordered pair (x, y) that makes both equations true.

Determine whether $(2, -5)$ is a solution of the system.

$$\begin{cases} x + y = -3 \\ 2x - 3y = 19 \end{cases}$$

Replace x with 2 and y with -5 in both equations.

$$x + y = -3 \qquad\qquad 2x - 3y = 19$$
$$2 + (-5) = -3 \qquad\qquad 2(2) - 3(-5) = 19$$
$$-3 = -3 \quad \text{True.} \qquad\qquad 4 + 15 = 19$$
$$\qquad\qquad 19 = 19 \quad \text{True.}$$

$(2, -5)$ is a solution of the system.

Geometrically, a solution of a system in two variables is a point common to the graphs of the equations.

Solve by graphing $\begin{cases} y = 2x - 1 \\ x + 2y = 13 \end{cases}$

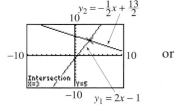

$$y_2 = -\tfrac{1}{2}x + \tfrac{13}{2}$$
$$y_1 = 2x - 1$$

or

the solution is $(3, 5)$.

A system of equations with at least one solution is a **consistent system.** A system that has no solution is an **inconsistent system.**

If the graphs of two linear equations are identical, the equations are **dependent.**

If their graphs are different, the equations are **independent.**

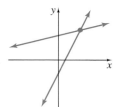

One solution;
consistent and
independent

No solution;
inconsistent and
independent

(continued)

DEFINITIONS AND CONCEPTS	EXAMPLES

SECTION 5.1   SOLVING SYSTEMS OF LINEAR EQUATIONS IN TWO VARIABLES

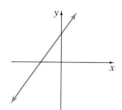

Infinite number
of solutions; consistent
and dependent

To solve a system of linear equations by the **substitution method:**

Step 1. Solve one equation for a variable.

Step 2. Substitute the expression for the variable into the other equation.

Step 3. Solve the equation from *step 2* to find the value of one variable.

Step 4. Substitute the value from *step 3* in either original equation to find the value of the other variable.

Step 5. Check the solution in both equations.

Solve by substitution:

$$\begin{cases} y = x + 2 \\ 3x - 2y = -5 \end{cases}$$

Substitute $x + 2$ for y in the second equation.

$$3x - 2y = -5$$
$$3x - 2(x + 2) = -5$$
$$3x - 2x - 4 = -5$$
$$x - 4 = -5 \qquad \text{Simplify.}$$
$$x = -1 \qquad \text{Add 4.}$$

To find y, let $x = -1$ in $y = x + 2$, so $y = -1 + 2 = 1$. The solution set $\{(-1, 1)\}$ checks.

To solve a system of linear equations by the **elimination method:**

Step 1. Rewrite each equation in standard form: $Ax + By = C$.

Step 2. Multiply one or both equations by a nonzero number so that the coefficients of a variable are opposites.

Step 3. Add the equations.

Step 4. Find the value of one variable by solving the resulting equation.

Step 5. Substitute the value from *step 4* into either original equation to find the value of the other variable.

Step 6. Check the solution in both equations.

Solve by elimination:

$$\begin{cases} x - 3y = -3 \\ -2x + y = 6 \end{cases}$$

Multiply both sides of the first equation by 2.

$$2x - 6y = -6$$
$$\underline{-2x + y = 6}$$
$$-5y = 0 \qquad \text{Add.}$$
$$y = 0 \qquad \text{Divide by } -5.$$

To find x, let $y = 0$ in an original equation.

$$x - 3y = -3$$
$$x - 3 \cdot 0 = -3$$
$$x = -3$$

The solution set $\{(-3, 0)\}$ checks.   *(continued)*

DEFINITIONS AND CONCEPTS	EXAMPLES

SECTION 5.2 SOLVING SYSTEMS OF LINEAR EQUATIONS IN THREE VARIABLES

A **solution** of an equation in three variables x, y, and z is an **ordered triple** (x, y, z) that makes the equation a true statement.

Verify that $(-2, 1, 3)$ is a solution of
$2x + 3y - 2z = -7$.

Replace x with -2, y with 1, and z with 3.

$$2(-2) + 3(1) - 2(3) = -7$$
$$-4 + 3 - 6 = -7$$
$$-7 = -7 \qquad \text{True.}$$

$(-2, 1, 3)$ is a solution.

To solve a system of three linear equations by the elimination method:

Step 1. Write each equation in standard form:
$Ax + By + Cz = D$.

Step 2. Choose a pair of equations and use the equations to eliminate a variable.

Step 3. Choose any other pair of equations and eliminate the same variable.

Step 4. Solve the system of two equations in two variables from *steps 1* and *2*.

Step 5. Solve for the third variable by substituting the values of the variables from *step 4* into any of the original equations.

Solve:

$$\begin{cases} 2x + y - z = 0 & (1) \\ x - y - 2z = -6 & (2) \\ -3x - 2y + 3z = -22 & (3) \end{cases}$$

1. Each equation is written in standard form.

2.
$$\begin{array}{ll} 2x + y - z = 0 & (1) \\ \underline{x - y - 2z = -6} & (2) \\ 3x - 3z = -6 & (4) \quad \text{Add.} \end{array}$$

3. Eliminate y from equations (1) and (3) also.

$$\begin{array}{lll} 4x + 2y - 2z = 0 & & \text{Multiply equation} \\ \underline{-3x - 2y + 3z = -22} & (3) & \text{(1) by 2.} \\ x + z = -22 & (5) & \text{Add.} \end{array}$$

4. Solve:

$$\begin{cases} 3x - 3z = -6 & (4) \\ x + z = -22 & (5) \end{cases}$$

$$\begin{array}{ll} x - z = -2 & \text{Divide equation (4) by 3.} \\ \underline{x + z = -22} & (5) \\ 2x = -24 & \\ x = -12 & \end{array}$$

To find z, use equation (5).

$$x + z = -22$$
$$-12 + z = -22$$
$$z = -10$$

5. To find y, use equation (1).

$$2x + y - z = 0$$
$$2(-12) + y - (-10) = 0$$
$$-24 + y + 10 = 0$$
$$y = 14$$

The solution set is $\{(-12, 14, -10)\}$. *(continued)*

DEFINITIONS AND CONCEPTS	EXAMPLES

SECTION 5.3   SYSTEMS OF EQUATIONS AND PROBLEM SOLVING

1. UNDERSTAND the problem.

Two numbers have a sum of 11. Twice one number is 3 less than 3 times the other. Find the numbers.

1. Read and reread.

2. ASSIGN.

2. x = one number

    y = other number

3. ILLUSTRATE.

3. No illustration needed.

4. TRANSLATE.

4. In words: | sum of numbers | is | 11 |

Translate: $x + y$ =   11

In words: | twice one number | is | 3 less than 3 times the other number |

Translate: $2x$ = $3y - 3$

5. COMPLETE.

5. Solve the system $\begin{cases} x + y = 11 \\ 2x = 3y - 3 \end{cases}$.

In the first equation, $x = 11 - y$. Substitute into the other equation.

$$2x = 3y - 3$$
$$2(11 - y) = 3y - 3$$
$$22 - 2y = 3y - 3$$
$$-5y = -25$$
$$y = 5$$

Replace y with 5 in the equation $x = 11 - y$. Then $x = 11 - 5 = 6$. The solution is $(6, 5)$.

6. INTERPRET.

6. To *check*, see that $6 + 5 = 11$, the required sum, and that twice 6 is 3 times 5 less 3. *State:* The numbers are 6 and 5.

SECTION 5.4   SOLVING SYSTEMS OF EQUATIONS BY MATRICES

A **matrix** is a rectangular array of numbers.

Matrices:

$$\begin{bmatrix} -7 & 0 & 3 \\ 1 & 2 & 4 \end{bmatrix} \quad \begin{bmatrix} a & b & c \\ d & e & f \\ g & h & i \end{bmatrix}$$

The **augmented matrix of the system** is obtained by writing a matrix composed of the coefficients of the variables and the constants of the system.

The augmented matrix of the system $\begin{cases} x - y = 1 \\ 2x + y = 11 \end{cases}$ is $\begin{bmatrix} 1 & -1 & | & 1 \\ 2 & 1 & | & 11 \end{bmatrix}$

(continued)

DEFINITIONS AND CONCEPTS	EXAMPLES

SECTION 5.4   SOLVING SYSTEMS OF EQUATIONS BY MATRICES

The following **row operations** can be performed on matrices, and the result is an equivalent matrix.

Elementary row operations:

1. Interchange any two rows.
2. Multiply (or divide) the elements of one row by the same nonzero number.
3. Multiply (or divide) the elements of one row by the same nonzero number and add to its corresponding elements in any other row.

Solve $\begin{cases} x - y = 1 \\ 2x + y = 11 \end{cases}$ using matrices.

The augmented matrix is

$$\left[\begin{array}{cc|c} 1 & -1 & 1 \\ 2 & 1 & 11 \end{array} \right]$$

Use row operations to write an equivalent matrix with 1's along the main diagonal and 0's below each 1 in the main diagonal. Multiply row 1 by -2 and add to row 2. Change row 2 only.

$$\left[\begin{array}{cc|c} 1 & -1 & 1 \\ -2(1) + 2 & -2(-1) + 1 & -2(1) + 11 \end{array} \right]$$

simplifies to $\left[\begin{array}{cc|c} 1 & -1 & 1 \\ 0 & 3 & 9 \end{array} \right]$

Divide row 2 by 3.

$$\left[\begin{array}{cc|c} 1 & -1 & 1 \\ \frac{0}{3} & \frac{3}{3} & \frac{9}{3} \end{array} \right] \text{ simplifies to } \left[\begin{array}{cc|c} 1 & -1 & 1 \\ 0 & 1 & 3 \end{array} \right]$$

This matrix corresponds to the system

$$\begin{cases} x - y = 1 \\ y = 3 \end{cases}$$

Let $y = 3$ in the first equation.

$$x - 3 = 1$$
$$x = 4$$

The solution set is $\{(4, 3)\}$.

SECTION 5.5   SOLVING SYSTEMS OF EQUATIONS BY DETERMINANTS

A **square matrix** is a matrix with the same number of rows and columns.

Square Matrices:

$$\left[\begin{array}{cc} -2 & 1 \\ 6 & 8 \end{array} \right] \quad \left[\begin{array}{ccc} 4 & -1 & 6 \\ 0 & 2 & 5 \\ 1 & 1 & 2 \end{array} \right]$$

A **determinant** is a real number associated with a square matrix. To denote the determinant, place vertical bars about the array of numbers.

The determinant of $\left[\begin{array}{cc} -2 & 1 \\ 6 & 8 \end{array} \right]$ is $\left| \begin{array}{cc} -2 & 1 \\ 6 & 8 \end{array} \right|$.

(continued)

DEFINITIONS AND CONCEPTS	EXAMPLES

SECTION 5.5   SOLVING SYSTEMS OF EQUATIONS BY DETERMINANTS

The determinant of a 2×2 matrix is

$$\begin{vmatrix} a & b \\ c & d \end{vmatrix} = ad - bc$$

$$\begin{vmatrix} -2 & 1 \\ 6 & 8 \end{vmatrix} = -2 \cdot 8 - 1 \cdot 6 = -22$$

Cramer's rule for two linear equations in two variables:

Use Cramer's rule to solve

The solution of the system $\begin{cases} ax + by = h \\ cx + dy = k \end{cases}$ is given by

$$x = \frac{\begin{vmatrix} h & b \\ k & d \end{vmatrix}}{\begin{vmatrix} a & b \\ c & d \end{vmatrix}} = \frac{D_x}{D}; \qquad y = \frac{\begin{vmatrix} a & h \\ c & k \end{vmatrix}}{\begin{vmatrix} a & b \\ c & d \end{vmatrix}} = \frac{D_y}{D}$$

as long as $D = ad - bc$ is not 0.

$$\begin{cases} 3x + 2y = 8 \\ 2x - y = -11 \end{cases}$$

$$D = \begin{vmatrix} 3 & 2 \\ 2 & -1 \end{vmatrix} = 3(-1) - 2(2) = -7$$

$$D_x = \begin{vmatrix} 8 & 2 \\ -11 & -1 \end{vmatrix} = 8(-1) - 2(-11) = 14$$

$$D_y = \begin{vmatrix} 3 & 8 \\ 2 & -11 \end{vmatrix} = 3(-11) - 8(2) = -49$$

$$x = \frac{D_x}{D} = \frac{14}{-7} = -2, \quad y = \frac{D_y}{D} = \frac{-49}{-7} = 7$$

The solution set is $\{(-2, 7)\}$.

CHAPTER 5 REVIEW

(5.1) *Solve each system of equations in two variables by each of three methods: (1) graphing, (2) substitution, and (3) elimination.*

1. $\begin{cases} 3x + 10y = 1 \\ x + 2y = -1 \end{cases}$

2. $\begin{cases} y = \dfrac{1}{2}x + \dfrac{2}{3} \\ 4x + 6y = 4 \end{cases}$

3. $\begin{cases} 2x - 4y = 22 \\ 5x - 10y = 16 \end{cases}$

4. $\begin{cases} 3x - 6y = 12 \\ 2y = x - 4 \end{cases}$

5. $\begin{cases} \dfrac{1}{2}x - \dfrac{3}{4}y = -\dfrac{1}{2} \\ \dfrac{1}{8}x + \dfrac{3}{4}y = \dfrac{19}{8} \end{cases}$

6. The revenue equation for a certain style of backpack is

$$y = 32x$$

where x is the number of backpacks sold and y is the income in dollars for selling x backpacks. The cost equation for these units is

$$y = 15x + 25,500$$

where x is the number of backpacks manufactured and y is the cost in dollars for manufacturing x backpacks. Find the number of units to be sold for the company to break even.

(5.2) *Solve each system of equations in three variables.*

7. $\begin{cases} x + z = 4 \\ 2x - y = 4 \\ x + y - z = 0 \end{cases}$

8. $\begin{cases} 2x + 5y = 4 \\ x - 5y + z = -1 \\ 4x - z = 11 \end{cases}$

9. $\begin{cases} 4y + 2z = 5 \\ 2x + 8y = 5 \\ 6x + 4z = 1 \end{cases}$

10. $\begin{cases} 5x + 7y = 9 \\ 14y - z = 28 \\ 4x + 2z = -4 \end{cases}$

11. $\begin{cases} 3x - 2y + 2z = 5 \\ -x + 6y + z = 4 \\ 3x + 14y + 7z = 20 \end{cases}$

12. $\begin{cases} x + 2y + 3z = 11 \\ y + 2z = 3 \\ 2x + 2z = 10 \end{cases}$

13. $\begin{cases} 7x - 3y + 2z = 0 \\ 4x - 4y - z = 2 \\ 5x + 2y + 3z = 1 \end{cases}$

14. $\begin{cases} x - 3y - 5z = -5 \\ 4x - 2y + 3z = 13 \\ 5x + 3y + 4z = 22 \end{cases}$

(5.3) *Use systems of equations to solve the following applications.*

15. The sum of three numbers is 98. The sum of the first and second is two more than the third number, and the second is four times the first. Find the numbers.

16. Alice's coin purse has 95 coins in it—all dimes, nickels, and pennies—worth $4.03 total. There are twice as many pennies as dimes and one fewer nickel than dimes. Find how many of each type of coin the purse contains.

17. One number is 3 times a second number, and twice the sum of the numbers is 168. Find the numbers.

18. Sue is 16 years older than Pat. In 15 years Sue will be twice as old as Pat is then. How old is each now?

19. Two cars leave Chicago, one traveling east and the other west. After 4 hours they are 492 miles apart. If one car is traveling 7 mph faster than the other, find the speed of each.

20. The foundation for a rectangular Hardware Warehouse has a length three times the width and is 296 feet around. Find the dimensions of the building.

21. James has available a 10% alcohol solution and a 60% alcohol solution. Find how many liters of each solution he should mix to make 50 liters of a 40% alcohol solution.

22. An employee at See's Candy Store needs a special mixture of candy. She has creme-filled chocolates that sell for $3.00 per pound, chocolate-covered nuts that sell for $2.70 per pound, and chocolate-covered raisins that sell for $2.25 per pound. She wants to have twice as many raisins as nuts in the mixture. Find how many pounds of each she should use to make 45 pounds worth $2.80 per pound.

23. Chris has $2.77 in his coin jar—all in pennies, nickels, and dimes. If he has 53 coins in all and four more nickels than dimes, find how many of each type of coin he has.

24. If $10,000 and $4000 are invested such that $1250 is earned in one year, and if the rate of interest on the larger investment is 2% more than that of the smaller investment, find the rates of interest.

25. The perimeter of an isosceles (two sides equal) triangle is 73 centimeters. If two sides are of equal length and the third side is 7 centimeters longer than the others, find the lengths of the three sides.

26. The sum of three numbers is 295. One number is five more than a second and twice the third. Find the numbers.

(5.4) *Use matrices to solve each system.*

27. $\begin{cases} 3x + 10y = 1 \\ x + 2y = -1 \end{cases}$

28. $\begin{cases} 3x - 6y = 12 \\ 2y = x - 4 \end{cases}$

29. $\begin{cases} 3x - 2y = -8 \\ 6x + 5y = 11 \end{cases}$

30. $\begin{cases} 6x - 6y = -5 \\ 10x - 2y = 1 \end{cases}$

31. $\begin{cases} 3x - 6y = 0 \\ 2x + 4y = 5 \end{cases}$

32. $\begin{cases} 5x - 3y = 10 \\ -2x + y = -1 \end{cases}$

33. $\begin{cases} 0.2x - 0.3y = -0.7 \\ 0.5x + 0.3y = 1.4 \end{cases}$

34. $\begin{cases} 3x + 2y = 8 \\ 3x - y = 5 \end{cases}$

35. $\begin{cases} x + z = 4 \\ 2x - y = 0 \\ x + y - z = 0 \end{cases}$

36. $\begin{cases} 2x + 5y = 4 \\ x - 5y + z = -1 \\ 4x - z = 11 \end{cases}$

37. $\begin{cases} 3x - y = 11 \\ x + 2z = 13 \\ y - z = -7 \end{cases}$

38. $\begin{cases} 5x + 7y + 3z = 9 \\ 14y - z = 28 \\ 4x + 2z = -4 \end{cases}$

39. $\begin{cases} 7x - 3y + 2z = 0 \\ 4x - 4y - z = 2 \\ 5x + 2y + 3z = 1 \end{cases}$

40. $\begin{cases} x + 2y + 3z = 14 \\ y + 2z = 3 \\ 2x - 2z = 10 \end{cases}$

(5.5) *Evaluate.*

41. $\begin{vmatrix} -1 & 3 \\ 5 & 2 \end{vmatrix}$

42. $\begin{vmatrix} 3 & -1 \\ 2 & 5 \end{vmatrix}$

43. $\begin{vmatrix} 2 & -1 & -3 \\ 1 & 2 & 0 \\ 3 & -2 & 2 \end{vmatrix}$

44. $\begin{vmatrix} -2 & 3 & 1 \\ 4 & 4 & 0 \\ 1 & -2 & 3 \end{vmatrix}$

Use Cramer's rule to solve each system of equations.

45. $\begin{cases} 3x - 2y = -8 \\ 6x + 5y = 11 \end{cases}$

46. $\begin{cases} 6x - 6y = -5 \\ 10x - 2y = 1 \end{cases}$

47. $\begin{cases} 3x + 10y = 1 \\ x + 2y = -1 \end{cases}$

48. $\begin{cases} y = \dfrac{1}{2}x + \dfrac{2}{3} \\ 4x + 6y = 4 \end{cases}$

53. $\begin{cases} x + 3y - z = 5 \\ 2x - y - 2z = 3 \\ x + 2y + 3z = 4 \end{cases}$

54. $\begin{cases} 2x \quad\;\; - z = 1 \\ 3x - y + 2z = 3 \\ x + y + 3z = -2 \end{cases}$

49. $\begin{cases} 2x - 4y = 22 \\ 5x - 10y = 16 \end{cases}$

50. $\begin{cases} 3x - 6y = 12 \\ 2y = x - 4 \end{cases}$

51. $\begin{cases} x \quad\;\; + z = 4 \\ 2x - y \quad\;\; = 0 \\ x + y - z = 0 \end{cases}$

52. $\begin{cases} 2x + 5y \quad\;\; = 4 \\ x - 5y + z = -1 \\ 4x \quad\;\; - z = 11 \end{cases}$

55. $\begin{cases} x + 2y + 3z = 14 \\ y + 2z = 3 \\ 2x \quad\;\; - 2z = 10 \end{cases}$

56. $\begin{cases} 5x + 7y \quad\;\; = 9 \\ 14y - z = 28 \\ 4x \quad\;\; + 2z = -4 \end{cases}$

CHAPTER 5 TEST

Evaluate each determinant.

1. $\begin{vmatrix} 4 & -7 \\ 2 & 5 \end{vmatrix}$

2. $\begin{vmatrix} 4 & 0 & 2 \\ 1 & -3 & 5 \\ 0 & -1 & 2 \end{vmatrix}$

Solve each system of equations graphically and then solve by the elimination method or the substitution method.

3. $\begin{cases} 2x - y = -1 \\ 5x + 4y = 17 \end{cases}$

4. $\begin{cases} 7x - 14y = 5 \\ x \quad\;\; = 2y \end{cases}$

Solve each system.

5. $\begin{cases} 4x - 7y = 29 \\ 2x + 5y = -11 \end{cases}$

6. $\begin{cases} 15x + 6y = 15 \\ 10x + 4y = 10 \end{cases}$

7. $\begin{cases} 2x - 3y \quad\;\; = 4 \\ 3y + 2z = 2 \\ x \quad\;\; - z = -5 \end{cases}$

8. $\begin{cases} 3x - 2y - z = -1 \\ 2x - 2y \quad\;\; = 4 \\ 2x \quad\;\; - 2z = -12 \end{cases}$

9. $\begin{cases} \dfrac{x}{2} + \dfrac{y}{4} = -\dfrac{3}{4} \\ x + \dfrac{3}{4}y = -4 \end{cases}$

Use Cramer's rule to solve each system.

10. $\begin{cases} 3x - y = 7 \\ 2x + 5y = -1 \end{cases}$

11. $\begin{cases} 4x - 3y = -6 \\ -2x + y = 0 \end{cases}$

12. $\begin{cases} x + y + z = 4 \\ 2x + 5y \quad\;\; = 1 \\ x - y - 2z = 0 \end{cases}$

13. $\begin{cases} 3x + 2y + 3z = 3 \\ x \quad\;\; - z = 9 \\ 4y + z = -4 \end{cases}$

Use matrices to solve each system.

14. $\begin{cases} x - y = -2 \\ 3x - 3y = -6 \end{cases}$

15. $\begin{cases} x + 2y = -1 \\ 2x + 5y = -5 \end{cases}$

16. $\begin{cases} x - y - z = 0 \\ 3x - y - 5z = -2 \\ 2x + 3y \quad\;\; = -5 \end{cases}$

17. $\begin{cases} 2x - y + 3z = 4 \\ 3x \quad\;\; - 3z = -2 \\ -5x + y \quad\;\; = 0 \end{cases}$

18. Frame Masters, Inc., recently purchased $5500 worth of new equipment in order to offer a new style of eyeglass frame. The marketing department of Frame Masters estimates that the cost of producing this new frame is $18 and that the frame will be sold to stores for $38. Find the number of frames that must be sold in order to break even.

19. A motel in New Orleans charges $90 per day for double occupancy and $80 per day for single occupancy. If 80 rooms are occupied for a total of $6930, how many rooms of each kind are there?

20. The research department of a company that manufactures children's fruit drinks is experimenting with a new flavor. A 17.5% fructose solution is needed, but only 10% and 20% solutions are available. How many gallons of a 10% fructose solution should be mixed with a 20% fructose solution in order to obtain 20 gallons of a 17.5% fructose solution?

CHAPTER 5   CUMULATIVE REVIEW

1. Write each sentence using mathematical symbols.

   a. The sum of 5 and y is greater than or equal to 7.

   b. 11 is not equal to z.

   c. 20 is less than the difference of 5 and twice x.

2. Use the cumulative property of addition to write an expression equivalent to $7x + 5$.

3. Use a calculator to evaluate each expression when $x = -7.1$ and $y = 2.9$.

   a. $x + 3y$ **b.** $2x - 5y$ **c.** $3y^2$

4. Solve for y: $\dfrac{y}{3} - \dfrac{y}{4} = \dfrac{1}{6}$.

5. Find the domain and range of each relation. Determine whether the relation is also a function.

   a.

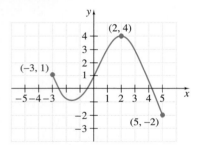

   b.

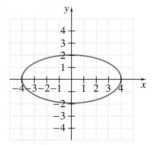

   c.   **d.**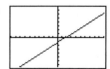

   e.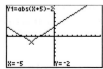

6. Find the slope of the line $x = -5$.

7. Find an equation of the line through points $(4, 0)$ and $(-4, -5)$. Write the equation using function notation.

8. Solve for x: $3x + 4 \geq 2x - 6$. Graph the solution set and write it in interval notation.

9. Solve $2 < 4 - x < 7$.

10. Solve $|5x + 3| = 7$.

11. Solve for x: $|x - 3| \geq 7$.

12. Solve for x: $\left| \dfrac{2(x + 1)}{3} \right| \leq 0$.

13. Graph $3x \geq y$.

14. Graph $2x - y < 6$.

15. Use the substitution method to solve the system

$$\begin{cases} 2x + 4y = -6 \\ x - 2y = -5 \end{cases}$$

16. Use the elimination method to solve the system

$$\begin{cases} 3x + \dfrac{y}{2} = 2 \\ 6x + y = 5 \end{cases}$$

17. Solve the system

$$\begin{cases} 3x - y + z = -15 \\ x + 2y - z = 1 \\ 2x + 3y - 2z = 0 \end{cases}$$

18. Two cars leave Indianapolis, one traveling east and the other west. After 3 hours they are 297 miles apart. If one car is traveling 5 mph faster than the other, what is the speed of each?

19. Use matrices to solve the system:

$$\begin{cases} x + 2y + z = 2 \\ -2x - y + 2z = 5 \\ x + 3y - 2z = -8 \end{cases}$$

20. Evaluate by expanding by minors.

$$\begin{vmatrix} 0 & 5 & 1 \\ 1 & 3 & -1 \\ -2 & 2 & 4 \end{vmatrix}$$

EXPONENTS, POLYNOMIALS, AND POLYNOMIAL FUNCTIONS

FINDING THE LARGEST AREA

Picture framers must often find either the optimal or the most pleasing dimensions of a frame or mat to fit the area of the item to be framed. If a standard-sized frame won't fit, the framer will need to build a custom frame. Sometimes custom framing requires working with a limited amount of framing or matting material.

IN THE CHAPTER GROUP ACTIVITY ON PAGE 373, YOU WILL HAVE THE OPPORTUNITY TO FIND THE DIMENSIONS OF A PICTURE FRAME MADE FROM A FIXED AMOUNT OF MATERIAL THAT WILL ALLOW THE LARGEST AMOUNT OF INTERIOR DISPLAY AREA.

6.1 EXPONENTS AND SCIENTIFIC NOTATION

O B J E C T I V E S

1 Use the product rule for exponents.

2 Evaluate a raised to the 0 power.

3 Use the quotient rule for exponents.

4 Define a raised to the negative nth power.

5 Write numbers in scientific notation.

6 Convert numbers from scientific notation to standard notation.

TAPE IAG 6.1

1 Recall that exponents may be used to write repeated factors in a more compact form. As we have seen in the previous chapters, exponents can be used when the repeated factor is a number or a variable. For example,

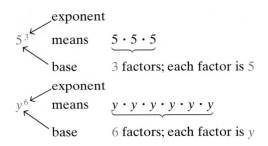

Expressions such as 5^3 and y^6 that contain exponents are called **exponential expressions**.

Exponential expressions can be multiplied, divided, added, subtracted, and themselves raised to powers. In this section, we review operations on exponential expressions.

DISCOVER THE CONCEPT

a. Use a calculator to evaluate each pair of expressions. Then compare the results.

$2^5 \cdot 2^4$ and 2^9

$(3.7)^2 \cdot (3.7)^3$ and $(3.7)^5$

$(-5)^4 \cdot (-5)^2$ and $(-5)^6$

b. From the results of part (a), write a pattern you observed for multiplying exponential expressions with the same base.

c. Use your findings to simplify the product

$x^2 \cdot x^3$.

```
2^5*2^4
            512
2^9
            512
```

```
3.7^2*3.7^3
        693.43957
3.7^5
        693.43957
```

```
(-5)^4*(-5)^2
          15625
(-5)^6
          15625
```

In part (a) above, we see that each pair of expressions are equal since they simplify to the same number as shown in the screens to the left. It appears that the product of exponential expressions with the same base is the base raised to the sum of the exponents. From this, we reason that

$$x^2 \cdot x^3 = x^{2+3} = x^5$$

To check, use the definition of a^n:

$$x^2 \cdot x^3 = \underbrace{(x \cdot x)(x \cdot x \cdot x)}_{x \text{ is a factor } 5 \text{ times}} = x^5$$

This suggests the following.

PRODUCT RULE FOR EXPONENTS

If m and n are positive integers and a is a real number, then

$$a^m \cdot a^n = a^{m+n}$$

In other words, the *product* of exponential expressions with a common base is the common base raised to a power equal to the *sum* of the exponents of the factors.

EXAMPLE 1   Use the product rule to simplify.

    a. $2^2 \cdot 2^5$ **b.** $x^7 x^3$ **c.** $y \cdot y^2 \cdot y^4$

Solution: **a.** $2^2 \cdot 2^5 = 2^{2+5} = 2^7$

                  b. $x^7 x^3 = x^{7+3} = x^{10}$

                  c. $y \cdot y^2 \cdot y^4 = (y^1 \cdot y^2) \cdot y^4$

                                       $= y^3 \cdot y^4$

                                       $= y^7$

EXAMPLE 2   Use the product rule to multiply.

    a. $(3x^6)(5x)$ **b.** $(-2x^3p^2)(4xp^{10})$

Solution: Here, we use properties of multiplication to group together like bases.

    a. $(3x^6)(5x) = 3(5)x^6 x^1 = 15x^7$

    b. $(-2x^3p^2)(4xp^{10}) = -2(4)x^3 x^1 p^2 p^{10} = -8x^4 p^{12}$

2   The definition of a^n does not include the possibility that n might be 0. But if it did, then, by the product rule,

$$\underbrace{a^0 \cdot a^n}_{} = a^{0+n} = a^n = \underbrace{1 \cdot a^n}_{}.$$

From this, we reasonably define that $a^0 = 1$, as long as a does not equal 0.

ZERO EXPONENT

If a does not equal 0, then $a^0 = 1$.

EXAMPLE 3   Evaluate the following.

    a. 7^0 **b.** -7^0 **c.** $(2x + 5)^0$ **d.** $2x^0$

Solution: **a.** $7^0 = 1$

    b. Without parentheses, only 7 is raised to the 0 power.

$$-7^0 = -(7^0) = -(1) = -1$$

    c. $(2x + 5)^0 = 1$

    d. $2x^0 = 2(1) = 2$

3   Next, we discover a pattern that occurs when dividing exponential expressions with a common base.

DISCOVER THE CONCEPT

a. Use a calculator to evaluate each pair of expressions. Then compare the results.

$$\frac{7^5}{7^2} \text{ and } 7^3$$

$$\frac{(-3)^9}{(-3)^4} \text{ and } (-3)^5$$

b. From the results of part (a), write a pattern you observed for dividing exponential expressions with the same base.

c. Use your findings to simplify the quotient $\dfrac{x^9}{x^2}$.

```
7^5/7^2
            343
7^3
            343
```

```
(-3)^9/(-3)^4
           -243
(-3)^5
           -243
```

In part (a), we see that each pair of expressions are equal since they simplify to the same number as shown in the screens to the left. It appears that the quotient of exponential expressions with the same base is the base raised to the difference of the exponents. From this, we reason that

$$\frac{x^9}{x^2} = x^{9-2} = x^7$$

To check, we again begin with the definition of a^n to simplify $\dfrac{x^9}{x^2}$.

$$\frac{x^9}{x^2} = \frac{\boxed{x} \cdot \boxed{x} \cdot x \cdot x \cdot x \cdot x \cdot x \cdot x \cdot x}{\boxed{x} \cdot \boxed{x}} = x^7$$

(Assume that denominators containing variables are not 0.)
    This suggests the following.

QUOTIENT RULE FOR EXPONENTS

If a is a nonzero real number and n and m are integers, then

$$\frac{a^m}{a^n} = a^{m-n}$$

In other words, the *quotient* of exponential expressions with a common base is the common base raised to a power equal to the *difference* of the exponents.

EXAMPLE 4   Use the quotient rule to simplify.

  a. $\dfrac{x^7}{x^4}$ **b.** $\dfrac{5^8}{5^2}$ **c.** $\dfrac{20x^6}{4x^5}$ **d.** $\dfrac{12y^{10}z^7}{14y^8z^7}$

Solution: **a.** $\dfrac{x^7}{x^4} = x^{7-4} = x^3$

  b. $\dfrac{5^8}{5^2} = 5^{8-2} = 5^6$

  c. $\dfrac{20x^6}{4x^5} = 5x^{6-5} = 5x^1$,   or $5x$

  d. $\dfrac{12y^{10}z^7}{14y^8z^7} = \dfrac{6}{7}y^{10-8} \cdot z^{7-7} = \dfrac{6}{7}y^2z^0 = \dfrac{6}{7}y^2$, or $\dfrac{6y^2}{7}$

---

4   When the exponent of the denominator is larger than the exponent of the numerator, applying the quotient rule yields a negative exponent. For example,

$$\frac{x^3}{x^5} = x^{3-5} = x^{-2}$$

Using the definition of a^n, though, gives us

$$\frac{x^3}{x^5} = \frac{\boxed{x} \cdot \boxed{x} \cdot \boxed{x}}{\boxed{x} \cdot \boxed{x} \cdot \boxed{x} \cdot x \cdot x} = \frac{1}{x^2}$$

From this, we reasonably define $x^{-2} = \dfrac{1}{x^2}$ or, in general, $a^{-n} = \dfrac{1}{a^n}$.

NEGATIVE EXPONENTS

If a is a real number other than 0 and n is a positive integer, then

$$a^{-n} = \frac{1}{a^n}$$

EXAMPLE 5

Use only positive exponents to write the following. Simplify if possible. Use a calculator to check parts (a) and (f).

a. 5^{-2} **b.** $2x^{-3}$ **c.** $(3x)^{-1}$ **d.** $\dfrac{m^5}{m^{15}}$ **e.** $\dfrac{3^3}{3^6}$ **f.** $2^{-1} + 3^{-2}$ **g.** $\dfrac{1}{t^{-5}}$

Solution:

a. $5^{-2} = \dfrac{1}{5^2} = \dfrac{1}{25}$

b. Without parentheses, only x is raised to the -3 power.

$$2x^{-3} = 2 \cdot \frac{1}{x^3} = \frac{2}{x^3}$$

c. With parentheses, both 3 and x are raised to the -1 power.

$$(3x)^{-1} = \frac{1}{(3x)^1} = \frac{1}{3x}$$

d. $\dfrac{m^5}{m^{15}} = m^{5-15} = m^{-10} = \dfrac{1}{m^{10}}$

e. $\dfrac{3^3}{3^6} = 3^{3-6} = 3^{-3} = \dfrac{1}{3^3} = \dfrac{1}{27}$

f. $2^{-1} + 3^{-2} = \dfrac{1}{2^1} + \dfrac{1}{3^2} = \dfrac{1}{2} + \dfrac{1}{9} = \dfrac{9}{18} + \dfrac{2}{18} = \dfrac{11}{18}$

g. $\dfrac{1}{t^{-5}} = \dfrac{1}{\dfrac{1}{t^5}} = 1 \div \dfrac{1}{t^5} = 1 \cdot \dfrac{t^5}{1} = t^5$

```
5^-2▸Frac
            1/25
2^-1+3^-2▸Frac
            11/18
```

A calculator check of parts (a) and (f) is to the left.

REMINDER Notice that, when a factor containing an exponent is moved from the numerator to the denominator or from the denominator to the numerator, the sign of its exponent changes.

$$x^{-3} = \frac{1}{x^3}, \qquad 5^{-2} = \frac{1}{5^2} = \frac{1}{25}$$

$$\frac{1}{y^{-4}} = y^4, \qquad \frac{1}{2^{-3}} = 2^3 = 8$$

EXAMPLE 6   Simplify each expression. Use positive exponents to write answers.

a. $\dfrac{x^{-9}}{x^2}$ b. $\dfrac{p^4}{p^{-3}}$ c. $\dfrac{2^{-3}}{2^{-1}}$ d. $\dfrac{2x^{-7}y^2}{10xy^{-5}}$ e. $\dfrac{(3x^{-3})(x^2)}{x^6}$

Solution: a. $\dfrac{x^{-9}}{x^2} = x^{-9-2} = x^{-11} = \dfrac{1}{x^{11}}$

b. $\dfrac{p^4}{p^{-3}} = p^{4-(-3)} = p^7$

c. $\dfrac{2^{-3}}{2^{-1}} = 2^{-3-(-1)} = 2^{-2} = \dfrac{1}{2^2} = \dfrac{1}{4}$

d. $\dfrac{2x^{-7}y^2}{10xy^{-5}} = \dfrac{x^{-7-1} \cdot y^{2-(-5)}}{5} = \dfrac{x^{-8}y^7}{5} = \dfrac{y^7}{5x^8}$

e. Simplify the numerator first.

$\dfrac{(3x^{-3})(x^2)}{x^6} = \dfrac{3x^{-3+2}}{x^6} = \dfrac{3x^{-1}}{x^6} = 3x^{-1-6} = 3x^{-7} = \dfrac{3}{x^7}$

EXAMPLE 7   Simplify. Assume that a and t are nonzero integers and that x is not 0.

a. $x^{2a} \cdot x^3$ b. $\dfrac{x^{2t-1}}{x^{t-5}}$

Solution: a. $x^{2a} \cdot x^3 = x^{2a+3}$ Use the product rule.

b. $\dfrac{x^{2t-1}}{x^{t-5}} = x^{(2t-1)-(t-5)}$ Use the quotient rule.

$= x^{2t-1-t+5} = x^{t+4}$

Very large and very small numbers occur frequently in nature. For example, the distance between the Earth and the Sun is approximately 150,000,000 kilometers. A helium atom has a diameter of 0.000 000 022 centimeters. It can be tedious to write these very large and very small numbers in standard notation like this. **Scientific notation** is a convenient shorthand notation for writing very large and very small numbers.

TECHNOLOGY NOTE

To determine the scientific notation capabilities of your graphing utility, consult your owner's manual.

SCIENTIFIC NOTATION

A positive number is written in **scientific notation** if it is written as the product of a number a, where $1 \le a < 10$, and an integer power r of 10:
$a \times 10^r$.

The following are examples of numbers written in scientific notation:

2.03×10^2, 7.362×10^7, 8.1×10^{-5}

To write the approximate distance between the Earth and the Sun in scientific notation, move the decimal point to the left until the number is between 1 and 10.

150,000,000.

The decimal point was moved 8 places to the left, so

$$150,000,000 = 1.5 \times 10^8$$

Next, to write the diameter of a helium atom in scientific notation, again move the decimal point until we have a number between 1 and 10.

0. 000 000 022

The decimal point was moved 8 places to the right, so

$$0.\,000\,000\,022 = 2.2 \times 10^{-8}$$

TO WRITE A NUMBER IN SCIENTIFIC NOTATION

Step 1.   Move the decimal point in the original number until the new number has a value between 1 and 10.

Step 2.   Count the number of decimal places the decimal point was moved in *step 1*. If the decimal point was moved to the left, the count is positive. If the decimal point was moved to the right, the count is negative.

Step 3.   Multiply the new number in *step 1* by 10 raised to an exponent equal to the count found in *step 2*.

EXAMPLE 8   Write each number in scientific notation. Use a calculator to check the results.

   a. 730,000   **b.** 0.00000104

Solution: **a.** *Step 1.*   Move the decimal point until the number is between 1 and 10.

730,000.

          Step 2.   The decimal point is moved 5 places to the left, so the count is positive 5.

          Step 3. $730,000 = 7.3 \times 10^5$.

    b. *Step 1.*   Move the decimal point until the number is between 1 and 10.

0.00000104

```
730000
          7.3E5
.00000104
       1.04E-6
```

          Step 2.   The decimal point is moved 6 places to the right, so the count is −6.

          Step 3. $0.00000104 = 1.04 \times 10^{-6}$.

The screen to the left verifies the results of Example 8. This screen was generated with the calculator in scientific notation mode.

6 To write a scientific notation number in standard form, we reverse the preceding steps.

> **TO WRITE A SCIENTIFIC NOTATION NUMBER IN STANDARD NOTATION**
>
> Move the decimal point in the number the same number of places as the exponent on 10. If the exponent is positive, move the decimal point to the right. If the exponent is negative, move the decimal point to the left.

EXAMPLE 9   Write each number in standard notation.

a. 7.7×10^8 **b.** 1.025×10^{-3}

Solution: **a.** Since the exponent is positive, move the decimal point 8 places to the right. Add zeros as needed.

$$7.7 \times 10^8 = 770,000,000$$

b. Since the exponent is negative, move the decimal point 3 places to the left. Add zeros as needed.

$$1.025 \times 10^{-3} = 0.001025$$

MENTAL MATH

Use positive exponents to state each expression.

1. $5x^{-1}y^{-2}$ **2.** $7xy^{-4}$ **3.** $a^2b^{-1}c^{-5}$ **4.** $a^{-4}b^2c^{-6}$ **5.** $\dfrac{y^{-2}}{x^{-4}}$ **6.** $\dfrac{x^{-7}}{z^{-3}}$

EXERCISE SET 6.1

Use the product rule to simplify each expression. See Examples 1 and 2.

1. $4^2 \cdot 4^3$ **2.** $3^3 \cdot 3^5$

3. $x^5 \cdot x^3$ **4.** $a^2 \cdot a^9$

5. $-7x^3 \cdot 20x^9$ **6.** $-3y \cdot -9y^4$

7. $(4xy)(-5x)$ **8.** $(7xy)(7aby)$

9. $(-4x^3p^2)(4y^3x^3)$ **10.** $(-6a^2b^3)(-3ab^3)$

Evaluate the following. See Example 3.

11. -8^0 **12.** $(-9)^0$ **13.** $(4x + 5)^0$

14. $8x^0 + 1$ **15.** $(5x)^0 + 5x^0$ **16.** $4y^0 - (4y)^0$

17. Explain why $(-5)^0$ simplifies to 1 but -5^0 simplifies to -1.

18. Explain why both $4x^0 - 3y^0$ and $(4x - 3y)^0$ simplify to 1.

Find each quotient. See Example 4.

19. $\dfrac{a^5}{a^2}$ **20.** $\dfrac{x^9}{x^4}$ **21.** $\dfrac{x^9y^6}{x^8y^6}$ **22.** $\dfrac{a^{12}b^2}{a^9b}$

23. $-\dfrac{26z^{11}}{2z^7}$

24. $\dfrac{16x^5}{8x}$

25. $\dfrac{-36a^5b^7c^{10}}{6ab^3c^4}$

26. $\dfrac{49a^3bc^{14}}{-7abc^8}$

Simplify each expression. Write answers with positive exponents. See Examples 5 and 6.

27. 4^{-2} **28.** 2^{-3} **29.** $\dfrac{x^7}{x^{15}}$ **30.** $\dfrac{z}{z^3}$

31. $5a^{-4}$ **32.** $10b^{-1}$ **33.** $\dfrac{x^{-2}}{x^5}$ **34.** $\dfrac{y^{-6}}{y^{-9}}$

35. $\dfrac{8r^4}{2r^{-4}}$ **36.** $\dfrac{3s^3}{15s^{-3}}$ **37.** $\dfrac{x^{-9}x^4}{x^{-5}}$ **38.** $\dfrac{y^{-7}y}{y^8}$

Simplify. Assume that variables in the exponent represent nonzero integers and that x, y, and z are not 0. See Example 7.

39. $x^5 \cdot x^{7a}$ **40.** $y^{2p} \cdot y^{9p}$ **41.** $\dfrac{x^{3t-1}}{x^t}$

42. $\dfrac{y^{4p-2}}{y^{3p}}$ **43.** $x^{4a} \cdot x^7$ **44.** $x^{9y} \cdot x^{-7y}$

45. $\dfrac{z^{6x}}{z^7}$ **46.** $\dfrac{y^6}{y^{4z}}$ **47.** $\dfrac{x^{3t} \cdot x^{4t-1}}{x^t}$

48. $\dfrac{z^{5x} \cdot z^{x-7}}{z^x}$

Simplify the following. Write answers with positive exponents.

49. $4^{-1} + 3^{-2}$ **50.** $1^{-3} - 4^{-2}$ **51.** $4x^0 + 5$

52. $-5x^0$ **53.** $x^7 \cdot x^8$ **54.** $y^6 \cdot y$

55. $2x^3 \cdot 5x^7$ **56.** $-3z^4 \cdot 10z^7$ **57.** $\dfrac{z^{12}}{z^{15}}$

58. $\dfrac{x^{11}}{x^{20}}$ **59.** $\dfrac{y^{-3}}{y^{-7}}$ **60.** $\dfrac{z^{-12}}{z^{10}}$ **61.** $3x^{-1}$

62. $(4x)^{-1}$ **63.** $3^0 - 3t^0$ **64.** $4^0 + 4x^0$ **65.** $\dfrac{r^4}{r^{-4}}$

66. $\dfrac{x^{-5}}{x^3}$ **67.** $\dfrac{x^{-7}y^{-2}}{x^2y^2}$ **68.** $\dfrac{a^{-5}b^7}{a^{-2}b^{-3}}$ **69.** $\dfrac{2a^{-6}b^2}{18ab^{-5}}$

70. $\dfrac{18ab^{-6}}{3a^{-3}b^6}$ **71.** $\dfrac{(24x^8)(x)}{20x^{-7}}$ **72.** $\dfrac{(30z^2)(z^5)}{55z^{-4}}$

Write each number in scientific notation. See Example 8.

73. 31,250,000 **74.** 678,000 **75.** 0.016

76. 0.007613 **77.** 67,413 **78.** 36,800,000

79. 0.0125 **80.** 0.00084 **81.** 0.000053

82. 98,700,000,000

Write each number in standard notation, without exponents. See Example 9.

83. 3.6×10^{-9} **84.** 2.7×10^{-5} **85.** 9.3×10^7

86. 6.378×10^8 **87.** 1.278×10^6 **88.** 7.6×10^4

89. 7.35×10^{12} **90.** 1.66×10^{-5} **91.** 4.03×10^{-7}

92. 8.007×10^8

93. Explain how to convert a number from standard notation to scientific notation.

94. Explain how to convert a number from scientific notation to standard notation.

Express each number in scientific notation.

95. The approximate distance between Jupiter and Earth is 918,000,000 kilometers.

96. The number of millimeters in 50 kilometers is 50,000,000.

97. A computer can perform 48,000,000,000 arithmetic operations in one minute.

98. The temperature at the center of the Sun is about 27,000,000°F.

99. A pulsar is a rotating neutron star that gives off sharp, regular pulses of radio waves. For one particular pulsar, the rate of pulses is one every 0.001 second.

100. To convert from cubic inches to cubic meters, multiply by 0.0000164.

Review Exercises

Evaluate. See Section 1.3.

101. $(5 \cdot 2)^2$ **102.** $5^2 \cdot 2^2$ **103.** $\left(\dfrac{3}{4}\right)^3$ **107.** $(2^{-1})^4$ **108.** $(2^4)^{-1}$

104. $\dfrac{3^3}{4^3}$ **105.** $(2^3)^2$ **106.** $(2^2)^3$

6.2 | More Work with Exponents and Scientific Notation

O B J E C T I V E S

1 Use the power rules for exponents.

2 Use exponent rules and definitions to simplify exponential expressions.

3 Compute, using scientific notation.

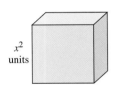

Tape IAG 6.2

1

The volume of the cube shown whose side measures x^2 units is $(x^2)^3$ cubic units. To simplify an expression such as $(x^2)^3$, we use the definition of a^n. Then

$$(x^2)^3 = \underbrace{(x^2)(x^2)(x^2)}_{x^2 \text{ is a factor 3 times}} = x^{2+2+2} = x^6$$

Notice that the result is exactly the same if the exponents are multiplied.

$$(x^2)^3 = x^{2 \cdot 3} = x^6$$

This suggests that the power of an exponential expression raised to a power is the product of the exponents. To discover two additional power rules for exponents, see below.

DISCOVER THE CONCEPT

a. Use a calculator to evaluate each pair of expressions. Then compare the results.

$(-2 \cdot 3)^7$ and $(-2)^7 \cdot 3^7$

$(3 \cdot 5)^{-2}$ and $3^{-2} \cdot 5^{-2}$

$\left(\dfrac{6}{7}\right)^4$ and $\dfrac{6^4}{7^4}$

$\left(\dfrac{3}{4}\right)^{-5}$ and $\dfrac{3^{-5}}{4^{-5}}$

b. From the results of part (a), write a pattern you observed for raising a product to a power and for raising a quotient to a power.

c. Use your findings to simplify $(xy)^3$ and $\left(\dfrac{r}{s}\right)^7$.

The results of part (a) are below.

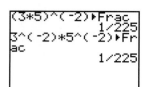

It appears that a product raised to a power is equal to the product of each factor raised to the power. Also, a quotient raised to a power is equal to the quotient of the numerator raised to the power and the denominator raised to the power. From this, we reason that

$$(xy)^3 = x^3 y^3 \quad \text{and} \quad \left(\frac{r}{s}\right)^7 = \frac{r^7}{s^7}.$$

These power rules suggested above are given in the following box.

POWER RULES FOR EXPONENTS

If a and b are real numbers and m and n are integers, then

$(a^m)^n = a^{m \cdot n}$ Power of a power.

$(ab)^m = a^m b^m$ Power of a product.

$\left(\dfrac{a}{b}\right)^n = \dfrac{a^n}{b^n} \ (b \neq 0)$ Power of a quotient.

EXAMPLE 1   Use the power of a power rule to simplify the following expressions. Use positive exponents to write all results.

   a. $(x^5)^7$ **b.** $(2^2)^3$ **c.** $(5^{-1})^2$ **d.** $(y^{-3})^{-4}$

Solution: **a.** $(x^5)^7 = x^{5 \cdot 7} = x^{35}$

   b. $(2^2)^3 = 2^{2 \cdot 3} = 2^6 = 64$

   c. $(5^{-1})^2 = 5^{-1 \cdot 2} = 5^{-2} = \dfrac{1}{5^2} = \dfrac{1}{25}$

   d. $(y^{-3})^{-4} = y^{-3(-4)} = y^{12}$

EXAMPLE 2   Use the power rules to simplify the following. Use positive exponents to write all results.

   a. $(5x^2)^3$ **b.** $\left(\dfrac{2}{3}\right)^3$ **c.** $\left(\dfrac{3p^4}{q^5}\right)^2$ **d.** $\left(\dfrac{2^{-3}}{y}\right)^{-2}$ **e.** $(x^{-5}y^2z^{-1})^7$

Solution: **a.** $(5x^2)^3 = 5^3 \cdot (x^2)^3 = 5^3 \cdot x^{2 \cdot 3} = 125x^6$

b. $\left(\dfrac{2}{3}\right)^3 = \dfrac{2^3}{3^3} = \dfrac{8}{27}$

c. $\left(\dfrac{3p^4}{q^5}\right)^2 = \dfrac{(3p^4)^2}{(q^5)^2} = \dfrac{3^2 \cdot (p^4)^2}{(q^5)^2} = \dfrac{9p^8}{q^{10}}$

d. $\left(\dfrac{2^{-3}}{y}\right)^{-2} = \dfrac{(2^{-3})^{-2}}{y^{-2}}$

$= \dfrac{2^6}{y^{-2}} = 64y^2$ Use the negative exponent rule.

e. $(x^{-5}y^2z^{-1})^7 = (x^{-5})^7 \cdot (y^2)^7 \cdot (z^{-1})^7$

$= x^{-35}y^{14}z^{-7} = \dfrac{y^{14}}{x^{35}z^7}$

2   In the next few examples, we practice the use of several of the rules and definitions for exponents. The following is a summary of these rules and definitions.

SUMMARY OF RULES FOR EXPONENTS

If a and b are real numbers and m and n are integers, then

Product rule for exponents	$a^m \cdot a^n = a^{m+n}$	
Zero exponent	$a^0 = 1$	$(a \neq 0)$
Negative exponent	$a^{-n} = \dfrac{1}{a^n}$	$(a \neq 0)$
Quotient rule	$\dfrac{a^m}{a^n} = a^{m-n}$	$(a \neq 0)$
Power rules	$(a^m)^n = a^{m \cdot n}$	
	$(ab)^m = a^m \cdot b^m$	
	$\left(\dfrac{a}{b}\right)^m = \dfrac{a^m}{b^m}$	$(b \neq 0)$

EXAMPLE 3   Simplify each expression. Use positive exponents to write answers.

a. $(2x^0y^{-3})^{-2}$ **b.** $\left(\dfrac{x^{-5}}{x^{-2}}\right)^{-3}$ **c.** $\left(\dfrac{2}{7}\right)^{-2}$ **d.** $\dfrac{5^{-2}x^{-3}y^{11}}{x^2y^{-5}}$

Solution:

a. $(2x^0y^{-3})^{-2} = 2^{-2}(x^0)^{-2}(y^{-3})^{-2}$

$= 2^{-2}x^0y^6$

$= \dfrac{1(y^6)}{2^2}$ Write x^0 as 1.

$= \dfrac{y^6}{4}$

b. $\left(\dfrac{x^{-5}}{x^{-2}}\right)^{-3} = \dfrac{(x^{-5})^{-3}}{(x^{-2})^{-3}} = \dfrac{x^{15}}{x^6} = x^{15-6} = x^9$

c. $\left(\dfrac{2}{7}\right)^{-2} = \dfrac{2^{-2}}{7^{-2}} = \dfrac{7^2}{2^2} = \dfrac{49}{4}$

d. $\dfrac{5^{-2}x^{-3}y^{11}}{x^2y^{-5}} = \left(5^{-2}\right)\left(\dfrac{x^{-3}}{x^2}\right)\left(\dfrac{y^{11}}{y^{-5}}\right) = 5^{-2}x^{-3-2}y^{11-(-5)} = 5^{-2}x^{-5}y^{16}$

$= \dfrac{y^{16}}{5^2x^5} = \dfrac{y^{16}}{25x^5}$

EXAMPLE 4   Simplify each expression. Use positive exponents to write answers.

a. $\left(\dfrac{3x^2y}{y^{-9}z}\right)^{-2}$ **b.** $\left(\dfrac{3a^2}{2x^{-1}}\right)^3\left(\dfrac{x^{-3}}{4a^{-2}}\right)^{-1}$

Solution: There is often more than one way to simplify exponential expressions. Here, we will simplify inside parentheses if possible before we apply power rules for exponents.

a. $\left(\dfrac{3x^2y}{y^{-9}z}\right)^{-2} = \left(\dfrac{3x^2y^{10}}{z}\right)^{-2} = \dfrac{3^{-2}x^{-4}y^{-20}}{z^{-2}} = \dfrac{z^2}{3^2x^4y^{20}} = \dfrac{z^2}{9x^4y^{20}}$

b. $\left(\dfrac{3a^2}{2x^{-1}}\right)^3\left(\dfrac{x^{-3}}{4a^{-2}}\right)^{-1} = \dfrac{27a^6}{8x^{-3}} \cdot \dfrac{x^3}{4^{-1}a^2}$

$= \dfrac{27 \cdot 4 \cdot a^6x^3x^3}{8 \cdot a^2} = \dfrac{27a^4x^6}{2}$

EXAMPLE 5   Simplify the expression. Assume that a and b are integers and that x and y are not 0.

a. $x^{-b}(2x^b)^2$ **b.** $\dfrac{(y^{3a})^2}{y^{a-6}}$

Solution: **a.** $x^{-b}(2x^b)^2 = x^{-b}2^2x^{2b} = 4x^{-b+2b} = 4x^b$

b. $\dfrac{(y^{3a})^2}{y^{a-6}} = \dfrac{y^{2(3a)}}{y^{a-6}} = \dfrac{y^{6a}}{y^{a-6}} = y^{6a-(a-6)} = y^{6a-a+6} = y^{5a+6}$

3   To perform operations on numbers written in scientific notation, we use properties of exponents.

EXAMPLE 6   Perform the indicated operations. Write each result in scientific notation. Use a calculator to check the results.

$$\textbf{a.}\ (8.1 \times 10^5)(5 \times 10^{-7}) \qquad \textbf{b.}\ \frac{1.2 \times 10^4}{3 \times 10^{-2}}$$

Solution: **a.** $(8.1 \times 10^5)(5 \times 10^{-7}) = 8.1 \times 5 \times 10^5 \times 10^{-7}$

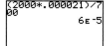

$$= 40.5 \times 10^{-2}$$
$$= (4.05 \times 10^1) \times 10^{-2}$$
$$= 4.05 \times 10^{-1}$$

$$\textbf{b.}\ \frac{1.2 \times 10^4}{3 \times 10^{-2}} = \left(\frac{1.2}{3}\right)\left(\frac{10^4}{10^{-2}}\right) = 0.4 \times 10^{4-(-2)}$$
$$= 0.4 \times 10^6 = (4 \times 10^{-1}) \times 10^6 = 4 \times 10^5$$

A calculator check is shown to the left.

EXAMPLE 7   Use scientific notation to simplify $\dfrac{2000 \times 0.000021}{700}$. Use a calculator to check the results.

Solution: $$\frac{2000 \times 0.000021}{700} = \frac{(2 \times 10^3)(2.1 \times 10^{-5})}{7 \times 10^2} = \frac{2(2.1)}{7} \cdot \frac{10^3 \cdot 10^{-5}}{10^2}$$

$$= 0.6 \times 10^{-4}$$
$$= (6 \times 10^{-1}) \times 10^{-4}$$
$$= 6 \times 10^{-5}$$

MENTAL MATH

Simplify. See Example 1.

1. $(x^4)^5$ **2.** $(5^6)^2$ **3.** $x^4 \cdot x^5$ **4.** $x^7 \cdot x^8$ **5.** $(y^6)^7$

6. $(x^3)^4$ **7.** $(z^4)^5$ **8.** $(z^3)^7$ **9.** $(z^{-6})^{-3}$ **10.** $(y^{-4})^{-2}$

EXERCISE SET 6.2

Simplify. Use positive exponents to write each answer.
See Examples 1 and 2.

12. $(6x^{-6}y^7z^0)^{-2}$ **13.** $\left(\dfrac{x^7y^{-3}}{z^{-4}}\right)^{-5}$ **14.** $\left(\dfrac{a^{-2}b^{-5}}{c^{-11}}\right)^{-6}$

1. $(3^{-1})^2$ **2.** $(2^{-2})^2$ **3.** $(x^4)^{-9}$ **4.** $(y^7)^{-3}$

5. $(y)^{-5}$ **6.** $(z^{-1})^{10}$ **7.** $(3x^2y^3)^2$ **8.** $(4x^3yz)^2$

Simplify. Use positive exponents to write each answer.
See Examples 3 and 4.

9. $\left(\dfrac{2x^5}{y^{-3}}\right)^4$ **10.** $\left(\dfrac{3a^{-4}}{b^7}\right)^3$ **11.** $(a^2bc^{-3})^{-6}$

15. $\left(\dfrac{a^{-4}}{a^{-5}}\right)^{-2}$ **16.** $\left(\dfrac{x^{-9}}{x^{-4}}\right)^{-3}$ **17.** $\left(\dfrac{2a^{-2}b^5}{4a^2b^7}\right)^{-2}$

18. $\left(\dfrac{5x^7y^4}{10x^3y^{-2}}\right)^{-3}$ **19.** $\dfrac{4^{-1}x^2yz}{x^{-2}yz^3}$ **20.** $\dfrac{8^{-2}x^{-3}y^{11}}{x^2y^{-5}}$

21. Is there a number a such that $a^{-1} = a^1$? If so, give the value of a.

22. Is there a number a such that a^{-2} is a negative number? If so, give the value of a.

Simplify. Use positive exponents to write each answer.

23. $(5^{-1})^3$ **24.** $(8^2)^{-1}$ **25.** $(x^7)^{-9}$ **26.** $(y^{-4})^5$

27. $\left(\dfrac{7}{8}\right)^3$ **28.** $\left(\dfrac{4}{3}\right)^2$ **29.** $(4x^2)^2$ **30.** $(-8x^3)^2$

31. $(-2^{-2}y)^3$ **32.** $(-4^{-6}y^{-6})^{-4}$ **33.** $\left(\dfrac{4^{-4}}{y^3x}\right)^{-2}$

34. $\left(\dfrac{7^{-3}}{ab^2}\right)^{-2}$ **35.** $\left(\dfrac{6p^6}{p^{12}}\right)^2$ **36.** $\left(\dfrac{4p^6}{p^9}\right)^3$

37. $(-8y^3xa^{-2})^{-3}$ **38.** $(-xy^0x^2a^3)^{-3}$ **39.** $\left(\dfrac{x^{-2}y^{-2}}{a^{-3}}\right)^{-7}$

40. $\left(\dfrac{x^{-1}y^{-2}}{5^{-3}}\right)^{-5}$ **41.** $\left(\dfrac{3x^5}{6x^4}\right)^4$ **42.** $\left(\dfrac{8^{-3}}{y^2}\right)^{-2}$

43. $\left(\dfrac{1}{4}\right)^{-3}$ **44.** $\left(\dfrac{1}{8}\right)^{-2}$ **45.** $\dfrac{(y^3)^{-4}}{y^3}$

46. $\dfrac{2(y^3)^{-3}}{y^{-3}}$ **47.** $\dfrac{8p^7}{4p^9}$ **48.** $\left(\dfrac{2x^4}{x^2}\right)^3$

49. $(4x^6y^5)^{-2}(6x^4y^3)$ **50.** $(5xy)^3(z^{-2})^{-3}$

51. $x^6(x^6bc)^{-6}$ **52.** $2(y^2b)^{-4}$ **53.** $\dfrac{2^{-3}x^2y^{-5}}{5^{-2}x^7y^{-1}}$

54. $\dfrac{7^{-1}a^{-3}b^5}{a^2b^{-2}}$ **55.** $\left(\dfrac{2x^2}{y^4}\right)^3 \cdot \left(\dfrac{2x^5}{y}\right)^{-2}$ **56.** $\left(\dfrac{3z^{-2}}{y}\right)^2 \cdot \left(\dfrac{9y^{-4}}{z^{-3}}\right)^{-1}$

Simplify the following. Assume that variables in the exponents represent nonzero integers and that all other variables are not 0. See Example 5.

57. $(x^{3a+6})^3$ **58.** $(x^{2b+7})^2$ **59.** $\dfrac{x^{4a}(x^{4a})^3}{x^{4a-2}}$ **60.** $\dfrac{x^{-5y+2}x^{2y}}{x}$

61. $(b^{5x-2})^{2x}$ **62.** $(c^{2a+3})^3$ **63.** $\dfrac{(y^{2a})^8}{y^{a-3}}$ **64.** $\dfrac{(y^{4a})^7}{y^{2a-1}}$

65. $\left(\dfrac{2x^{3t}}{x^{2t-1}}\right)^4$ **66.** $\left(\dfrac{3y^{5a}}{y^{-a+1}}\right)^2$

Perform indicated operations. Write each result in scientific notation. Use a calculator to check the results. See Examples 6 and 7.

67. $(5 \times 10^{11})(2.9 \times 10^{-3})$ **68.** $(3.6 \times 10^{-12})(6 \times 10^9)$

69. $(2 \times 10^5)^3$ **70.** $(3 \times 10^{-7})^3$ **71.** $\dfrac{3.6 \times 10^{-4}}{9 \times 10^2}$

72. $\dfrac{1.2 \times 10^9}{2 \times 10^{-5}}$ **73.** $\dfrac{0.0069}{0.023}$ **74.** $\dfrac{0.00048}{0.0016}$

75. $\dfrac{18{,}200 \times 100}{91{,}000}$ **76.** $\dfrac{0.0003 \times 0.0024}{0.0006 \times 20}$

77. $\dfrac{6000 \times 0.006}{0.009 \times 400}$ **78.** $\dfrac{0.00016 \times 300}{0.064 \times 100}$

79. $\dfrac{0.00064 \times 2000}{16{,}000}$ **80.** $\dfrac{0.00072 \times 0.003}{0.00024}$

81. $\dfrac{66{,}000 \times 0.001}{0.002 \times 0.003}$ **82.** $\dfrac{0.0007 \times 11{,}000}{0.001 \times 0.0001}$

83. $\dfrac{1.25 \times 10^{15}}{(2.2 \times 10^{-2})(6.4 \times 10^{-5})}$

84. $\dfrac{(2.6 \times 10^{-3})(4.8 \times 10^{-4})}{1.3 \times 10^{-12}}$

Solve.

85. A fast computer can add two numbers in about 10^{-8} second. Express in scientific notation how long it would take this computer to do this task 200,000 times.

86. The density D of an object is equivalent to the quotient of its mass M and volume V. Thus $D = \dfrac{M}{V}$. Express in scientific notation the density of an object whose mass is 500,000 pounds and whose volume is 250 cubic feet.

87. The density of ordinary water is 3.12×10^{-2} ton per cubic foot. The volume of water in the largest of the Great Lakes, Lake Superior, is 4.269×10^{14} cubic feet. Use the formula $D = \dfrac{M}{V}$ (see Exercise 86) to find the mass (in tons) of the water in Lake Superior. Express your answer in scientific notation. (*Source:* National Ocean Service)

88. The population of the United States in 1994 was 2.603×10^8. The land area of the United States is 3.536×10^6 square miles. Find the population density (number of people per square mile) for the United States in 1994. Round to the nearest whole. (*Source:* U.S. Bureau of the Census)

89. Each side of the cube shown is $\dfrac{2x^{-2}}{y}$ meters. Find its volume.

$\dfrac{2x^{-2}}{y}$ meters

90. The lot shown is in the shape of a parallelogram with base $\dfrac{3x^{-1}}{y^{-3}}$ feet and height $5x^{-7}$ feet. Find its area.

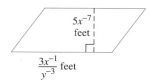

91. To convert from square inches to square meters, multiply by 6.452×10^{-4}. The area of the following square is 4×10^{-2} square inch. Convert this area to square meters.

4×10^{-2}
square inch

92. To convert from cubic inches to cubic meters, multiply by 1.64×10^{-5}. A grain of salt is in the shape of a cube. If an average size of a grain of salt is 3.8×10^{-6} cubic inch, convert this volume to cubic meters.

 93. Explain whether 0.4×10^{-5} is written in scientific notation.

Review Exercises

Simplify each expression. See Section 1.4.

94. $-5y + 4y - 18 - y$

95. $12m - 14 - 15m - 1$

96. $-3x - (4x - 2)$

97. $-9y - (5 - 6y)$

98. $3(z - 4) - 2(3z + 1)$

99. $5(x - 3) - 4(2x - 5)$

6.3 │ POLYNOMIALS AND POLYNOMIAL FUNCTIONS

TAPE IAG 6.3

O B J E C T I V E S

1. Define *term, constant, polynomial, monomial, binomial,* and *trinomial.*
2. Identify the degree of a term and of a polynomial.
3. Define polynomial functions.
4. Add polynomials.
5. Subtract polynomials.

---

1 A **term** is a number or the product of a number and one or more variables raised to powers. The **numerical coefficient,** or simply the **coefficient,** is the numerical factor of a term.

TERM	NUMERICAL COEFFICIENT
$-12x^5$	-12
x^3y	1
$-z$	-1
2	2

If a term contains only a number, it is called a **constant term,** or simply a **constant.**

A **polynomial** is a finite sum of terms in which all variables are raised to non-negative integer powers and no variables appear in the denominator.

POLYNOMIALS   NOT POLYNOMIALS

$4x^5y + 7xz$ $5x^{-3} + 2x$ (negative integer exponent)

$-5x^3 + 2x + \dfrac{2}{3}$ $\dfrac{6}{x^2} - 5x + 1$ (variable in denominator)

A polynomial that contains only one variable is called a **polynomial in one variable.** For example, $3x^2 - 2x + 7$ is a **polynomial in x.** This polynomial in x is written in *descending order* since the terms are listed in descending order of the variable's exponents. (The term 7 can be thought of as $7x^0$.) The following examples are polynomials in one variable written in **descending order.**

$$4x^3 - 7x^2 + 5, \qquad y^2 - 4, \qquad 8a^4 - 7a^2 + 4a$$

A **monomial** is a polynomial consisting of one term. A **binomial** is a polynomial consisting of two terms. A **trinomial** is a polynomial consisting of three terms.

MONOMIALS	BINOMIALS	TRINOMIALS
ax^2	$x + y$	$x^2 + 4xy + y^2$
$-3x$	$6y^2 - 2$	$-x^4 + 3x^3 + 1$
4	$\dfrac{5}{7}z^3 - 2z$	$8y^2 - 2y - 10$

By definition, all monomials, binomials, and trinomials are also polynomials.

 Each term of a polynomial has a **degree.**

> **DEGREE OF A TERM**
>
> The **degree of a term** is the sum of the exponents on the *variables* contained in the term.

EXAMPLE 1   Find the degree of each term.

a. $3x^2$ **b.** -2^3x^5 **c.** y **d.** $12x^2yz^3$ **e.** 5

Solution: **a.** The exponent on x is 2, so the degree of the term is 2.

   b. The exponent on x is 5, so the degree of the term is 5. (Recall that the degree is the sum of the exponents on only the *variables*.)

   c. The degree of y, or y^1, is 1.

   d. The degree is the sum of the exponents on the variables, or $2 + 1 + 3 = 6$.

   e. The degree of 5, which can be written as $5x^0$, is 0.

From the preceding example, we can say that the degree of a constant is 0. Also, the term 0 has no degree.

Each polynomial also has a degree.

> **DEGREE OF A POLYNOMIAL**
>
> The **degree of a polynomial** is the largest degree of all its terms.

EXAMPLE 2   Find the degree of each polynomial and indicate whether the polynomial is a monomial, binomial, trinomial, or none of these.

    a. $7x^3 - 3x + 2$ **b.** $-xyz$ **c.** $x^2 - 4$ **d.** $2xy + x^2y^2 - 5x^2 - 6$

Solution: **a.** The degree of the trinomial $7x^3 - 3x + 2$ is 3, the largest degree of any term.

    b. The degree of the monomial $-xyz$, or $-x^1y^1z^1$, is 3.

    c. The degree of the binomial $x^2 - 4$ is 2.

    d. The degree of each term of the polynomial $2xy + x^2y^2 - 5x^2 - 6$ is

TERM	DEGREE
$2xy$, or $2x^1y^1$	$1 + 1 = 2$
x^2y^2	$2 + 2 = 4$
$-5x^2$	2
-6	0

The largest degree of any term is 4, so the degree of this polynomial is 4.

3   At times, it is convenient to use function notation to represent polynomials. For example, we may write $P(x)$ to represent the polynomial $3x^2 - 2x - 5$. In symbols, this is

$$P(x) = 3x^2 - 2x - 5$$

This function is called a **polynomial function** because the expression $3x^2 - 2x - 5$ is a polynomial.

> R E M I N D E R   Recall that the symbol $P(x)$ **does not mean** P times x. It is a special symbol used to denote a function.

    If $P(x) = 3x^2 - 2x - 5$, let's evaluate $P(2)$. Below is a review of methods that we have used to evaluate functions at given values.

$P(x) = 3(x)^2 - 2x - 5$
$P(2) = 3 \cdot 2^2 - 2 \cdot 2 - 5$
$P(2) = 3$

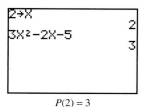

$P(2) = 3$

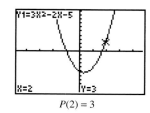

$P(2) = 3$

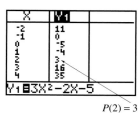

$P(2) = 3$

EXAMPLE 3   If $P(x) = 3x^2 - 2x - 5$, find each function value. Use a graph to check part (a) and a table to check part (b).

    a. $P(1)$ **b.** $P(-2)$

Solution: **a.** Substitute 1 for x in $P(x) = 3x^2 - 2x - 5$ and simplify.

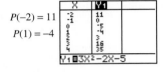

$$P(x) = 3x^2 - 2x - 5$$
$$P(1) = 3(1)^2 - 2(1) - 5 = -4$$

b. $P(x) = 3x^2 - 2x - 5$
$$P(-2) = 3(-2)^2 - 2(-2) - 5 = 11$$

The checks are shown to the left.

Many real-world phenomena are modeled by polynomial functions. If the polynomial function model is given, we can often find the solution of a problem by evaluating the function at a certain value.

EXAMPLE 4   The Royal Gorge suspension bridge in Colorado, the world's highest bridge, is 1053 feet above the Arkansas River. An object is dropped from the top of this bridge. Neglecting air resistance, the height of the object at time x seconds is given by the polynomial function $P(x) = -16x^2 + 1053$.

    a. Use a table to find the height of the object when $x = 3$ seconds and when $x = 7$ seconds.

    b. Approximate, to the nearest second, the time when the object hits the ground.

    c. Approximate, to the nearest tenth of a second, the time when the object hits the ground.

Solution: Enter the polynomial function $y_1 = -16x^2 + 1053$ in your graphing utility, where x is time in seconds and y_1 is height in feet.

a. Since we are asked to find height at 3 seconds and 7 seconds, we may set table start at 3 and table increment at 1.

X	Y1	
3	909	
4	797	
5	653	
6	477	
7	269	
8	29	
9	-243	

$P(3) = 909$

$P(7) = 269$

$Y1 \blacksquare -16X^2+1053$

Because $P(3) = 909$, the height of the object at 3 seconds is 909 feet. Because $P(7) = 269$, the height of the object at 7 seconds is 269 feet.

b. The object will hit the ground when $P(x)$ or $y = 0$. Since

$$P(8) = 29 \text{ and } P(9) = -243,$$

the object hits the ground between 8 and 9 seconds. Since 29 is closer to 0 than -243, the time to the nearest second is 8 seconds.

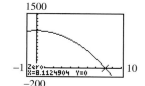

c. Given the table above, graph y_1 in an appropriate window. We graph y_1 in a $[-1, 10, 1]$ by $[-200, 1500, 100]$ window. Recall that the object hits the ground when $P(x)$ or $y = 0$. Thus, use a zero or root function of your graphing utility to find the x-intercept of the graph in the first quadrant. To the nearest tenth, the object hits the ground at 8.1 seconds.

Notice from the table or the first quadrant portion of the graph that as time x increases the height of the object decreases.

4   Before we add polynomials, recall that terms are considered to be **like terms** if they contain exactly the same variables raised to exactly the same powers.

LIKE TERMS	UNLIKE TERMS
$-5x^2, -x^2$	$4x^2, 3x$
$7xy^3z, -2xzy^3$	$12x^2y^3, -2xy^3$

To simplify a polynomial, **combine like terms** by using the distributive property. For example, by the distributive property,

$$5x + 7x = (5 + 7)x = 12x$$

EXAMPLE 5   Simplify by combining like terms.

a. $-12x^2 + 7x^2 - 6x$ **b.** $3xy - 2x + 5xy - x$

Solution: By the distributive property,

a. $-12x^2 + 7x^2 - 6x = (-12 + 7)x^2 - 6x = -5x^2 - 6x$

b. Use the associative and commutative properties to group together like terms; then combine.

$$3xy - 2x + 5xy - x = 3xy + 5xy - 2x - x$$
$$= (3 + 5)xy + (-2 - 1)x$$
$$= 8xy - 3x$$

Now we have reviewed the necessary skills to add polynomials.

To Add Polynomials

Combine all like terms.

EXAMPLE 6   Add.

a. $(7x^3y - xy^3 + 11) + (6x^3y - 4)$ **b.** $(3a^3 - b + 2a - 5) + (a + b + 5)$

Solution: **a.** To add, remove the parentheses and group like terms.

$$(7x^3y - xy^3 + 11) + (6x^3y - 4)$$
$$= 7x^3y - xy^3 + 11 + 6x^3y - 4$$
$$= 7x^3y + 6x^3y - xy^3 + 11 - 4 \qquad \text{Group like terms.}$$
$$= 13x^3y - xy^3 + 7 \qquad \text{Combine like terms.}$$

b. $(3a^3 - b + 2a - 5) + (a + b + 5)$
$$= 3a^3 - b + 2a - 5 + a + b + 5$$
$$= 3a^3 - b + b + 2a + a - 5 + 5 \qquad \text{Group like terms.}$$
$$= 3a^3 + 3a \qquad \text{Combine like terms.}$$

EXAMPLE 7   Add $11x^3 - 12x^2 + x - 3$ and $x^3 - 10x + 5$.

Solution: $(11x^3 - 12x^2 + x - 3) + (x^3 - 10x + 5)$
$$= 11x^3 + x^3 - 12x^2 + x - 10x - 3 + 5 \qquad \text{Group like terms.}$$
$$= 12x^3 - 12x^2 - 9x + 2 \qquad \text{Combine like terms.}$$

Sometimes it is more convenient to add polynomials vertically. To do this, line up like terms beneath one another and add like terms.

EXAMPLE 8   Add $11x^3 - 12x^2 + x - 3$ and $x^3 - 10x + 5$ vertically.

Solution:

$$11x^3 - 12x^2 + \quad x - 3$$
$$\underline{x^3 \qquad\quad - 10x + 5} \qquad \text{Line up like terms.}$$
$$12x^3 - 12x^2 - \quad 9x + 2 \qquad \text{Combine like terms.}$$

This example is the same as Example 7, only here we added vertically.

5   The definition of subtraction of real numbers can be extended to apply to polynomials. To subtract a number, we add its opposite.

$$a - b = a + (-b)$$

To subtract a polynomial, we add its opposite. In other words, if P and Q are polynomials, then

$$P - Q = P + (-Q)$$

The polynomial $-Q$ is the **opposite,** or **additive inverse,** of the polynomial Q. We can find $-Q$ by changing the sign of each term of Q.

To Subtract Polynomials

To subtract polynomials, change the signs of the terms of the polynomial being subtracted; then add.

For example,

To subtract, change the signs; then

Add

$$(3x^2 + 4x - 7) - (3x^2 - 2x - 5) = (3x^2 + 4x - 7) + (-3x^2 + 2x + 5)$$
$$= 3x^2 + 4x - 7 - 3x^2 + 2x + 5$$
$$= 6x - 2 \qquad \text{Combine like terms.}$$

EXAMPLE 9   Subtract: $(12z^5 - 12z^3 + z) - (-3z^4 + z^3 + 12z)$.

Solution: To subtract, change the sign of each term of the second polynomial and add the result to the first polynomial.

$$(12z^5 - 12z^3 + z) - (-3z^4 + z^3 + 12z)$$
$$= 12z^5 - 12z^3 + z + 3z^4 - z^3 - 12z \qquad \text{Change signs and add.}$$
$$= 12z^5 + 3z^4 - 12z^3 - z^3 + z - 12z \qquad \text{Group like terms.}$$
$$= 12z^5 + 3z^4 - 13z^3 - 11z \qquad \text{Combine like terms.}$$

EXAMPLE 10 Subtract $4x^3y^2 - 3x^2y^2 + 2y^2$ from $10x^3y^2 - 7x^2y^2$.

Solution: If we subtract 2 from 8, the difference is $8 - 2 = 6$. Notice the order of the numbers, and then write "Subtract $4x^3y^2 - 3x^2y^2 + 2y^2$ from $10x^3y^2 - 7x^2y^2$" as a mathematical expression.

$$(10x^3y^2 - 7x^2y^2) - (4x^3y^2 - 3x^2y^2 + 2y^2)$$
$$= 10x^3y^2 - 7x^2y^2 - 4x^3y^2 + 3x^2y^2 - 2y^2 \qquad \text{Remove parentheses.}$$
$$= 6x^3y^2 - 4x^2y^2 - 2y^2 \qquad \text{Combine like terms.}$$

EXAMPLE 11 Perform the subtraction $(10x^3y^2 - 7x^2y^2) - (4x^3y^2 - 3x^2y^2 + 2y^2)$ vertically.

Solution: Add the opposite of the second polynomial.

$$\begin{array}{l} 10x^3y^2 - 7x^2y^2 \\ - (4x^3y^2 - 3x^2y^2 + 2y^2) \end{array} \quad \text{is equivalent to} \quad \begin{array}{l} 10x^3y^2 - 7x^2y^2 \\ -4x^3y^2 + 3x^2y^2 - 2y^2 \\ \hline 6x^3y^2 - 4x^2y^2 - 2y^2 \end{array}$$

Polynomial functions, like polynomials, can be added, subtracted, multiplied, and divided. For example, if

$$P(x) = x^2 + x + 1$$

then

$$2P(x) = 2(x^2 + x + 1) = 2x^2 + 2x + 2$$

Also, if $Q(x) = 5x^2 - 1$, then $P(x) + Q(x) = (x^2 + x + 1) + (5x^2 - 1) = 6x^2 + x$.

A useful business and economics application of subtracting polynomial functions is finding the profit function $P(x)$ when given a revenue function $R(x)$ and a cost function $C(x)$. In business, it is true that

$$\text{profit} = \text{revenue} - \text{cost, or}$$
$$P(x) = R(x) - C(x)$$

For example, if the revenue function is $R(x) = 7x$ and the cost function is $C(x) = 2x + 5000$, then the profit function is

$$P(x) = R(x) - C(x)$$

or

$$P(x) = 7x - (2x + 5000) \qquad \text{Substitute } R(x) = 7x \text{ and } C(x) = 2x + 5000.$$
$$P(x) = 5x - 5000$$

Problem-solving exercises involving profit are in the exercise set.

In this section, we reviewed how to find the degree of a polynomial. Knowing the degree of a polynomial can help us recognize the graph of the related polynomial function. For example, we know from Sections 3.2 and 3.3 that the graph of the quadratic function $f(x) = x^2$ is the parabola shown at left.

The polynomial x^2 has degree 2. The graphs of all polynomial functions of degree 2 will have this same general shape, a parabola, opening upward, as shown, or

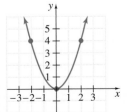

downward. Graphs of polynomial functions of degree 2 or 3 will, in general, resemble one of the graphs shown next.

Degree 2

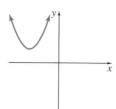

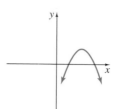

Coefficient of x^2
is a positive number.

Coefficient of x^2
is a negative number.

Degree 3

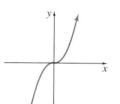

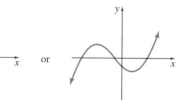

or

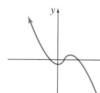

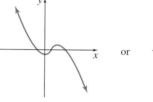

or

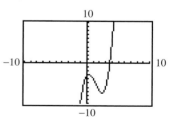

Coefficient of x^3
is a positive number.

Coefficient of x^3
is a negative number.

General Shapes of Graphs of Polynomial Functions

In addition to determining the general shape of the graph of a function from the degree of the related polynomial, we also can find the y-intercept from the constant term. That is, if $p(x) = x^2 + 3x - 2$, then $p(0) = -2$, so the point $(0, -2)$ is the y-intercept point.

EXAMPLE 12   Match each graph with its equation and give the coordinates of the y-intercept point of each graph.
  a. $f(x) = x^2 + x - 5$ **b.** $g(x) = x^3 - x^2 + x + 2$
  c. $h(x) = x^3 - 4x^2 + 2x - 3$ **d.** $F(x) = -x^3 + x - 6$
  e. $H(x) = -x^2 + 5x + 1$

1.

2.

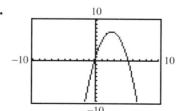

3.

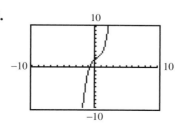

4.

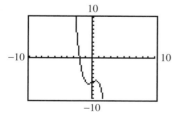

5.

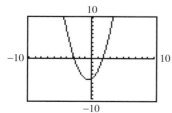

Solution: **a.** The polynomial function $f(x) = x^2 + x - 5$ has degree 2, so the shape of its graph is a parabola. Since the coefficient of x^2 is the positive number 1, the parabola opens upward and has the shape of option 4. Its y-intercept point is $(0, -5)$.

b. The polynomial function $g(x) = x^3 - x^2 + x + 2$ has degree 3, so the shape of its graph is option 1, 3, or 5. Since the coefficient of x^3 is the positive number 1, it has the shape of option 3 or 5. The y-intercept point of $g(x)$ is $(0, 2)$, so the graph resembles option 5.

c. The graph of $h(x)$ resembles option 3. The y-intercept point of $h(x)$ is $(0, -3)$.

d. Since the coefficient of x^3 is -1, the graph of $F(x)$ resembles option 1. The y-intercept point is $(0, -6)$.

e. The graph of $H(x)$ is a parabola opening downward, since the coefficient of x^2 is -1. Its graph is option 2. The y-intercept point is $(0, 1)$.   ▬▬

EXERCISE SET 6.3

Find the degree of each term. See Example 1.

1. 4   **2.** 7   **3.** $5x^2$

4. $-z^3$ **5.** $-3xy^2$ **6.** $12x^3z$

Find the degree of each polynomial and indicate whether the polynomial is a monomial, binomial, trinomial, or none of these. See Example 2.

7. $6x + 3$ **8.** $7x - 8$

9. $3x^2 - 2x + 5$ **10.** $5x^2 - 3x^2y - 2x^3$

11. $-xyz$ **12.** -9

13. $x^2y - 4xy^2 + 5x + y$

14. $-2x^2y - 3y^2 + 4x + y^5$

◻ 15. In your own words, describe how to find the degree of a term.

◻ 16. In your own words, describe how to find the degree of a polynomial.

If $P(x) = x^2 + x + 1$ and $Q(x) = 5x^2 - 1$, find the following. See Example 3.

17. $P(7)$ **18.** $Q(4)$ **19.** $Q(-10)$

20. $P(-4)$

Solve. See Example 4.

21. Enter $y_1 = -16x^2 + 75$ on a graphing utility. Look at a table of values with the table start at 0 and incrementing by 1.

    a. Between what two x-values do the y-values change from positive to negative?

    b. Graph y_1 in a $[0, 5, 1]$ by $[-10, 100, 10]$ window. Between what two integers does the graph cross the x-axis?

22. Enter $y_1 = -16x^2 - 50x + 400$ on a graphing utility. Look at a table of values with the table start at 0 and incrementing by 1.

    a. Between what two x-values do the y-values change from positive to negative?

    b. Graph y_1 in a $[0, 5, 1]$ by $[-100, 500, 100]$ window. Between what two integers does the graph cross the x-axis?

23. An object is dropped off the top of a building that is 525 feet high. The distance, in feet, above the ground at x seconds is given by $P(x) = -16x^2 + 525$. Approximate to the nearest tenth of a second when the object hits the ground.

24. An object is dropped off the top of a building that is 350 feet high. The distance, in feet, above the ground at x seconds is given by $P(x) = -16x^2 + 350$. Approximate to the nearest tenth of a second when the object will pass a window that is 100 feet above ground level.

◻ 25. Two intermediate algebra students perform the following experiment. An egg is dropped from the top of a 100-foot building at the same instant that another egg is thrown down with a speed of 85 feet per second from a nearby building 200 feet high. The distance, in feet, above the ground at x seconds for the dropped egg is given by $P_1(x) =$

$-16x^2 + 100$, and for the egg thrown with a speed of 85 feet per second, it is $P_2(x) = -16x^2 - 85x + 200$.

a. Which egg hits the ground first?

b. Are the eggs ever the same distance above the ground at the same time? If yes, approximate your answer to the nearest second.

26. A stone is dropped off a bridge that is 150 feet above a river. The distance, in feet, above the ground at x seconds is given by the equation $P_1(x) = -16x^2 + 150$. One second later, another stone is thrown down with a speed of 80 feet per second. The distance, in feet, above the ground at x seconds is given by $P_2(x) = -16(x - 1)^2 - 80(x - 1) + 150$.

a. Which stone hits the water first?

b. Does the second stone ever catch up with the first stone so that they are the same distance above the water at the same time? If yes, approximate your answer to the nearest second.

Simplify by combining like terms. See Example 5.

27. $5y + y$

28. $-x + 3x$

29. $4x + 7x - 3$

30. $-8y + 9y + 4y^2$

31. $4xy + 2x - 3xy - 1$

32. $-8xy^2 + 4x - x + 2xy^2$

Perform indicated operations. See Examples 6 through 11.

33. $(9y^2 - 8) + (9y^2 - 9)$

34. $(x^2 + 4x - 7) + (8x^2 + 9x - 7)$

35. $(x^2 + xy - y^2)$ and $(2x^2 - 4xy + 7y^2)$

36. $(4x^3 - 6x^2 + 5x + 7)$ and $(2x^2 + 6x - 3)$

37.
$$\begin{array}{r} x^2 - 6x + 3 \\ + \quad (2x + 5) \\ \hline \end{array}$$

38.
$$\begin{array}{r} -2x^2 + 3x - 9 \\ + \quad (2x - 3) \\ \hline \end{array}$$

39. $(9y^2 - 7y + 5) - (8y^2 - 7y + 2)$

40. $(2x^2 + 3x + 12) - (5x - 7)$

41. $(6x^2 - 3x)$ from $(4x^2 + 2x)$

42. $(xy + x - y)$ from $(xy + x - 3)$

43.
$$\begin{array}{r} 3x^2 - 4x + 8 \\ - \quad (5x^2 - 7) \\ \hline \end{array}$$

44.
$$\begin{array}{r} -3x^2 - 4x + 8 \\ - \quad (5x + 12) \\ \hline \end{array}$$

45. $(5x - 11) + (-x - 2)$

46. $(3x^2 - 2x) + (5x^2 - 9x)$

47. $(7x^2 + x + 1) - (6x^2 + x - 1)$

48. $(4x - 4) - (-x - 4)$

49. $(7x^3 - 4x + 8) + (5x^3 + 4x + 8x)$

50. $(9xyz + 4x - y) + (-9xyz - 3x + y + 2)$

51. $(9x^3 - 2x^2 + 4x - 7) - (2x^3 - 6x^2 - 4x + 3)$

52. $(3x^2 + 6xy + 3y^2) - (8x^2 - 6xy - y^2)$

53. Add $(y^2 + 4yx + 7)$ and $(-19y^2 + 7yx + 7)$.

54. Subtract $(x - 4)$ from $(3x^2 - 4x + 5)$.

55. $(3x^3 - b + 2a - 6) + (-4x^3 + b + 6a - 6)$

56. $(5x^2 - 6) + (2x^2 - 4x + 8)$

57. $(4x^2 - 6x + 2) - (-x^2 + 3x + 5)$

58. $(5x^2 + x + 9) - (2x^2 - 9)$

59. $(-3x + 8) + (-3x^2 + 3x - 5)$

60. $(5y^2 - 2y + 4) + (3y + 7)$

61. $(-3 + 4x^2 + 7xy^2) + (2x^3 - x^2 + xy^2)$

62. $(-3x^2y + 4) - (-7x^2y - 8y)$

63.
$$\begin{array}{r} 6y^2 - 6y + 4 \\ -(-y^2 - 6y + 7) \\ \hline \end{array}$$

64.
$$\begin{array}{r} -4x^3 + 4x^2 - 4x \\ - \ (2x^3 - 2x^2 + 3x) \\ \hline \end{array}$$

65.
$$\begin{array}{r} 3x^2 + 15x + 8 \\ + \ (2x^2 + 7x + 8) \\ \hline \end{array}$$

66.
$$\begin{array}{r} 9x^2 + 9x - 4 \\ + \ (7x^2 - 3x - 4) \\ \hline \end{array}$$

67. Find the sum of $(5q^4 - 2q^2 - 3q)$ and $(-6q^4 + 3q^2 + 5)$.

68. Find the sum of $(5y^4 - 7y^2 + x^2 - 3)$ and $(-3y^4 + 2y^2 + 4)$.

69. Subtract $(3x + 7)$ from the sum of $(7x^2 + 4x + 9)$ and $(8x^2 + 7x - 8)$.

70. Subtract $(9x + 8)$ from the sum of $(3x^2 - 2x - x^3 + 2)$ and $(5x^2 - 8x - x^3 + 4)$.

71. Find the sum of $(4x^4 - 7x^2 + 3)$ and $(2 - 3x^4)$.

72. Find the sum of $(8x^4 - 14x^2 + 6)$ and $(-12x^6 - 21x^4 - 9x^2)$.

73. $(8x^{2y} - 7x^y + 3) + (-4x^{2y} + 9x^y - 14)$

74. $(14z^{5x} + 3z^{2x} + z) - (2z^{5x} - 10z^{2x} + 3z)$

75. The polynomial function $P(x) = 45x - 100,000$ models the relationship between the number of lamps x that Sherry's Lamp Shop sells and the profit the shop makes, $P(x)$. Find $P(4000)$, the profit from selling 4000 lamps.

76. The polynomial $P(t) = -32t + 500$ models the relationship between the length of time t in seconds a particle flies through space, beginning at a velocity of 500 feet per second, and its accrued velocity, $P(t)$. Find $P(3)$, the accrued velocity after 3 seconds.

77. The function $f(x) = 4.31x^2 + 29.17x + 343.44$ can be used to approximate spending for health care in the United States, where x is the number of years since 1970 and $f(x)$ is the amount of money spent per capita. (*Sources:* Office of National Health Statistics and U.S. Bureau of the Census)

   a. Approximate the amount of money spent on health care per capita in the year 1980.

   b. Approximate the amount of money spent on health care per capita in the year 1990.

   c. Use the given function to predict the amount of money spent on health care per capita in the year 2000.

   d. From parts a, b, and c, is the amount of money spent rising at a steady rate? Why or why not?

78. An object is thrown upward with an initial velocity of 50 feet per second from the top of the 350-foot-high City Hall in Milwaukee, Wisconsin. The height of the object at any time t can be described by the polynomial function $P(t) = -16t^2 + 50t + 350$. Find the height of the projectile when $t = 1$ second, $t = 2$ seconds, and $t = 3$ seconds. (*Source: World Almanac*)

79. The total cost (in dollars) for MCD, Inc., Manufacturing Company to produce x blank audiocassette tapes per week is given by the polynomial function $C(x) = 0.8x + 10,000$. Find the total cost in producing 20,000 tapes per week.

80. The total revenues (in dollars) for MCD, Inc., Manufacturing Company to sell x blank audiocassette tapes per week is given by the polynomial function $R(x) = 2x$. Find the total revenue in selling 20,000 tapes per week.

A projectile is fired upward from the ground with an initial velocity of 300 feet per second. Neglecting air resistance, the height of the projectile at any time t can be described by the polynomial function

$$P(t) = -16t^2 + 300t$$

Use the following table of values to answer Excercises 81–83.

X	Y1
0	0
2	536
4	944
6	1224
8	1376
10	1400
12	1296

$$Y_1 = -16X^2 + 300X$$

81. Find the height of the projectile at the given times.

   a. $t = 2$ seconds   **b.** $t = 6$ seconds

   c. $t = 10$ seconds   **d.** $t = 12$ seconds

82. Explain why the height increases and then decreases as time passes.

83. Generate the table above for positive integer x-values and approximate (to the nearest second) how long it takes the object to hit the ground.

If $P(x) = 3x + 3$, $Q(x) = 4x^2 - 6x + 3$, and $R(x) = 5x^2 - 7$, find the following.

84. $P(x) + Q(x)$ **85.** $R(x) + P(x)$

86. $Q(x) - R(x)$ **87.** $P(x) - Q(x)$

88. $2[Q(x)] - R(x)$ **89.** $-5[P(x)] - Q(x)$

90. Find the coordinates of the y-intercept of the graph of each function.

   a. $f(x) = 2x^3 - 5x^2 + 3x + 7$

   b. $g(x) = x^2 + x - 4$ **c.** $H(x) = 5x + 1$

91. Find the coordinates of the y-intercept point of the graph of each function.

   a. $h(x) = -3x^2 + 2x + 4$

   b. $F(x) = 0.5x^3 - 1.2x - 3.9$ **c.** $G(x) = -x^2 + 4x$

92. If the revenue function of a certain company is given by $R(x) = 5.5x$, where x is the number of packages of personalized stationery sold and the cost function is given by $C(x) = 3x + 3000$,

   a. Find the profit function. (Recall that revenue − cost = profit.)

   b. Find the profit when 2000 packages of stationery are sold.

Match each equation with its graph found on the next page, and state the coordinates of its y-intercept point. See Example 12.

93. $f(x) = 3x^2 - 2$ **94.** $h(x) = 5x^3 - 6x + 2$

95. $g(x) = -2x^3 - 3x^2 + 3x - 2$

96. $g(x) = -2x^2 - 6x + 2$

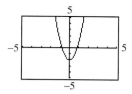

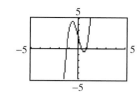

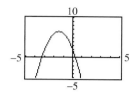

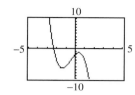

Find each perimeter.

97.

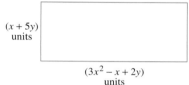

$(x + 5y)$
units

$(3x^2 - x + 2y)$
units

98.

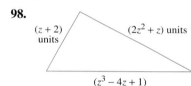

$(z + 2)$
units

$(2z^2 + z)$ units

$(z^3 - 4z + 1)$
units

Review Exercises

Multiply. See Section 1.2.

99. $5(3x - 2)$ **100.** $-7(2z - 6y)$

101. $-2(x^2 - 5x + 6)$ **102.** $5(-3y^2 - 2y + 7)$

A Look Ahead

EXAMPLE

If $P(x) = -3x + 5$, find the following:

a. $P(a)$ **b.** $P(-x)$ **c.** $P(x + h)$

Solution: **a.** $P(x) = -3x + 5$

$$P(a) = -3a + 5$$

b. $\quad P(x) = -3x + 5$

$$P(-x) = -3(-x) + 5$$
$$= 3x + 5$$

c. $\quad\quad P(x) = -3x + 5$

$$P(x + h) = -3(x + h) + 5$$
$$= -3x - 3h + 5$$

If $P(x)$ is the polynomial given, find $P(a)$, $P(-x)$, and $P(x + h)$. See the preceding example.

103. $P(x) = 2x - 3$ **104.** $P(x) = 8x + 3$

105. $P(x) = 4x$ **106.** $P(x) = -4x$

107. $P(x) = 4x - 1$ **108.** $P(x) = 3x - 2$

6.4   MULTIPLYING POLYNOMIALS

O B J E C T I V E S

 Multiply two polynomials.

2 Multiply binomials.

3 Square binomials.

4 Multiply the sum and difference of two terms.

TAPE IAG 6.4

Properties of real numbers and exponents are used continually in the process of multiplying polynomials. To multiply monomials, for example, we apply the commutative and associative properties of real numbers and the product rule for *exponents.*

EXAMPLE 1   Multiply.

    a. $(2x^3)(5x^6)$ **b.** $(7y^4z^4)(-xy^{11}z^5)$

Solution: Group like bases and apply the product rule for exponents.

    a. $(2x^3)(5x^6) = 2(5)(x^3)(x^6) = 10x^9$

    b. $(7y^4z^4)(-xy^{11}z^5) = 7(-1)x(y^4y^{11})(z^4z^5) = -7xy^{15}z^9$

To multiply a monomial by a polynomial other than a monomial, we use an expanded form of the distributive property.

$$a(b + c + d + \cdots + z) = ab + ac + ad + \cdots + az$$

Notice that the monomial a is multiplied by each term of the polynomial.

EXAMPLE 2   Multiply.

    a. $2x(5x - 4)$ **b.** $-3x^2(4x^2 - 6x + 1)$ **c.** $-xy(7x^2y + 3xy - 11)$

Solution: Apply the distributive property.

    a. $2x(5x - 4) = 2x(5x) + 2x(-4)$

                     $= 10x^2 - 8x$

    b. $-3x^2(4x^2 - 6x + 1) = -3x^2(4x^2) + (-3x^2)(-6x) + (-3x^2)(1)$

                                $= -12x^4 + 18x^3 - 3x^2$

    c. $-xy(7x^2y + 3xy - 11) = -xy(7x^2y) + (-xy)(3xy) + (-xy)(-11)$

                                $= -7x^3y^2 - 3x^2y^2 + 11xy$

In general, we have the following:

> To multiply any two polynomials, use the distributive property, and multiply each term of one polynomial by each term of the other polynomial. Then combine any like terms.

EXAMPLE 3   Multiply and simplify the product if possible.

    a. $(x + 3)(2x + 5)$ **b.** $(2x^3 - 3)(5x^2 - 6x + 7)$

Solution: **a.** Multiply each term of $(x + 3)$ by $(2x + 5)$.

$$(x + 3)(2x + 5) = x(2x + 5) + 3(2x + 5)$$ Apply the distributive property.

$$= 2x^2 + 5x + 6x + 15$$ Apply the distributive property again.

$$= 2x^2 + 11x + 15$$ Combine like terms.

b. Multiply each term of $(2x^3 - 3)$ by each term of $(5x^2 - 6x + 7)$.

$$(2x^3 - 3)(5x^2 - 6x + 7) = 2x^3(5x^2 - 6x + 7) + (-3)(5x^2 - 6x + 7)$$
$$= 10x^5 - 12x^4 + 14x^3 - 15x^2 + 18x - 21$$

Sometimes polynomials are easier to multiply vertically, in the same way we multiply real numbers. When multiplying vertically, line up like terms in the **partial products** vertically. This makes combining like terms easier.

EXAMPLE 4   Find the product of $(4x^2 + 7)$ and $(x^2 + 2x + 8)$.

Solution:

$$
\begin{array}{r}
x^2 + 2x + 8 \\
4x^2 + 7 \\
\hline
7x^2 + 14x + 56 \\
4x^4 + 8x^3 + 32x^2 \\
\hline
4x^4 + 8x^3 + 39x^2 + 14x + 56
\end{array}
$$

$7(x^2 + 2x + 8)$

$4x^2(x^2 + 2x + 8)$

Combine like terms.

When multiplying a binomial by a binomial, we can use a special order of multiplying terms, called the **FOIL** order. The letters of FOIL stand for "First–Outer–Inner–Last." To illustrate this method, multiply $(2x - 3)$ by $(3x + 1)$.

Multiply the **First** terms of each binomial. $(2x - 3)(3x + 1)$ **F** $2x(3x) = 6x^2$

Multiply the **Outer** terms of each binomial. $(2x - 3)(3x + 1)$ **O** $2x(1) = 2x$

Multiply the **Inner** terms of each binomial. $(2x - 3)(3x + 1)$ **I** $-3(3x) = -9x$

Multiply the **Last** terms of each binomial. $(2x - 3)(3x + 1)$ **L** $-3(1) = -3$

Combine like terms.

$$6x^2 + 2x - 9x - 3 = 6x^2 - 7x - 3$$

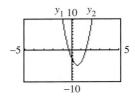

A graphing utility may be used to visualize operations on polynomials. To visualize a multiplication statement such as

$$(2x - 3)(3x + 1) = 6x^2 - 7x - 3,$$

graph $y_1 = (2x - 3)(3x + 1)$ and $y_2 = 6x^2 - 7x - 3$ on the same set of axes and see that their graphs coincide.

Note: If the graphs do not coincide, we can be sure that a mistake has been made in multiplying the polynomials or in entering keystrokes. If the graphs appear to coincide, we cannot be sure that our work is correct. This is because it is possible for the graphs to differ so slightly that we do not notice it.

The graphs of y_1 and y_2 are shown to the left. The graphs *appear* to coincide so the multiplication statement

$$(2x - 3)(3x + 1) = 6x^2 - 7x - 3$$

appears to be correct.

A table of values, such as the one to the left showing y_1 and y_2, can also help confirm operations on polynomials.

EXAMPLE 5   Multiply $(x - 1)(x + 2)$. Use the FOIL order.

Solution:

$$
\begin{array}{cccc}
\text{First} & \text{Outer} & \text{Inner} & \text{Last} \\
\downarrow & \downarrow & \downarrow & \downarrow
\end{array}
$$

$$(x - 1)(x + 2) = x \cdot x + 2 \cdot x + (-1)x + (-1)(2)$$
$$= x^2 + 2x - x - 2$$
$$= x^2 + x - 2 \qquad \text{Combine like terms.}$$

EXAMPLE 6   Multiply $(2x - 7)(3x - 4)$.

Solution:

$$
\begin{array}{cccc}
\text{First} & \text{Outer} & \text{Inner} & \text{Last} \\
\downarrow & \downarrow & \downarrow & \downarrow
\end{array}
$$

$$(2x - 7)(3x - 4) = 2x(3x) + 2x(-4) + (-7)(3x) + (-7)(-4)$$
$$= 6x^2 - 8x - 21x + 28$$
$$= 6x^2 - 29x + 28$$

 Certain products and powers of polynomials create striking patterns. For example, the **square of a binomial** is a special case of the product of two binomials. Find $(a + b)^2$ by the FOIL order for multiplying two binomials.

$$(a + b)^2 = (a + b)(a + b)$$

$$
\begin{array}{cccc}
\text{F} & \text{O} & \text{I} & \text{L}
\end{array}
$$
$$= a^2 + ab + ba + b^2$$
$$= a^2 + 2ab + b^2$$

This product can be visualized geometrically by analyzing areas. The area of the square shown next can be written as $(a + b)^2$ or as the sum of the areas of the smaller rectangles.

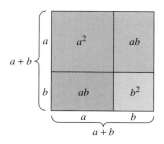

The sum of the areas of the smaller rectangles is $a^2 + ab + ab + b^2$, or $a^2 + 2ab + b^2$. Thus,

$$(a + b)^2 = a^2 + 2ab + b^2$$

We can use this pattern as the basis of a quick method for squaring a binomial. Two forms are shown for squaring a binomial: the sum of terms and the difference of terms. We call these **special products.**

SQUARE OF A BINOMIAL

$$(a + b)^2 = a^2 + 2ab + b^2, \quad (a - b)^2 = a^2 - 2ab + b^2$$

A binomial squared is the sum of the first term squared, twice the product of both terms, and the second term squared.

EXAMPLE 7

Multiply.

a. $(x + 5)^2$ **b.** $(x - 9)^2$ **c.** $(3x + 2z)^2$ **d.** $(4m^2 - 3n)^2$

Solution:

a. $(x + 5)^2 = x^2 + 2 \cdot x \cdot 5 + 5^2 = x^2 + 10x + 25$

b. $(x - 9)^2 = x^2 - 2 \cdot x \cdot 9 + 9^2 = x^2 - 18x + 81$

c. $(3x + 2z)^2 = (3x)^2 + 2(3x)(2z) + (2z)^2 = 9x^2 + 12xz + 4z^2$

d. $(4m^2 - 3n)^2 = (4m^2)^2 - 2(4m^2)(3n) + (3n)^2 = 16m^4 - 24m^2n + 9n^2$

TECHNOLOGY NOTE

To verify that $(x + 5)^2 \neq x^2 + 25$, for example, graph $y_1 = (x + 5)^2$ and $y_2 = x^2 + 25$ using the same standard window. Since the two graphs do not coincide, the expressions are not equivalent.

R E M I N D E R   Note that $(a + b)^2 = a^2 + 2ab + b^2$, **not** $a^2 + b^2$. Also, $(a - b)^2 = a^2 - 2ab + b^2$, **not** $a^2 - b^2$.

---

4   Another special product applies to the sum and difference of the same two terms. Multiply $(a + b)(a - b)$ to see a pattern.

$$(a + b)(a - b) = a^2 - ab + ba - b^2$$
$$= a^2 - b^2$$

PRODUCT OF THE SUM AND DIFFERENCE OF TWO TERMS

$$(a + b)(a - b) = a^2 - b^2$$

The product of the sum and difference of the same two terms is the difference of the first term squared and the second term squared.

EXAMPLE 8   Multiply.

a. $(x - 3)(x + 3)$ **b.** $(4y + 1)(4y - 1)$ **c.** $(x^2 + 2y)(x^2 - 2y)$

Solution: **a.** $(x - 3)(x + 3) = x^2 - 3^2 = x^2 - 9$
  b. $(4y + 1)(4y - 1) = (4y)^2 - 1^2 = 16y^2 - 1$
  c. $(x^2 + 2y)(x^2 - 2y) = (x^2)^2 - (2y)^2 = x^4 - 4y^2$

EXAMPLE 9   Multiply: $[3 + (2a + b)]^2$.

Solution: Think of 3 as the first term and $(2a + b)$ as the second term, and apply the method for squaring a binomial.

$$[3 + (2a + b)]^2 = \quad (3)^2 \quad + 2(3)(2a + b) + (2a + b)^2$$

$$\underset{\substack{\text{First term} \\ \text{squared}}}{} \quad \underset{\substack{\text{Twice the} \\ \text{product of} \\ \text{both terms}}}{} \quad \underset{\substack{\text{Last term} \\ \text{squared}}}{}$$

$$= 9 + 6(2a + b) + (2a + b)^2$$
$$= 9 + 12a + 6b + (2a)^2 + 2(2a)(b) + b^2 \quad \text{Square } (2a + b).$$
$$= 9 + 12a + 6b + 4a^2 + 4ab + b^2$$

EXAMPLE 10   Multiply: $[(5x - 2y) - 1][(5x - 2y) + 1]$.

Solution: Think of $(5x - 2y)$ as the first term and 1 as the second term, and apply the method for the product of the sum and difference of two terms.

$$[(5x - 2y) - 1][(5x - 2y) + 1] = (5x - 2y)^2 - 1^2$$

First term   Second term
squared   squared

$$= (5x - 2y)^2 - 1$$
$$= (5x)^2 - 2(5x)(2y) + (2y)^2 - 1 \text{ Square } (5x - 2y).$$
$$= 25x^2 - 20xy + 4y^2 - 1$$

Our work in multiplying polynomials is often useful in evaluating polynomial functions.

EXAMPLE 11   If $f(x) = x^2 + 5x - 2$, find $f(a + 1)$.

Solution: To find $f(a + 1)$, replace x with the expression $a + 1$ in the polynomial function $f(x)$.

$$f(x) = x^2 + 5x - 2$$
$$f(a + 1) = (a + 1)^2 + 5(a + 1) - 2$$
$$= a^2 + 2a + 1 + 5a + 5 - 2$$
$$= a^2 + 7a + 4$$

EXERCISE SET 6.4

Multiply. See Examples 1 through 4.

1. $(-4x^3)(3x^2)$ **2.** $(-6a)(4a)$ **3.** $3x(4x + 7)$

4. $5x(6x - 4)$ **5.** $-6xy(4x + y)$ **6.** $-8y(6xy + 4x)$

7. $-4ab(xa^2 + ya^2 - 3)$

8. $-6b^2z(z^2a + baz - 3b)$

9. $(x - 3)(2x + 4)$ **10.** $(y + 5)(3y - 2)$

11. $(2x + 3)(x^3 - x + 2)$ **12.** $(a + 2)(3a^2 - a + 5)$

13. $3x - 2$
$\underline{5x + 1}$

14. $2z - 4$
$\underline{6z - 2}$

15. $3m^2 + 2m - 1$
$\underline{5m + 2}$

16. $2x^2 - 3x - 4$
$\underline{x + 5}$

17. Explain how to multiply a polynomial by a polynomial.

18. Explain why $(3x + 2)^2$ does not equal $9x^2 + 4$.

Multiply the binomials. Use a graphing utility to visualize the product. See Examples 5 and 6.

19. $(x - 3)(x + 4)$ **20.** $(c - 3)(c + 1)$

21. $(5x + 8y)(2x - y)$ **22.** $(2n - 9m)(n - 7m)$

23. $(3x - 1)(x + 3)$ **24.** $(5d - 3)(d + 6)$

25. $\left(3x + \dfrac{1}{2}\right)\left(3x - \dfrac{1}{2}\right)$ **26.** $\left(2x - \dfrac{1}{3}\right)\left(2x + \dfrac{1}{3}\right)$

Multiply, using special product methods. See Examples 7 through 9.

27. $(x + 4)^2$ **28.** $(x - 5)^2$

29. $(6y - 1)(6y + 1)$ **30.** $(x - 9)(x + 9)$

31. $(3x - y)^2$ **32.** $(4x - z)^2$

33. $(3b - 6y)(3b + 6y)$ **34.** $(2x - 4y)(2x + 4y)$

Multiply, using special product methods. See Example 10.

35. $[3 + (4b + 1)]^2$ **36.** $[5 - (3b - 3)]^2$

37. $[(2s - 3) - 1][(2s - 3) + 1]$

38. $[(2y + 5) + 6][(2y + 5) - 6]$

39. $[(xy + 4) - 6]^2$ **40.** $[(2a^2 + 4a) + 1]^2$

41. Explain when the FOIL method can be used to multiply polynomials.

42. Explain why the product of $(a + b)$ and $(a - b)$ is not a trinomial.

Multiply.

43. $(3x + 1)(3x + 5)$ **44.** $(4x - 5)(5x + 6)$

45. $(2x^3 + 5)(5x^2 + 4x + 1)$

46. $(3y^3 - 1)(3y^3 - 6y + 1)$

47. $(7x - 3)(7x + 3)$ **48.** $(4x + 1)(4x - 1)$

49. $\begin{array}{r} 3x^2 + 4x - 4 \\ \underline{3x + 6} \end{array}$ **50.** $\begin{array}{r} 6x^2 + 2x - 1 \\ \underline{3x - 6} \end{array}$

51. $\left(4x + \dfrac{1}{3}\right)\left(4x - \dfrac{1}{2}\right)$ **52.** $\left(4y - \dfrac{1}{3}\right)\left(3y - \dfrac{1}{8}\right)$

53. $(6x + 1)^2$ **54.** $(4x + 7)^2$

55. $(x^2 + 2y)(x^2 - 2y)$ **56.** $(3x + 2y)(3x - 2y)$

57. $-6a^2b^2(5a^2b^2 - 6a - 6b)$

58. $7x^2y^3(-3ax - 4xy + z)$

59. $(a - 4)(2a - 4)$ **60.** $(2x - 3)(x + 1)$

61. $(7ab + 3c)(7ab - 3c)$

62. $(3xy - 2b)(3xy + 2b)$

63. $(m - 4)^2$ **64.** $(x + 2)^2$

65. $(3x + 1)^2$ **66.** $(4x + 6)^2$

67. $(y - 4)(y - 3)$ **68.** $(c - 8)(c + 2)$

69. $(x + y)(2x - 1)(x + 1)$

70. $(z + 2)(z - 3)(2z + 1)$

71. $(3x^2 + 2x - 1)^2$ **72.** $(4x^2 + 4x - 4)^2$

73. $(3x + 1)(4x^2 - 2x + 5)$

74. $(2x - 1)(5x^2 - x - 2)$

If $R(x) = x + 5$, $Q(x) = x^2 - 2$, and $P(x) = 5x$, find the following.

75. $P(x) \cdot R(x)$ **76.** $P(x) \cdot Q(x)$

77. $[Q(x)]^2$ **78.** $[R(x)]^2$

79. $R(x) \cdot Q(x)$ **80.** $P(x) \cdot R(x) \cdot Q(x)$

81. Perform the indicated operations. Explain the difference between the two problems.

   a. $(3x + 5) + (3x + 7)$ **b.** $(3x + 5)(3x + 7)$

82. Find the area of the circle. Do not approximate π.

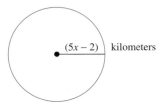

$(5x - 2)$ kilometers

83. Find the volume of the cylinder. Do not approximate π.

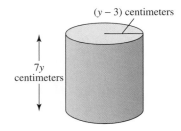

$(y - 3)$ centimeters

$7y$ centimeters

If $f(x) = x^2 - 3x$, find the following. See Example 11.

84. $f(a)$ **85.** $f(c)$

86. $f(a + h)$ **87.** $f(a + 5)$

88. $f(b - 2)$ **89.** $f(a - b)$

90. If $F(x) = x^2 + 3x + 2$, find **(a)** $F(a + h)$, **(b)** $F(a)$, **(c)** $F(a + h) - F(a)$.

91. If $g(x) = x^2 + 2x + 1$, find **(a)** $g(a + h)$, **(b)** $g(a)$, **(c)** $g(a + h) - g(a)$.

Multiply. Assume that variables represent positive integers.

92. $5x^2y^n(6y^{n+1} - 2)$ **93.** $-3yz^n(2y^3z^{2n} - 1)$

94. $(x^a + 5)(x^{2a} - 3)$ **95.** $(x^a + y^{2b})(x^a - y^{2b})$

Review Exercises

Use the slope–intercept form of a line, $y = mx + b$, to find the slope of each line. See Section 3.5.

96. $y = -2x + 7$ **97.** $y = \dfrac{3}{2}x - 1$

98. $3x - 5y = 14$ **99.** $x + 7y = 2$

Use the vertical line test to determine which of the following are graphs of functions. See Section 3.3.

100.

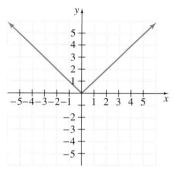

101.

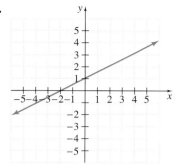

6.5 | THE GREATEST COMMON FACTOR AND FACTORING BY GROUPING

TAPE IAG 6.5

O B J E C T I V E S

1 Identify the GCF.

2 Factor out the GCF of a polynomial's terms.

3 Factor polynomials by grouping.

Factoring is the reverse process of multiplying. It is the process of writing a polynomial as a product.

$$\overbrace{6x^2 + 13x - 5 = (3x - 1)(2x + 5)}^{\text{factoring}}_{\text{multiplying}}$$

In the next few sections, we review techniques for factoring polynomials. These techniques are used at the end of this chapter to solve polynomial equations and to graph polynomial functions.

To factor a polynomial, we first factor out the greatest common factor (GCF) of its terms, using the distributive property. The GCF of a list of terms or monomials is the product of the GCF of the numerical coefficients and each GCF of the powers of a common variable.

TO FIND THE **GCF** OF A LIST OF **MONOMIALS**

Step 1. Find the GCF of the numerical coefficients.

Step 2. Find the GCF of the variable factors.

Step 3. The product of the factors found in *steps 1* and *2* is the GCF of the monomials.

EXAMPLE 1   Find the GCF of $20x^3y$, $10x^2y^2$, and $35x^3$.

Solution: The GCF of the numerical coefficients 20, 10, and 35 is 5, the largest integer that is a factor of each integer. The GCF of the variable factors x^3, x^2, and x^3 is x^2 because x^2 is the largest factor common to all three powers of x. The variable y is not a common factor because it does not appear in all three monomials. The GCF is thus

$$5 \cdot x^2, \quad \text{or} \quad 5x^2$$

2   A first step in factoring polynomials is to use the distributive property and write the polynomial as a product of the GCF of its monomial terms and a simpler polynomial. This is called **factoring out** the GCF.

EXAMPLE 2   Factor.

     a. $8x + 4$ **b.** $5y - 2z$ **c.** $6x^2 - 3x^3$

Solution: **a.** The GCF of terms $8x$ and 4 is 4.

$$8x + 4 = 4(2x) + 4(1) \qquad \text{Factor out 4 from each term.}$$
$$= 4(2x + 1) \qquad \text{Apply the distributive property.}$$

The factored form of $8x + 4$ is $4(2x + 1)$. To check, multiply $4(2x + 1)$ to see that the product is $8x + 4$.

b. There is no common factor of the terms $5y$ and $-2z$ other than 1 (or -1).

c. The greatest common factor of $6x^2$ and $-3x^3$ is $3x^2$. Thus,

$$6x^2 - 3x^3 = 3x^2(2) - 3x^2(x)$$
$$= 3x^2(2 - x)$$

> REMINDER   To verify that the GCF has been factored out correctly, multiply the factors together and see that their product is the original polynomial.

EXAMPLE 3   Factor $17x^3y^2 - 34x^4y^2$.

Solution: The GCF of the two terms is $17x^3y^2$, which we factor out of each term.

$$17x^3y^2 - 34x^4y^2 = 17x^3y^2(1) - 17x^3y^2(2x)$$
$$= 17x^3y^2(1 - 2x)$$

> REMINDER   If the GCF happens to be one of the terms in the polynomial, a factor of 1 will remain for this term when the GCF is factored out. For example, in the polynomial $21x^2 + 7x$, the GCF of $21x^2$ and $7x$ is $7x$, so
> $$21x^2 + 7x = 7x(3x) + 7x(1) = 7x(3x + 1)$$

EXAMPLE 4   Factor $-3x^3y + 2x^2y - 5xy$.

Solution: Two possibilities are shown for factoring this polynomial. First, the common factor xy is factored out.

$$-3x^3y + 2x^2y - 5xy = xy(-3x^2 + 2x - 5)$$

Also, the common factor $-xy$ can be factored out as shown.

$$-3x^3y + 2x^2y - 5xy = -xy(3x^2) + (-xy)(-2x) + (-xy)(5)$$
$$= -xy(3x^2 - 2x + 5)$$

Both of these alternatives are correct.

EXAMPLE 5   Factor $2(x - 5) + 3a(x - 5)$.

Solution: The GCF is the binomial factor $(x - 5)$. Thus,

$$2(x - 5) + 3a(x - 5) = (x - 5)(2 + 3a)$$

EXAMPLE 6   Factor $7x(x^2 + 5y) - (x^2 + 5y)$.

Solution: The GCF is the expression $(x^2 + 5y)$. Factor this from each term.

$$7x(x^2 + 5y) - (x^2 + 5y) = 7x(x^2 + 5y) - 1(x^2 + 5y) = (x^2 + 5y)(7x - 1)$$

Notice that we write $-(x^2 + 5y)$ as $-1(x^2 + 5y)$ to aid in factoring.

3   Sometimes it is possible to factor a polynomial by grouping the terms of the polynomial and looking for common factors in each group. This method of factoring is called **factoring by grouping.**

EXAMPLE 7   Factor $ab - 6a + 2b - 12$.

Solution: First, look for the GCF of all four terms. The GCF of all four terms is 1. Next, group the first two terms and the last two terms and factor out common factors from each group.

$$ab - 6a + 2b - 12 = (ab - 6a) + (2b - 12)$$

Factor a from the first group and 2 from the second group.

$$= a(b - 6) + 2(b - 6)$$

Now we see a GCF of $(b - 6)$. Factor out $(b - 6)$ to get

$$a(b - 6) + 2(b - 6) = (b - 6)(a + 2)$$

This factorization can be checked by multiplying $(b - 6)$ by $(a + 2)$ to verify that the product is the original polynomial.

> R E M I N D E R   Notice that the polynomial $a(b - 6) + 2(b - 6)$ is **not** in factored form. It is a **sum,** not a **product.** The factored form is $(b - 6)(a + 2)$.

EXAMPLE 8   Factor $m^2n^2 + m^2 - 2n^2 - 2$.

Solution: Once again, the GCF of all four terms is 1. Try grouping the first two terms together and the last two terms together.

$$m^2n^2 + m^2 - 2n^2 - 2 = (m^2n^2 + m^2) + (-2n^2 - 2)$$

Factor m^2 from the first group and 2 from the second group.

$$= m^2(n^2 + 1) + 2(-n^2 - 1)$$

Notice that the polynomial is not in factored form since it is still written as a sum and not a product.

There is no common factor in this resulting polynomial, but notice that $(n^2 + 1)$ and $(-n^2 - 1)$ are opposites. Try grouping the terms differently, as follows:

$$m^2n^2 + m^2 - 2n^2 - 2 = (m^2n^2 + m^2) - (2n^2 + 2) \qquad \text{Watch the signs!}$$

$$= m^2(n^2 + 1) - 2(n^2 + 1) \qquad \text{Factor common factors from the groups of terms.}$$

$$= (n^2 + 1)(m^2 - 2) \qquad \text{Factor out a GCF of } (n^2 + 1).$$

MENTAL MATH

Find the GCF of each list of monomials.

1. $6, 12$

2. $9, 27$

3. $15x, 10$

4. $9x, 12$

5. $13x, 2x$

6. $4y, 5y$

7. $7x, 14x$

8. $8z, 4z$

EXERCISE SET 6.5

Find the GCF of each list of monomials. See Example 1.

1. a^8, a^5, a^3

2. b^9, b^2, b^5

3. $x^2y^3z^3, y^2z^3, xy^2z^2$

4. $xy^2z^3, x^2y^2z^2, x^2y^3$

5. $6x^3y, 9x^2y^2, 12x^2y$

6. $4xy^2, 16xy^3, 8x^2y^2$

7. $10x^3yz^3, 20x^2z^5, 45xz^3$

8. $12y^2z^4, 9xy^3z^4, 15x^2y^2z^3$

Factor out the GCF in each polynomial. See Examples 2 through 6.

9. $18x - 12$

10. $21x + 14$

11. $4y^2 - 16xy^3$

12. $3z - 21xz^4$

13. $6x^5 - 8x^4 - 2x^3$

14. $9x + 3x^2 - 6x^3$

15. $8a^3b^3 - 4a^2b^2 + 4ab + 16ab^2$

16. $12a^3b - 6ab + 18ab^2 - 18a^2b$

17. $6(x + 3) + 5a(x + 3)$ **18.** $2(x - 4) + 3y(x - 4)$

19. $2x(z + 7) + (z + 7)$ **20.** $x(y - 2) + (y - 2)$

21. $3x(x^2 + 5) - 2(x^2 + 5)$

22. $4x(2y + 3) - 5(2y + 3)$

23. When $3x^2 - 9x + 3$ is factored, the result is $3(x^2 - 3x + 1)$. Explain why it is necessary to include the term 1 in this factored form.

24. Construct a trinomial whose GCF is $5x^2y^3$.

Factor each polynomial by grouping. See Examples 7 and 8.

25. $ab + 3a + 2b + 6$ **26.** $ab + 2a + 5b + 10$

27. $ac + 4a - 2c - 8$ **28.** $bc + 8b - 3c - 24$

29. $2xy - 3x - 4y + 6$ **30.** $12xy - 18x - 10y + 15$

31. $12xy - 8x - 3y + 2$ **32.** $20xy - 15 - 4y + 3$

33. The material needed to manufacture a tin can is given by the polynomial

$$2\pi r^2 + 2\pi rh$$

where the radius is r and height is h. Factor this expression.

34. The amount E of current in an electrical circuit is given by the formula

$$IR_1 + IR_2 = E$$

Write an equivalent equation by factoring the expression $IR_1 + IR_2$.

35. At the end of T years, the amount of money A in a savings account earning simple interest from an initial investment of P dollars at rate R is given by the formula

$$A = P + PRT$$

Write an equivalent equation by factoring the expression $P + PRT$.

36. An open-topped box has a square base and a height of 10 inches. If each of the bottom edges of the box has length x inches, find the amount of ma-

terial needed to construct the box. Write the answer in factored form.

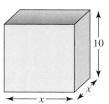

37. An object is thrown upward from the ground with an initial velocity of 64 feet per second. The height $h(t)$ of the object after t seconds is given by the polynomial function

$$h(t) = -16t^2 + 64t$$

a. Write an equivalent factored expression for the function $h(t)$ by factoring $-16t^2 + 64t$.

b. Find $h(1)$ by using $h(t) = -16t^2 + 64t$ and then by using the factored form of $h(t)$.

c. Explain why the values found in part (b) are the same.

38. An object is dropped from the gondola of a hot-air balloon at a height of 224 feet. The height $h(t)$ of the object after t seconds is given by the polynomial function

$$h(t) = -16t^2 + 224$$

a. Write an equivalent factored expression for the function $h(t)$ by factoring $-16t^2 + 224$.

b. Find $h(2)$ by using $h(t) = -16t^2 + 224$ and then by using the factored form of the function.

c. Explain why the values found in part (b) are the same.

Factor each polynomial.

39. $6x^3 + 9$

40. $6x^2 - 8$

41. $x^3 + 3x^2$

42. $x^4 - 4x^3$

43. $8a^3 - 4a$

44. $12b^4 + 3b^2$

45. $-20x^2y + 16xy^3$

46. $-18xy^3 + 27x^4y$

47. $10a^2b^3 + 5ab^2 - 15ab^3$ **48.** $10ef - 20e^2f^3 + 30e^3f$

49. $9abc^2 + 6a^2bc - 6ab + 3bc$

50. $4a^2b^2c - 6ab^2c - 4ac + 8a$

51. $4x(y - 2) - 3(y - 2)$ **52.** $8y(z + 8) - 3(z + 8)$

53. $6xy + 10x + 9y + 15$ **54.** $15xy + 20x + 6y + 8$

55. $xy + 3y - 5x - 15$ **56.** $xy + 4y - 3x - 12$

57. $6ab - 2a - 9b + 3$ **58.** $16ab - 8a - 6b + 3$

59. $12xy + 18x + 2y + 3$ **60.** $20xy + 8x + 5y + 2$

61. $2m(n - 8) - (n - 8)$ **62.** $3a(b - 4) - (b - 4)$

63. $15x^3y^2 - 18x^2y^2$ **64.** $12x^4y^2 - 16x^3y^3$

65. $2x^2 + 3xy + 4x + 6y$ **66.** $3x^2 + 12x + 4xy + 16y$

67. $5x^2 + 5xy - 3x - 3y$ **68.** $4x^2 + 2xy - 10x - 5y$

69. $x^3 + 3x^2 + 4x + 12$ **70.** $x^3 + 4x^2 + 3x + 12$

71. $x^3 - x^2 - 2x + 2$ **72.** $x^3 - 2x^2 - 3x + 6$

73. A factored polynomial can be in many forms. For example, a factored form of $xy - 3x - 2y + 6$ is $(x - 2)(y - 3)$. Which of the following is not a factored form of $xy - 3x - 2y + 6$?

a. $(2 - x)(3 - y)$ **b.** $(-2 + x)(-3 + y)$

c. $(x - 2)(y - 3)$ **d.** $(x + 2)(-y + 3)$

74. Consider the following sequence of algebraic steps:

$$x^3 - 6x^2 + 2x - 10 = (x^3 - 6x^2) + (2x - 10)$$
$$= x^2(x - 6) + 2(x - 5)$$

Explain whether the final result is the factored form of the original polynomial.

Review Exercises

Simplify the following. See Section 6.1.

75. $(5x^2)(11x^5)$ **76.** $(7y)(-2y^3)$

77. $(5x^2)^3$ **78.** $(-2y^3)^4$

Find each product by using the FOIL order of multiplying binomials. See Section 6.4.

79. $(x + 2)(x - 5)$ **80.** $(x - 7)(x - 1)$

81. $(x + 3)(x + 2)$ **82.** $(x - 4)(x + 2)$

83. $(y - 3)(y - 1)$ **84.** $(s + 8)(s + 10)$

A Look Ahead

EXAMPLE

Factor $x^{5a} - x^{3a} + x^{7a}$.

Solution:

The variable x is common to all three terms, and the power $3a$ is the smallest of the exponents. So factor out the common factor x^{3a}.

$$x^{5a} - x^{3a} + x^{7a} = x^{3a}(x^{2a}) - x^{3a}(1) + x^{3a}(x^{4a})$$
$$= x^{3a}(x^{2a} - 1 + x^{4a})$$

Factor. Assume that variables used as exponents represent positive integers.

85. $x^{3n} - 2x^{2n} + 5x^n$ **86.** $3y^n + 3y^{2n} + 5y^{8n}$

87. $6x^{8a} - 2x^{5a} - 4x^{3a}$ **88.** $3x^{5a} - 6x^{3a} + 9x^{2a}$

6.6 | FACTORING TRINOMIALS

TAPE IAG 6.6

OBJECTIVES

1 Factor trinomials of the form $x^2 + bx + c$.

2 Factor trinomials of the form $ax^2 + bx + c$.

3 Factor by substitution.

4 Factor trinomials by the AC method.

1 In the previous section, we used factoring by grouping to factor four-term polynomials. In this section, we present techniques for factoring trinomials. Since $(x - 2)(x + 5) = x^2 + 3x - 10$, we say that $(x - 2)(x + 5)$ is a factored form of

$x^2 + 3x - 10$. Taking a close look at how $(x - 2)$ and $(x + 5)$ are multiplied suggests a pattern for factoring trinomials of the form

$$x^2 + bx + c.$$

$$(x - 2)(x + 5) = x^2 + 3x - 10$$

The pattern for factoring is summarized next.

> **To Factor a Trinomial of the Form** $x^2 + bx + c$, find two numbers whose product is c and whose sum is b. The factored form of $x^2 + bx + c$ is
>
> **$(x + $ one number$)(x + $ other number$)$**

EXAMPLE 1   Factor $x^2 + 10x + 16$.

Solution: Look for two integers whose product is 16 and whose sum is 10. Since our integers must have a positive product and a positive sum, we look at only positive factors of 16.

Positive Factors of 16	Sum of Factors
1, 16	$1 + 16 = 17$
4, 4	$4 + 4 = 8$
2, 8	$2 + 8 = 10$

TECHNOLOGY NOTE

A graphing utility can be used to visualize the factorization of a polynomial. For example, to visualize the factorization to the right, graph $y_1 = x^2 + 10x + 16$ and $y_2 = (x + 2)(x + 8)$ and see that the graphs coincide.

The correct pair of numbers is 2 and 8 because their product is 16 and their sum is 10. Thus,

$$x^2 + 10x + 16 = (x + \text{one number})(x + \text{other number})$$

or

$$x^2 + 10x + 16 = (x + 2)(x + 8)$$

To check, see that $(x + 2)(x + 8) = x^2 + 10x + 16$.

EXAMPLE 2   Factor $x^2 - 12x + 35$.

Solution: Find two integers whose product is 35 and whose sum is -12. Since our integers must have a positive product and a negative sum, we consider only negative factors of 35. The numbers are -5 and -7.

$$x^2 - 12x + 35 = [x + (-5)][x + (-7)]$$
$$= (x - 5)(x - 7)$$

To check, see that $(x - 5)(x - 7) = x^2 - 12x + 35$.

EXAMPLE 3   Factor $5x^3 - 30x^2 - 35x$.

Solution: First, factor out a GCF of $5x$.

$$5x^3 - 30x^2 - 35x = 5x(x^2 - 6x - 7)$$

Next, factor $x^2 - 6x - 7$ by finding two numbers whose product is -7 and whose sum is -6. The numbers are 1 and -7.

$$5x^3 - 30x^2 - 35x = 5x(x^2 - 6x - 7)$$
$$= 5x(x + 1)(x - 7)$$

> **REMINDER**   If the polynomial to be factored contains a common factor that is factored out, don't forget to include that common factor in the final factored form of the original polynomial.

EXAMPLE 4   Factor $2n^2 - 38n + 80$.

Solution: The terms of this polynomial have a GCF of 2, which we factor out first.

$$2n^2 - 38n + 80 = 2(n^2 - 19n + 40)$$

Next, factor $n^2 - 19n + 40$ by finding two numbers whose product is 40 and whose sum is -19. Both numbers must be negative since their sum is -19. Possibilities are

$$-1 \text{ and } -40, \qquad -2 \text{ and } -20, \qquad -4 \text{ and } -10, \qquad -5 \text{ and } -8$$

None of the pairs has a sum of -19, so no further factoring with integers is possible. The factored form of $2n^2 - 38n + 80$ is

$$2n^2 - 38n + 80 = 2(n^2 - 19n + 40)$$

We call a polynomial such as $n^2 - 19n + 40$ that cannot be factored with integers a **prime polynomial.**

2

Next, we factor trinomials of the form $ax^2 + bx + c$, where the coefficient a of x^2 is not 1. Don't forget that the first step in factoring any polynomial is to factor out the GCF of its terms.

EXAMPLE 5   Factor $2x^2 + 11x + 15$.

Solution: Factors of $2x^2$ are $2x$ and x. Try these factors as first terms of the binomials.

$$2x^2 + 11x + 15 = (2x +\ \ )(x +\ \ )$$

Next, try combinations of factors of 15 until the correct middle term, $11x$, is obtained. We will try only positive factors of 15 since the coefficient of the middle term, 11, is positive. Positive factors of 15 are 1 and 15 and 3 and 5.

$(2x + 1)(x + 15)$ $\qquad\qquad$ $(2x + 15)(x + 1)$

$\qquad\quad 1x \qquad\qquad\qquad\qquad\qquad\quad 15x$

$\qquad\quad \dfrac{30x}{31x},$ incorrect middle term $\qquad \dfrac{2x}{17x},$ incorrect middle term

$(2x + 3)(x + 5)$ $\qquad\qquad$ $(2x + 5)(x + 3)$

$\qquad\quad 3x \qquad\qquad\qquad\qquad\qquad\quad 5x$

$\qquad\quad \dfrac{10x}{13x},$ incorrect middle term $\qquad \dfrac{6x}{11x},$ *correct* middle term

Thus, the factored form of $2x^2 + 11x + 15$ is $(2x + 5)(x + 3)$.

TO FACTOR A TRINOMIAL OF THE FORM $ax^2 + bx + c$

Step 1.   Write all pairs of factors of ax^2.

Step 2.   Write all pairs of factors of c, the constant term.

Step 3.   Try various combinations of these factors until the correct middle term bx is found.

Step 4.   If no combination exists, the polynomial is **prime.**

EXAMPLE 6   Factor $3x^2 - x - 4$.

Solution: Factors of $3x^2$: $3x \cdot x$.

Factors of -4: $-4 = -1 \cdot 4,\ -4 = 1 \cdot -4,\ -4 = -2 \cdot 2,\ -4 = 2 \cdot -2$.

Try possible combinations of these factors.

$$(3x - 1)(x + 4) \qquad\qquad (3x + 4)(x - 1)$$

$-1x$ $\qquad\qquad\qquad\qquad\qquad\qquad$ $4x$

$\dfrac{12x}{11x}$, incorrect middle term $\qquad\qquad$ $\dfrac{-3x}{1x}$, incorrect middle term

$$(3x - 4)(x + 1)$$

$-4x$

$\dfrac{3x}{-1x}$, *correct* middle term

Thus, $3x^2 - x - 4 = (3x - 4)(x + 1)$.

EXAMPLE 7   Factor $12x^3y - 22x^2y + 8xy$.

Solution: First, factor out the GCF of the terms of this trinomial, $2xy$.

$$12x^3y - 22x^2y + 8xy = 2xy(6x^2 - 11x + 4)$$

Now try to factor the trinomial $6x^2 - 11x + 4$.

Factors of $6x^2$: $6x^2 = 2x \cdot 3x,$ $6x^2 = 6x \cdot x$

Try $2x$ and $3x$.

$$2xy(6x^2 - 11x + 4) = 2xy(2x + \;)(3x + \;)$$

The constant term 4 is positive and the coefficient of the middle term -11 is negative, so factor 4 into negative factors only.

Negative factors of 4: $4 = -4(-1),$ $4 = -2(-2)$

Try -4 and -1.

$$2xy(2x - 4)(3x - 1)$$

$-12x$

$\dfrac{-2x}{-14x}$, incorrect middle term

This combination cannot be correct, because one of the factors $(2x - 4)$ has a common factor of 2. This cannot happen if the polynomial $6x^2 - 11x + 4$ has no common factors. Try -1 and -4.

$$2xy(2x - 1)(3x - 4)$$

$-3x$

$\dfrac{-8x}{-11x}$, *correct* middle term

If this combination had not worked, we would try -2 and -2 and then $6x$ and x as factors of $6x^2$.

$$12x^3y - 22x^2y + 8xy = 2xy(2x - 1)(3x - 4)$$

> R E M I N D E R   If a trinomial has no common factor (other than 1), then none of its binomial factors will contain a common factor (other than 1).

EXAMPLE 8   Factor $16x^2 + 24xy + 9y^2$.

Solution: No GCF can be factored out of this trinomial. Factors of $16x^2$ are

$$16x^2 = 16x \cdot x, \qquad 16x^2 = 8x \cdot 2x, \qquad 16x^2 = 4x \cdot 4x$$

Factors of $9y^2$ are

$$9y^2 = y \cdot 9y, \qquad 9y^2 = 3y \cdot 3y$$

Try possible combinations until the correct factorization is found.

$$16x^2 + 24xy + 9y^2 = (4x + 3y)(4x + 3y), \quad \text{or} \quad (4x + 3y)^2$$

The trinomial $16x^2 + 24xy + 9y^2$ in Example 8 is an example of a **perfect square trinomial** since its factors are two identical binomials. In the next section, we examine a special method for factoring perfect square trinomials.

> 3

A complicated looking polynomial may be a simpler trinomial "in disguise." Revealing the simpler trinomial is possible by substitution.

EXAMPLE 9   Factor $2(a + 3)^2 - 5(a + 3) - 7$.

Solution: The quantity $(a + 3)$ is in two of the terms of this polynomial. **Substitute** x for $(a + 3)$, and the result is the following simpler trinomial.

$$2(a + 3)^2 - 5(a + 3) - 7 \qquad \text{Original trinomial.}$$
$$\downarrow \qquad\qquad \downarrow$$
$$= \quad 2(x)^2 \quad - \quad 5(x) \quad - 7 \qquad \text{Substitute } x \text{ for } (a + 3).$$

Now factor $2x^2 - 5x - 7$.

$$2x^2 - 5x - 7 = (2x - 7)(x + 1)$$

But the quantity in the original polynomial was $(a + 3)$, not x. Thus, we need to reverse the substitution and replace x with $(a + 3)$.

$$(2x - 7)(x + 1) \qquad\qquad \text{Factored expression.}$$
$$= [2(a + 3) - 7][(a + 3) + 1] \qquad \text{Substitute } (a + 3) \text{ for } x.$$
$$= (2a + 6 - 7)(a + 3 + 1) \qquad \text{Remove inside parentheses.}$$
$$= (2a - 1)(a + 4) \qquad\qquad \text{Simplify.}$$

Thus, $2(a + 3)^2 - 5(a + 3) - 7 = (2a - 1)(a + 4)$.

EXAMPLE 10 Factor $5x^4 + 29x^2 - 42$.

Solution: Again, substitution may help us factor this polynomial more easily. Let $y = x^2$, so $y^2 = (x^2)^2$, or x^4. Then

$$5x^4 + 29x^2 - 42$$

becomes

$$5y^2 + 29y - 42$$

which factors as

$$5y^2 + 29y - 42 = (5y - 6)(y + 7)$$

Next, replace y with x^2 to get

$$(5x^2 - 6)(x^2 + 7)$$

 There is another method we can use when factoring trinomials of the form $ax^2 + bx + c$: Write the trinomial as a four-term polynomial, and then factor by grouping. The method is called the AC method.

TO FACTOR A TRINOMIAL OF THE FORM $ax^2 + bx + c$ BY THE AC METHOD

Step 1. Find two numbers whose product is $a \cdot c$ and whose sum is b.

Step 2. Write the term bx as a sum by using the factors found in *step 1*.

Step 3. Factor by grouping.

EXAMPLE 11 Factor $6x^2 + 13x + 6$.

Solution: In this trinomial, $a = 6$, $b = 13$, and $c = 6$.

Step 1. Find two numbers whose product is $a \cdot c$, or $6 \cdot 6 = 36$, and whose sum is b, 13.

The two numbers are 4 and 9.

Step 2. Write the middle term $13x$ as the sum $4x + 9x$.

$$6x^2 + 13x + 6 = 6x^2 + 4x + 9x + 6$$

Step 3. Factor $6x^2 + 4x + 9x + 6$ by grouping.

$$(6x^2 + 4x) + (9x + 6) = 2x(3x + 2) + 3(3x + 2)$$
$$= (3x + 2)(2x + 3)$$

DISCOVER THE CONCEPT

a. The following polynomials are factored. Graph each related polynomial function shown in a standard window. Compare the x-intercepts of the graph and the factors of the polynomial.

POLYNOMIAL	FACTORS	RELATED POLYNOMIAL FUNCTION	X-INTERCEPTS
$x^2 - x - 12$	$(x - 4)(x + 3)$	$y = x^2 - x - 12$	
$x^2 + 10x + 16$	$(x + 2)(x + 8)$	$y = x^2 + 10x + 16$	
$x^2 - 8x + 15$	$(x - 3)(x - 5)$	$y = x^2 - 8x + 15$	
$2x^2 - 17x + 30$	$(2x - 5)(x - 6)$	$y = 2x^2 - 17x + 30$	

b. Find a connection between the factors and the x-intercepts above. If you know the x-intercepts of the related graph, can it help you factor the polynomial?

c. If a function has x-intercepts of 2 and -8, what are two factors of the related polynomial? Graph a related function in factored form and see if the x-intercepts are 2 and -8.

In the above discovery, we find that if the graph of a polynomial function has an x-intercept at a, then the related polynomial has a factor of $(x - a)$. Also, if a polynomial has a factor of $(x - a)$ the graph of the related polynomial function has an x-intercept at a.

MENTAL MATH

1. Find two numbers whose product is 10 and whose sum is 7.

2. Find two numbers whose product is 12 and whose sum is 8.

3. Find two numbers whose product is 24 and whose sum is 11.

4. Find two numbers whose product is 30 and whose sum is 13.

EXERCISE SET 6.6

Factor each trinomial. See Examples 1 through 4.

1. $x^2 + 9x + 18$ **2.** $x^2 + 9x + 20$

3. $x^2 - 12x + 32$ **4.** $x^2 - 12x + 27$

5. $x^2 + 10x - 24$ **6.** $x^2 + 3x - 54$

7. $x^2 - 2x - 24$ **8.** $x^2 - 9x - 36$

9. $3x^2 - 18x + 24$ **10.** $x^2y^2 + 4xy^2 + 3y^2$

11. $4x^2z + 28xz + 40z$ **12.** $5x^2 - 45x + 70$

13. $2x^2 + 30x - 108$ **14.** $3x^2 + 12x - 96$

15. Find all positive and negative integers b such that $x^2 + bx + 6$ factors.

16. Find all positive and negative integers b such that $x^2 + bx - 10$ factors.

A polynomial function is graphed on each screen. Use the graph to write factors of the related polynomial. Each tick mark on the x-axis is 1 unit, and it can be assumed that both x-intercepts are integer values.

17.

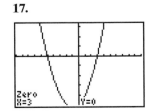

18.

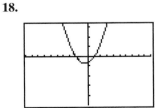

19. 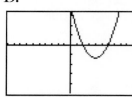 **20.**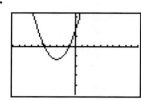

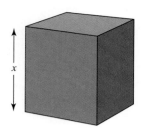

Factor each trinomial. See Examples 5 through 8 and 11.

21. $2x^2 - 11x + 12$

22. $3x^2 - 19x + 20$

23. $4x^2 - 12x + 9$

24. $25x^2 - 30x + 9$

25. $12x^2 + 10x - 50$

26. $12y^2 - 48y + 45$

27. $3y^4 - y^3 - 10y^2$

28. $2x^2z + 5xz - 12z$

29. $6x^3 + 8x^2 + 24x$

30. $18y^3 + 12y^2 + 2y$

31. $x^2 + 8xz + 7z^2$

32. $a^2 - 2ab - 15b^2$

33. $2x^2 - 5xy - 3y^2$

34. $6x^2 + 11xy + 4y^2$

35. $28y^2 + 22y + 4$

36. $24y^3 - 2y^2 - y$

Complete the following:

37. $x^2 - x - 20$ has factors of _____. The graph of the related equation has x-intercepts of _____.

38. $x^2 + 5x + 4$ has factors of _____. The graph of the related equation has x-intercepts of _____.

39. $x^2 - 11x + 28$ has factors of _____. The graph of the related equation has x-intercepts of _____.

40. $x^2 + 2x + 1$ has factors of _____. The graph of the related equation has x-intercepts of _____.

41. Find all positive and negative integers b such that $3x^2 + bx + 5$ factors.

42. Find all positive and negative integers b such that $2x^2 + bx + 7$ factors.

Use substitution to factor each polynomial completely. See Examples 9 and 10.

43. $x^4 + x^2 - 6$

44. $x^4 - x^2 - 20$

45. $(5x + 1)^2 + 8(5x + 1) + 7$

46. $(3x - 1)^2 + 5(3x - 1) + 6$

47. $x^6 - 7x^3 + 12$

48. $x^6 - 4x^3 - 12$

49. $(a + 5)^2 - 5(a + 5) - 24$

50. $(3c + 6)^2 + 12(3c + 6) - 28$

51. The volume $V(x)$ of a box in terms of its height x is given by the function $V(x) = 3x^3 - 2x^2 - 8x$. Factor this expression for $V(x)$.

52. Based on your results from Exercise 51, find the length and width of the box if the height is 5 inches and the dimensions of the box are whole numbers.

Factor each polynomial completely.

53. $x^2 - 24x - 81$

54. $x^2 - 48x - 100$

55. $x^2 - 15x - 54$

56. $x^2 - 15x + 54$

57. $3x^2 - 6x + 3$

58. $8x^2 - 8x + 2$

59. $3x^2 - 5x - 2$

60. $5x^2 - 14x - 3$

61. $8x^2 - 26x + 15$

62. $12x^2 - 17x + 6$

63. $18x^4 + 21x^3 + 6x^2$

64. $20x^5 + 54x^4 + 10x^3$

65. $3a^2 + 12ab + 12b^2$

66. $2x^2 + 16xy + 32y^2$

67. $x^2 + 4x + 5$

68. $x^2 + 6x + 8$

69. $2(x + 4)^2 + 3(x + 4) - 5$

70. $3(x + 3)^2 + 2(x + 3) - 5$

71. $6x^2 - 49x + 30$

72. $4x^2 - 39x + 27$

73. $x^4 - 5x^2 - 6$

74. $x^4 - 5x^2 + 6$

75. $6x^3 - x^2 - x$

76. $12x^3 + x^2 - x$

77. $12a^2 - 29ab + 15b^2$

78. $16y^2 + 6yx - 27x^2$

79. $9x^2 + 30x + 25$

80. $4x^2 + 6x + 9$

81. $3x^2y - 11xy + 8y$

82. $5xy^2 - 9xy + 4x$

83. $2x^2 + 2x - 12$

84. $3x^2 + 6x - 45$

85. $(x - 4)^2 + 3(x - 4) - 18$

86. $(x - 3)^2 - 2(x - 3) - 8$

87. $2x^6 + 3x^3 - 9$

88. $3x^6 - 14x^3 + 8$

89. $72xy^4 - 24xy^2z + 2xz^2$

90. $36xy^2 - 48xyz^2 + 16xz^4$

Recall that a graphing utility may be used to visualize the factorization of polynomials in one variable. For example, to see that

$$2x^3 - 9x^2 - 5x = x(2x + 1)(x - 5)$$

graph $y_1 = 2x^3 - 9x^2 - 5x$ and $y_2 = x(2x + 1)(x - 5)$. Then trace along both graphs or generate a table of values to see that they coincide. Factor the following and use this method to visualize your results.

91. $x^4 + 6x^3 + 5x^2$ **92.** $x^3 + 6x^2 + 8x$

93. $30x^3 + 9x^2 - 3x$ **94.** $-6x^4 + 10x^3 - 4x^2$

Review Exercises

Multiply the following. See Section 6.4.

95. $(x - 2)(x^2 + 2x + 4)$ **96.** $(y + 1)(y^2 - y + 1)$

If $P(x) = 3x^2 + 2x - 9$, find the following. See Section 6.3.

97. $P(0)$ **98.** $P(1)$ **99.** $P(-1)$ **100.** $P(-2)$

A Look Ahead

EXAMPLE

Factor $x^{2n} + 7x^n + 12$.

Solution:

Factors of x^{2n} are x^n and x^n, so $x^{2n} + 7x^n + 12 = (x^n + \text{one number})(x^n + \text{other number})$. Factors of 12 whose sum is 7 are 3 and 4. Thus

$$x^{2n} + 7x^n + 12 = (x^n + 4)(x^n + 3)$$

Factor. Assume that variables used as exponents represent positive integers. See the preceding example.

101. $x^{2n} + 10x^n + 16$ **102.** $x^{2n} - 7x^n + 12$

103. $x^{2n} - 3x^n - 18$ **104.** $x^{2n} + 7x^n - 18$

105. $2x^{2n} + 11x^n + 5$ **106.** $3x^{2n} - 8x^n + 4$

107. $4x^{2n} - 12x^n + 9$ **108.** $9x^{2n} + 24x^n + 16$

6.7 | FACTORING BY SPECIAL PRODUCTS AND FACTORING STRATEGIES

TAPE IAG 6.7

O B J E C T I V E S

  1   Factor a perfect square trinomial.

  2   Factor the difference of two squares.

  3   Factor the sum or difference of two cubes.

  4   Practice techniques for factoring polynomials.

1

In the previous section, we considered a variety of ways to factor trinomials of the form $ax^2 + bx + c$. In one particular example, we factored $16x^2 + 24xy + 9y^2$ as

$$16x^2 + 24xy + 9y^2 = (4x + 3y)^2$$

Recall that we called $16x^2 + 24xy + 9y^2$ a perfect square trinomial because its factors are two identical binomials. A perfect square trinomial can be factored quickly if you recognize the trinomial as a perfect square.

    A trinomial is a perfect square trinomial if it can be written so that its first term is the square of some quantity a, its last term is the square of some quantity b, and its middle term is twice the product of the quantities a and b. The following special formulas can be used to factor perfect square trinomials.

TECHNOLOGY NOTE

Below is the graph of $y = x^2 + 10x + 25$ or $y = (x + 5)^2$.

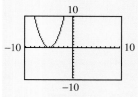

Notice that because the factors of the polynomial are the same, $(x + 5)$, only one x-intercept is generated, as shown by the graph.

PERFECT SQUARE TRINOMIALS

$$a^2 + 2ab + b^2 = (a + b)^2$$
$$a^2 - 2ab + b^2 = (a - b)^2$$

Notice that these equations are the same special products from Section 6.4 for the square of a binomial.

From $a^2 + 2ab + b^2 = (a + b)^2$, we see that

$$16x^2 + 24xy + 9y^2 = (4x)^2 + 2(4x)(3y) + (3y)^2 = (4x + 3y)^2$$

EXAMPLE 1   Factor $m^2 + 10m + 25$.

Solution: Notice that the first term is a square: $m^2 = (m)^2$,
the last term is a square: $25 = 5^2$,
and $10m = 2 \cdot 5 \cdot m$.
Thus,

$$m^2 + 10m + 25 = m^2 + 2(m)(5) + 5^2 = (m + 5)^2$$

EXAMPLE 2   Factor $3a^2x - 12abx + 12b^2x$.

Solution: The terms of this trinomial have a GCF of $3x$, which we factor out first.

$$3a^2x - 12abx + 12b^2x = 3x(a^2 - 4ab + 4b^2)$$

Now, the polynomial $a^2 - 4ab + 2b^2$ is a perfect square trinomial. Notice that
the first term is a square: $a^2 = (a)^2$,
the last term is a square: $4b^2 = (2b)^2$,
and $4ab = 2(a)(2b)$.
The factoring can now be completed as

$$3x(a^2 - 4ab + 4b^2) = 3x(a - 2b)^2$$

REMINDER   If you recognize a trinomial as a perfect square trinomial, use the special formulas to factor. However, methods for factoring trinomials in general from Section 6.6 will also result in the correct factored form.

---

2   We now factor special types of binomials, beginning with the **difference of two squares.** The special product pattern presented in Section 6.4 for the product of a sum and a difference of two terms is used again here. However, the emphasis is now on factoring rather than on multiplying.

DIFFERENCE OF TWO SQUARES
$$a^2 - b^2 = (a + b)(a - b)$$

Notice that a binomial is a difference of two squares when it is the difference of the square of some quantity a and the square of some quantity b.

EXAMPLE 3   Factor the following:

        a. $x^2 - 9$ **b.** $16y^2 - 9$ **c.** $50 - 8y^2$

Solution: **a.** $x^2 - 9 = x^2 - 3^2$ **b.** $16y^2 - 9 = (4y)^2 - 3^2$

                      $= (x + 3)(x - 3)$ $= (4y + 3)(4y - 3)$

     c. First factor out the common factor of 2.

           $50 - 8y^2 = 2(25 - 4y^2)$

                    $= 2(5 + 2y)(5 - 2y)$

       The binomial $x^2 + 9$ is a **sum of two squares** and cannot be factored by using real numbers. **In general, except for factoring out a GCF, the sum of two squares usually cannot be factored by using real numbers.**

> R E M I N D E R   The sum of two squares whose GCF is 1 usually cannot be factored by using real numbers.

EXAMPLE 4   Factor the following:

        a. $p^4 - 16$ **b.** $(x + 3)^2 - 36$

Solution: **a.** $p^4 - 16 = (p^2)^2 - 4^2$

                      $= (p^2 + 4)(p^2 - 4)$

       The binomial factor $p^2 + 4$ cannot be factored by using real numbers, but the binomial factor $p^2 - 4$ is a difference of squares.

         $(p^2 + 4)(p^2 - 4) = (p^2 + 4)(p + 2)(p - 2)$

     b. Factor $(x + 3)^2 - 36$ as the difference of squares.

         $(x + 3)^2 - 36 = (x + 3)^2 - 6^2$

                    $= [(x + 3) + 6][(x + 3) - 6]$ Factor.

                    $= [x + 3 + 6][x + 3 - 6]$ Remove parentheses.

                    $= (x + 9)(x - 3)$ Simplify.

    3   Although the sum of two squares usually cannot be factored, the sum of two cubes, as well as the difference of two cubes, can be factored as follows.

> **SUM AND DIFFERENCE OF TWO CUBES**
>
> $$a^3 + b^3 = (a + b)(a^2 - ab + b^2)$$
> $$a^3 - b^3 = (a - b)(a^2 + ab + b^2)$$

To check the first pattern, find the product of $(a + b)$ and $(a^2 - ab + b^2)$.

$$
\begin{array}{r}
a^2 - ab + b^2 \\
a + b \\
\hline
a^2b - ab^2 + b^3 \\
a^3 - a^2b + ab^2 \\
\hline
a^3 \qquad\qquad + b^3
\end{array}
$$

Then $a^3 + b^3 = (a + b)(a^2 - ab + b^2)$.

EXAMPLE 5   Factor $x^3 + 8$.

Solution: First, write the binomial in the form $a^3 + b^3$. Then, use the formula

$$a^3 + b^3 = (a + b)(a^2 - a \cdot b + b^2), \text{ where } a \text{ is } x \text{ and } b \text{ is } 2.$$

$$\downarrow \quad \downarrow \quad \downarrow \quad \downarrow \downarrow \quad \downarrow \downarrow \quad \downarrow$$

$$x^3 + 8 = x^3 + 2^3 = (x + 2)(x^2 - x \cdot 2 + 2^2)$$

Thus, $x^3 + 8 = (x + 2)(x^2 - 2x + 4)$.

EXAMPLE 6   Factor $p^3 + 27q^3$.

Solution: $p^3 + 27q^3 = p^3 + (3q)^3$
$$= (p + 3q)[p^2 - (p)(3q) + (3q)^2]$$
$$= (p + 3q)(p^2 - 3pq + 9q^2)$$

EXAMPLE 7   Factor $y^3 - 64$.

Solution: This, is a difference of cubes since $y^3 - 64 = y^3 - 4^3$.

From $a^3 - b^3 = (a - b)(a^2 + a \cdot b + b^2)$ we have that

$$\downarrow \quad \downarrow \quad \downarrow \quad \downarrow \downarrow \quad \downarrow \downarrow \quad \downarrow$$

$$y^3 - 4^3 = (y - 4)(y^2 + y \cdot 4 + 4^2)$$
$$= (y - 4)(y^2 + 4y + 16)$$

R E M I N D E R   When factoring sums or differences of cubes, be sure to notice the sign patterns.

same sign

$$x^3 + y^3 = (x + y)(x^2 - xy + y^2)$$

opposite sign   always positive

same sign

$$x^3 - y^3 = (x - y)(x^2 + xy + y^2)$$

opposite sign   always positive

EXAMPLE 8   Factor $125q^2 - n^3q^2$.

Solution: First, factor out a common factor of q^2.

$$125q^2 - n^3q^2 = q^2(125 - n^3)$$
$$= q^2(5^3 - n^3)$$

opposite sign   positive
$\downarrow$

$$= q^2(5 - n)[5^2 + (5)(n) + (n^2)]$$
$$= q^2(5 - n)(25 + 5n + n^2)$$

Thus, $125q^2 - n^3q^2 = q^2(5 - n)(25 + 5n + n^2)$. The trinomial $25 + 5n + n^2$ cannot be factored further.

EXAMPLE 9   Factor $x^2 + 4x + 4 - y^2$.

Solution: Factoring by grouping comes to mind since the sum of the first three terms of this polynomial is a perfect square trinomial.

$$x^2 + 4x + 4 - y^2 = (x^2 + 4x + 4) - y^2 \quad \text{Group the first three terms.}$$
$$= (x + 2)^2 - y^2 \qquad\quad \text{Factor the perfect square trinomial.}$$

This is not factored yet since we have a *difference,* not a *product.* Since $(x + 2)^2 - y^2$ is a difference of squares, we have

$$(x + 2)^2 - y^2 = [(x + 2) + y][(x + 2) - y]$$
$$= (x + 2 + y)(x + 2 - y)$$

4   The key to proficiency in factoring polynomials is to practice until you are comfortable with each technique. A strategy for factoring polynomials is given next.

TO FACTOR A POLYNOMIAL

Step 1.   Are there any common factors? If so, factor out the GCF.

Step 2.   How many terms are in the polynomial?
   a. If there are **two** terms, decide if one of the following formulas may be applied:
     i. Difference of two squares: $a^2 - b^2 = (a - b)(a + b)$.
     ii. Difference of two cubes: $a^3 - b^3 = (a - b)(a^2 + ab + b^2)$.
     iii. Sum of two cubes: $a^3 + b^3 = (a + b)(a^2 - ab + b^2)$.
   b. If there are **three** terms, try one of the following:
     i. Perfect square trinomial: $a^2 + 2ab + b^2 = (a + b)^2$.
     $$a^2 - 2ab + b^2 = (a - b)^2.$$
     ii. If not a perfect square trinomial, factor by using the methods presented in Section 6.6.
   c. If there are **four** or more terms, try factoring by grouping.

Step 3.   See if any factors in the factored polynomial can be factored further.

EXAMPLE 10   Factor each polynomial completely.

        a. $8a^2b - 4ab$ **b.** $36x^2 - 9$ **c.** $2x^2 - 5x - 7$

Solution: **a.** *Step 1.* The terms have a common factor of $4ab$, which we factor out.

$$8a^2b - 4ab = 4ab(2a - 1)$$

     Step 2. There are two terms, but the binomial $2a - 1$ is not the difference of two squares or the sum or difference of two cubes.

     Step 3. The factor $2a - 1$ cannot be factored further.

  b. *Step 1.* Factor out a common factor of 9.

$$36x^2 - 9 = 9(4x^2 - 1)$$

     Step 2. The factor $4x^2 - 1$ has two terms, and it is the difference of two squares.

$$9(4x^2 - 1) = 9(2x + 1)(2x - 1)$$

     Step 3. No factor with more than one term can be factored further.

  c. *Step 1.* The terms of $2x^2 - 5x - 7$ contain no common factor other than 1 or -1.

     Step 2. There are three terms. The trinomial is not a perfect square, so we factor by methods from Section 6.6.

$$2x^2 - 5x - 7 = (2x - 7)(x + 1)$$

     Step 3. No factor with more than one term can be factored further.

EXAMPLE 11   Factor each polynomial completely.

        a. $5p^2 + 5 + qp^2 + q$ **b.** $9x^2 + 24x + 16$ **c.** $y^2 + 25$

Solution: **a.** *Step 1.* There is no common factor of all terms of $5p^2 + 5 + qp^2 + q$.

     Step 2. The polynomial has four terms, so try factoring by grouping.

$$5p^2 + 5 + qp^2 + q = (5p^2 + 5) + (qp^2 + q) \qquad \text{Group the terms.}$$
$$= 5(p^2 + 1) + q(p^2 + 1)$$
$$= (p^2 + 1)(5 + q)$$

     Step 3. No factor can be factored further.

  b. *Step 1.* The terms of $9x^2 + 24x + 16$ contain no common factor other than 1 or -1.

     Step 2. The trinomial $9x^2 + 24x + 16$ is a perfect square trinomial, and $9x^2 + 24x + 16 = (3x + 4)^2$.

     Step 3. No factor can be factored further.

  c. *Step 1.* There is no common factor of $y^2 + 25$ other than 1.

     Step 2. This binomial is the sum of two squares and is prime.

     Step 3. The binomial $y^2 + 25$ cannot be factored further.

EXAMPLE 12   Factor each completely.

a. $27a^3 - b^3$ b. $3n^2m^4 - 48m^6$ c. $2x^2 - 12x + 18 - 2z^2$

d. $8x^4y^2 + 125xy^2$ e. $(x - 5)^2 - 49y^2$

Solution: **a.** This binomial is a difference of two cubes.

$$27a^3 - b^3 = (3a)^3 - b^3$$
$$= (3a - b)[(3a)^2 + (3a)(b) + b^2]$$
$$= (3a - b)(9a^2 + 3ab + b^2)$$

b. $3n^2m^4 - 48m^6 = 3m^4(n^2 - 16m^2)$ Factor out the GCF, $3m^4$.

$$= 3m^4(n + 4m)(n - 4m)$$ Factor the difference of squares.

c. $2x^2 - 12x + 18 - 2z^2 = 2(x^2 - 6x + 9 - z^2)$ The GCF is 2.

$$= 2[(x^2 - 6x + 9) - z^2]$$ Group the first three terms together.

$$= 2[(x - 3)^2 - z^2]$$ Factor the perfect square trinomial.

$$= 2[(x - 3) + z][(x - 3) - z]$$ Factor the difference of squares.

$$= 2(x - 3 + z)(x - 3 - z)$$

d. $8x^4y^2 + 125xy^2 = xy^2(8x^3 + 125)$ The GCF is xy^2.

$$= xy^2[(2x)^3 + 5^3]$$

$$= xy^2(2x + 5)[(2x)^2 - (2x)(5) + 5^2]$$ Factor the sum of cubes.

$$= xy^2(2x + 5)(4x^2 - 10x + 25)$$

e. This binomial is a difference of squares.

$$(x - 5)^2 - 49y^2 = (x - 5)^2 - (7y)^2$$

$$= [(x - 5) + 7y][(x - 5) - 7y]$$

$$= (x - 5 + 7y)(x - 5 - 7y)$$

EXERCISE SET 6.7

Factor the following. See Examples 1 and 2.

1. $x^2 + 6x + 9$ **2.** $x^2 - 10x + 25$

3. $4x^2 - 12x + 9$ **4.** $25x^2 + 10x + 1$

5. $3x^2 - 24x + 48$ **6.** $x^3 + 14x^2 + 49x$

7. $9y^2x^2 + 12yx^2 + 4x^2$ **8.** $32x^2 - 16xy + 2y^2$

Factor the following. See Examples 3 and 4.

9. $x^2 - 25$ **10.** $y^2 - 100$

11. $9 - 4z^2$ **12.** $16x^2 - y^2$

13. $(y + 2)^2 - 49$ **14.** $(x - 1)^2 - z^2$

15. $64x^2 - 100$ **16.** $4x^2 - 36$

Factor the following. See Examples 5 through 8.

17. $x^3 + 27$ **18.** $y^3 + 1$

19. $z^3 - 1$ **20.** $x^3 - 8$

21. $m^3 + n^3$ **22.** $r^3 + 125$

23. $x^3y^2 - 27y^2$ **24.** $64 - p^3$

25. $a^3b + 8b^4$ **26.** $8ab^3 + 27a^4$

27. $125y^3 - 8x^3$ **28.** $54y^3 - 128$

Factor the following. See Example 9.

29. $x^2 + 6x + 9 - y^2$ **30.** $x^2 + 12x + 36 - y^2$

31. $x^2 - 10x + 25 - y^2$ **32.** $x^2 - 18x + 81 - y^2$

33. $4x^2 + 4x + 1 - z^2$ **34.** $9y^2 + 12y + 4 - x^2$

Factor each polynomial completely.

35. $9x^2 - 49$ **36.** $25x^2 - 4$

37. $x^4 - 81$ **38.** $x^4 - 256$

39. $x^2 + 8x + 16 - 4y^2$

40. $x^2 + 14x + 49 - 9y^2$

41. $(x + 2y)^2 - 9$ **42.** $(3x + y)^2 - 25$

43. $x^3 - 1$ **44.** $x^3 - 8$

45. $x^3 + 125$ **46.** $x^3 + 216$

47. $4x^2 + 25$ **48.** $16x^2 + 25$

49. $4a^2 + 12a + 9$ **50.** $9a^2 - 30a + 25$

51. $18x^2y - 2y$ **52.** $12xy^2 - 108x$

53. $x^6 - y^3$ **54.** $x^3 - y^6$

55. $x^2 + 16x + 64 - x^4$ **56.** $x^2 + 20x + 100 - x^4$

57. $3x^6y^2 + 81y^2$ **58.** $x^2y^9 + x^2y^3$

59. $(x + y)^3 + 125$ **60.** $(x + y)^3 + 27$

61. $(2x + 3)^3 - 64$ **62.** $(4x + 2)^3 - 125$

63. The manufacturer of Antonio's Metal Washers needs to determine the cross-sectional area of each washer. If the outer radius of the washer is R and the radius of the hole is r, express the area of the washer as a polynomial. Factor this polynomial completely.

64. Express the area of the shaded region as a polynomial. Factor the polynomial completely.

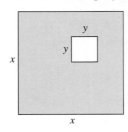

65. The manufacturer of Tootsie Roll Pops plans to change the size of its candy. To compute the new cost, the company needs a formula for the volume of the candy coating without the Tootsie Roll center. Given the diagram, express the volume as a polynomial. Factor this polynomial completely.

66. Express the area of the shaded region as a polynomial. Factor the polynomial completely.

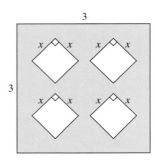

Factor completely. See Examples 10 through 12.

67. $x^2 - 8x + 16 - y^2$

68. $12x^2 - 22x - 20$

69. $x^4 - x$

70. $(2x + 1)^2 - 3(2x + 1) + 2$

71. $14x^2y - 2xy$ **72.** $24ab^2 - 6ab$

73. $4x^2 - 16$ **74.** $9x^2 - 81$

75. $3x^2 - 8x - 11$ **76.** $5x^2 - 2x - 3$

77. $4x^2 + 8x - 12$ **78.** $6x^2 - 6x - 12$

79. $4x^2 + 36x + 81$ **80.** $25x^2 + 40x + 16$

81. $8x^3 + 27y^3$ **82.** $125x^3 + 8y^3$

83. $64x^2y^3 - 8x^2$ **84.** $27x^5y^4 - 216x^2y$

85. $(x + 5)^3 + y^3$ **86.** $(y - 1)^3 + 27x^3$

87. $(5a - 3)^2 - 6(5a - 3) + 9$

88. $(4r + 1)^2 + 8(4r + 1) + 16$

Find a value of c that makes each trinomial a perfect square trinomial.

89. $x^2 + 6x + c$ **90.** $y^2 + 10y + c$

91. $m^2 - 14m + c$ **92.** $n^2 - 2n + c$

93. $x^2 + cx + 16$ **94.** $x^2 + cx + 36$

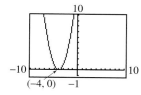
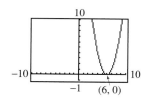

95. Factor $x^6 - 1$ completely, using the following methods from this chapter.

   a. Factor the expression by treating it as the difference of two squares, $(x^3)^2 - 1^2$.

   b. Factor the expression treating it as the difference of two cubes, $(x^2)^3 - 1^3$.

   c. Are the answers to parts (a) and (b) the same? Why or why not?

Review Exercises

Solve the following equations. See Section 2.1.

96. $x - 5 = 0$ **97.** $x + 7 = 0$

98. $3x + 1 = 0$ **99.** $5x - 15 = 0$

100. $-2x = 0$ **101.** $3x = 0$

102. $-5x + 25 = 0$ **103.** $-4x - 16 = 0$

A Look Ahead

EXAMPLE

Factor $x^{2n} - 100$.

Solution:

This binomial is a difference of squares.

$$x^{2n} - 100 = (x^n)^2 - 10^2$$
$$= (x^n + 10)(x^n - 10)$$

Factor each expression. Assume that variables used as exponents represent positive integers. See the preceding example.

104. $x^{2n} - 25$ **105.** $x^{2n} - 36$

106. $36x^{2n} - 49$ **107.** $25x^{2n} - 81$

108. $x^{4n} - 16$ **109.** $x^{4n} - 625$

6.8 | SOLVING POLYNOMIAL EQUATIONS ALGEBRAICALLY AND GRAPHICALLY

TAPE IAG 6.8

O B J E C T I V E S

1 Solve polynomial equations by factoring.

2 Solve problems that can be modeled by quadratic equations.

In this section, your efforts to learn factoring start to pay off. We use factoring to solve polynomial equations, which in turn helps us solve problems that can be modeled by polynomial equations.

A *polynomial equation* is the result of setting two polynomials equal to each other. Examples of polynomial equations are

$$3x^3 - 2x^2 = x^2 + 2x - 1 \qquad 2.6x + 7 = -1.3 \qquad -5x^2 - 5 = -9x^2 - 2x + 1$$

A polynomial equation is in *standard form* if one side of the equation is 0. In standard form, the polynomial equations above become

$$3x^3 - 3x^2 - 2x + 1 = 0 \qquad 2.6x + 8.3 = 0 \qquad 4x^2 + 2x - 6 = 0$$

The degree of a simplified polynomial equation in standard form is the same as the highest degree of any of its terms. A polynomial equation of degree 2 is also called a **quadratic equation.**

A solution of a polynomial equation in one variable is a value of the variable that makes the equation true. The method presented in this section for solving polynomial equations is called the *factoring method.* This method is based on the **zero-factor property.**

ZERO-FACTOR PROPERTY

If a and b are real numbers and $a \cdot b = 0$, then $a = 0$ or $b = 0$.
This property is true for three or more factors also.

In other words, if the product of two or more real numbers is zero, then at least one number must be zero.

EXAMPLE 1 Solve $(x + 2)(x - 6) = 0$.

Solution: By the zero-factor property, $(x + 2)(x - 6) = 0$ only if $x + 2 = 0$ or $x - 6 = 0$.

$$x + 2 = 0 \quad \text{or} \quad x - 6 = 0 \qquad \text{Apply the zero-factor property.}$$
$$x = -2 \quad \text{or} \qquad x = 6 \qquad \text{Solve each linear equation.}$$

To check, let $x = -2$ and then let $x = 6$ in the original equation as shown in the screen to the left.

Both -2 and 6 check, and the solution set is $\{-2, 6\}$.

```
-2→X:(X+2)(X-6)
                 0
6→X:(X+2)(X-6)
                 0
```

EXAMPLE 2 Solve $2x^2 + 9x - 5 = 0$.

Solution: To use the zero-factor property, one side of the equation must be 0, and the other side must be in factored form.

$$2x^2 + 9x - 5 = 0$$
$$(2x - 1)(x + 5) = 0 \qquad \text{Factor.}$$
$$2x - 1 = 0 \quad \text{or} \quad x + 5 = 0 \qquad \text{Set each factor equal to zero.}$$
$$2x = 1$$
$$x = \frac{1}{2} \quad \text{or} \qquad x = -5 \qquad \text{Solve each linear equation.}$$

TECHNOLOGY NOTE

We now have a variety of ways to check a solution to an equation. We can check numerically as above or graphically. We will only check each equation one way, but remember that you can select the method that you prefer.

```
1/2→X:2X²+9X-5
               0
-5→X:2X²+9X-5
               0
```

The solution set is $\left\{-5, \dfrac{1}{2}\right\}$. To check, let $x = \dfrac{1}{2}$ in the original equation; then let $x = -5$ in the original equation as shown in the screen to the left. ▬▬▬

TO SOLVE POLYNOMIAL EQUATIONS BY FACTORING

Step 1.   Write the equation in standard form so that one side of the equation is 0.

Step 2.   Factor the polynomial completely.

Step 3.   Set each factor containing a variable equal to 0.

Step 4.   Solve the resulting equations.

Step 5.   Check each solution in the original equation.

Since it is not always possible to factor a polynomial, not all polynomial equations can be solved by factoring. Other methods of solving polynomial equations are presented in Chapter 9.

EXAMPLE 3   Solve $x(2x - 7) = 4$.

Solution: First, write the equation in standard form; then factor.

```
-1/2→X:X(2X-7)
               4
4→X:X(2X-7)
               4
```

$$x(2x - 7) = 4$$
$$2x^2 - 7x = 4 \qquad \text{Apply the distributive property.}$$
$$2x^2 - 7x - 4 = 0 \qquad \text{Write in standard form.}$$
$$(2x + 1)(x - 4) = 0 \qquad \text{Factor.}$$
$$2x + 1 = 0 \quad \text{or} \quad x - 4 = 0 \qquad \text{Set each factor equal to zero.}$$
$$2x = -1 \quad \text{or} \qquad x = 4 \qquad \text{Solve.}$$
$$x = -\frac{1}{2} \quad \text{or} \qquad x = 4$$

The solution set is $\left\{-\dfrac{1}{2}, 4\right\}$. Check both solutions in the original equation, as shown.

▬▬▬

R E M I N D E R   To apply the zero-factor property, one side of the equation must be 0, and the other side of the equation must be factored. To solve the equation $x(2x - 7) = 4$, for example, you may **not** set each factor equal to 4.

EXAMPLE 4   Solve $3(x^2 + 4) + 5 = -6(x^2 + 2x) + 13$.

Solution: Rewrite the equation so that one side is 0.

$$3(x^2 + 4) + 5 = -6(x^2 + 2x) + 13$$

$$3x^2 + 12 + 5 = -6x^2 - 12x + 13 \qquad \text{Apply the distributive property.}$$

$$9x^2 + 12x + 4 = 0 \qquad \begin{array}{l}\text{Rewrite the equation so that} \\ \text{one side is 0.}\end{array}$$

$$(3x + 2)(3x + 2) = 0 \qquad \text{Factor.}$$

$$3x + 2 = 0 \quad \text{or} \quad 3x + 2 = 0 \qquad \text{Set each factor equal to 0.}$$

$$3x = -2 \quad \text{or} \qquad 3x = -2$$

$$x = -\frac{2}{3} \quad \text{or} \qquad x = -\frac{2}{3} \qquad \text{Solve each equation.}$$

Notice that the identical factors $(3x + 2)$, led to a single solution, $-\frac{2}{3}$. The solution set is $\left\{-\frac{2}{3}\right\}$.

We choose to check this time by graphing and using the intersection-of-graphs method as shown in the screen to the left. The intersection is at $x \approx -0.66667$ which is an approximation for $-\frac{2}{3}$.

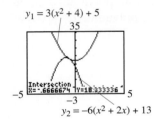

$y_1 = 3(x^2 + 4) + 5$

$y_2 = -6(x^2 + 2x) + 13$

If the equation contains fractions, we clear the equation of fractions as a first step.

EXAMPLE 5   Solve $2x^2 = \frac{17}{3}x + 1$.

Solution:

$$2x^2 = \frac{17}{3}x + 1$$

$$3(2x^2) = 3\left(\frac{17}{3}x + 1\right) \qquad \text{Clear the equation of fractions.}$$

$$6x^2 = 17x + 3 \qquad \text{Apply the distributive property.}$$

$$6x^2 - 17x - 3 = 0 \qquad \begin{array}{l}\text{Rewrite the equation in standard} \\ \text{form.}\end{array}$$

$$(6x + 1)(x - 3) = 0 \qquad \text{Factor.}$$

$$6x + 1 = 0 \quad \text{or} \quad x - 3 = 0 \qquad \text{Set each factor equal to zero.}$$

$$6x = -1$$

$$x = -\frac{1}{6} \quad \text{or} \qquad x = 3 \qquad \text{Solve each equation.}$$

The two graphs intersect twice. Verify the solutions of $x = -\frac{1}{6}$ and $x = 3$.

$\left(-\frac{1}{6}, \frac{1}{18}\right)$ $(3, 18)$

The solution set is $\left\{-\frac{1}{6}, 3\right\}$. To check, graph $y_1 = 2x^2$ and $y_2 = \frac{17}{3}x + 1$. Notice that the graph of y_1 is a parabola and the graph of y_2 is a line.

EXAMPLE 6   Solve $x^3 = 4x$.

Solution:

$$x^3 = 4x$$

$$x^3 - 4x = 0$$
Rewrite the equation so that one side is 0.

$$x(x^2 - 4) = 0$$
Factor out the GCF, x.

$$x(x + 2)(x - 2) = 0$$
Factor the difference of squares.

$x = 0$ or $x + 2 = 0$ or $x - 2 = 0$ Set each factor equal to 0.

$x = 0$ or $x = -2$ or $x = 2$ Solve each equation.

The solution set is $\{-2, 0, 2\}$.

Notice that the *third*-degree equation of Example 6 yielded *three* solutions. To see why, graph $y_1 = x^3$ and $y_2 = 4x$.

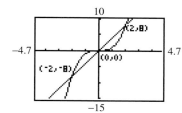

The graphs intersect in three points representing the three solutions of the equation. The solutions of the equation $x^3 = 4x$ are $x = -2, x = 0,$ and $x = 2$.

EXAMPLE 7   Solve $x^3 + 5x^2 = x + 5$.

Solution: First, write the equation so that one side is 0.

$$x^3 + 5x^2 - x - 5 = 0$$

$$(x^3 - x) + (5x^2 - 5) = 0$$
Factor by grouping.

$$x(x^2 - 1) + 5(x^2 - 1) = 0$$

$$(x^2 - 1)(x + 5) = 0$$

$$(x + 1)(x - 1)(x + 5) = 0$$
Factor the difference of squares.

$x + 1 = 0$ or $x - 1 = 0$ or $x + 5 = 0$ Set each factor equal to 0.

$x = -1$ or $x = 1$ or $x = -5$ Solve each equation.

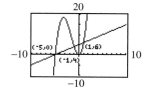

The solution set is $\{-5, -1, 1\}$. The screen to the left verifies the solutions.

Some problems may be modeled by polynomial equations. To solve these problems, we use the same problem-solving steps that were introduced in Section 2.2. When solving these problems, keep in mind that a solution of an equation that models a problem is not always a solution to the problem. For example, a person's weight or the length of a side of a geometric figure is always a positive number. Discard solutions that do not make sense as solutions of the problem.

EXAMPLE 8

An Alpha III model rocket is launched from the ground with an A8–3 engine. Without a parachute the height of the rocket $h(t)$ at time t seconds is approximated by the function

$$h(t) = -16t^2 + 144t$$

Find how long it takes the rocket to return to the ground.

Solution:

1. UNDERSTAND.   Read and reread the problem. The function $h(t) = -16t^2 + 144t$ models the height of the rocket. Familiarize yourself with this function by finding a few function values.

> When $t = 1$ second, the height of the rocket is
> $h(1) = -16(1)^2 + 144(1) = 128$ feet.

> When $t = 2$ seconds, the height of the rocket is
> $h(2) = -16(2)^2 + 144(2) = 224$ feet.

Since we have been given the needed function, we proceed to step 4.

4. TRANSLATE.   To find how long it takes the rocket to hit the ground, we want to know for what value of t is the height $h(t)$ equal to 0. That is, we want to solve $h(t) = 0$.

$$-16t^2 + 144t = 0$$

5. COMPLETE.   Solve the quadratic equation by factoring.

$$-16t^2 + 144t = 0$$
$$-16t(t - 9) = 0$$
$$-16t = 0 \text{ or } t - 9 = 0$$
$$t = 0 \quad \text{or} \quad t = 9$$

6. INTERPRET.   The height $h(t)$ is 0 feet at time 0 seconds (when the rocket is launched) and at time 9 seconds. To *check*, graph $y_1 = -16x^2 + 144x$ and see that $(9, 0)$ is an intercept point.

State: The rocket returns to the ground 9 seconds after it is launched.   ▬▬▬▬

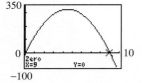

The rocket is at ground level at 0 seconds and 9 seconds.

Some of the exercises at the end of this section make use of the **Pythagorean theorem.** Before we review this theorem, recall that a **right triangle** is a triangle that contains a 90° angle, or right angle. The **hypotenuse** of a right triangle is the side opposite the right angle and is the longest side of the triangle. The **legs** of a right triangle are the other sides of the triangle.

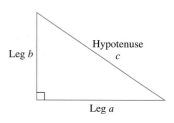

> **PYTHAGOREAN THEOREM**
>
> In a right triangle, the sum of the squares of the lengths of the two legs is equal to the square of the length of the hypotenuse.
>
> $$(\text{leg})^2 + (\text{leg})^2 = (\text{hypotenuse})^2 \quad \text{or} \quad a^2 + b^2 = c^2$$

EXAMPLE 9   When framing a new house, carpenters frequently make use of the Pythagorean theorem in order to determine whether a wall is "square"—that is, whether the wall forms a right angle with the floor. Often, they use a triangle whose sides are three consecutive integers. Find a right triangle whose sides are three consecutive integers.

Solution: **1.** UNDERSTAND.   Read and reread the problem.

2. ASSIGN.   Let $x, x + 1$, and $x + 2$ be three consecutive integers. Since these integers represent lengths of the sides of a right triangle, we have

$$x = \text{one leg}$$
$$x + 1 = \text{other leg}$$
$$x + 2 = \text{hypotenuse (longest side)}$$

3. ILLUSTRATE.   An illustration is to the left.

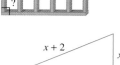

4. TRANSLATE. By the Pythagorean theorem, we have

In words: $(\text{leg})^2 + (\text{leg})^2 = (\text{hypotenuse})^2$
Translate: $(x)^2 + (x + 1)^2 = (x + 2)^2$

5. COMPLETE.   Solve the equation.

$$x^2 + (x + 1)^2 = (x + 2)^2$$
$$x^2 + x^2 + 2x + 1 = x^2 + 4x + 4 \qquad \text{Multiply.}$$
$$2x^2 + 2x + 1 = x^2 + 4x + 4$$
$$x^2 - 2x - 3 = 0 \qquad \text{Write in standard form.}$$
$$(x - 3)(x + 1) = 0$$
$$x - 3 = 0 \quad \text{or} \quad x + 1 = 0$$
$$x = 3 \quad \text{or} \qquad x = -1$$

6. INTERPRET.   Discard $x = -1$ since length cannot be negative. If $x = 3$, then $x + 1 = 4$ and $x + 2 = 5$.

Check: To check, see that $(\text{leg})^2 + (\text{leg})^2 = (\text{hypotenuse})^2$.

$$3^2 + 4^2 = 5^2$$

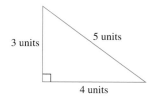

or

$$9 + 16 = 25 \qquad \text{True.}$$

State: The lengths of the sides of the right triangle are 3, 4, and 5 units.

Is this length 5 feet?

4 feet

3 feet

A carpenter uses this information, for example, by marking off lengths of 3 and 4 feet on the framing. If the diagonal length between these marks is 5 feet, the wall is "square." If not, adjustments must be made.

In Section 6.6, we found that if $(x - a)$ is a factor of a polynomial, then a is an x-intercept of the related polynomial function $p(x)$. Now we know that if $(x - a)$ is a factor of a polynomial, not only is a an x-intercept of the related polynomial function $p(x)$ but a is also a solution of the related equation $p(x) = 0$.

For example, if

$$p(x) = (x + 2)(x - 6),$$

then -2 and 6 are the x-intercepts of the graph of $p(x)$ and -2 and 6 are the solutions of

$$0 = (x + 2)(x - 6).$$

In Chapter 9, we explore graphs of quadratic functions such as $p(x)$ more fully.

EXAMPLE 10   Match each function with its graph:

$$f(x) = (x - 3)(x + 2), \quad g(x) = x(x + 2)(x - 2), \quad h(x) = (x - 2)(x + 2)(x - 1)$$

A 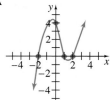   B 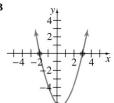   C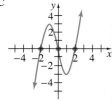

Solution: The graph of the function $f(x) = (x - 3)(x + 2)$ has two x-intercept points, $(3, 0)$ and $(-2, 0)$, because the equation $0 = (x - 3)(x + 2)$ has two solutions, 3 and -2.

The graph of $f(x)$ is graph B.

The graph of the function $g(x) = x(x + 2)(x - 2)$ has three x-intercept points, $(0, 0)$, $(-2, 0)$, and $(2, 0)$, because the equation $0 = x(x + 2)(x - 2)$ has three solutions, 0, -2, and 2.

The graph of $g(x)$ is graph C.

The graph of the function $h(x) = (x - 2)(x + 2)(x - 1)$ has three x-intercept points, $(-2, 0)$, $(1, 0)$, and $(2, 0)$, because the equation $0 = (x - 2)(x + 2)(x - 1)$ has three solutions, -2, 1, and 2.

The graph of $h(x)$ is graph A.

MENTAL MATH

Solve each equation for the variable. See Example 1.

1. $(x - 3)(x + 5) = 0$

2. $(y + 5)(y + 3) = 0$

3. $(z - 3)(z + 7) = 0$

4. $(c - 2)(c - 4) = 0$

5. $x(x - 9) = 0$

6. $w(w + 7) = 0$

EXERCISE SET 6.8

Solve each equation. See Example 1.

1. $(x + 3)(3x - 4) = 0$ **2.** $(5x + 1)(x - 2) = 0$

3. $3(2x - 5)(4x + 3) = 0$ **4.** $8(3x - 4)(2x - 7) = 0$

Solve each equation. See Examples 2 through 5.

5. $x^2 + 11x + 24 = 0$ **6.** $y^2 - 10y + 24 = 0$

7. $12x^2 + 5x - 2 = 0$ **8.** $3y^2 - y - 14 = 0$

9. $z^2 + 9 = 10z$ **10.** $n^2 + n = 72$

11. $x(5x + 2) = 3$ **12.** $n(2n - 3) = 2$

13. $x^2 - 6x = x(8 + x)$ **14.** $n(3 + n) = n^2 + 4n$

15. $\dfrac{z^2}{6} - \dfrac{z}{2} - 3 = 0$ **16.** $\dfrac{c^2}{20} - \dfrac{c}{4} + \dfrac{1}{5} = 0$

17. $\dfrac{x^2}{2} + \dfrac{x}{20} = \dfrac{1}{10}$ **18.** $\dfrac{y^2}{30} = \dfrac{y}{15} + \dfrac{1}{2}$

19. $\dfrac{4t^2}{5} = \dfrac{t}{5} + \dfrac{3}{10}$ **20.** $\dfrac{5x^2}{6} - \dfrac{7x}{2} + \dfrac{2}{3} = 0$

Solve each equation. See Examples 6 and 7.

21. $(x + 2)(x - 7)(3x - 8) = 0$

22. $(4x + 9)(x - 4)(x + 1) = 0$

23. $y^3 = 9y$ **24.** $n^3 = 16n$

25. $x^3 - x = 2x^2 - 2$ **26.** $m^3 = m^2 + 12m$

27. Explain how solving $2(x - 3)(x - 1) = 0$ differs from solving $2x(x - 3)(x - 1) = 0$.

28. Explain why the zero-factor property works for more than two numbers whose product is 0.

Solve each equation.

29. $(2x + 7)(x - 10) = 0$ **30.** $(x + 4)(5x - 1) = 0$

31. $3x(x - 5) = 0$ **32.** $4x(2x + 3) = 0$

33. $x^2 - 2x - 15 = 0$ **34.** $x^2 + 6x - 7 = 0$

35. $12x^2 + 2x - 2 = 0$ **36.** $8x^2 + 13x + 5 = 0$

37. $w^2 - 5w = 36$ **38.** $x^2 + 32 = 12x$

39. $25x^2 - 40x + 16 = 0$ **40.** $9n^2 + 30n + 25 = 0$

41. $2r^3 + 6r^2 = 20r$ **42.** $-2t^3 = 108t - 30t^2$

43. $z(5z - 4)(z + 3) = 0$ **44.** $2r(r + 3)(5r - 4) = 0$

45. $2z(z + 6) = 2z^2 + 12z - 8$

46. $3c^2 - 8c + 2 = c(3c - 8)$

47. $(x - 1)(x + 4) = 24$ **48.** $(2x - 1)(x + 2) = -3$

49. $\dfrac{x^2}{4} - \dfrac{5}{2}x + 6 = 0$ **50.** $\dfrac{x^2}{18} + \dfrac{x}{2} + 1 = 0$

51. $y^2 + \dfrac{1}{4} = -y$ **52.** $\dfrac{x^2}{10} + \dfrac{5}{2} = x$

53. $y^3 + 4y^2 = 9y + 36$ **54.** $x^3 + 5x^2 = x + 5$

55. $2x^3 = 50x$ **56.** $m^5 = 36m^3$

57. $x^2 + (x + 1)^2 = 61$ **58.** $y^2 + (y + 2)^2 = 34$

59. $m^2(3m - 2) = m$ **60.** $x^2(5x + 3) = 26x$

61. $3x^2 = -x$ **62.** $y^2 = -5y$

63. $x(x - 3) = x^2 + 5x + 7$

64. $z^2 - 4z + 10 = z(z - 5)$

65. $3(t - 8) + 2t = 7 + t$

66. $7c - 2(3c + 1) = 5(4 - 2c)$

67. $-3(x - 4) + x = 5(3 - x)$

68. $-4(a + 1) - 3a = -7(2a - 3)$

69. Describe two ways in which a linear equation differs from a quadratic equation.

70. Is the following step correct? Why or why not?

$$x(x - 3) = 5$$
$$x = 5 \text{ or } x - 3 = 5$$

Solve. See Examples 8 and 9.

71. One number exceeds another by five, and their product is 66. Find the numbers.

72. If the sum of two numbers is 4 and their product is $\frac{15}{4}$, find the numbers.

73. An electrician needs to run a cable from the top of a 60-foot tower to a transmitter box located 45 feet from the base of the tower. Find how long he should cut the cable.

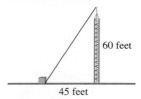

60 feet

45 feet

74. A stereo-system installer needs to run speaker wire along the two diagonals of a rectangular room whose dimensions are 40 feet by 75 feet. Find how much speaker wire she needs.

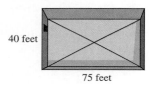

40 feet

75 feet

75. If the cost, $C(x)$, for manufacturing x units of a certain product is given by $C(x) = x^2 - 15x + 50$, find the number of units manufactured at a cost of $9500.

76. Determine whether any three consecutive integers represent the lengths of the sides of a right triangle.

77. The shorter leg of a right triangle is 3 centimeters less than the other leg. Find the length of the two legs if the hypotenuse is 15 centimeters.

78. Marie Mulroney has a rectangular board 12 inches by 16 inches around which she wants to put a uniform border of shells. If she has enough shells for a border whose area is 128 square inches, determine the width of the border.

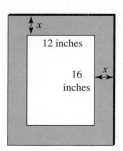

x

12 inches

16 inches

x

79. A gardener has a rose garden that measures 30 feet by 20 feet. He wants to put a uniform border of pine bark around the outside of the garden. Find how wide the border should be if he has enough pine bark to cover 336 square feet.

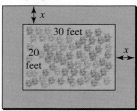

x

30 feet

20 feet

x

80. While hovering near the top of Ribbon Falls in Yosemite National Park at 1600 feet, a helicopter pilot accidentally drops his sunglasses. The height $h(t)$ of the sunglasses after t seconds is given by the polynomial function

$$h(t) = -16t^2 + 1600$$

When will the sunglasses hit the ground?

81. After t seconds, the height $h(t)$ of a model rocket launched from the ground into the air is given by the function

$$h(t) = -16t^2 + 80t$$

Find how long it takes the rocket to reach a height of 96 feet.

Match each polynomial function with its graph (A–F). See Example 10.

82. $f(x) = (x - 2)(x + 5)$ **83.** $g(x) = (x + 1)(x - 6)$

84. $h(x) = x(x + 3)(x - 3)$

85. $F(x) = (x + 1)(x - 2)(x + 5)$

86. $G(x) = 2x^2 + 9x + 4$ **87.** $H(x) = 2x^2 - 7x - 4$

A.

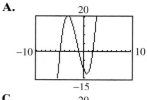

B.

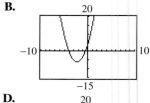

C.

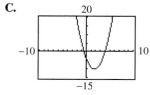

D.

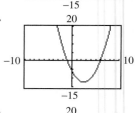

E.

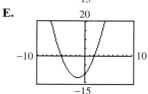

F.

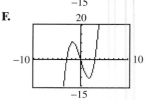

GROUP ACTIVITY

FINDING THE LARGEST AREA

MATERIALS:
- Calculator, graphing calculator (optional)

A picture framer has a piece of wood that measures 1 inch wide × 50 inches long with which she would like to make a picture frame with the largest possible interior area. Complete the following activity to help her determine the dimensions of the frame that she should use to achieve her goal.

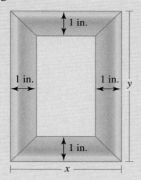

1. Use the situation given in the figure to write an equation in x and y for the *outer* perimeter of the frame. (Remember that the perimeter will equal 50 inches.)

2. Complete the following table, letting $A(x)$ be the interior area. From the table, what appears to be the largest interior area? Which exterior dimensions of the frame provide this area?*

3. Use the table and write $A(x)$ as a function of x alone. (*Hint*: Use the equation from Question 1.)

4. Graph the function $A(x)$. Locate and label the point from the table that represents the maximum interior area. Describe the location of the point in relation to the rest of the graph.

		FRAME'S INTERIOR DIMENSIONS		
x	y	Interior Width	Interior Height	$A(x)$ Interior Area
2.0				
2.5				
3.0				
3.5				
4.0				
4.5				
5.0				
5.5				
6.0				
6.5				
7.0				
7.5				
8.0				
8.5				
9.0				
9.5				
10.0				
10.5				
11.0				
11.5				
12.0				
12.5				
13.0				
13.5				
14.0				
14.5				
15.0				

* If using a graphing calculator, find the first few entries in the table to write $A(x)$. Then use the features Table Set and Δ Table to complete the table.

CHAPTER 6   HIGHLIGHTS

DEFINITIONS AND CONCEPTS	**EXAMPLES**
SECTION 6.1 EXPONENTS AND SCIENTIFIC NOTATION	

Product rule: $a^m \cdot a^n = a^{m+n}$

Zero exponent: $a^0 = 1, a \neq 0$

Quotient rule: $\dfrac{a^m}{a^n} = a^{m-n}$

Negative exponent: $a^{-n} = \dfrac{1}{a^n}$

$x^2 \cdot x^3 = x^5$

$7^0 = 1, \ (-10)^0 = 1$

$\dfrac{y^{10}}{y^4} = y^{10-4} = y^6$

$3^{-2} = \dfrac{1}{3^2} = \dfrac{1}{9}, \ \dfrac{x^{-5}}{x^{-7}} = x^{-5-(-7)} = x^2$

A positive number is written in **scientific notation** if it is written as the product of a number a, where $1 \leq a < 10$, and an integer power of 10: $a \times 10^r$.

Numbers written in scientific notation:
$$568,000 = 5.68 \times 10^5$$
$$0.0002117 = 2.117 \times 10^{-4}$$

| **SECTION 6.2   MORE WORK WITH EXPONENTS AND SCIENTIFIC NOTATION** | |

Power rules: $(a^m)^n = a^{m \cdot n}$

$\qquad\qquad\quad (ab)^m = a^m b^m$

$\qquad\qquad\quad \left(\dfrac{a}{b}\right)^n = \dfrac{a^n}{b^n}$

$(7^8)^2 = 7^{16}$

$(2y)^3 = 2^3 y^3 = 8y^3$

$\left(\dfrac{5x^{-3}}{x^2}\right)^{-2} = \dfrac{5^{-2} x^6}{x^{-4}}$

$\qquad\qquad = 5^{-2} \cdot x^{6-(-4)}$

$\qquad\qquad = \dfrac{x^{10}}{5^2}, \ \text{ or } \ \dfrac{x^{10}}{25}$

| **SECTION 6.3   POLYNOMIALS AND POLYNOMIAL FUNCTIONS** | |

A **polynomial** is a finite sum of terms in which all variables have exponents raised to nonnegative integer powers and no variables appear in the denominator.

Polynomials

$\qquad 1.3x^2 \qquad$ (monomial)

$\qquad -\dfrac{1}{3}y + 5 \qquad$ (binomial)

$\qquad 6z^2 - 5z + 7 \qquad$ (trinomial)

A function P is a **polynomial function** if $P(x)$ is a polynomial.

For the polynomial function
$$P(x) = -x^2 + 6x - 12, \text{ find } P(-2).$$
$$P(-2) = -(-2)^2 + 6(-2) - 12 = -28$$

To add polynomials, combine all like terms.

Add:
$$(3y^2x - 2yx + 11) + (-5y^2x - 7)$$
$$= -2y^2x - 2yx + 4$$

To subtract polynomials, change the signs of the terms of the polynomial being subtracted; then add.

Subtract:
$$(-2z^3 - z + 1) - (3z^3 + z - 6)$$
$$= -2z^3 - z + 1 - 3z^3 - z + 6$$
$$= -5z^3 - 2z + 7$$

DEFINITIONS AND CONCEPTS	EXAMPLES

SECTION 6.4   MULTIPLYING POLYNOMIALS

To multiply two polynomials, use the distributive property and multiply each term of one polynomial by each term of the other polynomial; then combine like terms.

Multiply:
$$(x^2 - 2x)(3x^2 - 5x + 1)$$
$$= 3x^4 - 5x^3 + x^2 - 6x^3 + 10x^2 - 2x$$
$$= 3x^4 - 11x^3 + 11x^2 - 2x$$

Special products:
$$(a + b)^2 = a^2 + 2ab + b^2$$
$$(a - b)^2 = a^2 - 2ab + b^2$$
$$(a + b)(a - b) = a^2 - b^2$$

$$(3m + 2n)^2 = 9m^2 + 12mn + 4n^2$$
$$(z^2 - 5)^2 = z^4 - 10z^2 + 25$$
$$(7y + 1)(7y - 1) = 49y^2 - 1$$

The FOIL method may be used when multiplying two binomials.

Multiply:
$$(x^2 + 5)(2x^2 - 9)$$
$$\quad\quad F\quad\quad O\quad\quad I\quad\quad L$$
$$= x^2(2x^2) + x^2(-9) + 5(2x^2) + 5(-9)$$
$$= 2x^4 - 9x^2 + 10x^2 - 45$$
$$= 2x^4 + x^2 - 45$$

SECTION 6.5   THE GREATEST COMMON FACTOR AND FACTORING BY GROUPING

The greatest common factor (GCF) of the terms of a polynomial is the product of the GCF of the numerical coefficients and the GCF of the variable factors.

Factor $14xy^3 - 2xy^2 = 2 \cdot 7 \cdot x \cdot y^3 - 2 \cdot x \cdot y^2$.
The GCF is $2 \cdot x \cdot y^2$, or $2xy^2$.
$$14xy^3 - 2xy^2 = 2xy^2(7y - 1)$$

To factor a polynomial by grouping, group the terms so that each group has a common factor. Factor out these common factors. Then see if the new groups have a common factor.

Factor: $x^4y - 5x^3 + 2xy - 10$.
$$= x^3(xy - 5) + 2(xy - 5)$$
$$= (xy - 5)(x^3 + 2)$$

SECTION 6.6   FACTORING TRINOMIALS

To factor $ax^2 + bx + c$:

Step 1.   Write all pairs of factors of ax^2.

Step 2.   Write all pairs of factors of c.

Step 3.   Try combinations of these factors until the middle term bx is found.

Factor $28x^2 - 27x - 10$.
Factors of $28x^2$: $28x$ and x, $2x$ and $14x$, $4x$ and $7x$.
Factors of -10: -2 and 5, 2 and -5, -10 and 1, 10 and -1.
$$28x^2 - 27x - 10 = (7x + 2)(4x - 5)$$

SECTION 6.7   FACTORING BY SPECIAL PRODUCTS AND FACTORING STRATEGIES

Perfect square trinomial:
$$a^2 + 2ab + b^2 = (a + b)^2$$
$$a^2 - 2ab + b^2 = (a - b)^2$$

Factor:
$$25x^2 + 30x + 9 = (5x + 3)^2$$
$$49z^2 - 28z + 4 = (7z - 2)^2$$

(continued)

DEFINITIONS AND CONCEPTS	EXAMPLES
SECTION 6.7 FACTORING BY SPECIAL PRODUCTS AND FACTORING STRATEGIES	

Difference of two squares

$$a^2 - b^2 = (a + b)(a - b)$$

$$36x^2 - y^2 = (6x + y)(6x - y)$$

Sum and difference of two cubes

$$a^3 + b^3 = (a + b)(a^2 - ab + b^2)$$
$$a^3 - b^3 = (a - b)(a^2 + ab + b^2)$$

$$8y^3 + 1 = (2y + 1)(4y^2 - 2y + 1)$$
$$27p^3 - 64q^3 = (3p - 4q)(9p^2 + 12pq + 16q^2)$$

To factor a polynomial,

Step 1. Factor out the GCF.

Step 2. If the polynomial is a binomial, see if it is a difference of two squares or a sum or difference of two cubes. If it is a trinomial, see if it is a perfect square trinomial. If not, try factoring by methods of Section 6.6. If it is a polynomial with four or more terms, try factoring by grouping.

Step 3. See if any factors can be factored further.

Factor $10x^4y + 5x^2y - 15y$.

$$= 5y(2x^4 + x^2 - 3)$$
$$= 5y(2x^2 + 3)(x^2 - 1)$$
$$= 5y(2x^2 + 3)(x + 1)(x - 1)$$

SECTION 6.8 SOLVING POLYNOMIAL EQUATIONS ALGEBRAICALLY AND GRAPHICALLY	

To solve polynomial equations by factoring:

Step 1. Write the equation so that one side is 0.

Step 2. Factor the polynomial completely.

Step 3. Set each factor equal to 0.

Step 4. Solve the resulting equations.

Step 5. Check each solution.

Solve:

$$2x^3 - 5x^2 = 3x$$
$$2x^3 - 5x^2 - 3x = 0$$
$$x(2x + 1)(x - 3) = 0$$
$$x = 0 \quad \text{or} \quad 2x + 1 = 0 \quad \text{or} \quad x - 3 = 0$$
$$x = 0 \quad \text{or} \quad x = -\frac{1}{2} \quad \text{or} \quad x = 3$$

CHAPTER 6 REVIEW

(6.1) *Evaluate.*

1. $(-2)^2$ **2.** $(-3)^4$ **3.** -2^2

4. -3^4 **5.** 8^0 **6.** -9^0

7. -4^{-2} **8.** $(-4)^{-2}$

Simplify each expression. Use only positive exponents.

9. $-xy^2 \cdot y^3 \cdot xy^2z$ **10.** $(-4xy)(-3xy^2b)$

11. $a^{-14} \cdot a^5$ **12.** $\dfrac{a^{16}}{a^{17}}$ **13.** $\dfrac{x^{-7}}{x^4}$

14. $\dfrac{9a(a^{-3})}{18a^{15}}$ **15.** $\dfrac{y^{6p-3}}{y^{6p+2}}$

Write in scientific notation.

16. 36,890,000 **17.** -0.000362

Write each number without exponents.

18. 1.678×10^{-6} **19.** 4.1×10^5

(6.2) *Simplify. Use only positive exponents.*

20. $(8^5)^3$ **21.** $\left(\dfrac{a}{4}\right)^2$ **22.** $(3x)^3$ **23.** $(-4x)^{-2}$

24. $\left(\dfrac{6x}{5}\right)^2$ **25.** $(8^6)^{-3}$ **26.** $\left(\dfrac{4}{3}\right)^{-2}$ **27.** $(-2x^3)^{-3}$

28. $\left(\dfrac{8p^6}{4p^4}\right)^{-2}$ **29.** $(-3x^{-2}y^2)^3$ **30.** $\left(\dfrac{x^{-5}y^{-3}}{z^3}\right)^{-5}$

31. $\dfrac{4^{-1}x^3yz}{x^{-2}yx^4}$ **32.** $(5xyz)^{-4}(x^{-2})^{-3}$ **33.** $\dfrac{2(3yz)^{-3}}{y^{-3}}$

Simplify each expression.

34. $x^{4a}(3x^{5a})^3$ **35.** $\dfrac{4y^{3x-3}}{2y^{2x+4}}$

Use scientific notation to find the quotient. Express each quotient in scientific notation.

36. $\dfrac{(0.00012)(144,000)}{0.0003}$ **37.** $\dfrac{(-0.00017)(0.00039)}{3000}$

Simplify. Use only positive exponents.

38. $\dfrac{27x^{-5}y^5}{18x^{-6}y^2} \cdot \dfrac{x^4y^{-2}}{x^{-2}y^3}$ **39.** $\dfrac{3x^5}{y^{-4}} \cdot \dfrac{(3xy^{-3})^{-2}}{(z^{-3})^{-4}}$

40. $\dfrac{(x^w)^2}{(x^{w-4})^{-2}}$

(6.3)   *Find the degree of each polynomial.*

41. $x^2y - 3xy^3z + 5x + 7y$

42. $3x + 2$

Simplify by combining like terms.

43. $4x + 8x - 6x^2 - 6x^2y$

44. $-8xy^3 + 4xy^3 - 3x^3y$

Add or subtract as indicated.

45. $(3x + 7y) + (4x^2 - 3x + 7) + (y - 1)$

46. $(4x^2 - 6xy + 9y^2) - (8x^2 - 6xy - y^2)$

47. $(3x^2 - 4b + 28) + (9x^2 - 30) - (4x^2 - 6b + 20)$

48. Add $(9xy + 4x^2 + 18)$ and $(7xy - 4x^3 - 9x)$.

49. Subtract $(x - 7)$ from the sum of $(3x^2y - 7xy - 4)$ and $(9x^2y + x)$.

50. $x^2 - 5x + 7$ **51.** $x^3 \quad\ + 2xy^2 - y$
$\underline{- \quad (x + 4)}$ $\underline{+ \quad (x - 4xy^2 \quad\ - 7)}$

If $P(x) = 9x^2 - 7x + 8$, find the following.

52. $P(6)$ **53.** $P(-2)$ **54.** $P(-3)$

If $P(x) = 2x - 1$ and $Q(x) = x^2 + 2x - 5$, find the following.

55. $P(x) + Q(x)$ **56.** $2[P(x)] - Q(x)$

57. Find the perimeter of the rectangle.

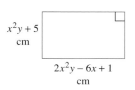

$x^2y + 5$ cm

$2x^2y - 6x + 1$ cm

(6.4)   *Multiply.*

58. $-6x(4x^2 - 6x + 1)$ **59.** $-4ab^2(3ab^3 + 7ab + 1)$

60. $(x - 4)(2x + 9)$ **61.** $(-3xa + 4b)^2$

62. $(9x^2 + 4x + 1)(4x - 3)$

63. $(5x - 9y)(3x + 9y)$ **64.** $\left(x - \dfrac{1}{3}\right)\left(x + \dfrac{2}{3}\right)$

65. $(x^2 + 9x + 1)^2$

Multiply, using special products.

66. $(3x - y)^2$ **67.** $(4x + 9)^2$

68. $(x + 3y)(x - 3y)$

69. $[4 + (3a - b)][4 - (3a - b)]$

70. If $P(x) = 2x - 1$ and $Q(x) = x^2 + 2x - 5$, find $P(x) \cdot Q(x)$.

71. Find the area of the rectangle.

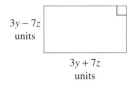

$3y - 7z$ units

$3y + 7z$ units

Multiply. Assume that all variable exponents represent integers.

72. $4a^b(3a^{b+2} - 7)$ **73.** $(4xy^z - b)^2$

74. $(3x^a - 4)(3x^a + 4)$

(6.5)   *Factor out the greatest common factor.*

75. $16x^3 - 24x^2$ **76.** $36y - 24y^2$

77. $6ab^2 + 8ab - 4a^2b^2$

78. $14a^2b^2 - 21ab^2 + 7ab$

79. $6a(a + 3b) - 5(a + 3b)$

80. $4x(x - 2y) - 5(x - 2y)$

81. $xy - 6y + 3x - 18$ **82.** $ab - 8b + 4a - 32$

83. $pq - 3p - 5q + 15$ **84.** $x^3 - x^2 - 2x + 2$

85. A smaller square is cut from a larger rectangle. Write the area of the shaded region as a factored polynomial.

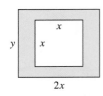

(6.6) *Completely factor each polynomial.*

86. $x^2 - 14x - 72$ **87.** $x^2 + 16x - 80$

88. $2x^2 - 18x + 28$ **89.** $3x^2 + 33x + 54$

90. $2x^3 - 7x^2 - 9x$ **91.** $3x^2 + 2x - 16$

92. $6x^2 + 17x + 10$ **93.** $15x^2 - 91x + 6$

94. $4x^2 + 2x - 12$ **95.** $9x^2 - 12x - 12$

96. $y^2(x + 6)^2 - 2y(x + 6)^2 - 3(x + 6)^2$

97. $(x + 5)^2 + 6(x + 5) + 8$

98. $x^4 - 6x^2 - 16$ **99.** $x^4 + 8x^2 - 20$

(6.7) *Factor each polynomial completely.*

100. $x^2 - 100$ **101.** $x^2 - 81$

102. $2x^2 - 32$ **103.** $6x^2 - 54$

104. $81 - x^4$ **105.** $16 - y^4$

106. $(y + 2)^2 - 25$ **107.** $(x - 3)^2 - 16$

108. $x^3 + 216$ **109.** $y^3 + 512$

110. $8 - 27y^3$ **111.** $1 - 64y^3$

112. $6x^4y + 48xy$ **113.** $2x^5 + 16x^2y^3$

114. $x^2 - 2x + 1 - y^2$

115. $x^2 - 6x + 9 - 4y^2$

116. $4x^2 + 12x + 9$

117. $16a^2 - 40ab + 25b^2$

118. The volume of the cylindrical shell is $\pi R^2 h - \pi r^2 h$ cubic units. Write this volume as a factored expression.

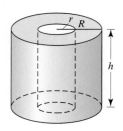

(6.8) *Solve each polynomial equation for the variable. Check each solution numerically or graphically.*

119. $(3x - 1)(x + 7) = 0$

120. $3(x + 5)(8x - 3) = 0$

121. $5x(x - 4)(2x - 9) = 0$

122. $6(x + 3)(x - 4)(5x + 1) = 0$

123. $2x^2 = 12x$

124. $4x^3 - 36x = 0$

125. $(1 - x)(3x + 2) = -4x$

126. $2x(x - 12) = -40$

127. $3x^2 + 2x = 12 - 7x$

128. $2x^2 + 3x = 35$

129. $x^3 - 18x = 3x^2$

130. $19x^2 - 42x = -x^3$

131. $12x = 6x^3 + 6x^2$

132. $8x^3 + 10x^2 = 3x$

133. The sum of a number and twice its square is 105. Find the number.

134. The length of a rectangular piece of carpet is 2 meters less than 5 times its width. Find the dimensions of the carpet if its area is 16 square meters.

135. A scene from an adventure film calls for a stunt dummy to be dropped from above the second-story platform of the Eiffel Tower, a distance of 400 feet. Its height $h(t)$ at the time t seconds is given by

$$h(t) = -16t^2 + 400$$

Determine how long before the stunt dummy reaches the ground.

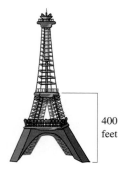

400 feet

CHAPTER 6 TEST

Simplify. Use positive exponents to write answers.

1. $(-9x)^{-2}$

2. $-3xy^{-2}(4xy^2)z$

3. $\dfrac{6^{-1}a^2b^{-3}}{3^{-2}a^{-5}b^2}$

4. $\left(\dfrac{-xy^{-5}z}{xy^3}\right)^{-5}$

Write in scientific notation.

5. 630,000,000

6. 0.01200

7. Write 5.0×10^{-6} without exponents.

8. Use scientific notation to find the quotient.

$$\frac{(0.0024)(0.00012)}{0.00032}$$

Perform the indicated operations.

9. $(4x^3 - 3x - 4) - (9x^3 + 8x + 5)$

10. $-3xy(4x + y)$

11. $(3x + 4)(4x - 7)$

12. $(5a - 2b)(5a + 2b)$

13. $(6m + n)^2$

14. $(2x - 1)(x^2 - 6x + 4)$

Factor each polynomial completely.

15. $16x^3y - 12x^2y^4$

16. $x^2 - 13x - 30$

17. $4y^2 + 20y + 25$

18. $6x^2 - 15x - 9$

19. $4x^2 - 25$

20. $x^3 + 64$

21. $3x^2y - 27y^3$

22. $6x^2 + 24$

23. $x^2y - 9y - 3x^2 + 27$

Solve the equation for the variable. Check each solution numerically or graphically.

24. $3(n - 4)(7n + 8) = 0$

25. $(x + 2)(x - 2) = 5(x + 4)$

26. $2x^3 + 5x^2 - 8x - 20 = 0$

27. Write the area of the shaded region as a factored polynomial.

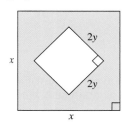

28. A pebble is hurled upward from the top of the Canada Trust Tower, which is 880 feet tall, with an initial velocity of 96 feet per second. Neglecting air resistance, the height $h(t)$ of the pebble after t seconds is given by the polynomial function

$$h(t) = -16t^2 + 96t + 880$$

  a. Find the height of the pebble when $t = 1$.

  b. Find the height of the pebble when $t = 5.1$.

  c. When will the pebble hit the ground?

CHAPTER 6 CUMULATIVE REVIEW

1. Translate each phrase to an algebraic expression. Use the variable x to represent each unknown number.

  a. Eight times a number

  b. Three more than eight times a number

  c. The quotient of a number and -7

  d. Negative one and six-tenths subtracted from twice a number

2. Find each absolute value.

  a. $|3|$ **b.** $|-5|$ **c.** $-|2|$

  d. $-|-8|$ **e.** $|0|$

3. Solve for $2(x - 3) = 5x - 9$.

4. Find the slope of the line whose equation is $f(x) = \frac{2}{3}x + 4$.

5. Determine whether the graphs of each pair of lines are parallel lines, perpendicular lines or neither.

  a. $3x + 7y = 21$ **b.** $-x + 3y = 2$ **c.** $2x - 3y = 12$
       $6x + 14y = 7$ $2x + 6y = 5$ $6x + 4y = 16$

6. Solve for x: $-(x - 3) + 2 < 3(2x - 5) + x$.

7. Graph the intersection of $x \geq 1$ and $y \geq 2x - 1$.

8. Graph the union of $x + \frac{1}{2}y \geq -4$ or $y \leq -2$.

9. Use the elimination method to solve the system.

$$\begin{cases} x - 5y = -12 \\ -x + y = 4 \end{cases}$$

10. Use matrices to solve the system.
$$\begin{cases} 2x - y = 3 \\ 4x - 2y = 5 \end{cases}$$

11. Evaluate.

a. $\begin{vmatrix} -1 & 2 \\ 3 & -4 \end{vmatrix}$ **b.** $\begin{vmatrix} 2 & 0 \\ 7 & -5 \end{vmatrix}$

12. Use the product rule to simplify.

a. $2^2 \cdot 2^5$ **b.** $x^7 x^3$ **c.** $y \cdot y^2 \cdot y^4$

13. Use the power of a power rule to simplify the following expressions. Use positive exponents to write all results.

a. $(x^5)^7$ **b.** $(2^2)^3$

c. $(5^{-1})^2$ **d.** $(y^{-3})^{-4}$

14. Find the degree of each polynomial and indicate whether the polynomial is a monomial, binomial, trinomial, or none of these.

a. $7x^3 - 3x + 2$

b. $-xyz$

c. $x^2 - 4$

d. $2xy + x^2y^2 - 5x^2 - 6$

15. Multiply.

a. $2x(5x - 4)$

b. $-3x^2(4x^2 - 6x + 1)$

c. $-xy(7x^2y + 3xy - 11)$

16. Factor $2(x - 5) + 3a(x - 5)$.

17. Factor $12x^3y - 22x^2y + 8xy$.

18. Factor $x^3 + 8$.

19. Solve $2x^2 = \frac{17}{3}x + 1$.

20. An Alpha III model rocket is launched from the ground with an A8-3 engine. Without a parachute the height of the rocket $h(t)$ at time t seconds is approximated by the function

$$h(t) = -16t^2 + 144t$$

Find how long it takes the rocket to return to the ground.

RATIONAL EXPRESSIONS

MODELING ELECTRICITY PRODUCTION

In 1994, energy produced by renewable sources, including hydroelectric, geothermal, wind and solar powers, accounted for over 8% of the United States' total production of energy (*Source*: U.S. Department of Energy). Wind energy can be harnessed by windmills to produce electricity, but it is the least utilized of these renewable energy sources. However, progressive communities are experimenting with fields of windmills for communal electricity needs, examining exactly how wind speed affects the amount of electricity produced.

IN THE CHAPTER GROUP ACTIVITY ON PAGE 449, YOU WILL HAVE THE OPPORTUNITY TO MODEL AND INVESTIGATE THE RELATIONSHIP BETWEEN WIND SPEED AND THE AMOUNT OF ELECTRICITY PRODUCED HOURLY BY A WINDMILL.

Polynomials are to algebra what integers are to arithmetic. We have added, subtracted, multiplied, and raised polynomials to powers, each operation yielding another polynomial, just as these operations on integers yield another integer. But when we divide one integer by another, the result may or may not be an integer. Likewise, when we divide one polynomial by another, we may or may not get a polynomial in return. The quotient $x \div (x + 1)$ is not a polynomial; it is a *rational expression* that can be written as $\dfrac{x}{x + 1}$.

In this chapter, we study these new algebraic forms known as rational expressions and the *rational functions* that they generate.

TAPE IAG 7.1

7.1   RATIONAL FUNCTIONS AND SIMPLIFYING RATIONAL EXPRESSIONS

O B J E C T I V E S

1 Define a rational expression and a rational function.

2 Find the domain of a rational function.

3 Write a rational expression in lowest terms.

4 Write a rational expression equivalent to a rational expression with a different denominator.

Recall that a *rational number,* or *fraction,* is a number that can be written as the quotient $\dfrac{p}{q}$ of two integers p and q as long as q is not 0. A **rational expression** is an expression that can be written as the quotient $\dfrac{P}{Q}$ of two polynomials P and Q as long as Q is not 0.

EXAMPLES OF RATIONAL EXPRESSIONS

$$\frac{3x + 7}{2} \qquad \frac{5x^2 - 3}{x - 1} \qquad \frac{7x - 2}{2x^2 + 7x + 6}$$

Rational expressions are sometimes used to describe functions. For example, we call the function $f(x) = \dfrac{x^2 + 2}{x - 3}$ a **rational function** since $\dfrac{x^2 + 2}{x - 3}$ is a rational expression.

EXAMPLE 1

FINDING UNIT COST

For the ICL Production Company, the rational function $C(x) = \dfrac{2.6x + 10,000}{x}$ describes the company's cost per disc of pressing x compact discs. Find the cost per disc for pressing:

a. 100 compact discs

b. 1000 compact discs

Solution: **a.** $C(100) = \dfrac{2.6(100) + 10{,}000}{100} = \dfrac{10{,}260}{100} = 102.6$

The cost per disc for pressing 100 compact discs is $102.60.

b. $C(1000) = \dfrac{2.6(1000) + 10{,}000}{1000} = \dfrac{12{,}600}{1000} = 12.6.$

The cost per disc for pressing 1000 compact discs is $12.60. Notice that as more compact discs are produced the cost per disc decreases.

2   As with fractions, a rational expression is **undefined** if the denominator is 0. If a variable in a rational expression is replaced with a number that makes the denominator 0, we say that the rational expression is **undefined** for this value of the variable. For example, the rational expression $\dfrac{x^2 + 2}{x - 3}$ is undefined when x is 3, because replacing x with 3 results in a denominator of 0. For this reason, we must exclude 3 from the domain of the function defined by $f(x) = \dfrac{x^2 + 2}{x - 3}$. The domain of f is then

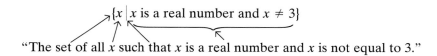

$$\{x \mid x \text{ is a real number and } x \neq 3\}$$

"The set of all x such that x is a real number and x is not equal to 3."

Unless told otherwise, we assume that the domain of a function described by an equation is the set of all real numbers for which the equation is defined.

EXAMPLE 2   Find the domain of each rational function.

a. $f(x) = \dfrac{8x^3 + 7x^2 + 20}{2}$ **b.** $g(x) = \dfrac{5x^2 - 3}{x - 1}$ **c.** $f(x) = \dfrac{7x - 2}{2x^2 + 7x + 6}$

Solution: The domain of each function will contain all real numbers except those values that make the denominator 0.

a. No matter what the value of x, the denominator of $f(x) = \dfrac{8x^3 + 7x^2 + 20}{2}$ is never 0, so the domain of f is $\{x \mid x \text{ is a real number}\}$.

b. To find the values of x that make the denominator of $g(x)$ equal to 0, we solve the equation "denominator $= 0$":

$$x - 1 = 0, \quad \text{or} \quad x = 1$$

The domain of $g(x)$ must exclude 1 since the rational expression is undefined when x is 1. The domain of g is $\{x \mid x \text{ is a real number and } x \neq 1\}$.

c. We find the domain by setting the denominator equal to 0.

$$2x^2 + 7x + 6 = 0 \qquad \text{Set the denominator equal to 0.}$$
$$(2x + 3)(x + 2) = 0 \qquad \text{Factor.}$$
$$2x + 3 = 0 \quad \text{or} \quad x + 2 = 0 \qquad \text{Set each factor equal to 0.}$$
$$x = -\frac{3}{2} \quad \text{or} \quad x = -2 \qquad \text{Solve.}$$

The domain of f is $\{x \mid x \text{ is a real number and } x \neq -\frac{3}{2} \text{ and } x \neq -2\}$. ▬▬▬

3 Recall that a fraction is in lowest terms or simplest form if the numerator and denominator have no common factors other than 1 (or −1). For example, $\frac{3}{13}$ is in lowest terms since 3 and 13 have no common factors other than 1 (or −1).

To **simplify** a rational expression, or to write it in lowest terms, we use the fundamental principle of rational expressions.

FUNDAMENTAL PRINCIPLE OF RATIONAL EXPRESSIONS

For any rational expression $\dfrac{P}{Q}$ and any polynomial R, $R \neq 0$,

$$\frac{P \cdot R}{Q \cdot R} = \frac{P}{Q}$$

Thus, the fundamental principle says that multiplying or dividing the numerator and denominator of a rational expression by the same nonzero polynomial yields an equivalent rational expression.

To simplify a rational expression such as $\dfrac{(x + 2)^2}{x^2 - 4}$, factor the numerator and the denominator and then use the fundamental principle of rational expressions to divide out common factors.

$$\frac{(x + 2)^2}{x^2 - 4} = \frac{(x + 2)(x + 2)}{(x + 2)(x - 2)}$$

$$= \frac{x + 2}{x - 2}$$

This means that the rational expression $\dfrac{(x + 2)^2}{x^2 - 4}$ has the same value as the rational expression $\dfrac{x + 2}{x - 2}$ for all values of x except 2 and −2. (Remember that when x is 2 the denominator of both rational expressions is 0 and that when x is −2 the original rational expression has a denominator of 0.) As we simplify rational expressions, we will assume that the simplified rational expression is equivalent to the original rational expression for all real numbers except those for which either denominator is 0.

In general, the following steps may be used to simplify rational expressions or to write a rational expression in lowest terms.

X	Y1	Y2
-3	.2	.2
-2	ERROR	0
-1	-.3333	-.3333
0	-1	-1
1	-3	-3
2	ERROR	ERROR
3	5	5

Y1▤(X+2)²/(X²−4)

In this table,

$$y_1 = \frac{(x + 2)^2}{x^2 - 4}$$

and $y_2 = \dfrac{x + 2}{x - 2}$.

Notice that both expressions are undefined at $x = 2$ (and an error message is given), and when $x = -2$, y_1 is not defined.

TO SIMPLIFY, OR WRITE A RATIONAL EXPRESSION IN LOWEST TERMS

Step 1.   Completely factor the numerator and denominator of the rational expression.

Step 2.   Use the fundamental principle of rational expressions to divide both the numerator and denominator by the GCF.

For now, we assume that variables in a rational expression do not represent values that make the denominator 0.

EXAMPLE 3   Write each rational expression in lowest terms.

a. $\dfrac{24x^6y^5}{8x^7y}$ **b.** $\dfrac{2x^2}{10x^3 - 2x^2}$

Solution: **a.** The GCF of the numerator and denominator is $8x^6y$.

$$\frac{24x^6y^5}{8x^7y} = \frac{(8x^6y)3y^4}{(8x^6y)x} \qquad \text{Factor the numerator and denominator.}$$

$$= \frac{3y^4}{x} \qquad \text{Apply the fundamental principle and divide out common factors.}$$

b. Factor out $2x^2$ from the denominator. Then divide the numerator and denominator by their GCF, $2x^2$.

$$\frac{2x^2}{10x^3 - 2x^2} = \frac{2x^2 \cdot 1}{2x^2(5x - 1)} = \frac{1}{5x - 1}$$

EXAMPLE 4   Write each rational expression in lowest terms.

a. $\dfrac{2 + x}{x + 2}$ **b.** $\dfrac{2 - x}{x - 2}$

Solution: **a.** By the commutative property of addition, $2 + x = x + 2$, so

$$\frac{2 + x}{x + 2} = \frac{x + 2}{x + 2} = 1$$

b. The terms in the numerator of $\dfrac{2 - x}{x - 2}$ differ by sign from the terms of the denominator, so the polynomials are opposites of each other and the expression simplifies to -1. To see this, factor out -1 from the numerator or the denominator. If -1 is factored from the numerator, then

$$\frac{2 - x}{x - 2} = \frac{-1(-2 + x)}{x - 2} = \frac{-1(x - 2)}{x - 2} = -1$$

> **REMINDER**
> When the numerator and the denominator of a rational expression are opposites of each other, the expression simplifies to -1.

a. Consider the expression $\dfrac{2x^2 - 18}{x^2 - 2x - 3}$. Graph the numerator as $y_1 = 2x^2 - 18$ and the denominator as $y_2 = x^2 - 2x - 3$ in the same window.

b. Completely factor the numerator and the denominator of the original expression.

c. What do you notice about the common factor in the expression and the point of intersection of the graphs of the numerator and denominator?

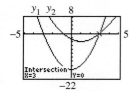

Graphs of $y_1 = 2x^2 - 18$ and $y_2 = x^2 - 2x - 3$ both intersect the x-axis at $x = 3$.

In the above discovery, we notice that both graphs intersect the x-axis at $x = 3$. In part (b), we find that $(x - 3)$ is a common factor of the numerator and denominator.

$$\frac{2x^2 - 18}{x^2 - 2x - 3} = \frac{2(x^2 - 9)}{x^2 - 2x - 3} = \frac{2(x + 3)(x - 3)}{(x + 1)(x - 3)}$$

In general, if the graphs of the numerator and denominator of a rational expression share an x-intercept c, then $(x - c)$ is a common factor of the numerator and denominator of the rational expression.

EXAMPLE 5   Write each rational expression in lowest terms.

a. $\dfrac{x^3 + 8}{2 + x}$

b. $\dfrac{2y^2 + 2}{y^3 - 5y^2 + y - 5}$

Solution: **a.** $\dfrac{x^3 + 8}{2 + x} = \dfrac{(x + 2)(x^2 - 2x + 4)}{x + 2}$ Factor the sum of the two cubes.

$= x^2 - 2x + 4$ Divide out common factors.

b. First factor the denominator by grouping.

$$y^3 - 5y^2 + y - 5 = (y^3 - 5y^2) + (y - 5)$$
$$= y^2(y - 5) + 1(y - 5)$$
$$= (y - 5)(y^2 + 1)$$

Then

$$\frac{2y^2 + 2}{y^3 - 5y^2 + y - 5} = \frac{2(y^2 + 1)}{(y - 5)(y^2 + 1)} = \frac{2}{y - 5}$$

4   The fundamental principle of rational expressions also allows us to write a rational expression as an equivalent rational expression with a different denominator. It is often necessary to do so in adding and subtracting rational expressions.

EXAMPLE 6   Write each rational expression as an equivalent rational expression with the given denominator.

a. $\dfrac{3x}{2y}$,   denominator $10xy^3$

b. $\dfrac{3x + 1}{x - 5}$,   denominator $2x^2 - 11x + 5$

Solution: **a.** $\dfrac{3x}{2y} = \dfrac{?}{10xy^3}$

If the denominator $2y$ is multiplied by $5xy^2$, the result is the given denominator $10xy^3$.

$$2y(5xy^2) \;=\; 10xy^3$$

$$\underset{\substack{\text{original}\\ \text{denominator}}}{\uparrow} \qquad \underset{\substack{\text{given}\\ \text{denominator}}}{\uparrow}$$

Use the fundamental principle of rational expressions and multiply the numerator and the denominator of the original rational expression by $5xy^2$. Then

$$\frac{3x}{2y} = \frac{3x(5xy^2)}{2y(5xy^2)} = \frac{15x^2y^2}{10xy^3}$$

b. The factored form of the given denominator, $2x^2 - 11x + 5$, is $(x - 5)(2x - 1)$.

$$\frac{3x + 1}{x - 5} = \frac{?}{2x^2 - 11x + 5} = \frac{?}{(x - 5)(2x - 1)}$$

Use the fundamental principle of rational expressions and multiply the numerator and denominator of the original rational expression by $2x - 1$.

$$\frac{3x + 1}{x - 5} = \frac{(3x + 1)(2x - 1)}{(x - 5)(2x - 1)} = \frac{6x^2 - x - 1}{(x - 5)(2x - 1)}$$

To prepare for adding and subtracting rational expressions, we multiply the binomials in the numerator but leave the denominator in factored form.

EXAMPLE 7 Find the domain of $f(x) = \dfrac{7x + 2}{x - 3}$. Confirm your result by graphing $f(x)$.

Solution: Recall that the domain of $f(x) = \dfrac{7x + 2}{x - 3}$ is the set of all real numbers for which $\dfrac{7x + 2}{x - 3}$ is defined. Since this expression is not defined when x is 3, the domain of $f(x)$ is $\{x \mid x \text{ is a real number and } x \neq 3\}$.

To confirm this domain, graph $y_1 = \dfrac{7x + 2}{x - 3}$. The domain of $f(x)$ does not include 3, so the graph of $f(x)$ does not exist at $x = 3$. If we graph $f(x)$ in dot mode, the graph shows that the function is undefined at $x = 3$ and no ordered pair solutions exist with an x-value of 3.

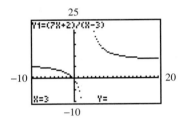

In dot mode the graphing utility plots only the ordered pairs that make the sentence true. Notice y_1 is undefined when $x = 3$ and thus a point does not exist on the graph of the function.

The graph of $f(x) = \dfrac{7x + 2}{x - 3}$ confirms that the domain is $\{x \mid x \text{ is a real number and } x \neq 3\}$.

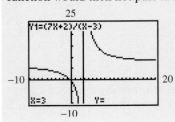

EXERCISE SET 7.1

Find each function value. See Example 1.

1. $f(x) = \dfrac{x + 8}{2x - 1}$; $f(2), f(0), f(-1)$

2. $f(y) = \dfrac{y - 2}{-5 + y}$; $f(-5), f(0), f(10)$

3. $g(x) = \dfrac{x^2 + 8}{x^3 - 25x}$; $g(3), g(-2), g(1)$

4. $s(t) = \dfrac{t^3 + 1}{t^2 + 1}$; $s(-1), s(1), s(2)$

Find the domain of each rational function. See Example 2.

5. $f(x) = \dfrac{5x - 7}{4}$

6. $g(x) = \dfrac{4 - 3x}{2}$

7. $s(t) = \dfrac{t^2 + 1}{2t}$

8. $v(t) = -\dfrac{5t + t^2}{3t}$

9. $f(x) = \dfrac{3x}{7 - x}$

10. $f(x) = \dfrac{-4x}{-2 + x}$

11. $g(x) = \dfrac{2 - 3x^2}{3x^2 + 9x}$

12. $C(x) = \dfrac{5 - 4x^2}{4x - 2x^2}$

13. $R(x) = \dfrac{3 + 2x}{x^3 + x^2 - 2x}$

14. $h(x) = \dfrac{5 - 3x}{2x^2 - 14x + 20}$

15. $C(x) = \dfrac{x + 3}{x^2 - 4}$

16. $R(x) = \dfrac{5}{x^2 - 7x}$

17. In your own words, explain how to find the domain of a rational function.

18. In your own words, explain how to simplify a rational expression or to write it in lowest terms.

Write each rational expression in lowest terms. See Examples 3–5.

19. $\dfrac{10x^3}{18x}$

20. $-\dfrac{48a^7}{16a^{10}}$

21. $\dfrac{9x^6y^3}{18x^2y^5}$

22. $\dfrac{10ab^5}{15a^3b^5}$

23. $\dfrac{8q^2}{16q^3 - 16q^2}$

24. $\dfrac{3y}{6y^2 - 30y}$

25. $\dfrac{x + 5}{5 + x}$

26. $\dfrac{x - 5}{5 - x}$

27. $\dfrac{x - 1}{1 - x^2}$

28. $\dfrac{10 + 5x}{x^2 + 2x}$

29. $\dfrac{7 - x}{x^2 - 14x + 49}$

30. $\dfrac{x^2 - 9}{2x^2 - 5x - 3}$

31. $\dfrac{4x - 8}{3x - 6}$

32. $\dfrac{12 - 6x}{30 - 15x}$

33. $\dfrac{2x - 14}{7 - x}$

34. $\dfrac{9 - x}{5x - 45}$

35. $\dfrac{x^2 - 2x - 3}{x^2 - 6x + 9}$

36. $\dfrac{x^2 + 10x + 25}{x^2 + 8x + 15}$

37. $\dfrac{2x^2 + 12x + 18}{x^2 - 9}$

38. $\dfrac{x^2 - 4}{2x^2 + 8x + 8}$

39. $\dfrac{3x + 6}{x^2 + 2x}$

40. $\dfrac{3x + 4}{9x^2 + 4}$

41. $\dfrac{x + 2}{x^2 - 4}$

42. $\dfrac{x^2 - 9}{x - 3}$

43. $\dfrac{2x^2 - x - 3}{2x^3 - 3x^2 + 2x - 3}$

44. $\dfrac{3x^2 - 5x - 2}{6x^3 + 2x^2 + 3x + 1}$

45. $\dfrac{x^4 - 16}{x^2 + 4}$

46. $\dfrac{x^2 + y^2}{x^4 - y^4}$

47. $\dfrac{x^2 + 6x - 40}{10 + x}$

48. $\dfrac{x^2 - 8x + 16}{4 - x}$

49. $\dfrac{2x^2 - 7x - 4}{x^2 - 5x + 4}$

50. $\dfrac{3x^2 - 11x + 10}{x^2 - 7x + 10}$

51. $\dfrac{x^3 - 125}{5 - x}$

52. $\dfrac{4x + 4}{2x^3 + 2}$

53. $\dfrac{8x^3 - 27}{4x - 6}$

54. $\dfrac{9x^2 - 15x + 25}{27x^3 + 125}$

55. $\dfrac{x + 5}{x^2 + 5}$

56. $\dfrac{5x}{5x^2 + 5x}$

Write each rational expression as an equivalent rational expression with the given denominator. See Example 6.

57. $\dfrac{5}{2y}$, $4y^3z$

58. $\dfrac{1}{z}$, $5z^5$

59. $\dfrac{3x}{2x - 1}$, $2x^2 + 9x - 5$

60. $\dfrac{5}{3x + 2}$, $3x^2 - 13x - 10$

61. $\dfrac{x - 2}{1}$, $x + 2$

62. $\dfrac{x - 5}{1}$, $x + 1$

63. $\dfrac{5}{m}$, $6m^3$

64. $\dfrac{3}{x}$, $3x$

65. $\dfrac{7}{m-2}$, $5m - 10$ **66.** $\dfrac{-2}{x+1}$, $10x + 10$

67. $\dfrac{y+4}{y-4}$, $y^2 - 16$ **68.** $\dfrac{5}{x-1}$, $x^2 - 1$

69. $\dfrac{12x}{x+2}$, $x^2 + 4x + 4$ **70.** $\dfrac{x+6}{x-4}$, $x^2 - 8x + 16$

71. $\dfrac{1}{x+2}$, $x^3 + 8$ **72.** $\dfrac{x}{3x-1}$, $27x^3 - 1$

73. $\dfrac{a}{a+2}$, $ab - 3a + 2b - 6$

74. $\dfrac{5}{x-y}$, $2x^2 + 5x - 2xy - 5y$

Find the domain of each rational function. Then use a graphing utility to graph each rational function and use the graph to confirm the domain. See Example 7.

75. $f(x) = \dfrac{x+1}{x^2 - 4}$ **76.** $g(x) = \dfrac{5x}{x^2 - 9}$

77. $h(x) = \dfrac{x^2}{2x^2 + 7x - 4}$ **78.** $f(x) = \dfrac{3x+2}{4x^2 - 19x - 5}$

79. Graph a portion of the function $f(x) = \dfrac{20x}{100 - x}$.

To do so, complete the given table, plot the points, and then connect the plotted points with a smooth curve. Verify the graph with a graphing utility.

x	0	10	30	50	70	90	95	99
y or $f(x)$								

80. The domain of the function $f(x) = \dfrac{1}{x}$ is all real numbers except 0. This means that the graph of this function will be in two pieces: one piece corresponding to x-values less than 0 and one piece corresponding to x-values greater than 0. Graph the function by completing the following tables, separately plotting the points and connecting each set of plotted points with a smooth curve. Verify the graph with a graphing utility.

x	$\frac{1}{4}$	$\frac{1}{2}$	1	2	4
y or $f(x)$					

x	-4	-2	-1	$-\frac{1}{2}$	$-\frac{1}{4}$
y or $f(x)$					

81. The function $f(x) = \dfrac{100{,}000x}{100 - x}$ models the cost in dollars for removing x percent of the pollutants from a bayou in which a nearby company dumped creosol.

  a. What is the domain of $f(x)$?

  b. Find the cost of removing 30% of the pollutants from the bayou. [*Hint:* Find $f(30)$.]

  c. Find the cost of removing 60% of the pollutants and then 80% of the pollutants.

  d. Find $f(90)$, then $f(95)$, and then $f(99)$. What happens to the cost as x approaches 100%?

82. The total revenue from the sale of a popular book is approximated by the rational function $R(x) = \dfrac{1000x^2}{x^2 + 4}$ where x is the number of years since publication and $R(x)$ is the total revenue in millions of dollars.

  a. Find the total revenue at the end of the first year.

  b. Find the total revenue at the end of the second year.

  c. Find the revenue during the second year only.

83. GROUP ACTIVITY. Frozen concentrated orange juice is diluted by mixing 1 can of concentrate to 3 cans water. The percent of water in the mixture is then $\dfrac{3 \text{ cans water}}{4 \text{ total cans}} = 0.75 = 75\%$. If we continue to add cans of water, x, the percent of water y is given by

$$y = \dfrac{3+x}{4+x}.$$

Graph this rational function and use the graph to determine the following.

  a. How much water should be added to the mixture to make it 80% water?

  b. How much water should be added to the mixture to make it 90% water?

  c. Will the mixture ever be 100% water? Discuss why or why not.

Review Exercises

Perform the indicated operations. See Section 1.3.

84. $\dfrac{6}{35} \cdot \dfrac{28}{9}$

85. $\dfrac{3}{8} \cdot \dfrac{4}{27}$

86. $\dfrac{8}{35} \div \dfrac{4}{5}$

87. $\dfrac{6}{11} \div \dfrac{2}{11}$

88. $\left(\dfrac{1}{2} \cdot \dfrac{1}{4}\right) \div \dfrac{3}{8}$

89. $\dfrac{3}{7} \cdot \left(\dfrac{1}{5} \div \dfrac{3}{10}\right)$

Recall that a **histogram** *is a special type of bar graph in which the width of the bar stands for a single number or a range of numbers and the height of each bar represents the number of occurrences. For example, the width of each bar in the following histogram represents a range of ages, and the height represents the number of occurrences, or percents. Use this graph to answer Exercises 90 through 92.*

90. What percent of people aged 45 and over have no health insurance?

91. Find the age group of people who have the lowest percent of no health insurance.

92. What percent of people aged 18 to 34 have no health insurance?

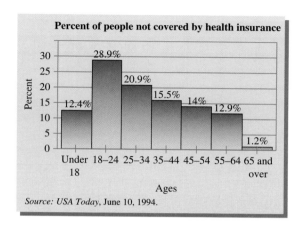

Percent of people not covered by health insurance

Source: *USA Today*, June 10, 1994.

7.2

MULTIPLYING AND DIVIDING RATIONAL EXPRESSIONS

TAPE IAG 7.2

O B J E C T I V E S

1. Multiply rational expressions.
2. Divide by a rational expression.

Arithmetic operations on rational expressions are performed in the same way as they are on rational numbers.

1

MULTIPLYING RATIONAL EXPRESSIONS

Let P, Q, R, and S be polynomials. Then

$$\frac{P}{Q} \cdot \frac{R}{S} = \frac{PR}{QS}$$

as long as $Q \neq 0$ and $S \neq 0$.

In other words, to multiply rational expressions, the product of their numerators is the numerator of the product and the product of their denominators is the denominator of the product.

EXAMPLE 1   Multiply.

a. $\dfrac{2x^3}{9y} \cdot \dfrac{y^2}{4x^3}$

b. $\dfrac{1 + 3n}{2n} \cdot \dfrac{2n - 4}{3n^2 - 2n - 1}$

Solution: **a.** $\dfrac{2x^3}{9y} \cdot \dfrac{y^2}{4x^3} = \dfrac{2x^3y^2}{36x^3y}$

To simplify, divide the numerator and the denominator by the common factor, $2x^3y$.

$$\dfrac{2x^3y^2}{36x^3y} = \dfrac{y\,(2x^3y)}{18\,(2x^3y)} = \dfrac{y}{18}$$

b. $\dfrac{1 + 3n}{2n} \cdot \dfrac{2n - 4}{3n^2 - 2n - 1} = \dfrac{1 + 3n}{2n} \cdot \dfrac{2(n - 2)}{(3n + 1)(n - 1)}$ Factor.

$$= \dfrac{(1 + 3n) \cdot 2\,(n - 2)}{2n\,(3n + 1)\,(n - 1)}$$ Multiply.

$$= \dfrac{n - 2}{n(n - 1)}$$ Divide out common factors.

When we multiply rational expressions, it is usually best to factor each numerator and denominator first. This will help when we apply the fundamental principle to write the product in lowest terms. The following steps may be used to multiply rational expressions.

TO MULTIPLY RATIONAL EXPRESSIONS

Step 1.   Completely factor the numerators and denominators.

Step 2.   Multiply the numerators and multiply the denominators.

Step 3.   Write the product in lowest terms by applying the fundamental principle of rational expressions and dividing both the numerator and the denominator by their GCF.

EXAMPLE 2   Multiply.

a. $\dfrac{2x^2 + 3x - 2}{-4x - 8} \cdot \dfrac{16x^2}{4x^2 - 1}$

b. $(ac - ad + bc - bd) \cdot \dfrac{a + b}{d - c}$

Solution: **a.** $\dfrac{2x^2 + 3x - 2}{-4x - 8} \cdot \dfrac{16x^2}{4x^2 - 1} = \dfrac{(2x - 1)(x + 2)}{-4(x + 2)} \cdot \dfrac{16x^2}{(2x + 1)(2x - 1)}$ Factor.

$$= \dfrac{4 \cdot 4x^2\,(2x - 1)(x + 2)}{-1 \cdot\ 4(x + 2)\ (2x + 1)\ (2x - 1)}$$ Multiply.

$$= -\dfrac{4x^2}{2x + 1}$$ Divide out common factors.

b. First factor $ac - ad + bc - bd$ by grouping.

$$ac - ad + bc - bd = (ac - ad) + (bc - bd)$$ Group terms.

$$= a(c - d) + b(c - d)$$

$$= (c - d)(a + b)$$

To multiply, write $(c - d)(a + b)$ as a fraction whose denominator is 1.

$$(ac - ad + bc - bd) \cdot \dfrac{a + b}{d - c} = \dfrac{(c - d)(a + b)}{1} \cdot \dfrac{a + b}{d - c}$$ Factor.

$$= \dfrac{(c - d)(a + b)(a + b)}{d - c}$$ Multiply.

Write $(c - d)$ as $-1(d - c)$ and simplify.

$$\dfrac{-1\,(d - c)\,(a + b)(a + b)}{d - c} = -(a + b)^2$$

To divide by a rational expression, multiply by its reciprocal. Recall that two numbers are reciprocals of each other if their product is 1. Similarly, if $\dfrac{P}{Q}$ is a rational expression, then $\dfrac{Q}{P}$ is its **reciprocal**, since

$$\dfrac{P}{Q} \cdot \dfrac{Q}{P} = \dfrac{P \cdot Q}{Q \cdot P} = 1$$

The following are examples of expressions and their reciprocals.

EXPRESSION	RECIPROCAL
$\dfrac{3}{x}$	$\dfrac{x}{3}$
$\dfrac{2 + x^2}{4x - 3}$	$\dfrac{4x - 3}{2 + x^2}$
x^3	$\dfrac{1}{x^3}$
0	no reciprocal

Division of rational expressions is defined as follows:

> **DIVIDING RATIONAL EXPRESSIONS**
> Let P, Q, R, and S be polynomials. Then
>
> $$\frac{P}{Q} \div \frac{R}{S} = \frac{P}{Q} \cdot \frac{S}{R} = \frac{PS}{QR}$$
>
> as long as $Q \neq 0$, $S \neq 0$, and $R \neq 0$.

Notice that division of rational expressions is the same as for rational numbers.

EXAMPLE 3 Divide.

a. $\dfrac{3x}{5y} \div \dfrac{9y}{x^5}$

b. $\dfrac{8m^2}{3m^2 - 12} \div \dfrac{40}{2 - m}$

Solution:

a. $\dfrac{3x}{5y} \div \dfrac{9y}{x^5} = \dfrac{3x}{5y} \cdot \dfrac{x^5}{9y}$ Multiply by the reciprocal of the divisor.

    $= \dfrac{x^6}{15y^2}$ Simplify.

b. $\dfrac{8m^2}{3m^2 - 12} \div \dfrac{40}{2 - m} = \dfrac{8m^2}{3m^2 - 12} \cdot \dfrac{2 - m}{40}$ Multiply by the reciprocal of the divisor.

    $= \dfrac{8m^2(2 - m)}{3(m + 2)(m - 2) \cdot 40}$ Factor and multiply.

    $= \dfrac{8\,m^2 \cdot -1(m - 2)}{3(m + 2)(m - 2) \cdot 8 \cdot 5}$ Write $(2 - m)$ as $-1(m - 2)$.

    $= -\dfrac{m^2}{15(m + 2)}$ Simplify.

> REMINDER   When dividing rational expressions, do not divide out common factors until the division problem is rewritten as a multiplication problem.

EXAMPLE 4 Divide: $\dfrac{8x^3 + 125}{x^4 + 5x^2 + 4} \div \dfrac{2x + 5}{2x^2 + 8}$

Solution: $\dfrac{8x^3 + 125}{x^4 + 5x^2 + 4} \div \dfrac{2x + 5}{2x^2 + 8} = \dfrac{8x^3 + 125}{x^4 + 5x^2 + 4} \cdot \dfrac{2x^2 + 8}{2x + 5}$

$$= \frac{(2x + 5)(4x^2 - 10x + 25) \cdot 2(x^2 + 4)}{(x^2 + 1)(x^2 + 4) \cdot (2x + 5)}$$

$$= \frac{2(4x^2 - 10x + 25)}{x^2 + 1}$$

As we have seen in earlier chapters, it is possible to add, subtract, and multiply functions. It is also possible to divide functions. Although we have not stated it as such, the sums, differences, products, and quotients of functions are themselves functions. For example, if $f(x) = \dfrac{3}{x}$ and $g(x) = \dfrac{x + 1}{5}$, their product, $f(x) \cdot g(x) = \dfrac{3}{x} \cdot \dfrac{x + 1}{5} = \dfrac{3(x + 1)}{5x}$, is a new function as long as the denominator is not 0. We can use the notation $(f \cdot g)(x)$ to denote this new function. Finding the sum, difference, product, and quotient of functions to generate new functions is called the **algebra of functions**.

This *algebra* of functions is defined next.

ALGEBRA OF FUNCTIONS

Let f and g be functions. Their **sum**, written $f + g$, is defined by

$$(f + g)(x) = f(x) + g(x)$$

Their **difference**, written as $f - g$, is defined by

$$(f - g)(x) = f(x) - g(x)$$

Their **product**, written as $f \cdot g$, is defined by

$$(f \cdot g)(x) = f(x) \cdot g(x)$$

Their **quotient**, written as $\dfrac{f}{g}$, is defined by

$$\left(\frac{f}{g}\right)(x) = \frac{f(x)}{g(x)}, \qquad g(x) \neq 0$$

EXAMPLE 5   If $f(x) = x - 1$ and $g(x) = 2x - 3$, find the following.

     a. $(f + g)(x)$ **b.** $(f - g)(x)$

     c. $(f \cdot g)(x)$ **d.** $\left(\dfrac{f}{g}\right)(x)$

Solution: Replace $f(x)$ by $x - 1$ and $g(x)$ by $2x - 3$. Then simplify.

     a. $(f + g)(x) = f(x) + g(x)$

$$= (x - 1) + (2x - 3)$$

$$= 3x - 4$$

     b. $(f - g)(x) = f(x) - g(x)$

$$= (x - 1) - (2x - 3)$$

$$= x - 1 - 2x + 3$$

$$= -x + 2$$

     c. $(f \cdot g)(x) = f(x) \cdot g(x)$

$$= (x - 1)(2x - 3)$$

$$= 2x^2 - 5x + 3$$

     d. $\left(\dfrac{f}{g}\right)(x) = \dfrac{f(x)}{g(x)} = \dfrac{x - 1}{2x - 3}, \quad x \neq \dfrac{3}{2}$

DISCOVER THE CONCEPT

Let $f(x) = \dfrac{1}{2}x + 2$ and $g(x) = \dfrac{1}{3}x^2 + 4$. Then

$$(f + g)(x) = f(x) + g(x)$$

$$= \left(\dfrac{1}{2}x + 2\right) + \left(\dfrac{1}{3}x^2 + 4\right)$$

$$= \dfrac{1}{3}x^2 + \dfrac{1}{2}x + 6.$$

To visualize this addition of functions with a graphing utility, graph

$$y_1 = \dfrac{1}{2}x + 2, \quad y_2 = \dfrac{1}{3}x^2 + 4, \quad \text{and} \quad y_3 = \dfrac{1}{3}x^2 + \dfrac{1}{2}x + 6 \text{ in a } [-4.7, 4.7, 1] \text{ by}$$

$[-1, 12, 1]$ window.

For a single x-value, trace to y_1, y_2, and y_3 to find corresponding y-values. For example, in the screen below, we choose the x-value 3 and trace to y_1, y_2, and y_3 to find corresponding y-values as shown. Repeat this process for other x-values until a pattern is discovered.

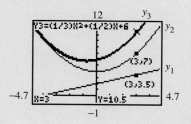

From the discovery above, we notice an interesting but not surprising relationship between the graphs of functions and the graph of their sum. For example, the graph of $(f + g)(x)$ can be found by adding the graph of $f(x)$ to the graph of $g(x)$. We add two graphs by adding corresponding y-values.

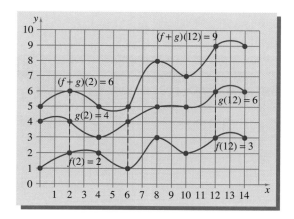

Similar statements can be made about the graphs of $(f - g)(x)$, $(f \cdot g)(x)$, and $\left(\dfrac{f}{g}\right)(x)$.

EXERCISE SET 7.2

Multiply as indicated. Write all answers in lowest terms. See Examples 1 and 2.

1. $\dfrac{4}{x} \cdot \dfrac{x^2}{8}$

2. $\dfrac{x}{3} \cdot \dfrac{9}{x^3}$

3. $\dfrac{2a^2b}{6ac} \cdot \dfrac{3c^2}{4ab}$

4. $\dfrac{5ab^4}{6abc} \cdot \dfrac{2bc^2}{10ab^2}$

5. $\dfrac{2x}{5} \cdot \dfrac{5x + 10}{6(x + 2)}$

6. $\dfrac{3x}{7} \cdot \dfrac{14 - 7x}{9(2 - x)}$

7. $\dfrac{2x - 4}{15} \cdot \dfrac{6}{2 - x}$

8. $\dfrac{10 - 2x}{7} \cdot \dfrac{14}{5x - 25}$

9. $\dfrac{18a - 12a^2}{4a^2 + 4a + 1} \cdot \dfrac{4a^2 + 8a + 3}{4a^2 - 9}$

10. $\dfrac{a - 5b}{a^2 + ab} \cdot \dfrac{b^2 - a^2}{10b - 2a}$

11. $\dfrac{x^2 - 6x - 16}{2x^2 - 128} \cdot \dfrac{x^2 + 16x + 64}{3x^2 + 30x + 48}$

12. $\dfrac{2x^2 + 12x - 32}{x^2 + 16x + 64} \cdot \dfrac{x^2 + 10x + 16}{x^2 - 3x - 10}$

13. $\dfrac{4x + 8}{x + 1} \cdot \dfrac{2 - x}{3x - 15} \cdot \dfrac{2x^2 - 8x - 10}{x^2 - 4}$

14. $\dfrac{3x - 15}{2 - x} \cdot \dfrac{x + 1}{4x + 8} \cdot \dfrac{x^2 - 4}{2x^2 - 8x - 10}$

Solve.

📋 **15.** Find the area of the rectangle.

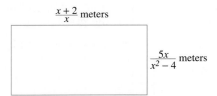

$\frac{x+2}{x}$ meters

$\frac{5x}{x^2-4}$ meters

📋 **16.** Find the area of the triangle.

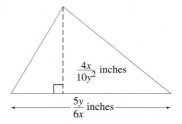

$\frac{4x}{10y^2}$ inches

$\frac{5y}{6x}$ inches

Divide as indicated. Write all answers in lowest terms. See Examples 3 and 4.

17. $\dfrac{4}{x} \div \dfrac{8}{x^2}$

18. $\dfrac{x}{3} \div \dfrac{x^3}{9}$

19. $\dfrac{4ab}{3c^2} \div \dfrac{2a^2b}{6ac}$

20. $\dfrac{6abc}{5ab^4} \div \dfrac{2bc^2}{10ab^2}$

21. $\dfrac{2x}{5} \div \dfrac{6x + 12}{5x + 10}$

22. $\dfrac{7}{3x} \div \dfrac{14 - 7x}{18 - 9x}$

23. $\dfrac{2(x + y)}{5} \div \dfrac{6(x + y)}{25}$

24. $\dfrac{4}{x - 2y} \div \dfrac{9}{3x - 6y}$

25. $\dfrac{x^2 - 6x + 9}{x^2 - x - 6} \div \dfrac{x^2 - 9}{4}$

26. $\dfrac{x^2 - 4}{3x + 6} \div \dfrac{2x^2 - 8x + 8}{x^2 + 4x + 4}$

27. $\dfrac{x^2 - 6x - 16}{2x^2 - 128} \div \dfrac{x^2 + 10x + 16}{x^2 + 16x + 64}$

28. $\dfrac{a^2 - a - 6}{a^2 - 81} \div \dfrac{a^2 - 7a - 18}{4a + 36}$

29. $\dfrac{14x^4}{y^5} \div \dfrac{2x^2}{y^7} \div \dfrac{2x^4}{7y^5}$

30. $\dfrac{x^2}{7y^2} \div \left(\dfrac{2x^2}{y^7} \div \dfrac{2x^2}{7y^5} \right)$

If $f(x) = x^2 + 1$ and $g(x) = 5x$, find the following. See Example 5.

31. $(f + g)(x)$

32. $(g \cdot f)(x)$

33. $(f - g)(x)$

34. $\left(\dfrac{f}{g} \right)(x)$

35. $\left(\dfrac{g}{f} \right)(x)$

36. $(g - f)(x)$

Perform the indicated operation. Write all answers in lowest terms.

37. $\dfrac{3xy^3}{4x^3y^2} \cdot \dfrac{-8x^3y^4}{9x^4y^7}$

38. $-\dfrac{2xyz^3}{5x^2z^2} \cdot \dfrac{10xy}{x^3}$

39. $\dfrac{8a}{3a^4b^2} \div \dfrac{4b^5}{6a^2b}$

40. $\dfrac{3y^3}{14x^4} \div \dfrac{8y^3}{7x}$

41. $\dfrac{a^2b}{a^2 - b^2} \cdot \dfrac{a + b}{4a^3b}$

42. $\dfrac{3ab^2}{a^2 - 4} \cdot \dfrac{a - 2}{6a^2b^2}$

43. $\dfrac{x^2 - 9}{4} \div \dfrac{x^2 - 6x + 9}{x^2 - x - 6}$

44. $\dfrac{a - 5b}{a^2 + ab} \div \dfrac{15b - 3a}{b^2 - a^2}$

45. $\dfrac{9x + 9}{4x + 8} \cdot \dfrac{2x + 4}{3x^2 - 3}$

46. $\dfrac{x^2 - 1}{10x + 30} \cdot \dfrac{12x + 36}{3x - 3}$

47. $\dfrac{a + b}{ab} \div \dfrac{a^2 - b^2}{4a^3b}$

48. $\dfrac{6a^2b^2}{a^2 - 4} \div \dfrac{3ab^2}{a - 2}$

49. $\dfrac{2x^2 - 4x - 30}{5x^2 - 40x - 75} \div \dfrac{x^2 - 8x + 15}{x^2 - 6x + 9}$

50. $\dfrac{4a + 36}{a^2 - 7a - 18} \div \dfrac{a^2 - a - 6}{a^2 - 81}$

51. $\dfrac{2x^3 - 16}{6x^2 + 6x - 36} \cdot \dfrac{9x + 18}{3x^2 + 6x + 12}$

52. $\dfrac{x^2 - 3x + 9}{5x^2 - 20x - 105} \cdot \dfrac{x^2 - 49}{x^3 + 27}$

53. $\dfrac{15b - 3a}{b^2 - a^2} \div \dfrac{a - 5b}{ab + b^2}$

54. $\dfrac{4x + 4}{x - 1} \div \dfrac{x^2 - 4x - 5}{x^2 - 1}$

55. $\dfrac{a^3 + a^2b + a + b}{a^3 + a} \cdot \dfrac{6a^2}{2a^2 - 2b^2}$

56. $\dfrac{a^2 - 2a}{ab - 2b + 3a - 6} \cdot \dfrac{8b + 24}{3a + 6}$

57. $\dfrac{5a}{12} \cdot \dfrac{2}{25a^2} \cdot \dfrac{15a}{2}$

58. $\dfrac{4a}{7} \div \dfrac{a^2}{14} \cdot \dfrac{3}{a}$

59. $\dfrac{3x - x^2}{x^3 - 27} \div \dfrac{x}{x^2 + 3x + 9}$

60. $\dfrac{x^2 - 3x}{x^3 - 27} \div \dfrac{2x}{2x^2 + 6x + 18}$

61. $\dfrac{4a}{7} \div \left(\dfrac{a^2}{14} \cdot \dfrac{3}{a} \right)$

62. $\dfrac{a^2}{14} \cdot \dfrac{3}{a} \div \dfrac{4a}{7}$

63. $\dfrac{8b + 24}{3a + 6} \div \dfrac{ab - 2b + 3a - 6}{a^2 - 4a + 4}$

64. $\dfrac{2a^2 - 2b^2}{a^3 + a^2b + a + b} \div \dfrac{6a^2}{a^3 + a}$

65. $\dfrac{4}{x} \div \dfrac{3xy}{x^2} \cdot \dfrac{6x^2}{x^4}$

66. $\dfrac{4}{x} \cdot \dfrac{3xy}{x^2} \div \dfrac{6x^2}{x^4}$

67. $\dfrac{3x^2 - 5x - 2}{y^2 + y - 2} \cdot \dfrac{y^2 + 4y - 5}{12x^2 + 7x + 1} \div \dfrac{5x^2 - 9x - 2}{8x^2 - 2x - 1}$

68. $\dfrac{x^2 + x - 2}{3y^2 - 5y - 2} \cdot \dfrac{12y^2 + y - 1}{x^2 + 4x - 5} \div \dfrac{8y^2 - 6y + 1}{5y^2 - 9y - 2}$

69. $\dfrac{5a^2 - 20}{3a^2 - 12a} \div \dfrac{a^3 + 2a^2}{2a^2 - 8a} \cdot \dfrac{9a^3 + 6a^2}{2a^2 - 4a}$

70. $\dfrac{5a^2 - 20}{3a^2 - 12a} \div \left(\dfrac{a^3 + 2a^2}{2a^2 - 8a} \cdot \dfrac{9a^3 + 6a^2}{2a^2 - 4a} \right)$

71. $\dfrac{5x^4 + 3x^2 - 2}{x - 1} \cdot \dfrac{x + 1}{x^4 - 1}$

72. $\dfrac{3x^4 - 10x^2 - 8}{x - 2} \cdot \dfrac{3x + 6}{15x^2 + 10}$

If $f(x) = -2x$, $g(x) = x^2 + 2$, and $h(x) = 4x + 3$, find the following.

73. $(f + g)(x)$

74. $(g - f)(x)$

75. $\left(\dfrac{f}{g} \right)(x)$

76. $\left(\dfrac{g}{f} \right)(x)$

77. $(g - h)(x)$

78. $(h \cdot g)(x)$

79. $(h + f)(x)$

80. $(h - g)(x)$

81. $f(a + b)$

82. $g(a + b)$

83. $\left(\dfrac{f}{h} \right)(x)$

84. $\left(\dfrac{h}{g} \right)(x)$

85. In our definition of division for

$$\dfrac{P}{Q} \div \dfrac{R}{S}$$

we stated that $Q \neq 0$, $S \neq 0$, and $R \neq 0$. Explain why R cannot equal 0.

86. Find the polynomial in the second numerator such that the following statement is true.

$$\dfrac{x^2 - 4}{x^2 - 7x + 10} \cdot \dfrac{?}{2x^2 + 11x + 14} = 1$$

87. A parallelogram has area $\dfrac{x^2 + x - 2}{x^3}$ square feet and height $\dfrac{x^2}{x - 1}$ feet. Express the length of its base as a rational expression in x. (*Hint:* Since $A = b \cdot h$, then $b = \dfrac{A}{h}$ or $b = A \div h$.)

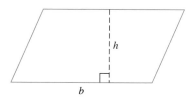

88. A lottery prize of $\dfrac{15x^3}{y^2}$ dollars is to be divided among $5x$ people. Express the amount of money each person is to receive as a rational expression in x and y.

89. Business people are concerned with cost functions, revenue functions, and profit functions. Recall that the profit $P(x)$ obtained from x units of a product is equal to the revenue $R(x)$ from selling the x units minus the cost $C(x)$ of manufacturing the x units. Write an equation expressing this relationship among $C(x)$, $R(x)$, and $P(x)$.

90. Suppose that the revenue $R(x)$ for x units of a product can be described by $R(x) = 25x$, and the cost $C(x)$ can be described by $C(x) = 50 + x^2 + 4x$. Find the profit $P(x)$ for x units.

Review Exercises

Perform the indicated operation. See Section 1.3.

91. $\dfrac{4}{5} + \dfrac{3}{5}$ **92.** $\dfrac{4}{10} - \dfrac{7}{10}$ **93.** $\dfrac{5}{28} - \dfrac{2}{21}$

94. $\dfrac{5}{13} + \dfrac{2}{7}$ **95.** $\dfrac{3}{8} + \dfrac{1}{2} - \dfrac{3}{16}$ **96.** $\dfrac{2}{9} - \dfrac{1}{6} + \dfrac{2}{3}$

A Look Ahead

EXAMPLE

Perform the following operation:

$$\frac{x^{2n} - 3x^n - 18}{x^{2n} - 9} \cdot \frac{3x^n + 9}{x^{2n}}$$

Solution:

$$\frac{x^{2n} - 3x^n - 18}{x^{2n} - 9} \cdot \frac{3x^n + 9}{x^{2n}}$$

$$= \frac{(x^n + 3)(x^n - 6) \cdot 3(x^n + 3)}{(x^n + 3)(x^n - 3) \cdot x^{2n}}$$

$$= \frac{3(x^n - 6)(x^n + 3)}{x^{2n}(x^n - 3)}$$

Perform the indicated operation. Write all answers in lowest terms. See the preceding example.

97. $\dfrac{x^{2n} - 4}{7x} \cdot \dfrac{14x^3}{x^n - 2}$

98. $\dfrac{x^{2n} + 4x^n + 4}{4x - 3} \cdot \dfrac{8x^2 - 6x}{x^n + 2}$

99. $\dfrac{y^{2n} + 9}{10y} \cdot \dfrac{y^n - 3}{y^{4n} - 81}$ **100.** $\dfrac{y^{4n} - 16}{y^{2n} + 4} \cdot \dfrac{6y}{y^n + 2}$

101. $\dfrac{y^{2n} - y^n - 2}{2y^n - 4} \div \dfrac{y^{2n} - 1}{1 + y^n}$

102. $\dfrac{y^{2n} + 7y^n + 10}{10} \div \dfrac{y^{2n} + 4y^n + 4}{5y^n + 25}$

7.3 | **ADDING AND SUBTRACTING RATIONAL EXPRESSIONS**

TAPE IAG 7.3

O B J E C T I V E S

1 Add or subtract rational expressions with common denominators.

2 Identify the least common denominator of two or more rational expressions.

3 Add or subtract rational expressions with unlike denominators.

1 Rational expressions, like rational numbers, can be added or subtracted. We define the sum or difference of rational expressions in the same way that we defined the sum or difference of rational numbers.

> **ADDING OR SUBTRACTING RATIONAL EXPRESSIONS WITH COMMON DENOMINATORS**
>
> If $\dfrac{P}{Q}$ and $\dfrac{R}{Q}$ are rational expressions, then
>
> $$\frac{P}{Q} + \frac{R}{Q} = \frac{P + R}{Q} \quad \text{and} \quad \frac{P}{Q} - \frac{R}{Q} = \frac{P - R}{Q}$$

To add or subtract rational expressions with common denominators, add or subtract the numerators and write the sum or difference over the common denominator.

EXAMPLE 1   Add or subtract.

a. $\dfrac{5}{7} + \dfrac{x}{7}$ b. $\dfrac{x}{4} + \dfrac{5x}{4}$

c. $\dfrac{x^2}{x + 7} - \dfrac{49}{x + 7}$ d. $\dfrac{x}{3y^2} - \dfrac{x + 1}{3y^2}$

Solution: The rational expressions have common denominators, so add or subtract their numerators and place the sum or difference over their common denominator.

a. $\dfrac{5}{7} + \dfrac{x}{7} = \dfrac{5 + x}{7}$

b. $\dfrac{x}{4} + \dfrac{5x}{4} = \dfrac{x + 5x}{4} = \dfrac{6x}{4} = \dfrac{3x}{2}$

c. $\dfrac{x^2}{x + 7} - \dfrac{49}{x + 7} = \dfrac{x^2 - 49}{x + 7}$

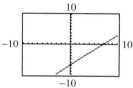

A graphing utility can be used to provide partial support of the result of operations on rational expressions. For example, the graphs of

$y_1 = \dfrac{x^2}{x + 7} - \dfrac{49}{x + 7}$

and $y_2 = x - 7$ appear to coincide. This provides partial support for the algebraic solution in part (c) of Example 1. The trace feature can be used to verify that the two graphs coincide (except when x is -7).

Next, write this rational expression in lowest terms.

$$\dfrac{x^2 - 49}{x + 7} = \dfrac{(x + 7)\,(x - 7)}{x + 7} = x - 7$$

d. $\dfrac{x}{3y^2} - \dfrac{x + 1}{3y^2} = \dfrac{x - (x + 1)}{3y^2}$ Subtract numerators.

$\qquad\qquad = \dfrac{x - x - 1}{3y^2}$ Apply the distributive property.

$\qquad\qquad = -\dfrac{1}{3y^2}$ Simplify.

2 To add or subtract rational expressions with unlike denominators, first write the rational expressions as equivalent rational expressions with common denominators.

The **least common denominator (LCD)** is usually the easiest common denominator to work with. The LCD of a list of rational expressions is a polynomial of least degree whose factors include the denominator factors in the list.

Use the following steps to find the LCD.

TO FIND THE LEAST COMMON DENOMINATOR (LCD)

Step 1. Factor each denominator completely.

Step 2. The LCD is the product of all unique factors formed in *step 1*, each raised to a power equal to the greatest number of times that the factor appears in any one factored denominator.

EXAMPLE 2 Find the LCD of the rational expressions in each list.

a. $\dfrac{2}{15x^5y^2}, \dfrac{3z}{5xy^3}$

b. $\dfrac{7}{z+1}, \dfrac{z}{z-1}$

c. $\dfrac{m-1}{m^2-25}, \dfrac{2m}{2m^2-9m-5}, \dfrac{7}{m^2-10m+25}$

Solution: **a.** Factor each denominator.

$$15x^5y^2 = 3 \cdot 5 \cdot x^5 \cdot y^2$$
$$5xy^3 = 5 \cdot x \cdot y^3$$

The unique factors are 3, 5, x, and y.

The greatest number of times that 3 appears in one denominator is 1.
The greatest number of times that 5 appears in one denominator is 1.
The greatest number of times that x appears in one denominator is 5.
The greatest number of times that y appears in one denominator is 3.
The LCD is the product of $3^1 \cdot 5^1 \cdot x^5 \cdot y^3$, or $15x^5y^3$.

b. The denominators $z + 1$ and $z - 1$ do not factor further. Each factor appears once so the

$$\text{LCD} = (z + 1)(z - 1)$$

c. First, factor each denominator.

$$m^2 - 25 = (m + 5)(m - 5)$$
$$2m^2 - 9m - 5 = (2m + 1)(m - 5)$$
$$m^2 - 10m + 25 = (m - 5)(m - 5)$$

The LCD $= (m + 5)(2m + 1)(m - 5)^2$, which is the product of each unique factor raised to a power equal to the greatest number of times it appears in any one factored denominator.

3

To add or subtract rational expressions with unlike denominators, we write each rational expression as an equivalent rational expression so that their denominators are alike.

TO ADD OR SUBTRACT RATIONAL EXPRESSIONS WITH UNLIKE DENOMINATORS

Step 1.   Find the LCD of the rational expressions.

Step 2.   Write each rational expression as an equivalent rational expression whose denominator is the LCD found in *step 1*.

Step 3.   Add or subtract numerators, and write the sum or difference over the common denominator.

Step 4.   Simplify, or write the resulting rational expression in lowest terms.

EXAMPLE 3   Perform the indicated operation.

$$\textbf{a. } \frac{2}{x^2y} + \frac{5}{3x^3y} \qquad \textbf{b. } \frac{3x}{x+2} + \frac{2x}{x-2} \qquad \textbf{c. } \frac{5k}{k^2-4} - \frac{2}{k^2+k-2}$$

Solution: **a.** The LCD is $3x^3y$. Write each fraction as an equivalent fraction with denominator $3x^3y$.

$$\frac{2}{x^2y} + \frac{5}{3x^3y} = \frac{2 \cdot 3x}{x^2y \cdot 3x} + \frac{5}{3x^3y}$$

$$= \frac{6x}{3x^3y} + \frac{5}{3x^3y}$$

$$= \frac{6x+5}{3x^3y} \qquad \text{Add the numerators.}$$

b. The LCD is the product of the two denominators: $(x+2)(x-2)$.

$$\frac{3x}{x+2} + \frac{2x}{x-2} = \frac{3x \cdot (x-2)}{(x+2) \cdot (x-2)} + \frac{2x \cdot (x+2)}{(x-2) \cdot (x+2)} \qquad \begin{array}{l}\text{Write equivalent}\\\text{rational expressions.}\end{array}$$

$$= \frac{3x(x-2) + 2x(x+2)}{(x+2)(x-2)} \qquad \text{Add the numerators.}$$

$$= \frac{3x^2 - 6x + 2x^2 + 4x}{(x+2)(x-2)} \qquad \begin{array}{l}\text{Apply the distributive}\\\text{property.}\end{array}$$

$$= \frac{5x^2 - 2x}{(x+2)(x-2)} \qquad \text{Simplify the numerator.}$$

$$\textbf{c. } \frac{5k}{k^2-4} - \frac{2}{k^2+k-2}$$

$$= \frac{5k}{(k+2)(k-2)} - \frac{2}{(k+2)(k-1)} \qquad \begin{array}{l}\text{Factor each}\\\text{denominator to}\\\text{find the LCD.}\end{array}$$

X	Y1	Y2
-3	10.2	10.2
-2	ERROR	ERROR
-1	-2.333	-2.333
0	0	0
1	-1	-1
2	ERROR	ERROR
3	7.8	7.8

Y1◻3X/(X+2)+2X/...

This table of

$$y_1 = \frac{3x}{x+2} + \frac{2x}{x-2}$$

and $y_2 = \dfrac{5x^2 - 2x}{(x+2)(x-2)}$

provides partial support for the result of part (b) in Example 3.

The LCD is $(k + 2)(k - 2)(k - 1)$. Write equivalent rational expressions with the LCD as denominators.

$$\frac{5k}{(k + 2)(k - 2)} - \frac{2}{(k + 2)(k - 1)} = \frac{5k(k - 1)}{(k + 2)(k - 2)(k - 1)} - \frac{2(k - 2)}{(k + 2)(k - 1)(k - 2)}$$

$$= \frac{5k(k - 1) - 2(k - 2)}{(k + 2)(k - 2)(k - 1)} \qquad \text{Subtract the numerators.}$$

$$= \frac{5k^2 - 5k - 2k + 4}{(k + 2)(k - 2)(k - 1)}$$

$$= \frac{5k^2 - 7k + 4}{(k + 2)(k - 2)(k - 1)} \qquad \text{Simplify the numerator.}$$

Since the numerator polynomial is prime, the numerator and denominator have no common factors and this rational expression is in lowest terms.

EXAMPLE 4   Perform the indicated operation.

a. $\dfrac{x}{x - 3} - 5$ 　　　　　**b.** $\dfrac{7}{x - y} + \dfrac{3}{y - x}$

Solution: **a.** $\dfrac{x}{x - 3} - 5 = \dfrac{x}{x - 3} - \dfrac{5}{1}$ 　　　Write 5 as $\dfrac{5}{1}$.

The LCD is $x - 3$.

$$= \frac{x}{x - 3} - \frac{5 \cdot (x - 3)}{1 \cdot (x - 3)}$$

$$= \frac{x - 5(x - 3)}{x - 3} \qquad \text{Subtract.}$$

$$= \frac{x - 5x + 15}{x - 3} \qquad \text{Apply the distributive property.}$$

$$= \frac{-4x + 15}{x - 3} \qquad \text{Simplify.}$$

b. Notice that the denominators $x - y$ and $y - x$ are opposites of one another. To write equivalent rational expressions with the LCD, write one denominator such as $y - x$ as $-1(x - y)$.

$$\frac{7}{x - y} + \frac{3}{y - x} = \frac{7}{x - y} + \frac{3}{-1(x - y)}$$

$$= \frac{7}{x - y} + \frac{-3}{x - y}$$

$$= \frac{4}{x - y} \qquad \text{Add the numerators.}$$

EXAMPLE 5   Add:

$$\frac{2x - 1}{2x^2 - 9x - 5} + \frac{x + 3}{6x^2 - x - 2}$$

Solution:

$$\frac{2x - 1}{2x^2 - 9x - 5} + \frac{x + 3}{6x^2 - x - 2} = \frac{2x - 1}{(2x + 1)(x - 5)} + \frac{x + 3}{(2x + 1)(3x - 2)} \qquad \text{Factor the denominators.}$$

The LCD is $(2x + 1)(x - 5)(3x - 2)$.

$$= \frac{(2x - 1) \cdot (3x - 2)}{(2x + 1)(x - 5) \cdot (3x - 2)} + \frac{(x + 3) \cdot (x - 5)}{(2x + 1)(3x - 2) \cdot (x - 5)}$$

$$= \frac{6x^2 - 7x + 2}{(2x + 1)(x - 5)(3x - 2)} + \frac{x^2 - 2x - 15}{(2x + 1)(x - 5)(3x - 2)}$$

$$= \frac{6x^2 - 7x + 2 + x^2 - 2x - 15}{(2x + 1)(x - 5)(3x - 2)} \qquad \text{Add.}$$

$$= \frac{7x^2 - 9x - 13}{(2x + 1)(x - 5)(3x - 2)} \qquad \text{Simplify the numerator.}$$

The rational expression is in lowest terms.

EXAMPLE 6   Perform the indicated operations.

$$\frac{7}{x - 1} + \frac{2(x + 1)}{(x - 1)^2} - \frac{2}{x^2}$$

Solution: The LCD is $x^2(x - 1)^2$.

$$\frac{7}{x - 1} + \frac{2(x + 1)}{(x - 1)^2} - \frac{2}{x^2} = \frac{7 \cdot x^2(x - 1)}{(x - 1) \cdot x^2(x - 1)} + \frac{2(x + 1) \cdot x^2}{(x - 1)^2 \cdot x^2} - \frac{2 \cdot (x - 1)^2}{x^2 \cdot (x - 1)^2}$$

$$= \frac{7x^3 - 7x^2}{x^2(x - 1)^2} + \frac{2x^3 + 2x^2}{x^2(x - 1)^2} - \frac{2x^2 - 4x + 2}{x^2(x - 1)^2}$$

$$= \frac{7x^3 - 7x^2 + 2x^3 + 2x^2 - 2x^2 + 4x - 2}{x^2(x - 1)^2} \qquad \text{Watch your signs!}$$

$$= \frac{9x^3 - 7x^2 + 4x - 2}{x^2(x - 1)^2} \qquad \text{Add or subtract like terms in the numerator.}$$

Since the numerator and denominator have no common factors, this rational expression is in lowest terms.

EXERCISE SET 7.3

Perform the indicated operation. Write each answer in lowest terms. See Example 1.

1. $\dfrac{2}{x} - \dfrac{5}{x}$

2. $\dfrac{4}{x^2} + \dfrac{2}{x^2}$

3. $\dfrac{2}{x-2} + \dfrac{x}{x-2}$

4. $\dfrac{x}{5-x} + \dfrac{2}{5-x}$

5. $\dfrac{x^2}{x+2} - \dfrac{4}{x+2}$

6. $\dfrac{4}{x-2} - \dfrac{x^2}{x-2}$

7. $\dfrac{2x-6}{x^2+x-6} + \dfrac{3-3x}{x^2+x-6}$

8. $\dfrac{5x+2}{x^2+2x-8} + \dfrac{2-4x}{x^2+2x-8}$

9. Find the perimeter and the area of the square.

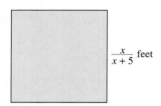

$\dfrac{x}{x+5}$ feet

10. Find the perimeter of the quadrilateral.

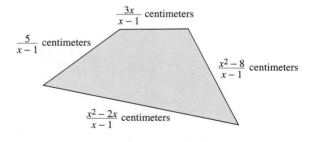

$\dfrac{3x}{x-1}$ centimeters

$\dfrac{5}{x-1}$ centimeters

$\dfrac{x^2-8}{x-1}$ centimeters

$\dfrac{x^2-2x}{x-1}$ centimeters

Find the LCD of the rational expressions in the list. See Example 2.

11. $\dfrac{2}{7}, \dfrac{3}{5x}$

12. $\dfrac{4}{5y}, \dfrac{3}{4y^2}$

13. $\dfrac{3}{x}, \dfrac{2}{x+1}$

14. $\dfrac{5}{2x}, \dfrac{7}{2+x}$

15. $\dfrac{12}{x+7}, \dfrac{8}{x-7}$

16. $\dfrac{1}{2x-1}, \dfrac{x}{2x+1}$

17. $\dfrac{5}{3x+6}, \dfrac{2x}{2x-4}$

18. $\dfrac{2}{3a+9}, \dfrac{5}{5a-15}$

19. $\dfrac{2a}{a^2-b^2}, \dfrac{1}{a^2-2ab+b^2}$

20. $\dfrac{2a}{a^2+8a+16}, \dfrac{7a}{a^2+a-12}$

21. When is the LCD of two rational expressions equal to the product of their denominators? (*Hint:* What is the LCD of $\dfrac{1}{x}$ and $\dfrac{7}{x+5}$?)

22. When is the LCD of two rational expressions with different denominators equal to one of the denominators? [*Hint:* What is the LCD of $\dfrac{3x}{x+2}$ and $\dfrac{7x+1}{(x+2)^3}$?]

Perform the indicated operation. Write each answer in the lowest terms. See Example 3.

23. $\dfrac{4}{3x} + \dfrac{3}{2x}$

24. $\dfrac{10}{7x} - \dfrac{5}{2x}$

25. $\dfrac{5}{2y^2} - \dfrac{2}{7y}$

26. $\dfrac{4}{11x^4y} - \dfrac{1}{4x^2y^3}$

27. $\dfrac{x-3}{x+4} - \dfrac{x+2}{x-4}$

28. $\dfrac{x-1}{x-5} - \dfrac{x+2}{x+5}$

29. $\dfrac{1}{x-5} + \dfrac{x}{x^2-x-20}$

30. $\dfrac{x+1}{x^2-x-20} - \dfrac{2}{x+4}$

Perform the indicated operation. Write each answer in lowest terms. See Example 4.

31. $\dfrac{1}{a-b} + \dfrac{1}{b-a}$

32. $\dfrac{1}{a-3} - \dfrac{1}{3-a}$

33. $x+1+\dfrac{1}{x-1}$

34. $5 - \dfrac{1}{x-1}$

35. $\dfrac{5}{x-2}+\dfrac{x+4}{2-x}$ **36.** $\dfrac{3}{5-x}+\dfrac{x+2}{x-5}$

Perform the indicated operation. Write each answer in lowest terms. See Examples 5 and 6.

37. $\dfrac{y+1}{y^2-6y+8}-\dfrac{3}{y^2-16}$

38. $\dfrac{x+2}{x^2-36}-\dfrac{x}{x^2+9x+18}$

39. $\dfrac{x+4}{3x^2+11x+6}+\dfrac{x}{2x^2+x-15}$

40. $\dfrac{x+3}{5x^2+12x+4}+\dfrac{6}{x^2-x-6}$

41. $\dfrac{7}{x^2-x-2}+\dfrac{x}{x^2+4x+3}$

42. $\dfrac{a}{a^2+10a+25}+\dfrac{4}{a^2+6a+5}$

43. $\dfrac{2}{x+1}-\dfrac{3x}{3x+3}+\dfrac{1}{2x+2}$

44. $\dfrac{5}{3x-6}-\dfrac{x}{x-2}+\dfrac{3+2x}{5x-10}$

45. $\dfrac{3}{x+3}+\dfrac{5}{x^2+6x+9}-\dfrac{x}{x^2-9}$

46. $\dfrac{x+2}{x^2-2x-3}+\dfrac{x}{x-3}-\dfrac{4}{x+1}$

Add or subtract as indicated. Write each answer in lowest terms.

47. $\dfrac{4}{3x^2y^3}+\dfrac{5}{3x^2y^3}$ **48.** $\dfrac{7}{2xy^4}+\dfrac{1}{2xy^4}$

49. $\dfrac{x-5}{2x}-\dfrac{x+5}{2x}$ **50.** $\dfrac{x+4}{4x}-\dfrac{x-4}{4x}$

51. $\dfrac{3}{2x+10}+\dfrac{8}{3x+15}$ **52.** $\dfrac{10}{3x-3}+\dfrac{1}{7x-7}$

53. $\dfrac{-2}{x^2-3x}-\dfrac{1}{x^3-3x^2}$ **54.** $\dfrac{-3}{2a+8}-\dfrac{8}{a^2+4a}$

55. $\dfrac{ab}{a^2-b^2}+\dfrac{b}{a+b}$ **56.** $\dfrac{x}{25-x^2}+\dfrac{2}{3x-15}$

57. $\dfrac{5}{x^2-4}-\dfrac{3}{x^2+4x+4}$ **58.** $\dfrac{3z}{z^2-9}-\dfrac{2}{3-z}$

59. $\dfrac{2}{a^2+2a+1}+\dfrac{3}{a^2-1}$

60. $\dfrac{9x+2}{3x^2-2x-8}+\dfrac{7}{3x^2+x-4}$

61. In your own words, explain how to add rational expressions with different denominators.

62. In your own words, explain how to multiply rational expressions.

63. In your own words, explain how to divide rational expressions.

64. In your own words, explain how to subtract rational expressions with different denominators.

Perform the indicated operation. Write each answer in lowest terms.

65. $\left(\dfrac{2}{3}-\dfrac{1}{x}\right)\cdot\left(\dfrac{3}{x}+\dfrac{1}{2}\right)$ **66.** $\left(\dfrac{2}{3}-\dfrac{1}{x}\right)\div\left(\dfrac{3}{x}+\dfrac{1}{2}\right)$

67. $\left(\dfrac{1}{x}+\dfrac{2}{3}\right)-\left(\dfrac{1}{x}-\dfrac{2}{3}\right)$ **68.** $\left(\dfrac{1}{2}+\dfrac{2}{x}\right)-\left(\dfrac{1}{2}-\dfrac{1}{x}\right)$

69. $\left(\dfrac{2a}{3}\right)^2\div\left(\dfrac{a^2}{a+1}-\dfrac{1}{a+1}\right)$

70. $\left(\dfrac{x+2}{2x}-\dfrac{x-2}{2x}\right)\cdot\left(\dfrac{5x}{4}\right)^2$

71. $\left(\dfrac{2x}{3}\right)^2\div\left(\dfrac{x}{3}\right)^2$ **72.** $\left(\dfrac{2x}{3}\right)^2\cdot\left(\dfrac{3}{x}\right)^2$

73. $\dfrac{x}{x^2-9}+\dfrac{3}{x^2-6x+9}-\dfrac{1}{x+3}$

74. $\dfrac{3}{x^2-9}-\dfrac{x}{x^2-6x+9}+\dfrac{1}{x+3}$

75. $\left(\dfrac{x}{x+1}-\dfrac{x}{x-1}\right)\div\dfrac{x}{2x+2}$

76. $\dfrac{x}{2x+2}\div\left(\dfrac{x}{x+1}+\dfrac{x}{x-1}\right)$

77. $\dfrac{4}{x}\cdot\left(\dfrac{2}{x+2}-\dfrac{2}{x-2}\right)$ **78.** $\dfrac{1}{x+1}\cdot\left(\dfrac{5}{x}+\dfrac{2}{x-3}\right)$

79. A student simplified $\dfrac{2x^2 + 3x}{4x^2 - 9} + \dfrac{6}{2x - 3}$ and got a result of $\dfrac{x + 6}{2x - 3}$.

a. Complete the following table. Does the table provide partial support for the student's result?

b. Verify this result algebraically.

x	−1	0	1	2	3
$\dfrac{2x^2 + 3x}{4x^2 - 9} + \dfrac{6}{2x - 3}$					
$\dfrac{x + 6}{2x - 3}$					

Use a graphing utility to provide partial support for the results of each exercise with a graph.

80. Exercise 3 **81.** Exercise 4

82. Exercise 27 **83.** Exercise 28

Use a graphing utility to provide partial support for the results of each exercise with a table.

84. Exercise 5 **85.** Exercise 6

86. Exercise 51 **87.** Exercise 52

If $f(x) = \dfrac{3x}{x^2 - 4}$ *and* $g(x) = \dfrac{6}{x^2 + 2x}$, *find the following:*

☐ 88. $(f + g)(x)$ **☐ 89.** $(f \cdot g)(x)$

☐ 90. $\left(\dfrac{f}{g}\right)(x)$ **☐ 91.** $(f - g)(x)$

Review Exercises

Use the distributive property to multiply the following. See Section 1.2.

92. $12\left(\dfrac{2}{3} + \dfrac{1}{6}\right)$ **93.** $14\left(\dfrac{1}{7} + \dfrac{3}{14}\right)$

94. $x^2\left(\dfrac{4}{x^2} + 1\right)$ **95.** $5y^2\left(\dfrac{1}{y^2} - \dfrac{1}{5}\right)$

Find each root. See Section 1.3.

96. $\sqrt{100}$ **97.** $\sqrt{25}$ **98.** $\sqrt[3]{8}$

99. $\sqrt[3]{27}$ **100.** $\sqrt[4]{81}$ **101.** $\sqrt[4]{16}$

Use the Pythagorean theorem to find each unknown length of a right triangle. See Section 6.8.

102.

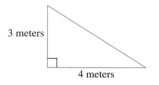

3 meters

4 meters

103.

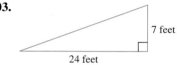

7 feet

24 feet

A Look Ahead

EXAMPLE

Add $x^{-1} + 3x^{-2}$.

Solution:

$$x^{-1} + 3x^{-2} = \frac{1}{x} + \frac{3}{x^2}$$

$$= \frac{1 \cdot x}{x \cdot x} + \frac{3}{x^2}$$

$$= \frac{x}{x^2} + \frac{3}{x^2}$$

$$= \frac{x + 3}{x^2}$$

Perform the indicated operation. See the preceding example.

104. $x^{-1} + (2x)^{-1}$ **105.** $3y^{-1} + (4y)^{-1}$

106. $4x^{-2} - 3x^{-1}$ **107.** $(4x)^{-2} - (3x)^{-1}$

108. $x^{-3}(2x + 1) - 5x^{-2}$ **109.** $4x^{-3} + x^{-4}(5x + 7)$

7.4   SIMPLIFYING COMPLEX FRACTIONS

O B J E C T I V E S

TAPE IAG 7.4

 1   Identify complex fractions.

2   Simplify complex fractions by simplifying the numerator and denominator and then dividing.

3   Simplify complex fractions by multiplying by a common denominator.

4   Simplify expressions with negative exponents.

---

1   A rational expression whose numerator, denominator, or both contain one or more rational expressions is called a **complex rational expression** or a **complex fraction**.

EXAMPLES OF COMPLEX FRACTIONS

$$\dfrac{\dfrac{1}{a}}{\dfrac{b}{2}} \qquad \dfrac{\dfrac{x}{2y^2}}{\dfrac{6x-2}{9y}} \qquad \dfrac{x+\dfrac{1}{y}}{y+1}$$

The parts of a complex fraction are

$$\dfrac{\left.\dfrac{x}{y+2}\right\}}{\left.7+\dfrac{1}{y}\right\}}$$

$\leftarrow$ Numerator of complex fraction.

$\leftarrow$ Main fraction bar.

$\leftarrow$ Denominator of complex fraction.

2   Our goal in this section is to simplify complex fractions. A complex fraction is simplified when it is in the form $\dfrac{P}{Q}$, where P and Q are polynomials that have no common factors. Two methods of simplifying complex fractions are introduced. The first method evolves from the definition of a fraction as a quotient.

TO SIMPLIFY A COMPLEX FRACTION: METHOD I

Step 1.   Simplify the numerator and the denominator of the complex fraction so that each is a single fraction.

Step 2.   Perform the indicated division by multiplying the numerator of the complex fraction by the reciprocal of the denominator of the complex fraction.

Step 3.   Simplify if possible.

EXAMPLE 1 Simplify each complex fraction.

a. $\dfrac{\dfrac{2x}{27y^2}}{\dfrac{6x^2}{9}}$ **b.** $\dfrac{\dfrac{5x}{x+2}}{\dfrac{10}{x-2}}$ **c.** $\dfrac{x+\dfrac{1}{y}}{y+\dfrac{1}{x}}$

Solution: **a.** The numerator of the complex fraction is already a single fraction, and so is the denominator. Perform the indicated division by multiplying the numerator, $\dfrac{2x}{27y^2}$, by the reciprocal of the denominator, $\dfrac{6x^2}{9}$. Then simplify.

$$\dfrac{\dfrac{2x}{27y^2}}{\dfrac{6x^2}{9}} = \dfrac{2x}{27y^2} \div \dfrac{6x^2}{9}$$

$$= \dfrac{2x}{27y^2} \cdot \dfrac{9}{6x^2} \qquad \text{Multiply by the reciprocal of } \dfrac{6x^2}{9}.$$

$$= \dfrac{2x \cdot 9}{27y^2 \cdot 6x^2}$$

$$= \dfrac{1}{9xy^2}$$

b. $\dfrac{\dfrac{5x}{x+2}}{\dfrac{10}{x-2}} = \dfrac{5x}{x+2} \cdot \dfrac{x-2}{10} \qquad \text{Multiply by the reciprocal of } \dfrac{10}{x-2}.$

$$= \dfrac{5x(x-2)}{2 \cdot 5(x+2)}$$

$$= \dfrac{x(x-2)}{2(x+2)} \qquad \text{Simplify.}$$

c. First simplify the numerator and the denominator of the complex fraction separately so that each is a single fraction.

$$\dfrac{x+\dfrac{1}{y}}{y+\dfrac{1}{x}} = \dfrac{\dfrac{x \cdot y}{1 \cdot y}+\dfrac{1}{y}}{\dfrac{y \cdot x}{1 \cdot x}+\dfrac{1}{x}} \qquad \begin{array}{l} \text{The LCD is } y. \\[2mm] \text{The LCD is } x. \end{array}$$

$$= \dfrac{\dfrac{xy+1}{y}}{\dfrac{yx+1}{x}} \qquad \begin{array}{l} \text{Add.} \\[2mm] \text{Add.} \end{array}$$

$$= \frac{xy + 1}{y} \cdot \frac{x}{xy + 1} \qquad \text{Multiply by the reciprocal of } \frac{yx + 1}{x}.$$

$$= \frac{x\,(xy + 1)}{y\,(xy + 1)}$$

$$= \frac{x}{y}$$

3 Next we look at another method of simplifying complex fractions. With this method we multiply the numerator and the denominator of the complex fraction by the LCD of all fractions in the complex fraction.

TO SIMPLIFY A COMPLEX FRACTION: METHOD II

Step 1. Multiply the numerator and the denominator of the complex fraction by the LCD of the fractions in both the numerator and the denominator.

Step 2. Simplify.

EXAMPLE 2 Simplify each complex fraction.

a. $\dfrac{\dfrac{5x}{x + 2}}{\dfrac{10}{x - 2}}$

b. $\dfrac{x + \dfrac{1}{y}}{y + \dfrac{1}{x}}$

Solution: **a.** The least common denominator of $\dfrac{5x}{x + 2}$ and $\dfrac{10}{x - 2}$ is $(x + 2)(x - 2)$. Multiply both the numerator, $\dfrac{5x}{x + 2}$, and the denominator, $\dfrac{10}{x - 2}$, by the LCD.

$$\frac{\dfrac{5x}{x + 2}}{\dfrac{10}{x - 2}} = \frac{\left(\dfrac{5x}{x + 2}\right) \cdot (\boxed{x + 2})(x - 2)}{\left(\dfrac{10}{x - 2}\right) \cdot (x + 2)(\boxed{x - 2})} \qquad \begin{array}{l}\text{Multiply numerator and}\\ \text{denominator by the LCD.}\end{array}$$

$$= \frac{\boxed{5}\,x \cdot (x - 2)}{2 \cdot \boxed{5} \cdot (x + 2)} \qquad \text{Simplify.}$$

$$= \frac{x(x - 2)}{2(x + 2)} \qquad \text{Simplify.}$$

b. The least common denominator of $\dfrac{1}{y}$ and $\dfrac{1}{x}$ is xy.

$$\dfrac{x + \dfrac{1}{y}}{y + \dfrac{1}{x}} = \dfrac{\left(x + \dfrac{1}{y}\right) \cdot xy}{\left(y + \dfrac{1}{x}\right) \cdot xy}$$

Multiply numerator and denominator by the LCD.

$$= \dfrac{x \cdot xy + \dfrac{1}{y} \cdot x\ y}{y \cdot xy + \dfrac{1}{x} \cdot x\ y}$$

Apply the distributive property.

$$= \dfrac{x^2 y + x}{xy^2 + y}$$

Simplify.

$$= \dfrac{x\,(xy + 1)}{y\,(xy + 1)}$$

Factor.

$$= \dfrac{x}{y}$$

Simplify.

4   If an expression contains negative exponents, write the expression as an equivalent expression with positive exponents.

EXAMPLE 3   Simplify.

$$\dfrac{x^{-1} + 2xy^{-1}}{x^{-2} - x^{-2}y^{-1}}$$

Solution: This fraction does not appear to be a complex fraction. If we write it by using only positive exponents, however, we see that it is a complex fraction.

$$\dfrac{x^{-1} + 2xy^{-1}}{x^{-2} - x^{-2}y^{-1}} = \dfrac{\dfrac{1}{x} + \dfrac{2x}{y}}{\dfrac{1}{x^2} - \dfrac{1}{x^2 y}}$$

The LCD of $\dfrac{1}{x}, \dfrac{2x}{y}, \dfrac{1}{x^2}$, and $\dfrac{1}{x^2 y}$ is $x^2 y$. Multiply both the numerator and denominator by $x^2 y$.

$$= \dfrac{\left(\dfrac{1}{x} + \dfrac{2x}{y}\right) \cdot x^2 y}{\left(\dfrac{1}{x^2} - \dfrac{1}{x^2 y}\right) \cdot x^2 y}$$

$$= \frac{\dfrac{1}{x} \cdot x^2 y + \dfrac{2x}{y} \cdot x^2 y}{\dfrac{1}{x^2} \cdot x^2 y - \left(\dfrac{1}{x^2 y}\right) \cdot x^2 y}$$ Apply the distributive property.

$$= \frac{xy + 2x^3}{y - 1}$$ Simplify.

Exercise Set 7.4

Simplify each complex fraction. See Examples 1 and 2.

1. $\dfrac{\dfrac{1}{3}}{\dfrac{2}{5}}$

2. $\dfrac{\dfrac{3}{5}}{\dfrac{4}{5}}$

3. $\dfrac{\dfrac{4}{x}}{\dfrac{5}{2x}}$

4. $\dfrac{\dfrac{5}{2x}}{\dfrac{4}{x}}$

5. $\dfrac{\dfrac{10}{3x}}{\dfrac{5}{6x}}$

6. $\dfrac{\dfrac{15}{2x}}{\dfrac{5}{6x}}$

7. $\dfrac{1 + \dfrac{2}{5}}{2 + \dfrac{3}{5}}$

8. $\dfrac{2 + \dfrac{1}{7}}{3 - \dfrac{4}{7}}$

9. $\dfrac{\dfrac{4}{x-1}}{\dfrac{x}{x-1}}$

10. $\dfrac{\dfrac{x}{x+2}}{\dfrac{2}{x+2}}$

11. $\dfrac{1 - \dfrac{2}{x}}{x - \dfrac{4}{9x}}$

12. $\dfrac{5 - \dfrac{3}{x}}{x + \dfrac{2}{3x}}$

13. $\dfrac{\dfrac{1}{x+1} - 1}{\dfrac{1}{x-1} + 1}$

14. $\dfrac{1 + \dfrac{1}{x-1}}{1 - \dfrac{1}{x+1}}$

Simplify. See Example 3.

15. $\dfrac{x^{-1}}{x^{-2} + y^{-2}}$

16. $\dfrac{a^{-3} + b^{-1}}{a^{-2}}$

17. $\dfrac{2a^{-1} + 3b^{-2}}{a^{-1} - b^{-1}}$

18. $\dfrac{x^{-1} + y^{-1}}{3x^{-2} + 5y^{-2}}$

19. $\dfrac{1}{x - x^{-1}}$

20. $\dfrac{x^{-2}}{x + 3x^{-1}}$

Simplify.

21. $\dfrac{\dfrac{x+1}{7}}{\dfrac{x+2}{7}}$

22. $\dfrac{\dfrac{y}{10}}{\dfrac{x+1}{10}}$

23. $\dfrac{\dfrac{1}{2} - \dfrac{1}{3}}{\dfrac{3}{4} + \dfrac{2}{5}}$

24. $\dfrac{\dfrac{5}{6} - \dfrac{1}{2}}{\dfrac{1}{3} + \dfrac{1}{8}}$

25. $\dfrac{\dfrac{x+1}{3}}{\dfrac{2x-1}{6}}$

26. $\dfrac{\dfrac{x+3}{12}}{\dfrac{4x-5}{15}}$

27. $\dfrac{\dfrac{x}{3}}{\dfrac{2}{x+1}}$

28. $\dfrac{\dfrac{x-1}{5}}{\dfrac{3}{x}}$

29. $\dfrac{\dfrac{2}{x} + 3}{\dfrac{4}{x^2} - 9}$

30. $\dfrac{2 + \dfrac{1}{x}}{4x - \dfrac{1}{x}}$

31. $\dfrac{1 - \dfrac{x}{y}}{\dfrac{x^2}{y^2} - 1}$

32. $\dfrac{1 - \dfrac{2}{x}}{x - \dfrac{4}{x}}$

33. $\dfrac{\dfrac{-2x}{x-y}}{\dfrac{y}{x^2}}$

34. $\dfrac{\dfrac{7y}{x^2 + xy}}{\dfrac{y^2}{x^2}}$

35. $\dfrac{\dfrac{2}{x} + \dfrac{1}{x^2}}{\dfrac{y}{x^2}}$

36. $\dfrac{\dfrac{5}{x^2} - \dfrac{2}{x}}{\dfrac{1}{x} + 2}$

37. $\dfrac{\dfrac{5}{9} - \dfrac{1}{x}}{1 + \dfrac{3}{x}}$

38. $\dfrac{\dfrac{x}{4} - \dfrac{4}{x}}{1 - \dfrac{4}{x}}$

39. $\dfrac{\dfrac{x-1}{x^2 - 4}}{1 + \dfrac{1}{x-2}}$

40. $\dfrac{\dfrac{2}{x+5} + \dfrac{4}{x+3}}{\dfrac{3x+13}{x^2 + 8x + 15}}$

41. $\dfrac{\dfrac{4}{5-x} + \dfrac{5}{x-5}}{\dfrac{2}{x} + \dfrac{3}{x-5}}$

42. $\dfrac{\dfrac{3}{x-4} - \dfrac{2}{4-x}}{\dfrac{2}{x-4} - \dfrac{2}{x}}$

43. $\dfrac{\dfrac{x+2}{x}-\dfrac{2}{x-1}}{\dfrac{x+1}{x}+\dfrac{x+1}{x-1}}$

44. $\dfrac{\dfrac{5}{a+2}-\dfrac{1}{a-2}}{\dfrac{3}{2+a}+\dfrac{6}{2-a}}$

45. $\dfrac{\dfrac{x-2}{x+2}+\dfrac{x+2}{x-2}}{\dfrac{x-2}{x+2}-\dfrac{x+2}{x-2}}$

46. $\dfrac{\dfrac{x-1}{x+1}-\dfrac{x+1}{x-1}}{\dfrac{x-1}{x+1}+\dfrac{x+1}{x-1}}$

47. $\dfrac{\dfrac{2}{y^2}-\dfrac{5}{xy}-\dfrac{3}{x^2}}{\dfrac{2}{y^2}+\dfrac{7}{xy}+\dfrac{3}{x^2}}$

48. $\dfrac{\dfrac{2}{x^2}-\dfrac{1}{xy}-\dfrac{1}{y^2}}{\dfrac{1}{x^2}-\dfrac{3}{xy}+\dfrac{2}{y^2}}$

49. $\dfrac{a^{-1}+1}{a^{-1}-1}$

50. $\dfrac{a^{-1}-4}{4+a^{-1}}$

51. $\dfrac{3x^{-1}+(2y)^{-1}}{x^{-2}}$

52. $\dfrac{5x^{-2}-3y^{-1}}{x^{-1}+y^{-1}}$

53. $\dfrac{2a^{-1}+(2a)^{-1}}{a^{-1}+2a^{-2}}$

54. $\dfrac{a^{-1}+2a^{-2}}{2a^{-1}+(2a)^{-1}}$

55. $\dfrac{5x^{-1}+2y^{-1}}{x^{-2}y^{-2}}$

56. $\dfrac{x^{-2}y^{-2}}{5x^{-1}+2y^{-1}}$

57. $\dfrac{5x^{-1}-2y^{-1}}{25x^{-2}-4y^{-2}}$

58. $\dfrac{3x^{-1}+3y^{-1}}{4x^{-2}-9y^{-2}}$

59. $(x^{-1}+y^{-1})^{-1}$

60. $\dfrac{xy}{x^{-1}+y^{-1}}$

61. $\dfrac{x}{1-\dfrac{1}{1+\dfrac{1}{x}}}$

62. $\dfrac{x}{1-\dfrac{1}{1-\dfrac{1}{x}}}$

In the study of calculus, the difference quotient $\dfrac{f(a+h)-f(a)}{h}$ is often found and simplified. Find and simplify this quotient for each function $f(x)$ by following steps a through d.

a. Find $f(a+h)$.

b. Find $f(a)$.

c. Use steps a and b to find $\dfrac{f(a+h)-f(a)}{h}$.

d. Simplify the result of step c.

◻ 63. $f(x)=\dfrac{1}{x}$

◻ 64. $f(x)=\dfrac{5}{x}$

◻ 65. $\dfrac{3}{x+1}$

◻ 66. $\dfrac{2}{x^2}$

Review Exercises

Simplify. See Sections 6.1 and 6.2.

67. $\dfrac{3x^3y^2}{12x}$

68. $\dfrac{-36xb^3}{9xb^2}$

69. $\dfrac{144x^5y^5}{-16x^2y}$

70. $\dfrac{48x^3y^2}{-4xy}$

Solve the following. See Sections 4.3 and 4.4.

71. $|x-5|=9$

72. $|2y+1|=1$

73. $|x-5|<9$

74. $|2x+1|\ge 1$

A Look Ahead

EXAMPLE

Simplify.

$$\dfrac{2(a+b)^{-1}-5(a-b)^{-1}}{4(a^2-b^2)^{-1}}$$

Solution:

$$\dfrac{2(a+b)^{-1}-5(a-b)^{-1}}{4(a^2-b^2)^{-1}}=\dfrac{\dfrac{2}{a+b}-\dfrac{5}{a-b}}{\dfrac{4}{a^2-b^2}}$$

$$=\dfrac{\left(\dfrac{2}{a+b}-\dfrac{5}{a-b}\right)\cdot(a+b)(a-b)}{\left[\dfrac{4}{(a+b)(a-b)}\right]\cdot(a+b)(a-b)}$$

$$=\dfrac{\dfrac{2}{a+b}\cdot(a+b)(a-b)-\dfrac{5}{a-b}\cdot(a+b)(a-b)}{\dfrac{4(a+b)(a-b)}{(a+b)(a-b)}}$$

$$=\dfrac{2(a-b)-5(a+b)}{4}$$

$$=\dfrac{-3a-7b}{4}\text{ or }-\dfrac{3a+7b}{4}$$

Simplify. See the preceding example.

75. $\dfrac{1}{1-(1-x)^{-1}}$

76. $\dfrac{1}{1+(1+x)^{-1}}$

77. $\dfrac{(x+2)^{-1}+(x-2)^{-1}}{(x^2-4)^{-1}}$

78. $\dfrac{(y-1)^{-1}-(y+4)^{-1}}{(y^2+3y-4)^{-1}}$

79. $\dfrac{3(a+1)^{-1}+4a^{-2}}{(a^3+a^2)^{-1}}$

80. $\dfrac{9x^{-1}-5(x-y)^{-1}}{4(x-y)^{-1}}$

7.5 | DIVIDING POLYNOMIALS

TAPE IAG 7.5

O B J E C T I V E S

1. Divide a polynomial by a monomial.
2. Divide by a polynomial.

Recall that a rational expression is a quotient of polynomials. An equivalent form of a rational expression can be obtained by performing the indicated division. For example, the rational expression $\dfrac{3x^5y^2 - 15x^3y - 6x}{6x^2y^3}$ can be thought of as the polynomial $3x^5y^2 - 15x^3y - 6x$ divided by the monomial $6x^2y^3$. To perform this division of a polynomial by a monomial (which we do in Example 3), recall the following addition fact for fractions with a common denominator:

$$\frac{a}{c} + \frac{b}{c} = \frac{a + b}{c}$$

If a, b, and c are monomials, we might read this equation from right to left and gain insight into dividing a polynomial by a monomial.

TO DIVIDE A POLYNOMIAL BY A MONOMIAL

Divide each term in the polynomial by the monomial:

$$\frac{a + b}{c} = \frac{a}{c} + \frac{b}{c}, \qquad c \neq 0$$

EXAMPLE 1   Divide $10x^2 - 5x + 20$ by 5.

Solution: Divide each term of $10x^2 - 5x + 20$ by 5 and simplify.

$$\frac{10x^2 - 5x + 20}{5} = \frac{10x^2}{5} - \frac{5x}{5} + \frac{20}{5} = 2x^2 - x + 4$$

To check, see that (quotient) (divisor) = dividend, or

$$(2x^2 - x + 4)\ (5) \quad = 10x^2 - 5x + 20$$

EXAMPLE 2   Find the quotient: $\dfrac{7a^2b - 2ab^2}{2ab^2}$.

Solution: Divide each term of the polynomial in the numerator by $2ab^2$.

$$\frac{7a^2b - 2ab^2}{2ab^2} = \frac{7a^2b}{2ab^2} - \frac{2ab^2}{2ab^2} = \frac{7a}{2b} - 1$$

EXAMPLE 3 Find the quotient: $\dfrac{3x^5y^2 - 15x^3y - 6x}{6x^2y^3}$

Solution: Divide each term in the numerator by $6x^2y^3$.

$$\frac{3x^5y^2 - 15x^3y - 6x}{6x^2y^3} = \frac{3x^5y^2}{6x^2y^3} - \frac{15x^3y}{6x^2y^3} - \frac{6x}{6x^2y^3} = \frac{x^3}{2y} - \frac{5x}{2y^2} - \frac{1}{xy^3}$$

2 To divide a polynomial by a polynomial other than a monomial, we use **long division.** Polynomial long division is similar to long division of real numbers. We review long division of real numbers by dividing 7 into 296.

$$
\begin{array}{r}
42 \\
\text{Divisor: } 7\overline{)296} \\
-28 \\
\hline
16 \\
-14 \\
\hline
2
\end{array}
$$

$4(7) = 28.$
Subtract and bring down the next digit in the dividend.
$2(7) = 14.$
Subtract. The remainder is 2.

The quotient is $42\dfrac{2 \text{ (remainder)}}{7 \text{ (divisor)}}$. To check, notice that

$$42(7) + 2 = 296, \qquad \text{the dividend.}$$

This same division process can be applied to polynomials, as shown next.

EXAMPLE 4 Divide $2x^2 - x - 10$ by $x + 2$.

Solution: $2x^2 - x - 10$ is the dividend, and $x + 2$ is the divisor.

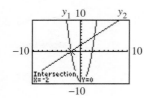

By graphing $y_1 = 2x^2 - x - 10$ and $y_2 = x + 2$ in the same window, we see that the graphs share the x-intercept $x = -2$. We know this means that $x + 2$ is a common factor of $\dfrac{2x^2 - x - 10}{x + 2}$. In other words, $x + 2$ divides into $2x^2 - x - 10$ evenly without a remainder.

Step 1. Divide $2x^2$ by x.

$$
\begin{array}{r}
2x \\
x + 2\overline{)2x^2 - x - 10}
\end{array}
$$

$\dfrac{2x^2}{x} = 2x$, so $2x$ is the first term of the quotient.

Step 2. Multiply $2x(x + 2)$.

$$
\begin{array}{r}
2x \\
x + 2\overline{)2x^2 - x - 10} \\
2x^2 + 4x
\end{array}
$$

$2x(x + 2)$
Like terms are lined up vertically.

Step 3. Subtract $(2x^2 + 4x)$ from $(2x^2 - x)$ by changing the signs of $(2x^2 + 4x)$ and adding.

$$
\begin{array}{r}
2x \\
x + 2\overline{)\;2x^2 - x - 10} \\
\underline{-2x^2 - 4x} \\
-5x
\end{array}
$$

Step 4.   Bring down the next term, -10, and start the process over.

$$
\begin{array}{r}
2x \\
x + 2\overline{)2x^2 - x - 10} \\
\underline{-2x^2 - 4x} \\
-5x - 10
\end{array}
$$

Step 5.   Divide $-5x$ by x.

$$
\begin{array}{r}
2x - 5 \\
x + 2\overline{)2x^2 - x - 10} \\
\underline{-2x^2 - 4x} \\
-5x - 10
\end{array}
$$
$\dfrac{-5x}{x} = -5$

So -5 is the second term of the quotient.

Step 6.   Multiply $-5(x + 2)$.

$$
\begin{array}{r}
2x - 5 \\
x + 2\overline{)2x^2 - x - 10} \\
\underline{-2x^2 - 4x} \\
-5x - 10 \\
\underline{-5x - 10}
\end{array}
$$
$-5(x + 2)$

Like terms are lined up vertically.

Step 7.   Subtract $(-5x - 10)$ from $(-5x - 10)$.

$$
\begin{array}{r}
2x - 5 \\
x + 2\overline{)2x^2 - x - 10} \\
\underline{-2x^2 - 4x} \\
-5x - 10 \\
\underline{+5x + 10} \\
0
\end{array}
$$

Then $\dfrac{2x^2 - x - 10}{x + 2} = 2x - 5$. There is no remainder.

Check this result by multiplying $2x - 5$ by $x + 2$. Their product is
$$(2x - 5)(x + 2) = 2x^2 - x - 10, \quad \text{the dividend}$$

EXAMPLE 5   Find the quotient: $\dfrac{6x^2 - 19x + 12}{3x - 5}$.

Solution:

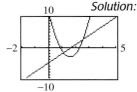

Notice that the graphs of $y_1 = 6x^2 - 19x + 12$ and $y_2 = 3x - 5$ do not have a common x-intercept. Therefore, $3x - 5$ does not divide into $6x^2 - 19x + 12$ evenly, and there will be a remainder.

$$
\begin{array}{r}
2x \\
3x - 5\overline{)6x^2 - 19x + 12} \\
\underline{6x^2 - 10x} \\
-9x + 12
\end{array}
$$

Divide: $\dfrac{6x^2}{3x} = 2x$.

Multiply: $2x(3x - 5)$.

Subtract by changing the signs of $6x^2 - 10x$ and adding. Bring down the next term, $+ 12$.

$$
\begin{array}{r}
2x - 3 \\
3x - 5\overline{)6x^2 - 19x + 12} \\
\underline{6x^2 - 10x} \\
-9x + 12 \\
\underline{-9x + 15} \\
-3
\end{array}
$$

Divide: $\dfrac{-9x}{3x} = -3$.

Multiply: $-3(3x - 5)$.

Subtract.

When checking, we call the **divisor** polynomial $3x - 5$. The **quotient** polynomial is $2x - 3$. The **remainder** polynomial is -3. See that

$$\text{dividend} = \text{divisor} \cdot \text{quotient} + \text{remainder}$$

or

$$6x^2 - 19x + 12 = (3x - 5)(2x - 3) + (-3)$$
$$= 6x^2 - 19x + 15 - 3$$
$$= 6x^2 - 19x + 12$$

The division checks, so

$$\frac{6x^2 - 19x + 12}{3x - 5} = 2x - 3 - \frac{3}{3x - 5}$$

EXAMPLE 6   Divide $2x^3 + 3x^4 - 8x + 6$ by $x^2 - 1$.

Solution: Before dividing, we write both the divisor and the dividend in descending order of exponents. Any "missing powers" can be represented by the product of 0 and the variable raised to the missing power. There is no x^2 term in the dividend, so include $0x^2$ to represent the missing term. Also, there is no x term in the divisor, so include $0x$ in the divisor.

$$
\begin{array}{r}
3x^2 + 2x + 3 \\
x^2 + 0x - 1 \overline{) 3x^4 + 2x^3 + 0x^2 - 8x + 6} \\
\underline{3x^4 + 0x^3 - 3x^2} \\
2x^3 + 3x^2 - 8x \\
\underline{2x^3 + 0x^2 - 2x} \\
3x^2 - 6x + 6 \\
\underline{3x^2 + 0x - 3} \\
-6x + 9
\end{array}
$$

$\dfrac{3x^4}{x^2} = 3x^2$.

$3x^2(x^2 + 0x - 1)$.
Subtract. Bring down $-8x$.
$2x^3/x^2 = 2x$, a term of the quotient.
$2x(x^2 + 0x - 1)$.
Subtract. Bring down 6.
$3x^2/x^2 = 3$, a term of the quotient.
$3(x^2 + 0x - 1)$.
Subtract.

The division process is finished when the degree of the remainder polynomial is less than the degree of the divisor. Thus,

$$\frac{3x^4 + 2x^3 - 8x + 6}{x^2 - 1} = 3x^2 + 2x + 3 + \frac{-6x + 9}{x^2 - 1}$$

To check, see that

$$3x^4 + 2x^3 - 8x + 6 = (x^2 - 1)(3x^2 + 2x + 3) + (-6x + 9)$$

EXAMPLE 7 Divide $27x^3 + 8$ by $2 + 3x$.

Solution: Write both the divisor and the dividend in descending order of exponents. Replace the missing terms in the dividend with $0x^2$ and $0x$.

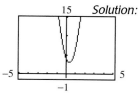

$$
\begin{array}{r}
9x^2 - 6x + 4 \\
3x + 2\overline{)27x^3 + 0x^2 + 0x + 8} \\
\underline{27x^3 + 18x^2} \qquad\qquad 9x^2(3x+2). \\
-18x^2 + 0x \qquad \text{Subtract. Bring down } 0x. \\
\underline{-18x^2 - 12x} \qquad -6x(3x+2). \\
12x + 8 \qquad \text{Subtract. Bring down } 8. \\
\underline{12x + 8} \qquad 4(3x+2).
\end{array}
$$

The graphs of
$y_1 = \dfrac{27x^3 + 8}{3x + 2}$ and
$y_2 = 9x^2 - 6x + 4$ appear to coincide. Notice that a third-degree polynomial divided by a linear polynomial is a second-degree polynomial whose graph is a parabola.

Thus, $\dfrac{27x^3 + 8}{3x + 2} = 9x^2 - 6x + 4.$

EXERCISE SET 7.5

Find each quotient. See Examples 1 through 3.

1. Divide $4a^2 + 8a$ by $2a$.

2. Divide $6x^4 - 3x^3$ by $3x^2$.

3. $\dfrac{12a^5b^2 + 16a^4b}{4a^4b}$

4. $\dfrac{4x^3y + 12x^2y^2 - 4xy^3}{4xy}$

5. $\dfrac{4x^2y^2 + 6xy^2 - 4y^2}{2y^2}$

6. $\dfrac{6x^5 + 74x^4 + 24x^3}{2x^3}$

7. $\dfrac{4x^2 + 8x + 4}{4}$

8. $\dfrac{15x^3 - 5x^2 + 10x}{5x^2}$

9. A board of length $3x^4 + 6x^2 - 18$ meters is to be cut into three pieces of the same length. Find the length of each piece.

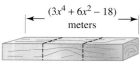

(3x⁴ + 6x² − 18) meters

10. The perimeter of a regular hexagon is given to be $12x^5 - 48x^3 + 3$ miles. Find the length of each side.

Find each quotient. See Examples 4 through 7.

11. $\dfrac{x^2 + 3x + 3}{x + 2}$

12. $\dfrac{y^2 + 7y + 10}{y + 5}$

13. $\dfrac{2x^2 - 6x - 8}{x + 1}$

14. $\dfrac{3x^2 + 19x + 20}{x + 5}$

15. Divide $2x^2 + 3x - 2$ by $2x + 4$.

16. Divide $6x^2 - 17x - 3$ by $3x - 9$.

17. $\dfrac{4x^3 + 7x^2 + 8x + 20}{2x + 4}$

18. $\dfrac{18x^3 + x^2 - 90x - 5}{9x^2 - 45}$

19. If the area of the rectangle is $15x^2 - 29x - 14$ square inches, and its length is $5x + 2$ inches, find its width.

(5x + 2) inches

20. If the area of a parallelogram is $2x^2 - 17x + 35$ square centimeters and its base is $2x - 7$ centimeters, find its height.

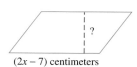

(2x − 7) centimeters

Use a graphing utility to decide if there will be a remainder when the indicated division is performed.

21. $\dfrac{3x^2 - 14x + 16}{x - 2}$ **22.** $\dfrac{3x^2 - 4x + 6}{x - 2}$

Find each quotient.

23. Divide $25a^2b^{12}$ by $10a^5b^7$.

24. Divide $12a^2b^3$ by $8a^7b$.

25. $\dfrac{x^6y^6 - x^3y^3}{x^3y^3}$

26. $\dfrac{25xy^2 + 75xyz + 125x^2yz}{-5x^2y}$

27. $\dfrac{a^2 + 4a + 3}{a + 1}$

28. $\dfrac{2x^2 + x - 10}{x - 2}$ **29.** $\dfrac{x^2 - 7x + 12}{x - 5}$

30. Divide $-16y^3 + 24y^4$ by $-4y^2$.

31. Divide $-20a^2b + 12ab^2$ by $-4ab$.

32. $\dfrac{2x^2 + 13x + 15}{x - 5}$ **33.** $\dfrac{2x^2 + 13x + 5}{2x + 3}$

34. $\dfrac{20x^2y^3 + 6xy^4 - 12x^3y^5}{2xy^3}$

35. $\dfrac{3x^2y + 6x^2y^2 + 3xy}{3xy}$ **36.** $\dfrac{6x^2 + 16x + 8}{3x + 2}$

37. $\dfrac{x^2 - 25}{x + 5}$ **38.** $\dfrac{2y^2 + 7y - 15}{2y - 3}$

39. Divide $4x^2 - 9$ by $2x - 3$.

40. Divide $8x^2 + 6x - 27$ by $4x + 9$.

41. Divide $2x^3 + 6x - 4$ by $x + 4$.

42. Divide $4x^3 - 5x$ by $2x - 1$.

43. Divide $3x^2 - 4$ by $x - 1$.

44. Divide $x^2 - 9$ by $x + 4$.

45. $\dfrac{-13x^3 + 2x^4 + 16x^2 - 9x + 20}{5 - x}$

46. $\dfrac{5x^2 - 5x + 2x^3 + 20}{4 + x}$

47. Divide $3x^5 - x^3 + 4x^2 - 12x - 8$ by $x^2 - 2$.

48. Divide $-8x^3 + 2x^4 + 19x^2 - 33x + 15$ by $x^2 - x + 5$.

49. $\dfrac{3x^3 - 5}{3x^2}$ **50.** $\dfrac{14x^3 - 2}{7x - 1}$

51. Find $P(1)$ for the polynomial function $P(x) = 3x^3 + 2x^2 - 4x + 3$. Next divide $3x^3 + 2x^2 - 4x + 3$ by $x - 1$. Compare the remainder with $P(1)$.

52. Find $P(-2)$ for the polynomial function $P(x) = x^3 - 4x^2 - 3x + 5$. Next divide $x^3 - 4x^2 - 3x + 5$ by $x + 2$. Compare the remainder with $P(-2)$.

53. Find $P(-3)$ for the polynomial $P(x) = 5x^4 - 2x^2 + 3x - 6$. Next, divide $5x^4 - 2x^2 + 3x - 6$ by $x + 3$. Compare the remainder with $P(-3)$.

54. Find $P(2)$ for the polynomial function $P(x) = -4x^4 + 2x^3 - 6x + 3$. Next, divide $-4x^4 + 2x^3 - 6x + 3$ by $x - 2$. Compare the remainder with $P(2)$.

55. Write down any patterns you noticed from Exercises 51–54.

56. Explain how to check polynomial long division.

57. Try performing the following division without changing the order of the terms. Describe why this makes the process more complicated. Then perform the division again after putting the terms in the dividend in descending order of exponents.

$$\dfrac{4x^2 - 12x - 12 + 3x^3}{x - 2}$$

Review Exercises

Insert $<$, $>$, or $=$ to make each statement true. See Section 1.3.

58. 3^2 ____ $(-3)^2$ **59.** $(-5)^2$ ____ 5^2

60. -2^3 ____ $(-2)^3$ **61.** 3^4 ____ $(-3)^4$

Solve each inequality. See Section 4.4.

62. $|x + 5| < 4$

63. $|x - 1| \le 8$

64. $|2x + 7| \ge 9$

65. $|4x + 2| > 10$

A Look Ahead

EXAMPLE   Divide $x^2 - \dfrac{7}{2}x + 4$ by $x + 2$.

Solution:

$$
\begin{array}{r}
x - \dfrac{11}{2} \\
x + 2 \overline{)\, x^2 - \dfrac{7}{2}x + 4 } \\
\underline{x^2 + 2x} \\
-\dfrac{11}{2}x + 4 \\
\underline{-\dfrac{11}{2}x - 11} \\
15
\end{array}
$$

The quotient is $x - \dfrac{11}{2} + \dfrac{15}{x + 2}$.

Find each quotient. See the preceding example.

66. $\dfrac{x^4 + \dfrac{2}{3}x^3 + x}{x - 1}$

67. $\dfrac{2x^3 + \dfrac{9}{2}x^2 - 4x - 10}{x + 2}$

68. $\dfrac{3x^4 - x - x^3 + \dfrac{1}{2}}{2x - 1}$

69. $\dfrac{2x^4 + \dfrac{1}{2}x^3 + x^2 + x}{x - 2}$

70. $\dfrac{5x^4 - 2x^2 + 10x^3 - 4x}{5x + 10}$

71. $\dfrac{9x^5 + 6x^4 - 6x^2 - 4x}{3x + 2}$

7.6 | ## SYNTHETIC DIVISION

TAPE IAG 7.6

O B J E C T I V E S

 Use synthetic division to divide a polynomial by a binomial.

Use the remainder theorem to evaluate polynomials.

When a polynomial is to be divided by a binomial of the form $x - c$, a shortcut process called **synthetic division** may be used. On the left is an example of long division, and on the right the same example showing the coefficients of the variables only.

$$
\begin{array}{r}
2x^2 + 5x + 2 \\
x - 3 \overline{)\, 2x^3 - x^2 - 13x + 1} \\
\underline{2x^3 - 6x^2} \\
5x^2 - 13x \\
\underline{5x^2 - 15x} \\
2x + 1 \\
\underline{2x - 6} \\
7
\end{array}
\qquad
\begin{array}{r}
2 \quad 5 \quad 2 \\
1 - 3 \overline{)\, 2 - 1 - 13 + 1} \\
\underline{2 - 6} \\
5 - 13 \\
\underline{5 - 15} \\
2 + 1 \\
\underline{2 - 6} \\
7
\end{array}
$$

Notice that, as long as we keep coefficients of powers of x in the same column, we can perform division of polynomials by performing algebraic operations on the coefficients only. This shortcut process of dividing with coefficients only in a special format is called synthetic division. To find $(2x^3 - x^2 - 13x + 1) \div (x - 3)$ by synthetic division, follow the next example.

EXAMPLE 1   Use synthetic division to divide $2x^3 - x^2 - 13x + 1$ by $x - 3$.

Solution: To use synthetic division, the divisor must be in the form $x - c$. Since we are dividing by $x - 3$, c is 3. Write down 3 and the coefficients of the dividend.

c

$$\begin{array}{r|rrrr} 3 & 2 & -1 & -13 & 1 \\ & \downarrow & & & \\ \hline & 2 & & & \end{array}$$

Next, draw a line and bring down the first coefficient of the dividend.

$$\begin{array}{r|rrrr} 3 & 2 & -1 & -13 & 1 \\ & & 6 & & \\ \hline & 2 & & & \end{array}$$

Multiply $3 \cdot 2$ and write down the product, 6.

$$\begin{array}{r|rrrr} 3 & 2 & -1 & -13 & 1 \\ & & 6 & & \\ \hline & 2 & 5 & & \end{array}$$

Add $-1 + 6$. Write down the sum, 5.

$$\begin{array}{r|rrrr} 3 & 2 & -1 & -13 & 1 \\ & & 6 & 15 & \\ \hline & 2 & 5 & 2 & \end{array}$$

$3 \cdot 5 = 15$

$-13 + 15 = 2$

$$\begin{array}{r|rrrr} 3 & 2 & -1 & -13 & 1 \\ & & 6 & 15 & 6 \\ \hline & 2 & 5 & 2 & 7 \end{array}$$

$3 \cdot 2 = 6$

$1 + 6 = 7$

The quotient is found in the bottom row. The numbers 2, 5, and 2 are the coefficients of the quotient polynomial, and the number 7 is the remainder. The degree of the quotient polynomial is one less than the degree of the dividend. In our example, the degree of the dividend is 3, so the degree of the quotient polynomial is 2. As we found when we performed the long division, the quotient is

$$2x^2 + 5x + 2, \qquad \text{remainder 7}$$

or

$$2x^2 + 5x + 2 + \frac{7}{x - 3}$$

EXAMPLE 2   Use synthetic division to divide $x^4 - 2x^3 - 11x^2 + 5x + 34$ by $x + 2$.

Solution: The divisor is $x + 2$, which we write in the form $x - c$ as $x - (-2)$. Thus, c is -2. The dividend coefficients are 1, -2, -11, 5, and 34.

c

$$\begin{array}{r|rrrrr} -2 & 1 & -2 & -11 & 5 & 34 \\ & & -2 & 8 & 6 & -22 \\ \hline & 1 & -4 & -3 & 11 & 12 \end{array}$$

The dividend is a fourth-degree polynomial, so the quotient polynomial is a third-degree polynomial. The quotient is $x^3 - 4x^2 - 3x + 11$ with a remainder of 12. Thus,

$$\frac{x^4 - 2x^3 - 11x^2 + 5x + 34}{x + 2} = x^3 - 4x^2 - 3x + 11 + \frac{12}{x + 2}$$

REMINDER   Before dividing by synthetic division, write the dividend in descending order of variable exponents. Any "missing powers" of the variable should be represented by 0 times the variable raised to the missing power.

DISCOVER THE CONCEPT

Let $P(x) = 2x^3 - 4x^2 + 5$

a. Find $P(2)$ by substitution.

b. Use synthetic division to find the remainder when $P(x)$ is divided by $x - 2$.

c. Compare the results of parts (a) and (b).

In the preceding discovery, we find that $P(2) = 5$ and that the remainder when $P(x)$ is divided by $x - 2$ is 5.

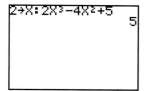

$P(2) = 5$

$$
\begin{array}{r}
c \quad\searrow \\
2\,\rvert \quad 2 \quad -4 \quad 0 \quad 5 \\
\underline{\quad\quad 4 \quad 0 \quad 0} \\
2 \quad 0 \quad 0 \quad 5 \quad\quad \text{remainder}
\end{array}
$$

This is no accident. This illustrates the **remainder theorem.**

REMAINDER THEOREM

If a polynomial $P(x)$ is divided by $x - c$, then the remainder R is $P(c)$.

EXAMPLE 3 Use the remainder theorem and synthetic division to find $P(4)$ if

$$P(x) = 4x^6 - 25x^5 + 35x^4 + 17x^2$$

Solution: To find $P(4)$ by the remainder theorem, we divide $P(x)$ by $x - 4$. The coefficients of $P(x)$ are 4, -25, 35, 0, 17, 0, and 0. Also, c is 4.

$$
\begin{array}{c}
c \\
\searrow \\
\underline{4|} \quad 4 \quad -25 \quad 35 \quad 0 \quad 17 \quad 0 \quad 0 \\
\quad\quad\quad 16 \quad -36 \quad -4 \quad -16 \quad 4 \quad 16 \\
\hline
\quad 4 \quad -9 \quad -1 \quad -4 \quad 1 \quad 4 \quad 16 \quad\text{ remainder}
\end{array}
$$

Thus, $P(4) = 16$, the remainder.

EXERCISE SET 7.6

Use synthetic division to find each quotient. See Examples 1 through 3.

1. $\dfrac{x^2 + 3x - 40}{x - 5}$

2. $\dfrac{x^2 - 14x + 24}{x - 2}$

3. $\dfrac{x^2 + 5x - 6}{x + 6}$

4. $\dfrac{x^2 + 12x + 32}{x + 4}$

5. $\dfrac{x^3 - 7x^2 - 13x + 5}{x - 2}$

6. $\dfrac{x^3 + 6x^2 - 4x - 7}{x + 5}$

7. $\dfrac{4x^2 - 9}{x - 2}$

8. $\dfrac{3x^2 - 4}{x - 1}$

For the given polynomial $P(x)$ and the given c, find $P(c)$ by (a) direct substitution and (b) the remainder theorem. See Example 3.

9. $P(x) = 3x^2 - 4x - 1$; $P(2)$

10. $P(x) = x^2 - x + 3$; $P(5)$

11. $P(x) = 4x^4 + 7x^2 + 9x - 1$; $P(-2)$

12. $P(x) = 8x^5 + 7x + 4$; $P(-3)$

13. $P(x) = x^5 + 3x^4 + 3x - 7$; $P(-1)$

14. $P(x) = 5x^4 - 4x^3 + 2x - 1$; $P(-1)$

Use synthetic division to find each quotient.

15. $\dfrac{x^3 - 3x^2 + 2}{x - 3}$

16. $\dfrac{x^2 + 12}{x + 2}$

17. $\dfrac{6x^2 + 13x + 8}{x + 1}$

18. $\dfrac{x^3 - 5x^2 + 7x - 4}{x - 3}$

19. $\dfrac{2x^4 - 13x^3 + 16x^2 - 9x + 20}{x - 5}$

20. $\dfrac{3x^4 + 5x^3 - x^2 + x - 2}{x + 2}$

21. $\dfrac{3x^2 - 15}{x + 3}$

22. $\dfrac{3x^2 + 7x - 6}{x + 4}$

23. $\dfrac{3x^3 - 6x^2 + 4x + 5}{x - \dfrac{1}{2}}$

24. $\dfrac{8x^3 - 6x^2 - 5x + 3}{x + \dfrac{3}{4}}$

25. $\dfrac{3x^3 + 2x^2 - 4x + 1}{x - \dfrac{1}{3}}$

26. $\dfrac{9y^3 + 9y^2 - y + 2}{y + \dfrac{2}{3}}$

27. $\dfrac{7x^2 - 4x + 12 + 3x^3}{x + 1}$

28. $\dfrac{x^4 + 4x^3 - x^2 - 16x - 4}{x - 2}$

29. $\dfrac{x^3 - 1}{x - 1}$

30. $\dfrac{y^3 - 8}{y - 2}$

31. $\dfrac{x^2 - 36}{x + 6}$

32. $\dfrac{4x^3 + 12x^2 + x - 12}{x + 3}$

For the given polynomial $P(x)$ and the given c, use the remainder theorem to find $P(c)$.

33. $P(x) = x^3 + 3x^2 - 7x + 4$; 1

34. $P(x) = x^3 + 5x^2 - 4x - 6$; 2

35. $P(x) = 3x^3 - 7x^2 - 2x + 5; -3$

36. $P(x) = 4x^3 + 5x^2 - 6x - 4; -2$

37. $P(x) = 4x^4 + x^2 - 2; -1$

38. $P(x) = x^4 - 3x^2 - 2x + 5; -2$

39. $P(x) = 2x^4 - 3x^2 - 2; \dfrac{1}{3}$

40. $P(x) = 4x^4 - 2x^3 + x^2 - x - 4; \dfrac{1}{2}$

41. $P(x) = x^5 + x^4 - x^3 + 3; \dfrac{1}{2}$

42. $P(x) = x^5 - 2x^3 + 4x^2 - 5x + 6; \dfrac{2}{3}$

43. Explain an advantage of using the remainder theorem instead of direct substitution.

44. Explain an advantage of using synthetic division instead of long division.

We say that 2 is a factor of 8 because 2 divides 8 evenly, or with a remainder of 0. In the same manner, the polynomial $x - 2$ is a factor of the polynomial $x^3 - 18x^2 + 24x$ because the remainder is 0 when $x^3 - 18x^2 + 24x$ is divided by $x - 2$. Use this information for Exercises 45 through 47.

45. Use synthetic division to show that $x + 3$ is a factor of $x^3 + 3x^2 + 4x + 12$.

46. Use synthetic division to show that $x - 2$ is a factor of $x^3 - 2x^2 - 3x + 6$.

47. From the remainder theorem, the polynomial $x - c$ is a factor of a polynomial function $P(x)$ if $P(c)$ is what value?

48. If a polynomial is divided by $x - 5$, the quotient is $2x^2 + 5x - 6$ and the remainder is 3. Find the original polynomial.

49. If a polynomial is divided by $x + 3$, the quotient is $x^2 - x + 10$ and the remainder is -2. Find the original polynomial.

50. If the area of a parallelogram is $x^4 - 23x^2 + 9x - 5$ square centimeters and its base is $x + 5$ centimeters, find its height.

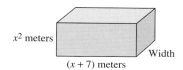

Height

$(x + 5)$ centimeters

51. If the volume of a box is $x^4 + 6x^3 - 7x^2$ cubic meters, its height is x^2 meters, and its length is $x + 7$ meters, find its width.

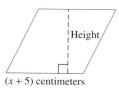

x^2 meters

Width

$(x + 7)$ meters

Review Exercises

Solve each equation for x. See Sections 2.1 and 6.8.

52. $7x + 2 = x - 3$

53. $4 - 2x = 17 - 5x$

54. $x^2 = 4x - 4$

55. $5x^2 + 10x = 15$

56. $\dfrac{x}{3} - 5 = 13$

57. $\dfrac{2x}{9} + 1 = \dfrac{7}{9}$

Factor the following. See Sections 6.5 and 6.7.

58. $x^3 - 1$

59. $8y^3 + 1$

60. $125z^3 + 8$

61. $a^3 - 27$

62. $xy + 2x + 3y + 6$

63. $x^2 - x + xy - y$

64. $x^3 - 9x$

65. $2x^3 - 32x$

7.7   SOLVING EQUATIONS CONTAINING RATIONAL EXPRESSIONS

O B J E C T I V E

1 Solve equations containing rational expressions.

1 To solve equations containing rational expressions algebraically, we first clear the equation of fractions by multiplying both sides of the equation by the LCD of all rational expressions.

EXAMPLE 1 Solve $\dfrac{8x}{5} + \dfrac{3}{2} = \dfrac{3x}{5}$ algebraically and check graphically.

Solution: The LCD of $\dfrac{8x}{5}, \dfrac{3}{2}$, and $\dfrac{3x}{5}$ is 10. Multiply both sides of the equation by 10.

$$\frac{8x}{5} + \frac{3}{2} = \frac{3x}{5}$$

$$10\left(\frac{8x}{5} + \frac{3}{2}\right) = 10\left(\frac{3x}{5}\right) \qquad \text{Multiply by the LCD.}$$

$$10 \cdot \frac{8x}{5} + 10 \cdot \frac{3}{2} = 10 \cdot \frac{3x}{5} \qquad \text{Apply the distributive property.}$$

$$16x + 15 = 6x \qquad \text{Simplify.}$$

$$15 = -10x \qquad \text{Subtract } 16x \text{ from both sides.}$$

$$-\frac{15}{10} = x \ \text{ or } \ x = -\frac{3}{2} \qquad \text{Solve.}$$

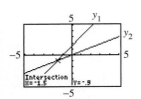

The point of intersection of $y_1 = \dfrac{8x}{5} + \dfrac{3}{2}$ and $y_2 = \dfrac{3x}{5}$ is $(-1.5, -0.9)$. Thus, the solution of the equation $\dfrac{8x}{5} + \dfrac{3}{2} = \dfrac{3x}{5}$ is $x = -1.5$ or $-\dfrac{3}{2}$.

To check this solution graphically, we use the intersection-of-graphs method shown to the left. The solution set is $\left\{-\dfrac{3}{2}\right\}$.

The important difference in the equations in this section is that the denominator of a rational expression may contain a variable. Recall that a rational expression is undefined for values of the variable that make the denominator 0. Thus, special precautions must be taken when an equation contains rational expressions with variables in the denominator. If a proposed solution makes the denominator 0, then it must be rejected as a solution. Such proposed solutions are called **extraneous solutions**.

EXAMPLE 2 Solve $\dfrac{3}{x} - \dfrac{x + 21}{3x} = \dfrac{5}{3}$ algebraically and check graphically.

Solution: The LCD of denominators x, $3x$, and 3 is $3x$. Multiply both sides by $3x$.

$$\frac{3}{x} - \frac{x + 21}{3x} = \frac{5}{3}$$

$$3x\left(\frac{3}{x} - \frac{x + 21}{3x}\right) = 3x\left(\frac{5}{3}\right)$$

$$3x\left(\frac{3}{x}\right) - 3x\left(\frac{x + 21}{3x}\right) = 3x\left(\frac{5}{3}\right) \qquad \text{Apply the distributive property.}$$

$$9 - (x + 21) = 5x \qquad \text{Simplify.}$$

$$9 - x - 21 = 5x$$

$$-12 = 6x$$

$$-2 = x \qquad \text{Solve.}$$

The proposed solution is -2. A check using a graphing utility is below.

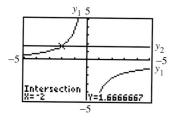

The intersection of $y_1 = \dfrac{3}{x} - \dfrac{x + 21}{3x}$ and $y_2 = \dfrac{5}{3}$ verifies that the solution of the equation

$\dfrac{3}{x} - \dfrac{x + 21}{3x} = \dfrac{5}{3}$ is $x = -2$.

The solution set is $\{-2\}$.

The following steps may be used to algebraically solve equations containing rational expressions.

TO SOLVE AN EQUATION CONTAINING RATIONAL EXPRESSIONS

Step 1.   Multiply both sides of the equation by the LCD of all rational expressions in the equation.

Step 2.   Simplify both sides.

Step 3.   Determine whether the equation is linear, quadratic, or higher degree and solve accordingly.

Step 4.   Check the solution in the original equation.

EXAMPLE 3   Solve $\dfrac{x + 6}{x - 2} = \dfrac{2(x + 2)}{x - 2}$.

Solution: First multiply both sides of the equation by the LCD, $x - 2$.

$$\frac{x + 6}{x - 2} = \frac{2(x + 2)}{x - 2}$$

$$(x - 2)\left(\frac{x + 6}{x - 2}\right) = (x - 2)\left[\frac{2(x + 2)}{x - 2}\right] \qquad \text{Multiply both sides by } x - 2.$$

This is a linear equation, so we solve by isolating the variable x.

$$x + 6 = 2(x + 2) \qquad \text{Simplify.}$$
$$x + 6 = 2x + 4 \qquad \text{Use the distributive property.}$$
$$x = 2 \qquad \text{Solve.}$$

Now check the proposed solution 2 in the *original equation.*

$$\frac{x + 6}{x - 2} = \frac{2(x + 2)}{x - 2}$$

$$\frac{2 + 6}{2 - 2} = \frac{2(2 + 2)}{2 - 2}$$

$$\frac{8}{0} = \frac{2(4)}{0}$$

The denominators are 0, since 2 is not in the domain of either rational expression in the equation. Therefore, 2 is an extraneous solution. There is no solution to the original equation. The solution set is $\emptyset$ or { }.

EXAMPLE 4   Solve $\dfrac{z}{2z^2 + 3z - 2} - \dfrac{1}{2z} = \dfrac{3}{z^2 + 2z}$.

Solution: Factor the denominators to find that the LCD is $2z(z + 2)(2z - 1)$. Multiply both sides by the LCD.

$$\frac{z}{2z^2 + 3z - 2} - \frac{1}{2z} = \frac{3}{z^2 + 2z}$$

$$\frac{z}{(2z - 1)(z + 2)} - \frac{1}{2z} = \frac{3}{z(z + 2)}$$

$$2z(z + 2)(2z - 1)\left[\frac{z}{(2z - 1)(z + 2)} - \frac{1}{2z}\right]$$

$$= 2z(z + 2)(2z - 1)\left[\frac{3}{z(z + 2)}\right]$$

$$2z(z + 2)(2z - 1)\left[\frac{z}{(2z - 1)(z + 2)}\right] - 2z(z + 2)(2z - 1)\left(\frac{1}{2z}\right)$$

$$= 2z(z + 2)(2z - 1)\left[\frac{3}{z(z + 2)}\right] \quad \begin{array}{l}\text{Apply the}\\\text{distributive}\\\text{property.}\end{array}$$

$$2z(z) - (z + 2)(2z - 1) = 3 \cdot 2(2z - 1) \qquad \text{Simplify.}$$

$$2z^2 - (2z^2 + 3z - 2) = 12z - 6$$

$$2z^2 - 2z^2 - 3z + 2 = 12z - 6$$

$$-3z + 2 = 12z - 6$$

$$-15z = -8$$

$$z = \frac{8}{15} \qquad \text{Solve.}$$

The proposed solution $\dfrac{8}{15}$ does not make any denominator 0 and a check is shown to the left. The solution set is $\left\{\dfrac{8}{15}\right\}$.

EXAMPLE 5   Solve $\dfrac{2x}{x-3} + \dfrac{6-2x}{x^2-9} = \dfrac{x}{x+3}$.

Solution: Factor the second denominator to find that the LCD is $(x+3)(x-3)$. Multiply both sides of the equation by $(x+3)(x-3)$. By the distributive property, this is the same as multiplying each term by $(x+3)(x-3)$.

$$\frac{2x}{x-3} + \frac{6-2x}{x^2-9} = \frac{x}{x+3}$$

$$(x+3)(x-3)\left(\frac{2x}{x-3}\right) + (x+3)(x-3)\left[\frac{6-2x}{(x+3)(x-3)}\right]$$

$$= (x+3)(x-3)\left(\frac{x}{x+3}\right)$$

$$2x(x+3) + (6-2x) = x(x-3) \qquad \text{Simplify.}$$
$$2x^2 + 6x + 6 - 2x = x^2 - 3x \qquad \text{Apply the distributive property.}$$

Next, we solve this quadratic equation by the factoring method. To do so, first write the equation so that one side is 0.

$$x^2 + 7x + 6 = 0$$

Next, see if the trinomial is factorable.

$$(x+6)(x+1) = 0 \qquad \text{Factor.}$$
$$x = -6 \quad \text{or} \quad x = -1 \qquad \text{Set each factor equal to 0.}$$

Neither -6 nor -1 makes any denominator 0. The solution set is $\{-6, -1\}$. ▬▬▬

EXERCISE SET 7.7

Solve each equation algebraically and check graphically. See Examples 1 and 2.

1. $\dfrac{x}{2} - \dfrac{x}{3} = 12$

2. $x = \dfrac{x}{2} - 4$

3. $\dfrac{x}{3} = \dfrac{1}{6} + \dfrac{x}{4}$

4. $\dfrac{x}{2} = \dfrac{21}{10} - \dfrac{x}{5}$

5. $\dfrac{2}{x} + \dfrac{1}{2} = \dfrac{5}{x}$

6. $\dfrac{5}{3x} + 1 = \dfrac{7}{6}$

7. $\dfrac{x+3}{x} = \dfrac{5}{x}$

8. $\dfrac{4-3x}{2x} = -\dfrac{8}{2x}$

Solve each equation. See Examples 3 through 5.

9. $\dfrac{x+5}{x+3} = \dfrac{8}{x+3}$

10. $\dfrac{5}{x-2} - \dfrac{2}{x+4} = -\dfrac{4}{x^2+2x-8}$

11. $\dfrac{1}{x-1} + \dfrac{1}{x+1} = \dfrac{2}{x^2-1}$

12. $\dfrac{1}{x-1} = \dfrac{2}{x+1}$

13. $\dfrac{6}{x+3} = \dfrac{4}{x-3}$

14. $\dfrac{1}{x-4} - \dfrac{3x}{x^2-16} = \dfrac{2}{x+4}$

15. $\dfrac{3}{2x+3} - \dfrac{1}{2x-3} = \dfrac{4}{4x^2-9}$

16. $\dfrac{1}{x-4} = \dfrac{8}{x^2-16}$

17. $\dfrac{2}{x^2-4} = \dfrac{1}{2x-4}$

18. $\dfrac{1}{x-2} - \dfrac{2}{x^2-2x} = 1$

19. $\dfrac{12}{3x^2+12x} = 1 - \dfrac{1}{x+4}$

Solve each equation.

20. $\dfrac{5}{x} = \dfrac{20}{12}$

21. $\dfrac{2}{x} = \dfrac{10}{5}$

22. $1 - \dfrac{4}{a} = 5$

23. $7 + \dfrac{6}{a} = 5$

24. $\dfrac{1}{2x} - \dfrac{1}{x + 1} = \dfrac{1}{3x^2 + 3x}$

25. $\dfrac{2}{x - 5} + \dfrac{1}{2x} = \dfrac{5}{3x^2 - 15x}$

26. $\dfrac{1}{x} - \dfrac{x}{25} = 0$

27. $\dfrac{x}{4} + \dfrac{5}{x} = 3$

28. $5 - \dfrac{2}{2y - 5} = \dfrac{3}{2y - 5}$

29. $1 - \dfrac{5}{y + 7} = \dfrac{4}{y + 7}$

30. $\dfrac{x - 1}{x + 2} = \dfrac{2}{3}$

31. $\dfrac{6x + 7}{2x + 9} = \dfrac{5}{3}$

32. $\dfrac{x + 3}{x + 2} = \dfrac{1}{x + 2}$

33. $\dfrac{2x + 1}{4 - x} = \dfrac{9}{4 - x}$

34. $\dfrac{1}{a - 3} + \dfrac{2}{a + 3} = \dfrac{1}{a^2 - 9}$

35. $\dfrac{12}{9 - a^2} + \dfrac{3}{3 + a} = \dfrac{2}{3 - a}$

36. $\dfrac{64}{x^2 - 16} + 1 = \dfrac{2x}{x - 4}$

37. $2 + \dfrac{3}{x} = \dfrac{2x}{x + 3}$

38. $\dfrac{-15}{4y + 1} + 4 = y$

39. $\dfrac{36}{x^2 - 9} + 1 = \dfrac{2x}{x + 3}$

40. $\dfrac{28}{x^2 - 9} + \dfrac{2x}{x - 3} + \dfrac{6}{x + 3} = 0$

41. $\dfrac{x^2 - 20}{x^2 - 7x + 12} = \dfrac{3}{x - 3} + \dfrac{5}{x - 4}$

42. $\dfrac{x + 2}{x^2 + 7x + 10} = \dfrac{1}{3x + 6} - \dfrac{1}{x + 5}$

43. $\dfrac{3}{2x - 5} + \dfrac{2}{2x + 3} = 0$

⃞ **44.** The average cost of producing x game disks for a computer is given by the function $f(x) = 3.3 + \dfrac{5400}{x}$.

Find the number of game disks that must be produced for the average cost to be $5.10.

⃞ **45.** The average cost of producing x electric pencil sharpeners is given by the function $f(x) = 20 + \dfrac{4000}{x}$. Find the number of electric pencil sharpeners that must be produced for the average cost to be $25.

Perform the indicated operation and simplify, or solve the equation for the variable.

46. $\dfrac{2}{x^2 - 4} = \dfrac{1}{x + 2} - \dfrac{3}{x - 2}$

47. $\dfrac{3}{x^2 - 25} = \dfrac{1}{x + 5} + \dfrac{2}{x - 5}$

48. $\dfrac{5}{x^2 - 3x} + \dfrac{4}{2x - 6}$

49. $\dfrac{5}{x^2 - 3x} \div \dfrac{4}{2x - 6}$

50. $\dfrac{x - 1}{x + 1} + \dfrac{x + 7}{x - 1} = \dfrac{4}{x^2 - 1}$

51. $\left(1 - \dfrac{y}{x}\right) \div \left(1 - \dfrac{x}{y}\right)$

52. $\dfrac{a^2 - 9}{a - 6} \cdot \dfrac{a^2 - 5a - 6}{a^2 - a - 6}$

53. $\dfrac{2}{a - 6} + \dfrac{3a}{a^2 - 5a - 6} - \dfrac{a}{5a + 5}$

54. $\dfrac{2x + 3}{3x - 2} = \dfrac{4x + 1}{6x + 1}$

55. $\dfrac{5x - 3}{2x} = \dfrac{10x + 3}{4x + 1}$

56. $\dfrac{a}{9a^2 - 1} + \dfrac{2}{6a - 2}$

57. $\dfrac{3}{4a - 8} - \dfrac{a + 2}{a^2 - 2a}$

58. $\dfrac{-3}{x^2} - \dfrac{1}{x} + 2 = 0$

59. $\dfrac{x}{2x + 6} + \dfrac{5}{x^2 - 9}$

60. $\dfrac{x - 8}{x^2 - x - 2} + \dfrac{2}{x - 2}$

61. $\dfrac{x - 8}{x^2 - x - 2} + \dfrac{2}{x - 2} = \dfrac{3}{x + 1}$

62. $\dfrac{3}{a} - 5 = \dfrac{7}{a} - 1$

63. $\dfrac{7}{3z - 9} + \dfrac{5}{z}$

Solve each equation. Begin by writing each equation with positive exponents only.

⃞ **64.** $x^{-2} - 19x^{-1} + 48 = 0$

⃞ **65.** $x^{-2} - 5x^{-1} - 36 = 0$

⃞ **66.** $p^{-2} + 4p^{-1} - 5 = 0$

⃞ **67.** $6p^{-2} - 5p^{-1} + 1 = 0$

Solve each equation. Round solutions to two decimal places.

68. $\dfrac{1.4}{x - 2.6} = \dfrac{-3.5}{x + 7.1}$

69. $\dfrac{-8.5}{x + 1.9} = \dfrac{5.7}{x - 3.6}$

70. $\dfrac{10.6}{y} - 14.7 = \dfrac{9.92}{3.2} + 7.6$

71. $\dfrac{12.2}{x} + 17.3 = \dfrac{9.6}{x} - 14.7$

Use a graphing utility to verify the solution of each given exercise.

72. Exercise 20

73. Exercise 21

74. Exercise 30

75. Exercise 31

Review Exercises

Write each sentence as an equation and solve. See Section 2.2.

76. Four more than 3 times a number is 19.

77. The sum of two consecutive integers is 147.

78. The length of a rectangle is 5 inches more than the width. Its perimeter is 50 inches. Find the length and width.

79. The sum of a number and its reciprocal is $\dfrac{5}{2}$.

Simplify the following. See Section 1.4.

80. $-|-6| - (-5)$

81. $\sqrt{49} - (10 - 6)^2$

82. $|4 - 8| + (4 - 8)$

83. $(-4)^2 - 5^2$

The following is from a survey of state prisons. Use this histogram to answer Exercises 84 through 88.

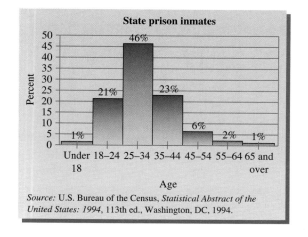

Source: U.S. Bureau of the Census, *Statistical Abstract of the United States: 1994,* 113th ed., Washington, DC, 1994.

84. What percent of state prison inmates are aged 45 to 54?

85. What percent of state prison inmates are 65 years old or older?

86. What age category shows the highest percent of prison inmates?

87. What percent of state prison inmates are 18 to 34 years old?

88. A state prison in Louisiana houses 2000 inmates. Approximately how many 25- to 34-year-old inmates might you expect this prison to hold?

A Look Ahead

EXAMPLE

Solve $\left(\dfrac{x}{x + 1}\right)^2 - 7\left(\dfrac{x}{x + 1}\right) + 10 = 0.$

Solution:

Let $u = \dfrac{x}{x + 1}$ and solve for u. Then substitute back and solve for x.

$$\left(\frac{x}{x + 1}\right)^2 - 7\left(\frac{x}{x + 1}\right) + 10 = 0$$
$$u^2 - 7u + 10 = 0 \quad \text{Let } u = \frac{x}{x + 1}.$$
$$(u - 5)(u - 2) = 0 \quad \text{Factor.}$$
$$u = 5 \quad \text{or} \quad u = 2 \quad \text{Solve.}$$

Since $u = \dfrac{x}{x + 1}$, we have that $5 = \dfrac{x}{x + 1}$ and $2 = \dfrac{x}{x + 1}$.

Thus, there are two rational equations to solve.

1.
$$5 = \frac{x}{x + 1}$$
$$5 \cdot (x + 1) = x$$
$$5x + 5 = x$$
$$5 = -4x$$
$$x = -\frac{5}{4}$$

2.
$$2 = \frac{x}{x + 1}$$
$$2 \cdot (x + 1) = x$$
$$2x + 2 = x$$
$$2 = -x$$
$$x = -2$$

Since neither $-\frac{5}{4}$ nor -2 makes the denominator 0, the solution set is $\left\{-\frac{5}{4}, -2\right\}$.

Solve each equation by substitution. See the preceding example.

89. $(x - 1)^2 + 3(x - 1) + 2 = 0$

90. $(4 - x)^2 - 5(4 - x) + 6 = 0$

91. $\left(\dfrac{3}{x - 1}\right)^2 + 2\left(\dfrac{3}{x - 1}\right) + 1 = 0$

92. $\left(\dfrac{5}{2 + x}\right)^2 + \left(\dfrac{5}{2 + x}\right) - 20 = 0$

7.8   RATIONAL EQUATIONS AND PROBLEM SOLVING

O B J E C T I V E S

 Solve an equation containing rational expressions for a specified variable.

2 Solve problems by writing equations containing rational expressions.

TAPE IAG 7.8

1 In Section 2.4 we solved equations for a specified variable. In this section, we continue practicing this skill by solving equations containing rational expressions for a specified variable. The steps given in Section 2.4 for solving equations for a specified variable are repeated here.

TO SOLVE EQUATIONS FOR A SPECIFIED VARIABLE

Step 1.   Clear the equation of fractions or rational expressions by multiplying each side of the equation by the least common denominator (LCD) of all denominators in the equation.

Step 2.   Use the distributive property to remove grouping symbols such as parentheses.

Step 3.   Combine like terms on the same side of the equation.

Step 4.   Use the addition property of equality to rewrite the equation as an equivalent equation with terms containing the specified variable on one side and all other terms on the other side.

Step 5.   Use the distributive property and the multiplication property of equality to isolate the specified variable.

EXAMPLE 1   Solve $\dfrac{1}{x} + \dfrac{1}{y} = \dfrac{1}{z}$ for x.

Solution: To clear this equation of fractions, multiply both sides of the equation by xyz, the LCD of $\dfrac{1}{x}, \dfrac{1}{y}$, and $\dfrac{1}{z}$.

$$\frac{1}{x} + \frac{1}{y} = \frac{1}{z}$$

$$xyz\left(\frac{1}{x} + \frac{1}{y}\right) = xyz\left(\frac{1}{z}\right) \qquad \text{Multiply both sides by } xyz.$$

$$xyz\left(\frac{1}{x}\right) + xyz\left(\frac{1}{y}\right) = xyz\left(\frac{1}{z}\right) \qquad \text{Apply the distributive property.}$$

$$yz + xz = xy \qquad \text{Simplify.}$$

Notice the two terms that contain the specified variable x.

Next subtract xz from both sides so that all terms containing the specified variable x are on one side of the equation and all other terms are on the other side.

$$yz = xy - xz$$

Use the distributive property to factor x from $xy - xz$ and then the multiplication property of equality to solve for x.

$$yz = x(y - z)$$

$$\frac{yz}{y - z} = x, \quad \text{or} \quad x = \frac{yz}{y - z} \qquad \text{Divide both sides by } y - z.$$

2   Problem solving sometimes involves modeling a described situation with an equation containing rational expressions. In Examples 2 through 5, we practice solving such problems and use the problem-solving steps first introduced in Section 2.2.

EXAMPLE 2   FINDING UNKNOWN NUMBERS
Find the number that, when subtracted from the numerator and added to the denominator of $\frac{9}{19}$, results in a fraction equivalent to $\frac{1}{3}$.

Solution: **1.** UNDERSTAND the problem. To do so, read and reread the problem and try guessing the solution. For example, if the unknown number is 3, we have the following:

$$\frac{9 - 3}{19 + 3} = \frac{1}{3}$$

To see if this is a true statement, simplify the fraction on the left.

$$\frac{6}{22} = \frac{1}{3}$$

or

$$\frac{3}{11} = \frac{1}{3} \qquad \text{False.}$$

when the fraction on the left is simplified further. Since this is not a true statement, 3 is not the correct number. Remember that the purpose of this step is not to guess the correct solution but to gain an understanding of the problem posed.

2. ASSIGN a variable. Let n be the number to be subtracted from the numerator and added to the denominator.

3. ILLUSTRATE. No illustration is needed.

4. TRANSLATE the problem.

In words:	when the number is subtracted from the numerator and added to the denominator of the fraction	this is equivalent to	$\frac{1}{3}$
Translate:	$\dfrac{9 - n}{19 + n}$	$=$	$\dfrac{1}{3}$

5. COMPLETE the work. Here we solve the equation for n.

$$\frac{9 - n}{19 + n} = \frac{1}{3}$$

To solve for n, begin by multiplying both sides by the LCD, $3(19 + n)$.

$$3(19 + n) \cdot \left(\frac{9 - n}{19 + n}\right) = 3(19 + n)\left(\frac{1}{3}\right) \qquad \text{Multiply by the LCD.}$$

$$3(9 - n) = 19 + n \qquad \text{Simplify.}$$

$$27 - 3n = 19 + n$$

$$8 = 4n$$

$$2 = n \qquad \text{Solve.}$$

6. INTERPRET the results. First, *check* the stated problem. If we subtract 2 from the numerator and add 2 to the denominator of $\frac{9}{19}$, we have $\frac{9 - 2}{19 + 2} = \frac{7}{21} = \frac{1}{3}$, and the problem checks. Next, *state* the conclusions. The unknown number is 2.

EXAMPLE 3   MEASURING LIGHT INTENSITY

The intensity of $I(x)$ of a light, as measured in foot-candles, that is x feet from its source is given by the rational function

$$I(x) = \frac{320}{x^2}$$

How far away is the source if the intensity of light is 5 foot-candles?

Solution: **1.** UNDERSTAND. Read and reread the problem. Since a function has been given that describes the relationship between $I(x)$ and x, we replace x with a few values to help us become familiar with the function.

To find the intensity $I(x)$ of light 1 foot from the source, we find $I(1)$.

$$I(1) = \frac{320}{1^2} = \frac{320}{1} = 320 \text{ foot-candles}$$

To find the intensity $I(x)$ of light 4 feet from the source, we find $I(4)$.

$$I(4) = \frac{320}{4^2} = \frac{320}{16} = 20 \text{ foot-candles}$$

Notice that as x increases $I(x)$ decreases. That is, as the number of feet from the light source increases, the intensity decreases, as expected. Steps 2 and 3 are not needed because a formula has been given.

4. TRANSLATE. We are given that the intensity $I(x)$ is 5 foot-candles, and we are asked to find how far away is the light source, x. To do so, let $I(x) = 5$.

$$I(x) = \frac{320}{x^2}$$

$$5 = \frac{320}{x^2} \qquad \text{Let } I(x) = 5.$$

5. COMPLETE. Here we solve the equation for x algebraically and verify graphically to the left.

$$5 = \frac{320}{x^2}$$

$$x^2(5) = x^2\left(\frac{320}{x^2}\right) \qquad \text{Multiply both sides by } x^2.$$

$$5x^2 = 320 \qquad \text{Simplify.}$$

$$x^2 = 64 \qquad \text{Divide both sides by 5.}$$

Then, since $8^2 = 64$ and also $(-8)^2 = 64$, we have that

$$x = 8 \qquad \text{or} \qquad x = -8$$

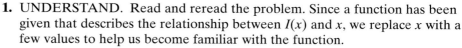

$y_1 = \frac{320}{x^2}$

$y_2 = 5$

Intersection X=8   Y=5

Since x represents feet, we choose a window showing positive x-values only. The solution for x is 8.

6. INTERPRET. Notice that the graph of y_1 clearly shows us that as distance x increases, the intensity of light y_1 decreases. Since x represents distance and distance cannot be negative, the proposed algebraic solution -8 must be rejected. *Check* the solution 8 feet in the given formula. Then *state* the solution: The source of light is 8 feet away when the intensity is 5 foot-candles. ▬▬▬▬▬

The following work example leads to an equation containing rational expressions.

EXAMPLE 4   **WORKING TOGETHER**

Melissa Scarlatti can clean the house in 4 hours, whereas her husband, Zack, can do the same job in 5 hours. They have agreed to clean together so that they can finish in time to watch a movie on TV that starts in 2 hours. How long will it take them to clean the house together? Can they finish before the movie starts?

Solution: **1.** UNDERSTAND. Read and reread the problem. The key idea here is the relationship between the *time* (in hours) it takes to complete the job and the *part of the job* completed in 1 unit of time (1 hour). For example, if the *time* it takes Melissa to complete the job is 4 hours, the *part of the job* she can complete in 1 hour is $\frac{1}{4}$. Similarly, Zack can complete $\frac{1}{5}$ of the job in 1 hour.

2. ASSIGN. Let t represent the *time* in hours it takes Melissa and Zack to clean the house together. Then $\frac{1}{t}$ represents the *part of the job* that they complete in 1 hour.

3. ILLUSTRATE. Here we summarize the given information on a chart.

	HOURS TO COMPLETE THE JOB	PART OF JOB COMPLETED IN 1 HOUR
MELISSA ALONE	4	$\frac{1}{4}$
ZACK ALONE	5	$\frac{1}{5}$
TOGETHER	t	$\frac{1}{t}$

4. TRANSLATE.

	part of job Melissa can complete in 1 hour	added to	part of job Zack can complete in 1 hour	is equal to	part of job they can complete together in 1 hour
In words:					
Translate:	$\frac{1}{4}$	$+$	$\frac{1}{5}$	$=$	$\frac{1}{t}$

5. COMPLETE. We solve the equation for x algebraically and verify graphically as shown to the left.

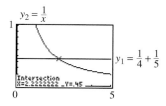

The solution of $\frac{1}{4} + \frac{1}{5} = \frac{1}{t}$ is $2.\overline{2}$ or $2\frac{2}{9}$.

$$\frac{1}{4} + \frac{1}{5} = \frac{1}{t}$$

$$20t\left(\frac{1}{4} + \frac{1}{5}\right) = 20t\left(\frac{1}{t}\right) \qquad \text{Multiply both sides by the LCD, } 20t.$$

$$5t + 4t = 20$$

$$9t = 20$$

$$t = \frac{20}{9} \quad \text{or} \quad 2\frac{2}{9} \qquad \text{Solve.}$$

6. INTERPRET. *Check:* The proposed solution is $2\frac{2}{9}$. That is, Melissa and Zack would take $2\frac{2}{9}$ hours to clean the house together. This proposed solution is reasonable since $2\frac{2}{9}$ hours is more than half of Melissa's time and less than half of Zack's time. Check this solution in the originally stated problem.

State: Can they finish before the movie starts? If they can clean the house together in $2\frac{2}{9}$ hours, they could not complete the job before the movie starts.

EXAMPLE 5

CALCULATING SPEED OF WATER CURRENT
In his boat, Steve Deitmer takes $1\frac{1}{2}$ times as long to go 72 miles upstream as he does to return. If the boat cruises at 30 mph in still water, what is the speed of the current?

Solution:

1. UNDERSTAND. Read and reread the problem. Next, guess a solution. Suppose that the current is 4 mph. The speed of the boat upstream is slowed down by the current: $30 - 4$, or 26 mph, and the speed of the boat downstream is speeded up by the current: $30 + 4$, or 34 mph. Next, let's find out how long it takes to travel 72 miles upstream and 72 miles downstream. To do so, we use the formula $d = r \cdot t$, or $\dfrac{d}{r} = t$.

UPSTREAM	DOWNSTREAM
$\dfrac{d}{r} = t$	$\dfrac{d}{r} = t$
$\dfrac{72}{26} = t$	$\dfrac{72}{34} = t$
$2\dfrac{10}{13} = t$	$2\dfrac{2}{17} = t$

Since the time upstream $\left(2\frac{10}{13} \text{ hours}\right)$ is not $1\frac{1}{2}$ times the time downstream $\left(2\frac{2}{17} \text{ hours}\right)$, our guess is not correct. We do, however, have a better understanding of the problem.

2. **ASSIGN.** Since we are asked to find the speed of the current, *let x represent the current's speed.* The speed of the boat traveling downstream is made faster by the current and is represented by $30 + x$. The speed of the boat traveling upstream is made slower by the current and is represented by $30 - x$.

3. **ILLUSTRATE.** This information is summarized in the following chart, where we use the formula $\dfrac{d}{r} = t$.

	DISTANCE	RATE	TIME $\left(\dfrac{d}{r}\right)$
UPSTREAM	72	$30 - x$	$\dfrac{72}{30 - x}$
DOWNSTREAM	72	$30 + x$	$\dfrac{72}{30 + x}$

4. **TRANSLATE.** Since the time spent traveling upstream is $1\frac{1}{2}$ times the time spent traveling downstream, we have

In words:

time upstream	is	$1\frac{1}{2}$	times	time downstream

Translate: $\dfrac{72}{30 - x} \quad = \quad \dfrac{3}{2} \quad \cdot \quad \dfrac{72}{30 + x}$

5. **COMPLETE.**

$$\frac{72}{30 - x} = \frac{3}{2} \cdot \frac{72}{30 + x}$$

Multiply both sides by the LCD, $2(30 + x)(30 - x)$.

$$2(30 + x)(30 - x)\left(\frac{72}{30 - x}\right) = 2(30 + x)(30 - x)\left(\frac{3}{2} \cdot \frac{72}{30 + x}\right)$$

$$72 \cdot 2(30 + x) = 3 \cdot 72 \cdot (30 - x) \qquad \text{Simplify.}$$

$$2(30 + x) = 3(30 - x) \qquad \text{Divide by 72.}$$

$$60 + 2x = 90 - 3x \qquad \text{Apply the distributive property.}$$

$$5x = 30$$

$$x = 6 \qquad \text{Solve.}$$

6. **INTERPRET.** *Check* this proposed solution of 6 mph in the originally stated problem. *State:* The current's speed is 6 mph.

EXERCISE SET 7.8

Solve each equation for the specified variable. See Example 1.

1. $F = \frac{9}{5}C + 32$; C

2. $V = \frac{1}{3}\pi r^2 h$; h

3. $\frac{1}{R} = \frac{1}{R_1} + \frac{1}{R_2}$; R

4. $\frac{1}{R} = \frac{1}{R_1} + \frac{1}{R_2}$; R_1

5. $S = \frac{n(a + L)}{2}$; n

6. $S = \frac{n(a + L)}{2}$; a

7. $A = \frac{h(a + b)}{2}$; b

8. $A = \frac{h(a + b)}{2}$; h

9. $\frac{P_1 V_1}{T_1} = \frac{P_2 V_2}{T_2}$; T_2

10. $H = \frac{kA(T_1 - T_2)}{L}$; T_2

11. $f = \frac{f_1 f_2}{f_1 + f_2}$; f_2

12. $I = \frac{E}{R + r}$; r

13. $\lambda = \frac{2L}{n}$; L

14. $S = \frac{a_1 - a_n r}{1 - r}$; a_1

15. $\frac{\theta}{\omega} = \frac{2L}{c}$; c

16. $F = \frac{-GMm}{r^2}$; M

Solve. See Example 2.

17. The sum of a number and 5 times its reciprocal is 6. Find the number(s).

18. The quotient of a number and 9 times its reciprocal is 1. Find the number(s).

19. If a number is added to the numerator of $\frac{12}{41}$ and twice the number is added to the denominator of $\frac{12}{41}$, the resulting fraction is equivalent to $\frac{1}{3}$. Find the number.

20. If a number is subtracted from the numerator of $\frac{13}{8}$ and added to the denominator of $\frac{13}{8}$, the resulting fraction is equivalent to $\frac{2}{5}$. Find the number.

In electronics, the relationship among the resistances r_1 and r_2 of two resistors wired in a parallel circuit and their combined resistance r is described by the formula $\frac{1}{r} = \frac{1}{r_1} + \frac{1}{r_2}$. Use this formula to solve Exercises 21 through 23. See Example 3.

21. If the combined resistance is 2 ohms and one of the two resistances is 3 ohms, find the other resistance.

22. Find the combined resistance of two resistors of 12 ohms each when they are wired in a parallel circuit.

23. The relationship among resistance of two resistors wired in a parallel circuit and their combined resistance may be extended to three resistors of resistances r_1, r_2, and r_3. Write an equation that you believe may describe the relationship, and use it to find the combined resistance if r_1 is 5, r_2 is 6, and r_3 is 2.

Solve. See Example 4.

24. Alan Cantrell can word process a research paper in 6 hours. With Steve Isaac's help, the paper can be processed in 4 hours. Find how long it takes Steve to word process the paper alone.

25. An experienced roofer can roof a house in 26 hours. A beginning roofer needs 39 hours to complete the same job. Find how long it takes for the two to do the job together.

26. A new printing press can print newspapers twice as fast as an old one. The old one can print the afternoon edition in 4 hours. Find how long it takes to print the afternoon edition if both printers are operating.

27. Three postal workers can sort a stack of mail in 20 minutes, 30 minutes, and 60 minutes, respectively. Find how long it takes to sort the mail if all three work together.

Solve. See Example 5.

28. An F-100 plane and a Toyota truck leave the same town at sunrise and head for a town 450 miles away. The speed of the plane is three times the speed of the truck, and the plane arrives 6 hours ahead of the truck. Find the speed of the truck.

29. Mattie Evans drove 150 miles in the same amount of time that it took a turbopropeller plane to travel 600 miles. The speed of the plane was 150 mph faster than the speed of the car. Find the speed of the plane.

30. The speed of a boat in still water is 24 mph. If the boat travels 54 miles upstream in the same time that it takes to travel 90 miles downstream, find the speed of the current.

31. The speed of Lazy River's current is 5 mph. If a boat travels 20 miles downstream in the same time that it takes to travel 10 miles upstream, find the speed of the boat in still water.

Solve.

32. The sum of the reciprocals of two consecutive odd integers is $\frac{20}{99}$. Find the two integers.

33. The sum of the reciprocals of two consecutive integers is $-\frac{15}{56}$. Find the two integers.

34. If Sarah Clark can do a job in 5 hours and Dick Belli and Sarah working together can do the same job in 2 hours, find how long it takes Dick to do the job alone.

35. One hose can fill a goldfish pond in 45 minutes, and two hoses can fill the same pond in 20 minutes. Find how long it takes the second hose alone to fill the pond.

36. The speed of a bicyclist is 10 mph faster than the speed of a walker. If the bicyclist travels 26 miles in the same amount of time that the walker travels 6 miles, find the speed of the bicyclist.

37. Two trains going in opposite directions leave at the same time. One train travels 15 mph faster than the other. In 6 hours the trains are 630 miles apart. Find the speed of each.

38. The numerator of a fraction is 4 less than the denominator. If both the numerator and the denominator are increased by 2, the resulting fraction is equivalent to $\frac{2}{3}$. Find the fraction.

39. Fabio Casartelli of Italy won the individual road race in cycling during the 1992 Summer Olympics. An amateur cyclist training for a road race rode the first 20-mile portion of his workout at a constant rate. For the 16-mile cooldown portion of his workout, he reduced his speed by 2 miles per hour. Each portion of the workout took equal time. Find the cyclist's rate during the first portion and his rate during the cooldown portion.

40. The denominator of a fraction is 1 more than the numerator. If both the numerator and the denominator are decreased by 3, the resulting fraction is equivalent to $\frac{4}{5}$. Find the fraction.

41. Moo Dairy has three machines to fill half-gallon milk cartons. The machines can fill the daily quota in 5 hours, 6 hours, and 7.5 hours, respectively. Find how long it takes to fill the daily quota if all three machines are running.

42. The inlet pipe of an oil tank can fill the tank in 1 hour 30 minutes. The outlet pipe can empty the tank in 1 hour. Find how long it takes to empty a full tank if both pipes are open.

43. A plane flies 465 miles with the wind and 345 miles against the wind in the same length of time. If the speed of the wind is 20 mph, find the speed of the plane in still air.

44. Two rockets are launched. The first travels at 9000 mph. Fifteen minutes later the second is launched at 10,000 mph. Find the distance at which both rockets are an equal distance from Earth.

45. Two joggers, one averaging 8 mph and one averaging 6 mph, start from a designated initial point. The slower jogger arrives at the end of the run a half-hour after the other jogger. Find the distance of the run.

46. A semi truck travels 300 miles through the flatland in the same amount of time that it travels 180 miles through the Great Smoky Mountains. The rate of the truck is 20 miles per hour slower in the mountains than in the flatland. Find both the flatland rate and mountain rate.

47. Smith Engineering is in the process of reviewing the salaries of their surveyors. During this review, the company found that an experienced surveyor surveys a roadbed in 4 hours. An apprentice surveyor needs 5 hours to survey the same stretch of road. If the two work together, find how long it takes them to complete the job.

48. An experienced bricklayer constructs a small wall in 3 hours. An apprentice completes the job in 6 hours. Find how long it takes if they work together.

49. A marketing manager travels 1080 miles in a corporate jet and then an additional 240 miles by car. If the car ride takes 1 hour longer, and if the rate of the jet is 6 times the rate of the car, find the time the manager travels by jet and find the time she travels by car.

50. Gary Marcus and Tony Alva work for Lombardo's Pipe and Concrete. Mr. Lombardo is preparing an estimate for a customer. He knows that Gary lays a slab of concrete in 6 hours. Tony lays the same size slab in 4 hours. If both work on the job and the cost of labor is $45.00 per hour, decide what the labor estimate should be.

51. In 2 minutes, a conveyor belt moves 300 pounds of recyclable aluminum cans from the delivery truck to a storage area. A smaller belt moves the same quantity of cans the same distance in 6 minutes. If both belts are used, find how long it takes to move the cans to the storage area.

52. Mr. Dodson can paint his house by himself in four days. His son needs an additional day to complete the job if he works by himself. If they work together, find how long it takes to paint the house.

53. While road testing a new make of car, the editor of a consumer magazine finds that she can go 10 miles into a 3-mile-per-hour wind in the same amount of time that she can go 11 miles with a 3-mile-per-hour wind behind her. Find the speed of the car in still air.

54. The world record for the largest white bass caught is held by Ronald Sprouse of Virginia. The bass weighed 6 pounds 13 ounces. If Ronald rows to his favorite fishing spot 9 miles downstream in the same amount of time that he rows 3 miles upstream and if the current is 6 miles per hour, find how long it takes him to cover the 12 miles.

Review Exercises

Solve the equation for x. See Section 2.1.

55. $\dfrac{x}{5} = \dfrac{x+2}{3}$

56. $\dfrac{x}{4} = \dfrac{x+3}{6}$

57. $\dfrac{x-3}{2} = \dfrac{x-5}{6}$

58. $\dfrac{x-6}{4} = \dfrac{x-2}{5}$

Use the circle graph to answer Exercises 59 through 64.

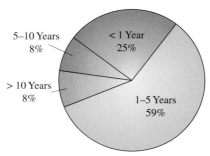

How long have you worn your hair in its current style?

5–10 Years 8%
< 1 Year 25%
> 10 Years 8%
1–5 Years 59%

Source: *Self Magazine*, June 1993.

59. What percent of those polled have worn their hair in its current style for less than 1 year?

60. What percent of those polled have worn their hair in its current style for more than 10 years?

61. What percent of those polled have worn their hair in its current style for 5 years or less?

62. What percent of those polled have worn their hair in its current style for 5 years or more?

 63. Poll your algebra class with this same question and compute the percents in each category.

 64. Use the results of Exercise 63 and construct a circle graph. [*Hint:* Recall that the number of degrees around a circle is 360°. Then, for example, the number of degrees in a sector representing 8% of a whole should be 0.08(360°) = 28.8°.]

7.9 VARIATION AND PROBLEM SOLVING

TAPE IAG 7.9

OBJECTIVES

1. Write an equation expressing direct variation.
2. Write an equation expressing inverse variation.
3. Write an equation expressing joint variation.

In this section, we solve problems that can be modeled by using the concepts of direct variation, inverse variation, or joint variation.

1

A very familiar example of direct variation is the relationship of the circumference C of a circle to its radius r. The formula $C = 2\pi r$ expresses that the circumference is always 2π times the radius. In other words, C is always a constant multiple (2π) of r. Because it is, we say that **C varies directly as r,** that **C varies directly with r,** or that **C is directly proportional to r.**

DIRECT VARIATION

y varies directly as x, or **y is directly proportional to x,** if there is a nonzero constant k such that

$$y = kx$$

The number k is called the **constant of variation** or the **constant of proportionality.**

Recall that the relationship described between x and y is linear. In other words, the graph of $y = kx$ is a line. The slope of the line is k, and the line passes through the origin.

The graph of the direct variation example $C = 2\pi r$ is shown next. The horizontal axis represents the radius r, and the vertical axis is the circumference C.

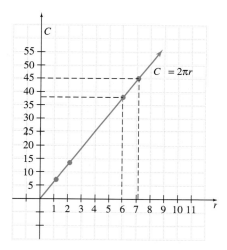

From the graph we can read that when the radius is 6 units the circumference is approximately 38 units. Also, when the circumference is 45 units, the radius is between 7 and 8 units. Notice that as the radius increases, the circumference increases.

EXAMPLE 1   Suppose that y varies directly as x. If y is 5 when x is 30, find the constant of variation. Also, find y when $x = 90$.

Solution: Since y varies directly as x, we write $y = kx$. If $y = 5$ when $x = 30$, we have that

$$y = kx$$
$$5 = k(30) \qquad \text{Replace } y \text{ with 5 and } x \text{ with 30.}$$
$$\frac{1}{6} = k \qquad \text{Solve for } k.$$

The constant of variation is $\frac{1}{6}$.

After finding the constant of variation k, the direct variation equation can be written as $y = \frac{1}{6}x$. Next, find y when x is 90.

$$y = \frac{1}{6}x$$
$$= \frac{1}{6}(90) \qquad \text{Let } x = 90.$$
$$= 15$$

When $x = 90$, y must be 15.

Notice that the direct variation equation $y = kx$ is not only a linear equation, but also y is a function of x. Thus, the notation $y = kx$ or $f(x) = kx$ may be used.

EXAMPLE 2   **USING HOOKE'S LAW**
Hooke's law states that the distance a spring stretches is directly proportional to the weight attached to the spring. If a 40-pound weight attached to the spring stretches the spring 5 inches, find the distance that a 65-pound weight attached to the spring stretches the spring.

Solution: 1. UNDERSTAND. Read and reread the problem. Notice we are given that the distance that a spring stretches is **directly proportional** to the weight attached.

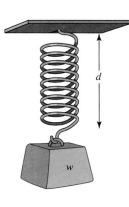

2. ASSIGN. Let d represent the distance stretched and let w represent the weight attached. The constant of variation is represented by k.

3. ILLUSTRATE. See the illustration.

4. TRANSLATE. Because d is directly proportional to w, we write

$$d = kw$$

When a weight of 40 pounds is attached, the spring stretches 5 inches. That is, when $w = 40$, $d = 5$.

$$5 = k(40) \qquad \text{Replace } d \text{ with 5 and } w \text{ with 40.}$$
$$\frac{1}{8} = k \qquad \text{Solve for } k.$$

Replace k with $\frac{1}{8}$ in the equation $d = kw$, and we have

$$d = \frac{1}{8}w$$

To find the stretch when a weight of 65 pounds is attached, replace w with 65; find d.

$$d = \frac{1}{8}(65)$$

5. COMPLETE.

$$d = \frac{1}{8}(65)$$

$$= \frac{65}{8} = 8\frac{1}{8}, \quad \text{or} \quad 8.125$$

6. INTERPRET. *Check* the proposed solution of 8.125 inches. *State:* The spring stretches 8.125 inches when a 65-pound weight is attached. ▬▬▬▬▬▬▬

2 When y is proportional to the *reciprocal* of another variable x, we say that **y varies inversely as x,** or that **y is inversely proportional to x.** An example of the inverse variation relationship is the relationship between the pressure that a gas exerts and the volume of its container. As the volume of a container decreases, the pressure of the gas that it contains increases.

INVERSE VARIATION

y varies inversely as x, or **y is inversely proportional to x,** if there is a nonzero constant k such that

$$y = \frac{k}{x}$$

The number k is called the **constant of variation** or the **constant of proportionality.**

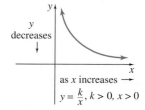

Notice that $y = \frac{k}{x}$, or $f(x) = \frac{k}{x}$, is a rational function. Its graph for $k > 0$ and $x > 0$ is shown. From the graph, we can see that as x increases, y decreases.

EXAMPLE 3 Suppose that u varies inversely as w. If u is 3 when w is 5, find u when w is 30.

Solution: Since u varies inversely as w, we have $u = \dfrac{k}{w}$. Let $u = 3$, $w = 5$, and solve for k.

$$u = \frac{k}{w}$$

$$3 = \frac{k}{5} \qquad \text{Let } u = 3 \text{ and } w = 5.$$

$$15 = k \qquad \text{Multiply both sides by 5 or cross-multiply.}$$

The constant of variation k is 15. This gives the following inverse variation equation:

$$u = \frac{15}{w}$$

Now find u when $w = 30$.

$$u = \frac{15}{30} \qquad \text{Let } w = 30.$$

$$= \frac{1}{2}$$

Thus, when $w = 30$, $u = \dfrac{1}{2}$.

EXAMPLE 4   **USING BOYLE'S LAW**

Boyle's law says that, if the temperature stays the same, the pressure P of a gas is inversely proportional to the volume V. If a cylinder in a steam engine has a pressure of 960 kilopascals when the volume is 1.4 cubic meters, find the pressure when the volume increases to 2.5 cubic meters.

Solution: **1.** UNDERSTAND. Read and reread the problem. Notice that we are given that the pressure of a gas is *inversely proportional* to the volume.

2. ASSIGN. Let P represent the pressure and let V represent the volume. The constant of variation is represented by k.

4. TRANSLATE. Because P is inversely proportional to V, we write

$$P = \frac{k}{V}$$

When $P = 960$ kilopascals, the volume $V = 1.4$ cubic meters. Use this information to find k.

$$960 = \frac{k}{1.4} \qquad \text{Let } P = 960 \text{ and } V = 1.4.$$

$$1344 = k \qquad \text{Cross-multiply.}$$

Thus, the value of k is 1344. Replace k with 1344 in the variation equation.

$$P = \frac{1344}{V}$$

Next, find P when V is 2.5 cubic meters.

$$P = \frac{1344}{2.5} \qquad \text{Let } V = 2.5.$$

5. COMPLETE.

$$P = \frac{1344}{2.5}$$

$$= 537.6$$

6. INTERPRET. *Check* the proposed solution. *State:* When the volume is 2.5 cubic meters, the pressure is 537.6 kilopascals.

Sometimes the ratio of a variable to the product of many other variables is constant. For example, the ratio of distance traveled to the product of speed and time traveled is constantly 1.

$$\frac{d}{rt} = 1, \quad \text{or} \quad d = rt$$

Such a relationship is called **joint variation**.

JOINT VARIATION

If the ratio of a variable y to the product of two or more variables is constant, then **y varies jointly as,** or **is jointly proportional to,** the other variables. If

$$y = kxz$$

then the number k is the **constant of variation** or the **constant of proportionality.**

EXAMPLE 5 CALCULATING SURFACE AREA

The surface area of a cylinder (neglecting bases) varies jointly as its radius and height. Express surface area S in terms of radius r and height h.

Solution: Because the surface area varies jointly as the radius r and the height h, we equate S to the constant multiple of r and h.

$$S = krh$$

EXERCISE SET 7.9

Write each statement as an equation. See Examples 1 through 5.

1. *A* is directly proportional to *B*.

2. *C* varies inversely as *D*.

3. *X* is inversely proportional to *Z*.

4. *G* varies directly with *M*.

5. *N* varies directly with the square of *P*.

6. *A* varies jointly with *D* and *E*.

7. *T* is inversely proportional to *R*.

8. *G* is inversely proportional to *H*.

9. *P* varies directly with *R*.

10. *T* is directly proportional to *S*.

Solve. See Examples 1 and 2.

11. *A* varies directly as *B*. If *A* is 60 when *B* is 12, find *A* when *B* is 9.

12. *C* varies directly as *D*. If *C* is 42 when *D* is 14, find *C* when *D* is 6.

13. Charles's law states that, if the pressure *P* stays the same, the volume *V* of a gas is directly proportional to its temperature *T*. If a balloon is filled with 20 cubic meters of a gas at a temperature of 300 K, find the new volume if the temperature rises to 360 K while the pressure stays the same.

14. The amount *P* of pollution varies directly with the population *N* of people. Kansas City has a population of 450,000 and produces 260,000 tons of pollutants. Find how many tons of pollution we should expect St. Louis to produce, if we know that its population is 980,000.

Solve. See Examples 3 and 4.

15. *H* is inversely proportional to *J*. If *H* is 4 when *J* is 5, find *H* when *J* is 2.

16. *D* varies inversely as *A*. If *D* is 16 when *A* is 2, find *D* when *A* is 8.

17. If the voltage *V* in an electric circuit is held constant, the current *I* is inversely proportional to the resistance *R*. If the current is 40 amperes when the resistance is 270 ohms, find the current when the resistance is 150 ohms.

18. Pairs of markings a set distance apart are made on highways so that police can detect drivers exceed-ing the speed limit. Over a fixed distance, the speed *R* varies inversely with the time *T*. In one particular pair of markings, *R* is 45 mph when *T* is 6 seconds. Find the speed of a car that travels the given distance in 5 seconds.

Write each statement as an equation. See Example 5.

19. *x* varies jointly as *y* and *z*.

20. *P* varies jointly as *R* and the square of *S*.

21. *r* varies jointly as *s* and the cube of *t*.

22. *a* varies jointly as *b* and *c*.

Solve.

23. *Q* is directly proportional to *R*. If *Q* is 4 when *R* is 20, find *Q* when *R* is 35.

24. *S* is directly proportional to *T*. If *S* is 4 when *T* is 16, find *S* when *T* is 40.

25. *M* varies directly with *P*. If *M* is 8 when *P* is 20, find *M* when *P* is 24.

26. *F* is directly proportional to *G*. If *F* is 18 when *G* is 10, find *F* when *G* is 16.

27. *B* is inversely proportional to *C*. If *B* is 12 when *C* is 3, find *B* when *C* is 18.

28. *U* varies inversely with *V*. If *U* is 14 when *V* is 4, find *U* when *V* is 7.

29. *W* varies inversely with *X*. If *W* is 18 when *X* is 6, find *W* when *X* is 40.

30. *Z* is inversely proportional to *Y*. If *Z* is 9 when *Y* is 6, find *Z* when *Y* is 24.

31. The weight of a synthetic ball varies directly with the cube of its radius. A ball with a radius of 2 inches weighs 1.20 pounds. Find the weight of a ball of the same material with a 3-inch radius.

32. At sea, the distance to the horizon is directly proportional to the square root of the elevation of the observer. If a person who is 36 feet above the water can see 7.4 miles, find how far a person 64 feet above the water can see.

33. Because it is more efficient to produce larger numbers of items, the cost of producing Dysan computer disks is inversely proportional to the number produced. If 4000 can be produced at a cost of $1.20 each, find the cost per disk when 6000 are produced.

34. The weight of an object on or above the surface of Earth varies inversely as the square of the distance between the object and Earth's center. If a person weighs 160 pounds on Earth's surface, find the individual's weight if he moves 200 miles above Earth. (Assume that Earth's radius is 4000 miles.)

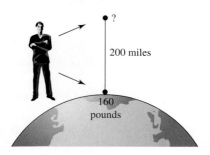

35. The number of cars manufactured on an assembly line at a General Motors plant varies jointly as the number of workers and the hours that they work. If 200 workers can produce 60 cars in 2 hours, find how many cars 240 workers should be able to make in 3 hours.

36. The volume of a cone varies jointly as the square of its radius and its height. If the volume of a cone is 32π when the radius is 4 inches and the height is 6 inches, find the volume of a cone when the radius is 3 inches and the height is 5 inches.

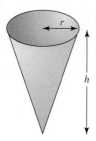

37. When a wind blows perpendicularly against a flat surface, its force is jointly proportional to the surface area and the speed of the wind. A sail whose surface area is 12 square feet experiences a 20-pound force when the wind speed is 10 miles per hour. Find the force on an 8-square-foot sail if the wind speed is 12 miles per hour.

38. The horsepower that can be safely transmitted to a shaft varies jointly as the shaft's angular speed of rotation (in revolutions per minute) and the cube of its diameter. A 2-inch shaft making 120 revolutions per minute safely transmits 40 horsepower. Find how much horsepower can be safely transmitted by a 3-inch shaft making 80 revolutions per minute.

39. A circular column has a *safe load* that is directly proportional to the fourth power of its diameter and inversely proportional to the square of its length. An 8-inch pillar 10 feet long can safely support a 16-ton load. Find the load that a 6-inch pillar made of the same material can support if it is 8 feet long.

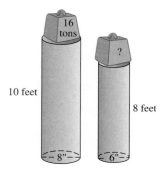

40. The maximum safe load for a rectangular beam varies jointly as its width and the square of its height and inversely as its length. If a beam 6 inches wide, 4 inches high, and 10 feet long supports 12 tons, find how much a similar beam can support if the beam is 8 inches wide, 5 inches high, and 16 feet long.

41. The area of a circle is directly proportional to the square of its radius. If the radius is tripled, determine how the area changes.

42. The horsepower to drive a boat varies directly as the cube of the speed of the boat. If the speed of the boat is to double, determine the corresponding increase in horsepower required.

43. The intensity I of light varies inversely as the square of the distance d from the light source. If the distance from the light source is doubled (see figure), determine what happens to the intensity of light at the new location.

44. The volume of a cylinder varies jointly as the height and the square of the radius. If the height is halved and the radius is doubled, determine what happens to the volume.

45. Suppose that y varies directly as x. If x is doubled, what is the effect on y?

46. Suppose that y varies directly as x^2. If x is doubled, what is the effect on y?

Complete the following table for the inverse variation $y = \dfrac{k}{x}$ over each given value of k. Plot the points on a rectangular coordinate system.

x	$\frac{1}{4}$	$\frac{1}{2}$	1	2	4
$y = \dfrac{k}{x}$					

47. $k = 1$ **48.** $k = 3$ **49.** $k = 5$ **50.** $k = \dfrac{1}{2}$

Review Exercises

Find the exact circumference and area of each circle. Leave results in terms of π.

51.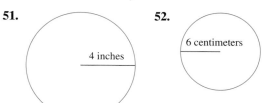

4 inches

52.

6 centimeters

53.

9 centimeters

54.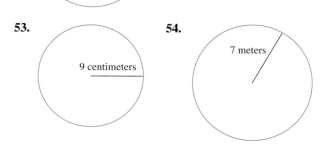

7 meters

Find the slope of the line containing each pair of points. See Section 3.5.

55. $(-5, -2), (0, 7)$ **56.** $(3, 6), (-2, 6)$
57. $(2, 1), (2, -3)$ **58.** $(4, -1), (5, -2)$

Graph each function. See Sections 3.2 and 3.3.

59. $f(x) = 2x - 3$ **60.** $f(x) = x + 4$
61. $g(x) = |x|$ **62.** $h(x) = |x| + 2$
63. $h(x) = x^2$ **64.** $f(x) = x^2 - 1$

GROUP ACTIVITY

MODELING ELECTRICITY PRODUCTION

MATERIALS:
• Calculator

A progressive community in California is experimenting with electricity generated by windmills. City engineers are analyzing the data that they have gathered about their field of windmills. The engineers are familiar with other research demonstrating that the amount of electricity that a windmill generates hourly (in watts per hour) is directly proportional to the cube of the wind speed (in miles per hour).

(continued)

1. The city engineers have documented that when the wind speed is exactly 10 miles per hour a windmill generates electricity at a rate of 15 watts per hour. Find a formula that models the relationship between the wind speed and the amount of electricity generated hourly by a windmill.

2. Use a calculator to complete the following table for the given wind speeds. Estimate from the table the wind speed required to obtain 400 watts per hour.

WIND SPEED (MILES PER HOUR)	ELECTRICITY (WATTS PER HOUR)
15	
17	
19	
21	
23	
25	
27	
29	
31	
33	
35	

3. Construct a bar graph for the data in the table. Describe the trend in the data.

4. The engineers' data show that for several days the wind speed was more or less steady at 20 miles per hour, and the windmills generated the expected 120 watts per hour. According to the weather forecast for the coming few days, wind speed will fluctuate wildly but will still average 20 miles per hour. Should the engineers still expect the windmills to generate 120 watts per hour? Demonstrate your reasoning with a numerical example.

5. During one three-day period, each windmill generated 150 watts per hour. The forecast predicts the wind speed for the coming few days will drop by half. How many watts per hour should the engineers now expect each windmill to generate? In general, if the wind speed yields c watts per hour, how many watts per hour does half the wind speed yield?

CHAPTER 7   HIGHLIGHTS

DEFINITIONS AND CONCEPTS	EXAMPLES
SECTION 7.1 RATIONAL FUNCTIONS AND SIMPLIFYING RATIONAL EXPRESSIONS	
A **rational expression** is the quotient $\dfrac{P}{Q}$ of two polynomials P and Q as long as Q is not 0.	Rational expressions: $$\dfrac{2x-6}{7}, \dfrac{t^2-3t+5}{t-1}$$
A **rational function** is a function described by a rational expression.	Rational functions: $$f(x)=\dfrac{2x-6}{7}, \quad h(t)=\dfrac{t^2-3t+5}{t-1}$$

(continued)

DEFINITIONS AND CONCEPTS	EXAMPLES

SECTION 7.1 RATIONAL FUNCTIONS AND SIMPLIFYING RATIONAL EXPRESSIONS

To simplify a rational expression:

Step 1. Completely factor the numerator and the denominator.

Step 2. Apply the fundamental principle.

Simplify:

$$\frac{2x^2 + 9x - 5}{x^2 - 25} = \frac{(2x - 1)\,(x + 5)}{(x - 5)\,(x + 5)}$$

$$= \frac{2x - 1}{x - 5}$$

SECTION 7.2 MULTIPLYING AND DIVIDING RATIONAL EXPRESSIONS

To multiply rational expressions:

Step 1. Completely factor numerators and denominators.

Step 2. Multiply the numerators and multiply the denominators.

Step 3. Apply the fundamental principle.

Multiply: $\dfrac{x^3 + 8}{12x - 18} \cdot \dfrac{14x^2 - 21x}{x^2 + 2x}$

$$= \frac{(x + 2)\,(x^2 - 2x + 4)}{6(2x - 3)} \cdot \frac{7x(2x - 3)}{x(x + 2)}$$

$$= \frac{7(x^2 - 2x + 4)}{6}$$

To divide rational expressions:

Multiply the first rational expression by the reciprocal of the second rational expression.

Divide: $\dfrac{x^2 + 6x + 9}{5xy - 5y} \div \dfrac{x + 3}{10y}$

$$= \frac{(x + 3)(x + 3)}{5y(x - 1)} \cdot \frac{2 \cdot 5y}{x + 3}$$

$$= \frac{2(x + 3)}{x - 1}$$

SECTION 7.3 ADDING AND SUBTRACTING RATIONAL EXPRESSIONS

To add or subtract rational expressions:

Step 1. Find the LCD.

Step 2. Write each rational expression as an equivalent rational expression whose denominator is the LCD.

Step 3. Add or subtract numerators and write the sum or difference over the common denominator.

Step 4. Write the result in lowest terms.

Subtract: $\dfrac{3}{x + 2} - \dfrac{x + 1}{x - 3}$

$$= \frac{3 \cdot (x - 3)}{(x + 2) \cdot (x - 3)} - \frac{(x + 1) \cdot (x + 2)}{(x - 3) \cdot (x + 2)}$$

$$= \frac{3(x - 3) - (x + 1)(x + 2)}{(x + 2)(x - 3)}$$

$$= \frac{3x - 9 - (x^2 + 3x + 2)}{(x + 2)(x - 3)}$$

$$= \frac{3x - 9 - x^2 - 3x - 2}{(x + 2)(x - 3)}$$

$$= \frac{-x^2 - 11}{(x + 2)(x - 3)}$$

DEFINITIONS AND CONCEPTS	EXAMPLES

SECTION 7.4 SIMPLIFYING COMPLEX FRACTIONS

Method I: Simplify the numerator and the denominator so that each is a single fraction. Then perform the indicated division and simplify if possible.

Simplify: $\dfrac{\dfrac{x+2}{x}}{x-\dfrac{4}{x}}$.

Method I: $\dfrac{\dfrac{x+2}{x}}{\dfrac{x\cdot x}{1\cdot x}-\dfrac{4}{x}} = \dfrac{\dfrac{x+2}{x}}{\dfrac{x^2-4}{x}}$

$= \dfrac{x+2}{x}\cdot\dfrac{x}{(x+2)(x-2)}$

$= \dfrac{1}{x-2}$

Method II: Multiply the numerator and the denominator of the complex fraction by the LCD of the fractions in both the numerator and the denominator. Then simplify.

Method II: $\dfrac{\left(\dfrac{x+2}{x}\right)\cdot x}{\left(x-\dfrac{4}{x}\right)\cdot x} = \dfrac{x+2}{x\cdot x-\dfrac{4}{x}\cdot x}$

$= \dfrac{x+2}{x^2-4} = \dfrac{x+2}{(x+2)(x-2)}$

$= \dfrac{1}{x-2}$

SECTION 7.5 DIVIDING POLYNOMIALS

To divide a polynomial by a monomial:
Divide each term in the polynomial by the monomial.

$\dfrac{12a^5b^3-6a^2b^2+ab}{6a^2b^2}$

$= \dfrac{12a^5b^3}{6a^2b^2} - \dfrac{6a^2b^2}{6a^2b^2} + \dfrac{ab}{6a^2b^2}$

$= 2a^3b - 1 + \dfrac{1}{6ab}$

To divide a polynomial by a polynomial, other than a monomial:
Use **long division.**

Divide $2x^3-x^2-8x-1$ by $x-2$.

$$
\begin{array}{r}
2x^2+3x-2 \\
x-2\overline{)2x^3-x^2-8x-1} \\
\underline{2x^3-4x^2} \\
3x^2-8x \\
\underline{3x^2-6x} \\
-2x-1 \\
\underline{-2x+4} \\
-5
\end{array}
$$

The quotient is $2x^2+3x-2-\dfrac{5}{x-2}$.

DEFINITIONS AND CONCEPTS	EXAMPLES

SECTION 7.6 SYNTHETIC DIVISION

A shortcut method called **synthetic division** may be used to divide a polynomial by a binomial of the form $x - c$.

Use synthetic division to divide $2x^3 - x^2 - 8x - 1$ by $x - 2$.

$$
\begin{array}{r|rrrr}
2 & 2 & -1 & -8 & -1 \\
 & & 4 & 6 & -4 \\
\hline
 & 2 & 3 & -2 & -5
\end{array}
$$

The quotient is $2x^2 + 3x - 2 - \dfrac{5}{x - 2}$.

SECTION 7.7 SOLVING EQUATIONS CONTAINING RATIONAL EXPRESSIONS

To algebraically solve an equation containing rational expressions: Multiply both sides of the equation by the LCD of all rational expressions. Then apply the distributive property and simplify. Solve the resulting equation and then check each proposed solution to see whether it makes a denominator 0. If so, it is an **extraneous solution.**

Solve $x - \dfrac{3}{x} = \dfrac{1}{2}$.

$$2x\left(x - \frac{3}{x}\right) = 2x\left(\frac{1}{2}\right) \qquad \text{The LCD is } 2x.$$

$$2x \cdot x - 2x\left(\frac{3}{x}\right) = 2x\left(\frac{1}{2}\right) \qquad \text{Distribute.}$$

$$2x^2 - 6 = x$$

$$2x^2 - x - 6 = 0 \qquad \text{Subtract } x.$$

$$(2x + 3)(x - 2) = 0 \qquad \text{Factor.}$$

$$x = -\frac{3}{2} \text{ or } x = 2$$

Both $-\dfrac{3}{2}$ and 2 check. The solution set is $\left\{2, -\dfrac{3}{2}\right\}$.

SECTION 7.8 RATIONAL EQUATIONS AND PROBLEM SOLVING

To solve an equation for a specified variable:
  Treat the specified variable as the only variable of the equation and solve as usual.

Solve for x.

$$A = \frac{2x + 3y}{5}$$

$$5A = 2x + 3y \qquad \text{Multiply by 5.}$$

$$5A - 3y = 2x \qquad \text{Subtract } 3y.$$

$$\frac{5A - 3y}{2} = x \qquad \text{Divide by 2.}$$

Problem-solving steps:

Jeanee and David Dillon volunteer every year to clean a strip of Lake Ponchatrain beach. Jeanee can clean all the trash in this area of beach in 6 hours; David takes 5 hours. How long will it take them to clean the area of beach together?

1. UNDERSTAND.

1. Read and reread the problem.

2. ASSIGN a variable.

2. Let $x =$ time in hours that it takes Jeanee and David to clean the beach together.　　(*continued*)

DEFINITIONS AND CONCEPTS	EXAMPLES

SECTION 7.8 RATIONAL EQUATIONS AND PROBLEM SOLVING

3. ILLUSTRATE the problem.

3.

	HOURS TO COMPLETE	PART COMPLETED IN 1 HOUR
JEANEE ALONE	6	$\frac{1}{6}$
DAVID ALONE	5	$\frac{1}{5}$
TOGETHER	x	$\frac{1}{x}$

4. TRANSLATE.

4. In words:

part Jeanee can complete in 1 hour	+	part David can complete in 1 hour	=	part they can complete together in 1 hour

Translate: $\frac{1}{6}$ + $\frac{1}{5}$ = $\frac{1}{x}$

5. COMPLETE.

5. $\frac{1}{6} + \frac{1}{5} = \frac{1}{x}$ Multiply by $30x$.

$$5x + 6x = 30$$
$$11x = 30$$
$$x = \frac{30}{11}, \text{ or } 2\frac{8}{11}$$

6. INTERPRET.

6. *Check* and then *state*. Together, they can clean the beach in $2\frac{8}{11}$ hours.

SECTION 7.9 VARIATION AND PROBLEM SOLVING

y **varies directly** as *x*, or *y* is **directly proportional** to *x*, if there is a nonzero constant *k* such that

$$y = kx$$

y **varies inversely** as *x*, or *y* is **inversely proportional** to *x*, if there is a nonzero constant *k* such that

$$y = \frac{k}{x}$$

The circumference of a circle *C* varies directly as its radius *r*.

$$C = \underset{k}{\underline{2\pi}} r$$

Pressure *P* varies inversely with volume *V*.

$$P = \frac{k}{V}$$

CHAPTER 7 REVIEW

(7.1) *Find the domain of each rational function.*

1. $f(x) = \dfrac{3 - 5x}{7}$

2. $g(x) = \dfrac{2x + 4}{11}$

3. $F(x) = \dfrac{-3x^2}{x - 5}$

4. $h(x) = \dfrac{4x}{3x - 12}$

5. $f(x) = \dfrac{x^3 + 2}{x^2 + 8x}$

6. $G(x) = \dfrac{20}{3x^2 - 48}$

Write each rational expression in lowest terms.

7. $\dfrac{15x^4}{45x^2}$

8. $\dfrac{x + 2}{2 + x}$

9. $\dfrac{18m^6 p^2}{10m^4 p}$

10. $\dfrac{x - 12}{12 - x}$

11. $\dfrac{5x - 15}{25x - 75}$

12. $\dfrac{22x + 8}{11x + 4}$

13. $\dfrac{2x}{2x^2 - 2x}$

14. $\dfrac{x + 7}{x^2 - 49}$

15. $\dfrac{2x^2 + 4x - 30}{x^2 + x - 20}$

16. $\dfrac{xy - 3x + 2y - 6}{x^2 + 4x + 4}$

17. The average cost of manufacturing x bookcases is given by the rational function

$$C(x) = \dfrac{35x + 4200}{x}$$

  a. Find the average cost per bookcase of manufacturing 50 bookcases.

  b. Find the average cost per bookcase of manufacturing 100 bookcases.

  c. As the number of bookcases increases, does the average cost per bookcase increase or decrease? (See parts (a) and (b).)

(7.2) *Perform the indicated operation. Write answers in lowest terms.*

18. $\dfrac{5}{x^3} \cdot \dfrac{x^2}{15}$

19. $\dfrac{3x^4 y z^3}{15x^2 y^2} \cdot \dfrac{10xy}{z^6}$

20. $\dfrac{4 - x}{5} \cdot \dfrac{15}{2x - 8}$

21. $\dfrac{x^2 - 6x + 9}{2x^2 - 18} \cdot \dfrac{4x + 12}{5x - 15}$

22. $\dfrac{a - 4b}{a^2 + ab} \cdot \dfrac{b^2 - a^2}{8b - 2a}$

23. $\dfrac{x^2 - x - 12}{2x^2 - 32} \cdot \dfrac{x^2 + 8x + 16}{3x^2 + 21x + 36}$

24. $\dfrac{2x^3 + 54}{5x^2 + 5x - 30} \cdot \dfrac{6x + 12}{3x^2 - 9x + 27}$

25. $\dfrac{3}{4x} \div \dfrac{8}{2x^2}$

26. $\dfrac{4x + 8y}{3} \div \dfrac{5x + 10y}{9}$

27. $\dfrac{5ab}{14c^3} \div \dfrac{10a^4 b^2}{6ac^5}$

28. $\dfrac{2}{5x} \div \dfrac{4 - 18x}{6 - 27x}$

29. $\dfrac{x^2 - 25}{3} \div \dfrac{x^2 - 10x + 25}{x^2 - x - 20}$

30. $\dfrac{a - 4b}{a^2 + ab} \div \dfrac{20b - 5a}{b^2 - a^2}$

31. $\dfrac{7x + 28}{2x + 4} \div \dfrac{x^2 + 2x - 8}{x^2 - 2x - 8}$

32. $\dfrac{3x + 3}{x - 1} \div \dfrac{x^2 - 6x - 7}{x^2 - 1}$

33. $\dfrac{2x - x^2}{x^3 - 8} \div \dfrac{x^2}{x^2 + 2x + 4}$

34. $\dfrac{5a^2 - 20}{a^3 + 2a^2 + a + 2} \div \dfrac{7a}{a^3 + a}$

35. $\dfrac{2a}{21} \div \dfrac{3a^2}{7} \cdot \dfrac{4}{a}$

36. $\dfrac{5x - 15}{3 - x} \cdot \dfrac{x + 2}{10x + 20} \cdot \dfrac{x^2 - 9}{x^2 - x - 6}$

37. $\dfrac{4a + 8}{5a^2 - 20} \cdot \dfrac{3a^2 - 6a}{a + 3} \div \dfrac{2a^2}{5a + 15}$

If $f(x) = x - 5$ and $g(x) = 2x + 1$, find:

38. $(f + g)(x)$

39. $(f - g)(x)$

40. $(f \cdot g)(x)$

41. $\left(\dfrac{g}{f}\right)(x)$

(7.3) *Find the LCD of the rational expressions in the list.*

42. $\dfrac{4}{9}, \dfrac{5}{2}$

43. $\dfrac{5}{4x^2 y^5}, \dfrac{3}{10x^2 y^4}, \dfrac{x}{6y^4}$

44. $\dfrac{5}{2x}, \dfrac{7}{x - 2}$

45. $\dfrac{3}{5x}, \dfrac{2}{x - 5}$

46. $\dfrac{1}{5x^3}, \dfrac{4}{x^2 + 3x - 28}, \dfrac{11}{10x^2 - 30x}$

Perform the indicated operation. Write answers in lowest terms.

47. $\dfrac{2}{15} + \dfrac{4}{15}$

48. $\dfrac{4}{x-4} + \dfrac{x}{x-4}$

49. $\dfrac{4}{3x^2} + \dfrac{2}{3x^2}$

50. $\dfrac{1}{x-2} - \dfrac{1}{4-2x}$

51. $\dfrac{2x+1}{x^2+x-6} + \dfrac{2-x}{x^2+x-6}$

52. $\dfrac{7}{2x} + \dfrac{5}{6x}$

53. $\dfrac{1}{3x^2y^3} - \dfrac{1}{5x^4y}$

54. $\dfrac{1}{10-x} + \dfrac{x-1}{x-10}$

55. $\dfrac{x-2}{x+1} - \dfrac{x-3}{x-1}$

56. $\dfrac{x}{9-x^2} - \dfrac{2}{5x-15}$

57. $2x + 1 - \dfrac{1}{x-3}$

58. $\dfrac{2}{a^2-2a+1} + \dfrac{3}{a^2-1}$

59. $\dfrac{x}{9x^2+12x+16} - \dfrac{3x+4}{27x^3-64}$

Perform the indicated operation. Write answers in lowest terms.

60. $\dfrac{2}{x-1} - \dfrac{3x}{3x-3} + \dfrac{1}{2x-2}$

61. $\dfrac{3}{2x} \cdot \left(\dfrac{2}{x+1} - \dfrac{2}{x-3} \right)$

62. $\left(\dfrac{2}{x} - \dfrac{1}{5} \right) \cdot \left(\dfrac{2}{x} + \dfrac{1}{3} \right)$

63. $\dfrac{2}{x^2-16} - \dfrac{3x}{x^2+8x+16} + \dfrac{3}{x+4}$

64. Find the perimeter of the heptagon (polygon with 7 sides).

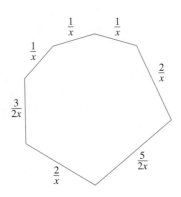

(7.4) *Simplify each complex fraction.*

65. $\dfrac{\frac{2}{5}}{\frac{3}{5}}$

66. $\dfrac{1 - \frac{3}{4}}{2 + \frac{1}{4}}$

67. $\dfrac{\frac{1}{x} - \frac{2}{3x}}{\frac{5}{2x} - \frac{1}{3}}$

68. $\dfrac{\frac{x^2}{15}}{\frac{x+1}{5x}}$

69. $\dfrac{\frac{3}{y^2}}{\frac{6}{y^3}}$

70. $\dfrac{\frac{x+2}{3}}{\frac{5}{x-2}}$

71. $\dfrac{2 - \frac{3}{2x}}{x - \frac{2}{5x}}$

72. $\dfrac{1 + \frac{x}{y}}{\frac{x^2}{y^2} - 1}$

73. $\dfrac{\frac{5}{x} + \frac{1}{xy}}{\frac{3}{x^2}}$

74. $\dfrac{\frac{x}{3} - \frac{3}{x}}{1 + \frac{3}{x}}$

75. $\dfrac{\frac{1}{x-1} + 1}{\frac{1}{x+1} - 1}$

76. $\dfrac{2}{1 - \frac{2}{x}}$

77. $\dfrac{1}{1 + \dfrac{2}{1 - \frac{1}{x}}}$

78. $\dfrac{\dfrac{x^2+5x-6}{(x+6)^2}}{\dfrac{4x+3}{8x+6}}$

79. $\dfrac{\dfrac{x-3}{x+3} + \dfrac{x+3}{x-3}}{\dfrac{x-3}{x+3} - \dfrac{x+3}{x-3}}$

80. $\dfrac{\dfrac{3}{x-1} - \dfrac{2}{1-x}}{\dfrac{2}{x-1} - \dfrac{2}{x}}$

81. If $f(x) = \dfrac{3}{x}$, find each of the following:

  a. $f(a+h)$ **b.** $f(a)$

  c. Use parts (a) and (b) to find $\dfrac{f(a+h) - f(a)}{h}$.

  d. Simplify the results of part (c).

(7.5) *Find each quotient. Use only positive exponents.*

82. $\dfrac{3x^5yb^9}{9xy^7}$

83. Divide $-9xb^4z^3$ by $-4axb^2$.

84. $\dfrac{4xy + 2x^2 - 9}{4xy}$

85. Divide $12xb^2 + 16xb^4$ by $4xb^3$.

Find each quotient.

86. $\dfrac{3x^4 - 25x^2 - 20}{x - 3}$

87. $\dfrac{-x^2 + 2x^4 + 5x - 12}{x - 3}$

88. $\dfrac{2x^4 - x^3 + 2x^2 - 3x + 1}{x - \dfrac{1}{2}}$

89. $\dfrac{x^3 + 3x^2 - 2x + 2}{x - \dfrac{1}{2}}$

90. $\dfrac{3x^4 + 5x^3 + 7x^2 + 3x - 2}{x^2 + x + 2}$

91. $\dfrac{9x^4 - 6x^3 + 3x^2 - 12x - 30}{3x^2 - 2x - 5}$

(7.6) *Use synthetic division to find each quotient.*

92. $\dfrac{3x^3 + 12x - 4}{x - 2}$

93. $\dfrac{3x^3 + 2x^2 - 4x - 1}{x + \dfrac{3}{2}}$

94. $\dfrac{x^5 - 1}{x + 1}$

95. $\dfrac{x^3 - 81}{x - 3}$

96. $\dfrac{x^3 - x^2 + 3x^4 - 2}{x - 4}$

97. $\dfrac{3x^4 - 2x^2 + 10}{x + 2}$

If $P(x) = 3x^5 - 9x + 7$, use the remainder theorem to find the following.

98. $P(4)$

99. $P(-5)$

100. $P\left(\dfrac{2}{3}\right)$

101. $P\left(-\dfrac{1}{2}\right)$

102. If the area of the rectangle is $x^4 - x^3 - 6x^2 - 6x + 18$ square miles and its width is $x - 3$ miles, find the length.

$$\boxed{\begin{array}{c} x^4 - x^3 - 6x^2 - 6x + 18 \\ \text{square miles} \end{array}} \quad \updownarrow \begin{array}{l} x - 3 \\ \text{miles} \end{array}$$

(7.7) *Solve each equation for x. Use a graphing utility to check Exercises 103 through 106.*

103. $\dfrac{2}{5} = \dfrac{x}{15}$

104. $\dfrac{3}{x} + \dfrac{1}{3} = \dfrac{5}{x}$

105. $4 + \dfrac{8}{x} = 8$

106. $\dfrac{2x + 3}{5x - 9} = \dfrac{3}{2}$

107. $\dfrac{1}{x - 2} - \dfrac{3x}{x^2 - 4} = \dfrac{2}{x + 2}$

108. $\dfrac{7}{x} - \dfrac{x}{7} = 0$

109. $\dfrac{x - 2}{x^2 - 7x + 10} = \dfrac{1}{5x - 10} - \dfrac{1}{x - 5}$

Solve the equations for x or perform the indicated operation. Simplify.

110. $\dfrac{5}{x^2 - 7x} + \dfrac{4}{2x - 14}$

111. $3 - \dfrac{5}{x} - \dfrac{2}{x^2} = 0$

112. $\dfrac{4}{3 - x} - \dfrac{7}{2x - 6} + \dfrac{5}{x}$

(7.8) *Solve the equation for the specified variable.*

113. $A = \dfrac{h(a + b)}{2}$, a

114. $\dfrac{1}{R} = \dfrac{1}{R_1} + \dfrac{1}{R_2}$, R_2

115. $I = \dfrac{E}{R + r}$, R

116. $A = P + Prt$, r

117. $H = \dfrac{kA(T_1 - T_2)}{L}$, A

Solve.

118. The sum of a number and twice its reciprocal is 3. Find the number(s).

119. If a number is added to the numerator of $\frac{3}{7}$, and twice that number is added to the denominator of $\frac{3}{7}$, the result is equivalent to $\frac{10}{21}$. Find the number.

120. The denominator of a fraction is 2 more than the numerator. If the numerator is decreased by 3 and the denominator is increased by 5, the resulting fraction is equivalent to $\frac{2}{3}$. Find the fraction.

121. The sum of the reciprocals of two consecutive even integers is $-\frac{9}{40}$. Find the two integers.

122. Three boys can paint a fence in 4 hours, 5 hours, and 6 hours, respectively. Find how long it will take all three boys to paint the fence.

123. If Sue Katz can type a certain number of mailing labels in 6 hours and Tom Neilson and Sue working together can type the same number of mailing labels in 4 hours, find how long it takes Tom alone to type mailing labels.

124. The inlet pipe of a water tank can fill the tank in 2 hours and 30 minutes. The outlet pipe can empty the tank in 2 hours. Find how long it takes to empty a full tank if both pipes are open.

125. Timmy Garnica drove 210 miles in the same amount of time that it took a DC 10 jet to travel 1715 miles. The speed of the jet was 430 mph faster than the speed of the car. Find the speed of the jet.

126. The combined resistance R of two resistors in parallel with resistances r_1 and r_2 is given by the formula $\dfrac{1}{R} = \dfrac{1}{r_1} + \dfrac{1}{r_2}$. If the combined resistance is $\dfrac{30}{11}$ ohms and the resistance of one of the two resistors is 5 ohms, find the resistance of the other resistor.

127. The speed of a Ranger boat in still water is 32 mph. If the boat travels 72 miles upstream in the same time that it takes to travel 120 miles downstream, find the current of the stream.

128. A B737 jet flies 445 miles with the wind and 355 miles against the wind in the same length of time. If the speed of the jet in still air is 400 mph, find the speed of the wind.

129. The speed of a jogger is 3 mph faster than the speed of a walker. If the jogger travels 14 miles in the same amount of time that the walker travels 8 miles, find the speed of the walker.

130. Two Amtrak trains traveling on parallel tracks leave Tucson at the same time. In 6 hours the faster train is 382 miles from Tucson and the trains are 112 miles apart. Find how fast each train is traveling.

(7.9) *Solve each proportion for x.*

131. $\dfrac{3x - 5}{14} = \dfrac{x}{4}$

132. $\dfrac{2 - 5x}{12} = \dfrac{x}{8}$

Solve each of the following variation problems.

133. A is directly proportional to B. If $A = 6$ when $B = 14$, find A when $B = 21$.

134. C is inversely proportional to D. If $C = 12$ when $D = 8$, find C when $D = 24$.

135. According to Boyle's law, the pressure exerted by a gas is inversely proportional to the volume, as long as the temperature stays the same. If a gas exerts a pressure of 1250 pounds per square inch when the volume is 2 cubic feet, find the volume when the pressure is 800 pounds per square inch.

136. The surface area of a sphere varies directly as the square of its radius. If the surface area is 36 square inches when the radius is 3 inches, find the surface area when the radius is 4 inches.

CHAPTER 7 TEST

Find the domain of each rational function.

1. $f(x) = \dfrac{5x^2}{1 - x}$

2. $g(x) = \dfrac{9x^2 - 9}{x^2 + 4x + 3}$

Write each rational expression in lowest terms.

3. $\dfrac{5x^7}{3x^4}$

4. $\dfrac{7x - 21}{24 - 8x}$

5. $\dfrac{x^2 - 4x}{x^2 + 5x - 36}$

Perform the indicated operation. Write answers in lowest terms.

6. $\dfrac{x}{x - 2} \cdot \dfrac{x^2 - 4}{5x}$

7. $\dfrac{2x^3 + 16}{6x^2 + 12x} \cdot \dfrac{5}{x^2 - 2x + 4}$

8. $\dfrac{26ab}{7c} \div \dfrac{13a^2c^5}{14a^4b^3}$

9. $\dfrac{3x^2 - 12}{x^2 + 2x - 8} \div \dfrac{6x + 18}{x + 4}$

10. $\dfrac{4x - 12}{2x - 9} \div \dfrac{3 - x}{4x^2 - 81} \cdot \dfrac{x + 3}{5x + 15}$

11. $\dfrac{5}{4x^3} + \dfrac{7}{4x^3}$

12. $\dfrac{3 + 2x}{10 - x} + \dfrac{13 + x}{x - 10}$

13. $\dfrac{3}{x^2 - x - 6} + \dfrac{2}{x^2 - 5x + 6}$

14. $\dfrac{5}{x - 7} - \dfrac{2x}{3x - 21} + \dfrac{x}{2x - 14}$

15. $\dfrac{3x}{5} \cdot \left(\dfrac{5}{x} - \dfrac{5}{2x} \right)$

Simplify each complex fraction.

16. $\dfrac{\dfrac{4x}{13}}{\dfrac{20x}{13}}$

17. $\dfrac{\dfrac{5}{x} - \dfrac{7}{3x}}{\dfrac{9}{8x} - \dfrac{1}{x}}$

18. $\dfrac{\dfrac{x^2 - 5x + 6}{x + 3}}{\dfrac{x^2 - 4x + 4}{x^2 - 9}}$

Divide.

19. $\dfrac{4x^2 y + 9x + z}{3xz}$

20. $\dfrac{x^6 + 3x^5 - 2x^4 + x^2 - 3x + 2}{x - 2}$

21. Use synthetic division to divide $4x^4 - 3x^3 + 2x^2 - x - 1$ by $x + 3$.

22. If $P(x) = 4x^4 + 7x^2 - 2x - 5$, use the remainder theorem to find $P(-2)$.

If $f(x) = x$, $g(x) = x - 7$, and $h(x) = x^2 - 6x + 5$, find the following.

23. $(h - g)(x)$

24. $(h \cdot f)(x)$

Solve each equation for x. Use a graphing utility to check Exercise 25.

25. $\dfrac{5x + 3}{3x - 7} = \dfrac{19}{7}$

26. $\dfrac{5}{x - 5} + \dfrac{x}{x + 5} = -\dfrac{29}{21}$

27. $\dfrac{x}{x - 4} = 3 - \dfrac{4}{x - 4}$

28. Solve for x: $\dfrac{x + b}{a} = \dfrac{4x - 7a}{b}$

29. The product of one more than a number and twice the reciprocal of the number is $\frac{12}{5}$. Find the number.

30. If Jan can weed the garden in 2 hours and her husband can weed it in 1 hour and 30 minutes, find how long it takes them to weed the garden together.

31. Suppose that W is inversely proportional to V. If $W = 20$ when $V = 12$, find W when $V = 15$.

32. Suppose that Q is jointly proportional to R and the square of S. If $Q = 24$ when $R = 3$ and $S = 4$, find Q when $R = 2$ and $S = 3$.

33. When an anvil is dropped into a gorge, the speed with which it strikes the ground is directly proportional to the square root of the distance it falls. An anvil that falls 400 feet hits the ground at a speed of 160 feet per second. Find the height of a cliff over the gorge if a dropped anvil hits the ground at a speed of 128 feet per second.

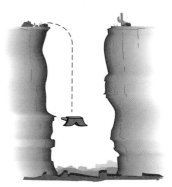

CHAPTER 7   CUMULATIVE REVIEW

1. Write each sentence using mathematical symbols.

   a. The sum of x and 5 is 20.

   b. Two times the sum of 3 and y amounts to 4.

   c. Subtract 8 from x, and the difference is the same as the product of 2 and x.

   d. The quotient of z and 9 is 3 times the difference of z and 5.

2. State the basic property that is being illustrated on the following calculator screens.

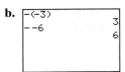

a.
```
11*(1/11)        1
-9*(-1/9)        1
5*5-1            1
```

b.
```
-(-3)            3
--6              6
```

c.
```
2(3+5)          16
2*3+2*5         16
```

3. A pennant in the shape of an isosceles triangle is to be constructed for the Slidell High School Athletic Club and sold as a fund-raiser. The company manufacturing the pennants charges according to perimeter. The club has determined that a perimeter of 149 centimeters each should make a nice profit. If each equal side of the triangle is 12 centimeters more than twice the length of the third side, find the lengths of the sides of each triangular pennant.

4. Find the midpoint of the line segment that joins points $P(-3, 3)$ and $Q(1, 0)$.

5. Graph $5x - 3y = 10$ in an integer window centered at the origin. Find the y-coordinate that makes the ordered pair $(8, ?)$ a solution for the given equation.

6. Find an equation of the line that contains the point $(4, 4)$ and is parallel to the line $2x + 3y = -6$. Write the equation in standard form.

7. Solve $\left|\dfrac{x}{2} - 1\right| = 11$.

8. Solve algebraically for x: $|5x + 1| + 1 \le 10$.

9. Solve the system:
$$\begin{cases} 2x - 4y + 8z = 2 \\ -x - 3y + z = 11 \\ x - 2y + 4z = 0 \end{cases}$$

10. Use the product rule to multiply.

a. $(3x^6)(5x)$

b. $(-2x^3p^2)(4xp^{10})$

11. Simplify each expression. Use positive exponents to write answers.

a. $\left(\dfrac{3x^2y}{y^{-9}z}\right)^{-2}$

b. $\left(\dfrac{3a^2}{2x^{-1}}\right)^3\left(\dfrac{x^{-3}}{4a^{-2}}\right)^{-1}$

12. Multiply $(2x - 7)(3x - 4)$.

13. Factor $2n^2 - 38n + 80$.

14. Factor $m^2n^2 + m^2 - 2n^2 - 2$.

15. Solve $x^3 = 4x$.

16. For the ICL Production Company, the rational function $C(x) = \dfrac{2.6x + 10{,}000}{x}$ describes the company's cost per disk of pressing x compact disks. Find the cost per disk for pressing:

a. 100 compact disks.

b. 1000 compact disks.

17. Multiply.

a. $\dfrac{2x^2 + 3x - 2}{-4x - 8} \cdot \dfrac{16x^2}{4x^2 - 1}$

b. $(ac - ad + bc - bd) \cdot \dfrac{a + b}{d - c}$

18. Simplify.
$$\dfrac{x^{-1} + 2xy^{-1}}{x^{-2} - x^{-2}y^{-1}}$$

19. Find the quotient: $\dfrac{7a^2b - 2ab^2}{2ab^2}$.

20. Solve $\dfrac{x + 6}{x - 2} = \dfrac{2(x + 2)}{x - 2}$.

21. The intensity $I(x)$ of light, in foot-candles, that is x feet from its source is given by the rational function

$$I(x) = \dfrac{320}{x^2}.$$

How far away is the source if the intensity of light is 5 foot-candles?

22. Suppose that y varies directly as x. If y is 5 when x is 30, find the constant of variation. Also, find y when $x = 90$.

RATIONAL EXPONENTS, RADICALS, AND COMPLEX NUMBERS

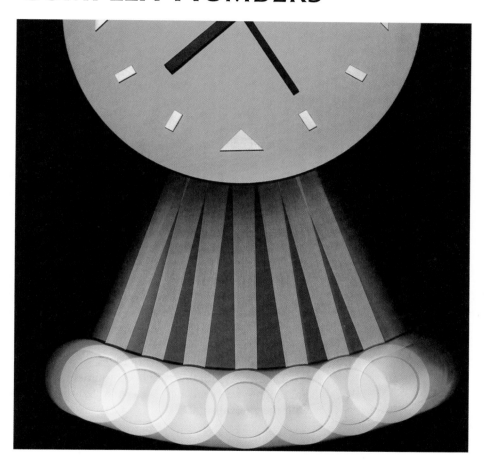

CALCULATING THE LENGTH AND PERIOD OF A PENDULUM

A simple pendulum, like the kind found in a clock, of a given length has a unique property. The time required to complete one full back-and-forth swing is the same regardless of the mass of the pendulum or the distance it travels. The time to complete one full swing *does*, however, depend on the pendulum's length.

IN THE CHAPTER GROUP ACTIVITY ON PAGE 507, YOU WILL HAVE THE OPPORTUNITY TO INVESTIGATE THE RELATIONSHIP BETWEEN THE LENGTH OF A PENDULUM AND ITS PERIOD.

I n this chapter, radical notation is reviewed, and then rational exponents are introduced. As the name implies, rational exponents are exponents that are rational numbers. We present an interpretation of rational exponents that is consistent with the meaning and rules already established for integer exponents, and we present two forms of notation for roots: radical and exponent. We conclude this chapter with complex numbers, a natural extension of the real number system.

8.1   RADICALS AND RADICAL FUNCTIONS

O B J E C T I V E S

1. Find square roots.
2. Find cube roots.
3. Find nth roots.
4. Graph square root functions.
5. Graph cube root functions.

TAPE IAG 8.1

Recall from Section 1.3 that if some number a is the square of some other number b, such that

$$a = b^2$$

then we also say that b is a square root of a.

Thus, because

$$25 = 5^2$$

we say that 5 is a square root of 25.

Also, because

$$25 = (-5)^2$$

we say that -5 is a square root of 25.

Like every other positive real number, 25 has two square roots—one positive, one negative. Recall that we symbolize the **nonnegative, or principal, square root** with the **radical sign:**

$$\sqrt{25} = 5$$

We denote the **negative square root** with the **negative radical sign:**

$$-\sqrt{25} = -5$$

An expression containing a radical sign is called a **radical expression.** An expression within, or "under," a radical sign is called a **radicand.**

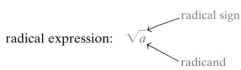

radical sign

radical expression: $\sqrt{a}$

radicand

> **SQUARE ROOT**
>
> The **principal square root** of a nonnegative number a is its nonnegative square root. The principal square root is written as $\sqrt{a}$. The **negative square root** of a is written as $-\sqrt{a}$. Thus,
>
> $$\sqrt{a} = b \text{ only if } b^2 = a \text{ and } b \geq 0$$

EXAMPLE 1   Simplify. Assume that all variables represent positive numbers.

a. $\sqrt{36}$ **b.** $\sqrt{0}$ **c.** $\sqrt{\dfrac{4}{49}}$ **d.** $\sqrt{0.25}$ **e.** $\sqrt{x^6}$ **f.** $\sqrt{9x^{10}}$ **g.** $-\sqrt{81}$

Solution: **a.** $\sqrt{36} = 6$ because $6^2 = 36$ and 6 is not negative.

b. $\sqrt{0} = 0$ because $0^2 = 0$ and 0 is not negative.

c. $\sqrt{\dfrac{4}{49}} = \dfrac{2}{7}$ because $\left(\dfrac{2}{7}\right)^2 = \dfrac{4}{49}$ and $\dfrac{2}{7}$ is not negative.

d. $\sqrt{0.25} = 0.5$ because $(0.5)^2 = 0.25$.

e. $\sqrt{x^6} = x^3$ because $(x^3)^2 = x^6$.

f. $\sqrt{9x^{10}} = 3x^5$ because $(3x^5)^2 = 9x^{10}$.

g. $-\sqrt{81} = -9$. The negative in front of the radical indicates the negative square root of 81.

TECHNOLOGY NOTE

Most graphing utilities will give a message such as the one below when taking the square root of a negative number when you are in real mode.

```
ERR:NONREAL ANS
1▶Quit
2:Goto
```

Can we find the square root of a negative number, say $\sqrt{-4}$? That is, can we find a real number whose square is -4? No, there is no real number whose square is -4, and we say that $\sqrt{-4}$ is not a real number. In general,

The square root of a negative number is not a real number.

Recall that numbers such as 1, 4, 9, and 25 are called **perfect squares,** since $1 = 1^2$, $4 = 2^2$, $9 = 3^2$, and $25 = 5^2$. Square roots of perfect square radicands simplify to rational numbers. What happens when we try to simplify a root such as $\sqrt{3}$? Since 3 is not a perfect square, $\sqrt{3}$ is not a rational number. It is called an **irrational number,** and we can use a calculator to find a decimal **approximation** of it. For example, an approximation for $\sqrt{3}$ is

$$\sqrt{3} \approx 1.732$$
$$\uparrow$$

approximation symbol

Finding roots can be extended to other roots such as cube roots. For example, since $2^3 = 8$, we call 2 the **cube root** of 8. In symbols, we write

$$\sqrt[3]{8} = 2$$

> **CUBE ROOT**
>
> The **cube root** of a real number a is written as $\sqrt[3]{a}$, and
>
> $$\sqrt[3]{a} = b \text{ only if } b^3 = a$$

From this definition, we have

$$\sqrt[3]{64} = 4 \text{ since } 4^3 = 64$$
$$\sqrt[3]{-27} = -3 \text{ since } (-3)^3 = -27$$

Notice that, unlike with square roots, **it is possible to have a negative radicand when finding a cube root.** This is so because the *cube* of a negative number is a negative number. Therefore, the *cube root* of a negative number is a negative number.

EXAMPLE 2   Find the cube roots.

a. $\sqrt[3]{1}$ b. $\sqrt[3]{-64}$ c. $\sqrt[3]{\dfrac{8}{125}}$ d. $\sqrt[3]{x^6}$ e. $\sqrt[3]{-8x^9}$

Solution:
a. $\sqrt[3]{1} = 1$ because $1^3 = 1$.

b. $\sqrt[3]{-64} = -4$ because $(-4)^3 = -64$.

c. $\sqrt[3]{\dfrac{8}{125}} = \dfrac{2}{5}$ because $\left(\dfrac{2}{5}\right)^3 = \dfrac{8}{125}$.

d. $\sqrt[3]{x^6} = x^2$ because $(x^2)^3 = x^6$.

e. $\sqrt[3]{-8x^9} = -2x^3$ because $(-2x^3)^3 = -8x^9$.

Just as we can raise a real number to powers other than 2 or 3, we can find roots other than square roots and cube roots. In fact, we can find the ***n*th root** of a number, where n is any natural number. In symbols, the nth root of a is written as $\sqrt[n]{a}$, where n is called the **index.** The index 2 is usually omitted for square roots.

> R E M I N D E R   If the index is even, such as $\sqrt{}, \sqrt[4]{}, \sqrt[6]{}$, and so on, the radicand must be nonnegative for the root to be a real number. For example,
>
> $$\sqrt[4]{16} = 2, \text{ but } \sqrt[4]{-16} \text{ is not a real number.}$$
> $$\sqrt[6]{64} = 2, \text{ but } \sqrt[6]{-64} \text{ is not a real number.}$$
>
> If the index is odd, such as $\sqrt[3]{}, \sqrt[5]{}$, and so on, the radicand may be any real number. For example,
>
> $$\sqrt[3]{64} = 4 \text{ and } \sqrt[3]{-64} = -4$$
> $$\sqrt[5]{32} = 2 \text{ and } \sqrt[5]{-32} = -2$$

A calculator can be used to check simplification of numerical radical expressions. See the margin for examples in this section and the next two sections.

EXAMPLE 3   Simplify the following expressions.

    a. $\sqrt[4]{81}$ **b.** $\sqrt[5]{-243}$ **c.** $-\sqrt{25}$ **d.** $\sqrt[4]{-81}$ **e.** $\sqrt[3]{64x^3}$

Solution: **a.** $\sqrt[4]{81} = 3$ because $3^4 = 81$ and 3 is positive.

```
4×⁴√81
                    3
5×⁵√-243
                   -3
-√(25)
                   -5
```

A check of parts (a), (b), and (c) of Example 3 using a calculator.

    b. $\sqrt[5]{-243} = -3$ because $(-3)^5 = -243$.

    c. $-\sqrt{25} = -5$ because -5 is the opposite of $\sqrt{25}$.

    d. $\sqrt[4]{-81}$ is not a real number. There is no real number that, when raised to the fourth power, is -81.

    e. $\sqrt[3]{64x^3} = 4x$ because $(4x)^3 = 64x^3$.

    Recall that the notation $\sqrt{a^2}$ indicates the positive square root of a^2 only. For example,

$$\sqrt{(-5)^2} = \sqrt{25} = 5$$

When variables are present in the radicand and it is unclear whether the variable represents a positive number or a negative number. Absolute value bars are sometimes needed to ensure that the result is a positive number. For example,

$$\sqrt{x^2} = |x|$$

This ensures that the result is positive. This same situation may occur when the index is any *even* positive integer. When the index is any *odd* positive integer, absolute value bars are not necessary.

> If n is an even positive integer, then $\sqrt[n]{a^n} = |a|$.
>
> If n is an odd positive integer, then $\sqrt[n]{a^n} = a$.

EXAMPLE 4   Simplify.

    a. $\sqrt{(-3)^2}$ **b.** $\sqrt{x^2}$ **c.** $\sqrt[4]{(x-2)^4}$ **d.** $\sqrt[3]{(-5)^3}$ **e.** $\sqrt[5]{(2x-7)^5}$

Solution: **a.** $\sqrt{(-3)^2} = |-3| = 3$ When the index is even, the absolute value bars ensure us

    b. $\sqrt{x^2} = |x|$ that our result is not negative.

    c. $\sqrt[4]{(x-2)^4} = |x-2|$

    d. $\sqrt[3]{(-5)^3} = -5$

    e. $\sqrt[5]{(2x-7)^5} = 2x - 7$ Absolute value bars are not needed when the index is odd.

4 Recall that an equation in x and y describes a function if each x-value is paired with exactly one y-value. With this in mind, does the equation

$$y = \sqrt{x}$$

describe a function? First, notice that replacement values for x must be nonnegative real numbers, since $\sqrt{x}$ is not a real number if $x < 0$. The notation $\sqrt{x}$ denotes the principal square root of x, so for every nonnegative number x, there is exactly one number, $\sqrt{x}$. Therefore, $y = \sqrt{x}$ describes a function, and we may write it as

$$f(x) = \sqrt{x}$$

Recall that the domain of a function in x is the set of all possible replacement values for x. This means that the domain of this function is the set of all nonnegative numbers, or $\{x \mid x \geq 0\}$.

We find function values for $f(x)$ as usual. For example,

$$f(0) = \sqrt{0} = 0$$
$$f(1) = \sqrt{1} = 1$$
$$f(4) = \sqrt{4} = 2$$
$$f(9) = \sqrt{9} = 3$$

Choosing perfect squares for x ensures us that $f(x)$ is a rational number, but it is important to stress that $f(x) = \sqrt{x}$ is defined for all nonnegative real numbers. For example,

$$f(3) = \sqrt{3} \approx 1.7$$

EXAMPLE 5 Identify the domain of the function $f(x) = \sqrt{x}$ and then graph the function.

Solution: Recall that the domain of this function is the set of all nonnegative numbers or $\{x \mid x \geq 0\}$. Below, we graph $y_1 = \sqrt{x}$ in a $[-3, 10, 1]$ by $[-2, 10, 1]$ window. The table below shows that a negative x-value gives an error message since $\sqrt{x}$ is not a real number when x is negative.

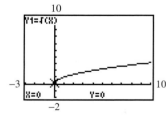

Notice that the graph of this function passes the vertical line test, as expected.

5 The equation $f(x) = \sqrt[3]{x}$ also describes a function. Here x may be any real number, so the domain of this function is the set of all real numbers. A few function values are given next.

$$f(0) = \sqrt[3]{0} = 0$$
$$f(1) = \sqrt[3]{1} = 1$$
$$f(-1) = \sqrt[3]{-1} = -1$$
$$f(6) = \sqrt[3]{6}$$
$$f(-6) = \sqrt[3]{-6}$$
$$f(8) = \sqrt[3]{8} = 2$$
$$f(-8) = \sqrt[3]{-8} = -2$$

} Here, the radicands are not perfect cubes. The radicals do not simplify to rational numbers.

EXAMPLE 6 Identify the domain of the function $f(x) = \sqrt[3]{x}$ and then graph the function.

Solution: The domain of this function is the set of all real numbers. Below, we graph $y_1 = \sqrt[3]{x}$ in a $[-10, 10, 1]$ by $[-5, 5, 1]$ window.

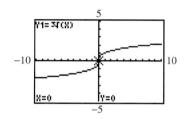

The graph of this function passes the vertical line test, as expected.

EXERCISE SET 8.1

Simplify. Assume that variables represent positive real numbers. See Example 1.

1. $\sqrt{100}$

2. $\sqrt{400}$

3. $\sqrt{\dfrac{1}{4}}$

4. $\sqrt{\dfrac{9}{25}}$

5. $\sqrt{0.0001}$

6. $\sqrt{0.04}$

7. $-\sqrt{36}$

8. $-\sqrt{9}$

9. $\sqrt{x^{10}}$

10. $\sqrt{x^{16}}$

11. $\sqrt{16y^6}$

12. $\sqrt{64y^{20}}$

Simplify. If the radical is not a real number, state so. Assume that variables represent positive real numbers. See Examples 1 through 3.

13. $\sqrt[3]{\dfrac{1}{8}}$

14. $\sqrt[3]{\dfrac{27}{64}}$

15. $-\sqrt[4]{16}$

16. $\sqrt[5]{-243}$

17. $\sqrt[4]{-16}$

18. $\sqrt{-16}$

19. $\sqrt[3]{-125x^9}$

20. $\sqrt[3]{a^{12}b^3}$

21. $\sqrt[5]{32x^5}$

22. $\sqrt[5]{-32y^{10}}$

Simplify each root. Assume that variables represent any real numbers. See Example 4.

23. $\sqrt{z^2}$

24. $\sqrt{y^{10}}$

25. $\sqrt[3]{x^3}$

26. $\sqrt[3]{z^6}$

27. $\sqrt{(x-5)^2}$

28. $\sqrt{(y-6)^2}$

29. $\sqrt[4]{(2z)^4}$

30. $\sqrt{(9x)^2}$

31. $\sqrt{100(2x-y)^6}$

32. $\sqrt[4]{81(y-10z)^{12}}$

33. When does a square root simplify to a rational number?

34. When does a cube root simplify to a rational number?

Simplify each radical. Assume that all variables represent positive real numbers.

35. $-\sqrt{121}$

36. $-\sqrt[3]{125}$

37. $\sqrt[3]{8x^3}$

38. $\sqrt{16x^8}$

39. $\sqrt{y^{12}}$

40. $\sqrt[3]{y^{12}}$

41. $\sqrt{25a^2b^{20}}$

42. $\sqrt{9x^4y^6}$

43. $\sqrt[3]{-27x^9}$

44. $\sqrt[3]{-8a^{21}b^6}$

45. $\sqrt[4]{a^{16}b^4}$

46. $\sqrt[4]{x^8y^{12}}$

47. $\sqrt[5]{-32x^{10}y^5}$

48. $\sqrt[5]{-243z^{15}}$

49. $\sqrt{\dfrac{25}{49}}$

50. $\sqrt{\dfrac{4}{81}}$

51. $\sqrt{\dfrac{x^2}{4y^2}}$

52. $\sqrt{\dfrac{y^{10}}{9x^6}}$

53. $-\sqrt[3]{\dfrac{z^{21}}{27x^3}}$

54. $-\sqrt[3]{\dfrac{64a^3}{b^9}}$

55. $\sqrt[4]{\dfrac{x^4}{16}}$

56. $\sqrt[4]{\dfrac{y^4}{81x^4}}$

*Determine whether each square root is rational or irrational. If it is rational, find its **exact value**. If it is irrational, use a calculator to write a three-decimal-place **approximation**.*

57. $\sqrt{9}$

58. $\sqrt{8}$

59. $\sqrt{37}$

60. $\sqrt{36}$

61. $\sqrt{169}$

62. $\sqrt{160}$

63. $\sqrt{4}$

64. $\sqrt{27}$

Simplify each root. Assume that variables represent any real numbers.

65. $\sqrt[3]{x^{15}}$

66. $\sqrt[3]{y^{12}}$

67. $\sqrt{x^{12}}$

68. $\sqrt{z^{16}}$

69. $\sqrt{81x^2}$

70. $\sqrt{100z^4}$

71. $-\sqrt{144y^{14}}$

72. $-\sqrt{121z^{22}}$

73. $\sqrt{x^2+4x+4}$

(*Hint:* Factor the polynomial first.)

74. $\sqrt{x^2-6x+9}$

(*Hint:* Factor the polynomial first.)

If $f(x) = \sqrt{2x+3}$ and $g(x) = \sqrt[3]{x-8}$, find the following function values.

75. $f(0)$

76. $g(0)$

77. $g(7)$

78. $f(-1)$

79. $g(-19)$

80. $f(3)$

81. $f(2)$

82. $g(1)$

Identify the domain of each function and then match the graphs with its equation. All graphs are in a $[-10, 10, 1]$ by $[-5, 5, 1]$ window.

A.

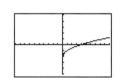

B.

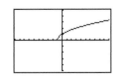

C.

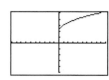

D.

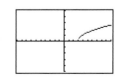

83. $f(x) = \sqrt{x} + 2$

84. $f(x) = \sqrt{x} - 2$

85. $f(x) = \sqrt{x-3}$

86. $f(x) = \sqrt{x+1}$

Identify the domain of each function and then match the graph with its equation. All graphs are in a $[-4.7, 4.7, 1]$ by $[-5, 5, 1]$ window.

87. $f(x) = \sqrt[3]{x} + 1$

88. $f(x) = \sqrt[3]{x} - 2$

89. $g(x) = \sqrt[3]{x - 1}$

90. $g(x) = \sqrt[3]{x + 1}$

A.

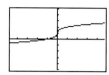

B.

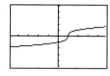

C.

D.

Review Exercises

Simplify each exponential expression. See Sections 6.1 and 6.2.

91. $(-2x^3y^2)^5$

92. $(4y^6z^7)^3$

93. $(-3x^2y^3z^5)(20x^5y^7)$

94. $(-14a^5bc^2)(2abc^4)$

95. $\dfrac{7x^{-1}y}{14(x^5y^2)^{-2}}$

96. $\dfrac{(2a^{-1}b^2)^3}{(8a^2b)^{-2}}$

8.2 | RATIONAL EXPONENTS

O B J E C T I V E S

TAPE IAG 8.2

1 Understand the meaning of a raised to the $\dfrac{1}{n}$-th power.

2 Understand the meaning of a raised to the $\dfrac{m}{n}$-th power.

3 Understand the meaning of a raised to the $-\dfrac{m}{n}$-th power.

4 Use rules for exponents to simplify expressions that contain rational exponents.

1 We introduce now an alternate, equivalent notation for the nth root of a number a. Recall that if b is the nth root of a we write with radicals

$$b = \sqrt[n]{a}$$

Alternately, we write with **rational exponents**

$$b = a^{1/n}$$

Thus, we define $a^{1/n}$ to be $\sqrt[n]{a}$. The two symbols have identical meaning and can be used interchangeably to denote the number whose nth power is a.

> **DEFINITION OF $a^{1/n}$**
>
> If n is a positive integer greater than 1 and $\sqrt[n]{a}$ is a real number, then $\sqrt[n]{a}$ can also be written as $a^{1/n}$:
>
> $$a^{1/n} = \sqrt[n]{a}$$

Notice that the denominator of the rational exponent corresponds to the index of the radical.

EXAMPLE 1 Use radical notation to write the following. Simplify if possible.

a. $4^{1/2}$ **b.** $64^{1/3}$ **c.** $x^{1/4}$ **d.** $0^{1/6}$ **e.** $-9^{1/2}$ **f.** $(81x^8)^{1/4}$ **g.** $(5y)^{1/3}$

Solution:

a. $4^{1/2} = \sqrt{4} = 2$

b. $64^{1/3} = \sqrt[3]{64} = 4$

c. $x^{1/4} = \sqrt[4]{x}$

d. $0^{1/6} = \sqrt[6]{0} = 0$

e. $-9^{1/2} = -\sqrt{9} = -3$

f. $(81x^8)^{1/4} = \sqrt[4]{81x^8} = 3x^2$

g. $(5y)^{1/3} = \sqrt[3]{5y}$

2 As we expand our use of exponents to include $\dfrac{m}{n}$, we define their meaning so that rules for exponents still hold true. For example, by properties of exponents,

$$8^{1/3} \cdot 8^{1/3} = 8^{1/3 + 1/3} = 8^{2/3}$$

For this to be true, we give meaning to the expression $8^{2/3}$ in such a way that the product and power rules do indeed hold true.

> **DEFINITION OF $a^{m/n}$**
>
> If m and n are positive integers greater than 1 with $\dfrac{m}{n}$ in lowest terms, then
>
> $$a^{m/n} = \sqrt[n]{a^m} = (\sqrt[n]{a})^m$$
>
> as long as $\sqrt[n]{a}$ is a real number.

Notice that the denominator n of the rational exponent corresponds to the index of the radical. The numerator m of the rational exponent indicates that the base is to be raised to the mth power.

This means that

$$8^{2/3} = \sqrt[3]{8^2} = \sqrt[3]{64} = 4 \quad \text{or}$$

$$8^{2/3} = \left(\sqrt[3]{8}\right)^2 = 2^2 = 8$$

Most of the time, $\left(\sqrt[n]{a}\right)^m$ will be easier to calculate.

EXAMPLE 2   Use radical notation to write the following. Then simplify if possible.

a. $4^{3/2}$ **b.** $-16^{3/4}$ **c.** $(-27)^{2/3}$ **d.** $\left(\dfrac{1}{9}\right)^{3/2}$ **e.** $(4x - 1)^{3/5}$

Solution: **a.** $4^{3/2} = (\sqrt{4})^3 = 2^3 = 8$
b. $-16^{3/4} = -(\sqrt[4]{16})^3 = -(2)^3 = -8$
c. $(-27)^{2/3} = (\sqrt[3]{-27})^2 = (-3)^2 = 9$
d. $\left(\dfrac{1}{9}\right)^{3/2} = \left(\sqrt{\dfrac{1}{9}}\right)^3 = \left(\dfrac{1}{3}\right)^3 = \dfrac{1}{27}$
e. $(4x - 1)^{3/5} = \sqrt[5]{(4x - 1)^3}$

> R E M I N D E R   The *denominator* of a rational exponent is the index of the corresponding radical. For example, $x^{1/5} = \sqrt[5]{x}$ and $z^{2/3} = \sqrt[3]{z^2}$, or $z^{2/3} = (\sqrt[3]{z})^2$.

3   The rational exponents we have given meaning to exclude negative rational numbers. To complete the set of definitions, we define $a^{-m/n}$.

DEFINITION OF $a^{-m/n}$

$$a^{-m/n} = \dfrac{1}{a^{m/n}}$$

as long as $a^{m/n}$ is a nonzero real number.

EXAMPLE 3   Write each expression with a positive exponent, and then simplify. Check using a calculator.

a. $16^{-3/4}$ **b.** $(-27)^{-2/3}$

Solution: **a.** $16^{-3/4} = \dfrac{1}{16^{3/4}} = \dfrac{1}{(\sqrt[4]{16})^3} = \dfrac{1}{2^3} = \dfrac{1}{8}$

b. $(-27)^{-2/3} = \dfrac{1}{(-27)^{2/3}} = \dfrac{1}{(\sqrt[3]{-27})^2} = \dfrac{1}{(-3)^2} = \dfrac{1}{9}$

```
16^(-3/4)▶Frac
            1/8
(-27)^(-2/3)▶Fra
c
            1/9
```

A calculator check for
Example 3.

REMINDER   Notice that the sign of the base is not affected by the sign of its exponent. For example,

$$9^{-3/2} = \dfrac{1}{9^{3/2}} = \dfrac{1}{(\sqrt{9})^3} = \dfrac{1}{27}$$

Also,

$$(-27)^{-1/3} = \dfrac{1}{(-27)^{1/3}} = -\dfrac{1}{3}$$

4   It can be shown that the properties of integer exponents hold for rational exponents. By using these properties and definitions, we can now simplify expressions that contain rational exponents.

These rules are repeated here for review.

SUMMARY OF EXPONENT RULES

If m and n are rational numbers, and a, b, and c are numbers for which the expressions below exist, then

Product rule for exponents:	$a^m \cdot a^n = a^{m+n}$
Power rule for exponents:	$(a^m)^n = a^{m \cdot n}$
Power rules for products and quotients:	$(ab)^n = a^n b^n$ and
	$\left(\dfrac{a}{c}\right)^n = \dfrac{a^n}{c^n}, \; c \neq 0$
Quotient rule for exponents:	$\dfrac{a^m}{a^n} = a^{m-n}, \; a \neq 0$
Zero exponent:	$a^0 = 1, \; a \neq 0$
Negative exponent:	$a^{-n} = \dfrac{1}{a^n}, \; a \neq 0$

EXAMPLE 4   Write with rational exponents. Then simplify if possible.

a. $\sqrt[5]{x}$ **b.** $\sqrt[3]{17x^2y^5}$ **c.** $\sqrt{x - 5a}$ **d.** $3\sqrt{2p} - 5\sqrt[3]{p^2}$

Solution: **a.** $\sqrt[5]{x} = x^{1/5}$

b. $\sqrt[3]{17x^2y^5} = (17x^2y^5)^{1/3}$. We can further simplify this expression by using a power rule for exponents.

$$(17x^2y^5)^{1/3} = 17^{1/3}x^{2/3}y^{5/3}$$

c. $\sqrt{x-5a} = (x-5a)^{1/2}$

d. $3\sqrt{2p} - 5\sqrt[3]{p^2} = 3(2p)^{1/2} - 5(p^2)^{1/3} = 3(2p)^{1/2} - 5p^{2/3}$

We can use rational exponents to help us simplify radicals.

EXAMPLE 5   Use rational exponents to simplify. Assume that variables represent positive numbers and give the result in radical form.

   a. $\sqrt[8]{x^4}$ **b.** $\sqrt[4]{r^2s^6}$

Solution: **a.** $\sqrt[8]{x^4} = x^{4/8} = x^{1/2} = \sqrt{x}$

   b. $\sqrt[4]{r^2s^6} = (r^2s^6)^{1/4} = r^{2/4}s^{6/4} = r^{1/2}s^{3/2} = (rs^3)^{1/2} = \sqrt{rs^3}$

EXAMPLE 6   Use properties of exponents to simplify the following expressions. Write results with only positive exponents.

   a. $x^{1/2}x^{1/3}$ **b.** $\dfrac{7^{1/3}}{7^{4/3}}$ **c.** $\dfrac{(2x^{2/5}y^{-1/3})^5}{x^2y}$

Solution: **a.** $x^{1/2}x^{1/3} = x^{(1/2+1/3)} = x^{3/6+2/6} = x^{5/6}$ **b.** $\dfrac{7^{1/3}}{7^{4/3}} = 7^{1/3-4/3} = 7^{-3/3} = 7^{-1} = \dfrac{1}{7}$

   c. We begin by using the power rule $(ab)^m = a^m b^m$ to simplify the numerator.

$$\frac{(2x^{2/5}y^{-1/3})^5}{x^2y} = \frac{2^5(x^{2/5})^5(y^{-1/3})^5}{x^2y} = \frac{32x^2y^{-5/3}}{x^2y}$$

$$= 32x^{2-2}y^{-5/3-3/3} \quad \text{Apply the quotient rule.}$$

$$= 32x^0y^{-8/3}$$

$$= \frac{32}{y^{8/3}}$$

EXAMPLE 7   Multiply.

   a. $z^{2/3}(z^{1/3} - z^5)$ **b.** $(x^{1/3} - 5)(x^{1/3} + 2)$

Solution: **a.** $z^{2/3}(z^{1/3} - z^5) = z^{2/3}z^{1/3} - z^{2/3}z^5$ Apply the distributive property.

   $= z^{(2/3+1/3)} - z^{(2/3+5)}$ Use the product rule.

   $= z^{3/3} - z^{(2/3+15/3)}$

   $= z - z^{17/3}$

   b. $(x^{1/3} - 5)(x^{1/3} + 2) = x^{2/3} + 2x^{1/3} - 5x^{1/3} - 10$

   $= x^{2/3} - 3x^{1/3} - 10$

EXAMPLE 8 Factor $x^{-1/2}$ from the expression $3x^{-1/2} - 7x^{5/2}$. Assume that all variables represent positive numbers.

Solution: $3x^{-1/2} - 7x^{5/2} = (x^{-1/2})(3) - (x^{-1/2})(7x^{6/2})$
$$= x^{-1/2}(3 - 7x^3)$$

To check, multiply $x^{-1/2}(3 - 7x^3)$ to see that the product is $3x^{-1/2} - 7x^{5/2}$.

EXERCISE SET 8.2

Use radical notation to write the following. Simplify if possible.

1. $49^{1/2}$

2. $64^{1/3}$

3. $27^{1/3}$

4. $8^{1/3}$

5. $\left(\dfrac{1}{16}\right)^{1/4}$

6. $\left(\dfrac{1}{64}\right)^{1/2}$

7. $169^{1/2}$

8. $81^{1/4}$

9. $2m^{1/3}$

10. $(2m)^{1/3}$

11. $(9x^4)^{1/2}$

12. $(16x^8)^{1/2}$

13. $(-27)^{1/3}$

14. $-64^{1/2}$

15. $-16^{1/4}$

16. $(-32)^{1/5}$

17. $16^{3/4}$

18. $4^{5/2}$

19. $(-64)^{2/3}$

20. $(-8)^{4/3}$

21. $(-16)^{3/4}$

22. $(-9)^{3/2}$

23. $(2x)^{3/5}$

24. $2x^{3/5}$

25. $(7x + 2)^{2/3}$

26. $(x - 4)^{3/4}$

27. $\left(\dfrac{16}{9}\right)^{3/2}$

28. $\left(\dfrac{49}{25}\right)^{3/2}$

Write with positive exponents. Simplify if possible. See Example 3.

29. $8^{-4/3}$

30. $64^{-2/3}$

31. $(-64)^{-2/3}$

32. $(-8)^{-4/3}$

33. $(-4)^{-3/2}$

34. $(-16)^{-5/4}$

35. $x^{-1/4}$

36. $y^{-1/6}$

37. $\dfrac{1}{a^{-2/3}}$

38. $\dfrac{1}{n^{-8/9}}$

39. $\dfrac{5}{7x^{-3/4}}$

40. $\dfrac{2}{3y^{-5/7}}$

41. Explain how writing x^{-7} with positive exponents is similar to writing $x^{-1/4}$ with positive exponents.

42. Explain how writing $2x^{-5}$ with positive exponents is similar to writing $2x^{-3/4}$ with positive exponents.

Write with rational exponents. Then simplify if possible. See Example 4.

43. $\sqrt{3}$

44. $\sqrt[3]{y}$

45. $\sqrt[3]{y^5}$

46. $\sqrt[4]{x^3}$

47. $\sqrt[5]{4y^7}$

48. $\sqrt[4]{11x^5}$

49. $\sqrt{(y + 1)^3}$

50. $\sqrt{(3 + y^2)^5}$

51. $2\sqrt{x} - 3\sqrt{y}$

52. $4\sqrt{2x} + \sqrt{xy}$

Use rational exponents to simplify each radical. (Variables represent positive numbers.) See Example 5.

53. $\sqrt[6]{x^3}$

54. $\sqrt[9]{a^3}$

55. $\sqrt[4]{16x^2}$

56. $\sqrt[8]{4y^2}$

57. $\sqrt[4]{(x + 3)^2}$

58. $\sqrt[9]{a^3b^6}$

59. $\sqrt[8]{x^4y^4}$

60. $\sqrt[9]{y^6z^3}$

Use properties of exponents to simplify each expression. Write with positive exponents. See Example 6.

61. $a^{2/3}a^{5/3}$

62. $b^{9/5}b^{8/5}$

63. $(4u^2v^{-6})^{3/2}$

64. $(32^{1/5}x^{2/3}y^{1/3})^3$

65. $\dfrac{b^{1/2}b^{3/4}}{-b^{1/4}}$

66. $\dfrac{a^{1/4}a^{-1/2}}{a^{2/3}}$

Multiply. See Example 7.

67. $y^{1/2}(y^{1/2} - y^{2/3})$

68. $x^{1/2}(x^{1/2} + x^{3/2})$

69. $x^{2/3}(2x - 2)$ **70.** $3x^{1/2}(x + y)$

71. $(2x^{1/3} + 3)(2x^{1/3} - 3)$ **72.** $(y^{1/2} + 5)(y^{1/2} + 5)$

Factor the common factor from the given expression. See Example 8.

73. $x^{8/3}; x^{8/3} + x^{10/3}$ **74.** $x^{3/2}; x^{5/2} - x^{3/2}$

75. $x^{1/5}; x^{2/5} - 3x^{1/5}$ **76.** $x^{2/7}; x^{3/7} - 2x^{2/7}$

77. $x^{-1/3}; 5x^{-1/3} + x^{2/3}$ **78.** $x^{-3/4}; x^{-3/4} + 3x^{1/4}$

Fill in the box with the correct expression.

79. $\boxed{} \cdot a^{2/3} = a^{3/3}$, or a **80.** $\boxed{} \cdot x^{1/8} = x^{4/8}$, or $x^{1/2}$

81. $\dfrac{\boxed{}}{x^{-2/5}} = x^{3/5}$ **82.** $\dfrac{\boxed{}}{y^{-3/4}} = y^{4/4}$, or y

Use a calculator to write a four-decimal-place approximation of each.

83. $8^{1/4}$ **84.** $20^{1/5}$

85. $18^{3/5}$ **86.** $76^{5/7}$

Review Exercises

Write each integer as a product of two integers such that one of the factors is a perfect square. For example, write 18 as 9 · 2, because 9 is a perfect square.

87. 75   **88.** 20   **89.** 48   **90.** 45

Write each integer as a product of two integers such that one of the factors is a perfect cube. For example, write 24 as 8 · 3, because 8 is a perfect cube.

91. 16   **92.** 56   **93.** 54   **94.** 80

8.3    SIMPLIFYING RADICAL EXPRESSIONS

TAPE IAG 8.3

O B J E C T I V E S

 Understand the product rule for radicals.

    Understand the quotient rule for radicals.

    Simplify radicals.

    Use rational exponents to simplify radical expressions.

 It is possible to simplify some radicals that do not evaluate to rational numbers. To do so, we use a product rule and a quotient rule for radicals. To discover the product rule, notice the following pattern:

$$\sqrt{9} \cdot \sqrt{4} = 3 \cdot 2 = 6$$
$$\sqrt{9 \cdot 4} = \sqrt{36} = 6$$

Since both expressions simplify to 6, it is true that

$$\sqrt{9} \cdot \sqrt{4} = \sqrt{9 \cdot 4}$$

This pattern suggests the following for radicals.

PRODUCT RULE FOR RADICALS

If $\sqrt[n]{a}$ and $\sqrt[n]{b}$ are real numbers, then

$$\sqrt[n]{a} \cdot \sqrt[n]{b} = \sqrt[n]{ab}$$

To see that this is true, notice that the product rule is the relationship $a^{1/n} \cdot b^{1/n} = (ab)^{1/n}$ stated in radical notation.

EXAMPLE 1 Multiply.

       a. $\sqrt{3} \cdot \sqrt{5}$ **b.** $\sqrt{21} \cdot \sqrt{x}$ **c.** $\sqrt[3]{4} \cdot \sqrt[3]{2}$ **d.** $\sqrt[4]{5y^2} \cdot \sqrt[4]{2x^3}$ **e.** $\sqrt{\dfrac{2}{a}} \cdot \sqrt{\dfrac{b}{3}}$

Solution: **a.** $\sqrt{3} \cdot \sqrt{5} = \sqrt{3 \cdot 5} = \sqrt{15}$

       b. $\sqrt{21} \cdot \sqrt{x} = \sqrt{21x}$

       c. $\sqrt[3]{4} \cdot \sqrt[3]{2} = \sqrt[3]{4 \cdot 2} = \sqrt[3]{8} = 2$

       d. $\sqrt[4]{5y^2} \cdot \sqrt[4]{2x^3} = \sqrt[4]{5y^2 \cdot 2x^3} = \sqrt[4]{10y^2x^3}$

       e. $\sqrt{\dfrac{2}{a}} \cdot \sqrt{\dfrac{b}{3}} = \sqrt{\dfrac{2}{a} \cdot \dfrac{b}{3}} = \sqrt{\dfrac{2b}{3a}}$

2 To discover a quotient rule for radicals, notice the following pattern:

$$\sqrt{\frac{4}{9}} = \frac{2}{3}$$

$$\frac{\sqrt{4}}{\sqrt{9}} = \frac{2}{3}$$

Since both expressions simplify to $\dfrac{2}{3}$, it is true that

$$\sqrt{\frac{4}{9}} = \frac{\sqrt{4}}{\sqrt{9}}$$

This pattern suggests the following quotient rule for radicals.

QUOTIENT RULE FOR RADICALS

If $\sqrt[n]{a}$ and $\sqrt[n]{b}$ are real numbers and $\sqrt[n]{b}$ is not zero, then

$$\sqrt[n]{\frac{a}{b}} = \frac{\sqrt[n]{a}}{\sqrt[n]{b}}$$

To see that this is true, notice that the quotient rule is the relationship $\left(\dfrac{a}{b}\right)^{1/n} = \dfrac{a^{1/n}}{b^{1/n}}$ stated in radical notation. We can use the quotient rule to simplify radical expressions by reading the rule from left to right or from right to left as shown in Example 2.

For the remainder of this chapter, we will assume that variables represent positive real numbers. If this is so, we need not insert absolute value bars when we simplify even roots.

EXAMPLE 2   Use the quotient rule to simplify.

a. $\sqrt{\dfrac{25}{49}}$ b. $\sqrt[3]{\dfrac{8}{27}}$ c. $\sqrt{\dfrac{x}{9}}$ d. $\sqrt[4]{\dfrac{3}{16y^4}}$

Solution: a. $\sqrt{\dfrac{25}{49}} = \dfrac{\sqrt{25}}{\sqrt{49}} = \dfrac{5}{7}$ b. $\sqrt[3]{\dfrac{8}{27}} = \dfrac{\sqrt[3]{8}}{\sqrt[3]{27}} = \dfrac{2}{3}$

c. $\sqrt{\dfrac{x}{9}} = \dfrac{\sqrt{x}}{\sqrt{9}} = \dfrac{\sqrt{x}}{3}$ d. $\sqrt[4]{\dfrac{3}{16y^4}} = \dfrac{\sqrt[4]{3}}{\sqrt[4]{16y^4}} = \dfrac{\sqrt[4]{3}}{2y}$ ▬

 Both the product and quotient rules can be used to simplify a radical. If the product rule is read from right to left, we have that $\sqrt[n]{ab} = \sqrt[n]{a} \cdot \sqrt[n]{b}$. This is used to simplify the following radicals.

EXAMPLE 3   Simplify the following.

a. $\sqrt{50}$ b. $\sqrt[3]{24}$ c. $\sqrt{26}$ d. $\sqrt[4]{32}$

Solution: a. Factor 50 such that one factor is the largest perfect square that divides 50. The largest perfect square factor of 50 is 25, so we write 50 as $25 \cdot 2$ and use the product rule for radicals to simplify.

$$\sqrt{50} = \sqrt{25 \cdot 2} = \sqrt{25} \cdot \sqrt{2} = 5\sqrt{2}$$

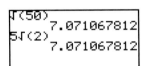

A calculator check for part (a) of Example 3.

b. Since the index is 3, we find the largest perfect cube factor of 24. We write 24 as $8 \cdot 3$ since 8 is the largest perfect cube factor of 24.

$$\sqrt[3]{24} = \sqrt[3]{8 \cdot 3} = \sqrt[3]{8} \cdot \sqrt[3]{3} = 2\sqrt[3]{3}$$

c. The largest perfect square factor of 26 is 1, so $\sqrt{26}$ cannot be simplified further.

d. $\sqrt[4]{32} = \sqrt[4]{16 \cdot 2} = \sqrt[4]{16} \cdot \sqrt[4]{2} = 2\sqrt[4]{2}$ ▬

After simplifying a radical such as a square root, always check the radicand to see that it contains no other perfect square factors. It may, if the largest perfect square factor of the radicand was not originally recognized. For example,

$$\sqrt{200} = \sqrt{4 \cdot 50} = \sqrt{4} \cdot \sqrt{50} = 2\sqrt{50}$$

Notice that the radicand 50 still contains the perfect square factor of 25. This is because 4 is not the largest perfect square factor of 200. We continue as follows:

$$2\sqrt{50} = 2\sqrt{25 \cdot 2} = 2 \cdot \sqrt{25} \cdot \sqrt{2} = 2 \cdot 5 \cdot \sqrt{2} = 10\sqrt{2}$$

The radical is now simplified since 2 contains no perfect square factors (other than 1).

In general, we say that a radicand of the form $\sqrt[n]{a}$ is simplified when a contains no factors that are perfect nth powers (other than 1 or -1).

EXAMPLE 4   Use the product rule to simplify.

a. $\sqrt{25x^3}$
b. $\sqrt[3]{54x^6y^8}$
c. $\sqrt[4]{81z^{11}}$

Solution:

a. $\sqrt{25x^3} = \sqrt{25x^2 \cdot x}$ Find the largest perfect square factor.

$= \sqrt{25x^2} \cdot \sqrt{x}$ Apply the product rule.

$= 5x\sqrt{x}$ Simplify.

b. $\sqrt[3]{54x^6y^8} = \sqrt[3]{27 \cdot 2 \cdot x^6 \cdot y^6 \cdot y^2}$ Factor the radicand and identify perfect cube factors.

$= \sqrt[3]{27x^6y^6 \cdot 2y^2}$

$= \sqrt[3]{27x^6y^6} \cdot \sqrt[3]{2y^2}$ Apply the product rule.

$= 3x^2y^2 \sqrt[3]{2y^2}$ Simplify.

c. $\sqrt[4]{81z^{11}} = \sqrt[4]{81 \cdot z^8 \cdot z^3}$ Factor the radicand and identify perfect fourth power factors.

$= \sqrt[4]{81z^8 \cdot z^3}$

$= \sqrt[4]{81z^8} \cdot \sqrt[4]{z^3}$ Apply the product rule.

$= 3z^2 \sqrt[4]{z^3}$ Simplify.

EXAMPLE 5   Use the quotient rule to divide, and simplify if possible.

a. $\dfrac{\sqrt{20}}{\sqrt{5}}$ b. $\dfrac{\sqrt{50x}}{2\sqrt{2}}$ c. $\dfrac{7\sqrt[3]{48x^4y^8}}{\sqrt[3]{6y^2}}$

Solution:

a. $\dfrac{\sqrt{20}}{\sqrt{5}} = \sqrt{\dfrac{20}{5}}$ Apply the quotient rule.

$= \sqrt{4}$ Simplify.

$= 2$ Simplify.

b. $\dfrac{\sqrt{50x}}{2\sqrt{2}} = \dfrac{1}{2} \cdot \sqrt{\dfrac{50x}{2}}$ Apply the quotient rule.

$= \dfrac{1}{2} \cdot \sqrt{25x}$ Simplify.

$= \dfrac{1}{2} \cdot \sqrt{25} \cdot \sqrt{x}$ Factor $25x$.

$= \dfrac{1}{2} \cdot 5 \cdot \sqrt{x}$ Simplify.

$= \dfrac{5}{2}\sqrt{x}$

c. $\dfrac{7\sqrt[3]{48x^4y^8}}{\sqrt[3]{6y^2}} = 7\cdot\sqrt[3]{\dfrac{48x^4y^8}{6y^2}}$ Apply the quotient rule.

$\qquad = 7\cdot\sqrt[3]{8x^4y^6}$ Simplify.

$\qquad = 7\sqrt[3]{8x^3y^6\cdot x}$ Factor.

$\qquad = 7\cdot\sqrt[3]{8x^3y^6}\cdot\sqrt[3]{x}$ Apply the product rule.

$\qquad = 7\cdot 2xy^2\cdot\sqrt[3]{x}$ Simplify.

$\qquad = 14xy^2\sqrt[3]{x}$

Rational exponents may be useful when multiplying radicands with different indices.

EXAMPLE 6   Use rational exponents to write $\sqrt{5}\cdot\sqrt[3]{2}$ as a single radical. Check using a calculator.

Solution: Notice that the indices 2 and 3 have a least common multiple of 6. Use rational exponents to write each radical as a sixth root.

$$\sqrt{5} = 5^{1/2} = 5^{3/6} = \sqrt[6]{5^3} = \sqrt[6]{125}$$

$$\sqrt[3]{2} = 2^{1/3} = 2^{2/6} = \sqrt[6]{2^2} = \sqrt[6]{4}$$

Thus,

$$\sqrt{5}\cdot\sqrt[3]{2} = \sqrt[6]{125}\cdot\sqrt[6]{4} = \sqrt[6]{500}$$

A calculator check for Example 6.

```
√(5)*³√(2)
       2.817269114
6×√500
       2.817269114
```

Exercise Set 8.3

Use the product rule to multiply. See Example 1.

1. $\sqrt{7}\cdot\sqrt{2}$
2. $\sqrt{11}\cdot\sqrt{10}$
3. $\sqrt[3]{4}\cdot\sqrt[3]{9}$
4. $\sqrt[3]{10}\cdot\sqrt[3]{5}$
5. $\sqrt{2}\cdot\sqrt{3x}$
6. $\sqrt{3y}\cdot\sqrt{5x}$
7. $\sqrt{\dfrac{7}{x}}\cdot\sqrt{\dfrac{2}{y}}$
8. $\sqrt{\dfrac{6}{m}}\cdot\sqrt{\dfrac{n}{5}}$
9. $\sqrt[4]{4x^3}\cdot\sqrt[4]{5}$
10. $\sqrt[4]{ab^2}\cdot\sqrt[4]{27ab}$

Use the quotient rule to divide, and simplify if possible. See Example 5.

11. $\dfrac{\sqrt{14}}{\sqrt{7}}$
12. $\dfrac{\sqrt{45}}{\sqrt{9}}$
13. $\dfrac{\sqrt[3]{24}}{\sqrt[3]{3}}$
14. $\dfrac{\sqrt[3]{10}}{\sqrt[3]{2}}$
15. $\dfrac{\sqrt{x^5y^3}}{\sqrt{xy}}$
16. $\dfrac{\sqrt{a^7b^6}}{\sqrt{a^3b^2}}$
17. $\dfrac{8\sqrt[3]{54m^7}}{\sqrt[3]{2m}}$
18. $\dfrac{\sqrt[3]{128x^3}}{-3\sqrt[3]{2x}}$

Simplify. See Examples 2–4.

19. $\sqrt{32}$
20. $\sqrt{27}$
21. $\sqrt[3]{192}$
22. $\sqrt[3]{108}$
23. $5\sqrt{75}$
24. $3\sqrt{8}$
25. $\sqrt{\dfrac{6}{49}}$
26. $\sqrt{\dfrac{8}{81}}$
27. $\sqrt{20}$
28. $\sqrt{24}$
29. $\sqrt[3]{\dfrac{4}{27}}$
30. $\sqrt[3]{\dfrac{3}{64}}$
31. $\sqrt{\dfrac{2}{49}}$
32. $\sqrt{\dfrac{5}{121}}$
33. $\sqrt{100x^5}$
34. $\sqrt{64y^9}$
35. $\sqrt[3]{16y^7}$
36. $\sqrt[3]{64y^9}$
37. $\sqrt[4]{a^8b^7}$
38. $\sqrt[3]{32z^{12}}$
39. $\sqrt{y^5}$
40. $\sqrt[3]{y^5}$
41. $\sqrt{25a^2b^3}$
42. $\sqrt{9x^5y^7}$
43. $\sqrt[5]{-32x^{10}y}$
44. $\sqrt[3]{-243z^9}$
45. $\sqrt[3]{50x^{14}}$
46. $\sqrt[3]{40y^{10}}$
47. $-\sqrt{32a^8b^7}$
48. $-\sqrt{20ab^6}$
49. $\sqrt{\dfrac{5x^2}{4y^2}}$
50. $\sqrt{\dfrac{y^{10}}{9x^6}}$
51. $-\sqrt[3]{\dfrac{z^7}{27x^3}}$

52. $-\sqrt[3]{\dfrac{64a}{b^9}}$ **53.** $\sqrt[4]{\dfrac{x^7}{16}}$ **54.** $\sqrt[4]{\dfrac{y}{81x^4}}$

55. $\sqrt{9x^7y^9}$ **56.** $\sqrt{12r^9s^{12}}$ **57.** $\sqrt[3]{125r^9s^{12}}$

58. $\sqrt[3]{8a^6b^9}$ **59.** $\sqrt{\dfrac{x^2y}{100}}$ **60.** $\sqrt{\dfrac{y^2z}{36}}$

61. $\sqrt[4]{\dfrac{8}{x^8}}$ **62.** $\sqrt[4]{\dfrac{a^3}{81}}$

Use rational exponents to write each radical with the same index. Then multiply. See Example 6.

63. $\sqrt{2} \cdot \sqrt[3]{3}$ **64.** $\sqrt[3]{5} \cdot \sqrt{2}$

65. $\sqrt[5]{7} \cdot \sqrt[3]{y}$ **66.** $\sqrt[3]{x} \cdot \sqrt[4]{5}$

67. $\sqrt[3]{x} \cdot \sqrt{x}$ **68.** $\sqrt[4]{y} \cdot \sqrt[7]{y}$

69. $\sqrt{5r} \cdot \sqrt[3]{s}$ **70.** $\sqrt[5]{4a} \cdot \sqrt[3]{b}$

71. The formula for the surface area A of a cone with height h and radius r is given by

$$A = \pi r \sqrt{r^2 + h^2}$$

a. Find the surface area of a cone whose height is 3 centimeters and whose radius is 4 centimeters.

b. Approximate to two decimal places the surface area of a cone whose height is 7.2 feet and whose radius is 6.8 feet.

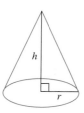

72. The formula for the radius r of a sphere, given the surface area A, is

$$r = \sqrt{\dfrac{A}{4\pi}}$$

a. The Safety First Company has decided to manufacture a new rubber ball for children. Both the marketing and production departments have determined that the surface area of the ball should be 300 square inches. Approximate the radius of the ball to two decimal places.

b. Why do you think that the marketing department of a company might be involved in the planning of a new product such as the one above?

73. Before Mount Vesuvius, a volcano in Italy, erupted violently in A.D. 79, its height was 4190 feet. Vesuvius was roughly cone-shaped, and its base had a radius of approximately 25,200 feet. Use the formula for the surface area of a cone, given in Exercise 71, to approximate the surface area this volcano had before it erupted. (*Source:* Global Volcanism Network)

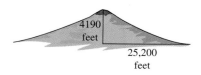

74. The surface area of Earth is approximately 1.97×10^8 square miles. Use the formula for the radius of a sphere, given in Exercise 72, to find the radius of Earth. Round to the nearest whole mile. (*Source:* American Museum of Natural History.)

75. The management team of GoodDay Tire Company has determined that the demand equation for their Premium GoodDay tires is given by

$$F(x) = \sqrt{130 - x}$$

where x is the unit price of the tire and $F(x)$ is the quantity demanded per week in thousands.

a. Approximate to two decimal places the quantity of Premium GoodDay tires demanded per week if the price per tire is $86.

b. Approximate to two decimal places the quantity of Premium GoodDay tires demanded per week if the tires go on sale and the price per tire is $79.99.

c. From the results of parts (a) and (b), explain why companies have sales.

76. The owner of Knightime Video has determined that the demand equation for renting older released tapes is given by the equation

$$F(x) = 0.6\sqrt{49 - x^2}$$

where x is the price in dollars per two-day rental and $F(x)$ is the quantity of the video demanded per week.

a. Approximate to one decimal place the demand per week of an older released video if the rental price is $3 per two-day rental.

b. Approximate to one decimal place the demand per week of an older released video if the rental price is $5 per two-day rental.

c. Explain how the owner of the video store can use this equation to predict the number of copies of each tape that should be in stock.

Review Exercises

Perform the following operations. See Sections 2.1 and 6.3.

77. $6x + 8x$

78. $(6x)(8x)$

79. $(2x + 3)(x - 5)$

80. $(2x + 3) + (x - 5)$

81. $9y^2 - 8y^2$

82. $(9y^2)(-8y^2)$

83. $-3(x + 5)$

84. $-3 + x + 5$

8.4 | ADDING, SUBTRACTING, AND MULTIPLYING RADICAL EXPRESSIONS

O B J E C T I V E S

 Add or subtract radical expressions.

 Multiply radical expressions.

We have learned that the sum or difference of like terms can be simplified. To simplify these sums or differences, we use the distributive property. For example,

$$2x + 3x = (2 + 3)x = 5x \quad \text{and} \quad 7x^2y - 4x^2y = (7 - 4)x^2y = 3x^2y$$

The distributive property can also be used to add **like radicals.**

LIKE RADICALS

Radicals with the same index and the same radicand are like radicals.

For example, $2\sqrt{7} + 3\sqrt{7} = (2 + 3)\sqrt{7} = 5\sqrt{7}$. Also,

$$5\sqrt{3x} - 7\sqrt{3x} = (5 - 7)\sqrt{3x} = -2\sqrt{3x}$$

The expression $2\sqrt{7} + 2\sqrt[3]{7}$ cannot be simplified since $2\sqrt{7}$ and $2\sqrt[3]{7}$ are not like radicals. A calculator can be used to check operations on numerical radical expressions. See the margin for examples in this section and the next.

EXAMPLE 1 Add or subtract. Assume that variables represent positive real numbers.

a. $\sqrt{20} + 2\sqrt{45}$ **b.** $\sqrt[3]{54} - 5\sqrt[3]{16} + \sqrt[3]{2}$ **c.** $\sqrt{27x} - 2\sqrt{9x} + \sqrt{72x}$

d. $\sqrt[3]{98} + \sqrt{98}$ **e.** $\sqrt[3]{48y^4} + \sqrt[3]{6y^4}$

Solution: First, simplify each radical. Then add or subtract any like radicals.

a. $\sqrt{20} + 2\sqrt{45} = \sqrt{4 \cdot 5} + 2\sqrt{9 \cdot 5}$

$\qquad\qquad\qquad = \sqrt{4} \cdot \sqrt{5} + 2 \cdot \sqrt{9} \cdot \sqrt{5}$

$\qquad\qquad\qquad = 2 \cdot \sqrt{5} + 2 \cdot 3 \cdot \sqrt{5}$

$\qquad\qquad\qquad = 2\sqrt{5} + 6\sqrt{5} = 8\sqrt{5}$

b. $\sqrt[3]{54} - 5\sqrt[3]{16} + \sqrt[3]{2} = \sqrt[3]{27} \cdot \sqrt[3]{2} - 5 \cdot \sqrt[3]{8} \cdot \sqrt[3]{2} + \sqrt[3]{2}$

$\qquad\qquad\qquad\qquad = 3 \cdot \sqrt[3]{2} - 5 \cdot 2 \cdot \sqrt[3]{2} + \sqrt[3]{2}$

$\qquad\qquad\qquad\qquad = 3\sqrt[3]{2} - 10\sqrt[3]{2} + \sqrt[3]{2}$

$\qquad\qquad\qquad\qquad = -6\sqrt[3]{2}$

c. $\sqrt{27x} - 2\sqrt{9x} + \sqrt{72x} = \sqrt{9} \cdot \sqrt{3x} - 2 \cdot \sqrt{9} \cdot \sqrt{x} + \sqrt{36} \cdot \sqrt{2x}$

$\qquad\qquad\qquad\qquad\qquad = 3 \cdot \sqrt{3x} - 2 \cdot 3 \cdot \sqrt{x} + 6 \cdot \sqrt{2x}$

$\qquad\qquad\qquad\qquad\qquad = 3\sqrt{3x} - 6\sqrt{x} + 6\sqrt{2x}$

d. We can simplify $\sqrt{98}$, but since the indexes are different, these radicals cannot be added.

$$\sqrt[3]{98} + \sqrt{98} = \sqrt[3]{98} + \sqrt{49} \cdot \sqrt{2} = \sqrt[3]{98} + 7\sqrt{2}$$

e. $\sqrt[3]{48y^4} + \sqrt[3]{6y^4} = \sqrt[3]{8y^3} \cdot \sqrt[3]{6y} + \sqrt[3]{y^3} \cdot \sqrt[3]{6y}$

$\qquad\qquad\qquad\quad = 2y\sqrt[3]{6y} + y\sqrt[3]{6y}$

$\qquad\qquad\qquad\quad = 3y\sqrt[3]{6y}$

A calculator check for Example 1, part (b).

The following may be used to simplify radical expressions.

TO SIMPLIFY RADICAL EXPRESSIONS

1. Write each radical term in simplest form.

2. Add or subtract any like radicals.

EXAMPLE 2 Simplify.

a. $\dfrac{\sqrt{45}}{4} - \dfrac{\sqrt{5}}{3}$ **b.** $\sqrt[3]{\dfrac{7x}{8}} + 2\sqrt[3]{7x}$

Solution: **a.** $\dfrac{\sqrt{45}}{4} - \dfrac{\sqrt{5}}{3} = \dfrac{3\sqrt{5}}{4} - \dfrac{\sqrt{5}}{3}$ To subtract, notice that the LCD is 12.

$\qquad\qquad\qquad\quad = \dfrac{3\sqrt{5} \cdot 3}{4 \cdot 3} - \dfrac{\sqrt{5} \cdot 4}{3 \cdot 4}$ Write each expression as an equivalent expression with a denominator of 12.

$\qquad\qquad\qquad\quad = \dfrac{9\sqrt{5}}{12} - \dfrac{4\sqrt{5}}{12}$ Multiply factors in the numerator and the denominator.

$\qquad\qquad\qquad\quad = \dfrac{5\sqrt{5}}{12}$ Subtract.

b. $\sqrt[3]{\dfrac{7x}{8}} + 2\sqrt[3]{7x} = \dfrac{\sqrt[3]{7x}}{\sqrt[3]{8}} + 2\sqrt[3]{7x}$ Apply the quotient property for radicals.

$\qquad\qquad\qquad = \dfrac{\sqrt[3]{7x}}{2} + 2\sqrt[3]{7x}$ Simplify.

$\qquad\qquad\qquad = \dfrac{\sqrt[3]{7x}}{2} + \dfrac{2\sqrt[3]{7x} \cdot 2}{2}$ Write each expression as an equivalent expression with a denominator of 2.

$\qquad\qquad\qquad = \dfrac{\sqrt[3]{7x}}{2} + \dfrac{4\sqrt[3]{7x}}{2}$

$\qquad\qquad\qquad = \dfrac{5\sqrt[3]{7x}}{2}$ Add.

2 Radical expressions are multiplied by using many of the same properties used to multiply polynomial expressions. For instance, to multiply $\sqrt{2}(\sqrt{6} - 3\sqrt{2})$, we use the distributive property and multiply $\sqrt{2}$ by each term inside the parentheses:

$$\sqrt{2}(\sqrt{6} - 3\sqrt{2}) = \sqrt{2}(\sqrt{6}) - \sqrt{2}(3\sqrt{2})$$
$$= \sqrt{12} - 3\sqrt{2 \cdot 2}$$
$$= \sqrt{4 \cdot 3} - 3 \cdot 2 \qquad \text{Apply the product rule for radicals.}$$
$$= 2\sqrt{3} - 6$$

EXAMPLE 3   Multiply.

a. $\sqrt{3}(5 + \sqrt{30})$ **b.** $(\sqrt{5} - \sqrt{6})(\sqrt{7} + 1)$ **c.** $(7\sqrt{x} + 5)(3\sqrt{x} - \sqrt{5})$
d. $(4\sqrt{3} - 1)^2$ **e.** $(\sqrt{2x} - 5)(\sqrt{2x} + 5)$

Solution: **a.** $\sqrt{3}(5 + \sqrt{30}) = \sqrt{3}(5) + \sqrt{3}(\sqrt{30})$

$\qquad\qquad\qquad\qquad = 5\sqrt{3} + \sqrt{3 \cdot 30}$

$\qquad\qquad\qquad\qquad = 5\sqrt{3} + \sqrt{3 \cdot 3 \cdot 10}$

$\qquad\qquad\qquad\qquad = 5\sqrt{3} + 3\sqrt{10}$

b. To multiply, we can use the FOIL method.

$$\qquad\qquad\qquad\qquad\quad \text{First}\quad\;\; \text{Outside}\quad \text{Inside}\quad\;\; \text{Last}$$
$$(\sqrt{5} - \sqrt{6})(\sqrt{7} + 1) = \sqrt{5} \cdot \sqrt{7} + \sqrt{5} \cdot 1 - \sqrt{6} \cdot \sqrt{7} - \sqrt{6} \cdot 1$$
$$= \sqrt{35} + \sqrt{5} - \sqrt{42} - \sqrt{6}$$

c. $(7\sqrt{x} + 5)(3\sqrt{x} - \sqrt{5}) = 7\sqrt{x}(3\sqrt{x}) - 7\sqrt{x}(\sqrt{5}) + 5(3\sqrt{x}) - 5(\sqrt{5})$
$$= 21x - 7\sqrt{5x} + 15\sqrt{x} - 5\sqrt{5}$$

d. $(4\sqrt{3} - 1)^2 = (4\sqrt{3} - 1)(4\sqrt{3} - 1)$
$$= 4\sqrt{3}(4\sqrt{3}) - 4\sqrt{3}(1) - 1(4\sqrt{3}) - 1(-1)$$
$$= 16 \cdot 3 - 4\sqrt{3} - 4\sqrt{3} + 1$$
$$= 48 - 8\sqrt{3} + 1$$
$$= 49 - 8\sqrt{3}$$

e. $(\sqrt{2x} - 5)(\sqrt{2x} + 5) = \sqrt{2x} \cdot \sqrt{2x} + 5\sqrt{2x} - 5\sqrt{2x} - 5 \cdot 5$
$$= 2x - 25$$

```
(4√(3)-1)²
          35.14359354
49-8√(3)
          35.14359354
```

A calculator check for Example 3, part (d).

MENTAL MATH

Simplify. Assume that all variables represent positive real numbers.

1. $2\sqrt{3} + 4\sqrt{3}$

2. $5\sqrt{7} + 3\sqrt{7}$

3. $8\sqrt{x} - 5\sqrt{x}$

4. $3\sqrt{y} + 10\sqrt{y}$

5. $7\sqrt[3]{x} + 5\sqrt[3]{x}$

6. $8\sqrt[3]{z} - 2\sqrt[3]{z}$

EXERCISE SET 8.4

Add or subtract. Assume that all variables represent positive numbers. See Examples 1 and 2.

1. $\sqrt{8} - \sqrt{32}$

2. $\sqrt{27} - \sqrt{75}$

3. $2\sqrt{2x^3} + 4x\sqrt{8x}$

4. $3\sqrt{45x^3} + x\sqrt{5x}$

5. $2\sqrt{50} - 3\sqrt{125} + \sqrt{98}$

6. $4\sqrt{32} - \sqrt{18} + 2\sqrt{128}$

7. $\sqrt[3]{16x} - \sqrt[3]{54x}$

8. $2\sqrt[3]{3a^4} - 3a\sqrt[3]{81a}$

9. $\sqrt{9b^3} - \sqrt{25b^3} + \sqrt{49b^3}$

10. $\sqrt{4x^7} + 9x^2\sqrt{x^3} - 5x\sqrt{x^5}$

11. $\dfrac{5\sqrt{2}}{3} + \dfrac{2\sqrt{2}}{5}$

12. $\dfrac{\sqrt{3}}{2} + \dfrac{4\sqrt{3}}{3}$

13. $\sqrt[3]{\dfrac{11}{8}} - \dfrac{\sqrt[3]{11}}{6}$

14. $\dfrac{2\sqrt[3]{4}}{7} - \dfrac{\sqrt[3]{4}}{14}$

15. $\dfrac{\sqrt{20x}}{9} + \sqrt{\dfrac{5x}{9}}$

16. $\dfrac{3x\sqrt{7}}{5} + \sqrt{\dfrac{7x^2}{100}}$

17. $7\sqrt{9} - 7 + \sqrt{3}$

18. $\sqrt{16} - 5\sqrt{10} + 7$

19. $2 + 3\sqrt{y^2} - 6\sqrt{y^2} + 5$

20. $3\sqrt{7} - \sqrt[3]{x} + 4\sqrt{7} - 3\sqrt[3]{x}$

21. $3\sqrt{108} - 2\sqrt{18} - 3\sqrt{48}$

22. $-\sqrt{75} + \sqrt{12} - 3\sqrt{3}$

23. $-5\sqrt[3]{625} + \sqrt[3]{40}$

24. $-2\sqrt[3]{108} - \sqrt[3]{32}$

25. $\sqrt{9b^3} - \sqrt{25b^3} + \sqrt{16b^3}$

26. $\sqrt{4x^7y^5} + 9x^2\sqrt{x^3y^5} - 5xy\sqrt{x^5y^3}$

27. $5y\sqrt{8y} + 2\sqrt{50y^3}$

28. $3\sqrt{8x^2y^3} - 2x\sqrt{32y^3}$

29. $\sqrt[3]{54xy^3} - 5\sqrt[3]{2xy^3} + y\sqrt[3]{128x}$

30. $2\sqrt[3]{24x^3y^4} + 4x\sqrt[3]{81y^4}$

31. $6\sqrt[3]{11} + 8\sqrt{11} - 12\sqrt{11}$

32. $3\sqrt[3]{5} + 4\sqrt{5}$

33. $-2\sqrt[4]{x^7} + 3\sqrt[4]{16x^7}$

34. $6\sqrt[3]{24x^3} - 2\sqrt[3]{81x^3} - x\sqrt[3]{3}$

35. $\dfrac{4\sqrt{3}}{3} - \dfrac{\sqrt{12}}{3}$

36. $\dfrac{\sqrt{45}}{10} + \dfrac{7\sqrt{5}}{10}$

37. $\dfrac{\sqrt[3]{8x^4}}{7} + \dfrac{3x\sqrt[3]{x}}{7}$

38. $\dfrac{\sqrt[4]{48}}{5x} - \dfrac{2\sqrt[3]{3}}{10x}$

39. $\sqrt{\dfrac{28}{x^2}} + \sqrt{\dfrac{7}{4x^2}}$

40. $\dfrac{\sqrt{99}}{5x} - \sqrt{\dfrac{44}{x^2}}$

41. $\sqrt[3]{\dfrac{16}{27}} - \dfrac{\sqrt[3]{54}}{6}$

42. $\dfrac{\sqrt[3]{3}}{10} + \sqrt[3]{\dfrac{24}{125}}$

43. $-\dfrac{\sqrt[3]{2x^4}}{9} + \sqrt[3]{\dfrac{250x^4}{27}}$

44. $\dfrac{\sqrt[3]{y^5}}{8} + \dfrac{5y\sqrt[3]{y^2}}{4}$

45. Find the perimeter of the trapezoid.

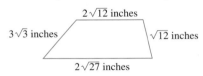

46. Find the perimeter of the triangle.

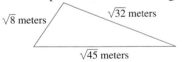

Multiply, and then simplify if possible. Assume that all variables represent positive real numbers. See Example 3.

47. $\sqrt{7}(\sqrt{5} + \sqrt{3})$

48. $\sqrt{5}(\sqrt{15} - \sqrt{35})$

49. $(\sqrt{5} - \sqrt{2})^2$

50. $(3x - \sqrt{2})(3x - \sqrt{2})$

51. $\sqrt{3x}(\sqrt{3} - \sqrt{x})$

52. $\sqrt{5y}(\sqrt{y} + \sqrt{5})$

53. $(2\sqrt{x} - 5)(3\sqrt{x} + 1)$ **54.** $(8\sqrt{y} + z)(4\sqrt{y} - 1)$

55. $(\sqrt[3]{a} - 4)(\sqrt[3]{a} + 5)$ **56.** $(\sqrt[3]{a} + 2)(\sqrt[3]{a} + 7)$

57. $6(\sqrt{2} - 2)$ **58.** $\sqrt{5}(6 - \sqrt{5})$

59. $\sqrt{2}(\sqrt{2} + x\sqrt{6})$ **60.** $\sqrt{3}(\sqrt{3} - 2\sqrt{5x})$

61. $(2\sqrt{7} + 3\sqrt{5})(\sqrt{7} - 2\sqrt{5})$

62. $(\sqrt{6} - 4\sqrt{2})(3\sqrt{6} + 1)$

63. $(\sqrt{x} - y)(\sqrt{x} + y)$

64. $(3\sqrt{x} + 2)(\sqrt{3x} - 2)$

65. $(\sqrt{3} + x)^2$ **66.** $(\sqrt{y} - 3x)^2$

67. $(\sqrt{5x} - 3\sqrt{2})(\sqrt{5x} - 3\sqrt{3})$

68. $(5\sqrt{3x} - \sqrt{y})(4\sqrt{x} + 1)$

69. $(\sqrt[3]{4} + 2)(\sqrt[3]{2} - 1)$

70. $(\sqrt[3]{3} + \sqrt[3]{2})(\sqrt[3]{9} - \sqrt[3]{4})$

71. $(\sqrt[3]{x} + 1)(\sqrt[3]{x} - 4\sqrt{x} + 7)$

72. $(\sqrt[3]{3x} + 3)(\sqrt[3]{2x} - 3x - 1)$

73. Baseboard needs to be installed around the perimeter of a rectangular room.

    a. Find how much baseboard should be ordered by finding the perimeter of the room.

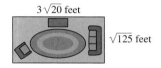

3 $\sqrt{20}$ feet

$\sqrt{125}$ feet

    b. Find the area of the room.

74. A border of wallpaper is to be used around the perimeter of the odd-shaped room shown.

    a. Find how much wallpaper border is needed by finding the perimeter of the room.

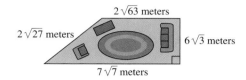

2 $\sqrt{63}$ meters

2 $\sqrt{27}$ meters

6 $\sqrt{3}$ meters

7 $\sqrt{7}$ meters

    b. Find the area of the room. (*Hint:* The area of a trapezoid is the product of half the height $6\sqrt{3}$ meters and the sum of the bases $2\sqrt{63}$ and $7\sqrt{7}$ meters.)

Review Exercises

Factor each numerator and denominator. Then simplify if possible. See Section 7.1.

75. $\dfrac{2x - 14}{2}$ **76.** $\dfrac{8x - 24y}{4}$

77. $\dfrac{7x - 7y}{x^2 - y^2}$ **78.** $\dfrac{x^3 - 8}{4x - 8}$

79. $\dfrac{6a^2b - 9ab}{3ab}$ **80.** $\dfrac{14r - 28r^2s^2}{7rs}$

81. $\dfrac{-4 + 2\sqrt{3}}{6}$ **82.** $\dfrac{-5 + 10\sqrt{7}}{5}$

8.5   RATIONALIZING NUMERATORS AND DENOMINATORS OF RADICAL EXPRESSIONS

TAPE IAG 8.5

O B J E C T I V E S

 Rationalize denominators.

2 Rationalize numerators.

Often in mathematics, it is helpful to write a radical expression such as $\dfrac{\sqrt{3}}{\sqrt{2}}$ either without a radical in the denominator or without a radical in the numerator. The process of writing this expression as an equivalent expression but without a radical in the denominator is called **rationalizing the denominator.** To rationalize the denominator of $\frac{\sqrt{3}}{\sqrt{2}}$, we use the fundamental principle of fractions and multiply the numerator and the denominator by $\sqrt{2}$. Recall that this is the same as multiplying by $\frac{\sqrt{2}}{\sqrt{2}}$, which simplifies to 1.

$$\frac{\sqrt{3}}{\sqrt{2}} = \frac{\sqrt{3} \cdot \sqrt{2}}{\sqrt{2} \cdot \sqrt{2}} = \frac{\sqrt{6}}{\sqrt{4}} = \frac{\sqrt{6}}{2}$$

EXAMPLE 1

Rationalize the denominator of each expression.

a. $\dfrac{\sqrt{27}}{\sqrt{5}}$
 b. $\dfrac{2\sqrt{16}}{\sqrt{9x}}$
 c. $\sqrt[3]{\dfrac{1}{2}}$

Solution:

a. First, we simplify $\sqrt{27}$; then we rationalize the denominator.

$$\frac{\sqrt{27}}{\sqrt{5}} = \frac{\sqrt{9 \cdot 3}}{\sqrt{5}} = \frac{3\sqrt{3}}{\sqrt{5}}$$

Recall that to rationalize the denominator we multiply the numerator and denominator by $\sqrt{5}$.

$$\frac{3\sqrt{3}}{\sqrt{5}} = \frac{3\sqrt{3} \cdot \sqrt{5}}{\sqrt{5} \cdot \sqrt{5}} = \frac{3\sqrt{15}}{5}$$

b. First, we simplify the radicals and then rationalize the denominator.

$$\frac{2\sqrt{16}}{\sqrt{9x}} = \frac{2(4)}{3\sqrt{x}} = \frac{8}{3\sqrt{x}}$$

To rationalize the denominator, multiply the numerator and denominator by $\sqrt{x}$. Then

$$\frac{8}{3\sqrt{x}} = \frac{8 \cdot \sqrt{x}}{3\sqrt{x} \cdot \sqrt{x}} = \frac{8\sqrt{x}}{3x}$$

c. $\sqrt[3]{\dfrac{1}{2}} = \dfrac{\sqrt[3]{1}}{\sqrt[3]{2}} = \dfrac{1}{\sqrt[3]{2}}$. Now we rationalize the denominator. Since $\sqrt[3]{2}$ is a cube root, we want to multiply by a value that will make the radicand 2 a perfect cube. If we multiply by $\sqrt[3]{2^2}$, we get $\sqrt[3]{2^3} = \sqrt[3]{8} = 2$.

$$\frac{1 \cdot \sqrt[3]{2^2}}{\sqrt[3]{2} \cdot \sqrt[3]{2^2}} = \frac{\sqrt[3]{4}}{\sqrt[3]{8}} = \frac{\sqrt[3]{4}}{2}$$

A different method is needed to rationalize a denominator that is a sum or difference of two terms. For example, to rationalize the denominator of the expression $\frac{5}{\sqrt{3} - 2}$, multiply both the numerator and denominator by $\sqrt{3} + 2$, the

conjugate of the denominator $\sqrt{3} - 2$. In general, the conjugate of $a + b$ is $a - b$. Then

$$\frac{5}{\sqrt{3} - 2} = \frac{5(\sqrt{3} + 2)}{(\sqrt{3} - 2)(\sqrt{3} + 2)}$$

$$= \frac{5(\sqrt{3} + 2)}{\sqrt{3} \cdot \sqrt{3} + 2\sqrt{3} - 2\sqrt{3} - 4}$$

$$= \frac{5(\sqrt{3} + 2)}{3 - 4}$$

$$= \frac{5(\sqrt{3} + 2)}{-1}$$

$$= -5(\sqrt{3} + 2), \quad \text{or} \quad -5\sqrt{3} - 10$$

EXAMPLE 2   Rationalize each denominator.

a. $\dfrac{2}{3\sqrt{2} + 4}$ **b.** $\dfrac{\sqrt{6} + 2}{\sqrt{5} - \sqrt{3}}$ **c.** $\dfrac{7\sqrt{y}}{\sqrt{12x}}$ **d.** $\dfrac{2\sqrt{m}}{3\sqrt{x} + \sqrt{m}}$

Solution: **a.** Multiply the numerator and denominator by the conjugate of the denominator, $3\sqrt{2} + 4$.

$$\frac{2}{3\sqrt{2} + 4} = \frac{2(3\sqrt{2} - 4)}{(3\sqrt{2} + 4)(3\sqrt{2} - 4)}$$

$$= \frac{2(3\sqrt{2} - 4)}{(3\sqrt{2})^2 - 4^2}$$

$$= \frac{2(3\sqrt{2} - 4)}{18 - 16}$$

$$= \frac{2(3\sqrt{2} - 4)}{2}, \quad \text{or} \quad 3\sqrt{2} - 4$$

```
2/(3√(2)+4)
        .2426406871
3√(2)-4
        .2426406871
```

A calculator check of Example 2, part (a). Notice the use of parentheses in the first expression.

It is often helpful to leave a numerator in factored form to help determine whether the expression can be simplified.

b. Multiply the numerator and denominator by the conjugate of $\sqrt{5} - \sqrt{3}$.

$$\frac{\sqrt{6} + 2}{\sqrt{5} - \sqrt{3}} = \frac{(\sqrt{6} + 2)(\sqrt{5} + \sqrt{3})}{(\sqrt{5} - \sqrt{3})(\sqrt{5} + \sqrt{3})}$$

$$= \frac{\sqrt{6}\sqrt{5} + \sqrt{6}\sqrt{3} + 2\sqrt{5} + 2\sqrt{3}}{(\sqrt{5})^2 - (\sqrt{3})^2}$$

$$= \frac{\sqrt{30} + \sqrt{18} + 2\sqrt{5} + 2\sqrt{3}}{5 - 3}$$

$$= \frac{\sqrt{30} + 3\sqrt{2} + 2\sqrt{5} + 2\sqrt{3}}{2}$$

c. Notice that the denominator of this example is *not the sum or difference of two terms.* For this reason, we simplify the radical expression and then multiply the numerator and denominator by a factor so that the resulting denominator is a rational expression:

$$\frac{7\sqrt{y}}{\sqrt{12x}} = \frac{7\sqrt{y}}{\sqrt{4}\sqrt{3x}} = \frac{7\sqrt{y} \cdot \sqrt{3x}}{2\sqrt{3x} \cdot \sqrt{3x}} = \frac{7\sqrt{3xy}}{2 \cdot 3x} = \frac{7\sqrt{3xy}}{6x}$$

d. Multiply by the conjugate of $3\sqrt{x} + \sqrt{m}$ to eliminate the radicals from the denominator.

$$\frac{2\sqrt{m}}{3\sqrt{x} + \sqrt{m}} = \frac{2\sqrt{m}(3\sqrt{x} - \sqrt{m})}{(3\sqrt{x} + \sqrt{m})(3\sqrt{x} - \sqrt{m})} = \frac{6\sqrt{mx} - 2m}{(3\sqrt{x})^2 - (\sqrt{m})^2}$$

$$= \frac{6\sqrt{mx} - 2m}{9x - m}$$

2   As mentioned earlier, it is also often helpful to write an expression such as $\frac{\sqrt{3}}{\sqrt{2}}$ as an equivalent expression without a radical in the numerator. This process is called **rationalizing the numerator.** To rationalize the numerator of $\frac{\sqrt{3}}{\sqrt{2}}$, we multiply the numerator and denominator by $\sqrt{3}$.

$$\frac{\sqrt{3}}{\sqrt{2}} = \frac{\sqrt{3} \cdot \sqrt{3}}{\sqrt{2} \cdot \sqrt{3}} = \frac{\sqrt{9}}{\sqrt{6}} = \frac{3}{\sqrt{6}}$$

EXAMPLE 3   Rationalize the numerator of each expression.

a. $\dfrac{\sqrt{28}}{\sqrt{45}}$

b. $\dfrac{\sqrt[3]{2x^2}}{\sqrt[3]{5y}}$

Solution: **a.** First, we simplify $\sqrt{28}$ and $\sqrt{45}$.

$$\frac{\sqrt{28}}{\sqrt{45}} = \frac{\sqrt{4 \cdot 7}}{\sqrt{9 \cdot 5}} = \frac{2\sqrt{7}}{3\sqrt{5}}$$

Next, we rationalize the numerator by multiplying the numerator and denominator by $\sqrt{7}$:

$$\frac{2\sqrt{7}}{3\sqrt{5}} = \frac{2\sqrt{7} \cdot \sqrt{7}}{3\sqrt{5} \cdot \sqrt{7}} = \frac{2 \cdot 7}{3\sqrt{5 \cdot 7}} = \frac{14}{3\sqrt{35}}$$

b. The numerator and the denominator of this expression are already simplified. To rationalize the numerator, $\sqrt[3]{2x^2}$, we multiply the numerator and denominator by a factor that will make the radicand a perfect cube. If we multiply $\sqrt[3]{2x^2}$ by $\sqrt[3]{4x}$, we get $\sqrt[3]{8x^3} = 2x$.

$$\frac{\sqrt[3]{2x^2}}{\sqrt[3]{5y}} = \frac{\sqrt[3]{2x^2} \cdot \sqrt[3]{4x}}{\sqrt[3]{5y} \cdot \sqrt[3]{4x}} = \frac{\sqrt[3]{8x^3}}{\sqrt[3]{20xy}} = \frac{2x}{\sqrt[3]{20xy}}$$

EXAMPLE 4   Rationalize the numerator of $\dfrac{\sqrt{x}+2}{5}$.

Solution: Multiply the numerator and denominator by the conjugate of $\sqrt{x}+2$, the numerator:

$$\frac{\sqrt{x}+2}{5} = \frac{(\sqrt{x}+2)(\sqrt{x}-2)}{5(\sqrt{x}-2)}$$

$$= \frac{x - 2\sqrt{x} + 2\sqrt{x} - 4}{5(\sqrt{x}-2)}$$

$$= \frac{x-4}{5(\sqrt{x}-2)}$$

EXERCISE SET 8.5

Simplify by rationalizing the denominator. Assume that all variables represent positive numbers. See Examples 1 and 2.

1. $\dfrac{1}{\sqrt{3}}$

2. $\dfrac{\sqrt{2}}{\sqrt{6}}$

3. $\sqrt{\dfrac{1}{5}}$

4. $\sqrt{\dfrac{1}{2}}$

5. $\dfrac{4}{\sqrt[3]{3}}$

6. $\dfrac{6}{\sqrt[3]{9}}$

7. $\dfrac{3}{\sqrt{8x}}$

8. $\dfrac{5}{\sqrt{27a}}$

9. $\dfrac{3}{\sqrt[3]{4x^2}}$

10. $\dfrac{5}{\sqrt[3]{3y}}$

11. $\dfrac{6}{2-\sqrt{7}}$

12. $\dfrac{3}{\sqrt{7}-4}$

13. $\dfrac{-7}{\sqrt{x}-3}$

14. $\dfrac{-8}{\sqrt{y}+4}$

15. $\dfrac{\sqrt{2}-\sqrt{3}}{\sqrt{2}+\sqrt{3}}$

16. $\dfrac{\sqrt{3}+\sqrt{4}}{\sqrt{2}+\sqrt{3}}$

17. $\dfrac{\sqrt{a}+1}{2\sqrt{a}-\sqrt{b}}$

18. $\dfrac{2\sqrt{a}-3}{2\sqrt{a}-\sqrt{b}}$

19. $\dfrac{9}{\sqrt{3a}}$

20. $\dfrac{x}{\sqrt{5}}$

21. $\dfrac{3}{\sqrt[3]{2}}$

22. $\dfrac{5}{\sqrt[3]{9}}$

23. $\dfrac{2\sqrt{3}}{\sqrt{7}}$

24. $\dfrac{-5\sqrt{2}}{\sqrt{11}}$

25. $\dfrac{8}{1+\sqrt{10}}$

26. $\dfrac{-3}{\sqrt{6}-2}$

27. $\dfrac{\sqrt{x}}{\sqrt{x}+\sqrt{y}}$

28. $\dfrac{2\sqrt{a}}{2\sqrt{x}-\sqrt{y}}$

Rationalize each numerator. See Examples 3 and 4.

29. $\sqrt{\dfrac{5}{3}}$

30. $\sqrt{\dfrac{3}{2}}$

31. $\sqrt{\dfrac{18}{5}}$

32. $\sqrt{\dfrac{12}{7}}$

33. $\dfrac{\sqrt{4x}}{7}$

34. $\dfrac{\sqrt{3x^5}}{6}$

35. $\dfrac{\sqrt[3]{5y^2}}{\sqrt[3]{4x}}$

36. $\dfrac{\sqrt[3]{4x}}{\sqrt[3]{z^4}}$

37. $\dfrac{2-\sqrt{7}}{-5}$

38. $\dfrac{\sqrt{5}+2}{\sqrt{2}}$

39. $\dfrac{\sqrt{x}+3}{\sqrt{x}}$

40. $\dfrac{5+\sqrt{2}}{\sqrt{2x}}$

41. $\sqrt{\dfrac{2}{5}}$

42. $\sqrt{\dfrac{3}{7}}$

43. $\dfrac{\sqrt{2x}}{11}$

44. $\dfrac{\sqrt{y}}{7}$

45. $\sqrt[3]{\dfrac{7}{8}}$

46. $\sqrt[3]{\dfrac{25}{2}}$

47. $\dfrac{2-\sqrt{11}}{6}$

48. $\dfrac{\sqrt{15}+1}{2}$

49. $\dfrac{\sqrt[3]{3x^5}}{10}$

50. $\sqrt[3]{\dfrac{9y}{7}}$

51. $\sqrt{\dfrac{18x^4y^6}{3z}}$

52. $\sqrt{\dfrac{8x^5y}{2z}}$

53. $\dfrac{\sqrt{2}-1}{\sqrt{2}+1}$

54. $\dfrac{\sqrt{8}-\sqrt{3}}{\sqrt{2}+\sqrt{3}}$

55. $\dfrac{\sqrt{x}+1}{\sqrt{x}-1}$

56. $\dfrac{\sqrt{x}+\sqrt{y}}{\sqrt{x}-\sqrt{y}}$

57. When rationalizing the denominator of $\dfrac{\sqrt{5}}{\sqrt{7}}$, explain why both the numerator and the denominator must be multiplied by $\sqrt{7}$.

58. When rationalizing the numerator of $\frac{\sqrt{5}}{\sqrt{7}}$, explain why both the numerator and the denominator must be multiplied by $\sqrt{5}$.

59. The formula of the radius of a sphere r with surface area A is given by the formula

$$r = \sqrt{\frac{A}{4\pi}}$$

Rationalize the denominator of the radical expression in this formula.

60. The formula for the radius of a cone r with height 7 centimeters and volume V is given by the formula

$$r = \sqrt{\frac{3V}{7\pi}}$$

Rationalize the numerator of the radical expression in this formula.

61. Explain why rationalizing the denominator does not change the value of the original expression.

62. Explain why rationalizing the numerator does not change the value of the original expression.

Review Exercises

Solve each equation. See sections 2.1 and 6.8.

63. $2x - 7 = 3(x - 4)$ **64.** $9x - 4 = 7(x - 2)$

65. $(x - 6)(2x + 1) = 0$ **66.** $(y + 2)(5y + 4) = 0$

67. $x^2 - 8x = -12$ **68.** $x^3 = x$

8.6 | RADICAL EQUATIONS AND PROBLEM SOLVING

TAPE IAG 8.6

O B J E C T I V E S

1 Solve equations that contain radical expressions.

2 Use the Pythagorean theorem to model problems.

1 In this section, we present both algebraic and graphical techniques to solve equations containing radical expressions such as

$$\sqrt{2x - 3} = 9$$

We use the power rule to help us solve these radical equations.

> **POWER RULE**
>
> If both sides of an equation are raised to the same power, **all** solutions of the original equation are **among** the solutions of the new equation.

This property *does not* say that raising both sides of an equation to a power yields an equivalent equation. A solution of the new equation *may or may not* be a solution of the original equation. Thus, *each solution of the new equation must be checked* to make sure it is a solution of the original equation.

EXAMPLE 1   Solve $\sqrt{2x - 3} = 9$ for x. Use a graphical approach to check.

Solution: Use the power rule to square both sides of the equation to eliminate the radical.

$$\sqrt{2x - 3} = 9$$
$$(\sqrt{2x - 3})^2 = 9^2$$
$$2x - 3 = 81$$
$$2x = 84$$
$$x = 42$$

Now check the solution using a graphical method. To check using the intersection-of-graphs method, graph $y_1 = \sqrt{2x - 3}$ and $y_2 = 9$ in a $[0, 95, 10]$ by $[-5, 15, 1]$ window. The intersection has x-value 42, the solution, as shown.

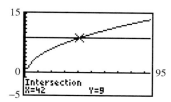

The solution checks, so we conclude that the solution set is {42}.

In the next example, we choose to solve the equation graphically, using the x-intercept method since the equation is written so that one side is 0.

EXAMPLE 2   Solve $\sqrt{-10x - 1} + 3x = 0$ for x.

Solution: Graph $y_1 = \sqrt{-10x - 1} + 3x$ in a decimal window.

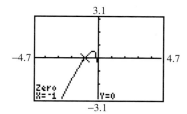

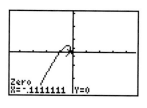

The equation has two solutions since there are two x-intercepts. The x-intercepts are $x = -1$ and $x = -0.11111\ldots$, a repeating decimal. Use the fraction command to write $-0.\overline{1}$ as the equivalent fraction $-\frac{1}{9}$.

Check the solution numerically as in the screen to the left.

The solution set is $\left\{-1, -\frac{1}{9}\right\}$.

To Solve a Radical Equation Algebraically

Step 1.   Isolate one radical so that it is by itself on one side of the equation.

Step 2.   Raise each side of the equation to a power equal to the index of the radical.

Step 3.   Simplify each side of the equation.

Step 4.   If the equation still contains a radical term, repeat steps 1–3 above.

Step 5.   Solve the equation.

Step 6.   Check all proposed solutions in the original equation for extraneous solutions to be eliminated from the solution set.

> R E M I N D E R   To solve a radical equation graphically, use the intersection-of-graphs method or the x-intercept method.

When do we solve a radical equation algebraically and when do we solve it graphically? If given a choice, it depends on the complexity of the equation itself. An equation that contains only one radical may be a good candidate for the algebraic process. Recall that by raising each side to a power, you are introducing the possibility of extraneous roots or solutions. For this reason, a check is mandatory. However, when solving graphically, you can visualize the numbers of solutions immediately. For the intersection-of-graphs method, the number of real solutions is equal to the number of intersections of the two graphs. For the x-intercept method, the number of real solutions is equal to the number of x-intercepts. We check after solving by a graphing method to confirm that the solutions are exact.

In Example 3, we choose to solve graphically by the intersection-of-graphs method and we confirm the solution algebraically.

EXAMPLE 3   Solve for x: $\sqrt[3]{x + 1} + 5 = 3$.

Solution: Graph $y_1 = \sqrt[3]{x + 1} + 5$ and $y_2 = 3$. The intersection of the two graphs has an x-value of -9, the solution of the equation.

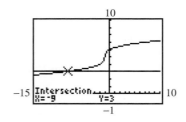

To check algebraically, first, isolate the radical by subtracting 5 from both sides of the equation.

$$\sqrt[3]{x + 1} + 5 = 3$$
$$\sqrt[3]{x + 1} = -2$$

Next, raise both sides of the equation to the third power to eliminate the radical.

$$(\sqrt[3]{x + 1})^3 = (-2)^3$$
$$x + 1 = -8$$
$$x = -9$$

The solution checks, so the solution set is $\{-9\}$.

In Example 4, we solve algebraically and check graphically. Notice that in solving algebraically, an extraneous solution is introduced. However, graphically we see that there is only one solution because there is only one point of intersection for the two graphs.

EXAMPLE 4   Solve $\sqrt{4 - x} = x - 2$ for x.

Solution:
$$\sqrt{4 - x} = x - 2$$
$$(\sqrt{4 - x})^2 = (x - 2)^2$$
$$4 - x = x^2 - 4x + 4$$
$$x^2 - 3x = 0 \qquad \text{Write the quadratic equation in standard form.}$$
$$x(x - 3) = 0 \qquad \text{Factor.}$$
$$x = 0 \quad \text{or} \quad x - 3 = 0$$
$$x = 3$$

Check the possible solutions.

$$\sqrt{4 - x} = x - 2 \qquad\qquad \sqrt{4 - x} = x - 2$$
$$\sqrt{4 - 0} = 0 - 2 \quad \text{Let } x = 0. \qquad \sqrt{4 - 3} = 3 - 2 \quad \text{Let } x = 3.$$
$$2 = -2 \qquad \text{False.} \qquad\qquad 1 = 1 \qquad \text{True.}$$

The proposed solution 3 checks, but 0 does not. When a proposed solution does not check, it is an **extraneous root or solution.** Since 0 is an extraneous solution, the solution set is $\{3\}$.

To check the results of Example 4, graph $y_1 = \sqrt{4 - x}$ and $y_2 = x - 2$. The x-value of the point of intersection is 3, as expected.

> **REMINDER** In Example 4, notice that $(x - 2)^2 = x^2 - 4x + 4$. Make sure binomials are squared correctly.

In Example 5 we solve graphically. Use the intersection-of-graphs method and confirm the solution algebraically.

EXAMPLE 5   Solve $\sqrt{2x+5} + \sqrt{2x} = 3$.

Solution: Graph $y_1 = \sqrt{2x+5} + \sqrt{2x}$ and $y_2 = 3$ in a $[-10, 10, 1]$ by $[-1, 10, 1]$ window. The intersection of the two graphs is $x = 0.22222\ldots$, a repeating decimal. Convert the value of x to fraction form using the fraction command. The solution is $\dfrac{2}{9}$.

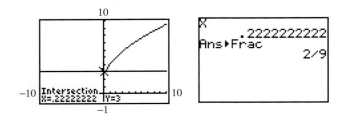

To check, we confirm the solution algebraically.

Isolate a radical by subtracting $\sqrt{2x}$ from both sides.

$$\sqrt{2x+5} + \sqrt{2x} = 3$$
$$\sqrt{2x+5} = 3 - \sqrt{2x}$$

Use the power rule to begin eliminating the radicals. Square both sides.

$$(\sqrt{2x+5})^2 = (3 - \sqrt{2x})^2$$
$$2x + 5 = 9 - 6\sqrt{2x} + 2x \qquad \text{Multiply } (3-\sqrt{2x})(3-\sqrt{2x}).$$

There is still a radical in the equation, so isolate the radical again. Then square both sides.

$$2x + 5 = 9 - 6\sqrt{2x} + 2x$$
$$6\sqrt{2x} = 4 \qquad\qquad \text{Isolate the radical.}$$
$$(6\sqrt{2x})^2 = 4^2 \qquad\quad \text{Square both sides of the equation}$$
$$36(2x) = 16 \qquad\qquad \text{to eliminate the radical.}$$
$$72x = 16 \qquad\qquad\quad \text{Multiply.}$$
$$x = \frac{16}{72} \qquad\qquad \text{Solve.}$$
$$x = \frac{2}{9} \qquad\qquad\;\; \text{Simplify.}$$

The proposed solution, $\dfrac{2}{9}$, does check, so the solution set is $\left\{\dfrac{2}{9}\right\}$.

REMINDER   Make sure expressions are squared correctly. In Example 5, we squared $(3 - \sqrt{2x})$.

$$(3 - \sqrt{2x})^2 = (3 - \sqrt{2x})(3 - \sqrt{2x})$$
$$= 3 \cdot 3 - 3\sqrt{2x} - 3\sqrt{2x} + \sqrt{2x} \cdot \sqrt{2x}$$
$$= 9 - 6\sqrt{2x} + 2x$$

2   Recall that the Pythagorean theorem states that in a right triangle the length of the hypotenuse squared equals the sum of the lengths of each of the legs squared.

PYTHAGOREAN THEOREM

If a and b are the lengths of the legs of a right triangle and c is the length of the hypotenuse, then $a^2 + b^2 = c^2$.

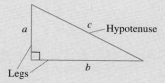

EXAMPLE 6   Find the length of the unknown leg of the following right triangle.

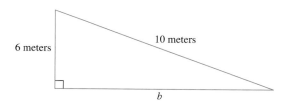

Solution: In the formula $a^2 + b^2 = c^2$, c is the hypotenuse. Let $c = 10$, the length of the hypotenuse, let $a = 6$, and solve for b. Then $a^2 + b^2 = c^2$ becomes

$$6^2 + b^2 = 10^2$$
$$36 + b^2 = 100$$
$$b^2 = 64 \qquad \text{Subtract 36 from both sides.}$$
$$b = 8 \quad \text{or} \quad b = -8 \qquad \text{Because } b^2 = 64.$$

Since we are solving for a length, we will list the positive solution only. The unknown leg of the triangle is 8 meters long.

EXAMPLE 7

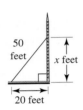

50 feet

75 feet

20 feet

DETERMINING PLACEMENT OF A SUPPORTING WIRE

A 50-foot supporting wire is to be attached to a 75-foot antenna. Because of surrounding buildings, sidewalks, and roadways, the wire must be anchored exactly 20 feet from the base of the antenna.

a. How high from the base of the antenna is the wire attached?

b. Local regulations require that a supporting wire be attached at a height no less than $\frac{3}{5}$ of the total height of the antenna. From part (a), have local regulations been met?

Solution:

50 feet

x feet

20 feet

1. UNDERSTAND. Read and reread the problem. From the diagram, notice that a right triangle is formed with hypotenuse 50 feet and one leg 20 feet.

2. ASSIGN. Let $x =$ the height from the base of the antenna to the attached wire.

3. ILLUSTRATE.

4. TRANSLATE. Use the Pythagorean theorem.

$$(\text{leg})^2 + (\text{leg})^2 = (\text{hypotenuse})^2$$
$$(20)^2 + x^2 = (50)^2$$

5. COMPLETE.
$$(20)^2 + x^2 = (50)^2$$
$$400 + x^2 = 2500$$
$$x^2 = 2100 \qquad \text{Subtract 400 from both sides.}$$
$$x = \sqrt{2100}$$

6. INTERPRET. *Check* the work and *state* the solution.

a. The wire is attached exactly $\sqrt{2100}$ feet from the base of the pole, or approximately 45.8 feet.

b. The supporting wire must be attached at a height no less than $\frac{3}{5}$ of the total height of the antenna. This height is $\frac{3}{5}(75 \text{ feet})$, or 45 feet. Since we know from part (a) that the wire is to be attached at a height of approximately 45.8 feet, local regulations have been met.

EXERCISE SET 8.6

Solve algebraically. Check the solution numerically. See Examples 1 and 2.

1. $\sqrt{2x} = 4$
2. $\sqrt{3x} = 3$
3. $\sqrt{x - 3} = 2$
4. $\sqrt{x + 1} = 5$
5. $\sqrt{2x} = -4$
6. $\sqrt{5x} = -5$
7. $\sqrt{4x - 3} - 5 = 0$
8. $\sqrt{x - 3} - 1 = 0$
9. $\sqrt{2x - 3} - 2 = 1$
10. $\sqrt{3x + 3} - 4 = 8$

Solve algebraically. Check the solution graphically. See Example 3.

11. $\sqrt[3]{6x} = -3$
12. $\sqrt[3]{4x} = -2$
13. $\sqrt[3]{x - 2} - 3 = 0$
14. $\sqrt[3]{2x - 6} - 4 = 0$

Solve by a graphical method. Check the solution numerically. See Examples 4 and 5.

15. $\sqrt{13 - x} = x - 1$

16. $\sqrt{2x - 3} = 3 - x$

17. $x - \sqrt{4 - 3x} = -8$

18. $2x + \sqrt{x + 1} = 8$

19. $\sqrt{y + 5} = 2 - \sqrt{y - 4}$

20. $\sqrt{x + 3} + \sqrt{x - 5} = 3$

21. $\sqrt{x - 3} + \sqrt{x + 2} = 5$

22. $\sqrt{2x - 4} - \sqrt{3x + 4} = -2$

Find the length of the unknown side in each triangle. See Example 6.

23.

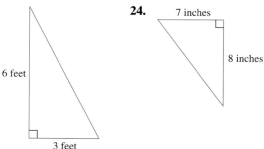

6 feet

3 feet

24.

7 inches

8 inches

25.

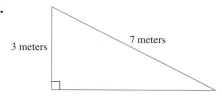

3 meters

7 meters

26.

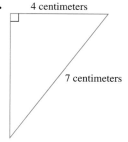

4 centimeters

7 centimeters

Solve by either an algebraic or graphical method. Check the solution using the other method.

27. $\sqrt{3x - 2} = 5$

28. $\sqrt{5x - 4} = 9$

29. $-\sqrt{2x} + 4 = -6$

30. $-\sqrt{3x + 9} = -12$

31. $\sqrt{3x + 1} + 2 = 0$

32. $\sqrt{3x + 1} - 2 = 0$

33. $\sqrt[4]{4x + 1} - 2 = 0$

34. $\sqrt[4]{2x - 9} - 3 = 0$

35. $\sqrt{4x - 3} = 5$

36. $\sqrt{3x + 9} = 12$

37. $\sqrt[3]{6x - 3} - 3 = 0$

38. $\sqrt[3]{3x} + 4 = 7$

39. $\sqrt[3]{2x - 3} - 2 = -5$

40. $\sqrt[3]{x - 4} - 5 = -7$

41. $\sqrt{x + 4} = \sqrt{2x - 5}$

42. $\sqrt{3y + 6} = \sqrt{7y - 6}$

43. $x - \sqrt{1 - x} = -5$

44. $x - \sqrt{x - 2} = 4$

45. $\sqrt[3]{-6x - 1} = \sqrt[3]{-2x - 5}$

46. $x + \sqrt{x + 5} = 7$

47. $\sqrt{5x - 1} - \sqrt{x + 2} = 3$

48. $\sqrt{2x - 1} - 4 = -\sqrt{x - 4}$

49. $\sqrt{2x - 1} = \sqrt{1 - 2x}$

50. $\sqrt{3x + 4} - 1 = \sqrt{2x + 1}$

51. Explain why proposed solutions of radical equations must be checked.

52. Consider the equations $\sqrt{2x} = 4$ and $\sqrt[3]{2x} = 4$.

    a. Explain the difference in solving these equations.

    b. Explain the similarity in solving these equations.

Find the length of the unknown side of each triangle. Give the exact length and a one-decimal-place approximation.

53.

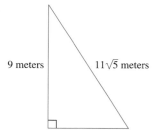

9 meters

$11\sqrt{5}$ meters

54.

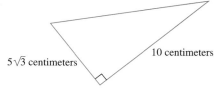

$5\sqrt{3}$ centimeters

10 centimeters

55.

7 millimeters

7.2 millimeters

56.

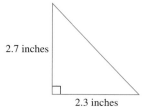

2.7 inches

2.3 inches

Use a graphing utility to solve each radical equation. Round all solutions to the nearest hundredth.

57. $\sqrt{x + 7} = x$

58. $\sqrt{3x + 5} = 2x$

59. $\sqrt{2x + 1} = \sqrt{2x + 2}$

60. $\sqrt{10x - 1} = \sqrt{-10x + 10} - 1$

61. $1.2x = \sqrt{3.1x + 5}$

62. $\sqrt{1.9x^2 - 2.2} = -0.8x + 3$

Solve. See Example 7. Give exact answers and two-decimal-place approximations where appropriate.

63. A wire is needed to support a vertical pole 15 feet high. The cable will be anchored to a stake 8 feet from the base of the pole. How much cable is needed?

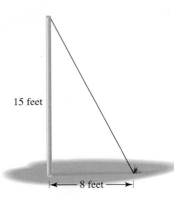

15 feet

8 feet

64. A furniture upholsterer wished to cut a strip from a piece of fabric that is 45 inches by 45 inches. The strip must be cut on the bias of the fabric. What is the longest strip that can be cut?

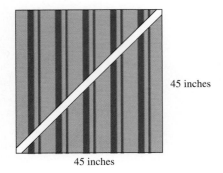

45 inches

45 inches

65. A spotlight is mounted on the eaves of a house 12 feet above the ground. A flower bed runs between the house and the sidewalk, so the closest the ladder can be placed to the house is 5 feet. How long a ladder is needed so that an electrician can reach the place where the light is mounted?

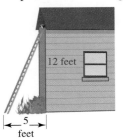

12 feet

5 feet

66. A wire is to be attached to support a telephone pole. Because of surrounding buildings, sidewalks, and roadway, the wire must be anchored exactly 15 feet from the base of the pole. Telephone company workers have only 30 feet of cable, and 2 feet of that must be used to attach the cable to the pole and to the stake on the ground. How high from the base of the pole can the wire be attached?

15 feet

67. The tallest structure in the United States is a TV tower in Blanchard, North Dakota. Its height is 2063 feet. A 2382-foot length of wire is to be used as a guy wire attached to the top of the tower. Approximate to the nearest foot how far from the base of the tower the guy wire must be anchored. (*Source:* U.S. Geological Survey)

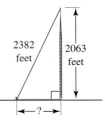

2382 feet

2063 feet

?

68. The radius of the Moon is 1080 miles. Use the formula for the radius r of a sphere given its surface area A,

$$r = \sqrt{\frac{A}{4\pi}}$$

to find the surface area of the Moon. Round to the nearest square mile. (*Source:* American Museum of Natural History)

69. The cost $C(x)$ in dollars per day to operate a small delivery service is given by $C(x) = 80\sqrt[3]{x} + 500$, where x is the number of deliveries per day. In July, the manager decides that it is necessary to keep delivery costs below $1620.00. Find the greatest number of deliveries this company can make per day and still keep overhead below $1620.00.

70. The formula $v = \sqrt{2gh}$ relates the velocity v, in feet per second, of an object after it falls h feet accelerated by gravity g, in feet per second squared. If g is approximately 32 feet per second squared, find how far an object has fallen if its velocity is 80 feet per second.

71. Two tractors are pulling a tree stump from a field. If two forces A and B pull at right angles (90°) to each other, the size of the resulting force R is given by the formula

$$R = \sqrt{A^2 + B^2}$$

If tractor A is exerting 600 pounds of force and the resulting force is 850 pounds, find how much force tractor B is exerting.

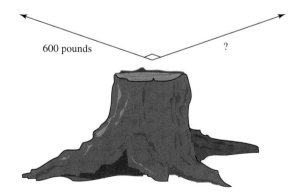

600 pounds   ?

72. Police departments find it very useful to be able to approximate the speed of a car when they are given the distance that the car skidded before it came to a stop. If the road surface is wet concrete, the function $S(x) = \sqrt{10.5x}$ is used, where $S(x)$ is the speed of the car in miles per hour and x is the distance skidded in feet. Find how fast a car was

moving if it skidded 280 feet on wet concrete. (See the accompanying photograph.)

73. The maximum distance $D(h)$ that a person can see from a height h kilometers above the ground is given by the function $D(h) = 111.7\sqrt{h}$. Find the height that would allow us to see 80 kilometers.

Review Exercises

Use the vertical line test to determine whether each graph represents the graph of a function. See Section 3.3.

74.
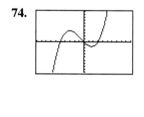

75.
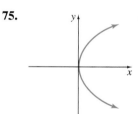

76.
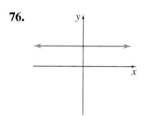

77.
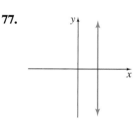

78.
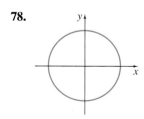

79.
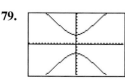

Simplify. See Section 7.4.

80. $\dfrac{\dfrac{x}{6}}{\dfrac{2x}{3} + \dfrac{1}{2}}$

81. $\dfrac{\dfrac{1}{y} + \dfrac{4}{5}}{\dfrac{-3}{20}}$

82. $\dfrac{\dfrac{z}{5} + \dfrac{1}{10}}{\dfrac{z}{20} - \dfrac{z}{5}}$

83. $\dfrac{\dfrac{1}{y} + \dfrac{1}{x}}{\dfrac{1}{y} - \dfrac{1}{x}}$

A Look Ahead

EXAMPLE

Solve $(t^2 - 3t) - 2\sqrt{t^2 - 3t} = 0$.

Solution:

Substitution can be used to make this problem somewhat simpler. Since $t^2 - 3t$ occurs more than once, let $x = t^2 - 3t$.

$$
\begin{aligned}
(t^2 - 3t) - 2\sqrt{t^2 - 3t} &= 0 \\
x - 2\sqrt{x} &= 0 \qquad \text{Let } x = t^2 - 3t.\\
x &= 2\sqrt{x} \\
x^2 &= (2\sqrt{x})^2 \\
x^2 &= 4x \\
x^2 - 4x &= 0 \\
x(x - 4) &= 0 \\
x = 0 \quad \text{or} \quad x - 4 &= 0 \\
x &= 4
\end{aligned}
$$

Now we "undo" the substitution.

$x = 0$ Replace x with $t^2 - 3t$.

$$
\begin{aligned}
t^2 - 3t &= 0 \\
t(t - 3) &= 0 \\
t = 0 \quad \text{or} \quad t - 3 &= 0 \\
t &= 3
\end{aligned}
$$

$x = 4$ Replace x with $t^2 - 3t$.

$$
\begin{aligned}
t^2 - 3t &= 4 \\
t^2 - 3t - 4 &= 0 \\
(t - 4)(t + 1) &= 0 \\
t - 4 = 0 \quad \text{or} \quad t + 1 &= 0 \\
t = 4 \quad \text{or} \quad t &= -1
\end{aligned}
$$

In this problem, we have four possible solutions: 0, 3, 4, and -1. All four solutions check in the original equation, so the solution set is $\{-1, 0, 3, 4\}$.

Solve. See the preceding example.

84. $3\sqrt{x^2 - 8x} = x^2 - 8x$

85. $\sqrt{(x^2 - x) + 7} = 2(x^2 - x) - 1$

86. $7 - (x^2 - 3x) = \sqrt{(x^2 - 3x) + 5}$

87. $x^2 + 6x = 4\sqrt{x^2 + 6x}$

8.7   COMPLEX NUMBERS

TAPE IAG 8.7

O B J E C T I V E S

1. Define imaginary and complex numbers.
2. Add or subtract complex numbers.
3. Multiply complex numbers.
4. Divide complex numbers.
5. Raise i to powers.

Our work with radical expressions has excluded expressions such as $\sqrt{-16}$ because $\sqrt{-16}$ is not a real number; there is no real number whose square is -16. In this section, we discuss a number system that includes roots of negative numbers. This number system is the **complex number system,** and it includes the set of real numbers as a subset. The complex number system allows us to solve equations such as $x^2 + 1 = 0$ that have no real number solutions. The set of complex numbers includes the **imaginary unit.**

TECHNOLOGY NOTE
Some graphing utilities have a complex number mode. Check your manual to see if your machine has this capability.

> ## IMAGINARY UNIT
>
> The imaginary unit, written i, is the number whose square is -1. That is,
>
> $$i^2 = -1 \quad \text{and} \quad i = \sqrt{-1}$$

To write the square root of a negative number in terms of i, use the property that if a is a positive number then

$$\sqrt{-a} = \sqrt{-1} \cdot \sqrt{a}$$
$$= i \cdot \sqrt{a}$$

Using i, we can write $\sqrt{-16}$ as

$$\sqrt{-16} = \sqrt{-1 \cdot 16} = \sqrt{-1} \cdot \sqrt{16} = i \cdot 4, \text{ or } 4i$$

Once a graphing utility is in complex mode, it can evaluate expressions such as $\sqrt{-16}$, whereas in real mode it gives an error message.

EXAMPLE 1   Write with i notation.

a. $\sqrt{-36}$

b. $\sqrt{-5}$

Solution: **a.** $\sqrt{-36} = \sqrt{-1 \cdot 36} = \sqrt{-1} \cdot \sqrt{36} = i \cdot 6, \text{ or } 6i$

b. $\sqrt{-5} = \sqrt{-1(5)} = \sqrt{-1} \cdot \sqrt{5} = i\sqrt{5}$. Since $\sqrt{5}i$ can easily be confused with $\sqrt{5i}$, we write $\sqrt{5}i$ as $i\sqrt{5}$.

Now that we have practiced working with the imaginary unit, complex numbers are defined.

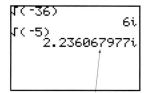

This is an approximate answer, whereas $i\sqrt{5}$ is an exact answer.

> ## COMPLEX NUMBERS
>
> A complex number is a number that can be written in the form $a + bi$, where a and b are real numbers.

The number a is the **real part** of $a + bi$, and the number b is the **imaginary part** of $a + bi$. Notice that the set of real numbers is a subset of the complex numbers since any real number can be written in the form of a complex number. For example,

$$16 = 16 + 0i$$

A calculator in complex number mode can be used to check operations on complex numbers.

In general, a complex number $a + bi$ is a real number if $b = 0$. Also, a complex number is called an **imaginary number** if $a = 0$. For example,

$$3i = 0 + 3i \quad \text{and} \quad i\sqrt{7} = 0 + i\sqrt{7}$$

are pure imaginary numbers.

The following diagram shows the relationship between complex numbers and their subsets.

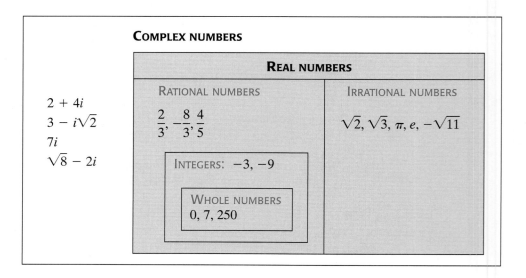

Two complex numbers $a + bi$ and $c + di$ are equal if and only if $a = c$ and $b = d$. Complex numbers can be added or subtracted by adding or subtracting their real parts and then adding or subtracting their imaginary parts.

SUM OR DIFFERENCE OF COMPLEX NUMBERS

If $a + bi$ and $c + di$ are complex numbers, then their sum is

$$(a + bi) + (c + di) = (a + c) + (b + d)i$$

Their difference is

$$(a + bi) - (c + di) = a + bi - c - di = (a - c) + (b - d)i$$

EXAMPLE 2   Add or subtract the complex numbers. Write the sum or difference in the form $a + bi$.

a. $(2 + 3i) + (-3 + 2i)$ **b.** $(5i) - (1 - i)$ **c.** $(-3 - 7i) - (-6)$

Solution: **a.** $(2 + 3i) + (-3 + 2i) = (2 - 3) + (3 + 2)i = -1 + 5i$

b. $5i - (1 - i) = 5i - 1 + i$

$$= -1 + (5 + 1)i$$

$$= -1 + 6i$$

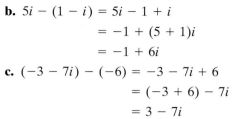

A calculator check for
Example 2.

c. $(-3 - 7i) - (-6) = -3 - 7i + 6$

$$= (-3 + 6) - 7i$$

$$= 3 - 7i$$

3   We will use the relationship $i^2 = -1$ to simplify when we multiply two complex numbers.

EXAMPLE 3   Multiply the complex numbers. Write the product in the form $a + bi$.

a. $(2 - 5i)(4 + i)$ **b.** $(2 - i)^2$ **c.** $(7 + 3i)(7 - 3i)$

Solution: Multiply complex numbers as though they were binomials.

a. $(2 - 5i)(4 + i) = 2(4) + 2(i) - 5i(4) - 5i(i)$

$$= 8 + 2i - 20i - 5i^2$$

$$= 8 - 18i - 5(-1) \qquad \text{Replace } i^2 = -1.$$

$$= 8 - 18i + 5$$

$$= 13 - 18i$$

b. $(2 - i)^2 = (2 - i)(2 - i)$

$$= 2(2) - 2(i) - 2(i) + i^2$$

$$= 4 - 4i + (-1) \qquad \text{Replace } i^2 = -1.$$

$$= 3 - 4i$$

c. $(7 + 3i)(7 - 3i) = 7(7) - 7(3i) + 3i(7) - 3i(3i)$

$$= 49 - 21i + 21i - 9i^2$$

$$= 49 - 9(-1) \qquad \text{Replace } i^2 = -1.$$

$$= 49 + 9$$

$$= 58$$

Notice that if you add, subtract, or multiply two complex numbers, just like real numbers, the result is a complex number.

4   From Example 3c, notice that the product of $7 + 3i$ and $7 - 3i$ is a real number. These two complex numbers are called **complex conjugates** of one another. In general, we have the following definition.

> ### COMPLEX CONJUGATES
>
> The complex numbers $(a + bi)$ and $(a - bi)$ are called **complex conjugates** of each other, and $(a + bi)(a - bi) = a^2 + b^2$.

To see that the product of a complex number $a + bi$ and its conjugate $a - bi$ is the real number $a^2 + b^2$, we multiply:

$$(a + bi)(a - bi) = a^2 - abi + abi - b^2i^2$$
$$= a^2 - b^2(-1)$$
$$= a^2 + b^2$$

We use complex conjugates to divide by a complex number.

EXAMPLE 4   Find each quotient. Write in the form $a + bi$.

a. $\dfrac{2 + i}{1 - i}$ **b.** $\dfrac{7}{3i}$

Solution: **a.** Multiply the numerator and denominator by the complex conjugate of $1 - i$ to eliminate the imaginary number in the denominator.

$$\frac{2 + i}{1 - i} = \frac{(2 + i)(1 + i)}{(1 - i)(1 + i)}$$
$$= \frac{2(1) + 2(i) + 1(i) + i^2}{1^2 - i^2}$$
$$= \frac{2 + 3i - 1}{1 + 1}$$
$$= \frac{1 + 3i}{2}, \quad \text{or} \quad \frac{1}{2} + \frac{3}{2}i$$

```
(2+i)/(1-i)▶Frac
           1/2+3/2i
7/(3i)▶Frac
             -7/3i
```

A check for Example 4. Notice the need for parentheses in the denominator when simplifying part (b).

To check that $\dfrac{2 + i}{1 - i} = \dfrac{1}{2} + \dfrac{3}{2}i$, multiply $\left(\dfrac{1}{2} + \dfrac{3}{2}i\right)(1 - i)$ to verify that the product is $2 + i$.

b. Multiply the numerator and denominator by the conjugate of $3i$. Note that $3i = 0 + 3i$, so its conjugate is $0 - 3i$ or $-3i$.

$$\frac{7}{3i} = \frac{7(-3i)}{(3i)(-3i)} = \frac{-21i}{-9i^2} = \frac{-21i}{-9(-1)} = \frac{-21i}{9} = \frac{-7i}{3}, \quad \text{or} \quad 0 - \frac{7}{3}i$$

The product rule for radicals does not necessarily hold true for imaginary numbers. *To multiply imaginary numbers, first write each number in terms of the imaginary unit i.* For example, to multiply $\sqrt{-4}$ and $\sqrt{-9}$, first write each number in the form bi.

$$\sqrt{-4}\sqrt{-9} = 2i(3i) = 6i^2 = -6$$

EXAMPLE 5   Multiply or divide the following as indicated.

a. $\sqrt{-3}\sqrt{-5}$ **b.** $\sqrt{-36}\sqrt{-1}$ **c.** $\sqrt{8}\sqrt{-2}$ **d.** $\dfrac{\sqrt{-125}}{\sqrt{5}}$

Solution: First, write each imaginary number in the form bi.

a. $\sqrt{-3}\,\sqrt{-5} = i\sqrt{3}\,(i\sqrt{5}) = i^2\sqrt{15} = -\sqrt{15}$

b. $\sqrt{-36}\,\sqrt{-1} = 6i(i) = 6i^2 = 6(-1) = -6$

c. $\sqrt{8}\,\sqrt{-2} = 2\sqrt{2}\,(i\sqrt{2}) = 2i(\sqrt{2}\,\sqrt{2}) = 2i(2) = 4i$

d. $\dfrac{\sqrt{-125}}{\sqrt{5}} = \dfrac{i\sqrt{125}}{\sqrt{5}} = i\sqrt{25} = 5i$

[5] We can use the fact that $i^2 = -1$ to find higher powers of i. To find i^3, we rewrite it as the product of i^2 and i.

$$i^3 = i^2 \cdot i = (-1)i = -i$$

Continue this process and find higher powers of i.

$$i^4 = i^2 \cdot i^2 = (-1)(-1) = 1$$
$$i^5 = i^4 \cdot i = 1 \cdot i = i$$

If we continue finding powers of i, we generate a pattern.

$i^1 = i$	$i^5 = i$	$i^9 = i$
$i^2 = -1$	$i^6 = -1$	$i^{10} = -1$
$i^3 = -i$	$i^7 = -i$	$i^{11} = -i$
$i^4 = 1$	$i^8 = 1$	$i^{12} = 1$

The values i, -1, $-i$, and 1 repeat as i is raised to higher and higher powers. This pattern allows us to find other powers of i.

EXAMPLE 6   Find the following powers of i.

a. i^7 b. i^{20} c. i^{46} d. i^{-12}

Solution: a. $i^7 = i^4 \cdot i^3 = 1(-i) = -i$ b. $i^{20} = (i^4)^5 = 1^5 = 1$

c. $i^{46} = i^{44} \cdot i^2 = (i^4)^{11} \cdot i^2 = 1^{11}(-1) = -1$

d. $i^{-12} = \dfrac{1}{i^{12}} = \dfrac{1}{(i^4)^3} = \dfrac{1}{(1)^3} = \dfrac{1}{1} = 1$

MENTAL MATH

Simplify. See Example 1.

1. $\sqrt{-81}$ 2. $\sqrt{-49}$ 3. $\sqrt{-7}$ 4. $\sqrt{-3}$

5. $-\sqrt{16}$ 6. $-\sqrt{4}$ 7. $\sqrt{-64}$ 8. $\sqrt{-100}$

EXERCISE SET 8.7

Write in terms of i. See Example 1.

1. $\sqrt{-24}$ **2.** $\sqrt{-32}$ **3.** $-\sqrt{-36}$

4. $-\sqrt{-121}$ **5.** $8\sqrt{-63}$ **6.** $4\sqrt{-20}$

7. $-\sqrt{54}$ **8.** $\sqrt{-63}$

Add or subtract. Write the sum or difference in the form a + bi. See Example 2.

9. $(4 - 7i) + (2 + 3i)$ **10.** $(2 - 4i) - (2 - i)$

11. $(6 + 5i) - (8 - i)$ **12.** $(8 - 3i) + (-8 + 3i)$

13. $6 - (8 + 4i)$ **14.** $(9 - 4i) - 9$

Multiply. Write the product in the form a + bi. See Example 3.

15. $6i(2 - 3i)$ **16.** $5i(4 - 7i)$

17. $(\sqrt{3} + 2i)(\sqrt{3} - 2i)$ **18.** $(\sqrt{5} - 5i)(\sqrt{5} + 5i)$

19. $(4 - 2i)^2$ **20.** $(6 - 3i)^2$

Write each quotient in the form a + bi. See Example 4.

21. $\dfrac{4}{i}$ **22.** $\dfrac{5}{6i}$ **23.** $\dfrac{7}{4 + 3i}$

24. $\dfrac{9}{1 - 2i}$ **25.** $\dfrac{3 + 5i}{1 + i}$ **26.** $\dfrac{6 + 2i}{4 - 3i}$

27. Describe how to find the conjugate of a complex number.

28. Explain why the product of a complex number and its complex conjugate is a real number.

Multiply or divide. See Example 5.

29. $\sqrt{-2} \cdot \sqrt{-7}$ **30.** $\sqrt{-11} \cdot \sqrt{-3}$

31. $\sqrt{-5} \cdot \sqrt{-10}$ **32.** $\sqrt{-2} \cdot \sqrt{-6}$

33. $\sqrt{16} \cdot \sqrt{-1}$ **34.** $\sqrt{3} \cdot \sqrt{-27}$

35. $\dfrac{\sqrt{-9}}{\sqrt{3}}$ **36.** $\dfrac{\sqrt{49}}{\sqrt{-10}}$

37. $\dfrac{\sqrt{-80}}{\sqrt{-10}}$ **38.** $\dfrac{\sqrt{-40}}{\sqrt{-8}}$

Find each power of i. See Example 6.

39. i^8 **40.** i^{10} **41.** i^{21} **42.** i^{15}

43. i^{11} **44.** i^{40} **45.** i^{-6} **46.** i^{-9}

Perform the indicated operation. Write the result in the form a + bi.

47. $(7i)(-9i)$ **48.** $(-6i)(-4i)$

49. $(6 - 3i) - (4 - 2i)$ **50.** $(-2 - 4i) - (6 - 8i)$

51. $(6 - 2i)(3 + i)$ **52.** $(2 - 4i)(2 - i)$

53. $(8 - \sqrt{-3}) - (2 + \sqrt{-12})$

54. $(8 - \sqrt{-4}) - (2 + \sqrt{-16})$

55. $(1 - i)(1 + i)$ **56.** $(6 + 2i)(6 - 2i)$

57. $\dfrac{16 + 15i}{-3i}$ **58.** $\dfrac{2 - 3i}{-7i}$

59. $(9 + 8i)^2$ **60.** $(4 - 7i)^2$

61. $\dfrac{2}{3 + i}$ **62.** $\dfrac{5}{3 - 2i}$

63. $(5 - 6i) - 4i$ **64.** $(6 - 2i) + 7i$

65. $\dfrac{2 - 3i}{2 + i}$ **66.** $\dfrac{6 + 5i}{6 - 5i}$

67. $(2 + 4i) + (6 - 5i)$ **68.** $(5 - 3i) + (7 - 8i)$

Review Exercises

Recall that the sum of the measures of the angles of a triangle is 180°. Find the unknown angle in each triangle. See Section 2.2.

69. 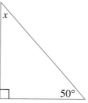   **70.**

Use synthetic division to divide the following. See Section 7.6.

71. $(x^3 - 6x^2 + 3x - 4) \div (x - 1)$

72. $(5x^4 - 3x^2 + 2) \div (x + 2)$

Thirty people were recently polled about their average monthly balance in their checking account. The results of this poll are shown in the following histogram. Use this graph to answer Exercises 73 through 78. See Section 1.1.

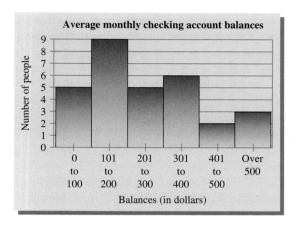

Average monthly checking account balances

Number of people (vertical axis)

Balances (in dollars)

0 to 100 | 101 to 200 | 201 to 300 | 301 to 400 | 401 to 500 | Over 500

73. How many people polled reported an average checking balance of $201 to $300?

74. How many people polled reported an average checking balance of 0 to $100?

75. How many people polled reported an average checking balance of $200 or less?

76. How many people polled reported an average checking balance of $301 or more?

77. What percent of people polled reported an average checking balance of $201 to $300?

78. What percent of people polled reported an average checking balance of 0 to $100?

GROUP ACTIVITY

CALCULATING THE LENGTH AND PERIOD OF A PENDULUM

MATERIALS:
- string (at least 1 meter long), weight, meter stick, stopwatch, calculator

Make a simple pendulum by securely tying the string to a weight.

The formula relating a pendulum's period T (in seconds) to its length l (in centimeters) is

$$T = 2\pi\sqrt{\frac{l}{980}}$$

The **period** of a pendulum is defined as the time it takes the pendulum to complete one full back-and-forth swing. In this activity, you will be measuring your simple pendulum's period with a stopwatch. Because the periods will be only a few seconds long, it will be more accurate for you to time a total of five complete swings and then find the average time of one complete swing.

(continued)

1. For each of the pendulum (string) lengths given in Table 1, measure the time required for 5 complete swings and record it in the appropriate column. Next, divide this value by 5 to find the measured period of the pendulum for the given length and record it in the Measured Period T_m column in the table. Use the given formula to calculate the theoretical period T for the same pendulum length and record it in the appropriate column. (Round to two decimal places.) Find and record in the last column the difference between the measured period and the theoretical period.

2. For each of the periods T given in Table 2, use the given formula and calculate the theoretical pendulum length l required to yield the given period. Record l in the appropriate column; round to one decimal place. Next, use this length l and measure and record the time for 5 complete swings. Divide this value by 5 to find the measured period T_m, and record it. Then find and record in the last column the difference between the theoretical period and the measured period.

3. Use the general trends you find in the tables to describe the relationship between a pendulum's period and its length.

4. Discuss the differences you found between the values of the theoretical period and the measured period. What factors contributed to these differences?

Table 1

Length l (centimeters)	Time for 5 Swings (seconds)	Measured Period T_m (seconds)	Theoretical Period T (seconds)	Difference $|T - T_m|$
30				
55				
70				

Table 2

Period T (seconds)	Theoretical Length l (centimeters)	Time for 5 Swings (seconds)	Measured Period T_m (seconds)	Difference $|T - T_m|$
1				
1.25				
2				

CHAPTER 8   HIGHLIGHTS

DEFINITIONS AND CONCEPTS	EXAMPLES
SECTION 8.1 RADICALS AND RADICAL FUNCTIONS	
The **nonnegative**, or **principal**, **square root** of a nonnegative number a is written as $\sqrt{a}$.	$\sqrt{36} = 6$ $\qquad$ $\sqrt{\dfrac{9}{100}} = \dfrac{3}{10}$
The **negative square root** of a is written as $-\sqrt{a}$. $\sqrt{a} = b$ only if $b^2 = a$ and $b \geq 0$	$-\sqrt{36} = -6$ $\qquad$ $\sqrt{0.04} = 0.2$

(continued)

DEFINITIONS AND CONCEPTS	EXAMPLES

SECTION 8.1   RADICALS AND RADICAL FUNCTIONS

The **cube root** of a real number a is written as $\sqrt[3]{a}$.

$$\sqrt[3]{a} = b \text{ only if } b^3 = a$$

If n is an even positive integer, then $\sqrt[n]{a^n} = |a|$.

If n is an odd positive integer, then $\sqrt[n]{a^n} = a$.

A **radical function** in x is a function defined by an expression containing a root of x.

$$\sqrt[3]{27} = 3 \qquad \sqrt[3]{-\frac{1}{8}} = -\frac{1}{2}$$

$$\sqrt[3]{y^6} = y^2 \qquad \sqrt[3]{64x^9} = 4x^3$$

$$\sqrt{(-3)^2} = |-3| = 3$$

$$\sqrt[3]{(-7)^3} = -7$$

If $f(x) = \sqrt{x} + 2$,

$$f(1) = \sqrt{1} + 2 = 1 + 2 = 3$$

$$f(3) = \sqrt{3} + 2 \approx 3.73$$

SECTION 8.2   RATIONAL EXPONENTS

$a^{1/n} = \sqrt[n]{a}$ if $\sqrt[n]{a}$ is a real number.

If m and n are positive integers, where $n \geq 2$ and $\sqrt[n]{a}$ is a real number, then

$$a^{m/n} = \left(a^{1/n}\right)^m = \left(\sqrt[n]{a}\right)^m$$

$a^{-m/n} = \dfrac{1}{a^{m/n}}$ as long as $a^{m/n}$ is a nonzero number.

Exponent rules are true for rational exponents.

$$81^{1/2} = \sqrt{81} = 9$$

$$(-8x^3)^{1/3} = \sqrt[3]{-8x^3} = -2x$$

$$4^{5/2} = \left(\sqrt{4}\right)^5 = 2^5 = 32$$

$$27^{2/3} = \left(\sqrt[3]{27}\right)^2 = 3^2 = 9$$

$$16^{-3/4} = \frac{1}{16^{3/4}} = \frac{1}{\left(\sqrt[4]{16}\right)^3} = \frac{1}{2^3} = \frac{1}{8}$$

$$x^{2/3} \cdot x^{-5/6} = x^{2/3 - 5/6} = x^{-1/6} = \frac{1}{x^{1/6}}$$

$$\left(8^{14}\right)^{1/7} = 8^2 = 64$$

$$\frac{a^{4/5}}{a^{-2/5}} = a^{4/5 - (-2/5)} = a^{6/5}$$

SECTION 8.3   SIMPLIFYING RADICAL EXPRESSIONS

Product and quotient rules:

If $\sqrt[n]{a}$ and $\sqrt[n]{b}$ are real numbers,

$$\sqrt[n]{a} \cdot \sqrt[n]{b} = \sqrt[n]{a \cdot b}$$

$$\frac{\sqrt[n]{a}}{\sqrt[n]{b}} = \sqrt[n]{\frac{a}{b}}, \text{ provided } \sqrt[n]{b} \neq 0$$

A radical of the form $\sqrt[n]{a}$ is **simplified** when a contains no factors that are perfect nth powers.

Multiply or divide as indicated:

$$\sqrt{11} \cdot \sqrt{3} = \sqrt{33}$$

$$\frac{\sqrt[3]{40x}}{\sqrt[3]{5x}} = \sqrt[3]{8} = 2$$

$$\sqrt{40} = \sqrt{4 \cdot 10} = 2\sqrt{10}$$

$$\sqrt{36x^5} = \sqrt{36x^4 \cdot x} = 6x^2\sqrt{x}$$

$$\sqrt[3]{24x^7y^3} = \sqrt[3]{8x^6y^3 \cdot 3x} = 2x^2y\sqrt[3]{3x}$$

(continued)

DEFINITIONS AND CONCEPTS	**EXAMPLES**

SECTION 8.4 ADDING, SUBTRACTING, AND MULTIPLYING RADICAL EXPRESSIONS

Radicals with the same index and the same radicand are **like radicals.**	$5\sqrt{6} + 2\sqrt{6} = (5 + 2)\sqrt{6} = 7\sqrt{6}$
The distributive property can be used to add like radicals.	$-\sqrt[3]{3x} - 10\sqrt[3]{3x} + 3\sqrt[3]{10x} = (-1 - 10)\sqrt[3]{3x}$ $+ 3\sqrt[3]{10x} = -11\sqrt[3]{3x} + 3\sqrt[3]{10x}$
Radical expressions are multiplied by using many of the same properties used to multiply polynomials.	Multiply: $$(\sqrt{5} - \sqrt{2x})(\sqrt{2} + \sqrt{2x})$$ $$= \sqrt{10} + \sqrt{10x} - \sqrt{4x} - 2x$$ $$= \sqrt{10} + \sqrt{10x} - 2\sqrt{x} - 2x$$ $$(2\sqrt{3} - \sqrt{8x})(2\sqrt{3} + \sqrt{8x})$$ $$= 4\sqrt{9} - 8x = 12 - 8x$$

SECTION 8.5 RATIONALIZING NUMERATORS AND DENOMINATORS OF RADICAL EXPRESSIONS

The **conjugate** of $a + b$ is $a - b$.	The conjugate of $\sqrt{7} + \sqrt{3}$ is $\sqrt{7} - \sqrt{3}$.
The process of writing the denominator of a radical expression without a radical is called **rationalizing the denominator.**	Rationalize each denominator: $$\frac{\sqrt{5}}{\sqrt{3}} = \frac{\sqrt{5} \cdot \sqrt{3}}{\sqrt{3} \cdot \sqrt{3}} = \frac{\sqrt{15}}{3}$$ $$\frac{6}{\sqrt{7} + \sqrt{3}} = \frac{6(\sqrt{7} - \sqrt{3})}{(\sqrt{7} + \sqrt{3})(\sqrt{7} - \sqrt{3})}$$ $$= \frac{6(\sqrt{7} - \sqrt{3})}{7 - 3}$$ $$= \frac{6(\sqrt{7} - \sqrt{3})}{4} = \frac{3(\sqrt{7} - \sqrt{3})}{2}$$
The process of writing the numerator of a radical expression without a radical is called **rationalizing the numerator.**	Rationalize each numerator: $$\frac{\sqrt[3]{9}}{\sqrt[3]{5}} = \frac{\sqrt[3]{9} \cdot \sqrt[3]{3}}{\sqrt[3]{5} \cdot \sqrt[3]{3}} = \frac{\sqrt[3]{27}}{\sqrt[3]{15}} = \frac{3}{\sqrt[3]{15}}$$ $$\frac{\sqrt{9} + \sqrt{3x}}{12} = \frac{(\sqrt{9} + \sqrt{3x})(\sqrt{9} - \sqrt{3x})}{12(\sqrt{9} - \sqrt{3x})}$$ $$= \frac{9 - 3x}{12(\sqrt{9} - \sqrt{3x})}$$ $$= \frac{3(3 - x)}{3 \cdot 4(3 - \sqrt{3x})} = \frac{3 - x}{4(3 - \sqrt{3x})}$$

SECTION 8.6 RADICAL EQUATIONS AND PROBLEM SOLVING

To solve a radical equation algebraically:	Solve: $x = \sqrt{4x + 9} + 3$.
Step 1. Write the equation so that one radical is by itself on one side of the equation.	1. $x - 3 = \sqrt{4x + 9}$
Step 2. Raise each side of the equation to a power equal to the index of the radical.	2. $(x - 3)^2 = 4x + 9$

(continued)

DEFINITIONS AND CONCEPTS	EXAMPLES

SECTION 8.6   RADICAL EQUATIONS AND PROBLEM SOLVING

Step 3. Simplify each side of the equation.

Step 4. If the equation still contains a radical, repeat *steps 1* through *3*.

Step 5. Solve the equation.

Step 6. Check proposed solutions in the original equation for extraneous solutions.

To solve a radical equation graphically, use the intersection-of-graphs method or the x-intercept method.

3. $x^2 - 6x + 9 = 4x + 9$

5. $x^2 - 10x = 0$
$x(x - 10) = 0$
$x = 0$ or $x = 10$

6. The proposed solution 10 checks, but 0 does not. The solution set is $\{10\}$.

Solve $x = \sqrt{4x + 9} + 3$ using the intersection-of-graphs method.

Graph $y_1 = x$ and $y_2 = \sqrt{4x + 9} + 3$ in a $[-5, 20, 5]$ by $[-10, 15, 5]$ window.

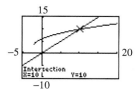

The intersection of the two graphs is at $x = 10$. The solution of the equation is 10.

See the algebraic solution above for a check.

SECTION 8.7   COMPLEX NUMBERS

A **complex number** is a number that can be written in the form $a + bi$, where a and b are real numbers.

$$i^2 = -1 \text{ and } i = \sqrt{-1}$$

Simplify $\sqrt{-9}$.
$$\sqrt{-9} = \sqrt{-1 \cdot 9} = \sqrt{-1} \cdot \sqrt{9} = i \cdot 3, \text{ or } 3i$$

COMPLEX NUMBERS	WRITTEN IN FORM $a + bi$
12	$12 + 0i$
$-5i$	$0 + (-5)i$
$-2 - 3i$	$-2 + (-3)i$

Multiply or divide as indicated.
$$\sqrt{-3} \cdot \sqrt{-7} = i\sqrt{3} \cdot i\sqrt{7}$$
$$= i^2\sqrt{21}$$
$$= -\sqrt{21}$$

To add or subtract complex numbers, add or subtract their real parts and then add or subtract their imaginary parts.

To multiply complex numbers, multiply as though they are binomials.

Perform the indicated operations.
$$(-3 + 2i) - (7 - 4i) = -3 + 2i - 7 + 4i$$
$$= -10 + 6i$$
$$(-7 - 2i)(6 + i) = -42 - 7i - 12i - 2i^2$$
$$= -42 - 19i - 2(-1)$$
$$= -42 - 19i + 2$$
$$= -40 - 19i$$

(continued)

DEFINITIONS AND CONCEPTS	EXAMPLES
SECTION 8.7 COMPLEX NUMBERS	
The complex numbers $(a + bi)$ and $(a - bi)$ are called **complex conjugates**.	The complex conjugate of $(3 + 6i)$ is $(3 - 6i)$
To divide complex numbers, multiply the numerator and the denominator by the conjugate of the denominator.	Divide: $\dfrac{4}{2 - i} = \dfrac{4(2 + i)}{(2 - i)(2 + i)}$ $= \dfrac{4(2 + i)}{4 - i^2}$ $= \dfrac{4(2 + i)}{5}$ $= \dfrac{8 + 4i}{5}$, or $\dfrac{8}{5} + \dfrac{4}{5}i$

CHAPTER 8 REVIEW

(8.1) *Take the root. Assume that all variables represent positive numbers.*

1. $\sqrt{81}$ **2.** $\sqrt[4]{81}$ **3.** $\sqrt[3]{-8}$ **4.** $\sqrt[4]{-16}$

5. $-\sqrt{\dfrac{1}{49}}$ **6.** $\sqrt{x^{64}}$ **7.** $-\sqrt{36}$ **8.** $\sqrt[3]{64}$

9. $\sqrt[3]{-a^6b^9}$ **10.** $\sqrt{16a^4b^{12}}$ **11.** $\sqrt[5]{32a^5b^{10}}$

12. $\sqrt[5]{-32x^{15}y^{20}}$ **13.** $\sqrt{\dfrac{x^{12}}{36y^2}}$ **14.** $\sqrt[3]{\dfrac{27y^3}{z^{12}}}$

Simplify. Use absolute value bars when necessary.

15. $\sqrt{(-x)^2}$ **16.** $\sqrt[4]{(x^2 - 4)^4}$ **17.** $\sqrt[3]{(-27)^3}$

18. $\sqrt[5]{(-5)^5}$ **19.** $-\sqrt[5]{x^5}$

20. $\sqrt[4]{16(2y + z)^{12}}$ **21.** $\sqrt{25(x - y)^{10}}$

22. $\sqrt[5]{-y^5}$ **23.** $\sqrt[9]{-x^9}$

Identify the domain and then graph each function.

24. $f(x) = \sqrt{x} + 3$

25. $g(x) = \sqrt[3]{x} - 3$

(8.2) *Evaluate the following. Check using a graphing utility.*

26. $\left(\dfrac{1}{81}\right)^{1/4}$ **27.** $\left(-\dfrac{1}{27}\right)^{1/3}$ **28.** $(-27)^{-1/3}$

29. $(-64)^{-1/3}$ **30.** $-9^{3/2}$ **31.** $64^{-1/3}$

32. $(-25)^{5/2}$ **33.** $\left(\dfrac{25}{49}\right)^{-3/2}$

34. $\left(\dfrac{8}{27}\right)^{-2/3}$ **35.** $\left(-\dfrac{1}{36}\right)^{-1/4}$

Write with rational exponents.

36. $\sqrt[3]{x^2}$ **37.** $\sqrt[5]{5x^2y^3}$

Write with radical notation.

38. $y^{4/5}$ **39.** $5(xy^2z^5)^{1/3}$ **40.** $(x + 2y)^{-1/2}$

Simplify each expression. Assume that all variables represent positive numbers. Write with only positive exponents.

41. $a^{1/3}a^{4/3}a^{1/2}$ **42.** $\dfrac{b^{1/3}}{b^{4/3}}$

43. $(a^{1/2}a^{-2})^3$ **44.** $(x^{-3}y^6)^{1/3}$

45. $\left(\dfrac{b^{3/4}}{a^{-1/2}}\right)^8$ **46.** $\dfrac{x^{1/4}x^{-1/2}}{x^{2/3}}$

47. $\left(\dfrac{49c^{5/3}}{a^{-1/4}b^{5/6}}\right)^{-1}$ **48.** $a^{-1/4}(a^{5/4} - a^{9/4})$

Use a calculator and write a three-decimal-place approximation.

49. $\sqrt{20}$ **50.** $\sqrt[3]{-39}$ **51.** $\sqrt[4]{726}$

52. $56^{1/3}$ **53.** $-78^{3/4}$ **54.** $105^{-2/3}$

(8.3) *Perform the indicated operations and then simplify if possible. For the remainder of this review, assume that variables represent positive numbers only.*

55. $\sqrt{3} \cdot \sqrt{8}$ **56.** $\sqrt[3]{7y} \cdot \sqrt[3]{x^2 z}$

57. $\dfrac{\sqrt{44x^3}}{\sqrt{11x}}$ **58.** $\dfrac{\sqrt[4]{a^6 b^{13}}}{\sqrt[4]{a^2 b}}$

Simplify.

59. $\sqrt{60}$ **60.** $-\sqrt{75}$

61. $\sqrt[3]{162}$ **62.** $\sqrt[3]{-32}$

63. $\sqrt{36x^7}$ **64.** $\sqrt[3]{24a^5 b^7}$

65. $\sqrt{\dfrac{p^{17}}{121}}$ **66.** $\sqrt[3]{\dfrac{y^5}{27x^6}}$

67. $\sqrt[4]{\dfrac{xy^6}{81}}$ **68.** $\sqrt{\dfrac{2x^3}{49y^4}}$

Use rational exponents to write each radical with the same index. Then multiply.

69. $\sqrt[3]{2} \cdot \sqrt{7}$ **70.** $\sqrt[3]{3} \cdot \sqrt[4]{x}$

71. The formula for the radius r of a circle of area A is

$$r = \sqrt{\dfrac{A}{\pi}}$$

  a. Find the exact radius of a circle whose area is 25 square meters.

  b. Approximate to two decimal places the radius of a circle whose area is 104 square inches.

(8.4) *Perform the indicated operation.*

72. $x\sqrt{75xy} - \sqrt{27x^3 y}$ **73.** $2\sqrt{32x^2 y^3} - xy\sqrt{98y}$

74. $\sqrt[3]{128} + \sqrt[3]{250}$ **75.** $3\sqrt[4]{32a^5} - a\sqrt[4]{162a}$

76. $\dfrac{5}{\sqrt[3]{4}} + \dfrac{\sqrt{3}}{3}$ **77.** $\sqrt{\dfrac{8}{x^2}} - \sqrt{\dfrac{50}{16x^2}}$

78. $2\sqrt{50} - 3\sqrt{125} + \sqrt{98}$

79. $2a\sqrt[4]{32b^5} - 3b\sqrt[4]{162a^4 b} + \sqrt[4]{2a^4 b^5}$

Multiply and then simplify if possible.

80. $\sqrt{3}(\sqrt{27} - \sqrt{3})$ **81.** $(\sqrt{x} - 3)^2$

82. $(\sqrt{5} - 5)(2\sqrt{5} + 2)$

83. $(2\sqrt{x} - 3\sqrt{y})(2\sqrt{x} + 3\sqrt{y})$

84. $(\sqrt{a} + 3)(\sqrt{a} - 3)$ **85.** $(\sqrt[3]{a} + 2)^2$

86. $(\sqrt[3]{5x} + 9)(\sqrt[3]{5x} - 9)$

87. $(\sqrt[3]{a} + 4)(\sqrt[3]{a^2} - 4\sqrt[3]{a} + 16)$

(8.5) *Rationalize each denominator.*

88. $\dfrac{3}{\sqrt{7}}$ **89.** $\sqrt{\dfrac{x}{12}}$

90. $\dfrac{5}{\sqrt[3]{4}}$ **91.** $\sqrt{\dfrac{24x^5}{3y^2}}$

92. $\sqrt[3]{\dfrac{15x^6 y^7}{z^2}}$ **93.** $\dfrac{5}{2 - \sqrt{7}}$

94. $\dfrac{3}{\sqrt{y} - 2}$ **95.** $\dfrac{\sqrt{2} - \sqrt{3}}{\sqrt{2} + \sqrt{3}}$

Rationalize each numerator.

96. $\dfrac{\sqrt{11}}{3}$ **97.** $\sqrt{\dfrac{18}{y}}$

98. $\dfrac{\sqrt[3]{9}}{7}$ **99.** $\sqrt{\dfrac{24x^5}{3y^2}}$

100. $\sqrt[3]{\dfrac{xy^2}{10z}}$ **101.** $\dfrac{\sqrt{x} + 5}{-3}$

(8.6) *Solve each equation either algebraically or graphically for the variable. Check the solution using the other method.*

102. $\sqrt{y - 7} = 5$ **103.** $\sqrt{2x + 10} = 4$

104. $\sqrt[3]{2x - 6} = 4$ **105.** $\sqrt{x + 6} = \sqrt{x + 2}$

106. $2x - 5\sqrt{x} = 3$ **107.** $\sqrt{x + 9} = 2 + \sqrt{x - 7}$

Find each unknown length.

108.

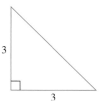

109.

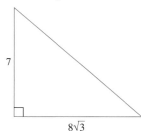

110. Beverly Hillis wants to determine the distance x across a pond on her property. She is able to measure the distances shown on the following diagram. Find how wide the lake is at the crossing point indicated by the triangle to the nearest tenth of a foot.

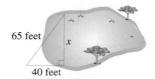

65 feet

x

40 feet

111. A pipefitter needs to connect two underground pipelines that are offset by 3 feet, as pictured in the diagram at the top of the next column. Neglecting the joints needed to join the pipes, find the length of the shortest possible connecting pipe rounded to the nearest hundredth of a foot.

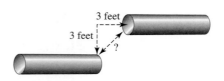

3 feet

3 feet

?

(8.7) *Perform the indicated operation and simplify. Write the result in the form a + bi.*

112. $\sqrt{-8}$ **113.** $-\sqrt{-6}$

114. $\sqrt{-4} + \sqrt{-16}$ **115.** $\sqrt{-2} \cdot \sqrt{-5}$

116. $(12 - 6i) + (3 + 2i)$ **117.** $(-8 - 7i) - (5 - 4i)$

118. $(\sqrt{3} + \sqrt{2}) + (3\sqrt{2} - \sqrt{-8})$

119. $2i(2 - 5i)$ **120.** $-3i(6 - 4i)$

121. $(3 + 2i)(1 + i)$ **122.** $(2 - 3i)^2$

123. $(\sqrt{6} - 9i)(\sqrt{6} + 9i)$

124. $\dfrac{2 + 3i}{2i}$ **125.** $\dfrac{1 + i}{-3i}$

CHAPTER 8 TEST

Raise to the power or take the root. Assume that all variables represent positive numbers. Write with only positive exponents.

1. $\sqrt{216}$ **2.** $-\sqrt[4]{x^{64}}$ **3.** $\left(\dfrac{1}{125}\right)^{1/3}$ **4.** $\left(\dfrac{1}{125}\right)^{-1/3}$

5. $\left(\dfrac{8x^3}{27}\right)^{2/3}$ **6.** $\sqrt[3]{-a^{18}b^9}$

7. $\left(\dfrac{64c^{4/3}}{a^{-2/3}b^{5/6}}\right)^{1/2}$ **8.** $a^{-2/3}(a^{5/4} - a^3)$

Take the root. Use absolute value bars when necessary.

9. $\sqrt[4]{(4xy)^4}$ **10.** $\sqrt[3]{(-27)^3}$

Rationalize the denominator. Assume that all variables represent positive numbers.

11. $\sqrt{\dfrac{9}{y}}$ **12.** $\dfrac{4 - \sqrt{x}}{4 + 2\sqrt{x}}$ **13.** $\dfrac{\sqrt[3]{ab}}{\sqrt[3]{ab^2}}$

14. Rationalize the numerator of $\dfrac{\sqrt{6} + x}{8}$ and simplify.

Perform the indicated operations. Assume that all variables represent positive numbers.

15. $\sqrt{125x^3} - 3\sqrt{20x^3}$ **16.** $\sqrt{3}(\sqrt{16} - \sqrt{2})$

17. $(\sqrt{x} + 1)^2$ **18.** $(\sqrt{2} - 4)(\sqrt{3} + 1)$

19. $(\sqrt{5} + 5)(\sqrt{5} - 5)$

Use a calculator to approximate each to three decimal places.

20. $\sqrt{561}$ **21.** $386^{-2/3}$

Solve either algebraically or graphically. Check using the other method.

22. $x = \sqrt{x - 2} + 2$ **23.** $\sqrt{x^2 - 7} + 3 = 0$

24. $\sqrt{x + 5} = \sqrt{2x - 1}$

Perform the indicated operation and simplify. Write the result in the form a + bi.

25. $\sqrt{-2}$ **26.** $-\sqrt{-8}$

27. $(12 - 6i) - (12 - 3i)$ **28.** $(6 - 2i)(6 + 2i)$

29. $(4 + 3i)^2$ **30.** $\dfrac{1 + 4i}{1 - i}$

31. Find x.

x

5

x

32. Identify the domain of $g(x) = \sqrt{x+2}$ and then graph $g(x)$.

Solve.

33. The function $V(r) = \sqrt{2.5r}$ can be used to estimate the maximum safe velocity, V, in miles per hour, at which a car can travel if it is driven along a curved road with a *radius of curvature*, r, in feet.

To the nearest whole number, find the maximum safe speed if a cloverleaf exit on an expressway has a radius of curvature of 300 feet.

34. Use the formula from Exercise 33 to find the radius of curvature if the safe velocity is 30 mph.

CHAPTER 8   CUMULATIVE REVIEW

1. Use the associative property of multiplication to write an expression equivalent to $4 \cdot (9y)$. Then simplify this equivalent expression.

2. Write the following as algebraic expressions. Then simplify.

   a. The sum of two consecutive integers, if x is the first integer.

   b. The perimeter of the triangle with sides x, $5x$, and $6x - 3$.

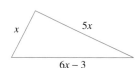

3. Find an equation of the line with no slope that contains the point $(2, 3)$.

4. Use a graphical approach to solve $\left| \dfrac{x}{3} - 1 \right| - 2 \geq 0$.

5. Use the elimination method to solve the system
$$\begin{cases} 3x - 2y = 10 \\ 4x - 3y = 15 \end{cases}$$

6. Lynn Pike, a pharmacist, needs 70 liters of 50% alcohol solution. She has available a 30% alcohol solution and an 80% alcohol solution. How many liters of each solution should she mix to obtain 70 liters of a 50% alcohol solution?

7. If $P(x) = 3x^2 - 2x - 5$, find the following:

   a. $P(1)$ **b.** $P(-2)$

8. Factor $ab - 6a + 2b - 12$.

9. Factor $2x^2 + 11x + 15$.

10. Factor each polynomial completely.

   a. $5p^2 + 5 + qp^2 + q$

   b. $9x^2 + 24x + 16$

   c. $y^2 + 25$

11. Write each rational expression in lowest terms.

   a. $\dfrac{24x^6y^5}{8x^7y}$ **b.** $\dfrac{2x^2}{10x^3 - 2x^2}$

12. Divide.

   a. $\dfrac{3x}{5y} \div \dfrac{9y}{x^5}$ **b.** $\dfrac{8m^2}{3m^2 - 12} \div \dfrac{40}{2 - m}$

13. Add or subtract.

   a. $\dfrac{5}{7} + \dfrac{x}{7}$ **b.** $\dfrac{x}{4} + \dfrac{5x}{4}$

   c. $\dfrac{x^2}{x+7} - \dfrac{49}{x+7}$ **d.** $\dfrac{x}{3y^2} - \dfrac{x+1}{3y^2}$

14. Simplify each complex fraction.

   a. $\dfrac{\dfrac{5x}{x+2}}{\dfrac{10}{x-2}}$ **b.** $\dfrac{x + \dfrac{1}{y}}{y + \dfrac{1}{x}}$

15. Find the quotient: $\dfrac{6x^2 - 19x + 12}{3x - 5}$.

16. Solve $\dfrac{8x}{5} + \dfrac{3}{2} = \dfrac{3x}{5}$ algebraically and check graphically.

17. Melissa Scarlatti can clean the house in 4 hours, and her husband, Zack, can do the same job in 5 hours. They have agreed to clean together so that they can finish in time to watch a movie on TV that starts in 2 hours. How long will it take them to clean the house together? Can they finish before the movie starts?

18. Simplify. Assume that all variables represent positive numbers.

a. $\sqrt{36}$ **b.** $\sqrt{0}$ **c.** $\sqrt{\dfrac{4}{49}}$

d. $\sqrt{0.25}$ **e.** $\sqrt{x^6}$ **f.** $\sqrt{9x^{10}}$

g. $-\sqrt{81}$

19. Use radical notation to write the following. Simplify if possible.

a. $4^{1/2}$ **b.** $64^{1/3}$ **c.** $x^{1/4}$

d. $0^{1/6}$ **e.** $-9^{1/2}$ **f.** $(81x^8)^{1/4}$

g. $(5y)^{1/3}$

20. Simplify the following.

a. $\sqrt{50}$ **b.** $\sqrt[3]{24}$ **c.** $\sqrt{26}$

d. $\sqrt[4]{32}$

21. Multiply.

a. $\sqrt{3}(5 + \sqrt{30})$

b. $(\sqrt{5} - \sqrt{6})(\sqrt{7} + 1)$

c. $(7\sqrt{x} + 5)(3\sqrt{x} - \sqrt{5})$

d. $(4\sqrt{3} - 1)^2$

e. $(\sqrt{2x} - 5)(\sqrt{2x} + 5)$

22. Rationalize the denominator of each expression.

a. $\dfrac{\sqrt{27}}{\sqrt{5}}$ **b.** $\dfrac{2\sqrt{16}}{\sqrt{9x}}$ **c.** $\sqrt[3]{\dfrac{1}{2}}$

23. Solve $\sqrt{4 - x} = x - 2$ algebraically and check graphically.

QUADRATIC EQUATIONS AND FUNCTIONS

MODELING THE POSITION OF A FREELY FALLING OBJECT

In physics, the position of an object in free fall is modeled by a quadratic function. Once an object is in free fall, its position depends on its initial position, its initial velocity, and the acceleration due to gravity that it experiences.

IN THE CHAPTER GROUP ACTIVITY ON PAGE 575, YOU WILL HAVE THE OPPORTUNITY TO MATCH A DESCRIPTION OF A FREELY FALLING OBJECT TO ITS MATHEMATICAL MODEL AND ITS GRAPH.

A
n important part of the study of algebra is learning to model and solve problems. Often, the model of a problem is a quadratic equation or a function containing a second-degree polynomial. In this chapter, we continue the work begun in Chapter 6, when we solved polynomial equations in one variable by factoring. Two additional methods of solving quadratic equations are analyzed, as well as methods of solving nonlinear inequalities in one variable.

9.1 SOLVING QUADRATIC EQUATIONS BY COMPLETING THE SQUARE

O B J E C T I V E S

1. Use the square root property to solve quadratic equations.
2. Write perfect square trinomials.
3. Solve quadratic equations by completing the square.
4. Use quadratic equations to solve problems.

TAPE IAG 9.1

1

In Chapter 6, we solved quadratic equations by factoring. Recall that a **quadratic, or second-degree, equation** is an equation that can be written in the form $ax^2 + bx + c = 0$, where a, b, and c are real numbers and a is not 0. To solve a quadratic equation such as $x^2 = 9$ by factoring, we use the zero-factor theorem. To use the zero-factor theorem, the equation must first be written in standard form, $ax^2 + bx + c = 0$.

$$x^2 = 9$$
$$x^2 - 9 = 0 \qquad \text{Subtract 9 from both sides.}$$
$$(x + 3)(x - 3) = 0 \qquad \text{Factor.}$$
$$x + 3 = 0 \quad \text{or} \quad x - 3 = 0 \qquad \text{Set each factor equal to 0.}$$
$$x = -3 \qquad\qquad x = 3 \qquad \text{Solve.}$$

The solution set is $\{-3, 3\}$, the positive and negative square roots of 9. Not all quadratic equations can be solved by factoring, so we need to explore other methods. Notice that the solutions of the equation $x^2 = 9$ are two numbers whose square is 9.

$$3^2 = 9 \quad \text{and} \quad (-3)^2 = 9$$

Thus, we can solve the equation $x^2 = 9$ by taking the square root of both sides. Be sure to include both $\sqrt{9}$ and $-\sqrt{9}$ as solutions since both $\sqrt{9}$ and $-\sqrt{9}$ are numbers whose square is 9.

$$x^2 = 9$$
$$x = \pm\sqrt{9}$$
$$x = \pm 3 \qquad \text{The notation } \pm 3 \text{ (read as plus or minus 3)}$$
$$\text{indicates the pair of numbers } +3 \text{ and } -3.$$

This illustrates the square root property.

TECHNOLOGY NOTE
Many graphing utilities have a complex mode. Check your manual, and if yours has it, set your graphing utility to this mode for this chapter. Recall that a real number is a complex number, so in complex mode the graphing utility will display real numbers and imaginary numbers.

SQUARE ROOT PROPERTY

If b is a real number and if $a^2 = b$, then $a = \pm\sqrt{b}$.

REMINDER   Complex Numbers

$a + bi$, a and b real numbers

Real Numbers   Imaginary Numbers
$a + bi$, $b = 0$ $a + bi$, $b \neq 0$

Examples of Real Numbers: -3, 7.6, $\sqrt{5}$, $\frac{13}{20}$
Examples of Imaginary Numbers: $2 - 6i$, $-1.2i$, $3 + \sqrt{11}i$

EXAMPLE 1   Use the square root property to solve $x^2 = 50$.

Solution:

$$x^2 = 50$$
$$x = \pm\sqrt{50} \qquad \text{Apply the square root property.}$$
$$x = \pm5\sqrt{2} \qquad \text{Simplify the radical.}$$

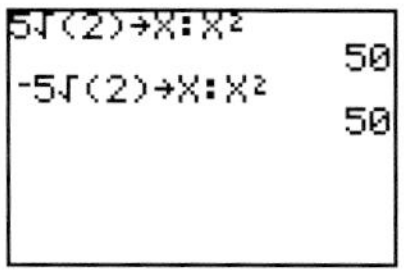

The solution set is $\{5\sqrt{2}, -5\sqrt{2}\}$.

A check is shown to the left.

```
5√(2)→X:X²
             50
-5√(2)→X:X²
             50
```

EXAMPLE 2   Use the square root property to solve $(x + 1)^2 = 12$ for x. Use a graphing utility to check.

Solution: By the square root property, we have

$$(x + 1)^2 = 12$$
$$x + 1 = \pm\sqrt{12} \qquad \text{Apply the square root property.}$$
$$x + 1 = \pm2\sqrt{3} \qquad \text{Simplify the radical.}$$
$$x = -1 \pm 2\sqrt{3} \qquad \text{Subtract 1 from both sides.}$$

The solution set is $\{-1 + 2\sqrt{3}, -1 - 2\sqrt{3}\}$.

To check graphically, we use the intersection-of-graphs method and graph

$$y_1 = (x + 1)^2 \quad \text{and} \quad y_2 = 12$$

The approximate points of intersection are shown below in the first two screens.

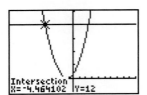 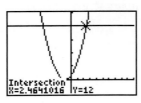 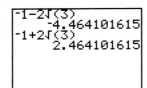

The approximate solutions are $x \approx -4.464$ and $x \approx 2.464$. The exact solutions are $x = -1 + 2\sqrt{3}$ and $x = -1 - 2\sqrt{3}$.

Next, see that the x-values of the points of intersection approximate the exact solutions. The screen above and to the right confirms this.

EXAMPLE 3   Solve $(2x - 5)^2 = -16$.

Solution:

$$(2x - 5)^2 = -16$$
$$2x - 5 = \pm\sqrt{-16} \qquad \text{Apply the square root property.}$$
$$2x - 5 = \pm 4i \qquad \text{Simplify the radical.}$$
$$2x = 5 \pm 4i \qquad \text{Add 5 to both sides.}$$
$$x = \frac{5 \pm 4i}{2} \qquad \text{Divide both sides by 2.}$$

The solution set is $\left\{ \dfrac{5 + 4i}{2}, \dfrac{5 - 4i}{2} \right\}$.

DISCOVER THE CONCEPT

From the example above, we know that the equation $(2x - 5)^2 = -16$ has two imaginary solutions. Use the intersection-of-graphs method to check.

a. Graph $y_1 = (2x - 5)^2$ and $y_2 = -16$.

b. Locate any points of intersection of the graphs.

c. Summarize the results of parts (a) and (b), and what you think occurred.

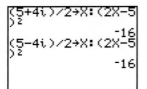

The x-axis and the y-axis of the rectangular coordinate system include real numbers only. Thus, coordinates of points of the associated plane are real numbers only. This means that the intersection-of-graphs method on the rectangular coordinate system gives real number solutions of a related equation only. Since the graphs above do not intersect, the equation has no real number solutions.

To check Example 3 numerically, see the screen to the left.

2   Notice from Examples 2 and 3 that, if we write a quadratic equation so that one side is the square of a binomial, we can solve algebraically by using the square root property. To write the square of a binomial, we write perfect square trinomials.

Recall that a perfect square trinomial is a trinomial that can be factored into two identical binomial factors.

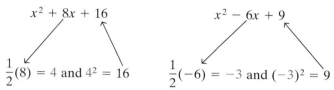

PERFECT SQUARE TRINOMIALS	FACTORED FORM
$x^2 + 8x + 16$	$(x + 4)^2$
$x^2 - 6x + 9$	$(x - 3)^2$
$x^2 + 3x + \dfrac{9}{4}$	$\left(x + \dfrac{3}{2}\right)^2$

Notice that for each perfect square trinomial **the constant term of the trinomial is the square of half the coefficient of the x-term.** For example,

$$x^2 + 8x + 16 \qquad\qquad x^2 - 6x + 9$$

$$\frac{1}{2}(8) = 4 \text{ and } 4^2 = 16 \qquad \frac{1}{2}(-6) = -3 \text{ and } (-3)^2 = 9$$

3 The process of writing a quadratic equation so that one side is a perfect square trinomial is called **completing the square.**

EXAMPLE 4   Solve $p^2 + 2p = 4$ by completing the square.

Solution: First, add the square of half the coefficient of p to both sides so that the resulting trinomial will be a perfect square trinomial. The coefficient of p is 2.

$$\frac{1}{2}(2) = 1 \quad \text{and} \quad 1^2 = 1$$

Add 1 to both sides of the original equation:

$$p^2 + 2p = 4$$
$$p^2 + 2p + 1 = 4 + 1 \qquad \text{Add 1 to both sides.}$$
$$(p + 1)^2 = 5 \qquad \text{Factor the trinomial; simplify the right side.}$$

We may now apply the square root property and solve for p:

$$p + 1 = \pm\sqrt{5} \qquad \text{Use the square root property.}$$
$$p = -1 \pm \sqrt{5} \qquad \text{Subtract 1 from both sides.}$$

Notice that there are two solutions: $-1 + \sqrt{5}$ and $-1 - \sqrt{5}$. The solution set is $\{-1 + \sqrt{5}, -1 - \sqrt{5}\}$. A numerical check is shown to the left.

```
-1+√(5)→P:P²+2P
              4
-1-√(5)→P:P²+2P
              4
```

EXAMPLE 5   Solve $2x^2 - 8x + 3 = 0$.

Solution: Our procedure for finding the constant term to complete the square works only if the coefficient of the squared variable term is 1. Therefore, to solve this equation, the first step is to divide both sides by 2, the coefficient of x^2.

$$2x^2 - 8x + 3 = 0$$

$$x^2 - 4x + \frac{3}{2} = 0 \qquad \text{Divide both sides by 2.}$$

$$x^2 - 4x = -\frac{3}{2} \qquad \text{Subtract } \frac{3}{2} \text{ from both sides.}$$

Next find the square of half of -4.

$$\frac{1}{2}(-4) = -2 \quad \text{and} \quad (-2)^2 = 4$$

Add 4 to both sides of the equation to complete the square.

$$x^2 - 4x + 4 = -\frac{3}{2} + 4$$

$$(x - 2)^2 = \frac{5}{2} \qquad \text{Factor the perfect square and simplify the right side.}$$

$$x - 2 = \pm\sqrt{\frac{5}{2}} \qquad \text{Apply the square root property.}$$

$$x - 2 = \pm\frac{\sqrt{10}}{2} \qquad \text{Rationalize the denominator.}$$

$$x = 2 \pm \frac{\sqrt{10}}{2} \qquad \text{Add 2 to both sides.}$$

$$= \frac{4}{2} \pm \frac{\sqrt{10}}{2} \qquad \text{Find the common denominator.}$$

$$= \frac{4 \pm \sqrt{10}}{2} \qquad \text{Simplify.}$$

The solution set is $\left\{ \dfrac{4 + \sqrt{10}}{2}, \dfrac{4 - \sqrt{10}}{2} \right\}$. Numerically or graphically, verify these solutions.

The following steps may be used to solve a quadratic equation such as $ax^2 + bx + c = 0$ by completing the square. This method may be used whether or not the polynomial $ax^2 + bx + c$ is factorable.

TO SOLVE A QUADRATIC EQUATION IN x BY COMPLETING THE SQUARE

Step 1. If the coefficient of x^2 is 1, go to *step 2*. Otherwise, divide both sides of the equation by the coefficient of x^2.

Step 2. Isolate all variable terms on one side of the equation.

Step 3. Complete the square for the resulting binomial by adding the square of half of the coefficient of x to both sides of the equation.

Step 4. Factor the resulting perfect square trinomial and write it as the square of a binomial.

Step 5. Apply the square root property to solve for x.

EXAMPLE 6   Solve $3x^2 - 9x + 8 = 0$ by completing the square.

Solution: $3x^2 - 9x + 8 = 0$

Step 1. $x^2 - 3x + \dfrac{8}{3} = 0$ Divide both sides of the equation by 3.

Step 2. $x^2 - 3x = -\dfrac{8}{3}$ Subtract $\dfrac{8}{3}$ from both sides.

Since $\dfrac{1}{2}(-3) = -\dfrac{3}{2}$ and $\left(-\dfrac{3}{2}\right)^2 = \dfrac{9}{4}$, we add $\dfrac{9}{4}$ to both sides of the equation.

Step 3. $x^2 - 3x + \dfrac{9}{4} = -\dfrac{8}{3} + \dfrac{9}{4}$

Step 4. $\left(x - \dfrac{3}{2}\right)^2 = -\dfrac{5}{12}$ Factor the perfect square trinomial.

Step 5. $x - \dfrac{3}{2} = \pm\sqrt{-\dfrac{5}{12}}$ Apply the square root property.

$\qquad x - \dfrac{3}{2} = \pm\dfrac{i\sqrt{5}}{2\sqrt{3}}$ Simplify the radical.

$\qquad x - \dfrac{3}{2} = \pm\dfrac{i\sqrt{15}}{6}$ Rationalize the denominator.

$\qquad x = \dfrac{3}{2} \pm \dfrac{i\sqrt{15}}{6}$ Add $\dfrac{3}{2}$ to both sides.

$\qquad = \dfrac{9}{6} \pm \dfrac{i\sqrt{15}}{6}$ Find a common denominator.

$\qquad = \dfrac{9 \pm i\sqrt{15}}{6}$ Simplify.

The solution set is $\left\{\dfrac{9 + i\sqrt{15}}{6}, \dfrac{9 - i\sqrt{15}}{6}\right\}$. Since the solutions are imaginary numbers, verify these solutions numerically.

4   Recall the **simple interest** formula $I = Prt$, where I is the interest earned, P is the principal, r is the rate of interest, and t is time. If \$100 is invested at a simple interest rate of 5% annually, at the end of 3 years the total interest I earned is

$$I = P \cdot r \cdot t$$

or

$$I = 100 \cdot 0.05 \cdot 3 = \$15$$

and the new principal is

$$\$100 + \$15 = \$115$$

Most of the time, the interest computed on money borrowed or money deposited is **compound interest.** Compound interest, unlike simple interest, is computed on original principal *and* on interest already earned. To see the difference between simple interest and compound interest, suppose that $100 is invested at a rate of 5% compounded annually. To find the total amount of money at the end of 3 years, we calculate as follows:

$$I \;=\; P \;\cdot\; r \;\cdot t$$

First year: Interest = $100 · 0.05 · 1 = $5.00

New principal = $100.00 + $5.00 = $105.00

Second year: Interest = $105.00 · 0.05 · 1 = $5.25

New principal = $105.00 + $5.25 = $110.25

Third year: Interest = $110.25 · 0.05 · 1 ≈ $5.51

New principal = $110.25 + $5.51 = $115.76

At the end of the third year, the total compound interest earned is $15.76, whereas the total simple interest earned is $15.

It is tedious to calculate compound interest as we did above, so we use a compound interest formula. The formula for calculating the total amount of money when interest is compounded annually is

$$A = P(1 + r)^t$$

where P is the original investment, r is the interest rate per compounding period, and t is the number of periods. For example, the amount of money A at the end of 3 years if $100 is invested at 5% compounded annually is

$$A = \$100(1 + 0.05)^3 \approx \$100(1.1576) = \$115.76$$

as expected.

EXAMPLE 7   **FINDING THE COMPOUND INTEREST RATE**
Find the interest rate r if $2000, compounded annually, grows to $2420 in 2 years.

Solution: **1. UNDERSTAND** the problem.   For this example, make sure that you understand the formula for compounding interest annually. Since a formula is known, we go to step 4.

4. TRANSLATE.   Here we substitute given values into the formula.

$$A = P(1 + r)^t$$
$$2420 = 2000(1 + r)^2 \qquad \text{Let } A = 2420, P = 2000, \text{ and } t = 2.$$

5. COMPLETE.   Solve the equation for r.

$$2420 = 2000(1 + r)^2$$

$$\frac{2420}{2000} = (1 + r)^2 \qquad\qquad \text{Divide both sides by 2000.}$$

$$\frac{121}{100} = (1 + r)^2 \qquad\qquad \text{Simplify the fraction.}$$

$$\pm\sqrt{\frac{121}{100}} = 1 + r \qquad\qquad \text{Apply the square root property.}$$

$$\pm\frac{11}{10} = 1 + r \qquad\qquad \text{Simplify.}$$

$$-1 \pm \frac{11}{10} = r$$

$$-\frac{10}{10} \pm \frac{11}{10} = r$$

$$\frac{1}{10} = r \quad \text{or} \quad -\frac{21}{10} = r$$

6. INTERPRET. The rate cannot be negative, so we reject $-\dfrac{21}{10}$ and *check*

$\dfrac{1}{10} = 0.10 = 10\%$ per year. To use the intersection-of-graphs method to check,

graph $y_1 = 2000 (1 + x)^2$ and $y_2 = 2420$. Here, x represents interest rate, which must be between 0% and 100% or between 0 and 1. Also, notice that we have a y-value of 2420. From these x- and y-values, we choose the window $[0, 1, 0.2]$ by $[2000, 3000, 500]$. The intersection shown to the left occurs at $x = 0.1$ or 10%.

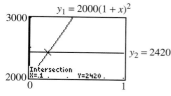

State: The interest rate is 10% compounded annually.

EXERCISE SET 9.1

Use the square root property to solve each equation. These equations have real number solutions. Check each solution graphically. See Examples 1 and 2.

1. $x^2 = 16$

2. $x^2 = 49$

3. $x^2 - 7 = 0$

4. $x^2 - 11 = 0$

5. $x^2 = 18$

6. $y^2 = 20$

7. $3z^2 - 30 = 0$

8. $2x^2 = 4$

9. $(x + 5)^2 = 9$

10. $(y - 3)^2 = 4$

11. $(z - 6)^2 = 18$

12. $(y + 4)^2 = 27$

13. $(2x - 3)^2 = 8$

14. $(4x + 9)^2 = 6$

Use the square root property to solve each equation. Check each solution numerically. See Examples 1 and 3.

15. $x^2 + 9 = 0$

16. $x^2 + 4 = 0$

17. $x^2 - 6 = 0$

18. $y^2 - 10 = 0$

19. $2z^2 + 16 = 0$

20. $3p^2 + 36 = 0$

21. $(x - 1)^2 = -16$

22. $(y + 2)^2 = -25$

23. $(z + 7)^2 = 5$

24. $(x + 10)^2 = 11$

25. $(x + 3)^2 = -8$

26. $(y - 4)^2 = -18$

Add the proper constant to each binomial so that the resulting trinomial is a perfect square trinomial. Then factor the trinomial.

27. $x^2 + 16x$

28. $y^2 + 2y$

29. $z^2 - 12z$

30. $x^2 - 8x$

31. $p^2 + 9p$

32. $n^2 + 5n$

33. $x^2 + x$

34. $y^2 - y$

Find two possible missing terms so that each is a perfect square trinomial.

35. $x^2 + \blacksquare + 16$

36. $y^2 + \blacksquare + 9$

37. $z^2 + \blacksquare + \dfrac{25}{4}$

38. $x^2 + \blacksquare + \dfrac{1}{4}$

Solve each equation by completing the square. These equations have real number solutions. Check each solution graphically. See Examples 4 through 6.

39. $x^2 + 8x = -15$

40. $y^2 + 6y = -8$

41. $x^2 + 6x + 2 = 0$

42. $x^2 - 2x - 2 = 0$

43. $x^2 + x - 1 = 0$

44. $x^2 + 3x - 2 = 0$

45. $x^2 + 2x - 5 = 0$

46. $y^2 + y - 7 = 0$

47. $3p^2 - 12p + 2 = 0$

48. $2x^2 + 14x - 1 = 0$

49. $4y^2 - 12y - 2 = 0$

50. $6x^2 - 3 = 6x$

51. $2x^2 + 7x = 4$

52. $3x^2 - 4x = 4$

53. $x^2 - 4x - 5 = 0$

54. $y^2 + 6y - 8 = 0$

55. $x^2 + 8x + 1 = 0$

56. $x^2 - 10x + 2 = 0$

57. $3y^2 + 6y - 4 = 0$

58. $2y^2 + 12y + 3 = 0$

59. $2x^2 - 3x - 5 = 0$

60. $5x^2 + 3x - 2 = 0$

Solve each equation by completing the square. Check each solution numerically or graphically. See Examples 4 through 6.

61. $y^2 + 2y + 2 = 0$

62. $x^2 + 4x + 6 = 0$

63. $x^2 - 6x + 3 = 0$

64. $x^2 - 7x - 1 = 0$

65. $2a^2 + 8a = -12$

66. $3x^2 + 12x = -14$

67. $5x^2 + 15x - 1 = 0$

68. $16y^2 + 16y - 1 = 0$

69. $2x^2 - x + 6 = 0$

70. $4x^2 - 2x + 5 = 0$

71. $x^2 + 10x + 28 = 0$

72. $y^2 + 8y + 18 = 0$

Use a graphing utility to solve each equation. Round all solutions to the nearest hundredth.

73. $x(x - 5) = 8$

74. $x(x + 2) = 5$

75. $x^2 + 0.5x = 0.3x + 1$

76. $x^2 - 2.6x = -2.2x + 3$

77. Use a graphing utility to solve Exercise 61. Compare the results. Are they the same? Why or why not?

78. What are the advantages and disadvantages of using a graphing utility to solve quadratic equations?

Use the formula $A = P(1 + r)^t$ to solve Exercises 79 through 82. See Example 7.

79. Find the rate r at which $3000 grows to $4320 in 2 years.

80. Find the rate r at which $800 grows to $882 in 2 years.

81. Find the rate at which $810 grows to $1000 in 2 years.

82. Find the rate at which $2000 grows to $2880 in 2 years.

83. In your own words, what is the difference between simple interest and compound interest?

84. If you are depositing money in an account that pays 4%, would you prefer the interest to be simple or compound?

85. If you are borrowing money at a rate of 10%, would you prefer the interest to be simple or compound?

Neglecting air resistance, the distance $s(t)$ traveled by a freely falling object is given by the function $s(t) = 16t^2$, where t is time in seconds. Use this formula to solve Exercises 86 through 89. Round answers to two decimal places.

86. The height of the Columbia Seafirst Center in Seattle is 954 feet. How long would it take an object to fall to the ground from the top of the building?

87. The height of the First Interstate World Center in Los Angeles is 1017 feet. How long would it take an object to fall from the top of the building to the ground?

88. The height of the John Hancock Center in Chicago is 1127 feet. How long would it take an object to fall from the top of the building to the ground?

89. The height of the Carew Tower in Cincinnati is 568 feet. How long would it take an object to fall from the top of the building to the ground?

Solve.

90. The area of a square room is 225 square feet. Find the dimensions of the room.

91. The area of a circle is 36π square inches. Find the radius of the circle.

92. An isosceles right triangle has legs of equal length. If the hypotenuse is 20 centimeters long, find the length of each leg.

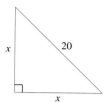

93. A 27-inch TV is advertised in the *Daily Sentry* newspaper. If 27 inches is the measure of the diagonal of the square picture tube, find the measure of the side of the picture tube.

A common equation used in business is a demand equation. It expresses the relationship between the unit price of some commodity and the quantity demanded. For Exercises 94 and 95, p represents the unit price and x represents the quantity demanded in thousands.

94. A manufacturing company has found that the demand equation for a certain type of scissors is

given by the equation $p = -x^2 + 47$. Find the demand for the scissors if the price is $11 per pair.

95. Acme, Inc., sells desk lamps and has found that the demand equation for a certain style of desk lamp is given by the equation $p = -x^2 + 15$. Find the demand for the desk lamp if the price is $7 per lamp.

Review Exercises

Simplify each expression. See Section 8.1.

96. $\dfrac{3}{4} - \sqrt{\dfrac{25}{16}}$

97. $\dfrac{3}{5} + \sqrt{\dfrac{16}{25}}$

98. $\dfrac{1}{2} - \sqrt{\dfrac{9}{4}}$

99. $\dfrac{9}{10} - \sqrt{\dfrac{49}{100}}$

Simplify each expression. See Section 8.5.

100. $\dfrac{6 + 4\sqrt{5}}{2}$

101. $\dfrac{10 - 20\sqrt{3}}{2}$

102. $\dfrac{3 - 9\sqrt{2}}{6}$

103. $\dfrac{12 - 8\sqrt{7}}{16}$

Evaluate $\sqrt{b^2 - 4ac}$ for each set of values. See Section 8.3.

104. $a = 2, b = 4, c = -1$

105. $a = 1, b = 6, c = 2$

106. $a = 3, b = -1, c = -2$

107. $a = 1, b = -3, c = -1$

9.2 | SOLVING QUADRATIC EQUATIONS BY THE QUADRATIC FORMULA

TAPE IAG 9.2

O B J E C T I V E S

1. Solve quadratic equations by using the quadratic formula.

2. Determine the number and type of solutions of a quadratic equation by using the discriminant.

3. Solve geometric problems that lead to quadratic equations.

Any quadratic equation can be solved by completing the square. Since the same sequence of steps is repeated each time that we complete the square, let's complete the square for a general quadratic equation, $ax^2 + bx + c = 0$. By doing so, we find a pattern for the solutions of a quadratic equation known as the **quadratic formula.**

Recall that, to complete the square for an equation such as $ax^2 + bx + c = 0$, we first divide both sides by the coefficient of x^2.

$$ax^2 + bx + c = 0$$

$$x^2 + \frac{b}{a}x + \frac{c}{a} = 0 \qquad \text{Divide both sides by } a, \text{ the coefficient of } x^2.$$

$$x^2 + \frac{b}{a}x = -\frac{c}{a} \qquad \text{Subtract the constant } \frac{c}{a} \text{ from both sides.}$$

Next, find the square of half $\dfrac{b}{a}$, the coefficient of x.

$$\frac{1}{2}\left(\frac{b}{a}\right) = \frac{b}{2a} \quad \text{and} \quad \left(\frac{b}{2a}\right)^2 = \frac{b^2}{4a^2}$$

Add this result to both sides of the equation:

$$x^2 + \frac{b}{a}x + \frac{b^2}{4a^2} = -\frac{c}{a} + \frac{b^2}{4a^2} \qquad \text{Add } \frac{b^2}{4a^2} \text{ to both sides.}$$

$$x^2 + \frac{b}{a}x + \frac{b^2}{4a^2} = \frac{-c \cdot 4a}{a \cdot 4a} + \frac{b^2}{4a^2} \qquad \substack{\text{Find a common denominator on the} \\ \text{right side.}}$$

$$x^2 + \frac{b}{a}x + \frac{b^2}{4a^2} = \frac{b^2 - 4ac}{4a^2} \qquad \text{Simplify the right side.}$$

$$\left(x + \frac{b}{2a}\right)^2 = \frac{b^2 - 4ac}{4a^2} \qquad \substack{\text{Factor the perfect square trinomial} \\ \text{on the left side.}}$$

$$x + \frac{b}{2a} = \pm\sqrt{\frac{b^2 - 4ac}{4a^2}} \qquad \text{Apply the square root property.}$$

$$x + \frac{b}{2a} = \pm\frac{\sqrt{b^2 - 4ac}}{2a} \qquad \text{Simplify the radical.}$$

$$x = -\frac{b}{2a} \pm \frac{\sqrt{b^2 - 4ac}}{2a} \qquad \text{Subtract } \frac{b}{2a} \text{ from both sides.}$$

$$= \frac{-b \pm \sqrt{b^2 - 4ac}}{2a} \qquad \text{Simplify.}$$

This equation identifies the solutions of the general quadratic equation in standard form and is called the quadratic formula. It can be used to solve any equation written in the standard form $ax^2 + bx + c = 0$ as long as a is not 0.

TECHNOLOGY NOTE

When evaluating the quadratic formula using a calculator, it is sometimes more convenient to evaluate the radicand separately. This prevents incorrect placement of parentheses.

QUADRATIC FORMULA

A quadratic equation written in the form $ax^2 + bx + c = 0$ has the solutions

$$x = \frac{-b \pm \sqrt{b^2 - 4ac}}{2a}$$

EXAMPLE 1   Solve $3x^2 + 16x + 5 = 0$ for x.

Solution: This equation is in standard form, so $a = 3$, $b = 16$, and $c = 5$. Substitute these values into the quadratic formula:

$$x = \frac{-b \pm \sqrt{b^2 - 4ac}}{2a} \qquad \text{Quadratic formula.}$$

$$= \frac{-16 \pm \sqrt{16^2 - 4(3)(5)}}{2(3)} \qquad \text{Let } a = 3, b = 16, \text{ and } c = 5.$$

$$= \frac{-16 \pm \sqrt{256 - 60}}{6}$$

$$= \frac{-16 \pm \sqrt{196}}{6} = \frac{-16 \pm 14}{6}$$

$$x = \frac{-16 + 14}{6} = -\frac{1}{3} \quad \text{or} \quad x = \frac{-16 - 14}{6} = -\frac{30}{6} = -5$$

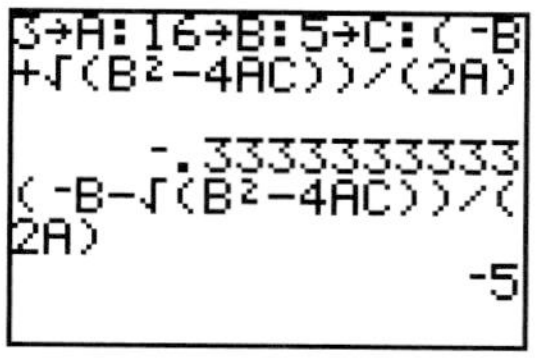

The solution set is $\left\{-\frac{1}{3}, -5\right\}$.

A numerical check is shown to the left. Notice the correct placement of parentheses.

```
3→A:16→B:5→C:(-B
+√(B²-4AC))/(2A)
            -.3333333333
(-B-√(B²-4AC))/(
2A)
                      -5
```

R E M I N D E R   To replace a, b, and c correctly in the quadratic formula, write the quadratic equation in standard form, $ax^2 + bx + c = 0$.

As usual, another way to check the solution of an equation is to use a graphical method such as the intersection-of-graphs method. Remember that this graphical method shows real number solutions only.

EXAMPLE 2   Solve $2x^2 - 4x = 3$.

Solution   First, write the equation in standard form by subtracting 3 from both sides.

$$2x^2 - 4x - 3 = 0$$

Now $a = 2$, $b = -4$, and $c = -3$. Substitute these values into the quadratic formula.

$$x = \frac{-b \pm \sqrt{b^2 - 4ac}}{2a}$$

$$= \frac{-(-4) \pm \sqrt{(-4)^2 - 4(2)(-3)}}{2(2)}$$

$$= \frac{4 \pm \sqrt{16 + 24}}{4}$$

$$= \frac{4 \pm \sqrt{40}}{4} = \frac{4 \pm 2\sqrt{10}}{4}$$

$$= \frac{2(2 \pm \sqrt{10})}{2 \cdot 2} = \frac{2 \pm \sqrt{10}}{2}$$

The solution set is $\left\{ \dfrac{2 + \sqrt{10}}{2}, \dfrac{2 - \sqrt{10}}{2} \right\}$.

Since the solutions are real numbers, let's check using the intersection-of-graphs method. To do so, graph $y_1 = 2x^2 - 4x$ and $y_2 = 3$. Find the approximate points of intersection as shown below, and see that the exact solutions have the same approximations.

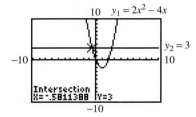
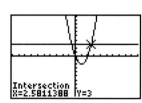

The screen to the right above shows that the exact solutions using the quadratic formula have the same approximations as those found by the intersection-of-graphs method.

R E M I N D E R   To simplify the expression $\dfrac{4 \pm 2\sqrt{10}}{4}$ in the preceding example, note that 2 is factored out of both terms of the numerator *before* simplifying.

$$\frac{4 \pm 2\sqrt{10}}{4} = \frac{2(2 \pm \sqrt{10})}{2 \cdot 2} = \frac{2 \pm \sqrt{10}}{2}$$

EXAMPLE 3   Solve $\dfrac{1}{4}m^2 - m + \dfrac{1}{2} = 0$.

Solution: We could use the quadratic formula with $a = \dfrac{1}{4}$, $b = -1$, and $c = \dfrac{1}{2}$. Instead, we find a simpler, equivalent standard form equation whose coefficients are not fractions.

Multiply both sides of the equation by 4 to clear fractions:

$$4\left(\frac{1}{4}m^2 - m + \frac{1}{2}\right) = 4 \cdot 0$$

$$m^2 - 4m + 2 = 0 \qquad \text{Simplify.}$$

Substitute $a = 1$, $b = -4$, and $c = 2$ into the quadratic formula and simplify:

$$m = \frac{-(-4) \pm \sqrt{(-4)^2 - 4(1)(2)}}{2(1)} = \frac{4 \pm \sqrt{16 - 8}}{2}$$

$$= \frac{4 \pm \sqrt{8}}{2} = \frac{4 \pm 2\sqrt{2}}{2} = \frac{2(2 \pm \sqrt{2})}{2} = 2 \pm \sqrt{2}$$

The solution set is $\{2 + \sqrt{2}, 2 - \sqrt{2}\}$. Check these real number solutions numerically or graphically.

EXAMPLE 4   Solve $p = -3p^2 - 3$.

Solution: The equation in standard form is $3p^2 + p + 3 = 0$. Thus, let $a = 3$, $b = 1$, and $c = 3$ in the quadratic formula:

```
(-1+i√(35))/6→P:
3P²+P+3
              0
(-1-i√(35))/6→P:
3P²+P+3
              0
```

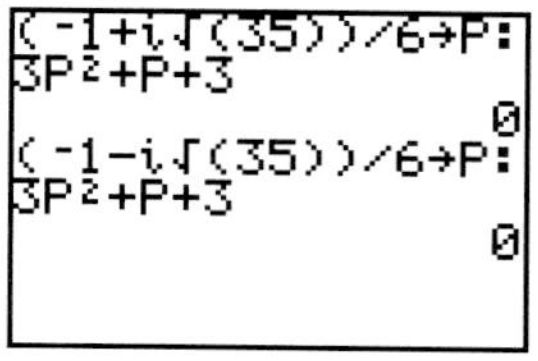

$$p = \frac{-1 \pm \sqrt{1^2 - 4(3)(3)}}{2 \cdot 3} = \frac{-1 \pm \sqrt{1 - 36}}{6} = \frac{-1 \pm \sqrt{-35}}{6} = \frac{-1 \pm i\sqrt{35}}{6}$$

The solution set is $\left\{\dfrac{-1 + i\sqrt{35}}{6}, \dfrac{-1 - i\sqrt{35}}{6}\right\}$. Since the solutions are imaginary numbers, we check numerically as shown to the left.

In the quadratic formula $x = \dfrac{-b \pm \sqrt{b^2 - 4ac}}{2a}$, the radicand $b^2 - 4ac$ is called the **discriminant** because, by knowing its value, we can **discriminate** among the possible number and type of solutions of a quadratic equation. Possible values of the discriminant and their meanings are summarized next.

DISCRIMINANT

The following table relates the discriminant, $b^2 - 4ac$, of a quadratic equation of the form $ax^2 + bx + c = 0$ to the number and type of solutions of the equation.

$b^2 - 4ac$	NUMBER AND TYPE OF SOLUTIONS
Positive	Two real solutions
Zero	One real solution
Negative	Two imaginary solutions

continued

DISCRIMINANT *continued*

To see the results of the discriminant graphically, study the screens below showing examples of graphs of $y = ax^2 + bx + c$.

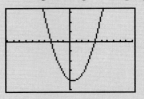
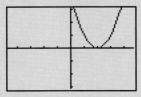
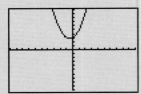

$b^2 - 4ac$ is positive
Two x-intercepts
Two real solutions

$b^2 - 4ac = 0$
One x-intercept
One real solution
(double root)

$b^2 - 4ac$ is negative
No x-intercept
Two imaginary solutions

EXAMPLE 5   Use the discriminant to determine the number and type of solutions of each quadratic equation. Confirm your answer by graphing the related equation.

a. $x^2 + 2x + 1 = 0$ **b.** $3x^2 + 2 = 0$ **c.** $2x^2 - 7x - 4 = 0$

Solution: **a.** In $x^2 + 2x + 1 = 0$, $a = 1$, $b = 2$, and $c = 1$. Thus,

$$b^2 - 4ac = 2^2 - 4(1)(1) = 0$$

Since $b^2 - 4ac = 0$, this quadratic equation has one real solution.

b. In this equation, $a = 3$, $b = 0$, and $c = 2$.

Then $b^2 - 4ac = 0 - 4(3)(2) = -24$.

Since $b^2 - 4ac$ is negative, the quadratic equation has two imaginary solutions.

c. In this equation, $a = 2$, $b = -7$, and $c = -4$. Then

$$b^2 - 4ac = 49 - 4(2)(-4) = 81$$

Since $b^2 - 4ac$ is positive, the quadratic equation has two real solutions.

To confirm graphically, see the results below.

a.

$y = x^2 + 2x + 1$ 3.1

−4.7   4.7

Zero
X=-1   Y=0

−3.1

One x-intercept indicates one real solution.

b.

10   $y = 3x^2 + 2$

−10   10

−10

No x-intercepts indicate no real solutions, but two imaginary solutions.

c.

10    $y = 2x^2 - 7x - 4$

−10   10

−10

Two x-intercepts indicate two real solutions.

3   The quadratic formula is useful in solving problems that are modeled by quadratic equations. In the example below, we review the Pythagorean Theorem.

EXAMPLE 6   **REVIEW OF THE PYTHAGOREAN THEOREM**
The hypotenuse of an isosceles right triangle is 2 centimeters longer than either of its legs. Find the perimeter of the triangle.

Solution: **1.** UNDERSTAND.   Read and reread the problem. Recall that an isosceles right triangle has legs of equal length. Also recall the Pythagorean theorem for right triangles:

$$a^2 + b^2 = c^2$$

2. ASSIGN. Let

$$x = \text{the length of each leg of the triangle}$$

so that

$$x + 2 = \text{the length of the hypotenuse of the triangle}$$

3. ILLUSTRATE.   Draw a triangle and label it with the assigned variables.

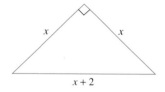

4. TRANSLATE.   By the Pythagorean theorem,

In words: $(\text{leg})^2 + (\text{leg})^2 = (\text{hypotenuse})^2$
Translate: $x^2 + x^2 = (x + 2)^2$

5. COMPLETE.   Solve the quadratic equation:

$$x^2 + x^2 = (x + 2)^2$$
$$x^2 + x^2 = x^2 + 4x + 4 \qquad \text{Square } (x + 2).$$
$$x^2 - 4x - 4 = 0 \qquad \text{Set the equation equal to 0.}$$

Next, substitute $a = 1$, $b = -4$, and $c = -4$ in the quadratic formula.

$$x = \frac{4 \pm \sqrt{16 - 4(1)(-4)}}{2} = \frac{4 \pm \sqrt{32}}{2} = \frac{4 \pm 4\sqrt{2}}{2}$$

$$= \frac{2 \cdot (2 \pm 2\sqrt{2})}{2} = 2 \pm 2\sqrt{2}$$

6. INTERPRET. Since the length of a side cannot be negative, we reject the solution $2 - 2\sqrt{2}$. The length of each leg is $(2 + 2\sqrt{2})$ centimeters. Since the hypotenuse is 2 centimeters longer than the legs, the hypotenuse is $(4 + 2\sqrt{2})$ centimeters. *Check* the lengths of the sides in the Pythagorean theorem. *State:* The perimeter of the triangle is $(2 + 2\sqrt{2})$ cm $+ (2 + 2\sqrt{2})$ cm $+ (4 + 2\sqrt{2})$ cm $= (8 + 6\sqrt{2})$ centimeters.

In Example 7 we choose to solve by the intersection-of-graphs method since it is convenient to use a graphing utility when the problem has several questions being asked and also the constant term contains a fraction.

EXAMPLE 7

DIMENSIONS OF A PARKING LOT

The length of a rectangular parking lot is 65 feet more than the width. The area of the parking lot is $12{,}808\frac{13}{16}$ square feet.

a. Find the dimensions of the parking lot.

b. Find the total cost of installing a chain link fence around the lot if the cost per foot of fencing is \$3.25.

Solution:

1. UNDERSTAND. Read and reread the problem. Notice that once we find the dimensions of the parking lot, we need to find the perimeter of the parking lot in order to determine the cost of installing a fence *around* it.

2. ASSIGN.

Let x = the width so that

$x + 65$ = length

3. ILLUSTRATE.

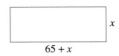

$65 + x$

4. TRANSLATE.

In Words:	The area of the rectangle	is	$12{,}808\frac{13}{16}$ square feet

Translate:	$x(65 + x)$	$=$	$12{,}808.8125$

5. COMPLETE. We choose to solve this equation by the intersection-of-graphs method. Graph $y_1 = x(65 + x)$ and $y_2 = 12{,}808.8125$ in a $[0, 100, 10]$ by $[0, 14{,}000, 100]$ window as shown in the screen below.

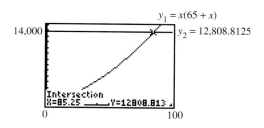

The intersection has an x-value of 85.25 so the width is 85.25 feet.

6. INTERPRET. **a.** *State:* The width of the parking lot is 85.25 feet and the length is therefore $x + 65 = 85.25 + 65$ or 150.25 feet.

b. To find the cost of installing a fence around the parking lot, notice that the perimeter of the lot is $2(150.25) + 2(85.25) = 471$ feet. Thus 471 feet of fencing is needed and the total cost is $471(\$3.25) = \$1530.75$ since the fencing is \$3.25 per foot. *State:* The total cost of fencing the parking lot is \$1530.75.

Check the dimensions of the parking lot from the originally stated problem.

EXERCISE SET 9.2

Use the quadratic formula to solve each equation. These equations have real number solutions and can be checked graphically. See Examples 1 through 3.

1. $m^2 + 5m - 6 = 0$

2. $p^2 + 11p - 12 = 0$

3. $2y = 5y^2 - 3$

4. $5x^2 - 3 = 14x$

5. $x^2 - 6x + 9 = 0$

6. $y^2 + 10y + 25 = 0$

7. $x^2 + 7x + 4 = 0$

8. $y^2 + 5y + 3 = 0$

9. $8m^2 - 2m = 7$

10. $11n^2 - 9n = 1$

11. $3m^2 - 7m = 3$

12. $x^2 - 13 = 5x$

13. $\dfrac{1}{2}x^2 - x - 1 = 0$

14. $\dfrac{1}{6}x^2 + x + \dfrac{1}{3} = 0$

15. $\dfrac{2}{5}y^2 + \dfrac{1}{5}y = \dfrac{3}{5}$

16. $\dfrac{1}{8}x^2 + x = \dfrac{5}{2}$

17. $\dfrac{1}{3}y^2 - y - \dfrac{1}{6} = 0$

18. $\dfrac{1}{2}y^2 = y + \dfrac{1}{2}$

19. Solve Exercise 1 by factoring. Explain the result.

20. Solve Exercise 2 by factoring. Explain the result.

Use the quadratic formula to solve each equation. Check numerically. See Example 4.

21. $6 = -4x^2 + 3x$

22. $9x^2 + x + 2 = 0$

23. $(x + 5)(x - 1) = 2$

24. $x(x + 6) = 2$

25. $10y^2 + 10y + 3 = 0$

26. $3y^2 + 6y + 5 = 0$

The solutions of the quadratic equation $ax^2 + bx + c = 0$ are

$$\frac{-b + \sqrt{b^2 - 4ac}}{2a} \quad and \quad \frac{-b - \sqrt{b^2 - 4ac}}{2a}$$

27. Show that the sum of these solutions is $\dfrac{-b}{a}$.

28. Show that the product of these solutions is $\dfrac{c}{a}$.

Use the discriminant to determine the number and types of solutions of each equation. Confirm your result by graphing the related equation of the form $y = ax^2 + bx + c$. See Example 5.

29. $9x - 2x^2 + 5 = 0$

30. $5 - 4x + 12x^2 = 0$

31. $4x^2 + 12x = -9$

32. $9x^2 + 1 = 6x$

33. $3x = -2x^2 + 7$

34. $3x^2 = 5 - 7x$

35. $6 = 4x - 5x^2$

36. $8x = 3 - 9x^2$

Use the quadratic formula to solve each equation. These equations have real number solutions.

37. $x^2 + 5x = -2$

38. $y^2 - 8 = 4y$

39. $(m + 2)(2m - 6) = 5(m - 1) - 12$

40. $7p(p - 2) + 2(p + 4) = 3$

41. $\dfrac{x^2}{3} - x = \dfrac{5}{3}$

42. $\dfrac{x^2}{2} - 3 = -\dfrac{9}{2}x$

43. $x(6x + 2) - 3 = 0$

44. $x(7x + 1) = 2$

Use the quadratic formula to solve each equation.

45. $x^2 + 6x + 13 = 0$

46. $x^2 + 2x + 2 = 0$

47. $\dfrac{2}{5}y^2 + \dfrac{1}{5}y + \dfrac{3}{5} = 0$

48. $\dfrac{1}{8}x^2 + x + \dfrac{5}{2} = 0$

49. $\dfrac{1}{2}y^2 = y - \dfrac{1}{2}$

50. $\dfrac{2}{3}x^2 - \dfrac{20}{3}x = -\dfrac{100}{6}$

51. $(n - 2)^2 = 15n$

52. $\left(p - \dfrac{1}{2}\right)^2 = \dfrac{p}{2}$

Given the following graphs of quadratic equations of the form $y = f(x)$, find the number of solutions of the related equation $f(x) = 0$ and whether the solutions are real or imaginary numbers.

53.

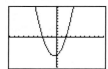

54.

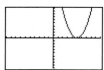

There is one
x-intercept.

55.

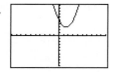

56.

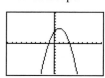

Solve. Find exact solutions and one decimal place approximations. See Example 6.

57. Uri Chechov's rectangular dog pen for his Irish setter must have an area of 400 square feet. Also, the length must be $2\frac{1}{2}$ times the width. Find the dimensions of the pen.

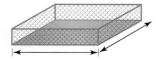

58. The hypotenuse of an isosceles right triangle is 5 inches longer than either of the legs. Find the length of the legs and the length of the hypotenuse.

59. The base of a triangle is twice its height. If the area of the triangle is 42 square centimeters, find its base and height.

60. The width of a rectangle is $\frac{1}{3}$ its length. If its area is 12 square inches, find its length and width.

61. An entry in the Peach Festival Poster Contest must be rectangular and have an area of 1200 square inches. Furthermore, its length must be $1\frac{1}{2}$ times its width. Find the dimensions each entry must have.

62. A holding pen for cattle must be square and have a diagonal length of 100 meters.

    a. Find the length of a side of the pen.

    b. Find the area of the pen.

63. A rectangle is three times longer than it is wide. It has a diagonal of length 50 centimeters.

    a. Find the dimensions of the rectangle.

    b. Find the perimeter of the rectangle.

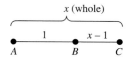

64. If a point B divides a line segment such that the smaller portion is to the larger portion as the larger is to the whole, the whole is the length of the *golden ratio*.

The golden ratio was thought by the Greeks to be the most pleasing to the eye, and many of their buildings contained numerous examples of the golden ratio. The value of the golden ratio is the positive solution of

$$\begin{matrix} \text{(smaller)} \\ \text{(larger)} \end{matrix} \quad \dfrac{x - 1}{1} = \dfrac{1}{x} \quad \begin{matrix} \text{(smaller)} \\ \text{(larger)} \end{matrix}$$

Find this value.

Solve each equation graphically. Approximate each solution to the nearest tenth.

65. $2x^2 - 6x + 3 = 0$

66. $5x^2 - 8x + 2 = 0$

67. $1.3x^2 - 2.5x - 7.9 = 0$

68. $3.6x^2 + 1.8x - 4.3 = 0$

A ball is thrown downward from the top of a 180-foot building with an initial velocity of -20 feet per second. The height of the ball $h(t)$ after t seconds is given by the function

$$h(t) = -16t^2 - 20t + 180$$

69. How long after the ball is thrown will it strike the ground? Round the result to the nearest tenth of a second.

70. How long after the ball is thrown will it be 50 feet from the ground? Round the result to the nearest tenth of a second.

Group Activity   The accompanying graph shows the daily low temperatures for one week in New Orleans, Louisiana. Use this graph for Exercises 71 through 74.

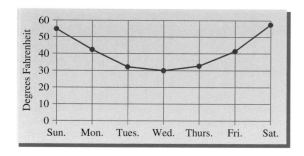

71. Which day of the week shows the greatest decrease in temperature low?

72. Which day of the week shows the greatest increase in temperature low?

73. Which day of the week had the lowest temperature?

74. Use the graph to estimate the low temperature on Thursday.

Notice that the shape of the temperature graph is similar to a parabola (see Section 3.2). In fact, this graph can be modeled by the quadratic function $f(x) = 3x^2 - 18x + 57$, where $f(x)$ is the temperature in degrees Fahrenheit and x is the number of days from Sunday. Use this function to answer Exercises 75 and 76.

75. Use the quadratic function given to approximate the temperature on Thursday. Does your answer agree with the graph above?

76. Use the function given and the quadratic formula to find when the temperature was 35°F. [*Hint:* Let $f(x) = 35$ and solve for x.] Round your answer to one decimal place and interpret your result. Does your answer agree with the graph above?

Review Exercises

Solve each equation. See Sections 7.7 and 8.6.

77. $\sqrt{5x - 2} = 3$

78. $\sqrt{y + 2} + 7 = 12$

79. $\dfrac{1}{x} + \dfrac{2}{5} = \dfrac{7}{x}$

80. $\dfrac{10}{z} = \dfrac{5}{z} - \dfrac{1}{3}$

Factor. See Section 6.7.

81. $x^4 + x^2 - 20$

82. $2y^4 + 11y^2 - 6$

83. $z^4 - 13z^2 + 36$

84. $x^4 - 1$

A Look Ahead

EXAMPLE

Solve $x^2 - 3\sqrt{2}x + 2 = 0$.

Solution:

In this equation, $a = 1$, $b = -3\sqrt{2}$, and $c = 2$. By the quadratic formula, we have

$$x = \frac{-b \pm \sqrt{b^2 - 4ac}}{2a}$$

$$= \frac{3\sqrt{2} \pm \sqrt{(-3\sqrt{2})^2 - 4(1)(2)}}{2(1)}$$

$$= \frac{3\sqrt{2} \pm \sqrt{18 - 8}}{2} = \frac{3\sqrt{2} \pm \sqrt{10}}{2}$$

The solution set is $\left\{ \dfrac{3\sqrt{2} + \sqrt{10}}{2}, \dfrac{3\sqrt{2} - \sqrt{10}}{2} \right\}$.

Use the quadratic formula to solve each quadratic equation. See the preceding example.

85. $3x^2 - \sqrt{12}x + 1 = 0$

86. $5x^2 + \sqrt{20}x + 1 = 0$

87. $x^2 + \sqrt{2}x + 1 = 0$

88. $x^2 - \sqrt{2}x + 1 = 0$

89. $2x^2 - \sqrt{3}x - 1 = 0$

90. $7x^2 + \sqrt{7}x - 2 = 0$

9.3 | SOLVING EQUATIONS BY USING QUADRATIC METHODS

O B J E C T I V E S

TAPE IAG 9.3

1   Solve various equations that are quadratic in form.

2   Solve problems that lead to quadratic equations.

In this section, we discuss various types of equations that can be solved in part by using the methods for solving quadratic equations.

Note that we have several methods for solving quadratic equations. To solve algebraically, we may solve by the factoring method if the related polynomial is factorable, by completing the square, or by the quadratic formula. To solve graphically, we may solve by the intersection-of-graphs method or the x-intercept method. Remember that these graphing methods give real number solutions only. When an exact solution is needed, it is wise to choose an algebraic method. When an approximation is sufficient and real number solutions are needed only, algebraic or graphical methods may be used.

The first example is a rational equation that simplifies to a quadratic equation. We solve algebraically by using the quadratic formula.

EXAMPLE 1   Solve $\dfrac{3x}{x-2} - \dfrac{x+1}{x} = \dfrac{6}{x(x-2)}$.

Solution: In this equation, x cannot be either 2 or 0, because these values cause denominators to equal zero. To solve for x, first multiply both sides of the equation by $x(x-2)$ to clear fractions. By the distributive property, this means that we multiply each term by $x(x-2)$.

$$x(x-2)\left(\frac{3x}{x-2}\right) - x(x-2)\left(\frac{x+1}{x}\right) = x(x-2)\left[\frac{6}{x(x-2)}\right]$$

$$3x^2 - (x-2)(x+1) = 6 \qquad \text{Simplify.}$$

$$3x^2 - (x^2 - x - 2) = 6 \qquad \text{Multiply.}$$

$$3x^2 - x^2 + x + 2 = 6$$

$$2x^2 + x - 4 = 0 \qquad \text{Simplify.}$$

$$x = \frac{-1 \pm \sqrt{1^2 - 4(2)(-4)}}{2 \cdot 2} \qquad \begin{array}{l}\text{Let } a = 2, b = 1, \text{ and}\\ c = -4 \quad \text{in the}\\ \text{quadratic formula.}\end{array}$$

$$= \frac{-1 \pm \sqrt{1 + 32}}{4} \qquad \text{Simplify.}$$

$$= \frac{-1 \pm \sqrt{33}}{4}$$

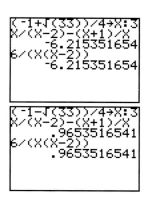

The solution set is $\left\{\dfrac{-1 + \sqrt{33}}{4}, \dfrac{-1 - \sqrt{33}}{4}\right\}$.

To check, we evaluate the left and right sides of the equation with the proposed solutions and see that a true statement results. Careful placement of parentheses is important, as shown to the left.

EXAMPLE 2   Solve $x^3 = 8$ for x.

Solution: Begin by subtracting 8 from both sides; then factor the resulting difference of cubes.

$$x^3 = 8$$
$$x^3 - 8 = 0$$
$$(x - 2)(x^2 + 2x + 4) = 0 \qquad \text{Factor.}$$
$$x - 2 = 0 \quad \text{or} \quad x^2 + 2x + 4 = 0 \qquad \text{Set each factor equal to 0.}$$

The solution of $x - 2 = 0$ is 2. Solve the second equation by the quadratic formula.

$$x = \frac{-2 \pm \sqrt{2^2 - 4(1)(4)}}{2 \cdot 1} = \frac{-2 \pm \sqrt{-12}}{2} \qquad \text{Let } a = 1, b = 2, c = 4.$$

$$= \frac{-2 \pm 2i\sqrt{3}}{2} = \frac{2(-1 \pm i\sqrt{3})}{2}$$

$$= -1 \pm i\sqrt{3}$$

The solution set is $\{2, -1 + i\sqrt{3}, -1 - i\sqrt{3}\}$.

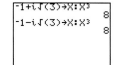

For this example, we can check the real solution 2 mentally since, from the original equation, we know that $2^3 = 8$. The imaginary solutions are checked to the left.

Recall that the graph of $y = x^3 - 8$ will reveal the real number solution of 2 only as shown.

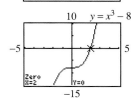

Notice that solutions of $x^3 = 8$ in Example 2 are numbers whose cube is 8. As expected, the real number 2 is a solution, since $2^3 = 8$. Unexpectedly, we also found two imaginary numbers whose cube is 8, $-1 + i\sqrt{3}$ and $-1 - i\sqrt{3}$.

EXAMPLE 3   Solve $p^4 - 3p^2 - 4 = 0$.

Solution: First, factor the trinomial.

$$p^4 - 3p^2 - 4 = 0$$

$$(p^2 - 4)(p^2 + 1) = 0 \qquad \text{Factor.}$$

$$(p - 2)(p + 2)(p^2 + 1) = 0 \qquad \text{Factor further.}$$

$$p - 2 = 0 \quad \text{or} \quad p + 2 = 0 \quad \text{or} \quad p^2 + 1 = 0 \qquad \text{Set each factor equal to 0.}$$

$$p = 2 \quad \text{or} \qquad p = -2 \quad \text{or} \qquad p^2 = -1$$

$$p = \pm\sqrt{-1} = \pm i$$

The solution set is $\{2, -2, i, -i\}$. Verify these solutions numerically.

EXAMPLE 4   Solve $(x - 3)^2 - 3(x - 3) - 4 = 0$.

Solution: Notice that the quantity $(x - 3)$ is repeated in this equation. Sometimes it is help-ful to substitute a variable (in this case other than x) for the repeated quantity. Let $y = x - 3$. Then

$$(x - 3)^2 - 3(x - 3) - 4 = 0$$

becomes

$$y^2 - 3y - 4 = 0 \qquad \text{Let } x - 3 = y.$$

$$(y - 4)(y + 1) = 0 \qquad \text{Factor.}$$

To solve, use the zero factor property.

$$y - 4 = 0 \quad \text{or} \quad y + 1 = 0 \qquad \text{Set each factor equal to 0.}$$

$$y = 4 \quad \text{or} \qquad y = -1 \qquad \text{Solve.}$$

To find values of x, substitute back. That is, let $y = x - 3$.

$$x - 3 = 4 \quad \text{or} \quad x - 3 = -1$$

$$x = 7 \quad \text{or} \qquad x = 2$$

The solution set is $\{2, 7\}$. Check to verify these solutions.

EXAMPLE 5   Solve $x^{2/3} - 5x^{1/3} + 6 = 0$.

Solution: The key to solving this equation algebraically is recognizing that $x^{2/3} = (x^{1/3})^2$. Replace $x^{1/3}$ with m so that

$$(x^{1/3})^2 - 5x^{1/3} + 6 = 0$$

becomes

$$m^2 - 5m + 6 = 0$$

Now solve by factoring.

$$m^2 - 5m + 6 = 0$$

$(m - 3)(m - 2) = 0$ Factor.

$m - 3 = 0$ or $m - 2 = 0$ Set each factor equal to 0.

$m = 3$ or $m = 2$

Since $m = x^{1/3}$, we have

$x^{1/3} = 3$ or $x^{1/3} = 2$

$x = 3^3 = 27$ or $x = 2^3 = 8$ Cube both sides of the equation.

The solution set is $\{8, 27\}$. To visualize these solutions, graph $y_1 = x^{2/3} - 5x^{1/3} + 6$ and find the x-intercepts of the graph.

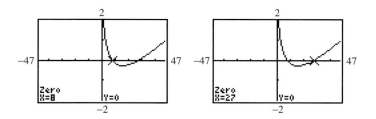

The x-intercepts are 8 and 27 and this verifies the solution set of $\{8, 27\}$. ▬▬▬▬

2   The next example is a work problem. This problem is modeled by a rational equation that simplifies to a quadratic equation.

EXAMPLE 6   **TIME MANAGEMENT**

Together, an experienced typist and an apprentice typist can process a document in 6 hours. Alone, the experienced typist can process the document 2 hours faster than the apprentice typist can. Find the time that it takes for each person to process the document alone. Round each time to the nearest tenth.

Solution: **1.** UNDERSTAND.   Read and reread the problem. The key idea here is the relationship between the *time* (hours) it takes to complete the job and the *part of the job* completed in one unit of time (hour). For example, because they can complete the job together in 6 hours, the *part of the job* that they can complete in 1 hour is $\frac{1}{6}$.

2. ASSIGN.   Let x represent the *time* in hours that it takes the apprentice typist to complete the job alone. Then $x - 2$ represents the time in hours that it takes the experienced typist to complete the job alone.

3. ILLUSTRATE. Here, we summarize in a chart the information discussed.

	TOTAL HOURS TO COMPLETE JOB	PART OF JOB COMPLETED IN 1 HOUR
APPRENTICE TYPIST	x	$\dfrac{1}{x}$
EXPERIENCED TYPIST	$x - 2$	$\dfrac{1}{x - 2}$
TOGETHER	6	$\dfrac{1}{6}$

4. TRANSLATE.

In words:	part of job completed by apprentice typist in 1 hour	added to	part of job completed by experienced typist in 1 hour	is equal to	part of job completed together in 1 hour
Translate:	$\dfrac{1}{x}$	$+$	$\dfrac{1}{x - 2}$	$=$	$\dfrac{1}{6}$

5. COMPLETE. Since we are interested in approximate real number solutions, we solve graphically. Graph $y_1 = \dfrac{1}{x} + \dfrac{1}{x - 2}$ and $y_2 = \dfrac{1}{6}$ in a $[-4, 18.8, 2]$ by $[-1, 1, 1]$ window. The approximate intersections are $x \approx 0.9$ and $x \approx 13.1$. A graph is shown below.

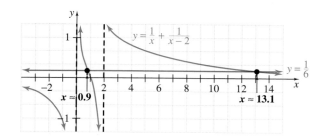

6. INTERPRET. *Check:* If the apprentice typist completes the job alone in 0.9 hours, the experienced typist completes the job alone in $x - 2 = 0.9 - 2 = -1.1$ hours. Since this is not possible, we reject the solution of 0.9. The approximate solution is thus 13.1 hours.

State: The apprentice typist can complete the job alone in approximately 13.1 hours, and the experienced typist completes the job alone in approximately $x - 2 = 13.1 - 2 = 11.1$ hours.

EXERCISE SET 9.3

Solve. See Example 1.

1. $\dfrac{2}{x} + \dfrac{3}{x-1} = 1$ **2.** $\dfrac{6}{x^2} = \dfrac{3}{x+1}$

3. $\dfrac{3}{x} + \dfrac{4}{x+2} = 2$ **4.** $\dfrac{5}{x-2} + \dfrac{4}{x+2} = 1$

5. $\dfrac{7}{x^2 - 5x + 6} = \dfrac{2x}{x-3} - \dfrac{x}{x-2}$

6. $\dfrac{11}{2x^2 + x - 15} = \dfrac{5}{2x-5} - \dfrac{x}{x+3}$

Solve. See Example 2.

7. $y^3 - 1 = 0$ **8.** $x^3 + 8 = 0$

9. $x^4 + 27x = 0$ **10.** $y^5 + y^2 = 0$

11. $z^3 = 64$ **12.** $z^3 = -125$

Solve. See Example 3.

13. $p^4 - 16 = 0$ **14.** $x^4 + 2x^2 - 3 = 0$

15. $4x^4 + 11x^2 = 3$ **16.** $z^4 = 81$

17. $z^4 - 13z^2 + 36 = 0$ **18.** $9x^4 + 5x^2 - 4 = 0$

Solve. See Examples 4 and 5.

19. $x^{2/3} - 3x^{1/3} - 10 = 0$ **20.** $x^{2/3} + 2x^{1/3} + 1 = 0$

21. $(5n + 1)^2 + 2(5n + 1) - 3 = 0$

22. $(m - 6)^2 + 5(m - 6) + 4 = 0$

23. $2x^{2/3} - 5x^{1/3} = 3$ **24.** $3x^{2/3} + 11x^{1/3} = 4$

25. $1 + \dfrac{2}{3t - 2} = \dfrac{8}{(3t - 2)^2}$

26. $2 - \dfrac{7}{x + 6} = \dfrac{15}{(x + 6)^2}$

27. $20x^{2/3} - 6x^{1/3} - 2 = 0$

28. $4x^{2/3} + 16x^{1/3} = -15$

29. Write a polynomial equation that has three solutions: 2, 5, and −7.

30. Write a polynomial equation that has three solutions: 0, 2*i*, and −2*i*.

Solve each equation. Check the solution numerically or graphically.

31. $a^4 - 5a^2 + 6 = 0$ **32.** $x^4 - 12x^2 + 11 = 0$

33. $\dfrac{2x}{x - 2} + \dfrac{x}{x + 3} = \dfrac{-5}{x + 3}$

34. $\dfrac{5}{x - 3} + \dfrac{x}{x + 3} = \dfrac{19}{x^2 - 9}$

35. $(p + 2)^2 = 9(p + 2) - 20$

36. $2(4m - 3)^2 - 9(4m - 3) = 5$

37. $x^3 + 64 = 0$ **38.** $y^3 - 27 = 0$

39. $x^{2/3} - 8x^{1/3} + 15 = 0$

40. $x^{2/3} - 2x^{1/3} - 8 = 0$

41. $y^3 + 9y - y^2 - 9 = 0$

42. $x^3 + x - 3x^2 - 3 = 0$

43. $2x^{2/3} + 3x^{1/3} - 2 = 0$

44. $6x^{2/3} - 25x^{1/3} - 25 = 0$

45. $x^{-2} - x^{-1} - 6 = 0$

46. $y^{-2} - 8y^{-1} + 7 = 0$

47. $2x^3 - 250 = 0$

48. $8y^3 + 8 = 0$

49. $\dfrac{x}{x - 1} + \dfrac{1}{x + 1} = \dfrac{2}{x^2 - 1}$

50. $\dfrac{x}{x - 5} + \dfrac{5}{x + 5} = \dfrac{-1}{x^2 - 25}$

51. $p^4 - p^2 - 20 = 0$

52. $x^4 - 10x^2 + 9 = 0$

53. $2x^3 = -54$ **54.** $y^3 - 216 = 0$

55. $1 = \dfrac{4}{x - 7} + \dfrac{5}{(x - 7)^2}$

56. $3 + \dfrac{1}{2p + 4} = \dfrac{10}{(2p + 4)^2}$

57. $27y^4 + 15y^2 = 2$ **58.** $8z^4 + 14z^2 = -5$

Solve. If appropriate, use a calculator to approximate each solution to the nearest tenth. See Example 6.

59. Bill Shaughnessy and his son Billy can clean the house together in 4 hours. When the son works alone, it takes him an hour longer to clean than it takes his dad alone. Find how long to the nearest hundredth hour it takes the son to clean alone.

60. Together, Scratchy and Freckles eat a 50-pound bag of dog food in 30 days. Scratchy by himself eats a 50-pound bag in 2 weeks less time than Freckles does by himself. How many days to the nearest whole day would a 50-pound bag of dog food last Freckles?

61. The product of a number and 4 less than the number is 96. Find the number.

62. A whole number increased by its square is two more than twice itself. Find the number.

63. An IBM computer and a Toshiba computer can complete a job together in 8 hours. The IBM computer alone can complete the job in 1 hour less time than the Toshiba computer alone. Find the time in which each computer can complete the job alone.

64. Two fax machines working together can fax a lengthy document in 75 minutes. Fax machine A alone can fax the document in 30 minutes less time than fax machine B alone. Find the time it take each machine to fax the document alone.

Review Exercises

Solve each inequality. See Section 4.1.

65. $\dfrac{5x}{3} + 2 \leq 7$

66. $\dfrac{2x}{3} + \dfrac{1}{6} \geq 2$

67. $\dfrac{y-1}{15} > \dfrac{-2}{5}$

68. $\dfrac{z-2}{12} < \dfrac{1}{4}$

Find the domain and range of each relation graphed. Decide which relations are also functions. See Section 3.3.

69.

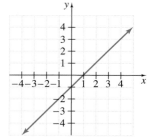

70.

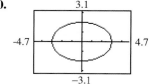

71.

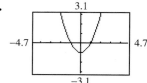

72.

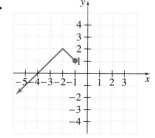

9.4 | NONLINEAR INEQUALITIES IN ONE VARIABLE

OBJECTIVES

1   Solve polynomial inequalities of degree 2 or greater.

2   Solve inequalities that contain rational expressions with variables in the denominator.

TAPE IAG 9.4

1

Just as we can solve linear inequalities in one variable, so we can also solve quadratic inequalities in one variable. A **quadratic inequality** is an inequality that can be written so that one side is a quadratic expression and the other side is 0. Here are examples of quadratic inequalities in one variable. Each is written in **standard form.**

$$x^2 - 10x + 7 \le 0 \qquad 3x^2 + 2x - 6 > 0$$
$$2x^2 + 9x - 2 < 0 \qquad x^2 - 3x + 11 \ge 0$$

A solution of a quadratic inequality in one variable is a value of the variable that makes the inequality a true statement.

DISCOVER THE CONCEPT

Graph the quadratic function $y = x^2 - 3x - 10$.

a. Find the values of x for which $y = 0$. (Recall that these are the x-intercepts of the graph.)

b. Find the x-values for which $y < 0$.

c. Find the x-values for which $y > 0$.

d. What is the solution of $y = 0$ or $x^2 - 3x - 10 = 0$?

e. What is the solution of $y < 0$ or $x^2 - 3x - 10 < 0$?

f. What is the solution of $y > 0$ or $x^2 - 3x - 10 > 0$?

Below is the graph of $y = x^2 - 3x - 10$.

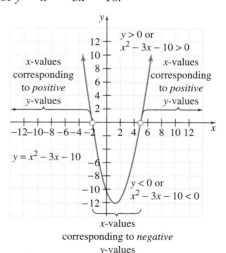

Notice that the x-values for which y or $x^2 - 3x - 10$ is positive are separated from the x values for which y or $x^2 - 3x - 10$ is negative by the values for which y or $x^2 - 3x - 10$ is 0, the x-intercepts. Thus, the solution set of $x^2 - 3x - 10 < 0$ consists of all real numbers from -2 to 5, or $(-2, 5)$. The solution set of $x^2 - 3x - 10 = 0$ is $\{-2, 5\}$; the solution set of $x^2 - 3x - 10 > 0$ is $(-\infty, -2) \cup (5, \infty)$.

It is not necessary to graph $y = x^2 - 3x - 10$ to solve a related inequality such as $x^2 - 3x - 10 < 0$. Instead, we can draw a number line representing the x-axis and keep the following in mind: **A region on the number line for which the value of $x^2 - 3x - 10$ is positive is separated from a region on the number line for which the value of $x^2 - 3x - 10$ is negative by a value for which the expression is 0.** Find these values for which the expression is 0 by solving the related equation.

$$x^2 - 3x - 10 = 0$$

$$(x - 5)(x + 2) = 0 \qquad \text{Factor.}$$

$$x - 5 = 0 \quad \text{or} \quad x + 2 = 0 \qquad \text{Set each factor equal to 0.}$$

$$x = 5 \quad \text{or} \qquad x = -2 \qquad \text{Solve.}$$

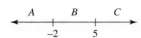

These two numbers divide the number line into three regions. We will call the regions A, B, and C. These regions are important because, if the value of $x^2 - 3x - 10$ is negative when a number from a region is substituted for x, then $x^2 - 3x - 10$ is negative when any number in the region is substituted for x. The same is true if the value of $x^2 - 3x - 10$ is positive for a particular value of x in a region.

To see whether the inequality $x^2 - 3x - 10 < 0$ is true or false in each region, choose a test point from each region and substitute its value for x in the inequality $x^2 - 3x - 10 < 0$ or $(x - 5)(x + 2) < 0$. If the resulting inequality is true, the region containing the test point is a solution region.

	TEST POINT VALUE	$(x - 5)(x + 2) < 0$	
REGION A	-3	$(-8)(-1) < 0$	False
REGION B	0	$(-5)(2) < 0$	True
REGION C	6	$(1)(8) < 0$	False

The values in region B satisfy the inequality. The numbers -2 and 5 are not included in the solution set since the inequality symbol is $<$. The solution set is $(-2, 5)$, as shown to the left.

EXAMPLE 1   Solve algebraically $(x + 3)(x - 3) < 0$.

Solution: First, solve the related equation $(x + 3)(x - 3) = 0$.

$$(x + 3)(x - 3) = 0$$

$$x + 3 = 0 \quad \text{or} \quad x - 3 = 0$$

$$x = -3 \quad \text{or} \qquad x = 3$$

The two numbers -3 and 3 separate the number line into three regions.

Substitute the value of a test point from each region. If the test value satisfies the inequality, every value in the region containing the test value is a solution.

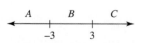

	TEST POINT VALUE	$(x + 3)(x - 3) < 0$	
REGION A	-4	$(-1)(-7) < 0$	False
REGION B	0	$(3)(-3) < 0$	True
REGION C	4	$(7)(1) < 0$	False

The values in region B satisfy the inequality. The numbers -3 and 3 are not included in the solution since the inequality symbol is $<$. The solution set in interval notation is $(-3, 3)$, and its graph is shown.

The following steps may be used to solve a polynomial inequality algebraically.

TO SOLVE A POLYNOMIAL INEQUALITY ALGEBRAICALLY

Step 1.   Write the inequality in standard form.

Step 2.   Solve the related equation.

Step 3.   Separate the number line into regions with the solutions from *step 2*.

Step 4.   For each region, choose a test point and determine whether its value satisfies the **original inequality.**

Step 5.   Write the solution set as the union of regions whose test point value is a solution.

EXAMPLE 2   Solve $x^2 - 4x \geq 0$.

Solution: This quadratic inequality is written in standard form, so first, solve the related equation $x^2 - 4x = 0$.

$$x^2 - 4x = 0$$
$$x(x - 4) = 0$$
$$x = 0 \quad \text{or} \quad x = 4$$

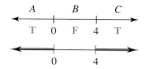

The numbers 0 and 4 separate the number line into three regions.

Check a test value in each region in the original inequality. Values in regions A and C satisfy the inequality. The numbers 0 and 4 are included in the solution since the inequality symbol is $\geq$. The solution set is $(-\infty, 0] \cup [4, \infty)$, and its graph is shown.

To solve polynomial inequalities graphically, we simply write the inequality so that the right side is 0 and then graph the related polynomial function. Instead of testing point values, we note the x-values where the graph is above the x-axis if the inequality symbol is $>$ or $\geq$ or we note the x-values where the graph is below the x-axis if the inequality symbol is $<$ or $\leq$.

Next, we solve the quadratic inequality of Example 2 graphically.

EXAMPLE 3   Use a graphical approach to solve $x^2 - 4x \geq 0$.

Solution: The x-intercepts of the graph of $y = x^2 - 4x$ are found by solving $x^2 - 4x = 0$. We found these intercepts earlier to be $x = 0$ or $x = 4$. Next, graph $y = x^2 - 4x$. Solutions of the quadratic inequality are x-values where $y \geq 0$, or where the graph is on or above the x-axis.

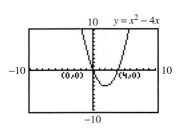

The x-values where the graph is on or above the x-axis are the x-intercepts of 0 and 4 and the x-values to the left of the x-intercept 0 and to the right of the x-intercept 4. In interval notation, the solution set is $(-\infty, 0] \cup [4, \infty)$.

EXAMPLE 4   Use a graphical approach to solve $(x + 2)(x - 1)(x - 5) < 0$.

Solution: The x-intercepts of the graph are found by solving $(x + 2)(x - 1)(x - 5) = 0$. These x-intercepts are -2, 1, and 5. Graph $y = (x + 2)(x - 1)(x - 5)$. To solve $y < 0$, find x-values where the graph lies below the x-axis.

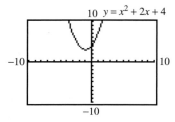

Again, the solution set of $y < 0$ contains x-values where the graph of y is below the x-axis. In interval notation, the solution set is $(-\infty, -2) \cup (1, 5)$.

EXAMPLE 5   Use a graphical approach to solve $x^2 + 2x > -4$.

Solution: In standard form, the quadratic inequality is $x^2 + 2x + 4 > 0$. The graph of $y = x^2 + 2x + 4$ is shown below. Notice that the entire graph lies above the x-axis, so $y > 0$ for all values of x. The solution set in interval notation is $(-\infty, \infty)$.

2   Inequalities containing rational expressions with variables in the denominator are solved algebraically by using a procedure similar to the one for polynomial inequalities.

EXAMPLE 6   Solve $\dfrac{x + 2}{x - 3} \le 0$.

Solution: First, find all values that make the denominator equal to 0. To do this, solve $x - 3 = 0$, or $x = 3$.

Next, solve the related equation $\dfrac{x + 2}{x - 3} = 0$.

$$\dfrac{x + 2}{x - 3} = 0 \qquad \text{Multiply both sides by the LCD, } x - 3.$$

$$x + 2 = 0$$

$$x = -2$$

Place these numbers on a number line and proceed as before, checking test point values in the original inequality.

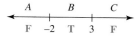

Choose -3 from region A Choose 0 from region B

$$\dfrac{x + 2}{x - 3} \le 0 \qquad\qquad\qquad \dfrac{x + 2}{x - 3} \le 0$$

$$\dfrac{-3 + 2}{-3 - 3} \le 0 \qquad\qquad\qquad \dfrac{0 + 2}{0 - 3} \le 0$$

$$\dfrac{-1}{-6} \le 0 \qquad\qquad\qquad\qquad -\dfrac{2}{3} \le 0 \qquad \text{True.}$$

$$\dfrac{1}{6} \le 0 \qquad \text{False.}$$

Choose 4 from region C

$$\dfrac{x + 2}{x - 3} \le 0$$

$$\dfrac{4 + 2}{4 - 3} \le 0$$

$$6 \le 0 \qquad \text{False.}$$

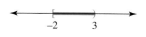

The solution set is $[-2, 3)$. This interval includes -2 because -2 satisfies the original inequality. This interval does not include 3, because 3 would make the denominator 0.

To visualize this graphically, graph

$$y = \dfrac{x + 2}{x - 3}$$

and find x-values where the graph is on or below the x-axis. The graph shown to the left is on the x-axis at $x = -2$. The graph is below the x-axis from -2 to the undefined value $x = 3$. Thus, we verify the solution set as $[-2, 3)$.

The following steps may be used to solve a rational inequality with variables in the denominator.

TO SOLVE A RATIONAL INEQUALITY ALGEBRAICALLY

Step 1.   Solve for values that make all denominators 0.

Step 2.   Solve the related equation.

Step 3.   Separate the number line into regions with the solutions from *steps 1* and *2*.

Step 4.   For each region, choose a test point and determine whether its value satisfies the **original inequality.**

Step 5.   Write the solution set as the union of regions whose test point value is a solution.

EXAMPLE 7   Solve $\dfrac{5}{x + 1} < -2$.

Solution: First, find values for x that make the denominator equal to 0.

$$x + 1 = 0$$
$$x = -1$$

Next, solve $\dfrac{5}{x + 1} = -2$.

$$(x + 1) \cdot \frac{5}{x + 1} = (x + 1) \cdot -2 \qquad \text{Multiply both sides by the LCD, } x + 1.$$
$$5 = -2x - 2 \qquad \text{Simplify.}$$
$$7 = -2x$$
$$-\frac{7}{2} = x$$

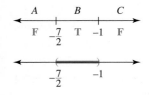

Use these two solutions to divide a number line into three regions and choose test points.

Only a test point value from region B satisfies the **original inequality.** The solution set is $\left(-\dfrac{7}{2}, -1\right)$.

EXERCISE SET 9.4

Solve each quadratic inequality. Write the solution set in interval notation. See Examples 1 through 5.

1. $(x + 1)(x + 5) > 0$ **2.** $(x + 1)(x + 5) \le 0$

3. $(x - 3)(x + 4) \le 0$ **4.** $(x + 4)(x - 1) > 0$

5. $x^2 - 7x + 10 \le 0$ **6.** $x^2 + 8x + 15 \ge 0$

7. $3x^2 + 16x < -5$ **8.** $2x^2 - 5x < 7$

9. $(x - 6)(x - 4)(x - 2) > 0$

10. $(x - 6)(x - 4)(x - 2) \le 0$

11. $x(x - 1)(x + 4) \le 0$ **12.** $x(x - 6)(x + 2) > 0$

13. $(x^2 - 9)(x^2 - 4) > 0$ **14.** $(x^2 - 16)(x^2 - 1) \le 0$

Solve each inequality. Write the solution set in interval notation. See Example 6.

15. $\dfrac{x + 7}{x - 2} < 0$ **16.** $\dfrac{x - 5}{x - 6} > 0$

17. $\dfrac{5}{x + 1} > 0$ **18.** $\dfrac{3}{y - 5} < 0$

19. $\dfrac{x + 1}{x - 4} \ge 0$ **20.** $\dfrac{x + 1}{x - 4} \le 0$

21. Explain why $\dfrac{x + 2}{x - 3} > 0$ and $(x + 2)(x - 3) > 0$ have the same solutions.

22. Explain why $\dfrac{x + 2}{x - 3} \ge 0$ and $(x + 2)(x - 3) \ge 0$ do not have the same solutions.

Solve each inequality. Write the solution set in interval notation. See Example 7.

23. $\dfrac{3}{x - 2} < 4$ **24.** $\dfrac{-2}{y + 3} > 2$

25. $\dfrac{x^2 + 6}{5x} \ge 1$ **26.** $\dfrac{y^2 + 15}{8y} \le 1$

Solve each inequality. Write the solution set in interval notation.

27. $(x - 8)(x + 7) > 0$ **28.** $(x - 5)(x + 1) < 0$

29. $(2x - 3)(4x + 5) \le 0$ **30.** $(6x + 7)(7x - 12) > 0$

31. $x^2 > x$ **32.** $x^2 < 25$

33. $(2x - 8)(x + 4)(x - 6) \le 0$

34. $(3x - 12)(x + 5)(2x - 3) \ge 0$

35. $6x^2 - 5x \ge 6$ **36.** $12x^2 + 11x \le 15$

37. $4x^3 + 16x^2 - 9x - 36 > 0$

38. $x^3 + 2x^2 - 4x - 8 < 0$

39. $x^4 - 26x^2 + 25 \ge 0$ **40.** $16x^4 - 40x^2 + 9 \le 0$

41. $(2x - 7)(3x + 5) > 0$ **42.** $(4x - 9)(2x + 5) < 0$

43. $\dfrac{x}{x - 10} < 0$ **44.** $\dfrac{x + 10}{x - 10} > 0$

45. $\dfrac{x - 5}{x + 4} \ge 0$ **46.** $\dfrac{x - 3}{x + 2} \le 0$

47. $\dfrac{x(x + 6)}{(x - 7)(x + 1)} \ge 0$ **48.** $\dfrac{(x - 2)(x + 2)}{(x + 1)(x - 4)} \le 0$

49. $\dfrac{-1}{x - 1} > -1$ **50.** $\dfrac{4}{y + 2} < -2$

51. $\dfrac{x}{x + 4} \le 2$ **52.** $\dfrac{4x}{x - 3} \ge 5$

53. $\dfrac{z}{z - 5} \ge 2z$ **54.** $\dfrac{p}{p + 4} \le 3p$

55. $\dfrac{(x + 1)^2}{5x} > 0$ **56.** $\dfrac{(2x - 3)^2}{x} < 0$

Find all numbers that satisfy each of the following.

57. A number minus its reciprocal is less than zero. Find the numbers.

58. Twice a number added to its reciprocal is nonnegative. Find the numbers.

59. The total profit function $P(x)$ for a company producing x thousand units is given by

$$P(x) = -2x^2 + 26x - 44$$

Find the values of x for which the company makes a profit. [*Hint:* The company makes a profit when $P(x) > 0$.]

60. A projectile is fired straight up from the ground with an initial velocity of 80 feet per second. Its height $s(t)$ in feet at any time t is given by the function

$$s(t) = -16t^2 + 80t$$

Find the interval of time for which the height of the projectile is greater than 96 feet.

Review Exercises

Recall that the graph of $f(x) + K$ is the same as the graph of $f(x)$ shifted $|K|$ units upward if $K > 0$ and $|K|$ units downward if $K < 0$. Use the graph of $f(x) = |x|$ below to sketch the graph of each function. (See Sections 3.2 and 3.3.)

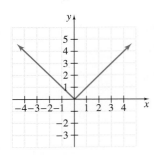

61. $g(x) = |x| + 2$

62. $H(x) = |x| - 2$

63. $F(x) = |x| - 1$

64. $h(x) = |x| + 5$

Use the graph of $f(x) = x^2$ below to sketch the graph of each function.

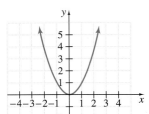

65. $F(x) = x^2 - 3$

66. $h(x) = x^2 - 4$

67. $H(x) = x^2 + 1$

68. $g(x) = x^2 + 3$

9.5   QUADRATIC FUNCTIONS AND THEIR GRAPHS

TAPE IAG 9.5

O B J E C T I V E S

1   Graph quadratic functions of the form $f(x) = x^2 + k$.

2   Graph quadratic functions of the form $f(x) = (x - h)^2$.

3   Graph quadratic functions of the form $f(x) = ax^2$.

4   Graph quadratic functions of the form $f(x) = a(x - h)^2 + k$.

We first graphed the quadratic equation $y = x^2$ in Section 3.2. In Section 3.3, we learned that this graph defines a function, and we wrote $y = x^2$ as $f(x) = x^2$. Throughout these sections, we discovered that the graph of a quadratic function is a parabola opening upward or downward. In this section, we continue our study of quadratic functions and their graphs.

First, let's recall the definition of a quadratic function.

QUADRATIC FUNCTION

A quadratic function is a function that can be written in the form $f(x) = ax^2 + bx + c$, where a, b, and c are real numbers and $a \neq 0$.

We know that an equation of the form $f(x) = ax^2 + bx + c$ may be written as $y = ax^2 + bx + c$. Thus, both $f(x) = ax^2 + bx + c$ and $y = ax^2 + bx + c$ define quadratic functions as long as a is not 0.

Recall the graph of the quadratic function defined by $f(x) = x^2$ below.

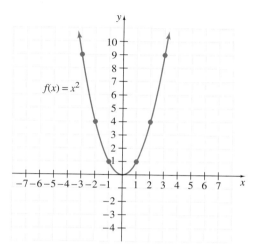

Notice that the graph passes the vertical line test, as it should since it is a function. Recall that this curve is called a **parabola.** The highest point on a parabola that opens downward or the lowest point on a parabola that opens upward is called the **vertex** of the parabola. The vertex of this parabola is $(0, 0)$, the lowest point on the graph. If we fold the graph along the y-axis, we can see that the two sides of the graph coincide. This means that this curve is symmetric about the y-axis, and the y-axis, or the line $x = 0$, is called the **axis of symmetry.** The graph of every quadratic function is a parabola and has an axis of symmetry: the vertical line that passes through the vertex of the parabola.

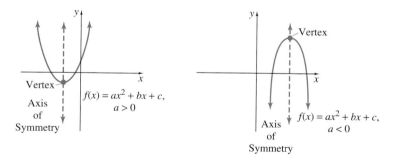

DISCOVER THE CONCEPT

Graph $f(x) = x^2$, $g(x) = x^2 + 3$ and $h(x) = x^2 - 5$. Compare the graphs of $g(x)$ and $h(x)$ to the graph of $f(x)$.

The graphs of $y_1 = x^2$, $y_2 = x^2 + 3$, $y_3 = x^2 - 5$ and corresponding tables are shown below.

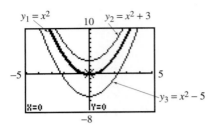

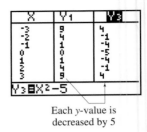

Each y-value is increased by 3

Each y-value is decreased by 5

Notice that the graph of $y_2 = x^2 + 3$ is the same as the graph of $y_1 = x^2$, but shifted *upward* 3 units. Also, the graph of $y_3 = x^2 - 5$ is the same as the graph of $y_1 = x^2$, but shifted *downward* 5 units. The axis of symmetry is the same for all graphs: $x = 0$.

In general, we have the following properties:

> **GRAPHING THE PARABOLA DEFINED BY $f(x) = x^2 + k$**
>
> If k is positive, the graph of $f(x) = x^2 + k$ is the graph of $y = x^2$ shifted upward k units.
>
> If k is negative, the graph of $f(x) = x^2 + k$ is the graph of $y = x^2$ shifted downward $|k|$ units.
>
> The vertex is $(0, k)$, and the axis of symmetry is the y-axis, or $x = 0$.

EXAMPLE 1   Sketch the graph of each function. Label the vertex and the axis of symmetry.

  a. $F(x) = x^2 + 2$ **b.** $g(x) = x^2 - 3$

Solution: **a.** The graph of $F(x) = x^2 + 2$ is obtained by shifting the graph of $y = x^2$ upward 2 units. The vertex is $(0, 2)$. The axis of symmetry is the y-axis, or $x = 0$.

  b. The graph of $g(x) = x^2 - 3$ is obtained by shifting the graph of $y = x^2$ downward 3 units. The vertex is $(0, -3)$. The axis of symmetry is the y-axis, or $x = 0$.

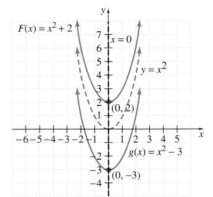

2 Next, we graph quadratic functions of the form $f(x) = (x - h)^2$.

DISCOVER THE CONCEPT

Graph $f(x) = x^2$, $g(x) = (x - 2)^2$, and $h(x) = (x + 4)^2$. Compare the graphs of $g(x)$ and $h(x)$ to the graph of $f(x)$.

The graphs of $y_1 = x^2$, $y_2 = (x - 2)^2$, and $y_3 = (x + 4)^2$ are shown below.

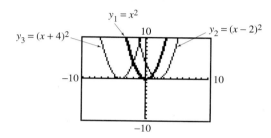

The graph of $y_2 = (x - 2)^2$ is the same as the graph of $y_1 = x^2$ shifted *to the right* 2 units. Also, the graph of $y_3 = (x + 4)^2$ is the same as the graph of $y_1 = x^2$ shifted *to the left* 4 units. Notice that the axes of symmetry have shifted also.

In general, we have the following properties:

GRAPHING THE PARABOLA DEFINED BY $f(x) = (x - h)^2$

If h is positive, the graph of $f(x) = (x - h)^2$ is the graph of $y = x^2$ shifted to the right h units.

If h is negative, the graph of $f(x) = (x - h)^2$ is the graph of $y = x^2$ shifted to the left $|h|$ units.

The vertex is $(h, 0)$, and the line of symmetry has the equation $x = h$.

EXAMPLE 2   Sketch the graph of each function. Label the vertex and the axis of symmetry.

    a. $G(x) = (x - 3)^2$ **b.** $F(x) = (x + 1)^2$

Solution: **a.** The graph of $G(x) = (x - 3)^2$ is obtained by shifting the graph of $y = x^2$ to the right 3 units. The vertex is $(3, 0)$, and the axis of symmetry has equation $x = 3$.

               b. The equation $F(x) = (x + 1)^2$ can be written as $F(x) = [x - (-1)]^2$. The graph of $F(x) = [x - (-1)]^2$ is obtained by shifting the graph of $y = x^2$ to the left 1 unit. The vertex is $(-1, 0)$, and the axis of symmetry has equation $x = -1$.

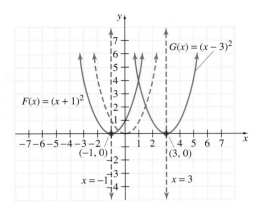

It is possible to combine vertical and horizontal shifts.

EXAMPLE 3 Sketch the graph of $F(x) = (x - 3)^2 + 1$. Label the vertex and the axis of symmetry.

Solution: The graph of $F(x) = (x - 3)^2 + 1$ is the graph of $y = x^2$ shifted 3 units to the right and 1 unit up. The vertex is then $(3, 1)$, and the axis of symmetry is $x = 3$. A few ordered pair solutions are plotted to aid in graphing.

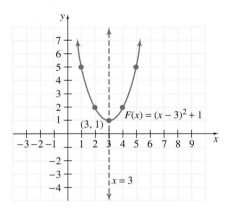

Next, we discover the change in the shape of the graph when the coefficient of x^2 is not 1.

DISCOVER THE CONCEPT

Graph $f(x) = ax^2$ for $a = 1, 3$ and $\frac{1}{2}$. Compare the other graphs to the graph of $f(x) = 1x^2$ or $f(x) = x^2$.

The graphs of $y_1 = x^2$, $y_2 = 3x^2$, and $y_3 = \frac{1}{2}x^2$ and corresponding tables are shown below.

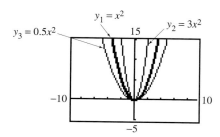

Compare the table of values. We see that for each x-value the corresponding value of y_2 is triple the corresponding value of y_1. Similarly, the value of y_3 is half the value of y_1. The result is that the graph of $y_2 = 3x^2$ is narrower than the graph of $f(x) = x^2$ and the graph of $y_3 = \frac{1}{2}x^2$ is wider. The vertex for each graph is $(0, 0)$, and the axis of symmetry is the y-axis.

GRAPHING THE PARABOLA DEFINED BY $f(x) = ax^2$

If a is positive, the parabola opens upward, and if a is negative, the parabola opens downward.

If $|a| > 1$, the graph of the parabola is narrower than the graph of $y = x^2$.

If $|a| < 1$, the graph of the parabola is wider than the graph of $y = x^2$.

EXAMPLE 4   Sketch the graph of $f(x) = -2x^2$. Label the vertex and the axis of symmetry.

Solution: Because $a = -2$, a negative value, this parabola opens downward. Since $|-2| = 2$ and $2 > 1$, the parabola is narrower than the graph of $y = x^2$. The vertex is $(0, 0)$, and the axis of symmetry is the y-axis. We verify this by plotting a few points.

x	$f(x) = -2x^2$
-2	-8
-1	-2
0	0
1	-2
2	-8

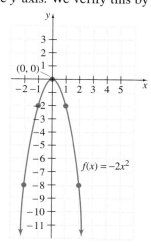

4   Next, we graph a quadratic function of the form $f(x) = a(x - h)^2 + k$.

EXAMPLE 5   Find the vertex and the axis of symmetry of the graph of $g(x) = \frac{1}{2}(x + 2)^2 + 5$. Use a graphing utility to verify the results.

Solution: The function $g(x) = \frac{1}{2}(x + 2)^2 + 5$ may be written as $g(x) = \frac{1}{2}[x - (-2)]^2 + 5$. Thus, this graph is the same as the graph of $y = x^2$ shifted 2 units to the left and 5 units up, and it is wider because a is $\frac{1}{2}$. The vertex is $(-2, 5)$, and the axis of symmetry is $x = -2$. The screen to the left shows the graph with the vertex indicated.

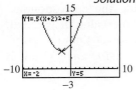

MENTAL MATH

State the vertex of the graph of each quadratic function.

1. $f(x) = x^2$ **2.** $f(x) = -5x^2$ **3.** $g(x) = (x - 2)^2$ **4.** $g(x) = (x + 5)^2$

5. $f(x) = 2x^2 + 3$ **6.** $h(x) = x^2 - 1$ **7.** $g(x) = (x + 1)^2 + 5$ **8.** $h(x) = (x - 10)^2 - 7$

EXERCISE SET 9.5

Sketch the graph of each quadratic function. Label the vertex and sketch and label the axis of symmetry. See Examples 1 through 3.

1. $f(x) = x^2 - 1$ **2.** $g(x) = x^2 + 3$

3. $h(x) = x^2 + 5$ **4.** $h(x) = x^2 - 4$

5. $g(x) = x^2 + 7$ **6.** $f(x) = x^2 - 2$

7. $f(x) = (x - 5)^2$ **8.** $g(x) = (x + 5)^2$

9. $h(x) = (x + 2)^2$ **10.** $H(x) = (x - 1)^2$

11. $G(x) = (x + 3)^2$ **12.** $f(x) = (x - 6)^2$

13. $f(x) = (x - 2)^2 + 5$ **14.** $g(x) = (x - 6)^2 + 1$

15. $h(x) = (x + 1)^2 + 4$ **16.** $G(x) = (x + 3)^2 + 3$

17. $g(x) = (x + 2)^2 - 5$ **18.** $h(x) = (x + 4)^2 - 6$

Sketch the graph of each quadratic function. Label the vertex, and sketch and label the axis of symmetry. See Example 4.

19. $g(x) = -x^2$ **20.** $f(x) = 5x^2$

21. $h(x) = \frac{1}{3}x^2$ **22.** $g(x) = -3x^2$

23. $H(x) = 2x^2$ **24.** $f(x) = -\frac{1}{4}x^2$

Sketch the graph of each quadratic function. Label the vertex, and sketch and label the axis of symmetry. See Example 5.

25. $f(x) = 2(x - 1)^2 + 3$

26. $g(x) = 4(x - 4)^2 + 2$

27. $h(x) = -3(x + 3)^2 + 1$

28. $f(x) = -(x - 2)^2 - 6$

29. $H(x) = \frac{1}{2}(x - 6)^2 - 3$

30. $G(x) = \frac{1}{5}(x + 4)^2 + 3$

Sketch the graph of each quadratic function. Label the vertex, and sketch and label the axis of symmetry. Use a graphing utility to verify the results.

31. $f(x) = -(x - 2)^2$ **32.** $g(x) = -(x + 6)^2$

33. $F(x) = -x^2 + 4$ **34.** $H(x) = -x^2 + 10$

35. $F(x) = 2x^2 - 5$ **36.** $g(x) = \dfrac{1}{2}x^2 - 2$

37. $h(x) = (x - 6)^2 + 4$ **38.** $f(x) = (x - 5)^2 + 2$

39. $F(x) = \left(x + \dfrac{1}{2}\right)^2 - 2$ **40.** $H(x) = \left(x + \dfrac{1}{4}\right)^2 - 3$

41. $F(x) = \dfrac{3}{2}(x + 7)^2 + 1$ **42.** $g(x) = -\dfrac{3}{2}(x - 1)^2 - 5$

43. $f(x) = \dfrac{1}{4}x^2 - 9$ **44.** $H(x) = \dfrac{3}{4}x^2 - 2$

45. $G(x) = 5\left(x + \dfrac{1}{2}\right)^2$ **46.** $F(x) = 3\left(x - \dfrac{3}{2}\right)^2$

47. $h(x) = -(x - 1)^2 - 1$

48. $f(x) = -3(x + 2)^2 + 2$

49. $g(x) = \sqrt{3}(x + 5)^2 + \dfrac{3}{4}$

50. $G(x) = \sqrt{5}(x - 7)^2 - \dfrac{1}{2}$

51. $h(x) = 10(x + 4)^2 - 6$

52. $h(x) = 8(x + 1)^2 + 9$

53. $f(x) = -2(x - 4)^2 + 5$

54. $G(x) = -4(x + 9)^2 - 1$

Write the equation of the parabola that has the same shape as $f(x) = 5x^2$, but with the following vertex.

❏ 55. $(2, 3)$ **❏ 56.** $(1, 6)$

❏ 57. $(-3, 6)$ **❏ 58.** $(4, -1)$

Use a graphing utility to graph the first function of each pair that follows. Then use its graph to predict the graph of the second function. Check your prediction by graphing both on the same set of axes.

59. $F(x) = \sqrt{x};\ G(x) = \sqrt{x} + 1$

60. $g(x) = x^3;\ H(x) = x^3 - 2$

61. $H(x) = |x|;\ f(x) = |x - 5|$

62. $h(x) = x^3 + 2;\ g(x) = (x - 3)^3 + 2$

63. $f(x) = |x + 4|;\ F(x) = |x + 4| + 3$

64. $G(x) = \sqrt{x} - 2;\ g(x) = \sqrt{x - 4} - 2$

The shifting properties covered in this section apply to the graphs of all functions. Given the accompanying graph of $y = f(x)$, sketch the graph of each of the following.

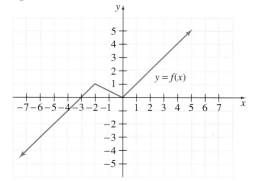

❏ 65. $y = f(x) + 1$ **❏ 66.** $y = f(x) - 2$

❏ 67. $y = f(x - 3)$ **❏ 68.** $y = f(x + 3)$

❏ 69. $y = f(x + 2) + 2$ **❏ 70.** $y = f(x - 1) + 1$

Review Exercises

Add the proper constant to each binomial so that the resulting trinomial is a perfect square trinomial. See Section 9.1.

71. $x^2 + 8x$ **72.** $y^2 + 4y$

73. $z^2 - 16z$ **74.** $x^2 - 10x$

75. $y^2 + y$ **76.** $z^2 - 3z$

Solve by completing the square. See Section 9.1.

77. $x^2 + 4x = 12$ **78.** $y^2 + 6y = -5$

79. $z^2 + 10z - 1 = 0$ **80.** $x^2 + 14x + 20 = 0$

81. $z^2 - 8z = 2$ **82.** $y^2 - 10y = 3$

9.6 | FURTHER GRAPHING OF QUADRATIC FUNCTIONS

TAPE IAG 9.6

O B J E C T I V E S

1. Write quadratic functions in the form $y = a(x - h)^2 + k$.
2. Derive a formula for finding the vertex of a parabola.
3. Use the vertex formula to find the vertex of a parabola.
4. Find the minimum or maximum value of a quadratic function.

1. We know that the graph of a quadratic function is a parabola. If a quadratic function is written in the form

$$f(x) = a(x - h)^2 + k$$

we can easily find the vertex (h, k) and graph the parabola. One way to write a quadratic function in this form is to complete the square on the variable x. See Section 9.1 for a review of completing the square.

EXAMPLE 1   Sketch the graph of $f(x) = x^2 - 4x - 12$ and check using a graphing utility.

Solution: The graph of this quadratic function is a parabola. To find the vertex of the parabola, we complete the square on the binomial $x^2 - 4x$. To simplify our work, let $f(x) = y$.

$$y = x^2 - 4x - 12 \qquad \text{Let } f(x) = y.$$
$$y + 12 = x^2 - 4x \qquad \text{Add 12 to both sides to isolate the } x\text{-variable terms.}$$

Now add the square of half of -4 to both sides.

$$\frac{1}{2}(-4) = -2 \quad \text{and} \quad (-2)^2 = 4$$

$$y + 12 + 4 = x^2 - 4x + 4 \qquad \text{Add 4 to both sides.}$$
$$y + 16 = (x - 2)^2 \qquad \text{Factor the trinomial.}$$
$$y = (x - 2)^2 - 16 \qquad \text{Subtract 16 from both sides.}$$
$$f(x) = (x - 2)^2 - 16 \qquad \text{Replace } y \text{ with } f(x).$$

From this equation, we can see that the vertex of the parabola is $(2, -16)$, and the axis of symmetry is the line $x = 2$.

Since $a > 0$, the parabola opens upward. This parabola opening upward with vertex $(2, -16)$ will have two x-intercepts. To find them, let $f(x)$ or $y = 0$.

$$0 = x^2 - 4x - 12$$
$$0 = (x - 6)(x + 2)$$
$$0 = x - 6 \quad \text{or} \quad 0 = x + 2$$
$$6 = x \qquad \text{or} \quad -2 = x$$

The two x-intercepts are 6 and -2. The sketch of $f(x) = x^2 - 4x - 12$ is shown. The graphing utility check is shown to the left.

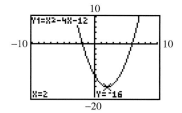

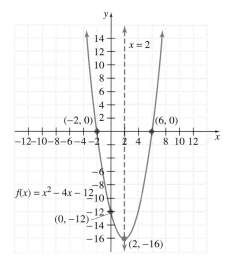

$f(x) = x^2 - 4x - 12$

2 Now that we have practiced completing the square, we introduce and derive a formula for finding the vertex of a parabola. The x-coordinate of the vertex of the graph of $f(x)$, or $y = ax^2 + bx + c$ can be found by the formula $x = \dfrac{-b}{2a}$. To see this, we complete the square on x and write the equation in the form $y = a(x - h)^2 + k$.

First, isolate the x-variable terms by subtracting c from both sides.

$$y = ax^2 + bx + c$$
$$y - c = ax^2 + bx$$

Next, factor a from the terms $ax^2 + bx$.

$$y - c = a\left(x^2 + \frac{b}{a}x\right)$$

Next, add the square of half of $\dfrac{b}{a}$, or $\left(\dfrac{b}{2a}\right)^2 = \dfrac{b^2}{4a^2}$, to the right side inside the parentheses. Because of the factor a, what we really added is $a\left(\dfrac{b^2}{4a^2}\right)$, and this must be added to the left side.

$$y - c + a\left(\frac{b^2}{4a^2}\right) = a\left(x^2 + \frac{b}{a}x + \frac{b^2}{4a^2}\right)$$

$$y - c + \frac{b^2}{4a} = a\left(x + \frac{b}{2a}\right)^2 \qquad \text{Simplify the left side and factor the right side.}$$

$$y = a\left(x + \frac{b}{2a}\right)^2 + c - \frac{b^2}{4a} \qquad \begin{array}{l}\text{Add } c \text{ to both sides and subtract} \\ \dfrac{b^2}{4a} \text{ from both sides.}\end{array}$$

Compare this form with $f(x)$ or $y = a(x - h)^2 + k$ and see that h is $\dfrac{-b}{2a}$, which means that the x-coordinate of the vertex of the graph of $f(x) = ax^2 + bx + c$ is $\dfrac{-b}{2a}$.

VERTEX FORMULA

The graph of $f(x) = ax^2 + bx + c$, when $a \neq 0$, is a parabola with vertex

$$\left(\frac{-b}{2a}, f\left(\frac{-b}{2a} \right) \right)$$

3

Let's practice using the vertex formula to find the vertex of a parabola algebraically.

EXAMPLE 2　　Use a graphing utility to graph $f(x) = 3x^2 + 3x + 1$. Find the vertex and any intercepts algebraically.

Solution:　　For $f(x) = 3x^2 + 3x + 1$, notice that $a = 3$, $b = 3$, and $c = 1$. Because $a > 0$, we know that the parabola opens upward. To algebraically find the vertex, we use the vertex formula.

$$x = \frac{-b}{2a} = \frac{-3}{2(3)} = \frac{-3}{6} = \frac{-1}{2} \quad \text{or} \quad -0.5$$

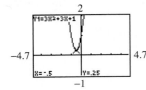

Also, $f(-0.5) = 3(-0.5)^2 + 3(-0.5) + 1 = 0.25$, so the vertex has coordinates $(-0.5, 0.25)$. Because the vertex is in the second quadrant and the parabola opens upward, the parabola has no x-intercepts, as shown in the graph to the left.

To find the y-intercept, let $x = 0$. Then

$$f(0) = 3(0)^2 + 3(0) + 1 = 1$$

The y-intercept point is $(0, 1)$.

EXAMPLE 3　　Use a graphing utility to graph $f(x) = 3 - 2x - x^2$. Find the vertex and any intercepts algebraically.

Solution:　　For $f(x) = 3 - 2x - x^2$, we have $a = -1$, $b = -2$, and $c = 3$. Since $a < 0$, the parabola opens downward. We use the vertex formula to find the vertex.

$$x = \frac{-b}{2a} = \frac{-(-2)}{2(-1)} = -1$$

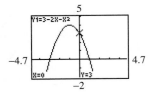

Also, $f(-1) = 4$, so the vertex has coordinates $(-1, 4)$. The vertex is in the second quadrant and the parabola opens downward, so the parabola has two x-intercepts and a y-intercept as shown to the left.

To find the x-intercepts, let y or $f(x) = 0$ and solve for x.

$$f(x) = 3 - 2x - x^2$$
$$0 = 3 - 2x - x^2$$
$$0 = (3 + x)(1 - x)$$
$$3 + x = 0 \quad \text{or} \quad 1 - x = 0$$
$$x = -3 \quad \text{or} \quad 1 = x$$

The x-intercept points are $(-3, 0)$ and $(1, 0)$.

To find the y-intercept, let $x = 0$.

$$f(0) = 3 - 2 \cdot 0 - 0^2 = 3$$

The y-intercept point is $(0, 3)$.

4 The quadratic function whose graph is a parabola that opens upward has a minimum value, and the quadratic function whose graph is a parabola that opens downward has a maximum value. The $f(x)$ or y-value of the vertex is the minimum or maximum value of the function.

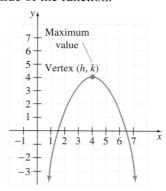 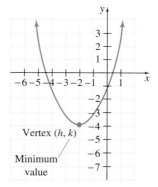

EXAMPLE 4   **FINDING MAXIMUM HEIGHT**
A rock is propelled upward from the ground. Its height in feet above ground after x seconds is given by the equation $f(x) = -16x^2 + 20x$. Find the maximum height of the rock and the number of seconds it took for the rock to reach its maximum height.

Solution: **1. UNDERSTAND.** The maximum height of the rock is the largest value of $f(x)$. Since the equation $f(x) = -16x^2 + 20x$ is a quadratic function, its graph is a parabola. It opens downward since $a = -16 < 0$. Thus, the maximum value of $f(x)$ is the $f(x)$ or y-value of the vertex of its graph.

4. TRANSLATE. To find the vertex, notice that, for $f(x) = -16x^2 + 20x$, $a = -16$, $b = 20$, and $c = 0$. Use these values and the vertex formula

$$\left(\frac{-b}{2a}, f\left(\frac{-b}{2a} \right) \right).$$

5. COMPLETE. $x = \dfrac{-b}{2a} = \dfrac{-20}{-32} = \dfrac{5}{8}$

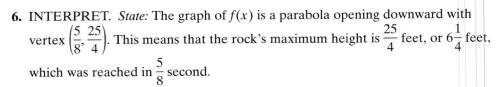

$$f\left(\frac{5}{8}\right) = -16\left(\frac{5}{8}\right)^2 + 20\left(\frac{5}{8}\right) = -16\left(\frac{25}{64}\right) + \frac{25}{2} = -\frac{25}{4} + \frac{50}{4} = \frac{25}{4}$$

Recall that $f\left(\dfrac{5}{8}\right)$ can also be found using a graphing utility, as shown in the screen to the left.

6. INTERPRET. *State:* The graph of $f(x)$ is a parabola opening downward with vertex $\left(\dfrac{5}{8}, \dfrac{25}{4}\right)$. This means that the rock's maximum height is $\dfrac{25}{4}$ feet, or $6\dfrac{1}{4}$ feet, which was reached in $\dfrac{5}{8}$ second.

To *check*, graph $y_1 = -16x^2 + 20x$ in a $[0, 2, 1]$ by $[0, 10, 1]$ window. The maximum height occurs when $x = 0.625$ or $\frac{5}{8}$ seconds and is $6\frac{1}{4}$ feet.

Exercise Set 9.6

Find the vertex of the graph of each quadratic function. See Example 1.

1. $f(x) = x^2 + 8x + 7$ **2.** $f(x) = x^2 + 6x + 5$

3. $f(x) = -x^2 + 10x + 5$ **4.** $f(x) = -x^2 - 8x + 2$

5. $f(x) = 5x^2 - 10x + 3$ **6.** $f(x) = -3x^2 + 6x + 4$

7. $f(x) = -x^2 + x + 1$ **8.** $f(x) = x^2 - 9x + 8$

Use a graphing utility to graph each function. Find the vertex, axis of symmetry, and any intercepts algebraically. See Examples 1 through 3.

9. $f(x) = x^2 + 4x - 5$ **10.** $f(x) = x^2 + 2x - 3$

11. $f(x) = -x^2 + 2x - 1$ **12.** $f(x) = -x^2 + 4x - 4$

13. $f(x) = x^2 - 4$ **14.** $f(x) = x^2 - 1$

15. $f(x) = 4x^2 + 4x - 3$ **16.** $f(x) = 2x^2 - x - 3$

17. $f(x) = x^2 + 8x + 15$ **18.** $f(x) = x^2 + 10x + 9$

19. $f(x) = x^2 - 6x + 5$ **20.** $f(x) = x^2 - 4x + 3$

21. $f(x) = x^2 - 4x + 5$ **22.** $f(x) = x^2 - 6x + 11$

23. $f(x) = 2x^2 + 4x + 5$ **24.** $f(x) = 3x^2 + 12x + 16$

25. $f(x) = -2x^2 + 12x$ **26.** $f(x) = -4x^2 + 8x$

27. $f(x) = x^2 + 1$ **28.** $f(x) = x^2 + 4$

29. $f(x) = x^2 - 2x - 15$ **30.** $f(x) = x^2 - 4x + 3$

31. $f(x) = -5x^2 + 5x$ **32.** $f(x) = 3x^2 - 12x$

33. $f(x) = -x^2 + 2x - 12$

34. $f(x) = -x^2 + 8x - 17$

35. $f(x) = 3x^2 - 12x + 15$

36. $f(x) = 2x^2 - 8x + 11$ **37.** $f(x) = x^2 + x - 6$

38. $f(x) = x^2 + 3x - 18$

39. $f(x) = -2x^2 - 3x + 35$

40. $f(x) = 3x^2 - 13x - 10$

Solve. See Example 4.

41. If a projectile is fired straight upward from the ground with an initial speed of 96 feet per second, then its height h in feet after t seconds is given by the equation

$$h(t) = -16t^2 + 96t$$

Find the maximum height of the projectile.

42. The cost C in dollars of manufacturing x bicycles at Holladay's Production Plant is given by the function $C(x) = 2x^2 - 800x + 92{,}000$.

   a. Find the number of bicycles that must be manufactured to minimize the cost.

   b. Find the minimum cost.

43. If a ball is propelled upward with an initial speed of 32 feet per second, then its height h in feet after t seconds is given by the equation

$$h(t) = -16t^2 + 32t$$

Find the maximum height of the ball.

44. The Utah Ski Club sells calendars to raise money. The profit P, in cents, from selling x calendars is given by the equation $P(x) = 360x - x^2$.

  a. Find how many calendars must be sold to maximize profit.

  b. Find the maximum profit.

45. Find two positive numbers whose sum is 60 and whose product is as large as possible. [*Hint:* Let x and $60 - x$ be the two positive numbers. Their product can be described by the function $f(x) = x(60 - x)$.]

46. The length and width of a rectangle must have a sum of 40. Find the dimensions of the rectangle that will have maximum area. (Use the hint for Exercise 45.)

Use a graphing utility to graph each quadratic function. Determine whether the graph opens upward or downward, find the vertex and y-intercept, and approximate the x-intercepts to one decimal place.

47. $f(x) = x^2 + 10x + 15$

48. $f(x) = x^2 - 6x + 4$

49. $f(x) = 3x^2 - 6x + 7$

50. $f(x) = 2x^2 + 4x - 1$

Use a graphing utility to find the maximum or minimum value of each function. Approximate to two decimal places.

51. $f(x) = 2.3x^2 - 6.1x + 3.2$

52. $f(x) = 7.6x^2 + 9.8x - 2.1$

53. $f(x) = -1.9x^2 + 5.6x - 2.7$

54. $f(x) = -5.2x^2 - 3.8x + 5.1$

Review Exercises

Sketch the graph of each function. See Section 9.5.

55. $f(x) = x^2 + 2$ **56.** $f(x) = (x - 3)^2$

57. $g(x) = x + 2$ **58.** $h(x) = x - 3$

59. $f(x) = (x + 5)^2 + 2$ **60.** $f(x) = 2(x - 3)^2 + 2$

61. $f(x) = 3(x - 4)^2 + 1$ **62.** $f(x) = (x + 1)^2 + 4$

63. $f(x) = -(x - 4)^2 + \dfrac{3}{2}$ **63.** $f(x) = -2(x + 7)^2 + \dfrac{1}{2}$

9.7 INTERPRETING DATA: RECOGNIZING LINEAR AND QUADRATIC MODELS

TAPE IAG 9.6

O B J E C T I V E S

1. Plot data points and find a linear model for the data.
2. Plot data points and find a quadratic model for the data.
3. Plot data points and recognize whether a linear or quadratic model best fits the data.

As we have seen thus far in our text, many situations arise that involve two related quantities. We can investigate and record certain facts about these situations, and these are called **data.** To *model* the data means to find an equation that describes the relationship between the given data quantities. Visually, the graph of the *best fit equation* should have a majority of the plotted ordered pair data points on the graph or close to it. In statistics, a technique called **regression** is used to determine an equation that best fits data that relates two quantities. In this text, we concentrate on modeling ordered pair data points with an equation that visually appears to best fit those data points.

Various graphing utilities have built-in capabilities for finding an equation that best fits a set of ordered pair data points. The equation is called a regression equation. We will concentrate on analyzing the data and finding an equation without these capabilities in order to better understand mathematically how models are generated on the graphing utility. Keep in mind that no model may fit the data exactly, but a model should fit the data closely enough so that useful predictions may be made using it.

 In this section, we will concentrate on two models, linear and quadratic. These models are shown below for your review.

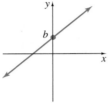

Linear Model
$y - y_1 = m(x - x_1)$
slope: m
y-intercept: b
point of line: (x_1, y_1)

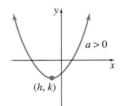

Quadratic Model
$y = a(x - h)^2 + k$
$a > 0$, parabola opens up
$a < 0$, parabola opens down
vertex: (h, k)

EXAMPLE 1   A LINEAR MODEL FOR PREDICTING SELLING PRICE
The following houses were sold in an Orange Park neighborhood in the past month. Listed below are each house's selling price and the number of square feet of living area in the house.

Total square feet	2,150	2,273	2,474	2,416
Selling price in dollars	120,000	134,500	158,900	149,800

a. Graph the data points.

b. Find a linear equation that models the data.

c. Graph the linear equation and use the graph or the equation to predict the approximate selling price for a house in the same neighborhood that has 2350 square feet of living area.

Solution: **a.** Enter the total number of square feet in a first list, L_1, and the corresponding selling price in a second list, L_2. Use the statistical plotting feature of your graphing utility to plot the points. From the plotting menu, select a scatterplot and indicate in which list you stored the data for the horizontal components of the graph, the x-values, and in which list you stored the data for the vertical components of the graph, the y-values.

In this text, values in L_1 will be x-values and values in L_2 will be y-values.

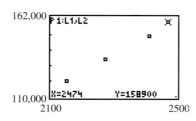

The x-values represent the number of square feet, and these data numbers are between 2100 and 2500. The y-values represent the selling price, and all these are between $110,000 and $162,000.

b. Notice that the plotted data points appear to lie approximately on a line. Take a straightedge and place it along the plotted points on the screen above. Find a line that joins at least two of the points and for which it appears that the remaining points are close to the line. We chose the points with coordinates (2150, 120,000) and (2273, 134,500). Recall from Chapter 3 that we can write the equation of line, given two points, by calculating its slope and using the point-slope form.

$$m = \frac{\Delta y}{\Delta x} = \frac{y_2 - y_1}{x_2 - x_1} = \frac{134,500 - 120,000}{2273 - 2150} = \frac{14,500}{123}$$

Using the point-slope form $y - y_1 = m(x - x_1)$ and the point with coordinates (2150, 120,000), we have

$$y - 120,000 = \frac{14,500}{123}(x - 2150)$$

or solving for y, we have

$$y = \frac{14,500}{123}(x - 2150) + 120,000$$

c. Graph this equation as y_1 and observe the fit of the graph to the data points that have been previously plotted.

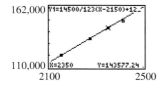

To predict an approximate selling price for a house in the same neighborhood that has 2350 square feet of living area, use the graph as shown to the left or the result of part (b) and find y when $x = 2350$.

$$y = \frac{14,500}{123}(x - 2150) + 120,000$$

$$= \frac{14,500}{123}(2350 - 2150) + 120,000$$

$$y \approx \$143,577$$

The predicted selling price is $143,577.

Notice from the graph that choosing different data points to write a linear model will probably result in a different model.

2   A quadratic model may be used when the plotted data points suggest a parabola.

EXAMPLE 2   **A QUADRATIC MODEL FOR FUEL CONSUMPTION**

The table below shows the average miles per gallon (MPG) for U.S. cars in the years shown. (*Source:* U.S. Federal Highway Administration, *Highway Statistics, annual.*)

Year	1940	1950	1960	1970	1980	1990
Average MPG	14.8	13.9	13.4	13.5	15.5	21.0

a. Graph the data points.

b. Find a quadratic equation that models the data.

c. Use the equation to predict the average MPG for the year 2002.

Solution: **a.** List the years in L_1 and the average MPG in L_2. Then draw a scatterplot as shown below.

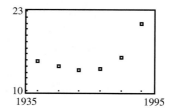

b. To find a quadratic model, we'll use the form $y = a(x - h)^2 + k$ and use the lowest point plotted for the vertex (h, k). The lowest data point has coordinates $(1960, 13.4)$. Thus far, we have

$$y = a(x - 1960)^2 + 13.4$$

To find a value for a, substitute the coordinates for another data point into the equation for x and y, and solve for a. We use the data point $(1990, 21.0)$.

$$21 = a(1990 - 1960)^2 + 13.4$$

$$21 = 900a + 13.4$$

$$\frac{7.6}{900} = a$$

Substituting for a, we have a quadratic model.

$$y = \frac{7.6}{900}(x - 1960)^2 + 13.4$$

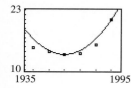

Graph this equation and observe the fit of the graph to the data points.

c. To predict the average MPG for the year 2002, let $x = 2002$ and solve for y.

$$y = \frac{7.6}{900}(2002 - 1960)^2 + 13.4$$

$$y \approx 28.3$$

The predicted mileage for automobiles in the year 2002 is 28.3 MPG.

As is true for linear models, choosing different data points for a quadratic model will lead to a different model.

A quadratic equation that best fits the data in Example 2 is

$$y = 0.0075x^2 - 29.372x + 28{,}770.797$$

To use this equation to predict MPG in the year 2002, we evaluate

$$y = 0.0075(2002)^2 - 29.372(2002) + 28{,}770.797$$
$$y \approx 28.1 \text{ MPG.}$$

Next, we practice deciding whether to choose a linear model or a quadratic model as an equation that best fits a data set.

EXAMPLE 3   CHOOSING AN APPROPRIATE MODEL
The table below shows the consumption of cigarettes for the years shown. (*Source:* U.S. Department of Agriculture.)

Year	1985	1987	1989	1991	1993
Annual consumption of cigarettes (billions of cigarettes)	594	575	540	510	485

a. Graph the data points.

b. Decide whether a linear model or a quadratic model best fits the data.

c. Find an equation that models the data.

d. Use the equation to predict the consumption of cigarettes in the year 2010.

Solution: **a.** List the years in L_1 and the consumption of cigarettes in L_2. Then draw a scatterplot as shown below.

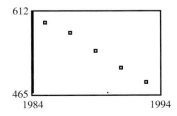

b. From the graph above, we choose to use a linear model to describe the data.

c. To write a linear model, choose two data points. We chose (1985, 594) and (1991, 510).

$$m = \frac{\Delta y}{\Delta x} = \frac{y_2 - y_1}{x_2 - x_1} = \frac{510 - 594}{1991 - 1985} = \frac{-84}{6} = -14$$

Using $y - y_1 = m(x - x_1)$ and the point (1985, 594), we have

$$y - 594 = -14(x - 1985)$$

or

$$y = -14(x - 1985) + 594$$

A graph of this model is shown to the left.

d. To predict the consumption of cigarettes in the year 2010, we assume that the linear trend continues, and let $x = 2010$ in the linear-model equation.

$$y = -14(2010 - 1985) + 594$$
$$= 244$$

Our prediction is that 244 billion cigarettes will be consumed in the year 2010.

EXERCISE SET 9.7

Given the following graphs of data points, tell whether the graph is likely to be best represented by a linear model, a quadratic model, or neither.

1.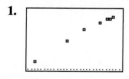

2.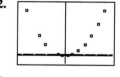

3.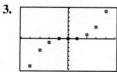

4.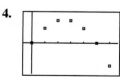

5.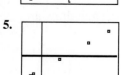

6.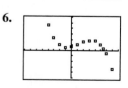

Plot the following data points. Then find an equation of a line that fits the data by using the data pairs indicated with an asterisk (*). See Example 1.

7.

x	*−3	−1	2	5	*7	9
y	−14	−7	1	9	16	18

8.

x	*−3	−1	3	5	*6	12
y	5	6	9	10	11	15

9.

x	−3	2	*3	4	5	*6
y	5	15	17	18	22	23

10.

x	−4	*−2	0	*2	4	6
y	10	−5	1	7	12	20

11.

x	−2	−1	*0	2	5	*6
y	−6	5	−5	−3	−3	−2

12.

x	2	3	4	5	6	8
y	1	4	9	14	17	24

Given the following data for four houses sold in comparable neighborhoods and their corresponding number of square feet, draw a scatter plot and find a linear equation representing a relationship between the number of square feet and the selling price of the house.

13.

Square feet	1,519	2,593	3,005	3,016
Selling price	$91,238	$220,000	$254,000	$269,000

14.

Square feet	1,754	2,151	2,587	2,671
Selling price	$91,238	$130,000	$155,000	$172,500

15. Use the equation found in Exercise 13 to predict the selling price of a house with 2200 square feet.

16. Use the equation found in Exercise 14 to predict the selling price of a house with 2400 square feet.

A salesperson's salary depends on the amount of sales for the week. Given the following amounts of sales and the corresponding weekly salaries, plot the data points and find a linear equation that describes the salary in terms of the amount of sales.

17.

Sales (in dollars)	100	300	500	800
Salary (in dollars)	158	174	190	214

18.

Sales (in dollars)	500	1000	2500	3000
Salary (in dollars)	200	225	300	325

19. The cost of a pizza depends on the diameter of the pizza. Given the following diameters and costs for the pizza, plot the data and find an equation for the cost y in terms of the diameter x.

Diameter (inches)	3	6	10	12
Cost (in dollars)	1.90	3.75	6.25	7.50

20. The following table gives the numbers of hours needed to cover the corresponding distances. Plot the data and find an equation for the distance y in terms of time x.

Time (hours)	2	4	7	9
Distance (miles)	110	215	385	495

State whether the relationship described can best be modeled by a linear function or a quadratic function.

21. The area of a square *and* the length of one side

22. A salary of $300 per week plus 5% commission on sales *and* weekly sales

23. A number of calculators *and* their cost, plus tax

24. The area of a circle *and* the radius of the circle

25. The average temperature of Cleveland, Ohio, *and* the months of March through October

26. An oven is turned on and set to reach 400 degrees. After it reaches the desired temperature, it is shut off. Temperatures are recorded every 2 minutes, from the time the oven is turned on until it cools to room temperature

27. The number of bacteria present in a person with respect to time when he is coming down with the flu, has the flu, and then recovers

28. The distance traveled during a recent trip with respect to the amount of time elapsed

Plot the following points and then find an equation that best models the data. See Example 2.

29. The table below shows the average temperature in Louisville, Kentucky, for the months of March through October.

March	April	May	June	July	August	Sept	Oct
45	57	65	74	78	76	70	58

30. The table below shows the average temperature in Wellington, New Zealand, for the months of March through October.

March	April	May	June	July	August	Sept	Oct
60	57	52	49	47	48	51	54

31. The prices of a scorecard in dollars at a particular Major League Baseball stadium for selected years from 1958 to 1995 is given in the table below. Plot the points and graph the equation of the quadratic model $y = 0.0015107x^2 - 5.89546x + 5751.7973$. Using the graph, approximate the cost of a scorecard in each given year.

a. 1967

b. 1999

1958	1963	1965	1968	1971	1975	1976	1983	1989	1995
0.15	0.25	0.30	0.45	0.60	0.90	1.00	1.75	2.00	3.00

32. The prices (in thousands of dollars) of a 30-second commercial spot on a local TV channel during a football bowl game for selected years from 1967 to 1996 are given in the table below. Plot the points and graph the equation of the linear model $y = 2.106x - 4123.631$. Using the graph, approximate the cost of a 30-second commercial in each given year.

a. 1982

b. 2000

1967	1970	1972	1975	1976	1979
5	20	35	38	46	50

1980	1985	1988	1990	1992	1996
55	60	63	67	70	72

33. Recall the quadratic equation from Example 2 that models the average miles per gallon (MPG) for U.S. cars:

$$y = \frac{7.6}{900}(x - 1960)^2 + 13.4$$

where x is the year. Use this model to fill in the predicted MPG below and compare with the actual given data. (Round predicted MPG to one decimal place.)

x (YEAR)	ACTUAL MPG	PREDICTED MPG	DIFFERENCE BETWEEN ACTUAL AND PREDICTED MPG
1990	21.0		
1991	21.7		
1992	21.7		
1993	21.6		

34. Recall the linear equation from Example 3 that models the annual consumption of cigarettes:

$$y = -14(x - 1985) + 594$$

where x is the year and y is in billions of cigarettes. Use this model to fill in the predicted consumption of cigarettes below and compare with the actual given data.

x (YEAR)	ACTUAL CONSUMPTION	PREDICTED CONSUMPTION	DIFFERENCE BETWEEN ACTUAL AND PREDICTED
1987	575		
1989	540		
1993	485		

The following sets of data from the U.S. Department of Commerce, Bureau of Labor Statistics, 1995, show the average weekly earnings in dollars for an employee in the following industries. Plot the data points and find an equation for the weekly earnings depending on the year, and then use this equation to predict the average weekly earnings in each industry for the year 2000.

35. MINING

1970	1980	1985	1990	1992	1993	1994
163	397	520	604	638	647	666

36. MANUFACTURING

1970	1980	1985	1990	1992	1993	1994
133	289	386	442	470	486	507

37. CONSTRUCTION

1970	1980	1985	1990	1992	1993	1994
195	368	464	526	538	551	570

38. RETAIL TRADE

1970	1980	1985	1990	1992	1993	1994
82	147	175	195	205	210	215

Plot the following data points. Then find an equation of the curve that models the data. See Example 3.

39.

x	0	1	3	4	6
y	1	6	10	9	1

40.

x	1	2	4	5	7
y	50	36	20	18	25

41.

x	1	3	4	5	7
y	−17	6	9	6	−15

42.

x	0	1	2	3	4
y	90	54	28	6	−6

43.

x	−4	−3	−1	0	2
y	11	12	8	3	−1

44.

x	−4	−2	0	2	4
y	11	15	11	−1	−21

45. The graph below shows the amounts given to charity (in billions of inflation-adjusted dollars) by corporations over the period 1984 through 1992. (*Source:* American Association of Fund Raising Council, Inc.)

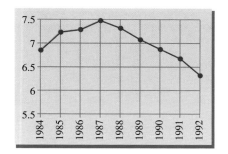

a. According to the graph, in what year was charitable giving by corporations the highest?

b. Plot the points using a graphing utility with the x-axis representing the number of years since 1980 and the y-axis representing the level of charitable giving in billions of dollars.

c. These data can be modeled by the quadratic function $f(x) = -0.045x^2 + 0.632x + 5.12$, where $f(x)$ is the level of charitable giving in billions of dollars and x is the number of years since 1980. Graph this function. Does this agree with the graph at the left?

d. Use the model to find the year in which giving was $7.296 billion. Does this agree with the graph at the left?

46. The graph below shows the depreciated (or decreased) value over 7 years for a copy machine recently purchased by the mathematics department. The school's budget needs to be projected for the next 7 years.

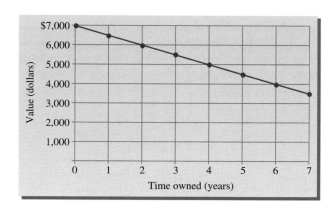

a. What was the purchase price of the machine?

b. What is the value of the machine after 7 years?

c. Find a linear model for the value of the machine in terms of time.

d. What does the slope of this line represent with respect to the value of the machine?

47. The 1988–1989 Chicago Bulls (other than Michael Jordan) had the following regular season totals.

Player	Brown	Corzin	Grant	Oakley	Paxson	Pippen	Sellers	Sparrow	Vincent
x (minutes played)	591	2328	1827	2816	1888	1650	2212	1044	1501
y (points)	197	804	622	1014	640	625	777	280	573

a. Plot the points using the statistical plotting feature of your graphing utility.

b. A line that best fits the data is the equation $y = 0.375x - 48.38$. Graph the equation and tell why this appears to be a reasonable line to use as a model.

c. If a player was to spend 1575 minutes playing, approximately how many points should we expect to see him score?

GROUP ACTIVITY

48. A tennis shoe manufacturer is studying the suggested selling price of a certain brand of tennis shoe. The suggested selling price cannot be too low or the cost of manufacturing the shoes will not be covered, and it cannot be too high or people will not purchase the shoes. Below are data collected for selling tennis shoes at various prices and the corresponding profit.

Price (in dollars)	39	49	65	75	85	95	120
Profit (in dollars)	9500	16,750	20,000	19,500	15,475	7500	1500

a. Plot the data points and tell why a quadratic model would seem to best represent the data.

b. A quadratic model for these data is Profit $= -6.77x^2 + 924.35x - 13,852$, where x is price of the tennis shoes. Graph this equation and check the fit of the equation to the graph of the data points.

c. Using the model, find the projected profit if the shoes sold for $80. For $60.

d. Use the table to find the profit if the shoes sold for $120. Is this a good selling price? Why or why not?

e. Using the quadratic model, find the selling price, to the nearest dollar, that will yield the maximum profit.

GROUP ACTIVITY

49. The local school board voted to give all its teachers an 8% raise in pay.

a. Find the new salary for teachers whose annual salaries at present are $26,500, $34,000, and $56,000.

b. Find a linear model for the projected salaries based on the present salaries of the faculty members.

50. The sales tax rate for Duval County is 7.5%.

a. Find the total purchase price for a sweater costing $29.95.

b. Write a linear equation for the total purchase price of an article y in terms of the price of the article x.

Review Exercises

Solve each of the following equations. See Sections 2.1 and 3.7.

51. $(x + 3)(x - 5) = 0$ **52.** $(x - 1)(x + 19) = 0$

53. $2x^2 - 7x - 15 = 0$ **54.** $6x^2 + 13x - 5 = 0$

55. $3(x - 4) + 2 = 5(x - 6)$

56. $4(2x + 7) = 2(x - 4)$

Find the y-intercept point of the graph of each equation. See Section 3.4.

57. $f(x) = x^3 + 3x^2 - 5x - 8$

58. $f(x) = 2x^3 + x^2 - 7x + 12$

59. $g(x) = x^2 - 3x + 5$ **60.** $g(x) = 3x^2 - 5x - 10$

GROUP ACTIVITY

MODELING THE POSITION OF A FREELY FALLING OBJECT

The quadratic model that describes the position of a freely falling object as a function of time (neglecting air resistance) is

$$s(t) = \frac{1}{2}gt^2 + v_0 t + s_0$$

where t is time in seconds, s_0 is the object's initial height or position in either meters or feet, v_0 is the object's initial velocity in either meters per second or feet per second, and g is the acceleration due to gravity. This function $s(t)$ is also known as a *position function*. Velocity is directional. If the velocity is directed upward, it has a positive value; if it is directed downward, it has a negative value. The acceleration due to gravity on Earth is -9.8 meters per second per second, or -32 feet per second per second. The acceleration due to gravity on the Moon is -1.617 meters per second per second, or -5.28 feet per second per second.

For each of the following descriptions of a freely falling object, (a) match the description to the position function that models it, (b) match the description to the graph that represents it, and (c) find how long the object is in the air (approximate to two decimal places).

1. A rock is thrown *down* a 60-foot-deep well at a velocity of 15 feet per second.

2. A golf ball is thrown *upward* into the air at a velocity of 100 feet per second from a height of 5 feet on the Moon.

3. A bottle rocket is *launched* from sea level at a velocity of 50 meters per second.

4. A forest ranger *drops* a bundle of mail from a 100-meter-tall fire tower.

5. A bottle rocket is *launched* at a velocity of 50 meters per second from the top of a 100-meter-tall fire tower.

6. A sandbag is thrown *toward the ground* with a velocity of 50 feet per second from a hot air balloon at a height of 400 feet.

(a) $s(t) = -4.9t^2 + 100$

(b) $s(t) = -16t^2 + 15t + 60$

(c) $s(t) = -4.9t^2 + 50t + 100$

(d) $s(t) = -4.9t^2 + 50t$

(e) $s(t) = -16t^2 - 50t + 400$

(f) $s(t) = -2.64t^2 + 100t + 5$

(g) $s(t) = -16t^2 - 15t + 60$

(h) $s(t) = -16t^2 + 50t + 400$

(i) $s(t) = -16t^2 + 100$

(i)

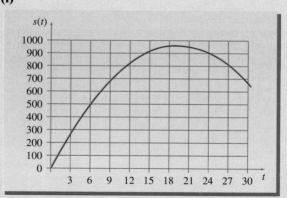

(continued)

(ii)

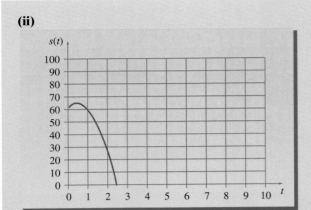

(iii)

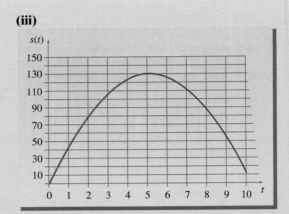

(iv)

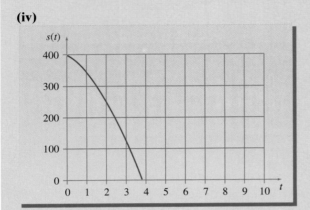

(v)

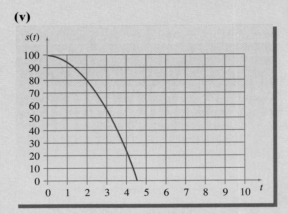

(vi)

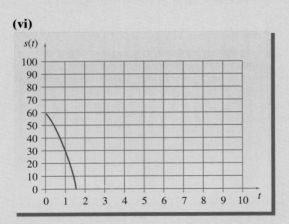

(vii)

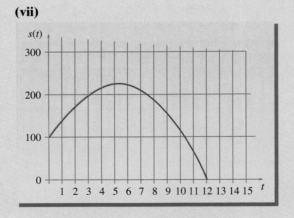

(viii)

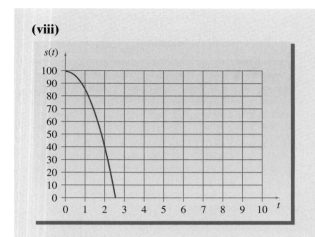

(ix)

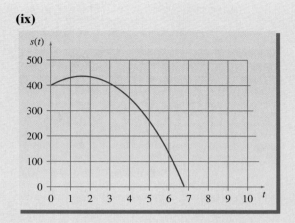

DEFINITIONS AND CONCEPTS	EXAMPLES
SECTION 9.1 SOLVING QUADRATIC EQUATIONS BY COMPLETING THE SQUARE	

Square root property:

If b is a real number and if $a^2 = b$, then $a = \pm\sqrt{b}$.

Solve: $(x + 3)^2 = 14$
$$x + 3 = \pm\sqrt{14}$$
$$x = -3 \pm \sqrt{14}$$

To solve a quadratic equation in x by completing the square:

Step 1. If the coefficient of x^2 is not 1, divide both sides of the equation by the coefficient of x^2.

Step 2. Isolate the variable terms.

Step 3. Complete the square by adding the square of half of the coefficient of x to both sides.

Step 4. Write the resulting trinomial as the square of a binomial.

Step 5. Apply the square root property and solve.

Solve $3x^2 - 12x - 18 = 0$.

1. $x^2 - 4x - 6 = 0$ Divide by 3.

2. $x^2 - 4x = 6$

3. $\dfrac{1}{2}(-4) = -2$ and $(-2)^2 = 4$

$$x^2 - 4x + 4 = 6 + 4$$

4. $(x - 2)^2 = 10$

5. $x - 2 = \pm\sqrt{10}$
$$x = 2 \pm \sqrt{10}$$

DEFINITIONS AND CONCEPTS	EXAMPLES

SECTION 9.2 SOLVING QUADRATIC EQUATIONS BY THE QUADRATIC FORMULA

A quadratic equation written in the form $ax^2 + bx + c = 0$ has solutions

$$x = \frac{-b \pm \sqrt{b^2 - 4ac}}{2a}$$

Solve $x^2 - x - 3 = 0$.

$$a = 1, b = -1, c = -3$$

$$x = \frac{-(-1) \pm \sqrt{(-1)^2 - 4(1)(-3)}}{2 \cdot 1}$$

$$x = \frac{1 \pm \sqrt{13}}{2}$$

SECTION 9.3 SOLVING EQUATIONS BY USING QUADRATIC METHODS

Substitution is often helpful in solving an equation that contains a repeated variable expression.

Solve $(2x + 1)^2 - 5(2x + 1) + 6 = 0$.

Let $m = 2x + 1$. Then

$$m^2 - 5m + 6 = 0 \quad \text{Let } m = 2x + 1.$$
$$(m - 3)(m - 2) = 0$$
$$m = 3 \quad \text{or} \quad m = 2$$
$$2x + 1 = 3 \quad \text{or} \quad 2x + 1 = 2 \quad \text{Substitute back.}$$
$$x = 1 \quad \text{or} \quad x = \frac{1}{2}$$

SECTION 9.4 NONLINEAR INEQUALITIES IN ONE VARIABLE

To solve a polynomial inequality algebraically:

Step 1. Write the inequality in standard form.

Step 2. Solve the related equation.

Step 3. Use solutions from *step 2* to separate the number line into regions.

Step 4. Use a test point to determine whether values in each region satisfy the original inequality.

Step 5. Write the solution set as the union of regions whose test point value is a solution.

Solve $x^2 \geq 6x$.

1. $x^2 - 6x \geq 0$

2. $x^2 - 6x = 0$

$$x(x - 6) = 0$$
$$x = 0 \quad \text{or} \quad x = 6$$

3.

	TEST POINT		
REGION	**VALUE**	$x^2 \geq 6x$	
A	-2	$(-2)^2 \geq 6(-2)$	True
B	1	$1^2 \geq 6(1)$	False
C	7	$7^2 \geq 6(7)$	True

4. (above)

5.

The solution set is $(-\infty, 0] \cup [6, \infty)$.

(continued)

DEFINITIONS AND CONCEPTS	**EXAMPLES**

SECTION 9.4 NONLINEAR INEQUALITIES IN ONE VARIABLE

To solve a polynomial inequality by graphing, graph the related polynomial function. Instead of testing point values, see where the graph lies above the x-axis for the inequality symbols $>$ or $\geq$. See where the graph lies below the x-axis for the inequality symbols $<$ or $\leq$.

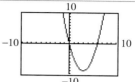

The solution of $x^2 - 6x \geq 0$ in interval notation is $(-\infty, 0] \cup [6, \infty)$. These are the x-values for which the graph lies on or above the x-axis.

SECTION 9.5 QUADRATIC FUNCTIONS AND THEIR GRAPHS

Graph of a quadratic function

The graph of a quadratic function written in the form $f(x) = a(x - h)^2 + k$ is a parabola with vertex (h, k). If $a > 0$, the parabola opens upward; if $a < 0$, the parabola opens downward. The axis of symmetry is the line whose equation is $x = h$.

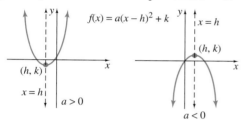

Graph $g(x) = 3(x - 1)^2 + 4$.

The graph is a parabola with vertex $(1, 4)$ and axis of symmetry $x = 1$. Since $a = 3$ is positive, the graph opens upward.

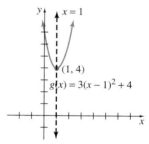

SECTION 9.6 FURTHER GRAPHING OF QUADRATIC FUNCTIONS

The graph of $f(x) = ax^2 + bx + c, a \neq 0$, is a parabola with vertex

$$\left(\frac{-b}{2a}, f\left(\frac{-b}{2a} \right) \right)$$

Graph $f(x) = x^2 - 2x - 8$. Find the vertex and x- and y-intercepts of $f(x) = x^2 - 2x - 8$.

$$\frac{-b}{2a} = \frac{-(-2)}{2 \cdot 1} = 1$$

$$f(1) = 1^2 - 2(1) - 8 = -9$$

The vertex is $(1, -9)$.

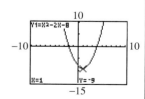

$$0 = x^2 - 2x - 8$$
$$0 = (x - 4)(x + 2)$$
$$x = 4 \quad \text{or} \quad x = -2$$

The x-intercept points are $(4, 0)$ and $(-2, 0)$.

$$f(0) = 0^2 - 2 \cdot 0 - 8 = -8$$

The y-intercept point is $(0, -8)$.

DEFINITIONS AND CONCEPTS	EXAMPLES

SECTION 9.7 INTERPRETING DATA: RECOGNIZING LINEAR AND QUADRATIC MODELS

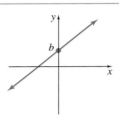

Linear Model:

$$y - y_1 = m(x - x_1)$$
Slope: m
y-intercept: b
Point of line: (x_1, y_1)

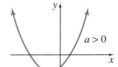

Quadratic Model:

$$y = a(x - h)^2 + k$$
$a > 0$, parabola opens up
$a < 0$, parabola opens down
Vertex: (h, k)

Given the following graphs of data points, tell whether the graph can best be modeled by a linear model or a quadratic model.

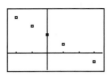 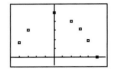

A linear model   A quadratic model

CHAPTER 9 REVIEW

(9.1) *Solve by factoring. Use a graphing utility to check.*

1. $x^2 - 15x + 14 = 0$

2. $x^2 - x - 30 = 0$

3. $10x^2 = 3x + 4$ **4.** $7a^2 = 29a + 30$

Solve by using the square root property. Check numerically.

5. $4m^2 = 196$ **6.** $9y^2 = 36$

7. $(9n + 1)^2 = 9$ **8.** $(5x - 2)^2 = 2$

Solve by completing the square. Check numerically.

9. $z^2 + 3z + 1 = 0$ **10.** $x^2 + x + 7 = 0$

11. $(2x + 1)^2 = x$ **12.** $(3x - 4)^2 = 10x$

13. Two ships leave a port at the same time and travel at the same speed. One ship is traveling due north and the other due east. In a few hours, the ships are 150 miles apart. How many miles has each ship traveled? Give an exact answer and a one-decimal-place approximation.

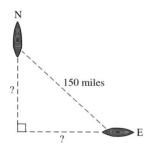

N

150 miles

?

?

E

(9.2) *If the discriminant of a quadratic equation has the given value, determine the number and type of solutions of the equation.*

14. -8 **15.** 48

16. 100 **17.** 0

Use the graph of the following quadratic equations to determine the number and type of solutions of the related equation.

18. 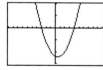   **19.**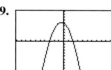

20. 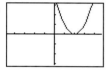   **21.**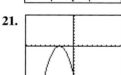

22. 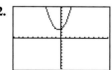   **23.**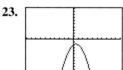

Solve by using the quadratic formula.

24. $x^2 - 16x + 64 = 0$ **25.** $x^2 + 5x = 0$

25. $x^2 + 11 = 0$ **27.** $2x^2 + 3x = 5$

28. $6x^2 + 7 = 5x$ **29.** $9a^2 + 4 = 2a$

30. $(5a - 2)^2 - a = 0$

31. $(2x - 3)^2 = x$

32. Cadets graduating from military school usually toss their hats high into the air at the end of the ceremony. One cadet threw his hat so that its dis-

tance $d(t)$ in feet above the ground t seconds after it was thrown was $d(t) = -16t^2 + 30t + 6$.

a. Find the distance above the ground of the hat 1 second after it was thrown.

b. Find the time it takes the hat to hit the ground. Give an exact time and a one-decimal-place approximation.

33. The hypotenuse of an isosceles right triangle is 6 centimeters longer than either of the legs. Find the length of the legs.

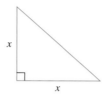

x

x

(9.3) *Solve each equation for the variable.*

34. $x^3 = 27$ **35.** $y^3 = -64$

36. $\dfrac{5}{x} + \dfrac{6}{x - 2} = 3$ **37.** $\dfrac{7}{8} = \dfrac{8}{x^2}$

38. $x^4 - 21x^2 - 100 = 0$

39. $5(x + 3)^2 - 19(x + 3) = 4$

40. $x^{2/3} - 6x^{1/3} + 5 = 0$ **41.** $x^{2/3} - 6x^{1/3} = -8$

42. $a^6 - a^2 = a^4 - 1$ **43.** $y^{-2} + y^{-1} = 20$

44. Two postal workers, Al MacDonell and Tim Bozik, can sort a stack of mail in 5 hours. Working alone, Tim can sort the mail in 1 hour less time than Al can. Find the time that each postal worker can sort the mail alone. Round the result to one decimal place.

45. A negative number decreased by its reciprocal is $-\dfrac{24}{5}$. Find the number.

46. The formula $A = P(1 + r)^2$ gives the amount A in an account paying interest rate r compounded annually after 2 years of P dollars were originally invested. Find the interest rate r such that $2500 increases to $2717 in 2 years. Round the result to the nearest hundredth of a percent.

(9.4) *Solve each inequality for x. Write each solution set in interval notation.*

47. $2x^2 - 50 \le 0$ **48.** $\dfrac{1}{4}x^2 < \dfrac{1}{16}$

49. $(2x - 3)(4x + 5) \geq 0$

50. $(x^2 - 16)(x^2 - 1) > 0$

51. $\dfrac{x - 5}{x - 6} < 0$ **52.** $\dfrac{x(x + 5)}{4x - 3} \geq 0$

53. $\dfrac{(4x + 3)(x - 5)}{x(x + 6)} > 0$

54. $(x + 5)(x - 6)(x + 2) \leq 0$

55. $x^3 + 3x^2 - 25x - 75 > 0$

56. $\dfrac{x^2 + 4}{3x} \leq 1$ **57.** $\dfrac{(5x + 6)(x - 3)}{x(6x - 5)} < 0$

58. $\dfrac{3}{x - 2} > 2$

(9.5) *Sketch the graph of each function. Label the vertex and the axis of symmetry.*

59. $f(x) = x^2 - 4$ **60.** $g(x) = x^2 + 7$

61. $H(x) = 2x^2$ **62.** $h(x) = -\dfrac{1}{3}x^2$

63. $F(x) = (x - 1)^2$ **64.** $G(x) = (x + 5)^2$

65. $f(x) = (x - 4)^2 - 2$

66. $f(x) = -3(x - 1)^2 + 1$

(9.6) *Use a graphing utility to graph each function. Find the vertex and the intercepts algebraically.*

67. $f(x) = x^2 + 10x + 25$ **68.** $f(x) = -x^2 + 6x - 9$

69. $f(x) = 4x^2 - 1$ **70.** $f(x) = -5x^2 + 5$

71. Use a graphing utility to graph the equation $f(x) = -3x^2 - 5x + 4$. Determine the vertex and

whether the graph opens upward or downward. Find the y-intercept, and approximate the x-intercepts to one decimal place.

72. The function $h(t) = -16t^2 + 120t + 300$ gives the height in feet of a projectile fired from the top of a building in t seconds.

    a. When will the object reach a height of 350 feet? Round your answer to one decimal place.

    b. Explain why part (a) has two answers.

73. Find two numbers whose product is as large as possible, given that their sum is 420.

74. Write an equation of a quadratic function whose graph is a parabola that has vertex $(-3, 7)$ and that passes through the origin.

(9.7)

75. According to the U.S. Bureau of the Census, Statistical Abstract of the United States, 1994, the U.S. population in millions is as given in the table below.

YEAR	1970	1980	1990	1993
Population (in millions)	203,302	226,548	248,710	257,908

    a. Plot the data using the statistical plotting feature of your graphing utility.

    b. Use the data given for the years 1980 and 1993 to find a linear equation that models the data. Graph this line with the data.

    c. Use this equation to predict the population of the United States in the year 2000.

CHAPTER 9 TEST

Solve each equation for the variable. Check each equation numerically or graphically.

1. $5x^2 - 2x = 7$ **2.** $(x + 1)^2 = 10$

3. $m^2 - m + 8 = 0$ **4.** $u^2 - 6u + 2 = 0$

5. $7x^2 + 8x + 1 = 0$

6. $a^2 - 3a = 5$

7. $\dfrac{4}{x + 2} + \dfrac{2x}{x - 2} = \dfrac{6}{x^2 - 4}$

8. $x^4 - 8x^2 - 9 = 0$

9. $x^6 + 1 = x^4 + x^2$

10. $(x + 1)^2 - 15(x + 1) + 56 = 0$

Solve the equation for the variable by completing the square.

11. $x^2 - 6x = -2$

12. $2a^2 + 5 = 4a$

Solve each inequality for x. Write the solution set in interval notation.

13. $2x^2 - 7x > 15$

14. $(x^2 - 16)(x^2 - 25) > 0$

15. $\dfrac{5}{x + 3} < 1$

16. $\dfrac{7x - 14}{x^2 - 9} \leq 0$

Sketch the graph of each function. Label the vertex.

17. $f(x) = 3x^2$

18. $G(x) = -2(x - 1)^2 + 5$

Graph each function. Find and label the vertex, y-intercept, and x-intercepts (if any) algebraically.

19. $h(x) = x^2 - 4x + 4$

20. $F(x) = 2x^2 - 8x + 9$

21. A 10-foot ladder is leaning against a house. The distance from the bottom of the ladder to the house is 4 feet less than the distance from the top of the ladder to the ground. Find how far the top of the ladder is from the ground. Give an exact answer and a one-decimal-place approximation.

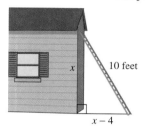

22. Dave and Sandy Hartranft can paint a room together in 4 hours. Working alone, Dave can paint the room in 2 hours less time than Sandy can. Find how long it takes Sandy to paint the room alone.

23. A stone is thrown upward from a bridge. The stone's height in feet, $s(t)$, above the water t seconds after the stone is thrown is a function given by the equation $s(t) = -16t^2 + 32t + 256$.

   a. Find the maximum height of the stone.

   b. Find the time it takes the stone to hit the water.

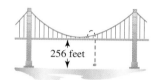

256 feet

24. According to the U.S. Bureau of the Census, Statistical Abstract of the United States, 1994, the population of the state of Florida in millions is as given in the table below.

YEAR	1970	1980	1990	1993
Population (in millions)	6791	9746	12,938	13,679

   a. Plot the data using the statistical plotting feature of your graphing utility.

   b. Use the data given for the years 1980 and 1993 to find a linear equation that models the data. Graph this line with the data.

   c. Use this equation to predict the population of Florida in the year 2000.

CHAPTER 9 CUMULATIVE REVIEW

1. Simplify each expression.

   a. $3xy - 2xy + 5 - 7 + xy$

   b. $7x^2 + 3 - 5(x^2 - 4)$

   c. $(2.1x - 5.6) - (-x - 5.3)$

2. Solve for x: $3x + 5 = 3(x + 2)$.

3. If $f(x) = 7x^2 - 3x + 1$, $g(x) = 3x - 2$, and $h(x) = x^2$, find the following.

   a. $f(1)$ **b.** $g(3)$ **c.** $h(-2)$

4. Solve $|2x| + 5 = 7$.

5. Simplify each expression. Write answers with positive exponents.

   a. $\dfrac{x^{-9}}{x^2}$ **b.** $\dfrac{p^4}{p^{-3}}$ **c.** $\dfrac{2^{-3}}{2^{-1}}$

   d. $\dfrac{2x^{-7}y^2}{10xy^{-5}}$ **e.** $\dfrac{(3x^{-3})(x^2)}{x^6}$

6. Simplify by combining like terms.

   a. $-12x^2 + 7x^2 - 6x$

   b. $3xy - 2x + 5xy - x$

7. Factor $16x^2 + 24xy + 9y^2$.

8. Solve $x^3 + 5x^2 = x + 5$.

9. Write each rational expression as an equivalent rational expression with the given denominator.

   a. $\dfrac{3x}{2y}$, denominator $10xy^3$

   b. $\dfrac{3x + 1}{x - 5}$, denominator $2x^2 - 11x + 5$

10. If $f(x) = x - 1$ and $g(x) = 2x - 3$, find the following.

    a. $(f + g)(x)$ b. $(f - g)(x)$

    c. $(f \cdot g)(x)$ d. $\left(\dfrac{f}{g}\right)(x)$

11. Perform the indicated operation.

    a. $\dfrac{x}{x - 3} - 5$ b. $\dfrac{7}{x - y} + \dfrac{3}{y - x}$

12. Divide $2x^2 - x - 10$ by $x + 2$.

13. Use synthetic division to divide $2x^3 - x^2 - 13x + 1$ by $x - 3$.

14. Simplify the following expressions.

    a. $\sqrt[4]{81}$ b. $\sqrt[5]{-243}$

    c. $-\sqrt{25}$ d. $\sqrt[4]{-81}$

    e. $\sqrt[3]{64x^3}$

15. Write with rational exponents.

    a. $\sqrt[5]{x}$ b. $\sqrt[3]{17x^2y^5}$

    c. $\sqrt{x - 5a}$

    d. $3\sqrt{2p} - 5\sqrt[3]{p^2}$

16. Simplify.

    a. $\dfrac{\sqrt{45}}{4} - \dfrac{\sqrt{5}}{3}$

    b. $\sqrt[3]{\dfrac{7x}{8}} + 2\sqrt[3]{7x}$

17. Rationalize the numerator of $\dfrac{\sqrt{x} + 2}{5}$.

18. Solve $\sqrt{2x + 5} + \sqrt{2x} = 3$.

19. A 50-foot supporting wire is to be attached to a 75-foot antenna. Because of surrounding buildings, sidewalks, and roadways, the wire must be anchored exactly 20 feet from the base of the antenna.

    a. How high from the base of the antenna is the wire attached?

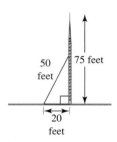

    b. Local regulations require that a supporting wire be attached at a height no less than $\frac{3}{5}$ of the total height of the antenna. From part (a), have local regulations been met?

20. Use the square root property to solve $(x + 1)^2 = 12$ for x. Use a graphing utility to check.

21. Solve $2x^2 - 4x = 3$.

22. Solve $p^4 - 3p^2 - 4 = 0$.

23. Solve $\dfrac{x + 2}{x - 3} \leq 0$.

24. Sketch the graph of $F(x) = (x - 3)^2 + 1$. Label the vertex and the axis of symmetry.

25. A rock is propelled upward from the ground. Its height in feet above ground after t seconds is given by the equation $f(x) = -16x^2 + 20x$. Find the maximum height of the rock and the number of seconds it took for the rock to reach that maximum height.

CONIC SECTIONS

MODELING CONIC SECTIONS

The shapes of conic sections are used in a variety of applications. They are used in architecture in the design of bridges, arches, and vaults. They are used in astronomy to model the orbits of planets, comets, and satellites. They are used also in engineering in the design of certain gears and reflectors. In blueprints or diagrams of these situations, the exact shape of the conic section involved must be depicted.

IN THE CHAPTER GROUP ACTIVITY ON PAGE 615, YOU WILL HAVE THE OPPORTUNITY TO DRAW THE SHAPES OF SEVERAL CONIC SECTIONS.

I n Chapter 9, we analyzed some of the important connections between a parabola and its equation. Parabolas are interesting in their own right, but are more interesting still because they are part of a collection of curves known as conic sections. This chapter is devoted to quadratic equations in two variables and their conic section graphs: the parabola, circle, ellipse, and hyperbola. ▬▬▬

10.1 | THE PARABOLA AND THE CIRCLE

O B J E C T I V E S

1. Graph parabolas of the forms $x = a(y - k)^2 + h$ and $y = a(x - h)^2 + k$.
2. Graph circles of the form $(x - h)^2 + (y - k)^2 = r^2$.
3. Write the equation of a circle given its center and radius.
4. Find the center and the radius of a circle, given its equation.

TAPE IAG 10.1

Conic sections are so named because each conic section is the intersection of a right circular cone and a plane. The circle, parabola, ellipse, and hyperbola are the conic sections.

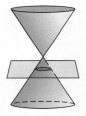 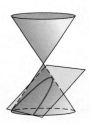 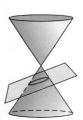 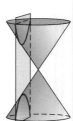

Circle   Parabola   Ellipse   Hyperbola

1 Thus far, we have seen that $f(x)$ or $y = a(x - h)^2 + k$ is the equation of a parabola that opens upward if $a > 0$ or downward if $a < 0$. Parabolas can also open left or right, or even on a slant. Equations of these parabolas are not functions of x, of course, since a parabola opening any way other than upward or downward fails the vertical line test. In this section, we introduce parabolas that open to the left and to the right. Parabolas opening on a slant will not be developed in this book.

Just as $y = a(x - h)^2 + k$ is the equation of a parabola that opens upward or downward, $x = a(y - k)^2 + h$ is the equation of a parabola that opens to the right or to the left. The parabola opens to the right if $a > 0$ and to the left if $a < 0$. The parabola has vertex (h, k), and its axis of symmetry is the line $y = k$.

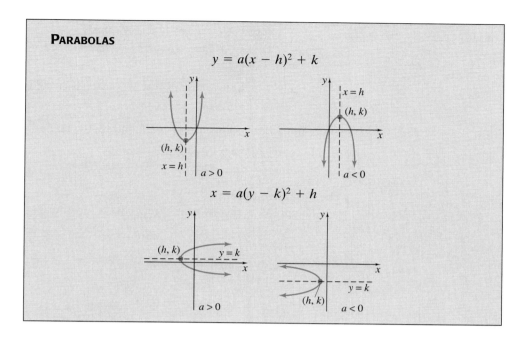

PARABOLAS

$$y = a(x - h)^2 + k$$

$$x = a(y - k)^2 + h$$

The forms $y = a(x - h)^2 + k$ and $x = a(y - k)^2 + h$ are called **standard forms**.

In this section, we graph parabolas and circles by hand and by using a graphing utility. Often a hand-drawn sketch is sufficient, but sometimes an exact graph drawn by a graphing utility is needed to further analyze the graph.

Let's first recall how we sketch the graph of a parabola opening upward or downward.

EXAMPLE 1   Find the vertex and the intercepts of the parabola described by $y = -x^2 - 2x + 15$. Sketch the graph.

Solution: In this equation, $a = -1$, $b = -2$, and $c = 15$. Since $a < 0$, this parabola opens downward. To find the vertex, first find $-\dfrac{b}{2a}$, the x-value of the vertex.

$$x = -\frac{b}{2a} = -\frac{-2}{2(-1)} = -1$$

The corresponding y-value is

$$y = -(-1)^2 - 2(-1) + 15 = 16$$

The vertex is $(-1, 16)$. Since the vertex is in the second quadrant and the parabola opens downward, the graph will have two x-intercepts and one y-intercept.

To find the x-intercepts, let $y = 0$.

$$0 = -x^2 - 2x + 15$$
$$0 = x^2 + 2x - 15 \qquad \text{Divide both sides by } -1.$$
$$0 = (x + 5)(x - 3) \qquad \text{Factor.}$$
$$x + 5 = 0 \quad \text{or} \quad x - 3 = 0 \qquad \text{Set each factor equal to 0.}$$
$$x = -5 \quad \text{or} \qquad x = 3 \qquad \text{Solve.}$$

The x-intercepts points are $(-5, 0)$, and $(3, 0)$. To find the y-intercept, let $x = 0$.

$$y = -0^2 - 2 \cdot 0 + 15$$
$$= 15$$

The y-intercept point is $(0, 15)$.

The graph is shown next, along with a few more ordered pair solutions.

x	y
-1	16
0	15
-2	15
1	12
-3	12
3	0
-5	0

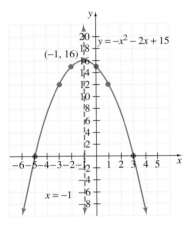

Notice that this graph is the graph of a function since it passes the vertical line test. To check using a graphing utility, graph $y_1 = -x^2 - 2x + 15$.

EXAMPLE 2 Sketch the parabola whose equation is $x = 2y^2$. Then use a graphing utility to check.

Solution: Written in standard form, the equation $x = 2y^2$ is $x = 2(y - 0)^2 + 0$ with $a = 2, h = 0$, and $k = 0$. Its graph is a parabola with vertex $(0, 0)$, and its axis of symmetry is the line $y = 0$. Since $a > 0$, this parabola opens to the right. The table shows a few more ordered pair solutions of $x = 2y^2$. Its graph is also shown.

x	y
8	-2
2	-1
0	0
2	1
8	2

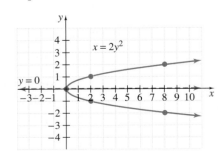

Notice that this parabola does not pass the vertical line test and is not the graph of a function. This means that we cannot enter a single function into the Y= editor to graph $x = 2y^2$. Instead, we solve for y and enter two functions separately, as shown below.

To graph $x = 2y^2$ using a graphing utility, solve the equation for y.

$$x = 2y^2$$

$$\frac{x}{2} = y^2$$

$$\pm\sqrt{\frac{x}{2}} = y$$

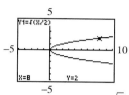

The graph of $y_1 = \sqrt{\dfrac{x}{2}}$ and $y_2 = -y_1$.

Both $y = \sqrt{\dfrac{x}{2}}$ and $y = -\sqrt{\dfrac{x}{2}}$ describe functions. Graph $y_1 = \sqrt{\dfrac{x}{2}}$ and $y_2 = -\sqrt{\dfrac{x}{2}}$ (or $y_1 = \sqrt{\dfrac{x}{2}}$ and $y_2 = -y_1$). The two graphs together form the graph of $x = 2y^2$, as shown to the left.

EXAMPLE 3   Graph the parabola $x = -3(y - 1)^2 + 2$.

Solution: The equation $x = -3(y - 1)^2 + 2$ is in the form $x = a(y - k)^2 + h$ with $a = -3$, $k = 1$, and $h = 2$. Since this is written in standard form, we sketch by hand. Since $a < 0$, the parabola opens to the left. The vertex (h, k) is $(2, 1)$, and the axis of symmetry is the line $y = 1$. When $y = 0$, $x = -1$, the x-intercept.

The parabola is sketched next. Also, a table containing a few ordered pair solutions is shown.

x	y
2	1
-1	0
-1	2

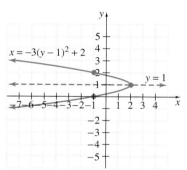

EXAMPLE 4   Find the vertex and the intercepts of the parabola described by $x = 2y^2 + 4y + 5$. Sketch the graph.

Solution: In this equation, $a = 2$, $b = 4$, and $c = 5$. Since $a > 0$, the parabola opens to the right. When a parabola opens to the left or to the right, the y-value of the vertex can be found by using the formula $-\dfrac{b}{2a}$.

$$y = -\frac{b}{2a} = -\frac{4}{2(2)} = -1$$

The corresponding x-value is

$$x = 2(-1)^2 + 4(-1) + 5 = 3$$

The vertex is $(3, -1)$. Since the vertex is in the first quadrant and opens to the right, there is one x-intercept and no y-intercepts. Let $y = 0$ and find that

$$x = 2 \cdot 0^2 + 4 \cdot 0 + 5 = 5$$

The x-intercept point is $(5, 0)$ and the graph is shown below.

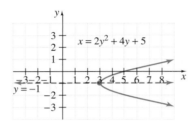

EXAMPLE 5 Use a graphing utility to graph $x = -y^2 - 4y - 1$.

Solution. Solve the equation for y. To do so, write the equation so that one side is 0 and then use the quadratic formula.

$$x = -y^2 - 4y - 1$$
$$y^2 + 4y + (x + 1) = 0$$

Since we are solving for y, $a = 1$, $b = 4$, and $c = x + 1$. By the quadratic formula,

$$y = \frac{-b \pm \sqrt{b^2 - 4ac}}{2a}$$

or

$$y = \frac{-4 \pm \sqrt{4^2 - 4(1)(x + 1)}}{2(1)}.$$

To enter these expressions on a graphing utility, it often is convenient to store the expression under the radical, $4^2 - 4(1)(x + 1)$, in y_1. Then deselect y_1, but use it to form y_2 and y_3 as shown below.

$$y_2 = \frac{-4 + \sqrt{y_1}}{2}$$

$$y_3 = \frac{-4 - \sqrt{y_1}}{2}$$

Enter y_2 and y_3 and graph these two equations in the standard window as shown below. (Don't forget to deselect y_1 so that its graph is not shown. The purpose of entering y_1 is to simplify the expressions for y_2 and y_3.)

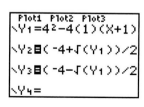

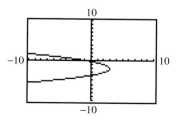

2

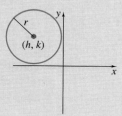

Another conic section is the **circle.** A circle is the set of all points in a plane that are the same distance from a fixed point called the **center.** The distance is called the **radius** of the circle. To find a standard equation for a circle, let (h, k) represent the center of the circle, and let (x, y) represent any point on the circle. The distance between (h, k) and (x, y) is defined to be the radius, r units. We can find this distance r by using the distance formula.

$$r = \sqrt{(x - h)^2 + (y - k)^2}$$
$$r^2 = (x - h)^2 + (y - k)^2 \qquad \text{Square both sides.}$$

CIRCLE

The graph of $(x - h)^2 + (y - k)^2 = r^2$ is a circle with center (h,k) and radius r.

The form $(x - h)^2 + (y - k)^2 = r^2$ is called **standard form.**
    If an equation can be written in the standard form

$$(x - h)^2 + (y - k)^2 = r^2$$

then its graph is a circle, which we can sketch by graphing the center (h, k) and using the radius r.

EXAMPLE 6 Graph $x^2 + y^2 = 4$.

Solution: The equation can be written in standard form as

$$(x - 0)^2 + (y - 0)^2 = 2^2$$

The center of the circle is $(0, 0)$, and the radius is 2. Its graph is shown at left.

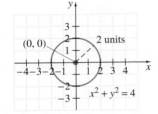

REMINDER Notice the difference between the equation of a circle and the equation of a parabola. The equation of a circle contains both x^2 and y^2 terms with equal coefficients. The equation of a parabola has either an x^2 term or a y^2 term but not both.

EXAMPLE 7 Graph $(x + 1)^2 + y^2 = 8$.

Solution: The equation can be written as $(x + 1)^2 + (y - 0)^2 = 8$ with $h = -1$, $k = 0$, and $r = \sqrt{8}$. The center is $(-1, 0)$, and the radius is $\sqrt{8} = 2\sqrt{2} \approx 2.8$.

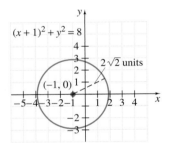

Just as for some parabolas, circles do not pass the vertical line test and are not functions. To graph a circle using a graphing utility, we solve for y and enter two functions separately, as shown below.

EXAMPLE 8 Use a graphing utility to graph $(x + 1)^2 + y^2 = 8$.

Solution: Solve the equation for y.

$$(x + 1)^2 + y^2 = 8$$
$$y^2 = 8 - (x + 1)^2$$
$$y = \pm\sqrt{8 - (x + 1)^2}$$

For the graph to appear circular, a square window must be used. Recall that a square window is one whose tick marks are equally spaced on the x- and y-axes.

Graph $y_1 = \sqrt{8 - (x + 1)^2}$ and $y_2 = -y_1$ in a decimal window as shown below.

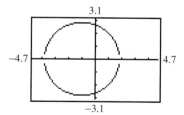

3 Since a circle is determined entirely by its center and radius, this information is all we need to write the equation of a circle.

EXAMPLE 9 Find an equation of the circle with the center $(-7, 3)$ and radius 10.

Solution: We are given that $h = -7$, $k = 3$, and $r = 10$. Using these values, we write the equation

$$(x - h)^2 + (y - k)^2 = r^2$$

or

$$[x - (-7)]^2 + (y - 3)^2 = 10^2$$

or

$$(x + 7)^2 + (y - 3)^2 = 100$$

4 To find the center and the radius of a circle from its equation, write the equation in standard form. To write the equation of a circle in standard form, we complete the square on both x and y.

EXAMPLE 10 Graph $x^2 + y^2 + 4x - 8y = 16$.

Solution: Since this equation contains x^2 and y^2 terms on the same side of the equation with equal coefficients, its graph is a circle. To write the equation in standard form, group the terms involving x and the terms involving y, and then complete the square on each variable.

$$(x^2 + 4x) + (y^2 - 8y) = 16$$

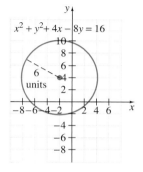

Thus, $\frac{1}{2}(4) = 2$ and $2^2 = 4$. Also, $\frac{1}{2}(-8) = -4$ and $(-4)^2 = 16$. Add 4 and then 16 to both sides.

$$(x^2 + 4x + 4) + (y^2 - 8y + 16) = 16 + 4 + 16$$
$$(x + 2)^2 + (y - 4)^2 = 36 \qquad \text{Factor.}$$

This circle has the center $(-2, 4)$ and radius 6, as shown.

MENTAL MATH

The graph of each equation is a parabola. Determine whether the parabola opens upward, downward, to the left, or to the right.

1. $y = x^2 - 7x + 5$

2. $y = -x^2 + 16$

3. $x = -y^2 - y + 2$

4. $x = 3y^2 + 2y - 5$

5. $y = -x^2 + 2x + 1$

6. $x = -y^2 + 2y - 6$

EXERCISE SET 10.1

The graph of each equation is a parabola. Find the vertex of the parabola and any intercepts. Then sketch its graph. See Examples 1 through 4.

1. $x = 3y^2$

2. $x = -2y^2$

3. $x = (y - 2)^2 + 3$

4. $x = (y - 4)^2 - 1$

5. $y = 3(x - 1)^2 + 5$

6. $x = -4(y - 2)^2 + 2$

7. $x = y^2 + 6y + 8$

8. $x = y^2 - 6y + 6$

9. $y = x^2 + 10x + 20$

10. $y = x^2 + 4x - 5$

11. $x = -2y^2 + 4y + 6$

12. $x = 3y^2 + 6y + 7$

The graph of each equation is a circle. Find the center and the radius, and then sketch. See Examples 6, 7, and 10.

13. $x^2 + y^2 = 9$

14. $x^2 + y^2 = 25$

15. $x^2 + (y - 2)^2 = 1$

16. $(x - 3)^2 + y^2 = 9$

17. $(x - 5)^2 + (y + 2)^2 = 1$

18. $(x + 3)^2 + (y + 3)^2 = 4$

19. $x^2 + y^2 + 6y = 0$

20. $x^2 + 10x + y^2 = 0$

21. $x^2 + y^2 + 2x - 4y = 4$

22. $x^2 + 6x - 4y + y^2 = 3$

Use a graphing utility to graph each conic section. If the conic section is a circle, graph using a square window. See Examples 5 and 8.

23. $x^2 + y^2 = 55$

24. $x^2 + y^2 = 20$

25. $x = 2 - 2y - y^2$

26. $x = -7 - 6y - y^2$

27. $(x - 4)^2 + (y + 2)^2 = 10$

28. $(x + 3)^2 + (y - 1)^2 = 15$

29. $x = 4y^2 - 16y + 11$

30. $x = 9y^2 - 6y + 4$

Write an equation of the circle with the given center and radius. See Example 9.

31. $(2, 3)$; 6

32. $(-7, 6)$; 2

33. $(0, 0)$; $\sqrt{3}$

34. $(0, -6)$; $\sqrt{2}$

35. $(-5, 4)$; $3\sqrt{5}$

36. the origin; $4\sqrt{7}$

37. If you are given a list of equations of circles and parabolas and none are in standard form, explain how you would determine which is an equation of a circle and which is an equation of a parabola. Explain also how you would distinguish the upward or downward parabolas from the left-opening or right-opening parabolas.

Sketch the graph of each equation. If the graph is a parabola, find its vertex. If the graph is a circle, find its center and radius.

38. $x = y^2 + 2$

39. $x = y^2 - 3$

40. $y = (x + 3)^2 + 3$

41. $y = (x - 2)^2 - 2$

42. $x^2 + y^2 = 49$

43. $x^2 + y^2 = 1$

44. $x = (y - 1)^2 + 4$

45. $x = (y + 3)^2 - 1$

46. $(x + 3)^2 + (y - 1)^2 = 9$

47. $(x - 2)^2 + (y - 2)^2 = 16$

48. $x = -2(y + 5)^2$

49. $x = -(y - 1)^2$

50. $x^2 + (y + 5)^2 = 5$

51. $(x - 4)^2 + y^2 = 7$

52. $y = 3(x - 4)^2 + 2$

53. $y = 5(x + 5)^2 + 3$

54. $2x^2 + 2y^2 = \dfrac{1}{2}$

55. $\dfrac{x^2}{8} + \dfrac{y^2}{8} = 2$

56. $y = x^2 - 2x - 15$

57. $y = x^2 + 7x + 6$

58. $x^2 + y^2 + 6x + 10y - 2 = 0$

59. $x^2 + y^2 + 2x + 12y - 12 = 0$

60. $x = y^2 + 6y + 2$ **61.** $x = y^2 + 8y - 4$

62. $x^2 + y^2 - 8y + 5 = 0$

63. $x^2 - 10y + y^2 + 4 = 0$

64. $x = -2y^2 - 4y$ **65.** $x = -3y^2 + 30y$

66. $\dfrac{x^2}{3} + \dfrac{y^2}{3} = 2$ **67.** $5x^2 + 5y^2 = 25$

68. $y = 4x^2 - 40x + 105$

69. $y = 5x^2 - 20x + 16$

Solve.

70. Two surveyors need to find the distance across a lake. They place a reference pole at point *A* in the diagram. Point *B* is 3 meters east and 1 meter north of the reference point *A*. Point *C* is 19 meters east and 13 meters north of point *A*. Find the distance across the lake, from *B* to *C*.

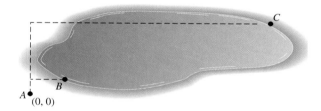

A •
(0, 0)

71. Determine whether the triangle with vertices $(2, 6)$, $(0, -2)$, and $(5, 1)$ is an isosceles triangle.

72. Cindy Brown, an architect, is drawing plans on grid paper for a circular pool with a fountain in the middle. The paper is marked off in centimeters, and each centimeter represents 1 foot. On the paper, the diameter of the "pool" is 20 centimeters, and the "fountain" is the point $(0, 0)$.

a. Sketch the architect's drawing. Be sure to label the axes.

b. Write an equation that describes the circular pool.

c. Cindy plans to place a circle of lights around the fountain such that each light is 5 feet from the fountain. Write an equation for the circle of lights and sketch the circle on your drawing.

73. A bridge constructed over a bayou has a supporting arch in the shape of a parabola. Find an equation of the parabolic arch if the length of the road over the arch is 100 meters and the maximum height of the arch is 40 meters.

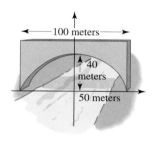

Review Exercises

Graph each equation. See Section 3.4.

74. $y = 2x + 5$ **75.** $y = -3x + 3$

76. $y = 3$ **77.** $x = -2$

Rationalize each denominator and simplify if possible. See Section 8.5

78. $\dfrac{1}{\sqrt{3}}$ **79.** $\dfrac{\sqrt{5}}{\sqrt{8}}$

80. $\dfrac{4\sqrt{7}}{\sqrt{6}}$ **81.** $\dfrac{10}{\sqrt{5}}$

10.2   THE ELLIPSE AND THE HYPERBOLA

TAPE IAG 10.2

O B J E C T I V E S

1 Define and graph an ellipse.

2 Define and graph a hyperbola.

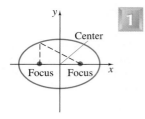

An **ellipse** can be thought of as the set of points in a plane such that the sum of the distances of those points from two fixed points is constant. Each of the two fixed points is called a **focus.** The plural of focus is **foci.** The point midway between the foci is called the **center.**

An ellipse may be drawn by hand by using two tacks, a piece of string, and a pencil. Secure the two tacks into a piece of cardboard, for example, and tie each end of the string to a tack. Use your pencil to pull the string tight and draw the ellipse. The two tacks are the foci of the drawn ellipse.

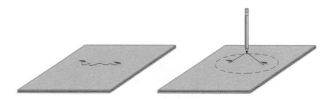

It can be shown that the **standard form** of an ellipse with center $(0, 0)$ is
$$\frac{x^2}{a^2} + \frac{y^2}{b^2} = 1.$$

ELLIPSE WITH CENTER (0, 0)

The graph of an equation of the form $\dfrac{x^2}{a^2} + \dfrac{y^2}{b^2} = 1$ is an ellipse with center $(0, 0)$. The x-intercepts are a and $-a$, and the y-intercepts are b and $-b$.

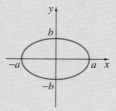

EXAMPLE 1   Graph $\dfrac{x^2}{9} + \dfrac{y^2}{16} = 1$.

Solution: The equation is of the form $\dfrac{x^2}{a^2} + \dfrac{y^2}{b^2} = 1$, with $a = 3$ and $b = 4$, so its graph is an ellipse with center $(0, 0)$, x-intercepts 3 and -3, and y-intercepts 4 and -4, as graphed next.

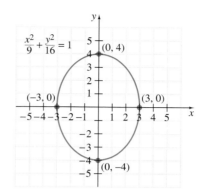

EXAMPLE 2   Graph the equation $4x^2 + 16y^2 = 64$. Use a graphing utility to check.

Solution: Although this equation contains a sum of squared terms in x and y on the same side of an equation, this is not the equation of a circle since the coefficients of x^2 and y^2 are not the same. When this happens, the graph is an ellipse. Since the standard form of the equation of an ellipse has 1 on one side, divide both sides of this equation by 64.

$$4x^2 + 16y^2 = 64$$

$$\frac{4x^2}{64} + \frac{16y^2}{64} = \frac{64}{64} \qquad \text{Divide both sides by 64.}$$

$$\frac{x^2}{16} + \frac{y^2}{4} = 1 \qquad \text{Simplify.}$$

We now recognize the equation of an ellipse with center $(0, 0)$, x-intercepts 4 and -4, and y-intercepts 2 and -2.

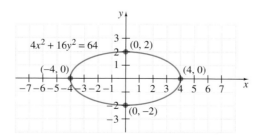

To check using a graphing utility, solve the equation for y.

$$4x^2 + 16y^2 = 64$$

$$16y^2 = 64 - 4x^2 \qquad \text{Subtract } 4x^2 \text{ from both sides.}$$

$$y^2 = \frac{64 - 4x^2}{16} \qquad \text{Divide both sides by 16.}$$

$$y = \pm\sqrt{\frac{64 - 4x^2}{16}} \qquad \text{Solve for } y.$$

$$y = \pm\frac{\sqrt{64 - 4x^2}}{4} \qquad \text{Simplify.}$$

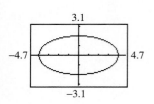

Graph $y_1 = \dfrac{\sqrt{64 - 4x^2}}{4}$ and $y_2 = -y_1$. The graph is shown to the left.

The center of an ellipse is not always $(0, 0)$, as shown in the next example.

EXAMPLE 3   Graph $\dfrac{(x + 3)^2}{25} + \dfrac{(y - 2)^2}{36} = 1$.

Solution: This ellipse has center $(-3, 2)$. Notice that $a = 5$ and $b = 6$. To find four points on the graph of the ellipse, first graph the center, $(-3, 2)$. Since $a = 5$, count 5 units right and then 5 units left of the point with coordinates $(-3, 2)$. Next, since $b = 6$, start at $(-3, 2)$ and count 6 units up and then 6 units down to find two more points on the ellipse.

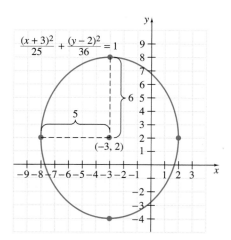

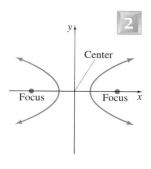

The final conic section discussed is the **hyperbola.** A hyperbola is the set of points in a plane such that the absolute value of the difference of the distance from two fixed points is constant. Each of the two fixed points is called a **focus.** The point midway between the foci is called the **center.**

Using the distance formula, we can show that the graph of $\dfrac{x^2}{a^2} - \dfrac{y^2}{b^2} = 1$ is a hyperbola with center $(0, 0)$ and x-intercepts a and $-a$. Also, the graph of $\dfrac{y^2}{b^2} - \dfrac{x^2}{a^2} = 1$ is a hyperbola with center $(0, 0)$ and y-intercepts b and $-b$. These equations are called **standard form** equations for a hyperbola.

HYPERBOLA WITH CENTER (0, 0)

The graph of an equation of the form $\dfrac{x^2}{a^2} - \dfrac{y^2}{b^2} = 1$ is a hyperbola with center $(0, 0)$ and x-intercepts a and $-a$.

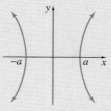

The graph of an equation of the form $\dfrac{y^2}{b^2} - \dfrac{x^2}{a^2} = 1$ is a hyperbola with center $(0, 0)$ and y-intercepts b and $-b$.

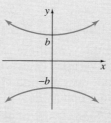

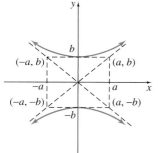

Graphing a hyperbola such as $\dfrac{y^2}{b^2} - \dfrac{x^2}{a^2} = 1$ is made easier by recognizing one of its important characteristics. Examining the figure to the left, notice how the sides of the branches of the hyperbola extend indefinitely and seem to approach the dashed lines in the figure. These dashed lines are called the **asymptotes** of the hyperbola.

To sketch these lines, or asymptotes, draw a rectangle with vertices (a, b), $(-a, b)$, $(a, -b)$, and $(-a, -b)$. The asymptotes of the hyperbola are the extended diagonals of this rectangle.

EXAMPLE 4   Sketch the graph of $\dfrac{x^2}{16} - \dfrac{y^2}{25} = 1$.

Solution: This equation has the form $\dfrac{x^2}{a^2} - \dfrac{y^2}{b^2} = 1$, with $a = 4$ and $b = 5$. Thus, its graph is a hyperbola with center $(0, 0)$ and x-intercepts 4 and -4. To aid in graphing the hyperbola, we first sketch its asymptotes. The extended diagonals of the rectangle with coordinates $(4, 5)$, $(4, -5)$, $(-4, 5)$, and $(-4, -5)$ are the asymptotes of the hyperbola. Then use the asymptotes to aid in sketching the hyperbola.

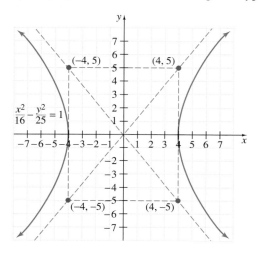

EXAMPLE 5   Sketch the graph of the equation $4y^2 - 9x^2 = 36$. Use a graphing utility to check.

Solution: Since this is a difference of squared terms in x and y on the same side of the equation, its graph is a hyperbola, as opposed to an ellipse or a circle. The standard form of the equation of a hyperbola has a 1 on one side, so divide both sides of the equation by 36.

$$4y^2 - 9x^2 = 36$$
$$\frac{4y^2}{36} - \frac{9x^2}{36} = \frac{36}{36} \qquad \text{Divide both sides by 36.}$$
$$\frac{y^2}{9} - \frac{x^2}{4} = 1 \qquad \text{Simplify.}$$

The equation is of the form $\dfrac{y^2}{b^2} - \dfrac{x^2}{a^2} = 1$, with $a = 2$ and $b = 3$, so the hyperbola is centered at $(0, 0)$ with y-intercepts 3 and -3. The sketch of the hyperbola is shown.

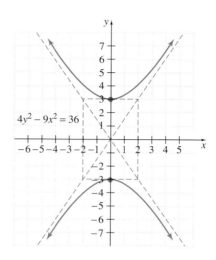

To check, solve the equation for y.

$$4y^2 - 9x^2 = 36$$

$$4y^2 = 36 + 9x^2 \qquad \text{Add } 9x^2 \text{ to both sides.}$$

$$y^2 = \frac{36 + 9x^2}{4} \qquad \text{Divide both sides by 4.}$$

$$y = \pm\sqrt{\frac{36 + 9x^2}{4}} \qquad \text{Solve for } y.$$

$$y = \pm\frac{\sqrt{36 + 9x^2}}{2} \qquad \text{Simplify.}$$

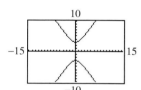

Graph $y_1 = \dfrac{\sqrt{36 + 9x^2}}{2}$ and $y_2 = -y_1$. The graph is shown to the left. ▬▬▬

The following box provides a summary of conic sections.

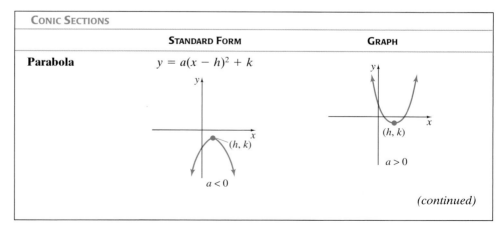

CONIC SECTIONS		
	STANDARD FORM	**GRAPH**
Parabola	$y = a(x - h)^2 + k$	

(continued)

CONIC SECTIONS (CONT.)

	STANDARD FORM	GRAPH
Parabola	$x = a(y - k)^2 + h$	

	STANDARD FORM	GRAPH
Circle	$(x - h)^2 + (y - k)^2 = r^2$	

	STANDARD FORM	GRAPH
Ellipse	$\dfrac{x^2}{a^2} + \dfrac{y^2}{b^2} = 1$	

	STANDARD FORM	GRAPH
Hyperbola	$\dfrac{x^2}{a^2} - \dfrac{y^2}{b^2} = 1$	

	STANDARD FORM	GRAPH
Hyperbola	$\dfrac{y^2}{b^2} - \dfrac{x^2}{a^2} = 1$	

EXERCISE SET 10.2

Sketch the graph of each equation. See Examples 1 and 2.

1. $\dfrac{x^2}{4} + \dfrac{y^2}{25} = 1$

2. $\dfrac{x^2}{9} + y^2 = 1$

3. $\dfrac{x^2}{16} + \dfrac{y^2}{9} = 1$

4. $x^2 + \dfrac{y^2}{4} = 1$

5. $9x^2 + 4y^2 = 36$

6. $x^2 + 4y^2 = 16$

7. $4x^2 + 25y^2 = 100$

8. $36x^2 + y^2 = 36$

Sketch the graph of each equation. See Example 3.

9. $\dfrac{(x+1)^2}{36} + \dfrac{(y-2)^2}{49} = 1$

10. $\dfrac{(x-3)^2}{9} + \dfrac{(y+3)^2}{16} = 1$

11. $\dfrac{(x-1)^2}{4} + \dfrac{(y-1)^2}{25} = 1$

12. $\dfrac{(x+3)^2}{16} + \dfrac{(y+2)^2}{4} = 1$

Sketch the graph of each equation. See Examples 4 and 5.

13. $\dfrac{x^2}{4} - \dfrac{y^2}{9} = 1$

14. $\dfrac{x^2}{36} - \dfrac{y^2}{36} = 1$

15. $\dfrac{y^2}{25} - \dfrac{x^2}{16} = 1$

16. $\dfrac{y^2}{25} - \dfrac{x^2}{49} = 1$

Sketch the graph of each equation. See Example 5.

17. $x^2 - 4y^2 = 16$

18. $4x^2 - y^2 = 36$

19. $16y^2 - x^2 = 16$

20. $4y^2 - 25x^2 = 100$

21. If you are given a list of equations of circles, parabolas, ellipses, and hyperbolas, explain how you could distinguish the different conic sections from their equations.

Identify whether each equation, when graphed, will be a parabola, circle, ellipse, or hyperbola. Sketch the graph of each equation.

22. $(x-7)^2 + (y-2)^2 = 4$

23. $y = x^2 + 4$

24. $y = x^2 + 12x + 36$

25. $\dfrac{x^2}{4} + \dfrac{y^2}{9} = 1$

26. $\dfrac{y^2}{9} - \dfrac{x^2}{9} = 1$

27. $\dfrac{x^2}{16} - \dfrac{y^2}{4} = 1$

28. $\dfrac{x^2}{16} + \dfrac{y^2}{4} = 1$

29. $x^2 + y^2 = 16$

30. $x = y^2 + 4y - 1$

31. $x = -y^2 + 6y$

32. $9x^2 - 4y^2 = 36$

33. $9x^2 + 4y^2 = 36$

34. $\dfrac{(x-1)^2}{49} + \dfrac{(y+2)^2}{25} = 1$ **35.** $y^2 = x^2 + 16$

36. $\left(x + \dfrac{1}{2}\right)^2 + \left(y - \dfrac{1}{2}\right)^2 = 1$

37. $y = -2x^2 + 4x - 3$

Solve.

38. A planet's orbit about the Sun can be described as an ellipse. Consider the Sun as the origin of a rectangular coordinate system. Suppose that the x-intercepts of the elliptical path of the planet are ±130,000,000 and that the y-intercepts are ±125,000,000. Write the equation of the elliptical path of the planet.

39. Comets orbit the Sun in elongated ellipses. Consider the Sun as the origin of a rectangular coordinate system. Suppose that the equation of the path of the comet is

$$\dfrac{(x - 1{,}782{,}000{,}000)^2}{(3.42)(10^{23})} + \dfrac{(y - 356{,}400{,}000)^2}{(1.368)(10^{22})} = 1$$

Find the center of the path of the comet.

Use a graphing utility to graph each conic section.

40. $10x^2 + y^2 = 32$

41. $20x^2 + 5y^2 = 100$

42. $2y^2 - 5x^2 = 10$

43. $7y^2 - 3x^2 = 21$

44. $7.3x^2 + 15.5y^2 = 95.2$

45. $18.8x^2 + 36.1y^2 = 205.8$

46. $4.6x^2 - 3.7y^2 = 70.2$

47. $4.5x^2 - 6.7y^2 = 50.7$

Tell whether the graph of the equation is an ellipse or a hyperbola. Find the coordinates of the center and sketch the graph of each equation.

48. $\dfrac{(x-1)^2}{4} - \dfrac{(y+1)^2}{25} = 1$

49. $\dfrac{(x+2)^2}{9} - \dfrac{(y-1)^2}{4} = 1$

50. $\dfrac{y^2}{16} - \dfrac{(x+3)^2}{9} = 1$ **51.** $\dfrac{(y+4)^2}{4} - \dfrac{x^2}{25} = 1$

52. $\dfrac{(x+5)^2}{16} + \dfrac{(y+2)^2}{25} = 1$

53. $\dfrac{(x-3)^2}{9} + \dfrac{(y-2)^2}{4} = 1$

Review Exercises

Solve each inequality. See Section 4.2.

54. $x < 5$ and $x < 1$

55. $x < 5$ or $x < 1$

56. $2x - 1 \geq 7$ or $-3x \leq -6$

57. $2x - 1 \geq 7$ and $-3x \leq -6$

Perform the indicated operations. See Sections 6.1 and 6.3.

58. $(2x^3)(-4x^2)$

59. $2x^3 - 4x^3$

60. $-5x^2 + x^2$

61. $(-5x^2)(x^2)$

10.3 | SOLVING NONLINEAR SYSTEMS OF EQUATIONS

TAPE IAG 10.3

O B J E C T I V E S

1 Solve a nonlinear system by substitution.

2 Solve a nonlinear system by elimination.

In Section 5.1, we used graphing, substitution, and elimination methods to find solutions of systems of linear equations in two variables. We now apply these same methods to nonlinear systems of equations in two variables. A **nonlinear system of equations** is a system of equations at least one of which is not linear. Since we will be graphing the equations in each system, we are interested in real number solutions only.

 First, nonlinear systems are solved by the substitution method.

EXAMPLE 1   Solve the system

$$\begin{cases} y = \sqrt{x} \\ x^2 + y^2 = 6 \end{cases}$$

Solution: This system is ideal for substitution since y is expressed in terms of x in the first equation. Notice that if $y = \sqrt{x}$, then both x and y must be nonnegative if they are real numbers. Substitute $\sqrt{x}$ for y in the second equation, and solve for x:

$$x^2 + y^2 = 6$$
$$x^2 + (\sqrt{x})^2 = 6 \qquad \text{Let } y = \sqrt{x}.$$
$$x^2 + x = 6$$
$$x^2 + x - 6 = 0$$
$$(x + 3)(x - 2) = 0$$
$$x = -3 \text{ or } x = 2$$

The solution -3 is discarded because we have noted that x must be nonnegative. To see this, let $x = -3$ in the first equation. Then let $x = 2$ in the first equation to find a corresponding y-value.

Let $x = -3$.   Let $x = 2$.

$y = \sqrt{x}$ $y = \sqrt{x}$

$y = \sqrt{-3}$ Not a real number. $y = \sqrt{2}$

Since we are interested in real number solutions only, the only solution is $(2, \sqrt{2})$. The solution set is $\{(2, \sqrt{2})\}$. Check to see that this solution satisfies both equations. To visualize this solution, the graph of each equation in this system is shown next.

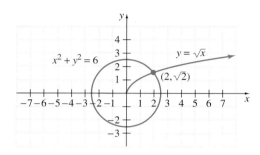

EXAMPLE 2   Solve the system

$$\begin{cases} x^2 - 3y = 1 \\ x - y = 1 \end{cases}$$

Use a graphical approach to check.

Solution: We can solve this system by substitution if we solve one equation for one of the variables. Solving the first equation for x is not the best choice since doing so introduces a radical. Also, solving for y in the first equation introduces a fraction. We solve the second equation for y.

$x - y = 1$ Second equation.

$x - 1 = y$ Solve for y.

Replace y with $x - 1$ in the first equation, and then solve for x:

$x^2 - 3y = 1$ First equation.

$x^2 - 3(x - 1) = 1$ Replace y with $x - 1$.

$x^2 - 3x + 3 = 1$

$x^2 - 3x + 2 = 0$

$(x - 2)(x - 1) = 0$

$x = 2$ or $x = 1$

Let $x = 2$ and then $x = 1$ in the equation $y = x - 1$ to find corresponding y-values.

$$\text{Let } x = 2. \qquad\qquad \text{Let } x = 1.$$
$$y = x - 1 \qquad\qquad y = x - 1$$
$$y = 2 - 1 = 1 \qquad\qquad y = 1 - 1 = 0$$

The solution set is $\{(2, 1), (1, 0)\}$. To check these solutions, graph each equation of the system. Since we are using a graphing utility, we first solve each equation for y.

$$x^2 - 3y = 1 \qquad\qquad\qquad\qquad x - y = 1$$
$$x^2 - 1 = 3y$$
$$y = \frac{x^2 - 1}{3} \qquad \text{Solve for } y. \qquad\qquad y = x - 1 \qquad \text{Solve for } y.$$

The graph of $y_1 = \dfrac{x^2 - 1}{3}$ and $y_2 = x - 1$ is shown below with the points of inter-section noted. Since the coordinates of the points of intersection are the same as the ordered pair solutions found above, the solutions check.

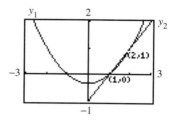

Some nonlinear systems may be solved by the elimination method.

EXAMPLE 3   Solve the system

$$\begin{cases} x^2 + 2y^2 = 10 \\ x^2 - y^2 = 1 \end{cases}$$

Solution: Use elimination, or addition, to solve this system. To eliminate x^2 when we add the two equations, multiply both sides of the second equation by -1. Then

$$\begin{cases} x^2 + 2y^2 = 10 \\ -1(x^2 - y^2) = -1 \cdot 1 \end{cases} \quad \begin{matrix} \text{is} \\ \text{equivalent} \\ \text{to} \end{matrix} \quad \begin{cases} x^2 + 2y^2 = 10 \\ -x^2 + y^2 = -1 \end{cases}$$

$$\overline{}$$
$$3y^2 = 9 \qquad \text{Add.}$$
$$y^2 = 3 \qquad \text{Divide by 3.}$$
$$y = \pm\sqrt{3}$$

To find the corresponding x-values, let $y = \sqrt{3}$ and $y = -\sqrt{3}$ in either original equation. We choose the second equation.

$$\text{Let } y = \sqrt{3}.$$
$$x^2 - y^2 = 1$$
$$x^2 - (\sqrt{3})^2 = 1$$
$$x^2 - 3 = 1$$
$$x^2 = 4$$
$$x = \pm\sqrt{4} = \pm 2$$

$$\text{Let } y = -\sqrt{3}.$$
$$x^2 - y^2 = 1$$
$$x^2 - (-\sqrt{3})^2 = 1$$
$$x^2 - 3 = 1$$
$$x^2 = 4$$
$$x = \pm\sqrt{4} = \pm 2$$

The solution set is $\{(2, \sqrt{3}), (-2, \sqrt{3}), (2, -\sqrt{3}), (-2, -\sqrt{3})\}$. Check all four ordered pairs in both equations of the system. To visualize these solutions, the graph of each equation in this system is shown next.

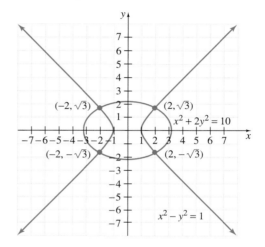

EXAMPLE 4   Solve the system

$$\begin{cases} x^2 + y^2 = 4 \\ x + y = 3 \end{cases}$$

Use a graphical approach to check the solution.

Solution: The elimination method is not a good choice here, since x and y are squared in the first equation but not in the second equation. Use the substitution method and solve the second equation for x.

$$x + y = 3 \qquad \text{Second equation.}$$
$$x = 3 - y$$

Let $x = 3 - y$ in the first equation:

$$x^2 + y^2 = 4 \qquad \text{First equation.}$$

$$(3 - y)^2 + y^2 = 4$$
$$9 - 6y + y^2 + y^2 = 4$$
$$2y^2 - 6y + 5 = 0$$

By the quadratic formula, where $a = 2$, $b = -6$, and $c = 5$, we have

$$y = \frac{6 \pm \sqrt{(-6)^2 - 4 \cdot 2 \cdot 5}}{2 \cdot 2} = \frac{6 \pm \sqrt{-4}}{4}$$

Since $\sqrt{-4}$ is not a real number, there is no solution.

To check graphically, notice that $x^2 + y^2 = 4$ solved for y is $y = \pm\sqrt{4 - x^2}$, and $x + y = 3$ solved for y is $y = 3 - x$. Graph $y_1 = \sqrt{4 - x^2}$, $y_2 = -y_1$, and $y_3 = 3 - x$ in a decimal window.

The graph of the circle and the line do not intersect, as expected. The system has no solution.

EXERCISE SET 10.3

Solve each nonlinear system of equations. See Examples 1 through 4.

1. $\begin{cases} x^2 + y^2 = 25 \\ 4x + 3y = 0 \end{cases}$

2. $\begin{cases} x^2 + y^2 = 25 \\ 3x + 4y = 0 \end{cases}$

3. $\begin{cases} x^2 + 4y^2 = 10 \\ y = x \end{cases}$

4. $\begin{cases} 4x^2 + y^2 = 10 \\ y = x \end{cases}$

5. $\begin{cases} y^2 = 4 - x \\ x - 2y = 4 \end{cases}$

6. $\begin{cases} x^2 + y^2 = 4 \\ x + y = -2 \end{cases}$

7. $\begin{cases} x^2 + y^2 = 9 \\ 16x^2 - 4y^2 = 64 \end{cases}$

8. $\begin{cases} 4x^2 + 3y^2 = 35 \\ 5x^2 + 2y^2 = 42 \end{cases}$

9. $\begin{cases} x^2 + 2y^2 = 2 \\ x - y = 2 \end{cases}$

10. $\begin{cases} x^2 + 2y^2 = 2 \\ x^2 - 2y^2 = 6 \end{cases}$

11. $\begin{cases} y = x^2 - 3 \\ 4x - y = 6 \end{cases}$

12. $\begin{cases} y = x + 1 \\ x^2 - y^2 = 1 \end{cases}$

13. $\begin{cases} y = x^2 \\ 3x + y = 10 \end{cases}$

14. $\begin{cases} 6x - y = 5 \\ xy = 1 \end{cases}$

15. $\begin{cases} y = 2x^2 + 1 \\ x + y = -1 \end{cases}$

16. $\begin{cases} x^2 + y^2 = 9 \\ x + y = 5 \end{cases}$

17. $\begin{cases} y = x^2 - 4 \\ y = x^2 - 4x \end{cases}$

18. $\begin{cases} x = y^2 - 3 \\ x = y^2 - 3y \end{cases}$

19. $\begin{cases} 2x^2 + 3y^2 = 14 \\ -x^2 + y^2 = 3 \end{cases}$

20. $\begin{cases} 4x^2 - 2y^2 = 2 \\ -x^2 + y^2 = 2 \end{cases}$

21. $\begin{cases} x^2 + y^2 = 1 \\ x^2 + (y + 3)^2 = 4 \end{cases}$

22. $\begin{cases} x^2 + 2y^2 = 4 \\ x^2 - y^2 = 4 \end{cases}$

23. $\begin{cases} y = x^2 + 2 \\ y = -x^2 + 4 \end{cases}$

24. $\begin{cases} x = -y^2 - 3 \\ x = y^2 - 5 \end{cases}$

25. $\begin{cases} 3x^2 + y^2 = 9 \\ 3x^2 - y^2 = 9 \end{cases}$

26. $\begin{cases} x^2 + y^2 = 25 \\ x = y^2 - 5 \end{cases}$

27. $\begin{cases} x^2 + 3y^2 = 6 \\ x^2 - 3y^2 = 10 \end{cases}$

28. $\begin{cases} x^2 + y^2 = 1 \\ y = x^2 - 9 \end{cases}$

29. $\begin{cases} x^2 + y^2 = 36 \\ y = \frac{1}{6}x^2 - 6 \end{cases}$

30. $\begin{cases} x^2 + y^2 = 16 \\ y = -\frac{1}{4}x^2 + 4 \end{cases}$

31. How many real solutions are possible for a system of equations whose graphs are a circle and a parabola?

32. How many real solutions are possible for a system of equations whose graphs are an ellipse and a line?

33. The sum of the squares of two numbers is 130. The difference of the squares of the two numbers is 32. Find the two numbers.

34. The sum of the squares of two numbers is 20. Their product is 8. Find the two numbers.

35. During the development stage of a new rectangular keypad for a security system, it was decided that the area of the rectangle should be 285 square centimeters and the perimeter should be 68 centimeters. Find the dimensions of the keypad.

36. A rectangular holding pen for cattle is to be designed so that its perimeter is 92 feet and its area is 525 feet. Find the dimensions of the holding pen.

Recall that in business a demand function *expresses the quantity of a commodity demanded as a function of the commodity's unit price. A* supply function *expresses the quantity of a commodity supplied as a function of the commodity's unit price. When the quantity produced and supplied is equal to the quantity demanded, then we have what is called* **market equilibrium.**

37. The demand function for a certain compact disc is given by the function

$$p = -0.01x^2 - 0.2x + 9$$

and the corresponding supply function is given by

$$p = 0.01x^2 - 0.1x + 3$$

where p is in dollars and x is in thousands of units. Find the equilibrium quantity and the corresponding price by solving the system consisting of the two given equations.

38. The demand function for a certain style of picture frame is given by the function

$$p = -2x^2 + 90$$

and the corresponding supply function is given by

$$p = 9x + 34$$

where p is in dollars and x is in thousands of units. Find the equilibrium quantity and the corresponding price by solving the system consisting of the two given equations.

Use a graphing utility to verify the results of each exercise.

39. Exercise 3.   **40.** Exercise 4.

41. Exercise 23.   **42.** Exercise 24.

Review Exercises

Graph each inequality in two variables. See Section 4.1.

43. $x > -3$ **44.** $y \le 1$

45. $y < 2x - 1$ **46.** $3x - y \le 4$

Find the perimeter of each geometric figure. See Section 6.3.

47.

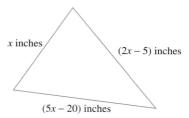

x inches $(2x - 5)$ inches

$(5x - 20)$ inches

48.

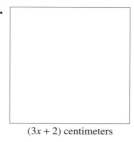

$(3x + 2)$ centimeters

49.

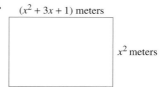

$(x^2 + 3x + 1)$ meters

x^2 meters

50.

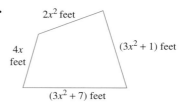

$2x^2$ feet

$4x$ feet

$(3x^2 + 1)$ feet

$(3x^2 + 7)$ feet

10.4   Nonlinear Inequalities and Systems of Inequalities

Tape IAG 10.4

O B J E C T I V E S

1 Sketch the graph of a nonlinear inequality.

2 Sketch the solution set of a system of nonlinear inequalities.

DISCOVER THE CONCEPT

Graph the circle defined by $x^2 + y^2 = 100$ in an integer window by graphing $y_1 = \sqrt{100 - x^2}$ and $y_2 = -\sqrt{100 - x^2}$ or $y_2 = -y_1$.

a. Trace to several points in the interior of the circle. For each ordered pair of numbers displayed, compare the value of the expression $x^2 + y^2$ to 100.

b. Trace to several points in the exterior of the circle. For each ordered pair of numbers displayed, compare the value of the expression $x^2 + y^2$ to 100.

c. What region do you think corresponds to ordered pair solutions of $x^2 + y^2 < 100$, and what region do you think corresponds to ordered pair solutions of $x^2 + y^2 > 100$?

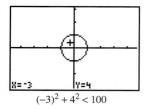

$(-3)^2 + 4^2 < 100$

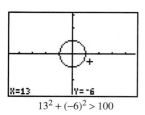

$13^2 + (-6)^2 > 100$

From the discovery above, we see that the circle divides the plane into two regions. All points in the interior of the circle correspond to ordered pair solutions of $x^2 + y^2 < 100$. All points in the exterior of the circles correspond to ordered pair solutions of $x^2 + y^2 > 100$. Don't forget that all points of the circle correspond to ordered pair solutions of $x^2 + y^2 = 100$.

In general, we can graph a nonlinear equality in two variables in a way similar to the way we graphed a linear inequality in two variables in Section 4.5. First, we graph the related equation. The graph of this equation is our boundary. Then, using test points, we determine and shade the region whose points satisfy the inequality.

EXAMPLE 1   Graph $\dfrac{x^2}{9} + \dfrac{y^2}{16} \le 1$.

Solution: First, graph the equation $\dfrac{x^2}{9} + \dfrac{y^2}{16} = 1$. Sketch a solid curve since the graph of $\dfrac{x^2}{9} + \dfrac{y^2}{16} \le 1$ includes the graph of $\dfrac{x^2}{9} + \dfrac{y^2}{16} = 1$. The graph is an ellipse, and it divides the plane into two regions, the inside and the outside of the ellipse. To determine which region contains the solutions, select a test point in either region and determine whether the coordinates of the point satisfy the inequality. We choose $(0, 0)$ as the test point.

$$\frac{x^2}{9} + \frac{y^2}{16} \le 1$$

$$\frac{0^2}{9} + \frac{0^2}{16} \le 1 \qquad \text{Let } x = 0 \text{ and } y = 0.$$

$$0 \le 1 \qquad \text{True.}$$

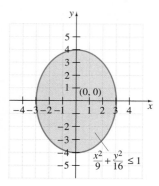

$\dfrac{x^2}{9} + \dfrac{y^2}{16} \le 1$

Since this statement is true, the solution set is the region containing $(0, 0)$. The graph of the solution set includes the points on and inside the ellipse, as shaded in the figure to the left.

EXAMPLE 2   Graph $4y^2 > x^2 + 16$.

Solution: The related equation is $4y^2 = x^2 + 16$, or $\dfrac{y^2}{4} - \dfrac{x^2}{16} = 1$, which is a hyperbola. Graph the hyperbola as a dashed curve since the graph of $4y^2 > x^2 + 16$ does *not* include the graph of $4y^2 = x^2 + 16$. The hyperbola divides the plane into three regions. Select a test point in each region—not on a boundary—to determine whether that region contains solutions of the inequality.

Test region A with $(0, 4)$	Test region B with $(0, 0)$	Test region C with $(0, -4)$
$4y^2 > x^2 + 16$	$4y^2 > x^2 + 16$	$4y^2 > x^2 + 16$
$4(4)^2 > 0^2 + 16$	$4(0)^2 > 0^2 + 16$	$4(-4)^2 > 0^2 + 16$
$64 > 16$ True.	$0 > 16$ False.	$64 > 16$ True.

The graph of the solution set includes the shaded regions only, not the boundary.

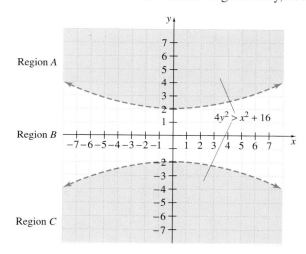

EXAMPLE 3   Use a graphing utility to graph $y \geq x^2 - 3$.

Solution: Graph the related equation $y_1 = x^2 - 3$ and shade the region above the boundary curve since the inequality symbol is $\geq$.

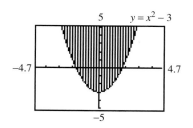

The solution region of the inequality $y \geq x^2 - 3$ is the region consisting of all points on the parabola itself and above the parabola.

2

In Section 4.5, we studied the intersection of graphs of inequalities in two variables. Although we did not identify them as such, we now can recognize these sets of inequalities as systems of inequalities. The graph of a system of inequalities is the intersection of the graphs of the inequalities.

EXAMPLE 4   Graph the system

$$\begin{cases} x \leq 1 - 2y \\ y \leq x^2 \end{cases}$$

Solution: Graph each inequality on the same set of axes. The intersection is the darkest shaded region along with its boundary lines. The coordinates of the points of intersection can be found by solving the related system

$$\begin{cases} x = 1 - 2y \\ y = x^2 \end{cases}$$

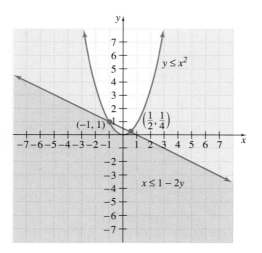

EXAMPLE 5   Graph the system

$$\begin{cases} x^2 + y^2 < 25 \\ \dfrac{x^2}{9} - \dfrac{y^2}{25} < 1 \\ y < x + 3 \end{cases}$$

Solution: Graph each inequality. The graph of $x^2 + y^2 < 25$ contains points inside the circle that has center $(0, 0)$ and radius 5. The graph of $\dfrac{x^2}{9} - \dfrac{y^2}{25} < 1$ is the region between the two branches of the hyperbola with x-intercepts -3 and 3 and center $(0, 0)$. The graph of $y < x + 3$ is the region below the line with the slope 1 and y-intercept 3. The graph of the solution set of the system is the intersection of all the graphs, the darkest shaded region shown. The boundary of this region is not part of the solution.

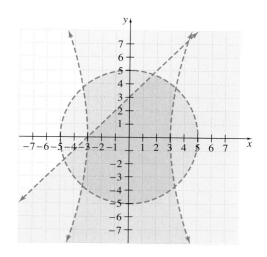

EXERCISE SET 10.4

Graph each inequality. See Examples 1 and 2.

1. $y < x^2$

2. $y < -x^2$

3. $x^2 + y^2 \geq 16$

4. $x^2 + y^2 < 36$

5. $\dfrac{x^2}{4} - y^2 < 1$

6. $x^2 - \dfrac{y^2}{9} \geq 1$

7. $y > (x - 1)^2 - 3$

8. $y > (x + 3)^2 + 2$

9. $x^2 + y^2 \leq 9$

10. $x^2 + y^2 > 4$

11. $y > -x^2 + 5$

12. $y < -x^2 + 5$

13. $\dfrac{x^2}{4} + \dfrac{y^2}{9} \leq 1$

14. $\dfrac{x^2}{25} + \dfrac{y^2}{4} \geq 1$

15. $\dfrac{y^2}{4} - x^2 \leq 1$

16. $\dfrac{y^2}{16} - \dfrac{x^2}{9} > 1$

Use a graphing utility to graph each inequality. See Example 3.

17. $y < (x - 2)^2 + 1$

18. $y > (x - 2)^2 + 1$

19. $y \leq x^2 + x - 2$

20. $y > x^2 + x - 2$

21. Discuss how graphing a linear inequality such as $x + y < 9$ is similar to graphing a nonlinear inequality such as $x^2 + y^2 < 9$.

22. Discuss how graphing a linear inequality such as $x + y < 9$ is different from graphing a nonlinear inequality such as $x^2 + y^2 < 9$.

Graph the solution of each system. See Examples 4 and 5.

23. $\begin{cases} 2x - y < 2 \\ y \le -x \end{cases}$

24. $\begin{cases} x - 2y > 4 \\ y > -x^2 \end{cases}$

25. $\begin{cases} 4x + 3y \ge 12 \\ x^2 + y^2 < 16 \end{cases}$

26. $\begin{cases} 3x - 4y \le 12 \\ x^2 + y^2 < 16 \end{cases}$

27. $\begin{cases} x^2 + y^2 \le 9 \\ x^2 + y^2 \ge 1 \end{cases}$

28. $\begin{cases} x^2 + y^2 \ge 9 \\ x^2 + y^2 \ge 16 \end{cases}$

29. $\begin{cases} y > x^2 \\ y \ge 2x + 1 \end{cases}$

30. $\begin{cases} y \le -x^2 + 3 \\ y \le 2x - 1 \end{cases}$

31. $\begin{cases} x > y^2 \\ y > 0 \end{cases}$

32. $\begin{cases} x < (y + 1)^2 + 2 \\ x + y \ge 3 \end{cases}$

33. $\begin{cases} x^2 + y^2 > 9 \\ \quad y > x^2 \end{cases}$

34. $\begin{cases} x^2 + y^2 \le 9 \\ \quad y < x^2 \end{cases}$

35. $\begin{cases} \dfrac{x^2}{4} + \dfrac{y^2}{9} \ge 1 \\ x^2 + y^2 \ge 4 \end{cases}$

36. $\begin{cases} x^2 + (y - 2)^2 \ge 9 \\ \dfrac{x^2}{4} + \dfrac{y^2}{25} < 1 \end{cases}$

37. $\begin{cases} x^2 - y^2 \ge 1 \\ \quad y \ge 0 \end{cases}$

38. $\begin{cases} x^2 - y^2 \ge 1 \\ \quad x \ge 0 \end{cases}$

39. $\begin{cases} x + y \ge 1 \\ 2x + 3y < 1 \\ x > -3 \end{cases}$

40. $\begin{cases} x - y < -1 \\ 4x - 3y > 0 \\ y > 0 \end{cases}$

41. $\begin{cases} x^2 - y^2 < 1 \\ \dfrac{x^2}{16} + y^2 \le 1 \\ x \ge -2 \end{cases}$

42. $\begin{cases} x^2 - y^2 \ge 1 \\ \dfrac{x^2}{16} + \dfrac{y^2}{4} \le 1 \\ y \ge 1 \end{cases}$

Review Exercises

Determine which graph is the graph of a function. See Section 3.3.

43.

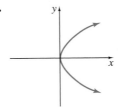

44.

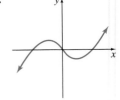

45.

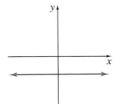

46.

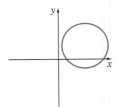

Find each function value if $f(x) = 3x^2 - 2$. See Section 3.3.

47. $f(-1)$

48. $f(-3)$

49. $f(a)$

50. $f(b)$

Group Activity

Modeling Conic Sections

Materials:
- Two thumbtacks (or nails), graph paper, cardboard, tape, string, pencil, ruler.

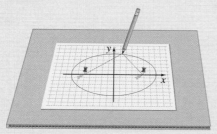

Figure 1

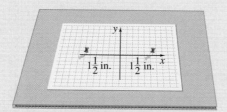

Figure 2

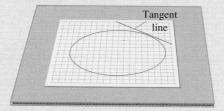

Figure 3

Try constructing the following model and investigating the results.

1. Draw an x-axis and a y-axis on the graph paper as shown in Figure 1.

2. Place the graph paper on the cardboard and use tape to attach.

3. Locate two points on the x-axis each about $1\frac{1}{2}$ inches from the origin and on opposite sides of the origin (see Figure 1). Insert thumbtacks (or nails) at each of these locations.

4. Fasten a 9-inch piece of string to the thumbtacks as shown in Figure 2. Use your pencil to draw and keep the string taut while you carefully move the pencil in a path all around the thumbtacks.

5. Using the grid of the graph paper as a guide, find an approximate equation of the ellipse that you drew.

6. Experiment by moving the tacks closer together or farther apart and drawing new ellipses. What do you observe?

7. Write a paragraph explaining why the figure drawn by the pencil is an ellipse. How might you use the same materials to draw a circle?

8. (Optional) Choose one of the ellipses that you drew with the string and pencil. Use a ruler to draw any six tangent lines to the ellipse. (A line is tangent to the ellipse if it intersects, or just touches, the ellipse at only one point. See Figure 3.) Extend the tangent lines to yield six points of intersection among the tangents. Use a straight edge to draw a line connecting each pair of opposite points of intersection. What do you observe? Repeat with a different ellipse. Can you make a conjecture about the relationship among the lines that connect opposite points of intersection?

CHAPTER 10   HIGHLIGHTS

DEFINITIONS AND CONCEPTS	**EXAMPLES**

SECTION 10.1   THE PARABOLA AND THE CIRCLE

Parabolas $y = a(x - h)^2 + k$

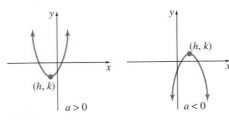

$x = a(y - k)^2 + h$

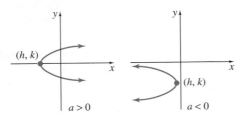

Graph $x = 3y^2 - 12y + 13$

$$x + 3(4) = 3(y^2 - 4y + 4) + 13$$
$$x = 3(y - 2)^2 + 1$$

Since $a = 3$, this parabola opens to the right with vertex $(1, 2)$. Its axis of symmetry is $y = 2$. The x-intercept is 13.

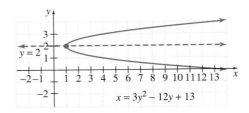

To use a graphing utility to graph an equation, we solve the equation for y.

Graph $x = 3y^2 - 12y + 13$.

$$0 = 3y^2 - 12y + (13 - x)$$
$$a = 3, \qquad b = -12, \qquad c = 13 - x$$

Substitute these values in the quadratic formula.

$$y = \frac{12 \pm \sqrt{144 - 4(3)(13 - x)}}{2(3)}$$

Let y_1 be the expression under the radical. Then *deselect* y_1 and graph y_2 and y_3 as shown.

$$y_1 = 144 - 4(3)(13 - x)$$
$$y_2 = \frac{12 + \sqrt{y_1}}{6}, \qquad y_3 = \frac{12 - \sqrt{y_1}}{6}$$

(continued)

DEFINITIONS AND CONCEPTS	**EXAMPLES**

SECTION 10.1 THE PARABOLA AND THE CIRCLE

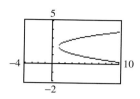

Circle

The graph of $(x - h)^2 + (y - k)^2 = r^2$ is a circle with center (h, k) and radius r.

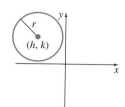

Graph $x^2 + (y + 3)^2 = 5$.

This equation can be written as

$$(x - 0)^2 + (y + 3)^2 = 5 \text{ with } h = 0, k = -3,$$
and $r = \sqrt{5}$.

The center of this circle is $(0, -3)$, and the radius is $\sqrt{5}$.

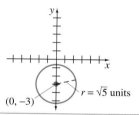

SECTION 10.2 THE ELLIPSE AND THE HYPERBOLA

Ellipse with center $(0, 0)$

The graph of an equation of the form $\dfrac{x^2}{a^2} + \dfrac{y^2}{b^2} = 1$

is an ellipse with center $(0, 0)$. The x-intercepts are a and $-a$, and the y-intercepts are b and $-b$.

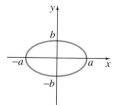

Graph $4x^2 + 9y^2 = 36$.

$$\frac{x^2}{9} + \frac{y^2}{4} = 1 \qquad \text{Divide by 36.}$$

$$\frac{x^2}{3^2} + \frac{y^2}{2^2} = 1$$

The ellipse has center $(0, 0)$, x-intercepts 3 and -3, and y-intercepts 2 and -2.

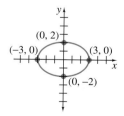

(continued)

DEFINITIONS AND CONCEPTS	EXAMPLES

SECTION 10.2 THE ELLIPSE AND THE HYPERBOLA

Hyperbola with center (0, 0)

The graph of an equation of the form $\dfrac{x^2}{a^2} - \dfrac{y^2}{b^2} = 1$ is a hyperbola with center $(0, 0)$ and x-intercepts a and $-a$.

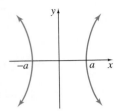

The graph of an equation of the form $\dfrac{y^2}{b^2} - \dfrac{x^2}{a^2} = 1$ is a hyperbola with center $(0, 0)$ and y-intercepts b and $-b$.

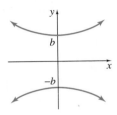

Graph $\dfrac{x^2}{9} - \dfrac{y^2}{4} = 1$. Here $a = 3$ and $b = 2$.

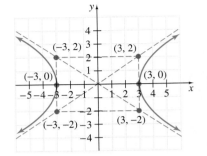

SECTION 10.3 SOLVING NONLINEAR SYSTEMS OF EQUATIONS

A **nonlinear system of equations** is a system of equations at least one of which is not linear. Both the substitution method and the elimination method may be used to solve a nonlinear system of equations.

Solve the nonlinear system $\begin{cases} y = x + 2 \\ 2x^2 + y^2 = 3. \end{cases}$

Substitute $x + 2$ for y in the second equation.

$$2x^2 + y^2 = 3$$
$$2x^2 + (x + 2)^2 = 3$$
$$2x^2 + x^2 + 4x + 4 = 3$$
$$3x^2 + 4x + 1 = 0$$
$$(3x + 1)(x + 1) = 0$$
$$x = -\frac{1}{3}, \qquad x = -1$$

(continued)

DEFINITIONS AND CONCEPTS	EXAMPLES
SECTION 10.3 SOLVING NONLINEAR SYSTEMS OF EQUATIONS	

If $x = -\dfrac{1}{3}$, $y = x + 2 = -\dfrac{1}{3} + 2 = \dfrac{5}{3}$.

If $x = -1$, $y = x + 2 = -1 + 2 = 1$.

The solution set is $\left\{\left(-\dfrac{1}{3}, \dfrac{5}{3}\right), (-1, 1)\right\}$.

SECTION 10.4 NONLINEAR INEQUALITIES AND SYSTEMS OF INEQUALITIES

The graph of a system of inequalities is the inter-section of the graphs of the inequalities.

Graph the system $\begin{cases} x \geq y^2 \\ x + y \leq 4 \end{cases}$.

The graph of the system is the darkest shaded region along with its boundary lines.

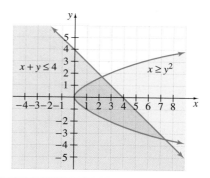

CHAPTER 10 REVIEW

(10.1) *Write an equation of the circle with the given center and radius.*

1. center $(-4, 4)$, *radius* 3

2. center $(5, 0)$, *diameter* 10

3. center $(-7, -9)$, *radius* $\sqrt{11}$

4. center $(0, 0)$, *diameter* 7

Sketch the graph of the equation. If the graph is a circle, find its center. If the graph is a parabola, find its vertex.

5. $x^2 + y^2 = 7$

6. $x = 2(y - 5)^2 + 4$

7. $x = -(y + 2)^2 + 3$

8. $(x - 1)^2 + (y - 2)^2 = 4$

9. $y = -x^2 + 4x + 10$

10. $x = -y^2 - 4y + 6$

11. $x = \dfrac{1}{2}y^2 + 2y + 1$

12. $y = -3x^2 + \dfrac{1}{2}x + 4$

13. $x^2 + y^2 + 2x + y = \dfrac{3}{4}$

14. $x^2 + y^2 + 3y = \dfrac{7}{4}$

15. $4x^2 + 4y^2 + 16x + 8y = 1$

16. $3x^2 + 6x + 3y^2 = 9$

17. $y = x^2 + 6x + 9$

18. $x = y^2 + 6y + 9$

19. Write an equation of the circle centered at $(5.6, -2.4)$ with diameter 6.2.

(10.2) *Sketch the graph of each equation.*

20. $x^2 + \dfrac{y^2}{4} = 1$ **21.** $x^2 - \dfrac{y^2}{4} = 1$

22. $\dfrac{y^2}{4} - \dfrac{x^2}{16} = 1$ **23.** $\dfrac{y^2}{4} + \dfrac{x^2}{16} = 1$

24. $\dfrac{x^2}{5} + \dfrac{y^2}{5} = 1$ **25.** $\dfrac{x^2}{5} - \dfrac{y^2}{5} = 1$

26. $-5x^2 + 25y^2 = 125$

27. $4y^2 + 9x^2 = 36$

28. $\dfrac{(x-2)^2}{4} + (y-1)^2 = 1$

29. $\dfrac{(x+3)^2}{9} + \dfrac{(y-4)^2}{25} = 1$

Use a graphing utility to graph each equation.

30. $x^2 - y^2 = 1$ **31.** $36y^2 - 49x^2 = 1764$

32. $y^2 = x^2 + 9$ **33.** $x^2 = 4y^2 - 16$

34. $100 - 25x^2 = 4y^2$

Sketch the graph of each equation.

35. $y = x^2 + 4x + 6$ **36.** $y^2 = x^2 + 6$

37. $y^2 + x^2 = 4x + 6$ **38.** $y^2 + 2x^2 = 4x + 6$

39. $x^2 + y^2 - 8y = 0$ **40.** $x - 4y = y^2$

41. $x^2 - 4 = y^2$ **42.** $x^2 = 4 - y^2$

43. $6(x-2)^2 + 9(y+5)^2 = 36$

44. $36y^2 = 576 + 16x^2$

45. $\dfrac{x^2}{16} - \dfrac{y^2}{25} = 1$

46. $3(x-7)^2 + 3(y+4)^2 = 1$

(10.3) *Solve each system of equations. Use a graphing utility to check.*

47. $\begin{cases} y = 2x - 4 \\ y^2 = 4x \end{cases}$ **48.** $\begin{cases} x^2 + y^2 = 4 \\ x - y = 4 \end{cases}$

49. $\begin{cases} y = x + 2 \\ y = x^2 \end{cases}$ **50.** $\begin{cases} y = x^2 - 5x + 1 \\ y = -x + 6 \end{cases}$

51. $\begin{cases} 4x - y^2 = 0 \\ 2x^2 + y^2 = 16 \end{cases}$ **52.** $\begin{cases} x^2 + 4y^2 = 16 \\ x^2 + y^2 = 4 \end{cases}$

53. $\begin{cases} x^2 + y^2 = 10 \\ 9x^2 + y^2 = 18 \end{cases}$ **54.** $\begin{cases} x^2 + 2y = 9 \\ 5x - 2y = 5 \end{cases}$

55. $\begin{cases} y = 3x^2 + 5x - 4 \\ y = 3x^2 - x + 2 \end{cases}$

56. $\begin{cases} x^2 - 3y^2 = 1 \\ 4x^2 + 5y^2 = 21 \end{cases}$

57. Find the length and the width of a room whose area is 150 square feet and whose perimeter is 50 feet.

58. What is the greatest number of real solutions possible for a system of two equations whose graphs are an ellipse and a hyperbola?

(10.4) *Graph the inequality or system of inequalities.*

59. $y \le -x^2 + 3$ **60.** $x^2 + y^2 < 9$

61. $x^2 - y^2 < 1$ **62.** $\dfrac{x^2}{4} + \dfrac{y^2}{9} \ge 1$

63. $\begin{cases} 2x \le 4 \\ x + y \ge 1 \end{cases}$ **64.** $\begin{cases} 3x + 4y \le 12 \\ x - 2y > 6 \end{cases}$

65. $\begin{cases} y > x^2 \\ x + y \ge 3 \end{cases}$ **66.** $\begin{cases} x^2 + y^2 \le 16 \\ x^2 + y^2 \ge 4 \end{cases}$

67. $\begin{cases} x^2 + y^2 < 4 \\ x^2 - y^2 \le 1 \end{cases}$ **68.** $\begin{cases} x^2 + y^2 < 4 \\ y \ge x^2 - 1 \\ x \ge 0 \end{cases}$

CHAPTER 10 TEST

Sketch the graph of each equation.

1. $x^2 + y^2 = 36$ **2.** $x^2 - y^2 = 36$

3. $16x^2 + 9y^2 = 144$ **4.** $y = x^2 - 8x + 16$

5. $x^2 + y^2 + 6x = 16$

6. $\dfrac{(x-4)^2}{16} + \dfrac{(y-3)^2}{9} = 1$

Use a graphing utility to graph each equation.

7. $y^2 - x^2 = 0$ **8.** $x = y^2 + 8y - 3$

Solve each system.

9. $\begin{cases} x^2 + y^2 = 169 \\ 5x + 12y = 0 \end{cases}$ **10.** $\begin{cases} x^2 + y^2 = 26 \\ x^2 - y^2 = 24 \end{cases}$

11. $\begin{cases} y = x^2 - 5x + 6 \\ y = 2x \end{cases}$ **12.** $\begin{cases} x^2 + 4y^2 = 5 \\ y = x \end{cases}$

Graph the solution of each system.

13. $\begin{cases} 2x + 5y \geq 10 \\ y \geq x^2 + 1 \end{cases}$

14. $\begin{cases} \dfrac{x^2}{4} + y^2 \leq 1 \\ x + y > 1 \end{cases}$

15. $\begin{cases} x^2 + y^2 > 1 \\ \dfrac{x^2}{4} - y^2 \geq 1 \end{cases}$

16. $\begin{cases} x^2 + y^2 \geq 4 \\ x^2 + y^2 < 16 \\ y \geq 0 \end{cases}$

17. Which graph best resembles the graph of $x = a(y - k)^2 + h$ if $a > 0, h < 0$, and $k > 0$?

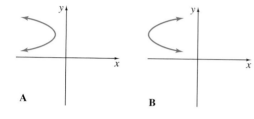

A

B

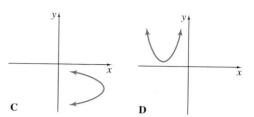

C

D

18. A bridge has an arch in the shape of half an ellipse. If the equation of the ellipse, measured in feet, is $100x^2 + 225y^2 = 22,500$, find the height of the arch from the road and the width of the arch.

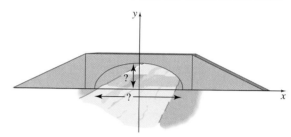

CHAPTER 10 CUMULATIVE REVIEW

1. Evaluate each expression.

a. 3^2

b. $\left(\dfrac{1}{2}\right)^4$

c. -5^2

d. $(-5)^2$

e. -5^3

f. $(-5)^3$

2. Solve for x: $\dfrac{x + 5}{2} + \dfrac{1}{2} = 2x - \dfrac{x - 3}{8}$.

3. Marial Callier just received an inheritance of $10,000 and plans to place all the money in a savings account that pays 5% compounded quarterly to help her son go to college in 3 years. How much money will be in the account in 3 years? The compound interest formula is $A = P\left(1 + \dfrac{r}{n}\right)^{nt}$.

4. The following graph shows the research and development expenditures by the Pharmaceutical Manufacturers Association as a function of time.

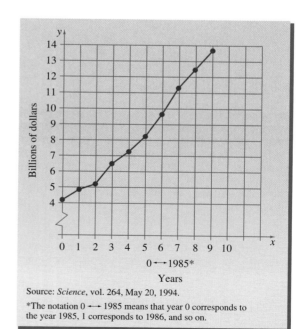

Source: *Science*, vol. 264, May 20, 1994.

*The notation 0 ⟷ 1985 means that year 0 corresponds to the year 1985, 1 corresponds to 1986, and so on.

a. Approximate the money spent on research and development in 1992.

b. In 1958, research and development expenditures were $200 million. Find the increase in expenditures from 1958 to 1994.

5. Write an equation of the line with y-intercept -3 and slope of $\frac{1}{4}$.

6. Solve $2|x| + 25 = 23$ algebraically and graphically.

7. Solve for x: $\left|2x - \frac{1}{10}\right| < -13$.

8. Use the elimination method to solve the system

$$\begin{cases} -\dfrac{x}{6} + \dfrac{y}{2} = \dfrac{1}{2} \\ \dfrac{x}{3} - \dfrac{y}{6} = -\dfrac{3}{4} \end{cases}$$

9. The measure of the largest angle of a triangle is $80°$ more than the measure of the smallest angle, and the measure of the remaining angle is $10°$ more than the measure of the smallest angle. Find the measure of each angle.

10. Evaluate the following.

a. 7^0 **b.** -7^0 **c.** $(2x + 5)^0$ **d.** $2x^0$

11. Simplify each expression. Write answers with positive exponents.

a. $(2x^0y^{-3})^{-2}$

b. $\left(\dfrac{x^{-5}}{x^{-2}}\right)^{-3}$

c. $\left(\dfrac{2}{7}\right)^{-2}$

d. $\dfrac{5^{-2}x^{-3}y^{11}}{x^2y^{-5}}$

12. Add $11x^3 - 12x^2 + x - 3$ and $x^3 - 10x + 5$.

13. Factor $x^2 - 12x + 35$.

14. Divide $\dfrac{8x^3 + 125}{x^4 + 5x^2 + 4} \div \dfrac{2x + 5}{2x^2 + 8}$.

15. Divide $10x^2 - 5x + 20$ by 5.

16. Find the cube roots.

a. $\sqrt[3]{1}$ **b.** $\sqrt[3]{-64}$ **c.** $\sqrt[3]{\dfrac{8}{125}}$

d. $\sqrt[3]{x^6}$ **e.** $\sqrt[3]{-8x^9}$

17. Rationalize the numerator of each expression.

a. $\dfrac{\sqrt{28}}{\sqrt{45}}$

b. $\dfrac{\sqrt[3]{2x^2}}{\sqrt[3]{5y}}$

18. Use the square root property to solve $x^2 = 50$.

19. Solve algebraically $(x + 3)(x - 3) < 0$.

20. Sketch the graph of each function. Label the vertex and the axis of symmetry.

a. $F(x) = x^2 + 2$

b. $g(x) = x^2 - 3$

21. Solve the system

$$\begin{cases} y = \sqrt{x} \\ x^2 + y^2 = 6 \end{cases}$$

Find real number solutions only.

Exponential and Logarithmic Functions

Modeling Temperature

When a cold object is placed in a warm room, the object's temperature gradually rises until it becomes, or nearly becomes, room temperature. Similarly, if a hot object is placed in a cooler room, the object's temperature gradually falls to room temperature. The way in which a cold or hot object warms up or cools off is modeled by an exact mathematical relationship.

In the chapter Group Activity on page 668, you will have the opportunity to investigate this model of cooling and warming.

I n this chapter, we discuss two closely related functions: exponential and logarithmic functions. These functions are vital in applications in economics, finance, engineering, the sciences, education, and other fields. Models of tumor growth and learning curves are two examples of the uses of exponential and logarithmic functions.

11.1 | COMPOSITE AND INVERSE FUNCTIONS

O B J E C T I V E S

 Compose functions.

2 Determine whether a function is a one-to-one function.

3 Use the horizontal line test to test whether a function is a one-to-one function.

4 Define the inverse of a function.

5 Find the equation of the inverse of a function.

1 Thus far in this text, we have seen several ways to combine functions and create new functions. Functions have been added, subtracted, multiplied, and divided, and the result each time has been a function.

Another way to combine functions is called **function composition.** To understand this new way of combining functions, study the tables below. They show degrees Fahrenheit converted to equivalent degrees Celsius, and then degrees Celsius converted to equivalent degrees Kelvin. (The Kelvin scale is a temperature scale devised by Lord Kelvin in 1848.)

x = DEGREES FAHRENHEIT (INPUT)	-31	-13	32	68	149	212
$C(x)$ = DEGREES CELSIUS (OUTPUT)	-35	-25	0	20	65	100

C = DEGREES CELSIUS (INPUT)	-35	-25	0	20	65	100
$K(C)$ = KELVINS (OUTPUT)	238.15	248.15	273.15	293.15	338.15	373.15

Suppose that we want a table that shows a direct conversion from degrees Fahrenheit to kelvins. In other words, suppose that a table is needed that shows kelvins as a function of degrees Fahrenheit. This can easily be done, because in the tables the output of the first table is the same as the input of the second table. The new table is as follows:

x = DEGREES FAHRENHEIT (INPUT)	-31	-13	32	68	149	212
$K(C(x))$ = KELVINS (OUTPUT)	238.15	248.15	273.15	293.15	338.15	373.15

Since the output of the first table is used as the input of the second table, we write the new function as $K(C(x))$. This new function is formed from the composition of the other two functions, and the mathematical symbol for this composition is $(K \circ C)(x)$. Thus, $(K \circ C)(x) = K(C(x))$.

It is possible to find an equation for the composition of the two functions $C(x)$ and $K(x)$. In other words, we can find a function that converts degrees Fahrenheit directly to kelvins. The function $C(x) = \frac{5}{9}(x - 32)$ converts degrees Fahrenheit to degrees Celsius, and the function $K(C) = C + 273.15$ converts degrees Celsius to kelvins. Thus,

$$(K \circ C)(x) = K(C(x)) = K\left(\frac{5}{9}(x - 32)\right) = \frac{5}{9}(x - 32) + 273.15$$

In general, the notation $f(g(x))$ means "f composed with g" and can be written as $(f \circ g)(x)$. Also, $g(f(x))$ or $(g \circ f)(x)$ means "g composed with f."

EXAMPLE 1   If $f(x) = x^2$ and $g(x) = x + 3$, find the following:

   a. $(f \circ g)(2)$ and $(g \circ f)(2)$ **b.** $(f \circ g)(x)$ and $(g \circ f)(x)$

Solution:

a. $(f \circ g)(2) = f(g(2))$

$\qquad\qquad\quad = f(5)$ Since $g(x) = x + 3$, then $g(2) = 2 + 3 = 5$.

$\qquad\qquad\quad = 5^2 = 25$

$(g \circ f)(2) = g(f(2))$

$\qquad\qquad\quad = g(4)$ Since $f(x) = x^2$, then $f(2) = 2^2 = 4$.

$\qquad\qquad\quad = 4 + 3 = 7$

b. $(f \circ g)(x) = f(g(x))$

$\qquad\qquad\quad = f(x + 3)$ Replace $g(x)$ with $x + 3$.

$\qquad\qquad\quad = (x + 3)^2$ $f(x + 3) = (x + 3)^2$.

$\qquad\qquad\quad = x^2 + 6x + 9$ Square $(x + 3)$.

$(g \circ f)(x) = g(f(x))$

$\qquad\qquad\quad = g(x^2)$ Replace $f(x)$ with x^2.

$\qquad\qquad\quad = x^2 + 3$ $g(x^2) = x^2 + 3$.

g(x)   f(x)

`2→X:X+3:Ans→X:X²` $f{\circ}g(2)$

`2→X:X²:Ans→X:X+3` $g{\circ}f(2)$

25

7

Using a graphing utility, we can evaluate $(f \circ g)(2)$ by first replacing the variable in $g(x)$ with 2 and evaluating the expression. We then take that answer and evaluate the $f(x)$ function.

EXAMPLE 2   If $f(x) = 5x$, $g(x) = x - 2$, and $h(x) = \sqrt{x}$, write each function as a composition with f, g, or h.

   a. $F(x) = \sqrt{x - 2}$ **b.** $G(x) = 5x - 2$

Solution: **a.** Notice the order in which the function F operates on an input value x. First, 2 is subtracted from x, and then the square root of that result is taken. This means that $F = h \circ g$. To check, find $h \circ g$:

$$(h \circ g)(x) = h(g(x))$$
$$= h(x - 2)$$
$$= \sqrt{x - 2}$$

b. Notice the order in which the function G operates on an input value x. First, x is multiplied by 5, and then 2 is subtracted from the result. This means that $G = g \circ f$. To check, find $g \circ f$:

$$(g \circ f)(x) = g(f(x))$$
$$= g(5x)$$
$$= 5x - 2$$

2 Study the following table. Determine whether the table describes a function.

STATE (INPUT)	ALABAMA	DELAWARE	MARYLAND	LOUISIANA	HAWAII	MONTANA
RECORD LOW TEMPERATURE IN DEGREES FAHRENHEIT (OUTPUT)	-27	-17	-40	-16	12	-70

Since each state (input) corresponds to exactly one record low temperature (output), this table of inputs and outputs does describe a function. Also notice that each output corresponds to a different input. This type of function is given a special name: a one-to-one function.

Does the set $f = \{(0, 1), (2, 2), (-3, 5), (7, 6)\}$ describe a one-to-one function? It is a function since each x-value corresponds to a unique y-value. For this particular function f, each y-value also corresponds to a unique x-value. Thus, this function is also a **one-to-one function.**

ONE-TO-ONE FUNCTION

For a one-to-one function, each x-value (input) corresponds to only one y-value (output), and each y-value (output) corresponds to only one x-value (input).

EXAMPLE 3 Determine whether each function described is one-to-one.

a. $f = \{(6, 2), (5, 4), (-1, 0), (7, 3)\}$
b. $g = \{(3, 9), (-4, 2), (-3, 9), (0, 0)\}$
c. $h = \{(1, 1), (2, 2), (10, 10), (-5, -5)\}$

d.

MINERAL (INPUT)	TALC	GYPSUM	DIAMOND	TOPAZ	STIBNITE
HARDNESS ON THE MOHS SCALE (OUTPUT)	1	2	10	8	2

e.

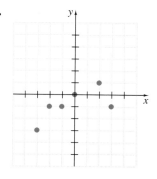

Solution: **a.** f is one-to-one since each y-value corresponds to only one x-value.

b. g is not one-to-one because the y-value 9 in $(3, 9)$ and $(-3, 9)$ corresponds to two different x-values.

c. h is a one-to-one function since each y-value corresponds to only one x-value.

d. This table does not describe a one-to-one function since the output 2 corresponds to two different inputs, gypsum and stibnite.

e. This graph does not describe a one-to-one function since the y-value -1 corresponds to three different x-values, -2, -1, and 3.

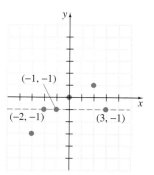

Recall that we recognize the graph of a function when it passes the vertical line test. Since every x-value of a function corresponds to exactly one y-value, each vertical line intersects the function's graph at most once. The graph shown next, for instance, is the graph of a function.

Is this function a *one-to-one* function? The answer is no. To see why not, notice that the y-value of the ordered pair $(-3, 3)$, for example, is the same as the y-value of the ordered pair $(3, 3)$. This function is therefore not one-to-one.

To test whether a graph is the graph of a one-to-one function, apply the vertical line test to see if it is a function, and then apply a similar **horizontal line test** to see if it is a one-to-one function.

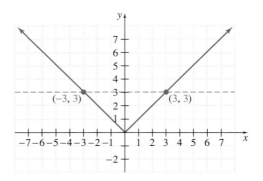

HORIZONTAL LINE TEST

If every horizontal line intersects the graph of a function at most once, then the function is a one-to-one function.

EXAMPLE 4 Determine whether each graph is the graph of a one-to-one function.

a.

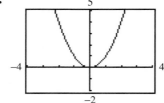

b.

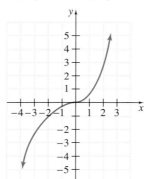

c.

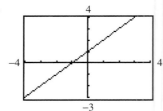

d.

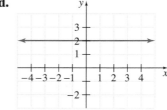

e.

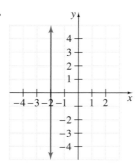

Solution: Graphs **a, b, c,** and **d** all pass the vertical line test, so only these graphs are graphs of functions. But, of these, only **b** and **c** pass the horizontal line test, so only **b** and **c** are graphs of one-to-one-functions.

> R E M I N D E R   All linear equations are one-to-one functions except those whose graphs are horizontal or vertical lines. A vertical line does not pass the vertical line test and hence is not the graph of a function. A horizontal line is the graph of a function but does not pass the horizontal line test, and hence it is not the graph of a one-to-one function.

One-to-one functions are special in that their graphs pass the vertical and horizontal line tests. They are special, too, in another sense: For each one-to-one function we can find its **inverse function** by switching the coordinates of the ordered pairs of the function, or the inputs and the outputs. For example, the inverse of the one-to-one function

STATE (INPUT)	**ALABAMA**	**DELAWARE**	**MARYLAND**	**LOUISIANA**	**HAWAII**	**MONTANA**
RECORD LOW TEMPERATURE (OUTPUT)	−27	−17	−40	−16	12	−70

is the function

RECORD LOW TEMPERATURE (INPUT)	**−27**	**−17**	**−40**	**−16**	**12**	**−70**
STATE (OUTPUT)	Alabama	Delaware	Maryland	Louisiana	Hawaii	Montana

Notice that the ordered pair (Alabama, −27) of the function f, for example, becomes the ordered pair (−27, Alabama) of its inverse.

Also, the inverse of the one-to-one function $f = \{(2, -3), (5, 10), (9, 1)\}$ is $\{(-3, 2), (10, 5), (1, 9)\}$. For a function f, we use the notation f^{-1}, read "f inverse," to denote its inverse function. Notice that since the coordinates of each ordered pair have been switched, the domain (set of inputs) of f is the range (set of outputs) of f^{-1}, and the range of f is the domain of f^{-1}.

INVERSE FUNCTION

The inverse of a one-to-one function f is the one-to-one function f^{-1} that consists of the set of all ordered pairs (y, x), where (x, y) belongs to f.

> REMINDER   The symbol f^{-1} is the single symbol used to denote the inverse of the function f. It is read as "f inverse." This symbol *does not mean* $\dfrac{1}{f}$.

5   If a one-to-one function f is defined as a set of ordered pairs, we can find f^{-1} by interchanging the x and y coordinates of the ordered pairs. If a one-to-one function f is given in the form of an equation, we can find f^{-1} by using a similar procedure.

TO FIND THE INVERSE OF A ONE-TO-ONE FUNCTION $f(x)$

Step 1.   Replace $f(x)$ with y.
Step 2.   Interchange x and y.
Step 3.   Solve the equation for y.
Step 4.   Replace y with the notation $f^{-1}(x)$.

EXAMPLE 5   Find the equation of the inverse of $f(x) = 3x - 5$.

Solution: First, let $y = f(x)$.

$$f(x) = 3x - 5$$
$$y = 3x - 5$$

Next, interchange x and y and solve for y:

$$x = 3y - 5 \qquad \text{Interchange } x \text{ and } y.$$
$$3y = x + 5$$
$$y = \frac{x + 5}{3} \qquad \text{Solve for } y.$$

Let $y = f^{-1}(x)$.

$$f^{-1}(x) = \frac{x + 5}{3}$$

Thus, the inverse of $f(x) = 3x - 5$ is
$$f^{-1}(x) = \frac{x + 5}{3}.$$

DISCOVER THE CONCEPT

a. Use your graphing utility to graph both $f(x) = 3x - 5$ as $y_1 = 3x - 5$ and

$f^{-1}(x) = \dfrac{x+5}{3}$ as $y_2 = (x + 5)/3$ using a square window.

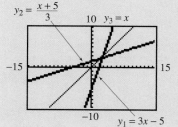

b. Trace to the point $(2, 1)$ on the graph of y_1. Is there a point $(1, 2)$ on y_2? Try this for a point on y_2. If (a, b) is a point of y_2, is (b, a) a point of y_1?

c. Next include the graph of $y_3 = x$ in the display. How are the graphs of $f(x) = 3x - 5$ and $f^{-1}(x) = \dfrac{x+5}{3}$ related to the graph of $y = x$?

In the above discovery, we notice that the graphs of f and f^{-1} are mirror images of each other, and the "mirror" is the dashed line $y = x$. This is true for every function and its inverse. For this reason, we say that **the graphs of f and f^{-1} are symmetric about the line $y = x$.**

X	Y1
0	-5
1	-2
2	1

Y1 ■ 3X-5

X	Y2
-5	0
-2	1
1	2

Y2 ■ (X+5)/3

Notice also in the above tables of values that $f(0) = -5$ and $f^{-1}(-5) = 0$, as expected. Also, for example, $f(1) = -2$ and $f^{-1}(-2) = 1$. In words, we say that for some input x, the function f^{-1} takes the output of x, called $f(x)$, back to x:

$$x \rightarrow f(x) \text{ and } f^{-1}(f(x)) \rightarrow x$$
$$\downarrow \quad \downarrow \qquad\qquad \downarrow \qquad \downarrow$$
$$f(0) = -5 \text{ and } f^{-1}(-5) = 0$$
$$f(1) = -2 \text{ and } f^{-1}(-2) = 1$$

In general,

If f is a one-to-one function, then the inverse of f is the function f^{-1} such that

$$(f^{-1} \circ f)(x) = x \quad \text{and} \quad (f \circ f^{-1})(x) = x$$

EXAMPLE 6   Show that if $f(x) = 3x + 2$, then $f^{-1}(x) = \dfrac{x - 2}{3}$.

Solution: To show this algebraically, we show that $f^{-1}(f(x)) = x$ and $f(f^{-1}(x)) = x$.

$$(f^{-1} \circ f)(x) = f^{-1}(f(x))$$
$$= f^{-1}(3x + 2) \qquad \text{Replace } f(x) \text{ with } 3x + 2.$$
$$= \frac{3x + 2 - 2}{3}$$
$$= \frac{3x}{3}$$
$$= x$$

$$(f \circ f^{-1})(x) = f(f^{-1}(x))$$
$$= f\left(\frac{x - 2}{3}\right) \qquad \text{Replace } f^{-1}(x) \text{ with } \frac{x - 2}{3}.$$
$$= 3\left(\frac{x - 2}{3}\right) + 2$$
$$= x - 2 + 2$$
$$= x$$

To visually see these results, graph $y_1 = 3x + 2$, $y_2 = \dfrac{x - 2}{3}$, and $y_3 = x$ using a square window, as shown below. Graphically, we see that the graphs of f and f^{-1} are mirror images of each other across the line $y = x$.

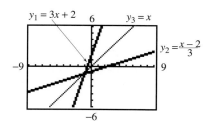

We can also make a table of values for f and f^{-1} and see that if $f(a) = b$ [the ordered pair (a, b)] then $f^{-1}(b) = a$ [the ordered pair (b, a)].

Notice that $f(0) = 2$

Notice that $f^{-1}(2) = 0$

EXERCISE SET 11.1

If $f(x) = x^2 - 6x + 2$, $g(x) = -2x$, and $h(x) = \sqrt{x}$, find the following. See Example 1.

1. $(f \circ g)(2)$

2. $(h \circ f)(-2)$

3. $(g \circ f)(-1)$

4. $(f \circ h)(1)$

5. $(g \circ h)(0)$

6. $(h \circ g)(0)$

Find $(f \circ g)(x)$, $(g \circ f)(x)$, and $(f \circ f)(x)$. See Example 1.

7. $f(x) = x^2 + 1$ $g(x) = 5x$

8. $f(x) = x - 3$ $g(x) = x^2$

9. $f(x) = 2x - 3$ $g(x) = x + 7$

10. $f(x) = x + 10$ $g(x) = 3x + 1$

Find $(f \circ g)(x)$ and $(g \circ f)(x)$. See Example 1.

11. $f(x) = x^3 + x - 2$ $g(x) = -2x$

12. $f(x) = -4x$ $g(x) = x^3 + x^2 - 6$

13. $f(x) = \sqrt{x}$ $g(x) = -5x + 2$

14. $f(x) = 7x - 1$ $g(x) = \sqrt[3]{x}$

If $f(x) = 3x$, $g(x) = \sqrt{x}$, and $h(x) = x^2 + 2$, write each of the following functions as a composition of f, g, and h. See Example 2.

15. $H(x) = \sqrt{x^2 + 2}$

16. $G(x) = \sqrt{3x}$

17. $F(x) = 9x^2 + 2$

18. $H(x) = 3x^2 + 6$

19. $G(x) = 3\sqrt{x}$

20. $F(x) = x + 2$

Determine whether each function is a one-to-one function. If it is one-to-one, list the inverse function by switching coordinates or inputs and outputs. See Example 3.

21. $f = \{(-1, -1), (1, 1), (0, 2), (2, 0)\}$

22. $g = \{(8, 6), (9, 6), (3, 4), (-4, 4)\}$

23. $h = \{(10, 10)\}$

24. $r = \{(1, 2), (3, 4), (5, 6), (6, 7)\}$

25. $f = \{(11, 12), (4, 3), (3, 4), (6, 6)\}$

26. $g = \{(0, 3), (3, 7), (6, 7), (-2, -2)\}$

27.

YEAR (INPUT)	1970	1975	1980	1985	1990
AVERAGE COMPOSITE SCORE ON THE ACT (OUTPUT)	19.9	18.6	18.5	18.6	20.6

28.

STATE (INPUT)	WASHINGTON	OHIO	GEORGIA	COLORADO	CALIFORNIA	ARIZONA
ELECTORAL VOTES (OUTPUT)	11	21	13	8	54	8

29.

STATE (INPUT)	CALIFORNIA	VERMONT	VIRGINIA	TEXAS	SOUTH DAKOTA
RANK IN POPULATION (OUTPUT)	1	49	12	2	45

30.

SHAPE (INPUT)	TRIANGLE	PENTAGON	QUADRILATERAL	HEXAGON	DECAGON
NUMBER OF SIDES (OUTPUT)	3	5	4	6	10

Given the one-to-one function $f(x) = x^3 + 2$, find the following:

31. a. $f(1)$ **b.** $f^{-1}(3)$

32. a. $f(0)$ **b.** $f^{-1}(2)$

33. a. $f(-1)$ **b.** $f^{-1}(1)$

34. a. $f(-2)$ **b.** $f^{-1}(-6)$

Determine whether the graph of each function is the graph of a one-to-one function. See Example 4.

35.

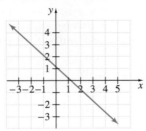

36.

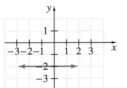

37.

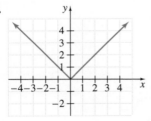

38.

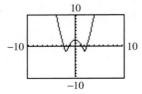

39.

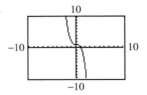

40.

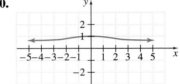

41.

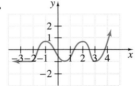

42.

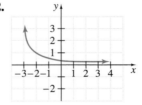

Each of the following functions is one-to-one. Find the inverse of each function and use a graphing utility to graph the function and its inverse in the same square window. See Example 5.

43. $f(x) = x + 4$ **44.** $f(x) = x - 5$

45. $f(x) = 2x - 3$ **46.** $f(x) = 4x + 9$

47. $f(x) = \dfrac{12x - 4}{3}$ **48.** $f(x) = \dfrac{7x + 5}{11}$

49. $f(x) = x^3$

50. $f(x) = x^3 - 1$

51. $f(x) = \dfrac{x - 2}{5}$

52. $f(x) = \dfrac{4x - 3}{2}$

53. $g(x) = x^2 + 5, x \geq 0$

54. $g(x) = x^2 - 2, x \geq 0$

Solve. See Example 6.

55. If $f(x) = 2x + 1$, show that $f^{-1}(x) = \dfrac{x - 1}{2}$.

56. If $f(x) = 3x - 10$, show that $f^{-1}(x) = \dfrac{x + 10}{3}$.

57. If $f(x) = x^3 + 6$, show that $f^{-1}(x) = \sqrt[3]{x - 6}$.

58. If $f(x) = x^3 - 5$, show that $f^{-1}(x) = \sqrt[3]{x + 5}$.

For Exercises 59 and 60,

a. Write the ordered pairs for $f(x)$ whose points are highlighted. (Include the points whose coordinates are given.)

b. Write the corresponding ordered pairs for f^{-1}, the inverse of f.

c. Graph the ordered pairs for f^{-1} found in part (b).

d. Graph $f^{-1}(x)$ by drawing a smooth curve through the plotted points.

59.

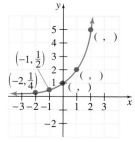

60.

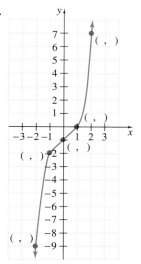

Find the inverse of each given one-to-one function. Then use a graphing utility to graph the function and its inverse on a square window.

61. $f(x) = 3x + 1$

62. $f(x) = -2x - 6$

63. $f(x) = \sqrt[3]{x + 3}$

64. $f(x) = x^3 - 3$

Review Exercises

Evaluate each of the following. See Section 8.2.

65. $25^{1/2}$

66. $49^{1/2}$

67. $16^{3/4}$

68. $27^{2/3}$

69. $9^{-3/2}$

70. $81^{-3/4}$

If $f(x) = 3^x$, find the following. In Exercises 73 and 74, give an exact answer and a two-decimal place approximation. See Sections 6.1 and 8.2.

71. $f(2)$

72. $f(0)$

73. $f\left(\tfrac{1}{2}\right)$

74. $f\left(\tfrac{2}{3}\right)$

11.2 EXPONENTIAL FUNCTIONS

O B J E C T I V E S

1. Identify exponential functions.
2. Graph exponential functions
3. Solve equations of the form $b^x = b^y$.

TAPE IAG 11.2

In earlier chapters, we gave meaning to exponential expressions such as 2^x, where x is a rational number. For example,

$$2^3 = 2 \cdot 2 \cdot 2 \qquad \text{Three factors, each factor is 2.}$$
$$2^{3/2} = (2^{1/2})^3 = \sqrt{2} \cdot \sqrt{2} \cdot \sqrt{2} \qquad \text{Three factors, each factor is } \sqrt{2}.$$

When x is an irrational number (for example, $\sqrt{3}$), what meaning can we give to $2^{\sqrt{3}}$?

It is beyond the scope of this book to give precise meaning to 2^x if x is irrational. We can confirm your intuition and say that $2^{\sqrt{3}}$ is a real number, and since $1 < \sqrt{3} < 2$, then $2^1 < 2^{\sqrt{3}} < 2^2$. We can also use a calculator and approximate $2^{\sqrt{3}}$: $2^{\sqrt{3}} \approx 3.321997$. In fact, as long as the base b is positive, b^x is a real number for all real numbers x. Finally, the rules of exponents apply whether x is rational or irrational, as long as b is positive. In this section, we are interested in functions of the form $f(x) = b^x$, where $b > 0$. A function of this form is called an **exponential function.**

EXPONENTIAL FUNCTION

A function of the form

$$f(x) = b^x$$

is called an exponential function if $b > 0$, b is not 1, and x is a real number.

2

Next, we practice graphing exponential functions.

EXAMPLE 1   Graph the exponential functions defined by $f(x) = 2^x$ and $g(x) = 3^x$ on the same set of axes.

Solution: Graph each function by plotting points. Set up a table of values for each of the two functions:

$f(x) = 2^x$

x	0	1	2	3	-1	-2
$f(x)$	1	2	4	8	$\dfrac{1}{2}$	$\dfrac{1}{4}$

$g(x) = 3^x$

x	0	1	2	3	-1	-2
$g(x)$	1	3	9	27	$\dfrac{1}{3}$	$\dfrac{1}{9}$

If each set of points is plotted and connected with a smooth curve, the following graphs result:

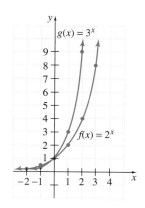

A number of things should be noted about these two graphs of exponential functions. First, the graphs show that $f(x) = 2^x$ and $g(x) = 3^x$ are one-to-one functions since each graph passes the vertical and horizontal line tests. The y-intercept of each graph is 1, but neither graph has an x-intercept. From the graph, we can also see that the domain of each function is all real numbers and that the range is $(0, \infty)$. We can also see that as x-values are increasing y-values are increasing also.

EXAMPLE 2   Graph the exponential functions $y = \left(\dfrac{1}{2}\right)^x$ and $y = \left(\dfrac{1}{3}\right)^x$ on the same set of axes.

Solution: As before, plot points and connect them with a smooth curve.

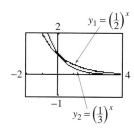

$y = \left(\dfrac{1}{2}\right)^x$

x	0	1	2	3	-1	-2
y	1	$\dfrac{1}{2}$	$\dfrac{1}{4}$	$\dfrac{1}{8}$	2	4

$y = \left(\dfrac{1}{3}\right)^x$

x	0	1	2	3	-1	-2
y	1	$\dfrac{1}{3}$	$\dfrac{1}{9}$	$\dfrac{1}{27}$	3	9

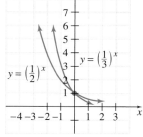

A graphing utility check is shown to the left.

To check, graph $y_1 = (1/2)^x$ and $y_2 = (1/3)^x$ using the same window. Compare the screens to the calculations done by hand.

Each function again is a one-to-one function. The y-intercept of both is 1. The domain is the set of all real numbers, and the range is $(0, \infty)$.

Notice the difference between the graphs of Example 1 and the graphs of Example 2. An exponential function is always increasing if the base is greater than 1. When the base is between 0 and 1, the graph is decreasing. The following figures summarize these characteristics of exponential functions.

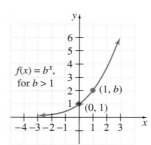

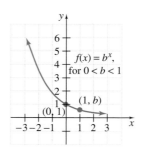

We have seen that an exponential function $y = b^x$ is a one-to-one function. Another way of stating this fact provides a property that we can use to solve exponential equations.

> **UNIQUENESS OF b^x**
>
> Let $b > 0$ and $b \neq 1$. Then $b^x = b^y$ is equivalent to $x = y$.

EXAMPLE 3   Solve each equation for x.

a. $2^x = 16$ **b.** $9^x = 27$ **c.** $4^{x+3} = 8^x$ **d.** $5^x = 10$

Solution: **a.** We write 16 as a power of 2 and then use the uniqueness of b^x to solve.

$$2^x = 16$$
$$2^x = 2^4$$

Since the bases are the same and are nonnegative, by the uniqueness of b^x, we then have that the exponents are equal. Thus,

$$x = 4$$

The solution set is $\{4\}$.

b. Notice that both 9 and 27 are powers of 3.

$$9^x = 27$$
$$(3^2)^x = 3^3 \qquad \text{Write 9 and 27 as powers of 3.}$$
$$3^{2x} = 3^3$$
$$2x = 3 \qquad \text{Apply the uniqueness of } b^x.$$
$$x = \frac{3}{2} \qquad \text{Divide by 2.}$$

To check, replace x with $\dfrac{3}{2}$ in the original expression, $9^x = 27$. The solution set is $\left\{\dfrac{3}{2}\right\}$.

c. Write both 4 and 8 as powers of 2.

$$4^{x+3} = 8^x$$
$$(2^2)^{x+3} = (2^3)^x$$
$$2^{2x+6} = 2^{3x}$$
$$2x + 6 = 3x \qquad \text{Apply the uniqueness of } b^x.$$
$$6 = x \qquad \text{Subtract } 2x \text{ from both sides.}$$

The solution set is {6}.

d. Notice that 5 and 10 cannot be easily written as powers of a common base. To solve this equation, we will approximate the solution graphically. Graph

$$y_1 = 5^x \qquad \text{(left side of equation)}$$
$$y_2 = 10 \qquad \text{(right side of equation)}$$

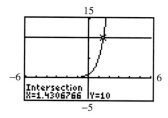

The point of intersection is approximately (1.431, 10), which indicates that the solution of the equation is $x \approx 1.431$. To check, replace x with 1.431 in the original equation.

$$5^x = 10$$
$$5^{1.431} = 10 \qquad \text{Let } x = 1.431.$$
$$10.00520695 \approx 10 \qquad \text{Approximately true.}$$

The solution set is approximately {1.431}.

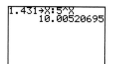

Use a calculator to evaluate $f(x) = 5^x$ at $x = 1.431$.

As we see in Example 3(d), often the two sides of an equation cannot easily be written as powers of a common base. We explore how to find exact solutions to an equation such as $5^x = 10$ with the help of **logarithms** later.

The world abounds with patterns that can be modeled by exponential functions. To make these applications realistic, we use numbers that warrant a calculator. The first application deals with exponential decay. When an application involves direct substitution into a given formula, problem-solving steps are not shown.

EXAMPLE 4

RADIOACTIVITY

As a result of the Chernobyl nuclear accident, radioactive debris was carried through the atmosphere. One immediate concern was the impact that debris had on the milk supply. The percent y of radioactive material in raw milk after t days is estimated by $y = 100 \, (2.7)^{-0.1t}$.

a. Estimate the expected percent of radioactive material in the milk after 30 days.

b. Make a table showing the percent of radioactive material in the milk every fifth day.

c. What percent of the radioactive material contaminated the milk after 5, 10, 15, and 20 days?

Solution:

a. Replace t with 30 in the given equation.

$$y = 100(2.7)^{-0.1t}$$
$$= 100(2.7)^{-0.1(30)} \qquad \text{Let } t = 30.$$
$$= 100(2.7)^{-3}$$

Use a calculator to approximate the percent y.

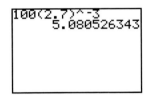

```
100(2.7)^-3
          5.080526343
```

Thus, approximately 5% of the radioactive material still contaminated the milk supply after 30 days.

b. Use a graphing utility to make a table.

X	Y1
0	100
5	60.858
10	37.037
15	22.54
20	13.717
25	8.3482
30	5.0805

Y1■100*2.7^(-.1...

Approximately 61% of the radioactive material contaminated the milk after 5 days.

c. We can read the table in part (b) to see that approximately 61% of the radioactive material contaminated the milk after 5 days, approximately 37% remained after 10 days, approximately 22.5% remained after 15 days, and approximately 14% remained after 20 days.

The exponential function defined by $A = P\left(1 + \dfrac{r}{n}\right)^{nt}$ models the pattern relating the dollars A accrued (or owed) after P dollars are invested (or loaned) at an annual rate of interest r compounded n times each year for t years.

EXAMPLE 5   REPAYING A LOAN

Find the amount owed at the end of 5 years if $1600 is loaned at a rate of 9% compounded monthly.

Solution: Use the formula $A = P\left(1 + \dfrac{r}{n}\right)^{nt}$, with the following values:

$P = \$1600$ (the amount of the loan)

$r = 9\% = 0.09$ (the annual rate of interest)

$n = 12$ (the number of times interest is compounded each year)

$t = 5$ (the duration of the loan, in years)

$A = P\left(1 + \dfrac{r}{n}\right)^{nt}$ Compound interest formula.

$\quad = 1600\left(1 + \dfrac{0.09}{12}\right)^{12(5)}$ Substitute known values.

$\quad = 1600(1.0075)^{60}$

Use a calculator to approximate A.

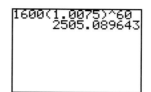

Thus, the amount A owed is $2505.09.

EXERCISE SET 11.2

Graph each exponential function. See Examples 1 and 2.

1. $y = 4^x$

2. $y = 5^x$

3. $y = 1 + 2^x$

4. $y = 3^x - 1$

5. $y = \left(\dfrac{1}{4}\right)^x$

6. $y = \left(\dfrac{1}{5}\right)^x$

7. $y = \left(\dfrac{1}{2}\right)^x - 2$

8. $y = \left(\dfrac{1}{3}\right)^x + 2$

9. $y = -2^x$

10. $y = -3^x$

11. $y = 3^x - 2$

12. $y = 2^x - 3$

13. $y = -\left(\dfrac{1}{4}\right)^x$

14. $y = -\left(\dfrac{1}{5}\right)^x$

15. $y = \left(\dfrac{1}{3}\right)^x + 1$

16. $y = \left(\dfrac{1}{2}\right)^x - 2$

17. Explain why the graph of an exponential function $y = b^x$ contains the point $(1, b)$.

18. Explain why an exponential function $y = b^x$ has a y-intercept of 1.

Solve each equation for x. See Example 3.

19. $3^x = 27$

20. $6^x = 36$

21. $16^x = 8$

22. $64^x = 16$

23. $32^{2x-3} = 2$

24. $9^{2x+1} = 81$

25. $\dfrac{1}{4} = 2^{3x}$

26. $\dfrac{1}{27} = 3^{2x}$

27. $5^x = 625$

28. $2^x = 64$

29. $4^x = 8$

30. $32^x = 4$

31. $27^{x+1} = 9$

32. $125^{x-2} = 25$

33. $81^{x-1} = 27^{2x}$

34. $4^{3x-7} = 32^{2x}$

Write an exponential equation defining the function whose graph is given.

35.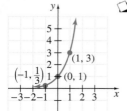

36.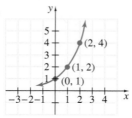

37.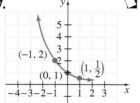

38.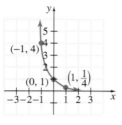

Solve. Unless otherwise indicated, round results to one decimal place. See Example 4.

39. One type of uranium has a daily radioactive decay rate of 0.4%. If 30 pounds of this uranium is available today, find how much will still remain after 50 days. Use $y = 30(2.7)^{-0.004t}$, and let t be 50.

40. The nuclear waste from an atomic energy plant decays at a rate of 3% each century. If 150 pounds of nuclear waste is disposed of, find how much of it will still remain after 10 centuries. Use $y = 150(2.7)^{-0.03t}$, and let t be 10.

41. The size of the rat population of a wharf area grows at a rate of 8% monthly. If there are 200 rats in January, find how many rats (rounded to the nearest whole number) should be expected by next January. Use $y = 200(2.7)^{0.08t}$.

42. National Park Service personnel are trying to increase the size of the bison population of Theodore Roosevelt National Park. If 260 bison currently live in the park, and if the population's rate of growth is 2.5% annually, find how many bison (rounded to the nearest whole number) there should be in 10 years. Use $y = 260(2.7)^{0.025t}$.

43. A rare isotope of a nuclear material is very unstable, decaying at a rate of 15% each second. Find how much isotope remains 10 seconds after 5 grams of the isotope is created. Use $y = 5(2.7)^{-0.15t}$.

44. An accidental spill of 75 grams of radioactive material in a local stream has led to the presence of radioactive debris decaying at a rate of 4% each day. Find how much debris still remains after 14 days. Use $y = 75(2.7)^{-0.04t}$.

45. Mexico City is growing at a rate of 0.7% annually. If there were 15,525,000 residents of Mexico City in 1994, predict how many (to the nearest ten-thousand) will be living in the city in 2000. Use $y = 15,525,000(2.7)^{0.007t}$.

46. An unusually wet spring has caused the size of the Cape Cod mosquito population to increase by 8% each day. If an estimated 200,000 mosquitoes are on Cape Cod on May 12, find how many thousands of mosquitoes will inhabit the Cape on May 25. Use $y = 200,000(2.7)^{0.08t}$.

Solve. Use $A = P\left(1 + \dfrac{r}{n}\right)^{nt}$. Round answers to two decimal places. See Example 5.

47. Find the amount Erica owes at the end of 3 years if $6000 is loaned to her at a rate of 8% compounded monthly.

48. Find the amount owed at the end of 5 years if $3000 is loaned at a rate of 10% compounded quarterly.

49. Find the total amount Janina has in a college savings account if $2000 was invested and earned 6% compounded semiannually for 12 years.

50. Find the amount accrued if $500 is invested and earns 7% compounded monthly for 4 years.

Use a graphing utility to solve. Estimate the result to two decimal places.

51. From Exercise 39, use a table to estimate the number of pounds of uranium that will be available after **(a)** 100 days, **(b)** 120 days, and **(c)** 150 days.

52. From Exercise 42, use a table to estimate the expected bison population for each year for the next 12 years.

53. Solve $7^x = 35$ for x.

54. Solve $3^{2x+1} = 14$ for x.

Review Exercises

Solve each equation. See Sections 2.1 and 6.8.

55. $5x - 2 = 18$

56. $3x - 7 = 11$

57. $3x - 4 = 3(x + 1)$

58. $2 - 6x = 6(1 - x)$

59. $x^2 + 6 = 5x$

60. $18 = 11x - x^2$

By inspection, find the value for x that makes each statement true.

61. $2^x = 8$

62. $3^x = 9$

63. $5^x = \dfrac{1}{5}$

64. $4^x = 1$

11.3 | LOGARITHMIC FUNCTIONS

TAPE IAG 11.3

O B J E C T I V E S

1. Write exponential equations with logarithmic notation and write logarithmic equations with exponential notation.

2. Solve logarithmic equations by using exponential notation.

3. Identify and graph logarithmic functions.

Since the exponential function $f(x) = 2^x$ is a one-to-one function, it has an inverse. We can create a table of values for f^{-1} by switching the coordinates in the accompanying table of values for $f(x) = 2^x$.

x	$y = f(x)$		x	$y = f^{-1}(x)$
-3	$\dfrac{1}{8}$		$\dfrac{1}{8}$	-3
-2	$\dfrac{1}{4}$		$\dfrac{1}{4}$	-2
-1	$\dfrac{1}{2}$		$\dfrac{1}{2}$	-1
0	1		1	0
1	2		2	1
2	4		4	2
3	8		8	3

The graphs of $f(x)$ and its inverse are shown next. Notice that the graphs of f and f^{-1} are symmetric about the line $y = x$, as expected.

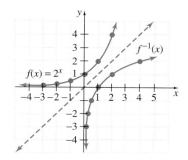

Finally, we would like to be able to write an equation for f^{-1}. To do so, we follow the steps for finding an inverse:

$$f(x) = 2^x$$

Step 1.	Replace $f(x)$ by y.	$y = 2^x$
Step 2.	Interchange x and y.	$x = 2^y$
Step 3.	Solve for y.	

At this point, we are stuck. To solve this equation for y, a new notation, the **logarithmic notation,** is needed. The symbol $\log_b x$ means "the power to which b is raised in order to produce a result of x."

$$\log_b x = y \quad \text{means} \quad b^y = x$$

We say that $\log_b x$ is "the logarithm of x to the base b" or "the log of x to the base b."

LOGARITHMIC DEFINITION

If $b > 0$ and $b \neq 1$, then

$$y = \log_b x \quad \text{means} \quad x = b^y$$

for every $x > 0$ and every real number y.

Before returning to the function $x = 2^y$ and solving it for y in terms of x, let's practice using the new notation $\log_b x$.

It is important to be able to write exponential equations with logarithmic notation, and vice versa. The following table shows examples of both forms.

TECHNOLOGY NOTE

A change of base formula, such as

$$\log_a b = \frac{\log b}{\log a}$$

can be used to evaluate logarithms on a calculator. (See also Section 11.5.)

LOGARITHMIC STATEMENT	CORRESPONDING EXPONENTIAL STATEMENT
$\log_3 9 = 2$	$3^2 = 9$
$\log_6 1 = 0$	$6^0 = 1$
$\log_2 8 = 3$	$2^3 = 8$
$\log_4 \dfrac{1}{16} = -2$	$4^{-2} = \dfrac{1}{16}$
$\log_8 2 = \dfrac{1}{3}$	$8^{1/3} = 2$

REMINDER   Notice that the base of the logarithmic notation is the base of the exponential notation.

EXAMPLE 1   Find the value of each logarithmic expression.

a. $\log_4 16$ **b.** $\log_{10} \dfrac{1}{10}$ **c.** $\log_9 3$

Solution: **a.** $\log_4 16 = 2$ because $4^2 = 16$.   **b.** $\log_{10} \dfrac{1}{10} = -1$ because $10^{-1} = \dfrac{1}{10}$.

         c. $\log_9 3 = \dfrac{1}{2}$ because $9^{1/2} = \sqrt{9} = 3$.

```
log(16)/log(4)
                    2
log(.1)/log(10)
                   -1
log(3)/log(9)▶Fr
ac
                  1/2
```

2   The ability to interchange the logarithmic and exponential forms of a statement is often the key to solving logarithmic equations.

EXAMPLE 2   Solve each equation for x.

a. $\log_4 \dfrac{1}{4} = x$ **b.** $\log_5 x = 3$ **c.** $\log_x 25 = 2$

d. $\log_3 1 = x$ **e.** $\log_b 1 = x$

Solution: **a.** $\log_4 \dfrac{1}{4} = x$ means $4^x = \dfrac{1}{4}$. Solve $4^x = \dfrac{1}{4}$ for x.

$$4^x = \dfrac{1}{4}$$
$$4^x = 4^{-1}$$

Since the bases are the same, by the uniqueness of b^x we have that

$$x = -1$$

The solution set is $\{-1\}$. To check, see that $\log_4 \dfrac{1}{4} = -1$, since $4^{-1} = \dfrac{1}{4}$.

b. $\log_5 x = 3$ means $5^3 = x$ or

$$x = 125$$

The solution set is $\{125\}$.

c. $\log_x 25 = 2$ means $x^2 = 25$ and $x > 0$ and $x \ne 1$.

$$x = 5$$

Even though $(-5)^2 = 25$, the base b of a logarithm must be positive. The solution set is $\{5\}$.

d. $\log_3 1 = x$ means $3^x = 1$. Either solve this equation by inspection or solve by writing 1 as 3^0 as shown:

$$3^x = 3^0 \qquad \text{Write 1 as } 3^0.$$
$$x = 0 \qquad \text{Apply the uniqueness of } b^x.$$

The solution set is $\{0\}$.

e. $\log_b 1 = x$ means $b^x = 1$ and $b > 0$ and $b \neq 1$.

$$b^x = b^0 \qquad \text{Write 1 as } b^0.$$

$$x = 0 \qquad \text{Apply the uniqueness of } b^x.$$

The solution set is $\{0\}$.

In Example 2(e) we proved an important property of logarithms. That is, $\log_b 1$ is always 0. This property as well as two important others are given next.

PROPERTIES OF LOGARITHMS

If b is a real number, $b > 0$ and $b \neq 1$, then

1. $\log_b 1 = 0$
2. $\log_b b^x = x$
3. $b^{\log_b x} = x$

To see that $\log_b b^x = x$, change the logarithmic form to exponential form. Then, $\log_b b^x = x$ means $b^x = b^x$. In exponential form, the statement is true, so in logarithmic form the statement is also true.

EXAMPLE 3   Simplify.

a. $\log_3 3^2$ **b.** $\log_7 7^{-1}$ **c.** $5^{\log_5 3}$ **d.** $2^{\log_2 6}$

Solution: **a.** From property 2, $\log_3 3^2 = 2$.
b. From property 2, $\log_7 7^{-1} = -1$.
c. From property 3, $5^{\log_5 3} = 3$.
d. From property 3, $2^{\log_2 6} = 6$.

Having gained proficiency with the notation $\log_b x$, we return to the function $f(x) = 2^x$ and write an equation for its inverse, $f^{-1}(x)$. Recall our earlier work.

$$f(x) = 2^x$$

Step 1.	Replace $f(x)$ by y.	$y = 2^x$
Step 2.	Interchange x and y.	$x = 2^y$
Step 3.	Solve for y.	$y = \log_2 x$
Step 4.	Replace y with $f^{-1}(x)$.	$f^{-1}(x) = \log_2 x$

Thus, $f^{-1}(x) = \log_2 x$ defines a function that is the inverse function of the function $f(x) = 2^x$. The function $f^{-1}(x)$ or $y = \log_2 x$ is called a **logarithmic function**.

LOGARITHMIC FUNCTION

If x is a positive real number, b is a constant positive real number, and b is not 1, then a **logarithmic function** is a function that can be defined by

$$f(x) = \log_b x$$

The domain of f is the set of positive real numbers, and the range of f is the set of real numbers.

We can explore logarithmic functions by graphing them.

EXAMPLE 4   Graph the logarithmic function $y = \log_2 x$.

Solution: Write the equation with exponential notation as $2^y = x$. Find some ordered pair solutions that satisfy this equation. Plot the points and connect them with a smooth curve. The domain of this function is $(0, \infty)$, and the range of the function is all real numbers, **R.** A graphing utility check is shown to the left.

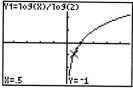

To graph $y = \log_2 x$, we use a change of base formula and enter the

function as $y = \dfrac{\log x}{\log 2}$.

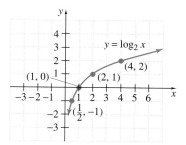

EXAMPLE 5   Graph the logarithmic function $f(x) = \log_{1/3} x$.

Solution: Replace $f(x)$ with y, and write the result with exponential notation:

$$f(x) = \log_{1/3} x$$
$$y = \log_{1/3} x \qquad \text{Replace } f(x) \text{ with } y.$$
$$\left(\frac{1}{3}\right)^y = x \qquad \text{Write in exponential form.}$$

Find ordered pair solutions that satisfy $\left(\frac{1}{3}\right)^y = x$, plot these points, and connect them with a smooth curve as shown in the figure.

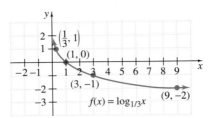

The domain of this function is $(0, \infty)$, and the range is the set of all real numbers.

The following figures summarize characteristics of logarithmic functions.

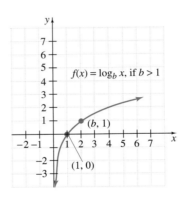

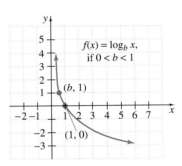

EXERCISE SET 11.3

Find the value of each logarithmic expression. See Example 1.

1. $\log_2 8$

2. $\log_4 64$

3. $\log_3 \dfrac{1}{9}$

4. $\log_2 \dfrac{1}{32}$

5. $\log_{25} 5$

6. $\log_8 \dfrac{1}{2}$

7. $\log_{1/2} 2$

8. $\log_{2/3} \dfrac{4}{9}$

9. $\log_7 1$

10. $\log_9 9$

11. $\log_2 2^4$

12. $\log_6 6^{-2}$

13. $\log_{10} 100$

14. $\log_{10} \dfrac{1}{10}$

15. $3^{\log_3 5}$

16. $5^{\log_5 7}$

17. $\log_3 81$

18. $\log_2 16$

19. $\log_4 \dfrac{1}{64}$

20. $\log_3 \dfrac{1}{9}$

21. Explain why negative numbers are not included as logarithmic bases.

22. Explain why 1 is not included as a logarithmic base.

Solve each equation for x. See Example 2.

23. $\log_3 9 = x$

24. $\log_2 8 = x$

25. $\log_3 x = 4$

26. $\log_2 x = 3$

27. $\log_x 49 = 2$

28. $\log_x 8 = 3$

29. $\log_2 \frac{1}{8} = x$

30. $\log_3 \frac{1}{81} = x$

Solve each equation for x.

31. $\log_3 \frac{1}{27} = x$

32. $\log_5 \frac{1}{125} = x$

33. $\log_8 x = \frac{1}{3}$

34. $\log_9 x = \frac{1}{2}$

35. $\log_4 16 = x$

36. $\log_2 16 = x$

37. $\log_{3/4} x = 3$

38. $\log_{2/3} x = 2$

39. $\log_x 100 = 2$

40. $\log_x 27 = 3$

Simplify. See Example 3.

41. $\log_5 5^3$

42. $\log_6 6^2$

43. $2^{\log_2 3}$

44. $7^{\log_7 4}$

45. $\log_9 9$

46. $\log_8 (8)^{-1}$

Graph each logarithmic function. Label any intercepts. See Example 4.

47. $y = \log_3 x$

48. $y = \log_2 x$

49. $f(x) = \log_{1/4} x$

50. $f(x) = \log_{1/2} x$

51. $f(x) = \log_5 x$

52. $f(x) = \log_6 x$

53. $f(x) = \log_{1/6} x$

54. $f(x) = \log_{1/5} x$

Graph each function and its inverse function on the same set of axes. Label any intercepts. See Example 5.

55. $y = 4^x$; $y = \log_4 x$

56. $y = 3^x$; $y = \log_3 x$

57. $y = \left(\frac{1}{3}\right)^x$; $y = \log_{1/3} x$

58. $y = \left(\frac{1}{2}\right)^x$; $y = \log_{1/2} x$

59. The formula $\log_{10}(1 - k) = \frac{-0.3}{H}$ models the relationship between the half-life H of a radioactive material and its rate of decay k. Find the rate of decay of the iodine isotope I-131 if its half-life is 8 days. Round to 4 decimal places.

60. Explain why the graph of the function $y = \log_b x$ contains the point $(1, 0)$ no matter what b is.

61. $\text{Log}_3 10$ is between which two integers? Explain your answer.

Review Exercises

Simplify each rational expression. See Section 7.1.

62. $\frac{x + 3}{3 + x}$

63. $\frac{x - 5}{5 - x}$

64. $\frac{x^2 - 8x + 16}{2x - 8}$

65. $\frac{x^2 - 3x - 10}{2 + x}$

Add or subtract as indicated. See Section 7.3.

66. $\frac{2}{x} + \frac{3}{x^2}$

67. $\frac{3x}{x + 3} + \frac{9}{x + 3}$

68. $\frac{m^2}{m + 1} - \frac{1}{m + 1}$

69. $\frac{5}{y + 1} - \frac{4}{y - 1}$

11.4   PROPERTIES OF LOGARITHMS

TAPE IAG 11.4

O B J E C T I V E S

1 Apply the product property of logarithms.

2 Apply the quotient property of logarithms.

3 Apply the power property of logarithms.

In the previous section we explored some basic properties of logarithms. We now introduce and explore additional properties. Because a logarithm is an exponent, logarithmic properties are just restatements of exponential properties.

1

The first of these properties is called the **product property of logarithms,** because it deals with the logarithm of a product.

PRODUCT PROPERTY OF LOGARITHMS

If x, y, and b are positive real numbers and $b \neq 1$, then

$$\log_b xy = \log_b x + \log_b y$$

To prove this, let $\log_b x = M$ and $\log_b y = N$. Now write each logarithm with exponential notation:

$$\log_b x = M \qquad \text{is equivalent to} \qquad b^M = x$$
$$\log_b y = N \qquad \text{is equivalent to} \qquad b^N = y$$

Multiply the left sides and the right sides of the exponential equations, and we have that

$$xy = (b^M)(b^N) = b^{M+N}$$

Now, write the equation $xy = b^{M+N}$ in equivalent logarithmic form:

$$xy = b^{M+N} \quad \text{is equivalent to} \quad \log_b xy = M + N$$

But since $M = \log_b x$ and $N = \log_b y$, we can write

$$\log_b xy = M + N$$

as

$$\log_b xy = \log_b x + \log_b y \qquad M = \log_b x \text{ and } N = \log_b y.$$

The logarithm of a product is the sum of the logarithms of the factors. This property is sometimes used to simplify logarithmic expressions.

EXAMPLE 1   Use the product property to write each sum as the logarithm of a single expression. Assume that variables represent positive numbers.

a. $\log_{11} 10 + \log_{11} 3$ **b.** $\log_3 \dfrac{1}{2} + \log_3 12$ **c.** $\log_2 (x + 2) + \log_2 x$

Solution: In each case, both terms have a common logarithmic base.

a. $\log_{11} 10 + \log_{11} 3 = \log_{11} (10 \cdot 3)$ Apply the product property.
$$= \log_{11} 30$$

b. $\log_3 \dfrac{1}{2} + \log_3 12 = \log_3 \left(\dfrac{1}{2} \cdot 12 \right) = \log_3 6$

c. $\log_2 (x + 2) + \log_2 x = \log_2 [(x + 2) \cdot x] = \log_2 (x^2 + 2x)$

2   The second property is the **quotient property of logarithms.**

QUOTIENT PROPERTY OF LOGARITHMS

If x, y, and b are positive real numbers and $b \neq 1$, then

$$\log_b \frac{x}{y} = \log_b x - \log_b y$$

The proof of the quotient property of logarithms is similar to the proof of the product rule. Notice that the quotient property says that the logarithm of a quotient is the difference of the logarithms of the dividend and divisor.

EXAMPLE 2   Use the quotient property to write each difference as the logarithm of a single expression. Assume that x represents positive numbers.

a. $\log_{10} 27 - \log_{10} 3$ **b.** $\log_5 8 - \log_5 x$ **c.** $\log_4 25 + \log_4 3 - \log_4 5$
d. $\log_3 (x^2 + 5) - \log_3 (x^2 + 1)$

Solution: All terms have a common logarithmic base.

a. $\log_{10} 27 - \log_{10} 3 = \log_{10} \dfrac{27}{3} = \log_{10} 9$

b. $\log_5 8 - \log_5 x = \log_5 \dfrac{8}{x}$

c. Use both the product and quotient properties:

$$\log_4 25 + \log_4 3 - \log_4 5 = \log_4 (25 \cdot 3) - \log_4 5 \qquad \text{Apply the product property.}$$
$$= \log_4 75 - \log_4 5 \qquad \text{Simplify.}$$
$$= \log_4 \frac{75}{5} \qquad \text{Apply the quotient property.}$$
$$= \log_4 15 \qquad \text{Simplify.}$$

d. $\log_3 (x^2 + 5) - \log_3 (x^2 + 1) = \log_3 \dfrac{x^2 + 5}{x^2 + 1} \qquad \text{Apply the quotient property.}$

---

3   The third and final property that we introduce is the **power property of logarithms.**

POWER PROPERTY OF LOGARITHMS

If x and b are positive real numbers, $b \neq 1$, and r is a real number, then

$$\log_b x^r = r \log_b x$$

For example,

$$\log_3 2^4 = 4 \log_3 2, \quad \log_5 x^3 = 3 \log_5 x, \quad \log_4 \sqrt{x} = \log_4 (x)^{1/2} = \frac{1}{2} \log_4 x$$

EXAMPLE 3   Write each sum as the logarithm of a single expression. Assume that x is a positive number.

a. $2 \log_5 3 + 3 \log_5 2$ **b.** $2 \log_9 x + \log_9 (x + 1)$

Solution: In each case, both terms have a common logarithmic base.

a. $2 \log_5 3 + 3 \log_5 2 = \log_5 3^2 + \log_5 2^3$ Apply the power property.

$\qquad\qquad\qquad\qquad\quad = \log_5 9 + \log_5 8$

$\qquad\qquad\qquad\qquad\quad = \log_5 (9 \cdot 8)$ Apply the product property.

$\qquad\qquad\qquad\qquad\quad = \log_5 72$

b. $2 \log_9 x + \log_9 (x + 1) = \log_9 x^2 + \log_9 (x + 1)$ Apply the power property.

$\qquad\qquad\qquad\qquad\qquad\quad = \log_9 [x^2(x + 1)]$ Apply the product property.

EXAMPLE 4   Use properties of logarithms to write each expression as a sum or difference of multiples of logarithms.

a. $\log_3 \dfrac{5 \cdot 7}{4}$ **b.** $\log_2 \dfrac{x^5}{y^2}$

Solution: **a.** $\log_3 \dfrac{5 \cdot 7}{4} = \log_3 (5 \cdot 7) - \log_3 4$ Apply the quotient property.

$\qquad\qquad\qquad = \log_3 5 + \log_3 7 - \log_3 4$ Apply the product property.

b. $\log_2 \dfrac{x^5}{y^2} = \log_2 (x^5) - \log_2 (y^2)$ Apply the quotient property.

$\qquad\qquad = 5 \log_2 x - 2 \log_2 y$ Apply the power property.

R E M I N D E R   Notice that we are not able to simplify further a logarithmic expression such as $\log_5 (2x - 1)$. None of the basic properties gives a way to write the logarithm of a difference in some equivalent form.

EXAMPLE 5   If $\log_b 2 = 0.43$ and $\log_b 3 = 0.68$, use the properties of logarithms to evaluate

a. $\log_b 6$ **b.** $\log_b 9$ **c.** $\log_b \sqrt{2}$

Solution: **a.** $\log_b 6 = \log_b (2 \cdot 3)$ Write 6 as $2 \cdot 3$.

$\qquad\qquad = \log_b 2 + \log_b 3$ Apply the product property.

$\qquad\qquad = 0.43 + 0.68$ Substitute given values.

$\qquad\qquad = 1.11$ Simplify.

b. $\log_b 9 = \log_b 3^2$ Write 9 as 3^2.

$\qquad\quad = 2 \log_b 3$

$\qquad\quad = 2(0.68)$ Substitute 0.68 for $\log_b 3$.

$\qquad\quad = 1.36$ Simplify.

c. First, recall that $\sqrt{2} = 2^{1/2}$. Then

$\log_b \sqrt{2} = \log_b 2^{1/2}$ Write $\sqrt{2}$ as $2^{1/2}$.

$\qquad\quad = \dfrac{1}{2} \log_b 2$ Apply the power property.

$\qquad\quad = \dfrac{1}{2}(0.43)$ Substitute the given value.

$\qquad\quad = 0.215$ Simplify.

A summary of the basic properties of logarithms that we have developed so far is given next.

PROPERTIES OF LOGARITHMS

If x, y, and b are positive real numbers, $b \neq 1$, and r is a real number, then:

1. $\log_b 1 = 0$

2. $\log_b b^x = x$

3. $b^{\log_b x} = x$

4. $\log_b xy = \log_b x + \log_b y$ Product property.

5. $\log_b \dfrac{x}{y} = \log_b x - \log_b y$ Quotient property.

6. $\log_b x^r = r \log_b x$ Power property.

EXERCISE SET 11.4

Write each sum as the logarithm of a single expression. Assume that variables represent positive numbers. See Example 1.

1. $\log_5 2 + \log_5 7$

2. $\log_3 8 + \log_3 4$

3. $\log_4 9 + \log_4 x$

4. $\log_2 x + \log_2 y$

5. $\log_{10} 5 + \log_{10} 2 + \log_{10} (x^2 + 2)$

6. $\log_6 3 + \log_6 (x + 4) + \log_6 5$

Write each of the following as the logarithm of a single expression. Assume that variables represent positive numbers. See Example 2.

7. $\log_5 12 - \log_5 4$

8. $\log_7 20 - \log_7 4$

9. $\log_2 x - \log_2 y$

10. $\log_3 12 - \log_3 z$

11. $\log_4 2 + \log_4 10 - \log_4 5$

12. $\log_6 18 + \log_6 2 - \log_6 9$

Write each of the following as the logarithm of a single expression. Assume that variables represent positive numbers. See Example 3.

13. $2 \log_2 5$

14. $3 \log_5 2$

15. $3 \log_5 x + 6 \log_5 z$

16. $2 \log_7 y + 6 \log_7 z$

17. $\log_{10} x - \log_{10} (x + 1) + \log_{10} (x^2 - 2)$

18. $\log_9 (4x) - \log_9 (x - 3) + \log_9 (x^3 + 1)$

Write each expression as a sum or difference of multiples of logarithms. Assume that variables represent positive numbers. See Example 4.

19. $\log_3 \dfrac{4y}{5}$

20. $\log_4 \dfrac{2}{9z}$

21. $\log_2 \dfrac{x^3}{y}$

22. $\log_5 \dfrac{x}{y^4}$

23. $\log_b \sqrt{7x}$

24. $\log_b \sqrt{\dfrac{3}{y}}$

If $\log_b 3 \approx 0.5$ and $\log_b 5 \approx 0.7$, approximate the following. See Example 5.

25. $\log_b \dfrac{5}{3}$

26. $\log_b 25$

27. $\log_b 15$

28. $\log_b \dfrac{3}{5}$

29. $\log_b \sqrt[3]{5}$

30. $\log_b \sqrt[4]{3}$

Write each of the following as the logarithm of a single expression. Assume that variables represent positive numbers.

31. $\log_4 5 + \log_4 7$

32. $\log_3 2 + \log_3 5$

33. $\log_3 8 - \log_3 2$

34. $\log_5 12 - \log_5 3$

35. $\log_7 6 + \log_7 3 - \log_7 4$

36. $\log_8 5 + \log_8 15 - \log_8 20$

37. $3 \log_4 2 + \log_4 6$

38. $2 \log_3 5 + \log_3 2$

39. $3 \log_2 x + \dfrac{1}{2} \log_2 x - 2 \log_2 (x + 1)$

40. $2 \log_5 x + \dfrac{1}{3} \log_5 x - 3 \log_5 (x + 5)$

41. $2 \log_8 x - \dfrac{2}{3} \log_8 x + 4 \log_8 x$

42. $5 \log_6 x - \dfrac{3}{4} \log_6 x + 3 \log_6 x$

Write each expression as a sum or difference of multiples of logarithms. Assume that variables represent positive numbers.

43. $\log_7 \dfrac{5x}{4}$

44. $\log_9 \dfrac{7}{y}$

45. $\log_5 x^3 (x + 1)$

46. $\log_2 y^3 z$

47. $\log_6 \dfrac{x^2}{x + 3}$

48. $\log_3 \dfrac{(x + 5)^2}{x}$

If $\log_b 2 = 0.43$ and $\log_b 3 = 0.68$, evaluate the following.

49. $\log_b 8$

50. $\log_b 81$

51. $\log_b \dfrac{3}{9}$

52. $\log_b \dfrac{4}{32}$

53. $\log_b \sqrt{\dfrac{2}{3}}$

54. $\log_b \sqrt{\dfrac{3}{2}}$

Answer the following true or false.

55. $\log_2 x^3 = 3 \log_2 x$

56. $\log_3 (x + y) = \log_3 x + \log_3 y$

57. $\dfrac{\log_7 10}{\log_7 5} = \log_7 2$

58. $\log_7 \dfrac{14}{8} = \log_7 14 - \log_7 8$

59. $\dfrac{\log_7 x}{\log_7 y} = (\log_7 x) - (\log_7 y)$

60. $(\log_3 6) \cdot (\log_3 4) = \log_3 24$

Review Exercises

61. Graph the functions $y = 10^x$ and $y = \log_{10} x$ on the same set of axes. See Section 11.3.

Evaluate each expression. See Section 11.3.

62. $\log_{10} 100$

63. $\log_{10} \dfrac{1}{10}$

64. $\log_7 7^2$

65. $\log_7 \sqrt{7}$

11.5 | COMMON LOGARITHMS, NATURAL LOGARITHMS, AND CHANGE OF BASE

TAPE IAG 11.5

O B J E C T I V E S

1. Identify common logarithms and approximate them by calculator.
2. Evaluate common logarithms of powers of 10.
3. Identify natural logarithms and approximate them by calculator.
4. Evaluate natural logarithms of powers of e.
5. Apply the change of base formula.

In this section we look closely at two particular logarithmic bases. These two logarithmic bases are used so frequently that logarithms to their bases are given special names. **Common logarithms** are logarithms to base 10. **Natural logarithms** are logarithms to base e, which we introduce in this section. Because of the wide availability and low cost of calculators today, the work in this section is based on the use of calculators, which typically have both the common "log" and the natural "log" keys.

Logarithms to base 10, common logarithms, are used frequently because our number system is a base 10 decimal system. The notation $\log x$ means the same as $\log_{10} x$.

> **COMMON LOGARITHMS**
>
> $$\log x \quad \text{means} \quad \log_{10} x$$

EXAMPLE 1

a. Use a calculator to approximate $\log 7$ to four decimal places.

b. Use a graphing utility to graph $f(x) = \log_{10} x$ in an appropriate window. From the graph, complete the ordered pair (4,).

Solution:

a. See the screen to the left. To four decimal places, $\log 7 \approx 0.8451$.

b. We know from Section 11.3 that the domain of f is the positive real numbers. Define the window to be $[-1, 9, 1]$ by $[-3, 3, 1]$.

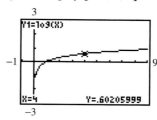

The approximate ordered pair (4, 0.602) is on the graph of $f(x) = \log_{10} x$.

> **2** To evaluate the common log of a power of 10, a calculator is not needed. According to the property of logarithms,
>
> $$\log_b b^x = x$$
>
> It follows that
>
> $$\log 10^x = x$$
>
> because the base of this logarithm is understood to be 10.

EXAMPLE 2   Find the exact value of each logarithm.

     a. $\log 10$ **b.** $\log 1000$ **c.** $\log \dfrac{1}{10}$ **d.** $\log \sqrt{10}$

Solution: **a.** $\log 10 = \log 10^1 = 1$ **b.** $\log 1000 = \log 10^3 = 3$

              c. $\log \dfrac{1}{10} = \log 10^{-1} = -1$ **d.** $\log \sqrt{10} = \log 10^{1/2} = \dfrac{1}{2}$

As we will soon see, equations containing common logs are useful models of many natural phenomena.

EXAMPLE 3   Solve $\log x = 1.2$ for x. Give an exact solution, and then approximate the solution to four decimal places.

Solution: Write the logarithmic equation with exponential notation. Keep in mind that the base of a common log is understood to be 10.

$$\log x = 1.2$$
$$10^{1.2} = x \qquad \text{Write with exponential notation.}$$

The exact solution is $10^{1.2}$. To approximate, use a calculator.

$$15.848932 \approx x$$

To four decimal places, $x \approx 15.8489$. We can check this solution with a graphing utility by graphing the left side and the right side of the original equation and finding the point of intersection.

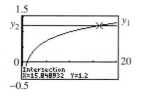

The point of intersection of $y_1 = \log x$ and $y_2 = 1.2$ is approximately (15.8489, 1.2). Thus, the solution of the equation $\log x = 1.2$ is $x \approx 15.8489$.

The Richter scale measures the intensity, or magnitude, of an earthquake. The formula for the magnitude R of an earthquake is $R = \log\left(\dfrac{a}{T}\right) + B$, where a is the amplitude in micrometers of the vertical motion of the ground at the recording station, T is the number of seconds between successive seismic waves, and B is an adjustment factor that takes into account the weakening of the seismic wave as the distance increases from the epicenter of the earthquake.

EXAMPLE 4   MEASURING EARTHQUAKE INTENSITY
Find an earthquake's magnitude on the Richter scale if a recording station measures an amplitude of 300 micrometers and 2.5 seconds between waves. Assume
that B is 4.2. Approximate the solution to the nearest tenth.

Solution: Substitute the known values into the formula for earthquake intensity.

$$R = \log\left(\frac{a}{T}\right) + B \qquad \text{Richter scale formula.}$$

$$= \log\left(\frac{300}{2.5}\right) + 4.2 \qquad \text{Let } a = 300, T = 2.5, \text{ and } B = 4.2.$$

$$= \log(120) + 4.2$$

$$\approx 2.1 + 4.2 \qquad \text{Approximate } \log 120 \text{ by } 2.1.$$

$$= 6.3$$

This earthquake had a magnitude of 6.3 on the Richter scale.

3   **Natural logarithms** are also frequently used, especially to describe natural events;
hence the label "natural logarithm." Natural logarithms are logarithms to the base
e, which is a constant approximately equal to 2.7183. The number e is an irrational
number, as is π. The notation $\log_e x$ is usually abbreviated to $\ln x$. (The abbrevia-
tion ln is read "el en.")

NATURAL LOGARITHMS

   $\ln x$ means $\log_e x$

The graph of $y = \ln x$ is shown next.

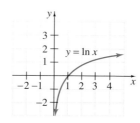

The screen below shows some values of the natural logarithm function.

EXAMPLE 5 Use a calculator to approximate ln 8 to four decimal places.

Solution: See the screen below.

```
ln(8)
        2.079441542
```

To four decimal places, ln 8 ≈ 2.0794.

4 As a result of the property $\log_b b^x = x$, we know that $\log_e e^x = x$, or $\ln e^x = x$.

EXAMPLE 6 Find the exact value of each natural logarithm.

a. $\ln e^3$ **b.** $\ln \sqrt[5]{e}$

Solution: **a.** $\ln e^3 = 3$ **b.** $\ln \sqrt[5]{e} = \ln e^{1/5} = \dfrac{1}{5}$

EXAMPLE 7 Solve the equation $\ln 3x = 5$ for x. Give an exact solution and then approximate the solution to four decimal places.

Solution: Write the equation with exponential notation. Keep in mind that the base of a natural logarithm is understood to be e.

$$\ln 3x = 5$$
$$e^5 = 3x \qquad \text{Write with exponential notation.}$$
$$\frac{e^5}{3} = x$$

The point of intersection of $y_1 = \ln(3x)$ and $y_2 = 5$ is approximately (49.4711, 5). Thus, the solution of the equation $\ln 3x = 5$ is $x \approx 49.4711$.

The exact solution is $\dfrac{e^5}{3}$. To four decimal places, $x \approx 49.4711$. To check this solution with a graphing utility, see the screen to the left.

Recall from Section 11.2 the formula $A = P\left(1 + \dfrac{r}{n}\right)^{nt}$ for compound interest, where n represents the number of compoundings per year. When interest is compounded continuously, the formula $A = Pe^{rt}$ is used, where r is the annual interest rate and interest is compounded continuously for t years.

EXAMPLE 8 **REPAYING A LOAN**
Find the amount owed at the end of 5 years if $1600 is loaned at a rate of 9% compounded continuously.

Solution: Use the formula $A = Pe^{rt}$, where

$$P = \$1600 \text{ (the size of the loan)}$$
$$r = 9\% = 0.09 \text{ (the rate of interest)}$$
$$t = 5 \text{ (the 5-year duration of the loan)}$$
$$A = Pe^{rt}$$
$$= 1600e^{0.09(5)} \qquad \text{Substitute in known values.}$$
$$= 1600e^{0.45}$$

Now use a calculator to approximate the solution.

$$A \approx 2509.30$$

The total amount of money owed is approximately \$2509.30.

5   Calculators are handy tools for approximating natural and common logarithms. Unfortunately, some calculators cannot be used to approximate logarithms to bases other than e or 10, at least not directly. In such cases, we use the change of base formula.

CHANGE OF BASE

If a, b, and c are positive real numbers and neither b nor c is 1, then

$$\log_b a = \frac{\log_c a}{\log_c b}$$

EXAMPLE 9   Approximate $\log_5 3$ to four decimal places.

Solution: Use the change of base property to write $\log_5 3$ as a quotient of logarithms to base 10:

$$\log_5 3 = \frac{\log 3}{\log 5} \qquad \text{Use the change of base property.}$$

$$\approx \frac{0.4771213}{0.69897} \qquad \text{Approximate logarithms by calculator.}$$

$$\approx 0.6826063 \qquad \text{Simplify by calculator.}$$

To four decimal places, $\log_5 3 \approx 0.6826$.

EXERCISE SET 11.5

Use a calculator to approximate each logarithm to four decimal places. See Examples 1 and 5.

1. log 8

2. log 6

3. log 2.31

4. log 4.86

5. ln 2

6. ln 3

7. ln 0.0716

8. ln 0.0032

9. log 12.6

10. log 25.9

11. ln 5

12. ln 7

13. log 41.5

14. ln 41.5

15. Use a calculator and try to approximate log 0. Describe what happens and explain why.

16. Use a calculator and try to approximate ln 0. Describe what happens and explain why.

Find the exact value. See Examples 2 and 6.

17. log 100

18. log 10,000

19. $\log\left(\dfrac{1}{1000}\right)$

20. $\log\left(\dfrac{1}{100}\right)$

21. $\ln e^2$

22. $\ln e^4$

23. $\ln \sqrt[4]{e}$

24. $\ln \sqrt[5]{e}$

25. $\log 10^3$

26. $\ln e^5$

27. $\ln e^2$

28. $\log 10^7$

29. log 0.0001

30. log 0.001

31. $\ln \sqrt{e}$

32. $\log \sqrt{10}$

33. Without using a calculator, explain which of log 50 or ln 50 must be larger.

34. Without using a calculator, explain which of $\log 50^{-1}$ or $\ln 50^{-1}$ must be larger.

Solve each equation for x. Give an exact solution and a four-decimal-place approximation. See Examples 3 and 7.

35. log x = 1.3

36. log x = 2.1

37. log 2x = 1.1

38. log 3x = 1.3

39. ln x = 1.4

40. ln x = 2.1

41. ln (3x − 4) = 2.3

42. ln (2x + 5) = 3.4

43. log x = 2.3

44. log x = 3.1

45. ln x = −2.3

46. ln x = −3.7

47. log (2x + 1) = −0.5

48. log (3x − 2) = −0.8

49. ln 4x = 0.18

50. ln 3x = 0.76

Approximate each logarithm to four decimal places. See Example 9.

51. $\log_2 3$

52. $\log_3 2$

53. $\log_{1/2} 5$

54. $\log_{1/3} 2$

55. $\log_4 9$

56. $\log_9 4$

57. $\log_3 \dfrac{1}{6}$

58. $\log_6 \dfrac{2}{3}$

59. $\log_8 6$

60. $\log_6 8$

Use the formula $R = \log\left(\dfrac{a}{T}\right) + B$ to find the intensity R on the Richter scale of the earthquakes that fit the de-

scriptions given. Round answers to one decimal place. See Example 4.

61. Amplitude *a* is 200 micrometers, time *T* between waves is 1.6 seconds, and *B* is 2.1.

62. Amplitude *a* is 150 micrometers, time *T* between waves is 3.6 seconds, and *B* is 1.9.

63. Amplitude *a* is 400 micrometers, time *T* between waves is 2.6 seconds, and *B* is 3.1.

64. Amplitude *a* is 450 micrometers, time *T* between waves is 4.2 seconds, and *B* is 2.7.

Use the formula $A = Pe^{rt}$ to solve. See Example 8.

65. Find how much money Dana Jones has after 12 years if $1400 is invested at 8% interest compounded continuously.

66. Determine the size of an account in which $3500 earns 6% interest compounded continuously for 1 year.

67. Find the amount of money Barbara Mack owes at the end of 4 years if 6% interest is compounded continuously on her $2000 debt.

68. Find the amount of money for which a $2500 certificate of deposit is redeemable if it has been paying 10% interest compounded continuously for 3 years.

Graph each function by finding ordered pair solutions, plotting the solutions, and then drawing a smooth curve through the plotted points. Use a graphing utility to check your results.

69. $f(x) = e^x$

70. $f(x) = e^{2x}$

71. $f(x) = e^{-3x}$

72. $f(x) = e^{-x}$

73. $f(x) = e^x + 2$

74. $f(x) = e^x - 3$

75. $f(x) = e^{x-1}$

76. $f(x) = e^{x+4}$

77. $f(x) = 3e^x$

78. $f(x) = -2e^x$

79. $f(x) = \ln x$

80. $f(x) = \log x$

81. $f(x) = -2 \log x$

82. $f(x) = 3 \ln x$

83. $f(x) = \log (x + 2)$

84. $f(x) = \log (x - 2)$

85. $f(x) = \ln x - 3$

86. $f(x) = \ln x + 3$

87. Graph $f(x) = e^x$ (Exercise 69), $f(x) = e^x + 2$ (Exercise 73), and $f(x) = e^x - 3$ (Exercise 74) on the same screen. Discuss any trends shown on the graphs.

88. Graph $f(x) = \ln x$ (Exercise 79), $f(x) = \ln x - 3$ (Exercise 85), and $f(x) = \ln x + 3$ (Exercise 86). Discuss any trends shown on the graphs.

Review Exercises

Solve each equation for x. See Sections 2.1 and 6.8.

89. $6x - 3(2 - 5x) = 6$

90. $2x + 3 = 5 - 2(3x - 1)$

91. $2x + 3y = 6x$ **92.** $4x - 8y = 10x$

93. $x^2 + 7x = -6$ **94.** $x^2 + 4x = 12$

Solve each system of equations. See Section 5.1.

95. $\begin{cases} x + 2y = -4 \\ 3x - y = 9 \end{cases}$ **96.** $\begin{cases} 5x + y = 5 \\ -3x - 2y = -10 \end{cases}$

11.6 | EXPONENTIAL AND LOGARITHMIC EQUATIONS AND PROBLEM SOLVING

TAPE IAG 11.6

O B J E C T I V E S

 Solve exponential equations.

2. Solve logarithmic equations.

3. Solve problems that can be modeled by exponential and logarithmic equations.

1 In Section 11.2 we solved exponential equations such as $2^x = 16$ by writing 16 as a power of 2 and applying the uniqueness of b^x:

$$2^x = 16$$
$$2^x = 2^4 \qquad \text{Write 16 as } 2^4.$$
$$x = 4 \qquad \text{Apply the uniqueness of } b^x.$$

To solve an equation such as $3^x = 7$, we use the fact that $f(x) = \log_b x$ is a one-to-one function. Another way of stating this fact is as a property of equality.

LOGARITHM PROPERTY OF EQUALITY

Let a, b, and c be real numbers such that $\log_b a$ and $\log_b c$ are real numbers and b is not 1. Then

$$\log_b a = \log_b c \quad \text{is equivalent to} \quad a = c$$

EXAMPLE 1   Solve $3^x = 7$.

Solution: To solve, use the logarithm property of equality and take the logarithm of both sides. For this example, we use the common logarithm.

$$3^x = 7$$
$$\log 3^x = \log 7 \qquad \text{Take the common log of both sides.}$$
$$x \log 3 = \log 7 \qquad \text{Apply the power property of logarithms.}$$
$$x = \frac{\log 7}{\log 3} \qquad \text{Divide both sides by log 3.}$$

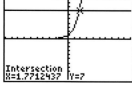

To verify the solution of the equation $3^x = 7$, graph $y_1 = 3^x$ and $y_2 = 7$. See that $x \approx 1.7712$.

The exact solution is $\dfrac{\log 7}{\log 3}$. If a decimal approximation is preferred,

$$\frac{\log 7}{\log 3} \approx \frac{0.845098}{0.4771213} \approx 1.7712 \text{ to four decimal places. The solution set is } \left\{\frac{\log 7}{\log 3}\right\}, \text{ or}$$

approximately $\{1.7712\}$.

2

By applying the appropriate properties of logarithms, we can solve a broad variety of logarithmic equations.

EXAMPLE 2 Solve $\log_4 (x - 2) = 2$ for x.

Solution: Notice that $x - 2$ must be positive, so x must be greater than 2. First, write the equation with exponential notation:

$$\log_4 (x - 2) = 2$$
$$4^2 = x - 2$$
$$16 = x - 2$$
$$18 = x \qquad \text{Add 2 to both sides.}$$

To check numerically, replace x with 18 in the *original equation*:

$$\log_4 (x - 2) = 2$$
$$\log_4 (18 - 2) = 2 \qquad \text{Let } x = 18.$$
$$\log_4 16 = 2$$
$$4^2 = 16 \qquad \text{True.}$$

The solution set is $\{18\}$.

To check graphically,

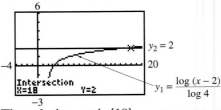

$y_1 = \dfrac{\log (x - 2)}{\log 4}$

The solution set is $\{18\}$.

EXAMPLE 3 Solve $\log_2 x + \log_2 (x - 1) = 1$ for x.

Solution: Notice that $x - 1$ must be positive, so x must be greater than 1. Apply the product rule to the left side of the equation:

$$\log_2 x + \log_2 (x - 1) = 1$$
$$\log_2 x(x - 1) = 1 \qquad \text{Use the product rule.}$$
$$\log_2 (x^2 - x) = 1$$

Next, write the equation with exponential notation and solve for x.

$$2^1 = x^2 - x$$
$$0 = x^2 - x - 2 \qquad \text{Subtract 2 from both sides.}$$
$$0 = (x - 2)(x + 1) \qquad \text{Factor.}$$
$$0 = x - 2 \quad \text{or} \quad 0 = x + 1 \qquad \text{Set each factor equal to 0.}$$
$$2 = x \qquad\qquad -1 = x$$

Verify that 2 satisfies the original equation.

Recall that -1 cannot be a solution because x must be greater than 1. If this is not noticed, -1 will be rejected after checking. To see this, replace x with -1 in the original equation.

$$\log_2 x + \log_2 (x - 1) = 1$$
$$\log_2 (-1) + \log_2 (-1 - 1) = 1 \qquad \text{Let } x = -1.$$

Because the logarithm of a negative number is undefined, -1 is an extraneous solution. The solution set is $\{2\}$.

EXAMPLE 4   Solve $\log (x + 2) - \log x = 2$ for x.

Solution: Apply the quotient property of logarithms to the left side of the equation.

$$\log (x + 2) - \log x = 2$$
$$\log \frac{x + 2}{x} = 2 \qquad \text{Apply the quotient property.}$$
$$10^2 = \frac{x + 2}{x} \qquad \text{Write with exponential notation.}$$
$$100 = \frac{x + 2}{x}$$
$$100x = x + 2 \qquad \text{Multiply both sides by } x.$$
$$99x = 2 \qquad \text{Subtract } x \text{ from both sides.}$$
$$x = \frac{2}{99} \qquad \text{Divide both sides by 99.}$$

```
2/99→X:log(X+2)-
log (X)
                2
```

To verify with a calculator that the solution set is $\left\{\dfrac{2}{99}\right\}$, see the screen at the left.

---

3   Throughout this chapter we have emphasized that logarithmic and exponential functions are used in a variety of scientific, technical, and business settings. A few examples follow.

EXAMPLE 5   ESTIMATING POPULATION SIZE
The population size y of a community of lemmings varies according to the relationship $y = y_0 e^{0.15t}$. In this formula, t is time in months, and y_0 is some initial population at time 0. Estimate the population after 6 months if there were originally 5000 lemmings.

Solution: Substitute 5000 for y_0 and 6 for t.

$$y = y_0 e^{0.15t}$$
$$= 5000 e^{0.15(6)} \qquad \text{Let } t = 6 \text{ and } y_0 = 5000.$$
$$= 5000 e^{0.9} \qquad \text{Multiply.}$$

Using a calculator, we find that $y \approx 12{,}298.016$. In 6 months the population will be approximately 12,300 lemmings.

EXAMPLE 6   FINDING DOUBLING TIME

How long does it take an investment of $2000 to double if it is invested at 5% interest compounded quarterly? The necessary formula is $A = P\left(1 + \dfrac{r}{n}\right)^{nt}$, where A is the accrued (or owed) amount, P is the principal invested, r is the annual rate of interest, n is the number of compounding periods per year, and t is the number of years.

Solution: We are given that $P = \$2000$ and $r = 5\% = 0.05$. Compounding quarterly means 4 times a year, so $n = 4$. The investment is to double, so A must be $4000. Substitute these values and solve for t.

$$A = P\left(1 + \frac{r}{n}\right)^{nt}$$

$$4000 = 2000\left(1 + \frac{0.05}{4}\right)^{4t} \qquad \text{Substitute in known values.}$$

$$4000 = 2000(1.0125)^{4t} \qquad \text{Simplify } 1 + \frac{0.05}{4}.$$

$$2 = (1.0125)^{4t} \qquad \text{Divide both sides by 2000.}$$

$$\log 2 = \log 1.0125^{4t} \qquad \text{Take the logarithm of both sides.}$$

$$\log 2 = 4t(\log 1.0125) \qquad \text{Apply the power property.}$$

$$\frac{\log 2}{4 \log 1.0125} = t \qquad \text{Divide both sides by 4 log 1.0125.}$$

$$13.949408 \approx t \qquad \text{Approximate by calculator.}$$

Thus, it takes nearly 14 years for the money to double in value.

EXAMPLE 7   TRACKING AN INVESTMENT

Suppose that you invest $1500 at an annual rate of 8% compounded monthly.

a. Make a table giving the value of the investment at the end of each year for the next 14 years.

b. How long does it take the investment to triple?

Solution: **a.** First, let $P = \$1500$, $r = 0.08$, and $n = 12$ (for monthly compounding) in the formula $A = P\left(1 + \dfrac{r}{n}\right)^{nt}$. Using the function $A = 1500\left(1 + \dfrac{0.08}{12}\right)^{12t}$, we can use a graphing utility to make the following table shown in two screens:

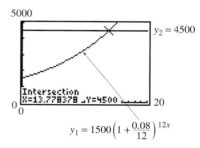

b. Notice that when the investment has tripled the accrued amount A is \$4500. Use a graphing utility to approximate the value of x that gives $A = 4500$. We can see from the table in part (a) that the value of the investment is \$4500 between $x = 13$ years and $x = 14$ years. Using this information, we can determine an appropriate viewing window, such as $[0, 20, 1]$ by $[0, 5000, 1000]$. Graph $y_1 = 1500(1 + 0.08/12)^{12x}$ and $y_2 = 4500$ in the same window. As usual, the point of intersection of the two curves is the solution.

The screen above shows that it takes $x \approx 13.78$ years, or 13 years and 9 months, for the original investment to triple in value to \$4500.

EXERCISE SET 11.6

Solve each equation. Give an exact solution, and also approximate the solution to four decimal places. Verify the result with a graphing utility. See Example 1.

1. $3^x = 6$

2. $4^x = 7$

3. $3^{2x} = 3.8$

4. $5^{3x} = 5.6$

5. $2^{x-3} = 5$

6. $8^{x-2} = 12$

7. $9^x = 5$

8. $3^x = 11$

9. $4^{x+7} = 3$

10. $6^{x+3} = 2$

11. $7^{3x-4} = 11$

12. $5^{2x-6} = 12$

13. $e^{6x} = 5$

14. $e^{2x} = 8$

Solve each equation. See Examples 2 through 4.

15. $\log_2 (x + 5) = 4$

16. $\log_6 (x^2 - x) = 1$

17. $\log_3 x^2 = 4$

18. $\log_2 x^2 = 6$

19. $\log_4 2 + \log_4 x = 0$

20. $\log_3 5 + \log_3 x = 1$

21. $\log_2 6 - \log_2 x = 3$

22. $\log_4 10 - \log_4 x = 2$

23. $\log_4 x + \log_4 (x + 6) = 2$

24. $\log_3 x + \log_3 (x + 6) = 3$

25. $\log_5 (x + 3) - \log_5 x = 2$

26. $\log_6 (x + 2) - \log_6 x = 2$

27. $\log_3 (x - 2) = 2$

28. $\log_2 (x - 5) = 3$

29. $\log_4 (x^2 - 3x) = 1$

30. $\log_8 (x^2 - 2x) = 1$

31. $\ln 5 + \ln x = 0$

32. $\ln 3 + \ln (x - 1) = 0$

33. $3 \log x - \log x^2 = 2$

34. $2 \log x - \log x = 3$

35. $\log_2 x + \log_2 (x + 5) = 1$

36. $\log_4 x + \log_4 (x + 7) = 1$

37. $\log_4 x - \log_4 (2x - 3) = 3$

38. $\log_2 x - \log_2 (3x + 5) = 4$

39. $\log_2 x + \log_2 (3x + 1) = 1$

40. $\log_3 x + \log_3 (x - 8) = 2$

Solve. See Example 5.

41. The size of the wolf population at Isle Royale National Park increases at a rate of 4.3% per year. If the size of the current population is 83 wolves, find how many there should be in 5 years. Use $y = y_0 e^{0.043t}$ and round to the nearest whole.

42. The number of victims of a flu epidemic is increasing at a rate of 7.5% per week. If 20,000 persons are currently infected, find in how many days we can expect 45,000 to have the flu. Use $y = y_0 e^{0.075t}$ and round to the nearest whole.

43. The size of the population of Senegal is increasing at a rate of 2.6% per year. If 9,000,000 people lived in Senegal in 1995, find how many inhabitants there will be by 2000. Round to the nearest ten thousand. Use $y = y_0 e^{0.026t}$.

44. In 1995, 937 million people were citizens of India. Find how long it will take India's population to reach a size of 1500 million (that is, 1.5 billion) if the population size is growing at a rate of 2.1% per year. Round to the nearest tenth. Use $y = y_0 e^{0.021t}$.

Use the formula $A = P\left(1 + \dfrac{r}{n}\right)^{nt}$ to solve these compound interest problems. Round money to the nearest hundredth and years to the nearest tenth. See Examples 6 and 7.

45. Suppose that you invest $600 at an annual interest rate of 7% compounded monthly.
    a. Make a table giving the value of the investment at the end of each year for the next 12 years.
    b. How long does it take the investment to double?

46. Suppose that you invest $600 at an annual interest rate of 12% compounded monthly.
    a. Make a table giving the value of the investment at the end of each year for the next 7 years.
    b. How long does it take the investment to double?

47. Find how long it takes a $1200 investment to earn $200 interest if it is invested at 9% interest compounded quarterly.

48. Find how long it takes a $1500 investment to earn $200 interest if it is invested at 10% compounded semiannually.

49. Find how long it takes $1000 to double if it is invested at 8% interest compounded semiannually.

50. Find how long it takes $1000 to double if it is invested at 8% interest compounded monthly.

The formula $w = 0.00185h^{2.67}$ is used to estimate the normal weight w of a boy h inches tall. Use this formula to solve the height–weight problems. Round to the nearest tenth.

51. Find the expected height of a boy who weighs 85 pounds.

52. Find the expected height of a boy who weighs 140 pounds.

The formula $P = 14.7e^{-0.21x}$ gives the average atmospheric pressure P, in pounds per square inch, at an altitude x, in miles above sea level. Use this formula to solve these pressure problems. Round answers to the nearest tenth.

53. Find the average atmospheric pressure of Denver, which is 1 mile above sea level.

54. Find the average atmospheric pressure of Pikes Peak, which is 2.7 miles above sea level.

55. Find the elevation of a Delta jet if the atmospheric pressure outside the jet is 7.5 lb/in.^{2}.

56. Find the elevation of a remote Himalayan peak if the atmospheric pressure atop the peak is 6.5 lb/in.^{2}.

Psychologists call the graph of the formula

$$t = \frac{1}{c} \ln \left(\frac{A}{A - N} \right)$$ *the learning curve, since the formula*

relates time t passed, in weeks, to a measure N of learning achieved, to a measure A of maximum learning possible, and to a measure c of an individual's learning style. Round to the nearest week.

57. Norman is learning to type. If he wants to type at a rate of 50 words per minute (N is 50) and his expected maximum rate is 75 words per minute (A is 75), find how many weeks it should take him to achieve his goal. Assume that c is 0.09.

58. In an experiment teaching chimpanzees sign language, a typical chimp can master a maximum of 65 signs. Find how many weeks it should take a chimpanzee to master 30 signs if c is 0.03.

59. Janine is working on her dictation skills. She wants to take dictation at a rate of 150 words per minute and believes that the maximum rate that she can hope for is 210 words per minute. Find how many weeks it should take her to achieve the 150-word level if c is 0.07.

60. A psychologist is measuring human capability to memorize nonsense syllables. Find how many weeks it should take a subject to learn 15 nonsense syllables if the maximum possible to learn is 24 syllables and c is 0.17.

Review Exercises

If $x = -2$, $y = 0$, and $z = 3$, find the value of each expression. See Section 1.4.

61. $\dfrac{x^2 - y + 2z}{3x}$

62. $\dfrac{x^3 - 2y + z}{2z}$

63. $\dfrac{3z - 4x + y}{x + 2z}$

64. $\dfrac{4y - 3x + z}{2x + y}$

Find the inverse function of each one-to-one function. See Section 11.1.

65. $f(x) = 5x + 2$

66. $f(x) = \dfrac{x - 3}{4}$

GROUP ACTIVITY

MODELING TEMPERATURE

METHOD 1 MATERIALS:
- A container of either cold or hot liquid, thermometer, stopwatch, grapher with curve-fitting capabilities (optional).

METHOD 2 MATERIALS:
- A container of either cold or hot liquid, TI-82, TI-83, or TI-85 graphing calculator with unit-to-unit link cable, TI-CBL (Calculator-Based Laboratory) unit with temperature probe.

Newton's law of cooling relates the temperature of an object to the time elapsed since its warming or cooling began. In this activity you will investigate experimental data to find a mathematical model for this relationship. You may collect the temperature data by using either Method 1 (stopwatch and thermometer) or Method 2 (CBL).

t	T
0	

Method 1:

a. Insert the thermometer into the liquid and allow a thermometer reading to register. Take a temperature reading T as you start the stopwatch (at $t = 0$) and record it in the accompanying data table.

b. Continue taking temperature readings at uniform intervals anywhere between 5 and 10 minutes long. At each reading use the stopwatch to measure the length of time that has elapsed *since the temperature readings started.* Record your time t and liquid temperature T in the data table. Gather data for six to twelve readings.

c. Plot the data from the data table. Plot t on the horizontal axis and T on the vertical axis.

Method 2:

a. Prepare the CBL unit and TI-82, TI-83, or TI-85 graphing calculator. Insert the temperature probe into the liquid.

b. Start the HEAT program on the TI graphing calculator and follow its instructions to begin collecting data. The program will collect 36 temperature readings in degrees Celsius and plot them in real time with t on the horizontal axis and T on the vertical axis.

1. Which of the following mathematical models best fits the data you collected? Explain your reasoning. (Assume $a > 0$.)

   a. $T = ab^t + c$
   b. $T = ab^{-t} + c$
   c. $T = -ab^{-t} + c$
   d. $T = \ln(-ax + b) + c$
   e. $T = -\ln(-ax + b) + c$

2. What does the constant c represent in the model you chose? What is the value of c in this activity?

3. (Optional) Subtract the value of c from each of your observations of T. Enter the new ordered pairs $(t, T - c)$ into a grapher. Use the exponential or logarithmic curve-fitting feature to find a model for your experimental data. Graph the ordered pairs $(t, T - c)$ with the model that you found. How well does the model fit the data? How does the model compare with your selection from question 1?

CHAPTER 11   HIGHLIGHTS

DEFINITIONS AND CONCEPTS	EXAMPLES

SECTION 11.1   COMPOSITE AND INVERSE FUNCTIONS

The notation $(f \circ g)(x)$ means "f composed with g":

$$(f \circ g)(x) = f(g(x))$$
$$(g \circ f)(x) = g(f(x))$$

Let $f(x) = x^2 + 1$ and let $g(x) = x - 5$.

$$(f \circ g)(x) = f(g(x))$$
$$= f(x - 5)$$
$$= (x - 5)^2 + 1$$
$$= x^2 - 10x + 26$$

If f is a function, then f is a **one-to-one function** only if each y-value (output) corresponds to only one x-value (input).

Determine whether each graph is a one-to-one function.

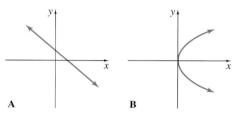

A   B

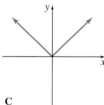

C

Horizontal line test:

If every horizontal line intersects the graph of a function at most once, then the function is a one-to-one function.

The **inverse** of a one-to-one function f is the one-to-one function f^{-1} that is the set of all ordered pairs (b, a) such that (a, b) belongs to f.

To find the inverse of a one-to-one function $f(x)$,

Step 1.   Replace $f(x)$ with y.

Step 2.   Interchange x and y.

Step 3.   Solve for y.

Step 4.   Replace y with $f^{-1}(x)$.

Graphs A and C pass the vertical line test, so only these are graphs of functions. Of graphs A and C, only graph A passes the horizontal line test, so only graph A is the graph of a one-to-one function.

Find the inverse of $f(x) = 2x + 7$.

$$y = 2x + 7 \qquad \text{Let } f(x) \text{ be } y.$$
$$x = 2y + 7 \qquad \text{Switch } x \text{ and } y.$$
$$2y = x - 7$$
$$y = \frac{x - 7}{2} \qquad \text{Solve for } y.$$
$$f^{-1}(x) = \frac{x - 7}{2} \qquad \text{Let } y \text{ be } f^{-1}(x).$$

The inverse of $f(x) = 2x + 7$ is $f^{-1}(x) = \dfrac{x - 7}{2}$.

DEFINITIONS AND CONCEPTS	**EXAMPLES**

SECTION 11.2 EXPONENTIAL FUNCTIONS

A function of the form $f(x) = b^x$ is an **exponential function,** where $b > 0, b \neq 1$, and x is a real number.

Graph the exponential function $y = 4^x$.

x	y
-2	$\dfrac{1}{16}$
-1	$\dfrac{1}{4}$
0	1
1	4
2	16

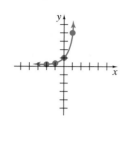

Solve $2^{x+5} = 8$.

Uniqueness of b^x

If $b > 0$ and $b \neq 1$, then $b^x = b^y$ is equivalent to $x = y$.

$2^{x+5} = 2^3$	Write 8 as 2^3.
$x + 5 = 3$	Apply uniqueness of b^x.
$x = -2$	Subtract 5.

SECTION 11.3 LOGARITHMIC FUNCTIONS

Logarithmic definition:

If $b > 0$ and $b \neq 1$, then

$$y = \log_b x \quad \text{means} \quad x = b^y$$

for positive x and real number y.

LOGARITHMIC FORM	CORRESPONDING EXPONENTIAL STATEMENT
$\log_5 25 = 2$	$5^2 = 25$
$\log_9 3 = \dfrac{1}{2}$	$9^{1/2} = 3$

Properties of logarithms:

If b is a real number, $b > 0$, and $b \neq 1$, then

$$\log_b 1 = 0, \quad \log_b b^x = x, \quad b^{\log_b x} = x$$

$\log_5 1 = 0, \quad \log_7 7^2 = 2, \quad 3^{\log_3 6} = 6$

Logarithmic function:

If $b > 0$ and $b \neq 1$, then a **logarithmic function** is a function that can be defined as

$$f(x) = \log_b x$$

The domain of f is the set of positive real numbers, and the range of f is the set of real numbers.

Graph $y = \log_3 x$.

Write $y = \log_3 x$ as $3^y = x$. Some ordered pair solutions are listed in the table.

x	y
3	1
1	0
$\dfrac{1}{3}$	-1
$\dfrac{1}{9}$	-2

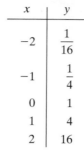

DEFINITIONS AND CONCEPTS	**EXAMPLES**

SECTION 11.4   PROPERTIES OF LOGARITHMS

Let x, y, and b be positive numbers and $b \neq 1$.

Product property:

$$\log_b xy = \log_b x + \log_b y$$

Quotient property:

$$\log_b \frac{x}{y} = \log_b x - \log_b y$$

Power property:

$$\log_b x^r = r \log_b x$$

Write as a single logarithm:

$2 \log_5 6 + \log_5 x - \log_5 (y + 2)$

$= \log_5 6^2 + \log_5 x - \log_5 (y + 2)$ Power property.

$= \log_5 36 \cdot x - \log_5 (y + 2)$ Product property.

$= \log_5 \dfrac{36x}{y + 2}$ Quotient property.

SECTION 11.5   COMMON LOGARITHMS, NATURAL LOGARITHMS, AND CHANGE OF BASE

Common logarithms:

$\log x$ means $\log_{10} x$

Natural logarithms:

$\ln x$ means $\log_e x$

Continuously compounded interest formula:

$$A = Pe^{rt}$$

where r is the annual interest rate for P dollars invested for t years.

$\log 5 = \log_{10} 5 \approx 0.69897$

$\ln 7 = \log_e 7 \approx 1.94591$

Find the amount in an account at the end of 3 years if \$1000 is invested at an interest rate of 4% compounded continuously.

Here, $t = 3$ years, $P = \$1000$, and $r = 0.04$:

$A = Pe^{rt}$

$= 1000e^{0.04(3)}$

$\approx \$1127.50$

SECTION 11.6   EXPONENTIAL AND LOGARITHMIC EQUATIONS AND PROBLEM SOLVING

Logarithm property of equality:

Let $\log_b a$ and $\log_b c$ be real numbers and $b \neq 1$. Then

$$\log_b a = \log_b c \quad \text{is equivalent to} \quad a = c$$

Solve $2^x = 5$.

$\log 2^x = \log 5$ Log property of equality.

$x \log 2 = \log 5$ Power property.

$x = \dfrac{\log 5}{\log 2}$ Divide by log 2.

$x \approx 2.3219$

CHAPTER 11 REVIEW

(11.1) If $f(x) = x^2 - 2$, $g(x) = x + 1$, and $h(x) = x^3 - x^2$, find the following.

1. $(f \circ g)(x)$

2. $(g \circ f)(x)$

3. $(h \circ g)(2)$

4. $(f \circ f)(x)$

5. $(f \circ g)(-1)$

6. $(h \circ h)(2)$

Determine whether each of the following functions is a one-to-one function. If it is one-to-one, list the elements of its inverse.

7. $h = \{(-9, 14), (6, 8), (-11, 12), (15, 15)\}$

8. $f = \{(-5, 5), (0, 4), (13, 5), (11, -6)\}$

9.

U.S. REGION (INPUT)	WEST	MIDWEST	SOUTH	NORTHEAST
RANK IN AUTOMOBILE THEFTS (OUTPUT)	2	4	1	3

10.

SHAPE (INPUT)	SQUARE	TRIANGLE	PARALLELOGRAM	RECTANGLE
NUMBER OF SIDES (OUTPUT)	4	3	4	4

Given that $f(x) = \sqrt{x + 2}$ is a one-to-one function, find the following.

11. a. $f(7)$ **b.** $f^{-1}(3)$

12. a. $f(-1)$ **b.** $f^{-1}(1)$

Determine whether each function is a one-to-one function.

13.

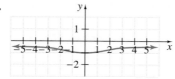

14.

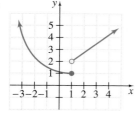

15.

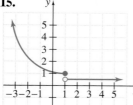

16.

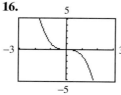

Find an equation defining the inverse function of the given one-to-one function.

17. $f(x) = 6x + 11$ **18.** $f(x) = 12x$

19. $q(x) = mx + b$ **20.** $g(x) = \dfrac{12x - 7}{6}$

21. $r(x) = \dfrac{13}{2}x - 4$

On the same set of axes, graph the given one-to-one function and its inverse.

22. $g(x) = \sqrt{x}$ **23.** $h(x) = 5x - 5$

24. Find the inverse of the one-to-one function $f(x) = 2x - 3$. Then use a graphing utility to graph both $f(x)$ and $f^{-1}(x)$ in a square window.

(11.2) *Solve each equation for x.*

25. $4^x = 64$ **26.** $3^x = \dfrac{1}{9}$

27. $2^{3x} = \dfrac{1}{16}$ **28.** $5^{2x} = 125$

29. $9^{x+1} = 243$ **30.** $8^{3x-2} = 4$

Graph each exponential function.

31. $y = 3^x$ **32.** $y = \left(\dfrac{1}{3}\right)^x$

33. $y = 4 \cdot 2^x$ **34.** $y = 2^x + 4$

Use the formula $A = P\left(1 + \dfrac{r}{n}\right)^{nt}$ to solve the interest problems. In this formula,

A = amount accrued (or owed)
P = principal invested (or loaned)
r = rate of interest
n = number of compounding periods per year
t = time in years

35. Find the amount accrued if $1600 is invested at 9% interest compounded semiannually for 7 years.

36. A total of $800 is invested in a 7% certificate of deposit for which interest is compounded quarterly. Use a table to find the value that this certificate will have at the end of each year for the next 5 years.

37. Use a graphing utility to verify the results of Exercise 33.

(11.3) *Write each equation with logarithmic notation.*

38. $49 = 7^2$

39. $2^{-4} = \dfrac{1}{16}$

Write each logarithmic equation with exponential notation.

40. $\log^{1/2} 16 = -4$

41. $\log_{0.4} 0.064 = 3$

Solve for x.

42. $\log_4 x = -3$

43. $\log_3 x = 2$

44. $\log_3 1 = x$

45. $\log_4 64 = x$

46. $\log_x 64 = 2$

47. $\log_x 81 = 4$

48. $\log_4 4^5 = x$

49. $\log_7 7^{-2} = x$

50. $5^{\log_5 4} = x$

51. $2^{\log_2 9} = x$

52. $\log_2 (3x - 1) = 4$

53. $\log_3 (2x + 5) = 2$

54. $\log_4 (x^2 - 3x) = 1$

55. $\log_8 (x^2 + 7x) = 1$

Graph each pair of equations on the same coordinate system.

56. $y = 2^x$ and $y = \log_2 x$

57. $y = \left(\dfrac{1}{2}\right)^x$ and $y = \log_{1/2} x$

(11.4) *Write each of the following as single logarithms.*

58. $\log_3 8 + \log_3 4$

59. $\log_2 6 + \log_2 3$

60. $\log_7 15 - \log_7 20$

61. $\log 18 - \log 12$

62. $\log_{11} 8 + \log_{11} 3 - \log_{11} 6$

63. $\log_5 14 + \log_5 3 - \log_5 21$

64. $2 \log_5 x - 2 \log_5 (x + 1) + \log_5 x$

65. $4 \log_3 x - \log_3 x + \log_3 (x + 2)$

Use properties of logarithms to write each expression as a sum or difference of multiples of logarithms.

66. $\log_3 \dfrac{x^3}{x + 2}$

67. $\log_4 \dfrac{x + 5}{x^2}$

68. $\log_2 \dfrac{3x^2 y}{z}$

69. $\log_7 \dfrac{yz^3}{x}$

If $\log_b 2 = 0.36$ and $\log_b 5 = 0.83$, find the following:

70. $\log_b 50$

71. $\log_b \dfrac{4}{5}$

(11.5) *Use a calculator to approximate the logarithm to four decimal places.*

72. $\log 3.6$

73. $\log 0.15$

74. $\ln 1.25$

75. $\ln 4.63$

Find the exact value.

76. $\log 1000$

77. $\log \dfrac{1}{10}$

78. $\ln \left(\dfrac{1}{e}\right)$

79. $\ln (e^4)$

Solve each equation for x.

80. $\ln (2x) = 2$

81. $\ln (3x) = 1.6$

82. $\ln (2x - 3) = -1$

83. $\ln (3x + 1) = 2$

Use the formula $\ln \dfrac{I}{I_0} = -kx$ to solve radiation problems. In this formula,

$x = $ depth in millimeters

$I = $ intensity of radiation

$I_0 = $ initial intensity

$k = $ a constant measure dependent on the material

Round answers to two decimal places.

84. Find the depth at which the intensity of the radiation passing through a lead shield is reduced to 3% of the original intensity if the value of k is 2.1.

85. If k is 3.2, find the depth at which 2% of the original radiation will penetrate.

Approximate the logarithm to four decimal places.

86. $\log_5 1.6$

87. $\log_3 4$

Use the formula $A = Pe^{rt}$ to solve the interest problems in which interest is compounded continuously. In this formula,

$A = $ amount accrued (or owed)

$P = $ principal invested (or loaned)

$r = $ rate of interest

$t = $ time in years

88. Bank of New York offers a 5-year, 6%, continuously compounded investment option. Find the amount accrued if $1450 is invested.

89. Find the amount to which a $940 investment grows if it is invested at 11% compounded continuously for 3 years.

(11.6) *Solve each exponential equation for x. Give an exact solution and also approximate the solution to four decimal places.*

90. $3^{2x} = 7$

91. $6^{3x} = 5$

92. $3^{2x+1} = 6$

93. $4^{3x+2} = 9$

94. $5^{3x-5} = 4$

95. $8^{4x-2} = 3$

96. $2 \cdot 5^{x-1} = 1$

97. $3 \cdot 4^{x+5} = 2$

Solve the equation for x.

98. $\log_5 2 + \log_5 x = 2$

99. $\log_3 x + \log_3 10 = 2$

100. $\log (5x) - \log (x + 1) = 4$

101. $\ln (3x) - \ln (x - 3) = 2$

102. $\log_2 x + \log_2 2x - 3 = 1$

103. $-\log_6 (4x + 7) + \log_6 x = 1$

Use the formula $y = y_0 e^{kt}$ to solve the population growth problems. In this formula,

$$y = \text{size of population}$$
$$y_0 = \text{initial count of population}$$
$$k = \text{rate of growth}$$
$$t = \text{time}$$

Round each answer to the nearest whole.

104. The population of mallard ducks in Nova Scotia is expected to grow at a rate of 6% per week during the spring migration. If 155,000 ducks are already in Nova Scotia, find how many are expected by the end of 4 weeks.

105. The population of Indonesia is growing at a rate of 1.7% per year. If the population in 1995 was 203,583,886, find the expected population by the year 2005.

106. Anaheim, California, is experiencing an annual growth rate of 3.16%. If 230,000 people now live in Anaheim, find how long it will take for the size of the population to be 500,000.

107. Memphis, Tennessee, is growing at a rate of 0.36% per year. Find how long it will take the population of Memphis to increase from 650,000 to 700,000.

108. Egypt's population is increasing at a rate of 2.1% per year. Find how long it will take for its 50,500,000-person population to double in size.

109. The greater Mexico City area had a population of 15.5 million in 1994. How long will it take the city to triple in population if its growth rate is 0.7% annually?

Use the compound interest equation $A = P\left(1 + \dfrac{r}{n}\right)^{nt}$ to solve the following. (See the directions for Exercises 35 and 36 for an explanation of this formula. Round answers to the nearest tenth.)

110. Find how long it will take a $5000 investment to grow to $10,000 if it is invested at 8% interest compounded quarterly.

111. An investment of $6000 has grown to $10,000 while the money was invested at 6% interest compounded monthly. Find how long it was invested.

Use a graphing utility to solve each equation. Round all solutions to two decimal places.

112. $e^x = 2$

113. $10^{0.3x} = 7$

CHAPTER 11 TEST

If $f(x) = x, g(x) = x - 7$, and $h(x) = x^2 - 6x + 5$, find the following.

1. $(f \circ h)(0)$

2. $(g \circ f)(x)$

3. $(g \circ h)(x)$

On the same set of axes, graph the given one-to-one function and its inverse.

4. $f(x) = 7x - 14$

Determine whether the given graph is the graph of a one-to-one function.

5.

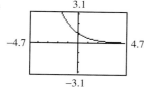

6.

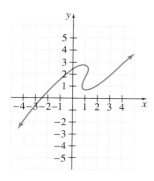

Determine whether each function is one-to-one. If it is one-to-one, find an equation or a set of ordered pairs that defines the inverse function of the given function.

7. $y = 6 - 2x$

8. $f = \{(0, 0), (2, 3), (-1, 5)\}$

9.

WORD (INPUT)	DOG	CAT	HOUSE	DESK	CIRCLE
FIRST LETTER OF WORD (OUTPUT)	d	c	h	d	c

Use the properties of logarithms to write each expression as a single logarithm.

10. $\log_3 6 + \log_3 4$

11. $\log_5 x + 3\log_5 x - \log_5 (x + 1)$

12. Write the expression $\log_6 \dfrac{2x}{y^3}$ as the sum or difference of multiples of logarithms.

13. If $\log_b 3 = 0.79$ and $\log_b 5 = 1.16$, find the value of $\log_b \left(\dfrac{3}{25}\right)$.

14. Approximate $\log_7 8$ to four decimal places.

15. Solve $8^{x-1} = \dfrac{1}{64}$ for x. Give an exact solution.

16. Solve $3^{2x+5} = 4$ for x. Give an exact solution, and also approximate the solution to four decimal places.

Solve each logarithmic equation for x. Give an exact solution.

17. $\log_3 x = -2$

18. $\ln \sqrt{e} = x$

19. $\log_8 (3x - 2) = 2$

20. $\log_5 x + \log_5 3 = 2$

21. $\log_4 (x + 1) - \log_4 (x - 2) = 3$

22. Solve $\ln (3x + 7) = 1.31$ accurate to four decimal places.

23. Graph $y = \left(\dfrac{1}{2}\right)^x + 1$.

24. Graph the functions $y = 3^x$ and $y = \log_3 x$ on the same coordinate system.

Use the formula $A = P\left(1 + \dfrac{r}{n}\right)^{nt}$ to solve Exercises 25 and 26.

25. Find the amount in the account if $4000 is invested for 3 years at 9% interest compounded monthly.

26. Find how long it will take $2000 to grow to $3000 if the money is invested at 7% interest compounded semiannually. Round to the nearest whole number.

Use the population growth formula $y = y_0 e^{kt}$ to solve Exercises 27 and 28.

27. The prairie dog population of the Grand Rapids area now stands at 57,000 animals. If the population is growing at a rate of 2.6% annually, use a table to estimate the prairie dog population for each year for the next 5 years.

28. In an attempt to save an endangered species of wood duck, naturalists would like to increase the wood duck population from 400 to 1000 ducks. If the annual population growth rate is 6.2%, find how long it will take the naturalists to reach their goal. Round to the nearest whole year.

29. The formula $\log(1 + k) = \dfrac{0.3}{D}$ relates the doubling time D, in days, and the growth rate k for a population of mice. Find the rate at which the population is increasing if the doubling time is 56 days. Round to the nearest tenth of a percent.

30. Use a graphing utility to approximate the solution of

$$e^{0.2x} = e^{-0.4x} + 2$$

to two decimal places.

CHAPTER 11 CUMULATIVE REVIEW

1. Write each value as an algebraic expression.

  a. Two numbers have a sum of 20. If one number is x, represent the other number as an expression in x.

  b. The older sister is 8 years older than her younger sister. If the age of the younger sister is x, represent the age of the older sister as an expression in x.

  c. Two angles are complementary if the sum of their measures is $90°$. If the measure of one angle is x degrees, represent the measure of the other angle as an expression in x.

  d. If x is the first of two consecutive integers, represent the second integer as an expression in x.

  e. A 2-foot section is to be cut from a board having a length of x feet. Express the length of the remaining board as an algebraic expression in x.

2. Use the graph of $y = 1500 + 0.1x$ to answer the questions that follow.

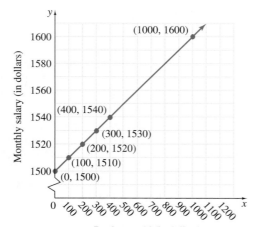

  a. If a salesperson has $800 of products sold for a particular month, what is the salary for that month?

  b. If the salesperson wants to make more than $1600 per month, what must be the total amount of products sold?

3. Graph $x = 2$.

4. Solve for x: $\dfrac{2}{5}(x - 6) \geq x - 1$.

5. Use the elimination method to solve the following system:

$$\begin{cases} -5x - 3y = 9 \\ 10x + 6y = -18 \end{cases}$$

6. Solve the system:

$$\begin{cases} 2x + 4y = 1 \\ 4x - 4z = -1 \\ y - 4z = -3 \end{cases}$$

7. Use the quotient rule to simplify.

  a. $\dfrac{x^7}{x^4}$

  b. $\dfrac{5^8}{5^2}$

  c. $\dfrac{20x^6}{4x^5}$

  d. $\dfrac{12y^{10}z^7}{14y^8z^7}$

8. Add $11x^3 - 12x^2 + x - 3$ and $x^3 - 10x + 5$ vertically.

9. Multiply:

  a. $(2x^3)(5x^6)$

  b. $(7y^4z^4)(-xy^{11}z^5)$

10. Find the product of $(4x^2 + 7)$ and $(x^2 + 2x + 8)$.

11. Factor $3a^2x - 12abx + 12b^2x$.

12. Solve $2x^2 + 9x - 5 = 0$.

13. Solve $\dfrac{3}{x} - \dfrac{x+21}{3x} = \dfrac{5}{3}$ algebraically and check graphically.

14. Write with radical notation. Then simplify if possible.

   a. $4^{3/2}$ **b.** $-16^{3/4}$

   c. $(-27)^{2/3}$ **d.** $\left(\dfrac{1}{9}\right)^{3/2}$

   e. $(4x - 1)^{3/5}$

15. Multiply the complex numbers. Write the product in the form $a + bi$.

   a. $(2 - 5i)(4 + i)$

   b. $(2 - i)^2$

   c. $(7 + 3i)(7 - 3i)$

16. Use a graphical approach to solve $(x + 2)(x - 1)(x - 5) < 0$.

17. Sketch the graph of $f(x) = -2x^2$. Label the vertex and the axis of symmetry.

18. Graph $x^2 + y^2 = 4$.

19. If $f(x) = x^2$ and $g(x) = x + 3$, find the following.

   a. $(f \circ g)(2)$ and $(g \circ f)(2)$

   b. $(f \circ g)(x)$ and $(g \circ f)(x)$

20. Solve each equation for x.

   a. $2^x = 16$ **b.** $9^x = 27$

   c. $4^{x+3} = 8^x$ **d.** $5^x = 10$

21. If $\log_b 2 = 0.43$ and $\log_b 3 = 0.68$, use the properties of logarithms to evaluate

   a. $\log_b 6$ **b.** $\log_b 9$

   c. $\log_b \sqrt{2}$

22. Solve the equation $\ln 3x = 5$ for x. Give an exact solution and then approximate the solution to four decimal places.

23. Solve $\log_2 x + \log_2 (x - 1) = 1$ for x.

Sequences, Series, and the Binomial Theorem

Modeling College Tuition

Annual college tuition has steadily increased over the past few decades. During the 1969–1970 academic year, the average annual tuition was $427 at a public four-year university and $1089 at a private four-year university. By the 1993–1994 academic year, the average annual tuitions had increased to $2543 and $10,994, respectively. (*Source:* National Center for Education Statistics.) Over the past several years, average annual tuition has been increasing at a rate greater than the annual rate of inflation.

In the chapter Group Activity on page 710, you will have the opportunity to model and investigate the trend in increasing tuition at public and private universities.

H aving explored in some depth the concept of function, we turn now in this final chapter to *sequences*. In one sense, a sequence is simply an ordered list of numbers. In another sense, a sequence is itself a function. Phenomena modeled by such functions are everywhere around us. The starting place for all mathematics is the sequence of natural numbers: 1, 2, 3, 4, and so on.

Sequences lead us to *series,* which are a sum of ordered numbers. Through series we gain new insight, for example, about the expansion of a binomial $(a + b)^n$, the concluding topic of this book.

12.1 SEQUENCES

TAPE IAG 12.1

O B J E C T I V E S

1 Write the terms of a sequence given its general term.

2 Find the general term of a sequence.

3 Solve applications that involve sequences.

Suppose that a town's present population of 100,000 is growing by 5% each year. After the first year, the town's population will be

$$100,000 + 0.05(100,000) = 105,000$$

After the second year, the town's population will be

$$105,000 + 0.05(105,000) = 110,250$$

After the third year, the town's population will be

$$110,250 + 0.05(110,250) \approx 115,763$$

If we continue to calculate, the town's yearly population can be written as the **infinite sequence** of numbers

$$105,000, 110,250, 115,763, \ldots$$

If we decide to stop calculating after a certain year (say, the fourth year), we obtain the **finite sequence**

$$105,000, 110,250, 115,763, 121,551$$

TECHNOLOGY NOTE

Some graphing calculators have a designated sequence mode. This mode allows the user to list terms of a sequence, investigate tables and graph sequences in terms of n, where n is a natural number. Consult your owner's manual for specific instructions on how to use this feature.

SEQUENCES

An infinite sequence is a function whose domain is the set of natural numbers $\{1, 2, 3, 4, \ldots\}$.

A finite sequence is a function whose domain is the set of natural numbers $\{1, 2, 3, 4, \ldots, n\}$, where n is some natural number.

Given the sequence 2, 4, 8, 16, . . . , we say that each number is a **term** of the sequence. Because a sequence is a function, we could describe it by writing $f(n) = 2^n$, where n is a natural number. Instead, we use the notation

$$a_n = 2^n$$

Some function values are

$$a_1 = 2^1 = 2 \qquad \text{First term of sequence.}$$
$$a_2 = 2^2 = 4 \qquad \text{Second term.}$$
$$a_3 = 2^3 = 8 \qquad \text{Third term.}$$
$$a_4 = 2^4 = 16 \qquad \text{Fourth term.}$$
$$a_{10} = 2^{10} = 1024 \qquad \text{Tenth term.}$$

The nth term of the sequence a^n is called the **general term.**

EXAMPLE 1   Write the first five terms of the sequence whose general term is given by

$$a_n = n^2 - 1$$

Solution: Evaluate a_n, where n is 1, 2, 3, 4, and 5.

n	$u(n)$
1	0
2	3
3	8
4	15
5	24
6	35
7	48

$u(n) \boxplus n^2 - 1$

Here we use a sequence mode and list the terms of the sequence in a table.

$$a_n = n^2 - 1$$
$$a_1 = 1^2 - 1 = 0 \qquad \text{Replace } n \text{ with 1}$$
$$a_2 = 2^2 - 1 = 3 \qquad \text{Replace } n \text{ with 2.}$$
$$a_3 = 3^2 - 1 = 8 \qquad \text{Replace } n \text{ with 3.}$$
$$a_4 = 4^2 - 1 = 15 \qquad \text{Replace } n \text{ with 4.}$$
$$a_5 = 5^2 - 1 = 24 \qquad \text{Replace } n \text{ with 5.}$$

Thus, the first five terms of the sequence $a_n = n^2 - 1$ are 0, 3, 8, 15, and 24.

EXAMPLE 2   If the general term of a sequence is given by $a_n = \dfrac{(-1)^n}{3n}$, find

a. the first term of the sequence

b. a_8

c. the one-hundredth term of the sequence

d. a_{15}

```
1→n: (-1)^n/(3n)▶
Frac
                    -1/3
8→n: (-1)^n/(3n)▶
Frac
                    1/24
```

Solution: **a.** $a_1 = \dfrac{(-1)^1}{3(1)} = -\dfrac{1}{3}$ Replace n with 1.

b. $a_8 = \dfrac{(-1)^8}{3(8)} = \dfrac{1}{24}$ Replace n with 8.

Single terms of a sequence may be found by evaluating the general term expression at given values.

c. $a_{100} = \dfrac{(-1)^{100}}{3(100)} = \dfrac{1}{300}$ Replace n with 100.

d. $a_{15} = \dfrac{(-1)^{15}}{3(15)} = -\dfrac{1}{45}$ Replace n with 15.

2   Suppose that we know the first few terms of a sequence and want to find a general term that fits the pattern of the first few terms.

EXAMPLE 3   Find a general term a_n of the sequence whose first few terms are given.

   a. $1, 4, 9, 16, \ldots$

   b. $\dfrac{1}{1}, \dfrac{1}{2}, \dfrac{1}{3}, \dfrac{1}{4}, \dfrac{1}{5}, \ldots$

   c. $-3, -6, -9, -12, \ldots$

   d. $\dfrac{1}{2}, \dfrac{1}{4}, \dfrac{1}{8}, \dfrac{1}{16}, \ldots$

Solution:

TECHNOLOGY NOTE

Another command that may be available on your calculator for listing the terms of a sequence is the sequence command. An advantage of this command is that it allows the terms to be written as fractions.

```
seq(n²,n,1,10,1)
{1 4 9 16 25 36…
seq(1/n,n,1,10,1
)▶Frac
{1 1/2 1/3 1/4 …
```

a. These numbers are the squares of the first four natural numbers, so a general term might be $a_n = n^2$.

b. These numbers are the reciprocals of the first five natural numbers, so a general term might be $a_n = \dfrac{1}{n}$.

c. These numbers are the product of -3 and the first four natural numbers, so a general term might be $a_n = -3n$.

d. Notice that the denominators double each time.

$$\dfrac{1}{2}, \quad \dfrac{1}{2 \cdot 2}, \quad \dfrac{1}{2(2 \cdot 2)}, \quad \dfrac{1}{2(2 \cdot 2 \cdot 2)}$$

or $\dfrac{1}{2}, \quad \dfrac{1}{2^2}, \quad \dfrac{1}{2^3}, \quad \dfrac{1}{2^4}$

We might then suppose that the general term is $a_n = \dfrac{1}{2^n}$.

Sequences model many phenomena of the physical world, as illustrated by the following example.

EXAMPLE 4   **FINDING PUPPY WEIGHT GAIN**

The amount of weight, in pounds, that a puppy gains in each month of its first year is modeled by a sequence whose general term is $a_n = n + 4$, where n is the number of the month. Write the first five terms of the sequence, and find how much weight the puppy should gain in its fifth month.

Solution: Evaluate $a_n = n + 4$ when n is 1, 2, 3, 4, and 5:

$$a_1 = 1 + 4 = 5$$
$$a_2 = 2 + 4 = 6$$
$$a_3 = 3 + 4 = 7$$
$$a_4 = 4 + 4 = 8$$
$$a_5 = 5 + 4 = 9$$

The puppy should gain 9 pounds in its fifth month.

EXERCISE SET 12.1

Write the first five terms of each sequence whose general term is given. See Example 1.

1. $a_n = n + 4$ **2.** $a_n = 5 - n$

3. $a_n = (-1)^n$ **4.** $a_n = (-2)^n$

5. $a_n = \dfrac{1}{n + 3}$ **6.** $a_n = \dfrac{1}{7 - n}$

7. $a_n = 2n$ **8.** $a_n = -6n$

9. $a_n = -n^2$ **10.** $a_n = n^2 + 2$

11. $a_n = 2^n$ **12.** $a_n = 3^{n-2}$

13. $a_n = 2n + 5$ **14.** $a_n = 1 - 3n$

15. $a^n = (-1)^n n^2$ **16.** $a_n = (-1)^{n+1}(n - 1)$

Find the indicated term for each sequence whose general term is given. See Example 2.

17. $a_n = 3n^2; a_5$ **18.** $a_n = -n^2; a_{15}$

19. $a_n = 6n - 2; a_{20}$ **20.** $a_n = 100 - 7n; a_{50}$

21. $a_n = \dfrac{n + 3}{n}; a_{15}$ **22.** $a_n = \dfrac{n}{n + 4}; a_{24}$

23. $a_n = (-3)^n; a_6$ **24.** $a_n = 5^{n+1}; a_3$

25. $a_n = \dfrac{n - 2}{n + 1}; a_6$ **26.** $a_n = \dfrac{n + 3}{n + 4}; a_8$

27. $a_n = \dfrac{(-1)^n}{n}; a_8$ **28.** $a_n = \dfrac{(-1)^n}{2n}; a_{100}$

29. $a_n = -n^2 + 5; a_{10}$ **30.** $a_n = 8 - n^2; a_{20}$

31. $a_n = \dfrac{(-1)^n}{n + 6}; a_{19}$ **32.** $a_n = \dfrac{n - 4}{(-2)^n}; a_6$

Find a general term a_n for each sequence whose first four terms are given. See Example 3.

33. 3, 7, 11, 15   **34.** 2, 7, 12, 17

35. $-2, -4, -8, -16$ **36.** $-4, 16, -64, 256$

37. $\dfrac{1}{3}, \dfrac{1}{9}, \dfrac{1}{27}, \dfrac{1}{81}$ **38.** $\dfrac{2}{5}, \dfrac{2}{25}, \dfrac{2}{125}, \dfrac{2}{625}$

Solve. See Example 4.

39. The distance, in feet, that a thermos dropped from a cliff falls in each consecutive second is modeled by a sequence whose general term is $a_n = 32n - 16$, where n is the number of seconds. Find the distance the thermos falls in the second, third, and fourth seconds.

40. The population size of a culture of bacteria triples every hour such that its size is modeled by the sequence $a_n = 50(3)^{n-1}$, where n is the number of the hour just beginning. Find the size of the culture at the beginning of the fourth hour and the size of the culture originally.

41. Mrs. Laser agrees to give her son Mark an allowance of $0.10 on the first day of his 14-day vacation, $0.20 on the second day, $0.40 on the third day, and so on. Write an equation of a sequence whose terms correspond to Mark's allowance. Use a graphing utility to find the allowance Mark will receive on the last day of his vacation.

42. A small theater has 10 rows with 12 seats in the first row, 15 seats in the second row, 18 seats in the third row, and so on. Write an equation of a sequence whose terms correspond to the seats in each row. Find the number of seats in the eighth row.

43. The number of cases of a new infectious disease is doubling every year such that the number of cases is modeled by a sequence whose general term is $a_n = 75(2)^{n-1}$, where n is the number of the year just beginning. Use a graphing utility to find how many cases there will be at the beginning of each of the first seven years.

44. A new college had an initial enrollment of 2700 students in 1995, and each year the enrollment increases by 150 students. Find the enrollment for each of 5 years, beginning with 1995.

45. An endangered species of sparrow had an estimated population of 800 in 1996, and scientists predict that its population will decrease by half each year. Estimate the population in 2000. Estimate the year the sparrow will be extinct.

46. A **Fibonacci sequence** is a special type of sequence in which the first two terms are 1 and each term thereafter is the sum of the two previous terms: 1, 1, 2, 3, 5, 8, ... Many plants and animals seem to grow according to a Fibonacci sequence, including pine cones, pineapple scales, nautilus shells, and certain flowers. Write the first 15 terms of the Fibonacci sequence.

Use a graphing utility to find the first five terms of each sequence. As necessary, round each term to four decimal places.

47. $a_n = \dfrac{1}{\sqrt{n}}$

48. $\dfrac{\sqrt{n}}{\sqrt{n}+1}$

49. $a_n = \left(1 + \dfrac{1}{n}\right)^n$

50. $a_n = \left(1 + \dfrac{0.05}{n}\right)^n$

Review Exercises

Sketch the graph of each quadratic function. See Section 9.5.

51. $f(x) = (x-1)^2 + 3$ **52.** $f(x) = (x-2)^2 + 1$

53. $f(x) = 2(x+4)^2 + 2$ **54.** $f(x) = 3(x-3)^2 + 4$

Find the distance between each pair of points. See Section 3.1.

55. $(-4, -1)$ and $(-7, -3)$ **56.** $(-2, -1)$ and $(-1, 5)$

57. $(2, -7)$ and $(-3, -3)$ **58.** $(10, -14)$ and $(5, -11)$

12.2 │ ARITHMETIC AND GEOMETRIC SEQUENCES

O B J E C T I V E S

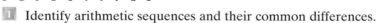

1   Identify arithmetic sequences and their common differences.

2   Identify geometric sequences and their common ratios.

TAPE IAG 12.2

1   Find the first four terms of the sequence whose general term is

$$a_n = 5 + (n-1)3$$
$$a_1 = 5 + (1-1)3 = 5 \qquad \text{Replace } n \text{ with 1.}$$
$$a_2 = 5 + (2-1)3 = 8 \qquad \text{Replace } n \text{ with 2.}$$
$$a_3 = 5 + (3-1)3 = 11 \qquad \text{Replace } n \text{ with 3.}$$
$$a_4 = 5 + (4-1)3 = 14 \qquad \text{Replace } n \text{ with 4.}$$

The first four terms are 5, 8, 11, and 14. Notice that the difference of any two successive terms is 3:

$$\begin{aligned} 8 - 5 &= 3 \\ 11 - 8 &= 3 \\ 14 - 11 &= 3 \\ &\vdots \\ \underset{\substack{\uparrow \\ n\text{th} \\ \text{term}}}{a_n} - \underset{\substack{\uparrow \\ \text{previous} \\ \text{term}}}{a_{n-1}} &= 3 \end{aligned}$$

Because the difference of any two successive terms is a constant, we call the sequence an **arithmetic sequence,** or an **arithmetic progression.** The constant difference d in successive terms is called the **common difference.** In this example, d is 3.

ARITHMETIC SEQUENCE AND COMMON DIFFERENCE

An **arithmetic sequence** is a sequence in which each term (after the first) differs from the preceding term by a constant amount d. The constant d is called the **common difference** of the sequence.

The sequence 2, 6, 10, 14, 18, . . . is an arithmetic sequence. Its common difference is 4. Given the first term, a_1, and the common difference, d, of an arithmetic sequence, we can find any term of the sequence.

EXAMPLE 1 Write the first five terms of the arithmetic sequence whose first term is 7 and whose common difference is 2.

Solution:

$$\begin{aligned} a_1 &= 7 \\ a_2 &= 7 + 2 = 9 \\ a_3 &= 9 + 2 = 11 \\ a_4 &= 11 + 2 = 13 \\ a_5 &= 13 + 2 = 15 \end{aligned}$$

The first five terms are 7, 9, 11, 13, 15.

Notice the general pattern of the terms in Example 1:

$$a_1 = 7$$
$$a_2 = 7 + 2 = 9 \qquad \text{or} \quad a_2 = a_1 + d$$
$$a_3 = 9 + 2 = 11 \qquad \text{or} \quad a_3 = a_2 + d = (a_1 + d) + d = a_1 + 2d$$
$$a_4 = 11 + 2 = 13 \qquad \text{or} \quad a_4 = a_3 + d = (a_1 + 2d) + d = a_1 + 3d$$
$$a_5 = 13 + 2 = 15 \qquad \text{or} \quad a_5 = a_4 + d = (a_1 + 3d) + d = a_1 + 4d$$

$\llcorner\rightarrow$ (subscript $- 1$) is multiplier $\longrightarrow\urcorner$

The pattern on the right suggests that the general term a_n of an arithmetic sequence is given by

$$a_n = a_1 + (n - 1)d$$

GENERAL TERM OF AN ARITHMETIC SEQUENCE

The general term a_n of an arithmetic sequence is given by

$$a_n = a_1 + (n - 1)d$$

where a_1 is the first term and d is the common difference.

EXAMPLE 2   Consider the arithmetic sequence whose first term is 3 and common difference is -5.

   a. Write an expression for the general term a_n.

   b. Find the twentieth term of this sequence.

Solution: **a.** Since this is an arithmetic sequence, the general term a_n is given by $a_n = a_1 + (n - 1)d$. Here, $a_1 = 3$ and $d = -5$, so

$$a_n = 3 + (n - 1)(-5) \qquad \text{Let } a_1 = 3 \text{ and } d = -5.$$
$$= 3 - 5n + 5 \qquad \text{Multiply.}$$
$$= 8 - 5n \qquad \text{Simplify.}$$

   b. $a_n = 8 - 5n$

$$a_{20} = 8 - 5 \cdot 20 \qquad \text{Let } n = 20.$$
$$= 8 - 100 = -92$$

n	$u(n)$	
20	-92	

u(n)🔲3+(n-1)-5

To check using a graphing utility, see the screen to the left.

EXAMPLE 3   Find the eleventh term of the arithmetic sequence whose first three terms are 2, 9, and 16.

Solution: Since the sequence is arithmetic, the eleventh term is

$$a_{11} = a_1 + (11 - 1)d = a_1 + 10d$$

We know a_1 is the first term of the sequence, so $a_1 = 2$. Also, d is the constant difference of terms, so $d = a_2 - a_1 = 9 - 2 = 7$. Thus,

$$a_{11} = a_1 + 10d$$
$$= 2 + 10 \cdot 7 \qquad \text{Let } a_1 = 2 \text{ and } d = 7.$$
$$= 72$$

EXAMPLE 4   If the third term of an arithmetic progression is 12 and the eighth term is 27, find the fifth term.

Solution: We need to find a_1 and d to write the general term, which then enables us to find a_5, the fifth term. The given facts about terms a_3 and a_8 lead to a system of linear equations:

$$\begin{cases} a_3 = a_1 + (3-1)d \\ a_8 = a_1 + (8-1)d \end{cases} \quad \text{or} \quad \begin{cases} 12 = a_1 + 2d \\ 27 = a_1 + 7d \end{cases}$$

Next, we solve the system $\begin{cases} 12 = a_1 + 2d \\ 27 = a_1 + 7d \end{cases}$ by elimination. Multiply both sides of the second equation by -1 so that

$$\begin{cases} 12 = a_1 + 2d \\ -1(27) = -1(a_1 + 7d) \end{cases} \quad \begin{matrix} \text{simplifies} \\ \text{to} \end{matrix} \quad \begin{cases} 12 = a_1 + 2d \\ -27 = -a_1 - 7d \end{cases} \quad \begin{matrix} \text{Add the} \\ \text{equations.} \end{matrix}$$

$$\begin{matrix} -15 = -5d \quad \begin{matrix} \text{Divide both} \\ \text{sides by } -5. \end{matrix} \\ 3 = d \end{matrix}$$

To find a_1, let $d = 3$ in $12 = a_1 + 2d$. Then

$$12 = a_1 + 2(3)$$
$$12 = a_1 + 6$$
$$6 = a_1$$

Thus, $a_1 = 6$ and $d = 3$, so

$$a_n = 6 + (n-1)(3)$$
$$= 6 + 3n - 3$$
$$= 3 + 3n$$
$$a_5 = 3 + 3 \cdot 5 = 18$$

EXAMPLE 5   **FINDING ANNUAL SALARIES**
Donna has an offer for a job starting at $20,000 per year and guaranteeing her a raise of $800 per year for the next 5 years. Write the general term for the arithmetic sequence that models Donna's potential annual salaries, and find her salary for the fourth year.

Solution: The first term $a_1 = 20,000$, and $d = 800$. So

$$a_n = 20,000 + (n - 1)(800) = 19,200 + 800n$$
$$a_4 = 19,200 + 800 \cdot 4 = 22,400$$

Her salary for the fourth year will be $22,400.

n	$u(n)$
1	20000
2	20800
3	21600
4	22400
5	23200
6	24000
7	24800

$u(n)\blacksquare 20000+(n-1\ldots$

Verify that her salary for the fourth year is $22,400.

2

We now investigate a **geometric sequence**, also called a **geometric progression**. In the sequence 5, 15, 45, 135, . . . , each term after the first is the *product* of 3 and the preceding term. This pattern of multiplying by a constant to get the next term defines a geometric sequence. The constant is called the **common ratio** because it is the ratio of any term (after the first) to its preceding term.

$$\frac{15}{5} = 3$$

$$\frac{45}{15} = 3$$

$$\frac{135}{45} = 3$$

$$\vdots$$

nth term $\rightarrow$
previous term $\rightarrow$ $\dfrac{a_n}{a_{n-1}} = 3$

GEOMETRIC SEQUENCE AND COMMON RATIO

A **geometric sequence** is a sequence in which each term (after the first) is obtained by multiplying the preceding term by a constant r. The constant r is called the **common ratio** of the sequence.

The sequence $12, 6, 3, \dfrac{3}{2}, \ldots$ is geometric since each term after the first is the product of the previous term and $\dfrac{1}{2}$.

EXAMPLE 6   Write the first five terms of a geometric sequence whose first term is 7 and whose common ratio is 2.

Solution:
$$a_1 = 7$$
$$a_2 = 7(2) = 14$$
$$a_3 = 14(2) = 28$$
$$a_4 = 28(2) = 56$$
$$a_5 = 56(2) = 112$$

The first five terms are 7, 14, 28, 56, and 112.

Notice the general pattern of the terms in Example 6:

$$a_1 = 7$$
$$a_2 = 7(2) = 14 \quad \text{or} \quad a_2 = a_1(r)$$
$$a_3 = 14(2) = 28 \quad \text{or} \quad a_3 = a_2(r) = (a_1 \cdot r) \cdot r = a_1 r^2$$
$$a_4 = 28(2) = 56 \quad \text{or} \quad a_4 = a_3(r) = (a_1 \cdot r^2) \cdot r = a_1 r^3$$
$$a_5 = 56(2) = 112 \quad \text{or} \quad a_5 = a_4(r) = (a_1 \cdot r^3) \cdot r = a_1 r^4 \leftarrow$$
$$\hookrightarrow (\text{subscript} - 1) \text{ is power} $$

The pattern on the right suggests that the general term of a geometric sequence is given by $a_n = a_1 r^{n-1}$.

GENERAL TERM OF A GEOMETRIC SEQUENCE

The general term a_n of a geometric sequence is given by

$$a_n = a_1 r^{n-1}$$

where a_1 is the first term and r is the common ratio.

EXAMPLE 7   Find the eighth term of the geometric sequence whose first term is 12 and whose common ratio is $\dfrac{1}{2}$.

Solution: Since this is a geometric sequence, the general term a_n is given by

$$a_n = a_1 r^{n-1}$$

Here $a_1 = 12$ and $r = \dfrac{1}{2}$, so $a_n = 12\left(\dfrac{1}{2}\right)^{n-1}$. Evaluate a_n for $n = 8$:

$$a_8 = 12\left(\frac{1}{2}\right)^{8-1} = 12\left(\frac{1}{2}\right)^7 = 12\left(\frac{1}{128}\right) = \frac{3}{32}$$

EXAMPLE 8   Find the fifth term of the geometric sequence whose first three terms are 2, −6, and 18.

Solution: Since the sequence is geometric and $a_1 = 2$, the fifth term must be $a_1 r^{5-1}$, or $2r^4$. We know that r is the common ratio of terms, so r must be $\dfrac{-6}{2}$, or −3. Thus,

$$a_5 = 2r^4$$
$$a_5 = 2(-3)^4 = 162$$

EXAMPLE 9   If the second term of a geometric sequence is $\frac{5}{4}$ and the third term is $\frac{5}{16}$, find the first term and the common ratio.

Solution: Notice that $\frac{5}{16} \div \frac{5}{4} = \frac{1}{4}$, so $r = \frac{1}{4}$. Then

$$a_2 = a_1\left(\frac{1}{4}\right)^{2-1}$$

$$\frac{5}{4} = a_1\left(\frac{1}{4}\right)^1, \text{ or } a_1 = 5 \qquad \text{Replace } a_2 \text{ with } \frac{5}{4}.$$

The first term is 5.

EXAMPLE 10   **PREDICTING BACTERIAL CULTURE SIZE**
The population size of a bacterial culture growing under controlled conditions is doubling each day. Predict how large the culture will be at the beginning of day 7 if it measures 10 units at the beginning of day 1.

Solution: Since the culture doubles in size each day, the population sizes are modeled by a geometric sequence. Here $a_1 = 10$ and $r = 2$. Thus,

$$a_n = a_1 r^{n-1} = 10(2)^{n-1} \quad \text{and} \quad a_7 = 10(2)^{7-1} = 640$$

The bacterial culture should measure 640 units at the beginning of day 7.

The bacterial culture measures 640 units at the beginning of day 7.

EXERCISE SET 12.2

Write the first five terms of the arithmetic or geometric sequence whose first term, a_1, and common difference, d, or common ratio, r, are given. See Examples 1 and 6.

1. $a_1 = 4; d = 2$

2. $a_1 = 3; d = 10$

3. $a_1 = 6; d = -2$

4. $a_1 = -20; d = 3$

5. $a_1 = 1; r = 3$

6. $a_1 = -2; r = 2$

7. $a_1 = 48; r = \frac{1}{2}$

8. $a_1 = 1; r = \frac{1}{3}$

Find the indicated term of each sequence. See Examples 2 and 7.

9. The eighth term of the arithmetic sequence whose first term is 12 and whose common difference is 3.

10. The twelfth term of the arithmetic sequence whose first term is 32 and whose common difference is −4.

11. The fourth term of the geometric sequence whose first term is 7 and whose common ratio is −5.

12. The fifth term of the geometric sequence whose first term is 3 and whose common ratio is 3.

13. The fifteenth term of the arithmetic sequence whose first term is −4 and whose common difference is −4.

14. The sixth term of the geometric sequence whose first term is 5 and whose common ratio is −4.

Find the indicated term of each sequence. See Examples 3 and 8.

15. The ninth term of the arithmetic sequence 0, 12, 24, ...

16. The thirteenth term of the arithmetic sequence −3, 0, 3, ...

17. The twenty-fifth term of the arithmetic sequence 20, 18, 16, ...

18. The ninth term of the geometric sequence 5, 10, 20, ...

19. The fifth term of the geometric sequence 2, –10, 50, ...

20. The sixth term of the geometric sequence $\frac{1}{2}$, $\frac{3}{2}$, $\frac{9}{2}$, ...

Find the indicated term of each sequence. See Examples 4 and 9.

21. The eighth term of the arithmetic sequence whose fourth term is 19 and whose fifteenth term is 52.

22. If the second term of an arithmetic sequence is 6 and the tenth term is 30, find the twenty-fifth term.

23. If the second term of an arithmetic progression is –1 and the fourth term is 5, find the ninth term. 20

24. If the second term of a geometric progression is 15 and the third term is 3, find a_1 and r.

25. If the second term of a geometric progression is $-\frac{4}{3}$ and the third term is $\frac{8}{3}$, find a_1 and r.

26. If the third term of a geometric sequence is 4 and the fourth term is –12, find a_1 and r.

27. Explain why 14, 10, 6 may be the first three terms of an arithmetic sequence when it appears that we are subtracting instead of adding to get the next term.

28. Explain why 80, 20, 5 may be the first three terms of a geometric sequence when it appears that we are dividing instead of multiplying to get the next term.

Given are the first three terms of a sequence that is either arithmetic or geometric. If the sequence is arithmetic, find a_1 and d. If a sequence is geometric, find a_1 and r.

29. 2, 4, 6

30. 8, 16, 24

31. 5, 10, 20

32. 2, 6, 18

33. $\frac{1}{2}, \frac{1}{10}, \frac{1}{50}$

34. $\frac{2}{3}, \frac{4}{3}, 2$

35. $x, 5x, 25x$

36. $y, -3y, 9y$

37. $p, p + 4, p + 8$

38. $t, t - 1, t - 2$

Find the indicated term of each sequence.

39. The twenty-first term of the arithmetic sequence whose first term is 14 and whose common difference is $\frac{1}{4}$.

40. The fifth term of the geometric sequence whose first term is 8 and whose common ratio is –3.

41. The fourth term of the geometric sequence whose first term is 3 and whose common ratio is $-\frac{2}{3}$.

42. The fourth term of the arithmetic sequence whose first term is 9 and whose common difference is 5.

43. The fifteenth term of the arithmetic sequence $\frac{3}{2}$, 2, $\frac{5}{2}$, ...

44. The eleventh term of the arithmetic sequence 2, $\frac{5}{3}$, $\frac{4}{3}$, ...

45. The sixth term of the geometric sequence 24, 8, $\frac{8}{3}$, ...

46. The eighteenth term of the arithmetic sequence 5, 2, –1, ...

47. If the third term of an arithmetic progression is 2 and the seventeenth term is –40, find the tenth term.

48. If the third term of a geometric sequence is –28 and the fourth term is –56, find a_1 and r.

Solve. See Examples 5 and 10.

49. An auditorium has 54 seats in the first row, 58 seats in the second row, 62 seats in the third row, and so on. Find the general term of this arithmetic sequence and the number of seats in the twentieth row.

50. A triangular display of cans in a grocery store has 20 cans in the first row, 17 cans in the next row, and so on, in an arithmetic sequence. Find the general term and the number of cans in the fifth row. Find how many rows there are in the display and how many cans are in the top row.

51. The initial size of a virus culture is 6 units, and it triples its size every day. Find the general term of the geometric sequence that models the culture's size. Use a graphing utility to find the size on the first five days.

52. A real-estate investment broker predicts that a certain property will increase in value 15% each year. Thus, the yearly property values can be modeled by a geometric sequence whose common ratio r is 1.15. If the initial property value was $500,000, write the first four terms of the sequence and predict the value at the end of the third year.

53. A rubber ball is dropped from a height of 486 feet, and it continues to bounce one-third the height from which it last fell. Write out the first five terms of this geometric sequence and find the general term. Find how many bounces it takes for the ball to rebound less than 1 foot.

54. On the first swing, the length of the arc through which a pendulum swings is 50 inches. The length of each successive swing is 80% of the preceding swing. Determine whether this sequence is arithmetic or geometric. Find the length of the fourth swing.

55. Jose takes a job that offers a monthly starting salary of $2000 and guarantees him a monthly raise of $125 during his first year of training. Find the general term of this arithmetic sequence and his monthly salary at the end of his training.

56. At the beginning of Claudia Schaffer's exercise program, she rides 15 minutes on the Lifecycle. Each week she increases her riding time by 5 minutes. Write the general term of this arithmetic sequence, and find her riding time after 7 weeks. Find how many weeks it takes her to reach a riding time of 1 hour.

57. If a radioactive element has a half-life of 3 hours, then x grams of the element dwindles to $\dfrac{x}{2}$ grams after 3 hours. If a nuclear reactor has 400 grams of this radioactive element, find the amount of radioactive material after 12 hours.

Use a graphing utility to write the first four terms of the arithmetic or geometric sequence whose first term, a_1, and common difference, d, or common ratio, r, are given.

58. $a_1 = \$3720, d = -\$268.50$

59. $a_1 = \$11,782.40, r = 0.5$

60. $a_1 = 26.8, r = 2.5$

61. $a_1 = 19.652; d = -0.034$

62. Describe a situation in your life that can be modeled by a geometric sequence. Write an equation for the sequence.

63. Describe a situation in your life that can be modeled by an arithmetic sequence. Write an equation for the sequence.

Review Exercises

Evaluate. See Section 1.4.

64. $5(1) + 5(2) + 5(3) + 5(4)$

65. $\dfrac{1}{3(1)} + \dfrac{1}{3(2)} + \dfrac{1}{3(3)}$

66. $2(2 - 4) + 3(3 - 4) + 4(4 - 4)$

67. $3^0 + 3^1 + 3^2 + 3^3$

68. $\dfrac{1}{4(1)} + \dfrac{1}{4(2)} + \dfrac{1}{4(3)}$

69. $\dfrac{8 - 1}{8 + 1} + \dfrac{8 - 2}{8 + 2} + \dfrac{8 - 3}{8 + 3}$

12.3 | SERIES

TAPE IAG 12.3

O B J E C T I V E S

1 Identify finite and infinite series.

2 Use summation notation.

3 Find partial sums.

1 A person who conscientiously saves money by saving first $100 and then saving $10 more each month than he saved the preceding month is saving money according to the arithmetic sequence

$$a_n = 100 + 10(n - 1)$$

Following this sequence, he can predict how much money he should save for any particular month. But if he also wants to know how much money *in total* he has saved, say, by the fifth month, he must find the *sum* of the first five terms of the sequence

$$\underbrace{100}_{a_1} + \underbrace{100 + 10}_{a_2} + \underbrace{100 + 20}_{a_3} + \underbrace{100 + 30}_{a_4} + \underbrace{100 + 40}_{a_5}$$

A sum of the terms of a sequence is called a **series** (the plural is also "series"). As our example here suggests, series are frequently used to model financial and natural phenomena.

A series is a **finite series** if it is the sum of only the first k terms of the sequence, for some natural number k. A series is an **infinite series** if it is the sum of all the terms of the sequence. For example,

SEQUENCE	SERIES	
$5, 9, 13$	$5 + 9 + 13$	Finite, k is 3
$5, 9, 13, \ldots$	$5 + 9 + 13 + \cdots$	Infinite
$4, -2, 1, -\dfrac{1}{2}, \dfrac{1}{4}$	$4 + (-2) + 1 + \left(-\dfrac{1}{2}\right) + \left(\dfrac{1}{4}\right)$	Finite, k is 5
$4, -2, 1, \ldots$	$4 + (-2) + 1 + \cdots$	Infinite
$3, 6, \ldots, 99$	$3 + 6 + \cdots + 99$	Finite, k is 33

2

A shorthand notation for denoting a series when the general term of the sequence is known is called **summation notation.** The Greek uppercase letter **sigma, Σ,** is used to mean "sum." The expression $\displaystyle\sum_{n=1}^{5}(3n + 1)$ is read "the sum of $3n + 1$ as n goes from 1 to 5"; this expression means the sum of the first five terms of the sequence whose general term is $a_n = 3n + 1$. Often, the variable i is used instead of n in summation notation: $\displaystyle\sum_{i=1}^{5}(3i + 1)$. Whether we use n, i, k, or some other variable, the variable is called the **index of summation.** The notation $i = 1$ below the symbol Σ indicates the beginning value of i, and the number 5 above the symbol Σ indicates the ending value of i. Thus, the terms of the sequence are found by successively replacing i with the natural numbers 1, 2, 3, 4, 5. To find the sum, we write out the terms and then add:

$$\sum_{i=1}^{5}(3i + 1) = (3 \cdot 1 + 1) + (3 \cdot 2 + 1) + (3 \cdot 3 + 1)$$
$$+ (3 \cdot 4 + 1) + (3 \cdot 5 + 1)$$
$$= 4 + 7 + 10 + 13 + 16 = 50$$

TECHNOLOGY NOTE

You can use a graphing utility to evaluate summations. Use a sum command to add terms of a specified sequence.

```
sum(seq(3n+1,n,1
,5,1))
               50
```

EXAMPLE 1   Evaluate.

a. $\displaystyle\sum_{i=0}^{6} \dfrac{i - 2}{2}$

b. $\displaystyle\sum_{i=3}^{5} 2^i$

Solution:

a. $\sum\limits_{i=0}^{6} \dfrac{i-2}{2} = \dfrac{0-2}{2} + \dfrac{1-2}{2} + \dfrac{2-2}{2} + \dfrac{3-2}{2} + \dfrac{4-2}{2} + \dfrac{5-2}{2} + \dfrac{6-2}{2}$

$$= (-1) + \left(-\dfrac{1}{2}\right) + 0 + \dfrac{1}{2} + 1 + \dfrac{3}{2} + 2$$

$$= \dfrac{7}{2}, \quad \text{or} \quad 3\dfrac{1}{2}$$

b. $\sum\limits_{i=3}^{5} 2^i = 2^3 + 2^4 + 2^5$

$$= 8 + 16 + 32$$

$$= 56$$

EXAMPLE 2   Write each series with summation notation.

a. $3 + 6 + 9 + 12 + 15$
b. $\dfrac{1}{2} + \dfrac{1}{4} + \dfrac{1}{8} + \dfrac{1}{16}$

Solution:

a. Since the *difference* of each term and the preceding term is 3, the terms correspond to the first five terms of the arithmetic sequence $a_n = a_1 + (n-1)d$ with $a_1 = 3$ and $d = 3$. So $a_n = 3 + (n-1)3$. Thus, in summation notation,

$$3 + 6 + 9 + 12 + 15 = \sum\limits_{i=1}^{5} 3 + (i-1)3$$

b. Since each term is the *product* of the preceding term and $\dfrac{1}{2}$, these terms correspond to the first four terms of the geometric sequence $a_n = a_1 r^{n-1}$. Here $a_1 = \dfrac{1}{2}$ and $r = \dfrac{1}{2}$, so $a_n = \left(\dfrac{1}{2}\right)\left(\dfrac{1}{2}\right)^{n-1} = \left(\dfrac{1}{2}\right)^{1+(n-1)} = \left(\dfrac{1}{2}\right)^n$. In summation notation,

$$\dfrac{1}{2} + \dfrac{1}{4} + \dfrac{1}{8} + \dfrac{1}{16} = \sum\limits_{i=1}^{4} \left(\dfrac{1}{2}\right)^i$$

3   The sum of the first n terms of a sequence is a finite series known as a **partial sum,** S_n. Thus, for the sequence $a_1, a_2, \ldots, a_n$, the first three partial sums are

$$S_1 = a_1$$
$$S_2 = a_1 + a_2$$
$$S_3 = a_1 + a_2 + a_3$$

and so on. In general, S_n is the sum of the first n terms of a sequence:

$$S_n = \sum\limits_{i=1}^{n} a_n$$

EXAMPLE 3   Find the sum of the first three terms of the sequence whose general term is $a_n = \dfrac{n+3}{2n}$.

Solution: $S_3 = \displaystyle\sum_{i=1}^{3} \dfrac{i+3}{2i} = \dfrac{1+3}{2 \cdot 1} + \dfrac{2+3}{2 \cdot 2} + \dfrac{3+3}{2 \cdot 3}$

$= 2 + \dfrac{5}{4} + 1 = 4\dfrac{1}{4}.$

The next example illustrates how these sums model real-life phenomena.

EXAMPLE 4   **SUMMING GORILLA BIRTHS**
The number of baby gorillas born at the San Diego Zoo is a sequence defined by $a_n = n(n-1)$, where n is the number of years the zoo has owned gorillas. Find the *total* number of baby gorillas born in the *first 4 years*.

Solution: To solve, find the sum

```
sum(seq(n(n-1),n
,1,4,1))
                20
```

$S_4 = \displaystyle\sum_{i=1}^{4} i(i-1)$

$= 1(1-1) + 2(2-1) + 3(3-1) + 4(4-1)$

$= 0 + 2 + 6 + 12 = 20$

There were 20 gorillas born in the first 4 years. A check is shown to the left

EXERCISE SET 12.3

Evaluate. See Example 1.

1. $\displaystyle\sum_{i=1}^{4} (i-3)$

2. $\displaystyle\sum_{i=1}^{5} (i+6)$

3. $\displaystyle\sum_{i=4}^{7} (2i+4)$

4. $\displaystyle\sum_{i=2}^{3} (5i-1)$

5. $\displaystyle\sum_{i=2}^{4} (i^2-3)$

6. $\displaystyle\sum_{i=3}^{5} i^3$

7. $\displaystyle\sum_{i=1}^{3} \dfrac{1}{i+5}$

8. $\displaystyle\sum_{i=2}^{4} \dfrac{2}{i+3}$

9. $\displaystyle\sum_{i=1}^{3} \dfrac{1}{6i}$

10. $\displaystyle\sum_{i=1}^{3} \dfrac{1}{3i}$

11. $\displaystyle\sum_{i=2}^{6} 3i$

12. $\displaystyle\sum_{i=3}^{6} -4i$

13. $\displaystyle\sum_{i=3}^{5} i(i+2)$

14. $\displaystyle\sum_{i=2}^{4} i(i-3)$

15. $\displaystyle\sum_{i=1}^{5} 2^i$

16. $\displaystyle\sum_{i=1}^{4} 3^{i-1}$

17. $\displaystyle\sum_{i=1}^{4} \dfrac{4i}{i+3}$

18. $\displaystyle\sum_{i=2}^{5} \dfrac{6-i}{6+i}$

Write each series with summation notation. See Example 2.

19. $1 + 3 + 5 + 7 + 9$

20. $4 + 7 + 10 + 13$

21. $4 + 12 + 36 + 108$

22. $5 + 10 + 20 + 40 + 80 + 160$

23. $12 + 9 + 6 + 3 + 0 + (-3)$

24. $5 + 1 + (-3) + (-7)$

25. $12 + 4 + \dfrac{4}{3} + \dfrac{4}{9}$

26. $80 + 20 + 5 + \dfrac{5}{4} + \dfrac{5}{16}$

27. $1 + 4 + 9 + 16 + 25 + 36 + 49$

28. $1 + (-4) + 9 + (-16)$

Find each partial sum. See Example 3.

29. Find the sum of the first two terms of the sequence whose general term is $a_n = (n + 2)(n - 5)$.

30. Find the sum of the first six terms of the sequence whose general term is $a_n = (-1)^n$.

31. Find the sum of the first two terms of the sequence whose general term is $a_n = n(n - 6)$.

32. Find the sum of the first seven terms of the sequence whose general term is $a_n = (-1)^{n-1}$.

33. Find the sum of the first four terms of the sequence whose general term is $a_n = (n + 3)(n + 1)$.

34. Find the sum of the first five terms of the sequence whose general term is $a_n = \dfrac{(-1)^n}{2n}$.

35. Find the sum of the first four terms of the sequence whose general term is $a_n = -2n$.

36. Find the sum of the first five terms of the sequence whose general term is $a_n = (n - 1)^2$.

37. Find the sum of the first three terms of the sequence whose general term is $a_n = -\dfrac{n}{3}$.

38. Find the sum of the first three terms of the sequence whose general term is $a_n = (n + 4)^2$.

Solve. See Example 4.

39. A gardener is making a triangular planting with 1 tree in the first row, 2 trees in the second row, 3 trees in the third row, and so on for 10 rows. Write the sequence that describes the number of trees in each row. Find the total number of trees planted.

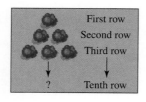

First row
Second row
Third row
?
Tenth row

40. Some surfers at the beach form a human pyramid with 2 surfers in the top row, 3 surfers in the second row, 4 surfers in the third row, and so on. If there are 6 rows in the pyramid, write the sequence that describes the number of surfers in each row of the pyramid. Find the total number of surfers.

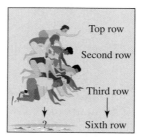

Top row
Second row
Third row
?
Sixth row

41. A culture of fungus starts with 6 units and doubles every day. Write the general term of the sequence that describes the growth of this fungus. Find the number of fungus units that there will be at the beginning of the fifth day.

42. A bacterial colony begins with 50 bacteria and doubles every 12 hours. Write the sequence that describes the growth of the bacteria. Find the number of bacteria that there will be after 48 hours.

43. A rubber ball bounces so that the number of feet it travels on each successive bounce forms a sequence whose general term is $a_n = 20(0.75)^{n-1}$. Use a graphing utility to find the total distance traveled in 15 bounces.

44. The number of otters born each year in a new aquarium forms a sequence whose general term is $a_n = (n - 1)(n + 3)$. Find the number of otters born in the third year, and find the total number of otters born in the first three years.

45. The number of opossums killed each month on a new highway describes the sequence whose general term is $a_n = (n + 1)(n + 2)$, where n is the number of the months. Find the number of opossums killed in the fourth month, and find the total number killed in the first four months.

46. In 1993 the population of an endangered fish was estimated by environmentalists to be decreasing each year. The size of the population in a given year is $24 - 4n$ thousand fish fewer than the previous year. Find the decrease in population in 1995, if year 1 is 1993. Find how many total fish died from 1993 through 1995.

47. The amount of decay in pounds of a radioactive isotope each year is given by the sequence whose general term is $a_n = 100(0.5)^n$, where n is the number of the year. Find the amount of decay in the fourth year, and find the total amount of decay in the first four years.

48. Susan has a choice between two job offers. Job A has an annual starting salary of $20,000 with guaranteed annual raises of $1200 for the next 4 years, whereas job B has an annual starting salary of $18,000 with guaranteed annual raises of $2500 for the next 4 years. Compare the fifth partial sums for each sequence to determine which job would pay Susan more money over the next 5 years.

49. A pendulum swings a length of 40 inches on its first swing. Each successive swing is $\frac{4}{5}$ of the preceding swing. Find the length of the fifth swing and the total length swung during the first five swings. (Round to the nearest tenth of an inch.)

50. Explain the difference between a sequence and a series.

51. a. Write the sum $\sum_{i=1}^{7} (i + i^2)$ without summation notation.

b. Write the sum $\sum_{i=1}^{7} i + \sum_{i=1}^{7} i^2$ without summation notation.

c. Compare the results of parts (a) and (b).

d. Do you think the following is true or false? Explain your answer.

$$\sum_{i=1}^{n} (a_n + b_n) = \sum_{i=1}^{n} a_n + \sum_{i=1}^{n} b_n$$

52. a. Write the sum $\sum_{i=1}^{6} 5i^3$ without summation notation.

b. Write the expression $5 \cdot \sum_{i=1}^{6} i^3$ without summation notation.

c. Compare the results of parts (a) and (b).

d. Do you think the following is true or false? Explain your answer.

$$\sum_{i=1}^{n} c \cdot a_n = c \cdot \sum_{i=1}^{n} a_n \text{ where } c \text{ is a constant}$$

Review Exercises

Evaluate. See Section 1.4.

53. $\dfrac{5}{1 - \dfrac{1}{2}}$

54. $\dfrac{-3}{1 - \dfrac{1}{7}}$

55. $\dfrac{\dfrac{1}{3}}{1 - \dfrac{1}{10}}$

56. $\dfrac{\dfrac{6}{11}}{1 - \dfrac{1}{10}}$

57. $\dfrac{3(1 - 2^4)}{1 - 2}$

58. $\dfrac{2(1 - 5^3)}{1 - 5}$

59. $\dfrac{10}{2}(3 + 15)$

60. $\dfrac{12}{2}(2 + 19)$

12.4 | PARTIAL SUMS OF ARITHMETIC AND GEOMETRIC SEQUENCES

TAPE IAG 12.4

O B J E C T I V E S

1 Find the partial sum of an arithmetic sequence.

2 Find the partial sum of a geometric sequence.

3 Find the sum of the terms of an infinite geometric sequence.

1

Partial sums S_n are relatively easy to find when n is small, that is, when the number of terms to add is small. But when n is large, finding S_n can be tedious. For large n, S_n is still relatively easy to find if the addends are terms of an arithmetic sequence or a geometric sequence.

For an arithmetic sequence, $a_n = a_1 + (n - 1)d$ for some first term a_1 and some common difference d. So S_n, the sum of the first n terms, is

$$S_n = a_1 + (a_1 + d) + (a_1 + 2d) + \cdots + [a_1 + (n - 1)d]$$

We might also find S_n by "working backward" from the nth term a_n, finding the preceding term a_{n-1}, by subtracting d each time:

$$S_n = a_n + (a_n - d) + (a_n - 2d) + \cdots + [a_n - (n - 1)d]$$

Now add the left sides of these two equations, and add the right sides:

$$2S_n = (a_1 + a_n) + (a_1 + a_n) + (a_1 + a_n) + \cdots + (a_1 + a_n)$$

The d terms subtract out, leaving n sums of the first term, a_1, and last term, a_n. Thus, we write

$$2S_n = n(a_1 + a_n)$$

or

$$S_n = \frac{n}{2}(a_1 + a_n)$$

PARTIAL SUM S_n OF AN ARITHMETIC SEQUENCE

The partial sum S_n of the first n terms of an arithmetic sequence is given by

$$S_n = \frac{n}{2}(a_1 + a_n)$$

where a_1 is the first term of the sequence and a_n is the nth term.

EXAMPLE 1 Use the partial sum formula to find the sum of the first six terms of the arithmetic sequence 2, 5, 8, 11, 14, 17, ...

Solution: Use the formula for S_n of an arithmetic sequence, replacing n with 6, a_1 with 2, and a_n with 17.

$$S_n = \frac{n}{2}(a_1 + a_n) = \frac{6}{2}(2 + 17) = 3(19) = 57$$

EXAMPLE 2   Find the sum of the first 30 positive integers.

Solution: Because $1, 2, 3, \ldots, 30$ is an arithmetic sequence, use the formula for S_n with $n = 30$, $a_1 = 1$, and $a_n = 30$. Thus,

$$S_n = \frac{n}{2}(a_1 + a_n) = \frac{30}{2}(1 + 30) = 15(31) = 465$$

EXAMPLE 3   **ANALYZING A CARPET DISPLAY**
Rolls of carpet are stacked in 20 rows with 3 rolls in the top row, 4 rolls in the next row, and so on, forming an arithmetic sequence. Find the total number of carpet rolls if there are 22 rolls in the bottom row.

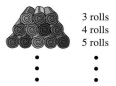

3 rolls
4 rolls
5 rolls

Solution: The list $3, 4, 5, \ldots, 22$ is the first 20 terms of an arithmetic sequence. Use the formula for S_n with $a_1 = 3$, $a_n = 22$, and $n = 20$ terms. Thus,

$$S_{20} = \frac{20}{2}(3 + 22) = 10(25) = 250$$

There are a total of 250 rolls of carpet in the display.

2

We can also derive a formula for the partial sum S_n of the first n terms of a geometric series. If $a_n = a_1 r^{n-1}$, then

$$S_n = a_1 + a_1 r + a_1 r^2 + \cdots + a_1 r^{n-1}$$
$$\quad\;\; \uparrow \quad\;\; \uparrow \quad\;\; \uparrow \qquad\qquad \uparrow$$
$$\quad\;\; \text{1st} \quad \text{2nd} \quad \text{3rd} \qquad\;\; n\text{th}$$
$$\quad\;\; \text{term term term} \qquad\;\; \text{term}$$

Multiply each side of the equation by $-r$:

$$-rS_n = -a_1 r - a_1 r^2 - a_1 r^3 - \cdots - a_1 r^n$$

Add the two equations:

$$S_n - rS_n = a_1 + (a_1 r - a_1 r) + (a_1 r^2 - a_1 r^2) + (a_1 r^3 - a_1 r^3) + \cdots - a_1 r^n$$
$$S_n - rS_n = a_1 - a_1 r^n$$

Now factor each side:

$$S_n(1 - r) = a_1(1 - r^n)$$

Solve for S_n by dividing both sides by $1 - r$. Thus,

$$S_n = \frac{a_1(1 - r^n)}{1 - r}$$

as long as r is not 1.

PARTIAL SUM S_n OF A GEOMETRIC SEQUENCE

The partial sum S_n of the first n terms of a geometric sequence is given by

$$S_n = \frac{a_1(1 - r^n)}{1 - r}$$

where a_1 is the first term of the sequence, r is the common ratio, and $r \neq 1$.

EXAMPLE 4   Find the sum of the first six terms of the geometric sequence 5, 10, 20, 40, 80, 160.

Solution: Use the formula for the partial sum S_n of the terms of a geometric sequence. Here, $n = 6$, the first term $a_1 = 5$, and the common ratio $r = 2$.

$$S_n = \frac{a_1(1 - r^n)}{1 - r}$$

$$S_6 = \frac{5(1 - 2^6)}{1 - 2} = \frac{5(-63)}{-1} = 315$$

EXAMPLE 5   TOTALING DONATIONS
A grant from an alumnus to a university specified that the university was to receive $800,000 during the first year and 75% of the preceding year's donation during each of the following 5 years. Find the total amount donated during the 6 years.

Solution: The donations are modeled by the first six terms of a geometric sequence. Evaluate S_n when $n = 6$, $a_1 = 800{,}000$, and $r = 0.75$.

$$S_6 = \frac{800{,}000[1 - (0.75)^6]}{1 - 0.75}$$

$$= \$2{,}630{,}468.75$$

The total amount donated during the 6 years is \$2,630,468.75.

3 Is it possible to find the sum of all the terms of an infinite sequence? Examine the partial sums of the geometric sequence $\frac{1}{2}, \frac{1}{4}, \frac{1}{8}, \ldots$.

$$S_1 = \frac{1}{2}$$

$$S_2 = \frac{1}{2} + \frac{1}{4} = \frac{3}{4}$$

$$S_3 = \frac{1}{2} + \frac{1}{4} + \frac{1}{8} = \frac{7}{8}$$

$$S_4 = \frac{1}{2} + \frac{1}{4} + \frac{1}{8} + \frac{1}{16} = \frac{15}{16}$$

$$S_5 = \frac{1}{2} + \frac{1}{4} + \frac{1}{8} + \frac{1}{16} + \frac{1}{32} = \frac{31}{32}$$

$$\vdots$$

$$S_{10} = \frac{1}{2} + \frac{1}{4} + \frac{1}{8} + \cdots + \frac{1}{2^{10}} = \frac{1023}{1024}$$

Even though each partial sum is larger than the preceding partial sum, we see that each partial sum is closer to 1 than the preceding partial sum. If n gets larger and larger, then S_n gets closer and closer to 1. We say that 1 is the **limit** of S_n and also that 1 is the sum of the terms of this infinite sequence. In general, if $|r| < 1$, the following formula gives the sum of the terms of an infinite geometric sequence.

SUM OF THE TERMS OF AN INFINITE GEOMETRIC SEQUENCE

The sum S_∞ of the terms of an infinite geometric sequence is given by

$$S_\infty = \frac{a_1}{1 - r}$$

where a is the first term of the sequence, r is the common ratio, and $|r| < 1$. If $|r| \geq 1$, S_∞ does not exist.

What happens for other values of r? For example, in the following geometric sequence, $r = 3$:

6, 18, 54, 162, . . .

Here, as n increases, the sum S_n increases also. This time, though, S_n does not get closer and closer to a fixed number but instead increases without bound.

EXAMPLE 6   Find the sum of the terms of the geometric sequence $2, \dfrac{2}{3}, \dfrac{2}{9}, \dfrac{2}{27}, \ldots$.

Solution: For this geometric sequence, $r = \dfrac{1}{3}$. Since $|r| < 1$, we may use the formula S_∞ of a geometric sequence with $a_1 = 2$ and $r = \dfrac{1}{3}$:

$$S_\infty = \frac{a_1}{1 - r} = \frac{2}{1 - \dfrac{1}{3}} = \frac{2}{\dfrac{2}{3}} = 3$$

The formula for the sum of the terms of an infinite geometric sequence can be used to write a repeating decimal as a fraction. For example,

$$0.33\overline{3} = \frac{3}{10} + \frac{3}{100} + \frac{3}{1000} + \cdots$$

This sum is the sum of the terms of an infinite geometric sequence whose first term $a_1 = \dfrac{3}{10}$ and whose common ratio $r = \dfrac{1}{10}$. Using the formula for S_∞,

$$S_\infty = \frac{a_1}{1 - r} = \frac{\dfrac{3}{10}}{1 - \dfrac{1}{10}} = \frac{1}{3}$$

So $0.33\overline{3} = \dfrac{1}{3}$.

EXAMPLE 7   **FINDING THE DISTANCE TRAVELED BY A PENDULUM**
On its first pass, a pendulum swings through an arc whose length is 24 inches. On each pass thereafter, the arc length is 75% of the arc length on the preceding pass. Find the total distance that the pendulum travels before it comes to rest.

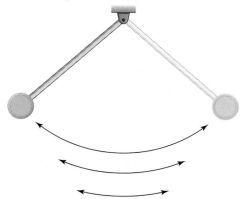

Solution: We must find the sum of the terms of an infinite geometric sequence whose first term, a_1, is 24 and whose common ratio, r, is 0.75. Since $|r| < 1$, we may use the formula for S_∞:

$$S_\infty = \frac{a_1}{1 - r} = \frac{24}{1 - 0.75} = \frac{24}{0.25} = 96$$

The pendulum travels a total distance of 96 inches before it comes to rest.

EXERCISE SET 12.4

Use the partial sum formula to find the partial sum of the given arithmetic or geometric sequence. See Examples 1 and 4.

1. Find the sum of the first six terms of the arithmetic sequence 1, 3, 5, 7, ...

2. Find the sum of the first seven terms of the arithmetic sequence $-7, -11, -15, \ldots$.

3. Find the sum of the first five terms of the geometric sequence 4, 12, 36, ...

4. Find the sum of the first eight terms of the geometric sequence $-1, 2, -4, \ldots$.

5. Find the sum of the first six terms of the arithmetic sequence 3, 6, 9, ...

6. Find the sum of the first four terms of the arithmetic sequence $-4, -8, -12, \ldots$.

7. Find the sum of the first four terms of the geometric sequence $2, \frac{2}{5}, \frac{2}{25}, \ldots$.

8. Find the sum of the first five terms of the geometric sequence $\frac{1}{3}, -\frac{2}{3}, \frac{4}{3}, \ldots$.

Solve. See Example 2.

9. Find the sum of the first ten positive integers.

10. Find the sum of the first eight negative integers.

11. Find the sum of the first four positive odd integers.

12. Find the sum of the first five negative odd integers.

Find the sum of the terms of each infinite geometric sequence. See Example 6.

13. 12, 6, 3, ...

14. 45, 15, 5, ...

15. $\frac{1}{10}, \frac{1}{100}, \frac{1}{1000}, \ldots$

16. $\frac{3}{5}, \frac{3}{20}, \frac{3}{80}, \ldots$

17. $-10, -5, -\frac{5}{2}, \ldots$

18. $-16, -4, -1, \ldots$

19. $2, -\frac{1}{4}, \frac{1}{32}, \ldots$

20. $-3, \frac{3}{5}, -\frac{3}{25}, \ldots$

21. $\frac{2}{3}, -\frac{1}{3}, \frac{1}{6}, \ldots$

22. $6, -4, \frac{8}{3}, \ldots$

Solve.

23. Find the sum of the first ten terms of the sequence $-4, 1, 6, \ldots$.

24. Find the sum of the first twelve terms of the sequence $-3, -13, -23, \ldots$.

25. Find the sum of the first seven terms of the sequence $3, \frac{3}{2}, \frac{3}{4}, \ldots$.

26. Find the sum of the first five terms of the sequence $-2, -6, -18, \ldots$.

27. Find the sum of the first five terms of the sequence $-12, 6, -3, \ldots$.

28. Find the sum of the first four terms of the sequence $-\frac{1}{4}, -\frac{3}{4}, -\frac{9}{4}, \ldots$.

29. Find the sum of the first twenty terms of the sequence $\frac{1}{2}, \frac{1}{4}, 0, \ldots$.

30. Find the sum of the first fifteen terms of the sequence $-5, -9, -13, \ldots$.

31. If a_1 is 8 and r is $-\frac{2}{3}$, find S_3.

32. If a_1 is 10 and d is $-\frac{1}{2}$, find S_{18}.

Solve. See Example 3.

33. Modern Car Company has come out with a new car model. Market analysts predict that 4000 cars will be sold in the first month and that sales will drop by 50 cars per month after that during the first year. Write out the first five terms of the sequence, and find the number of sold cars predicted for the twelfth month. Find the total predicted number of sold cars for the first year.

34. A company that sends faxes charges $3 for the first page sent and $0.10 less than the preceding page for each additional page sent. The cost per page forms an arithmetic sequence. Write the first five terms of this sequence, and use a partial sum to find the cost of sending a nine-page document.

35. Sal has two job offers: Firm A starts at $22,000 per year and guarantees raises of $1000 per year, whereas Firm B starts at $20,000 and guarantees raises of $1200 per year. Use a graphing utility to determine the more profitable offer over a 10-year period.

36. The game of pool uses 15 balls numbered 1 to 15. In the variety called rotation, a player who sinks a ball receives as many points as the number on the ball. Use an arithmetic series to find the score of a player who sinks all 15 balls.

Solve. See Example 5.

37. A woman made $30,000 during the first year she owned her business and made an additional 10% over the previous year in each subsequent year. Find how much she made during her fourth year of business. Find her total earnings during the first 4 years.

38. In free fall, a parachutist falls 16 feet during the first second, 48 feet during the second second, 80 feet during the third second, and so on. Find how far she falls during the eighth second. Find the total distance she falls during the first 8 seconds.

39. A trainee in a computer company takes 0.9 times as long to assemble each computer as he took to assemble the preceding computer. If it took him 30 minutes to assemble the first computer, find how long it takes him to assemble the fifth computer. Find the total time he takes to assemble the first five computers (round to the nearest minute).

40. On a gambling trip to Reno, Carol doubled her bet each time she lost. If her first losing bet was $5 and she lost six consecutive bets, find how much she lost on the sixth bet. Find the total amount lost on these six bets.

Solve. See Example 7.

41. A ball is dropped from a height of 20 feet and repeatedly rebounds to a height that is $\frac{4}{5}$ of its previous height. Find the total distance that the ball covers before it comes to rest.

42. A rotating flywheel coming to rest makes 300 revolutions in the first minute and in each minute thereafter makes $\frac{2}{5}$ as many revolutions as in the preceding minute. Find how many revolutions the wheel makes before it comes to rest.

Solve.

43. In the pool game of rotation, player A sinks balls numbered 1 to 9, and player B sinks the rest of the balls. Use arithmetic series to find each player's score (see Exercise 36).

44. A godfather deposited $250 in a savings account on the day his godchild was born. On each subsequent birthday he deposited $50 more than he deposited the previous year. Find how much money he deposited on his godchild's twenty-first birthday. Find the total amount deposited over the 21 years.

45. During the holiday rush a business can rent a computer system for $200 the first day, with the rental fee decreasing $5 for each additional day. Find the fee paid for 20 days during the holiday rush.

46. The spraying of a field with insecticide killed 6400 weevils the first day, 1600 the second day, 400 the third day, and so on. Find the total number of weevils killed during the first 5 days.

47. A college student humorously asks his parents to charge him room and board according to this geometric sequence: $0.01 for the first day of the month, $0.02 for the second day, $0.04 for the third day, and so on. Find the total room and board he would pay for 30 days.

48. Following its television advertising campaign, a bank attracted 80 new customers the first day, 120 the second day, 160 the third day, and so on, in an arithmetic sequence. Find how many new customers were attracted during the first 5 days following its television campaign.

49. Write $0.88\overline{8}$ as an infinite geometric series and use the formula for S_∞ to write it as a rational number.

50. Write $0.54\overline{54}$ as an infinite geometric series and use the formula S_∞ to write it as a rational number.

Review Exercises

Evaluate. See Section 12.2 and 12.4.

51. Explain whether the sequence 5, 5, 5, . . . is arithmetic, geometric, neither, or both.

52. Describe a situation in everyday life that can be modeled by an infinite geometric series.

Multiply. See Section 6.4.

53. $6 \cdot 5 \cdot 4 \cdot 3 \cdot 2 \cdot 1$

54. $8 \cdot 7 \cdot 6 \cdot 5 \cdot 4 \cdot 3 \cdot 2 \cdot 1$

55. $\dfrac{3 \cdot 2 \cdot 1}{2 \cdot 1}$

56. $\dfrac{5 \cdot 4 \cdot 3 \cdot 2 \cdot 1}{3 \cdot 2 \cdot 1}$

Multiply. See Section 5.4.

57. $(x + 5)^2$

58. $(x - 2)^2$

59. $(2x - 1)^3$

60. $(3x + 2)^3$

12.5 | THE BINOMIAL THEOREM

TAPE IAG 12.5

O B J E C T I V E S

1. Use Pascal's triangle to expand binomials.
2. Evaluate factorials.
3. Use the binomial theorem to expand binomials.
4. Find the nth term in the expansion of a binomial raised to a positive power.

In this section, we learn how to **expand** binomials of the form $(a + b)^n$ easily. Expanding a binomial such as $(a + b)^n$ means to write the factored form as a sum. First, we review the patterns in the expansions of $(a + b)^n$.

$(a + b)^0 = 1$	1 term
$(a + b)^1 = a + b$	2 terms
$(a + b)^2 = a^2 + 2ab + b^2$	3 terms
$(a + b)^3 = a^3 + 3a^2b + 3ab^2 + b^3$	4 terms
$(a + b)^4 = a^4 + 4a^3b + 6a^2b^2 + 4ab^3 + b^4$	5 terms
$(a + b)^5 = a^5 + 5a^4b + 10a^3b^2 + 10a^2b^3 + 5ab^4 + b^5$	6 terms

Notice the following patterns:

1. The expansion of $(a + b)^n$ contains $n + 1$ terms. For example, for $(a + b)^3$, $n = 3$, and the expansion contains $3 + 1$ terms, or 4 terms.
2. The first term of the expansion of $(a + b)^n$ is a^n, and the last term is b^n.
3. The powers of a decrease by 1 for each term, whereas the powers of b increase by 1 for each term.
4. For each term of the expansion of $(a + b)^n$, the sum of the exponents of a and b is n. (For example, the sum of the exponents of $5a^4b$ is $4 + 1$, or 5, and the sum of the exponents of $10a^3b^2$ is $3 + 2$, or 5.)

There are patterns in the coefficients of the terms as well. Written in a triangular array, the coefficients are called **Pascal's triangle.**

		Row
$(a + b)^0$:	1	Row 1
$(a + b)^1$:	1 1	Row 2
$(a + b)^2$:	1 2 1	Row 3
$(a + b)^3$:	1 3 3 1	Row 4
$(a + b)^4$:	1 4 6 4 1	Row 5
$(a + b)^5$:	1 5 10 10 5 1	Row 6

Each row in Pascal's triangle begins and ends with 1. Any other number in a row is the sum of the two closest numbers above it. Using this pattern, we can write the next row, the seventh row, by first writing the number 1. Then we can add the consecutive numbers in row 6 and write each sum "between and below" the pair. We complete the row by writing a 1.

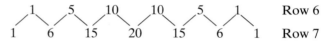

We can use Pascal's triangle and the patterns noted to expand $(a + b)^n$ without actually multiplying any terms.

EXAMPLE 1   Expand $(a + b)^6$.

Solution: Using the seventh row of Pascal's triangle as the coefficients and following the patterns noted, $(a + b)^6$ can be expanded as

$$a^6 + 6a^5b + 15a^4b^2 + 20a^3b^3 + 15a^2b^4 + 6ab^5 + b^6$$

For large n, the use of Pascal's triangle to find coefficients for $(a + b)^n$ can be tedious. An alternative method for determining these coefficients is based on the concept of a **factorial.**

The **factorial of n,** written $n!$ (read "n factorial"), is the product of the first n consecutive natural numbers.

> **FACTORIAL OF n: $n!$**
>
> If n is a natural number, then $n! = n(n - 1)(n - 2)(n - 3) \cdot \cdots \cdot 3 \cdot 2 \cdot 1$. The factorial of 0, written $0!$, is defined to be 1.

For example, $3! = 3 \cdot 2 \cdot 1 = 6$, $5! = 5 \cdot 4 \cdot 3 \cdot 2 \cdot 1 = 120$, and $0! = 1$.

EXAMPLE 2   Evaluate each expression.

a. $\dfrac{5!}{6!}$

b. $\dfrac{10!}{7!3!}$

c. $\dfrac{3!}{2!1!}$

d. $\dfrac{7!}{7!0!}$

Solution: **a.** $\dfrac{5!}{6!} = \dfrac{5 \cdot 4 \cdot 3 \cdot 2 \cdot 1}{6 \cdot 5 \cdot 4 \cdot 3 \cdot 2 \cdot 1} = \dfrac{1}{6}$

TECHNOLOGY NOTE

When evaluating facto-
rials using a graphing
utility, notice the need
for parentheses group-
ing the denominator.

```
5!/6!▶Frac
              1/6
10!/(7!3!)
              120
3!/(2!1!)
                3
```

b. $\dfrac{10!}{7!3!} = \dfrac{10 \cdot 9 \cdot 8 \cdot 7!}{7! \cdot 3 \cdot 2 \cdot 1} = \dfrac{10 \cdot 9 \cdot 8}{3 \cdot 2 \cdot 1} = 120$

c. $\dfrac{3!}{2!1!} = \dfrac{3 \cdot 2 \cdot 1}{2 \cdot 1 \cdot 1} = 3$

d. $\dfrac{7!}{7!0!} = \dfrac{7!}{7! \cdot 1} = 1$

R E M I N D E R   We can use a calculator with a factorial key to evaluate a factorial. A calculator uses scientific notation for large results.

3   It can be proved, although we won't do so here, that the coefficients of terms in
the expansion of $(a + b)^n$ can be expressed in terms of factorials. Following patterns 1 through 4 given earlier and using the factorial expressions of the coefficients, we have what is known as the **binomial theorem.**

BINOMIAL THEOREM

If n is a positive integer, then

$$(a + b)^n = a^n + \frac{n}{1!} a^{n-1}b^1 + \frac{n(n - 1)}{2!} a^{n-2}b^2$$

$$+ \frac{n(n - 1)(n - 2)}{3!} a^{n-3}b^3 + \cdots + b^n$$

We call the formula for $(a + b)^n$ given by the binomial theorem the **binomial formula.**

EXAMPLE 3 Use the binomial theorem to expand $(x + y)^{10}$.

Solution: Let $a = x$, $b = y$, and $n = 10$ in the binomial formula:

$$(x + y)^{10} = x^{10} + \frac{10}{1!}x^9y + \frac{10 \cdot 9}{2!}x^8y^2 + \frac{10 \cdot 9 \cdot 8}{3!}x^7y^3 + \frac{10 \cdot 9 \cdot 8 \cdot 7}{4!}x^6y^4$$

$$+ \frac{10 \cdot 9 \cdot 8 \cdot 7 \cdot 6}{5!}x^5y^5 + \frac{10 \cdot 9 \cdot 8 \cdot 7 \cdot 6 \cdot 5}{6!}x^4y^6$$

$$+ \frac{10 \cdot 9 \cdot 8 \cdot 7 \cdot 6 \cdot 5 \cdot 4}{7!}x^3y^7$$

$$+ \frac{10 \cdot 9 \cdot 8 \cdot 7 \cdot 6 \cdot 5 \cdot 4 \cdot 3}{8!}x^2y^8$$

$$+ \frac{10 \cdot 9 \cdot 8 \cdot 7 \cdot 6 \cdot 5 \cdot 4 \cdot 3 \cdot 2}{9!}xy^9 + y^{10}$$

$$= x^{10} + 10x^9y + 45x^8y^2 + 120x^7y^3 + 210x^6y^4 + 252x^5y^5 + 210x^4y^6$$

$$+ 120x^3y^7 + 45x^2y^8 + 10xy^9 + y^{10}$$

EXAMPLE 4 Use the binomial theorem to expand $(x + 2)^5$.

Solution: Let $a = x$ and $b = 2$ in the binomial formula:

$$(x + 2)^5 = x^5 + \frac{5}{1!}x^4(2) + \frac{5 \cdot 4}{2!}x^3(2)^2 + \frac{5 \cdot 4 \cdot 3}{3!}x^2(2)^3$$

$$+ \frac{5 \cdot 4 \cdot 3 \cdot 2}{4!}x(2)^4 + (2)^5$$

$$= x^5 + 10x^4 + 40x^3 + 80x^2 + 80x + 32$$

TECHNOLOGY NOTE

To visually check Example 4, graph $y_1 = (x + 2)^5$ and $y_2 = x^5 + 10x^4 + 40x^3\ 80x^2 + 80x + 32$ in the same viewing window. If the graphs do not coincide, you should double-check your bi-nomial expansion. In this case, the graphs appear to coincide, reinforcing the conclusion of Example 4.

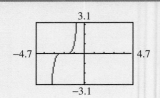

EXAMPLE 5 Use the binomial theorem to expand $(3m - n)^4$.

Solution: Let $a = 3m$ and $b = -n$ in the binomial formula:

$$(3m - n)^4 = (3m)^4 + \frac{4}{1!}(3m)^3(-n) + \frac{4 \cdot 3}{2!}(3m)^2(-n)^2$$

$$+ \frac{4 \cdot 3 \cdot 2}{3!}(3m)(-n)^3 + (-n)^4$$

$$= 81m^4 - 108m^3n + 54m^2n^2 - 12mn^3 + n^4$$

4 Sometimes it is convenient to find a specific term of a binomial expansion without writing out the entire expansion. By studying the expansion of binomials, a pattern forms for each term. This pattern is most easily stated for the $(r + 1)$st term.

(r + 1)ST TERM IN A BINOMIAL EXPANSION

The $(r + 1)$st term of the expansion of $(a + b)^n$ is $\dfrac{n!}{r!(n - r)!}a^{n-r}b$.

EXAMPLE 6   Find the eighth term in the expansion of $(2x - y)^{10}$.

Solution: Use the formula, with $n = 10$, $a = 2x$, $b = -y$, and $r + 1 = 8$. Notice that, since $r + 1 = 8$, $r = 7$:

$$\frac{n!}{r!(n - r)!}a^{n-r}b^r = \frac{10!}{7!3!}(2x)^3(-y)^7$$

$$= 120(8x^3)(-y)^7$$

$$= -960x^3y^7$$

EXERCISE SET 12.5

Use Pascal's triangle to expand the binomial. See Example 1.

1. $(m + n)^3$ **2.** $(x + y)^4$ **3.** $(c + d)^5$

4. $(a + b)^6$ **5.** $(y - x)^5$ **6.** $(q - r)^7$

7. Explain how to generate a row of Pascal's triangle.

8. Write the ninth row of Pascal's triangle.

Evaluate each expression. See Example 2.

9. $\dfrac{8!}{7!}$ **10.** $\dfrac{6!}{0!}$ **11.** $\dfrac{7!}{5!}$ **12.** $\dfrac{8!}{5!}$

13. $\dfrac{10!}{7!2!}$ **14.** $\dfrac{9!}{5!3!}$ **15.** $\dfrac{8!}{6!0!}$ **16.** $\dfrac{10!}{4!6!}$

Use the binomial formula to expand each binomial. See Examples 3 through 5.

17. $(a + b)^7$ **18.** $(x + y)^8$ **19.** $(a + 2b)^5$

20. $(x + 3y)^6$ **21.** $(q + r)^9$ **22.** $(b + c)^6$

23. $(4a + b)^5$ **24.** $(3m + n)^4$ **25.** $(5a - 2b)^4$

26. $(m - 4)^6$ **27.** $(2a + 3b)^3$ **28.** $(4 - 3x)^5$

29. $(x + 2)^5$ **30.** $(3 + 2a)^4$

Find the indicated term. See Example 6.

31. The fifth term of the expansion of $(c - d)^5$

32. The fourth term of the expansion of $(x - y)^6$

33. The eighth term of the expansion of $(2c + d)^7$

34. The tenth term of the expansion of $(5x - y)^9$

35. The fourth term of the expansion of $(2r - s)^5$

36. The first term of the expansion of $(3q - 7r)^6$

37. The third term of the expansion of $(x + y)^4$

38. The fourth term of the expansion of $(a + b)^8$

39. The second term of the expansion of $(a + 3b)^{10}$

40. The third term of the expansion of $(m + 5n)^7$

Review Exercises

Sketch the graph of each function. Decide whether each function is one-to-one. See Section 9.5.

41. $f(x) = |x|$ **42.** $g(x) = 3(x - 1)^2$

43. $H(x) = 2x + 3$ **44.** $F(x) = -2$

45. $f(x) = x^2 + 3$ **46.** $h(x) = -(x + 1)^2 - 4$

GROUP ACTIVITY

MODELING COLLEGE TUITION

MATERIALS:
• Newspapers, news magazines.

Annual tuitions at public and private universi-ties alike are increasing at a rate of 6% per year. During the 1995–1996 academic year, the average tuition was $2860 at a public four-year university and $12,432 at a private four-year university. (*Source: USA Today*, February 20, 1996)

1. Find the general term of the sequence that describes the pattern of average annual tuition for public universities. Find the general term of the sequence that describes the pattern of average annual tuition for private universities. In each case let $n = 1$ represent the 1995–1996 academic year.

2. Assuming that the rate of tuition increase remains the same, find the annual tuition for the 1999–2000 academic year for each type of university.

3. Use a graphing utility to find the tuition at each type of university for the next 10 years. Make a table of the results.

4. Use partial sums to find the average cost of a four-year college education at both a public university and a private university for a student starting college in the 1997–1998 academic year. Verify your result using the table from Question 3.

5. Find the year in which tuition at a public university will reach $4300 per year.

6. Use newspapers and/or news magazines to find a situation that can be modeled by a sequence. Briefly describe the situation and write several reasonable questions about the situation. Exchange your problem with another member of your group to solve. Compare your answers.

CHAPTER 12   HIGHLIGHTS

DEFINITIONS AND CONCEPTS	EXAMPLES
SECTION 12.1 SEQUENCES	
An **infinite sequence** is a function whose domain is the set of natural numbers $\{1, 2, 3, 4, \ldots\}$.	Infinite sequence: $\quad 2, 4, 6, 8, 10, \ldots$
A **finite sequence** is a function whose domain is the set of natural numbers $\{1, 2, 3, 4, \ldots, n\}$, where n is some natural number.	Finite sequence: $\quad 1, -2, 3, -4, 5, -6$

(continued)

DEFINITIONS AND CONCEPTS	**EXAMPLES**

SECTION 12.1   SEQUENCES

The notation a_n, where n is a natural number, is used to denote a sequence.

Write the first four terms of the sequence whose general term is $a_n = n^2 + 1$:

$$a_1 = 1^2 + 1 = 2$$
$$a_2 = 2^2 + 1 = 5$$
$$a_3 = 3^2 + 1 = 10$$
$$a_4 = 4^2 + 1 = 17$$

SECTION 12.2   ARITHMETIC AND GEOMETRIC SEQUENCES

An **arithmetic sequence** is a sequence in which each term differs from the preceding term by a constant amount d, called the **common difference.**

Arithmetic sequence:

$$5, 8, 11, 14, 17, 20, \ldots$$

Here, $a_1 = 5$ and $d = 3$.

The general term is

The **general term** a_n of an arithmetic sequence is given by

$$a_n = a_1 + (n - 1)d$$

where a_1 is the first term and d is the common difference.

$$a_n = a_1 + (n - 1)d \quad \text{or}$$
$$a_n = 5 + (n - 1)3$$

A **geometric sequence** is a sequence in which each term is obtained by multiplying the preceding term by a constant r, called the **common ratio.**

Geometric sequence:

$$12, -6, 3, -\frac{3}{2}, \ldots$$

The **general term** a_n of a geometric sequence is given by

$$a_n = a_1 r^{n-1}$$

where a_1 is the first term and r is the common ratio.

Here, $a_1 = 12$ and $r = -\frac{1}{2}$.

The general term is

$$a_n = a_1 r^{n-1} \quad \text{or}$$
$$a_n = 12\left(-\frac{1}{2}\right)^{n-1}$$

SECTION 12.3   SERIES

A sum of the terms of a sequence is called a **series.**

A shorthand notation for denoting a series is called **summation notation:**

$$\underset{\substack{\text{index of} \\ \text{summation}}}{\xrightarrow{\hspace{1cm}}} \sum_{i=1}^{4} \xrightarrow{\hspace{0.5cm}} \text{Greek letter sigma is used to mean sum}$$

SEQUENCE	SERIES	
3, 7, 11, 15	$3 + 7 + 11 + 15$	finite
3, 7, 11, 15, ...	$3 + 7 + 11 + 15 + \ldots$	infinite

$$\sum_{i=1}^{4} 3^i = 3^1 + 3^2 + 3^3 + 3^4$$
$$= 3 + 9 + 27 + 81$$
$$= 120$$

DEFINITIONS AND CONCEPTS	EXAMPLES

Section 12.4   Partial Sums of Arithmetic and Geometric Sequences

Partial sum, S_n, of the first n terms of an arithmetic sequence:

$$S_n = \frac{n}{2}(a_1 + a_n)$$

where a_1 is the first term and a_n is the nth term.

Partial sum, S_n, of the first n terms of a geometric sequence:

$$S_n = \frac{a_1(1 - r^n)}{1 - r}$$

where a_1 is the first term, r is the common ratio, and $r \neq 1$.

Sum of the terms of an infinite geometric sequence:

$$S_\infty = \frac{a_1}{1 - r}$$

where a_1 is the first term, r is the common ratio, and $|r| < 1$. If $|r| \geq 1$, S_∞ does not exist.

The sum of the first five terms of the arithmetic sequence

$$12, 24, 36, 48, 60, \ldots \quad \text{is}$$

$$S_n = \frac{5}{2}(12 + 60) = 180$$

The sum of the first five terms of the geometric sequence

$$15, 30, 60, 120, 240, \ldots \quad \text{is}$$

$$S_5 = \frac{15(1 - 2^5)}{1 - 2} = 465$$

The sum of the terms of the infinite geometric sequence

$$1, \frac{1}{3}, \frac{1}{9}, \frac{1}{27}, \ldots \quad \text{is}$$

$$S_\infty = \frac{1}{1 - \dfrac{1}{3}} = \frac{3}{2}$$

Section 12.5   The Binomial Theorem

The **factorial of n,** written $n!$, is the product of the first n consecutive natural numbers.

Binomial theorem:

If n is a positive integer, then

$$(a + b)^n = a^n + \frac{n}{1!}a^{n-1}b^1 + \frac{n(n-1)}{2!}a^{n-2}b^2$$

$$+ \frac{n(n-1)(n-2)}{3!}a^{n-3}b^3 + \cdots + b^n$$

$$5! = 5 \cdot 4 \cdot 3 \cdot 2 \cdot 1 = 120$$

Expand $(3x + y)^4$.

$$(3x + y)^4 = (3x)^4 + \frac{4}{1!}(3x)^3(y)^1$$

$$+ \frac{4 \cdot 3}{2!}(3x)^2(y)^2 + \frac{4 \cdot 3 \cdot 2}{3!}(3x)^1y^3 + y^4$$

$$= 81x^4 + 108x^3y + 54x^2y^2 + 12xy^3 + y^4$$

CHAPTER 12 REVIEW

(12.1)   *Find the indicated term(s) of the given sequence.*

1. The first five terms of the sequence $a_n = -3n^2$.

2. The first five terms of the sequence $a_n = n^2 + 2n$.

3. The one-hundredth term of the sequence $a_n = \frac{(-1)^n}{100}$.

4. The fiftieth term of the sequence $a_n = \frac{2n}{(-1)^2}$.

5. The general term a_n of the sequence $\frac{1}{6}, \frac{1}{12}, \frac{1}{18}, \ldots$

6. The general term a_n of the sequence $-1, 4, -9, 16, \ldots$

Solve the following applications.

7. The distance in feet that an olive falling from rest in a vacuum will travel during each second is given by an arithmetic sequence whose general term is $a_n = 32n - 16$, where n is the number of the second. Find the distance the olive will fall during the fifth, sixth, and seventh seconds.

8. A culture of yeast doubles every day in a geometric progression whose general term is $a_n = 100(2)^{n-1}$, where n is the number of the day just beginning. Find how many days it takes the yeast culture to measure at least 10,000. Find the original measure of the yeast culture.

9. The Centers for Disease Control and Prevention (CDC) reported that a new type of virus infected approximately 450 people during 1996, the year it was first discovered. The CDC predicts that during the next decade the virus will infect three times as many people each year as the year before. Write out the first five terms of this geometric sequence, and predict the number of infected people in 2000.

10. The first row of an amphitheater contains 50 seats, and each row thereafter contains 8 additional seats. Write the first ten terms of this arithmetic progression, and find the number of seats in the tenth row.

(12.2)

11. Find the first five terms of the geometric sequence whose first term is -2 and whose common ratio is $\frac{2}{3}$.

12. Find the first five terms for the arithmetic sequence whose first term is 12 and whose common difference is -1.5.

13. Find the thirtieth term of the arithmetic sequence whose first term is -5 and whose common difference is 4.

14. Find the eleventh term of the arithmetic sequence whose first term is 2 and whose common difference is $\frac{3}{4}$.

15. Find the twentieth term of the arithmetic sequence whose first three terms are 12, 7, and 2.

16. Find the sixth term of the geometric sequence whose first three terms are 4, 6, and 9.

17. If the fourth term of an arithmetic sequence is 18 and the twentieth term is 98, find the first term and the common difference.

18. If the third term of a geometric sequence is -48 and the fourth term is 192, find the first term and the common ratio.

19. Find the general term of the sequence $\frac{3}{10}, \frac{3}{100}, \frac{3}{1000}, \ldots$

20. Find a general term that satisfies the terms shown for the sequence 50, 58, 66, ...

Determine whether each of the following sequences is arithmetic, geometric, or neither. If a sequence is arithmetic, find a_1 and d. If a sequence is geometric, find a_1 and r.

21. $\frac{8}{3}, 4, 6, \ldots$

22. $-10.5, -6.1, -1.7$

23. $7x, -14x, 28x$

24. $3x^2, 9x^4, 81x^8, \ldots$

Solve the following applications.

25. To test the bounce of a racketball, the ball is dropped from a height of 8 feet. The ball is judged "good" if it rebounds at least 75% of its previous height with each bounce. Write out the first six terms of this geometric sequence (round to the nearest tenth). Determine if a ball is "good" that rebounds to a height of 2.5 feet after the fifth bounce.

26. A display of oil cans in an auto parts store has 25 cans in the bottom row, 21 cans in the next row, and so on, in an arithmetic progression. Find the general term and the number of cans in the top row.

27. Suppose that you save $1 the first day of a month, $2 the second day, $4 the third day, continuing to double your savings each day. Write the general term of this geometric sequence and find the amount that you will save on the tenth day. Estimate the amount that you will save on the thirtieth day of the month, and check your estimate with a calculator.

28. On the first swing, the length of an arc through which a pendulum swings is 30 inches. The length of the arc for each successive swing is 70% of the preceding swing. Find the length of the arc for the fifth swing.

29. Rosa takes a job that has a monthly starting salary of $900 and guarantees her a monthly raise of $150 during her 6-month training period. Find the general term of this sequence and her salary at the end of her training.

30. A sheet of paper is $\frac{1}{512}$-inch thick. By folding the sheet in half, the total thickness will be $\frac{1}{256}$ inch: A second fold produces a total thickness of $\frac{1}{128}$ inch. Estimate the thickness of the stack after 15 folds, and then check your estimate with a calculator.

(12.3) *Write out the terms and find the sum for each of the following.*

31. $\displaystyle\sum_{i=1}^{5} 2i - 1$

32. $\displaystyle\sum_{i=1}^{5} i(i + 2)$

33. $\displaystyle\sum_{i=2}^{4} \frac{(-1)^i}{2i}$

34. $\displaystyle\sum_{i=3}^{5} 5(-1)^{i-1}$

Find the partial sum of the given sequence.

35. S_4 of the sequence $a_n = (n - 3)(n + 2)$

36. S_6 of the sequence $a_n = n^2$

37. S_5 of the sequence $a_n = -8 + (n - 1)3$

38. S_3 of the sequence $a_n = 5(4)^{n-1}$

Write the sum with Σ notation.

39. $1 + 3 + 9 + 27 + 81 + 243$

40. $6 + 2 + (-2) + (-6) + (-10) + (-14) + (-18)$

41. $\dfrac{1}{4} + \dfrac{1}{16} + \dfrac{1}{64} + \dfrac{1}{256}$

42. $1 + \left(-\dfrac{3}{2}\right) + \dfrac{9}{4}$

Solve.

43. A yeast colony begins with 20 yeast cells and doubles every 8 hours. Write the sequence that describes the growth of the yeast, and find the total yeast cells after 48 hours. $a_n = 20(2)^n$; n represents the number of 8-hour periods.

44. The number of cranes born each year in a new aviary forms a sequence whose general term is $a_n = n^2 + 2n - 1$. Find the number of cranes born in the fourth year and the total number of cranes born in the first four years.

45. Harold has a choice between two job offers. Job A has an annual starting salary of $19,500 with guaranteed annual raises of $1100 for the next four years, whereas job B has an annual starting salary of $21,000 with guaranteed annual raises of $700 for the next four years. Compare the salaries for the fifth year under each job offer.

46. A sample of radioactive waste is decaying such that the amount decaying in kilograms during year n is $a_n = 200(0.5)^n$. Find the amount of decay in the third year, and the total amount of decay in the first three years.

(12.4) *Find the partial sum of the given sequence.*

47. The sixth partial sum of the sequence 15, 19, 23, ...

48. The ninth partial sum of the sequence 5, −10, 20, ...

49. The sum of the first 30 odd positive integers

50. The sum of the first 20 positive multiples of 7

51. The sum of the first 20 terms of the sequence 8, 5, 2, ...

52. The sum of the first eight terms of the sequence $\frac{3}{4}, \frac{9}{4}, \frac{27}{4}, \dots$

53. S_4 if $a_1 = 6$ and $r = 5$

54. S_{100} if $a_1 = -3$ and $d = -6$

Find the sum of each infinite geometric sequence.

55. $5, \dfrac{5}{2}, \dfrac{5}{4}, \dots$

56. $18, -2, \dfrac{2}{9}, \dots$

57. $-20, -4, -\dfrac{4}{5}, \dots$

58. $0.2, 0.02, 0.002, \dots$

Solve.

59. A frozen-yogurt store owner cleared $20,000 the first year he owned his business and made an additional 15% over the previous year in each subsequent year. Find how much he made during his fourth year of business. Find his total earnings during the first 4 years (round to the nearest dollar).

60. On his first morning in a television assembly factory, a trainee takes 0.8 times as long to assemble each television as he took to assemble the one before. If it took him 40 minutes to assemble the first television, find how long it takes him to assemble the fourth television. Find the total time he takes to assemble the first four televisions (round to the nearest minute).

61. During the harvest season a farmer can rent a combine machine for $100 the first day, with the rental fee decreasing $7 for each additional day. Find how much the farmer pays for the rental on the seventh day. Find how much total rent the farmer pays for 7 days.

62. A rubber ball is dropped from a height of 15 feet and rebounds 80% of its previous height after each bounce. Find the total distance the ball travels before it comes to rest.

63. After a pond was sprayed once with insecticide, 1800 mosquitoes were killed the first day, 600 the second day, 200 the third day, and so on. Find the total number of mosquitoes killed during the first 6 days after the spraying (round to the nearest unit).

64. See Exercise 63. Find the day on which the insecticide is no longer effective, and find the total number of mosquitoes killed (round to the nearest mosquito).

65. Use the formula S_∞ to write $0.55\overline{5}$ as a fraction.

66. A movie theater has 27 seats in the first row, 30 seats in the second row, 33 seats in the third row,

and so on. Find the total number of seats in the theater if there are 20 rows.

(12.5)   *Use Pascal's triangle to expand the binomial.*

67. $(x + z)^5$

68. $(y - r)^6$

69. $(2x + y)^4$

70. $(3y - z)^4$

Use the binomial formula to expand the following.

71. $(b + c)^8$

72. $(x - w)^7$

73. $(4m - n)^4$

74. $(p - 2r)^5$

Find the indicated term.

75. The fourth term of the expansion of $(a + b)^7$

76. The eleventh term of the expansion of $(y + 2z)^{10}$

CHAPTER 12 TEST

Find the indicated term(s) of the given sequence.

1. The first five terms of the sequence $a_n = \dfrac{(-1)^n}{n + 4}$.

2. The first five terms of the sequence $a_n = \dfrac{3}{(-1)^n}$.

3. The eightieth term of the sequence
$a_n = 10 + 3(n - 1)$.

4. The two-hundredth term of the sequence
$a_n = (n + 1)(n - 1)(-1)^n$.

5. The general term of the sequence $\dfrac{2}{5}, \dfrac{2}{25}, \dfrac{2}{125}, \ldots$

6. The general term of the sequence $-9, 18, -27, 36, \ldots$

Find the partial sum of the given sequence.

7. S_5 of the sequence $a_n = 5(2)^{n-1}$

8. S_{30} of the sequence $a_n = 18 + (n - 1)(-2)$

9. S_∞ of the sequence $a_1 = 24$ and $r = \dfrac{1}{6}$

10. S_∞ of the sequence $\dfrac{3}{2}, -\dfrac{3}{4}, \dfrac{3}{8}, \ldots$

11. $\displaystyle\sum_{i=1}^{4} i(i - 2)$

12. $\displaystyle\sum_{i=2}^{4} 5(2)^i(-1)^{i-1}$

Expand the binomial by using Pascal's triangle.

13. $(a - b)^6$

14. $(2x + y)^5$

Expand the binomial by using the binomial formula.

15. $(y + z)^8$

16. $(2p + r)^7$

Solve the following applications.

17. The population of a small town is growing yearly according to the sequence defined by $a_n = 250 + 75(n - 1)$, where n is the number of the year just beginning. Predict the population at the beginning of the tenth year. Find the town's initial population.

18. A gardener is making a triangular planting with one shrub in the first row, three shrubs in the second row, five shrubs in the third row, and so on, for eight rows. Write the finite series of this sequence, and find the total number of shrubs planted.

19. A pendulum swings through an arc of length 80 centimeters on its first swing. On each successive swing, the length of the arc is $\dfrac{3}{4}$ the length of the arc on the preceding swing. Find the length of the arc on the fourth swing, and find the total arc lengths for the first four swings.

20. See Exercise 19. Find the total arc lengths before the pendulum comes to rest.

21. A parachutist in free fall falls 16 feet during the first second, 48 feet during the second second, 80 feet during the third second, and so on. Find how far he falls during the tenth second. Find the total distance he falls during the first 10 seconds.

22. Use the formula S_∞ to write $0.42\overline{42}$ as a fraction.

CHAPTER 12 CUMULATIVE REVIEW

1. Solve for c: $25 - 3.5c = 6$.

2. Which of the following relations are also functions?

a. $\{(-2, 5), (2, 7), (-3, 5), (9, 9)\}$

b.

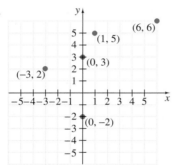

c.

Input:	Correspondence:	Output:
People in a certain city	Each person's age	Nonnegative integers

3. Write a function that describes the line that contains the point $(4, 4)$ and is perpendicular to the line $2x + 3y = -6$.

4. Solve $|x| < 7$.

5. A small manufacturing company manufactures and sells compact disc storage units. The revenue equation for these units is

$$y = 50x$$

where x is the number of units sold and y is the revenue, or income, in dollars for selling x units. The cost equation for these units is

$$y = 30x + 10,000$$

where x is the number of units sold and y is the total cost in dollars for manufacturing x units. Use these equations to find the number of units to be sold for the company to break even.

6. The first number is 4 less than a second number. Four times the first number is 6 more than twice the second. Find the numbers.

7. Write the following with only positive exponents. Simplify if possible.

a. 5^{-2} **b.** $2x^{-3}$ **c.** $(3x)^{-1}$ **d.** $\dfrac{m^5}{m^{15}}$

e. $\dfrac{3^3}{3^6}$ **f.** $2^{-1} + 3^{-2}$ **g.** $\dfrac{1}{t^{-5}}$

8. Subtract: $(12z^5 - 12z^3 + z) - (-3z^4 + z^3 + 12z)$.

9. Factor $y^3 - 64$.

10. Add:

$$\frac{2x - 1}{2x^2 - 9x - 5} + \frac{x + 3}{6x^2 - x - 2}$$

11. Divide $2x^3 + 3x^4 - 8x + 6$ by $x^2 - 1$.

12. Use the quotient rule to simplify.

a. $\sqrt{\dfrac{25}{49}}$ **b.** $\sqrt[3]{\dfrac{8}{27}}$

c. $\sqrt{\dfrac{x}{9}}$ **d.** $\sqrt[4]{\dfrac{3}{16y^4}}$

13. Solve $\sqrt{-10x - 1} + 3x = 0$ for x.

14. Solve $3x^2 + 16x + 5 = 0$ for x.

15. Use a graphing utility to graph $f(x) = 3x^2 + 3x + 1$. Find the vertex and any intercepts, algebraically.

16. Graph $\dfrac{x^2}{9} + \dfrac{y^2}{16} = 1$.

17. Solve the system

$$\begin{cases} x^2 + 2y^2 = 10 \\ x^2 - y^2 = 1 \end{cases}$$

18. Determine whether each function described is one-to-one.

a. $f = \{(6, 2), (5, 4), (-1, 0), (7, 3)\}$

b. $g = \{(3, 9), (-4, 2), (-3, 9), (0, 0)\}$

c. $h = \{(1, 1), (2, 2), (10, 10), (-5, -5)\}$

d.

MINERAL (INPUT)	HARDNESS ON THE MOHS SCALE (OUTPUT)
Talc	1
Gypsum	2
Diamond	10
Topaz	8
Stibnite	2

e.

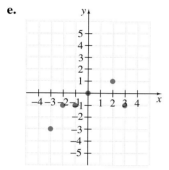

19. Find the amount owed at the end of 5 years if $1600 is loaned at a rate of 9% compounded monthly. Use the formula $A = P\left(1 + \dfrac{r}{n}\right)^{nt}$.

20. Solve each equation for x.

a. $\log_4 \dfrac{1}{4} = x$

b. $\log_5 x = 3$

c. $\log_x 25 = 2$

d. $\log_3 1 = x$

e. $\log_b 1 = x$

21. Solve $\log (x + 2) - \log x = 2$ for x.

22. If the general term of a sequence is given by $a_n = \dfrac{(-1)^n}{3n}$, find:

a. the first term of the sequence

b. a_8

c. the one-hundredth term of the sequence

d. a_{15}

23. Donna has an offer for a job starting at $20,000 per year and guaranteeing her a raise of $800 per year for the next 5 years. Write the general term for the arithmetic sequence that models Donna's potential annual salaries, and find her salary for the fourth year.

24. Use the binomial theorem to expand $(x + 2)^5$.

REVIEW OF ANGLES, LINES, AND SPECIAL TRIANGLES

The word **geometry** is formed from the Greek words, **geo,** meaning earth, and **metron,** meaning measure. Geometry literally means to measure the earth.

This section contains a review of some basic geometric ideas. It will be assumed that fundamental ideas of geometry such as point, line, ray, and angle are known. In this appendix, the notation $\angle 1$ is read "angle 1" and the notation $m\angle 1$ is read "the measure of angle 1."

We first review types of angles.

ANGLES

An angle whose measure is more than 0° but less than 90° is called an **acute angle**.

A **right angle** is an angle whose measure is 90°. A right angle can be indicated by a square drawn at the vertex of the angle, as shown below.

An angle whose measure is greater than 90° but less than 180° is called an **obtuse angle**.

An angle whose measure is 180° is called a **straight angle**.

Two angles are said to be **complementary** if the sum of their measures is 90°. Each angle is called the **complement** of the other.

Two angles are said to be **supplementary** if the sum of their measures is 180°. Each angle is called the **supplement** of the other.

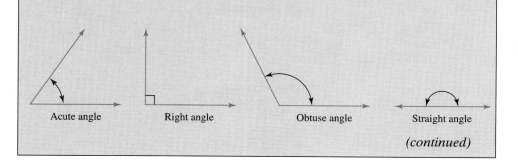

Acute angle   Right angle   Obtuse angle   Straight angle

(continued)

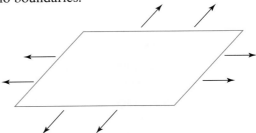

Complementary angles:
$m\angle 1 + m\angle 2 = 90°$

Supplementary angles:
$m\angle 3 + m\angle 4 = 180°$

EXAMPLE 1   If an angle measures 28°, find its complement.

Solution: Two angles are complementary if the sum of their measures is 90°. The complement of a 28° angle is an angle whose measure is $90° - 28° = 62°$. To check, notice that $28° + 62° = 90°$.

Plane is an undefined term that we will describe. A plane can be thought of as a flat surface with infinite length and width, but no thickness. A plane is two dimensional. The arrows in the following diagram indicate that a plane extends indefinitely and has no boundaries.

Figures that lie on a plane are called **plane figures.** (See the description of common plane figures in Appendix B.) Lines that lie in the same plane are called **coplanar.**

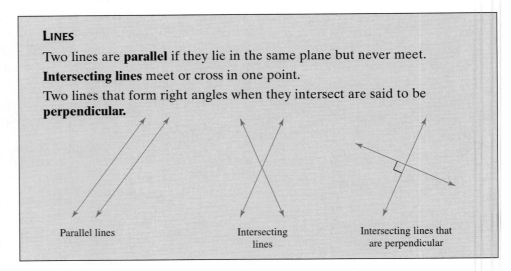

LINES

Two lines are **parallel** if they lie in the same plane but never meet.

Intersecting lines meet or cross in one point.

Two lines that form right angles when they intersect are said to be **perpendicular.**

Parallel lines

Intersecting
lines

Intersecting lines that
are perpendicular

Two intersecting lines form **vertical angles.** Angles 1 and 3 are vertical angles. Also angles 2 and 4 are vertical angles. It can be shown that **vertical angles have equal measures.**

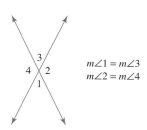

$$m\angle 1 = m\angle 3$$
$$m\angle 2 = m\angle 4$$

Adjacent angles have the same vertex and share a side. Angles 1 and 2 are adjacent angles. Other pairs of adjacent angles are angles 2 and 3, angles 3 and 4, and angles 4 and 1.

A **transversal** is a line that intersects two or more lines in the same plane. Line l is a transversal that intersects lines m and n. The eight angles formed are numbered and certain pairs of these angles are given special names.

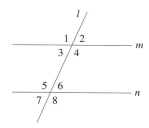

Corresponding angles: $\angle 1$ and $\angle 5$, $\angle 3$ and $\angle 7$, $\angle 2$ and $\angle 6$, and $\angle 4$ and $\angle 8$.

Exterior angles: $\angle 1$, $\angle 2$, $\angle 7$, and $\angle 8$.

Interior angles: $\angle 3$, $\angle 4$, $\angle 5$, and $\angle 6$.

Alternate interior angles: $\angle 3$ and $\angle 6$, $\angle 4$ and $\angle 5$.

These angles and parallel lines are related in the following manner.

PARALLEL LINES CUT BY A TRANSVERSAL

1. If two parallel lines are cut by a transversal, then
   a. corresponding angles are equal and
   b. alternate interior angles are equal.

2. If corresponding angles formed by two lines and a transversal are equal, then the lines are parallel.

3. If alternate interior angles formed by two lines and a transversal are equal, then the lines are parallel.

EXAMPLE 2   Given that lines m and n are parallel and that the measure of angle 1 is 100°, find the measures of angles 2, 3, and 4.

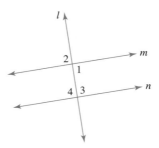

Solution: $m\angle 2 = 100°$,   since angles 1 and 2 are vertical angles.

$m\angle 4 = 100°$,   since angles 1 and 4 are alternate interior angles.

$m\angle 3 = 180° - 100° = 80°$, since angles 4 and 3 are supplementary angles.

A **polygon** is the union of three or more coplanar line segments that intersect each other only at each end point, with each end point shared by exactly two segments.

A **triangle** is a polygon with three sides. The sum of the measures of the three angles of a triangle is 180°. In the following figure, $m\angle 1 + m\angle 2 + m\angle 3 = 180°$.

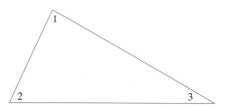

EXAMPLE 3   Find the measure of the third angle of the triangle shown.

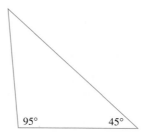

Solution: The sum of the measures of the angles of a triangle is 180°. Since one angle measures 45° and the other angle measures 95°, the third angle measures 180° − 45° − 95° = 40°.

Two triangles are **congruent** if they have the same size and the same shape. In congruent triangles, the measures of corresponding angles are equal and the lengths of corresponding sides are equal. The following triangles are congruent.

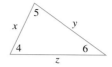 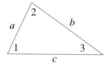

Corresponding angles are equal: $m\angle 1 = m\angle 4, m\angle 2 = m\angle 5$, and $m\angle 3 = m\angle 6$. Also, lengths of corresponding sides are equal: $a = x, b = y$, and $c = z$.

Any one of the following may be used to determine whether two triangles are congruent.

CONGRUENT TRIANGLES

1. If the measures of two angles of a triangle equal the measures of two angles of another triangle and the lengths of the sides between each pair of angles are equal, the triangles are congruent.

$$m\angle 1 = m\angle 3$$
$$m\angle 2 = m\angle 4$$
and
$$a = x$$

2. If the lengths of the three sides of a triangle equal the lengths of corresponding sides of another triangle, the triangles are congruent.

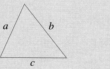

$$a = x$$
$$b = y$$
and
$$c = z$$

3. If the lengths of two sides of a triangle equal the lengths of corresponding sides of another triangle, and the measures of the angles between each pair of sides are equal, the triangles are congruent.

$$a = x$$
$$b = y$$
and
$$m\angle 1 = m\angle 2$$

Two triangles are **similar** if they have the same shape. In similar triangles, the measures of corresponding angles are equal and corresponding sides are in proportion. The following triangles are similar. (All similar triangles drawn in this appendix will be oriented the same.)

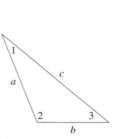
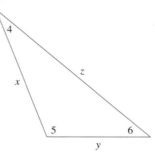

Corresponding angles are equal: $m\angle 1 = m\angle 4$, $m\angle 2 = m\angle 5$, and $m\angle 3 = m\angle 6$.

Also, corresponding sides are proportional: $\dfrac{a}{x} = \dfrac{b}{y} = \dfrac{c}{z}$.

Any one of the following may be used to determine whether two triangles are similar.

SIMILAR TRIANGLES

1. If the measures of two angles of a triangle equal the measures of two angles of another triangle, the triangles are similar.

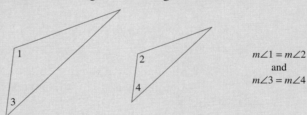

$m\angle 1 = m\angle 2$
and
$m\angle 3 = m\angle 4$

2. If three sides of one triangle are proportional to three sides of another triangle, the triangles are similar.

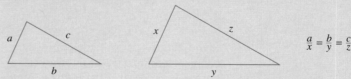

$\dfrac{a}{x} = \dfrac{b}{y} = \dfrac{c}{z}$

3. If two sides of a triangle are proportional to two sides of another triangle and the measures of the included angles are equal, the triangles are similar.

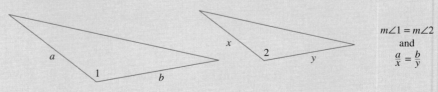

$m\angle 1 = m\angle 2$
and
$\dfrac{a}{x} = \dfrac{b}{y}$

EXAMPLE 4   Given that the following triangles are similar, find the missing length x.

Solution: Since the triangles are similar, corresponding sides are in proportion. Thus, $\dfrac{2}{3} = \dfrac{10}{x}$. To solve this equation for x, we multiply both sides by the LCD $3x$.

$$3x\left(\frac{2}{3}\right) = 3x\left(\frac{10}{x}\right)$$
$$2x = 30$$
$$x = 15$$

The missing length is 15 units.

A **right triangle** contains a right angle. The side opposite the right angle is called the **hypotenuse,** and the other two sides are called the **legs.** The **Pythagorean theorem** gives a formula that relates the lengths of the three sides of a right triangle.

THE PYTHAGOREAN THEOREM

If a and b are the lengths of the legs of a right triangle, and c is the length of the hypotenuse, then $a^2 + b^2 = c^2$.

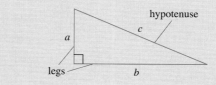

EXAMPLE 5   Find the length of the hypotenuse of a right triangle whose legs have lengths of 3 centimeters and 4 centimeters.

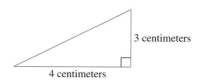

Solution: Because we have a right triangle, we use the Pythagorean theorem. The legs are 3 centimeters and 4 centimeters, so let $a = 3$ and $b = 4$ in the formula.

$$a^2 + b^2 = c^2$$
$$3^2 + 4^2 = c^2$$
$$9 + 16 = c^2$$
$$25 = c^2$$

Since c represents a length, we assume that c is positive. Thus, if c^2 is 25, c must be 5. The hypotenuse has a length of 5 centimeters.

APPENDIX A EXERCISE SET

Find the complement of each angle. See Example 1.

1. $19°$ **2.** $65°$ **3.** $70.8°$

4. $45\frac{2}{3}°$ **5.** $11\frac{1}{4}°$ **6.** $19.6°$

Find the supplement of each angle.

7. $150°$ **8.** $90°$ **9.** $30.2°$

10. $81.9°$ **11.** $79\frac{1}{2}°$ **12.** $165\frac{8}{9}°$

13. If lines m and n are parallel, find the measures of angles 1 through 7. See Example 2.

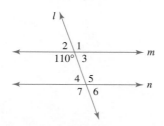

14. If lines m and n are parallel, find the measures of angles 1 through 5. See Example 2.

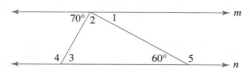

In each of the following, the measures of two angles of a triangle are given. Find the measure of the third angle. See Example 3.

15. $11°, 79°$ **16.** $8°, 102°$ **17.** $25°, 65°$

18. $44°, 19°$ **19.** $30°, 60°$ **20.** $67°, 23°$

In each of the following, the measure of one angle of a right triangle is given. Find the measures of the other two angles.

21. $45°$ **22.** $60°$ **23.** $17°$

24. $30°$ **25.** $39\frac{3}{4}°$ **26.** $72.6°$

Given that each of the following pairs of triangles is similar, find the missing lengths. See Example 4.

27.

28.

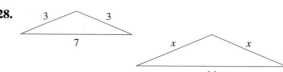

29.

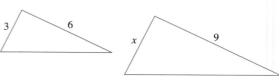

30.

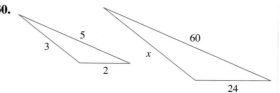

Use the Pythagorean Theorem to find the missing lengths in the right triangles. See Example 5.

31.

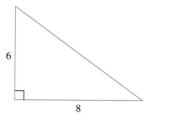

32.

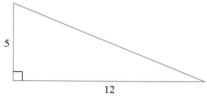

33.

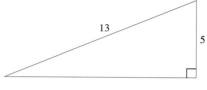

34.

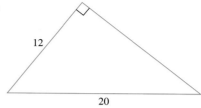

B | REVIEW OF GEOMETRIC FIGURES

PLANE FIGURES HAVE LENGTH AND WIDTH BUT NO THICKNESS OR DEPTH.		
NAME	**DESCRIPTION**	**FIGURE**
POLYGON	Union of three or more coplanar line segments that intersect with each other only at each end point, with each end point shared by two segments.	
TRIANGLE	Polygon with three sides (sum of measures of three angles is 180°).	
SCALENE TRIANGLE	Triangle with no sides of equal length.	
ISOSCELES TRIANGLE	Triangle with two sides of equal length.	
EQUILATERAL TRIANGLE	Triangle with all sides of equal length.	
RIGHT TRIANGLE	Triangle that contains a right angle.	leg, hypotenuse, leg
QUADRILATERAL	Polygon with four sides (sum of measures of four angles is 360°).	

PLANE FIGURES HAVE LENGTH AND WIDTH BUT NO THICKNESS OR DEPTH.		
NAME	**DESCRIPTION**	**FIGURE**
TRAPEZOID	Quadrilateral with exactly one pair of opposite sides parallel.	
ISOSCELES TRAPEZOID	Trapezoid with legs of equal length.	
PARALLELOGRAM	Quadrilateral with both pair of opposite sides parallel and equal in length.	
RHOMBUS	Parallelogram with all sides of equal length.	
RECTANGLE	Parallelogram with four right angles.	
SQUARE	Rectangle with all sides of equal length.	
CIRCLE	All points in a plane the same distance from a fixed point called the **center.**	

	SOLID FIGURES HAVE LENGTH, WIDTH, AND HEIGHT OR DEPTH.	
NAME	**DESCRIPTION**	**FIGURE**
RECTANGULAR SOLID	A solid with six sides, all of which are rectangles.	
CUBE	A rectangular solid whose six sides are squares.	
SPHERE	All points the same distance from a fixed point, called the **center.**	
RIGHT CIRCULAR CYLINDER	A cylinder consisting of two circular bases that are perpendicular to its altitude.	
RIGHT CIRCULAR CONE	A cone with a circular base that is perpendicular to its altitude.	

USING A GRAPHING UTILITY

This appendix is intended to give extra guidance, through examples and practice problems, on using technology as a mathematical problem-solving tool. The organization of material parallels the presentation of chapters in the text. You can use this appendix most effectively if you read it with your graphing utility at hand, trying examples and practice problems as you go.

REAL NUMBERS AND ALGEBRAIC EXPRESSIONS

Chapter 1 reviews operations on real numbers, the order of operations, and evaluating algebraic expressions for given numbers. It can be helpful to use a graphing utility to check any of these calculations or to assist when computing with "messy" numbers. A graphing utility follows the standard order of operations as well as implied mutiplication. For instance, when we see $5x$ or $2(6)$, we know that multiplication is implied even without the symbols $\cdot$, $\times$, or $*$. Likewise, a graphing utility will recognize that multiplication is implied when we enter $5x$ and will give a result of 12 when we enter $2(6)$. Let's practice entering and evaluating several expressions using a graphing utility.

EXAMPLE 1   Evaluate each expression.

a. $-6.3 + (-4.5) + 2.13 - (-12.6)$ **b.** $\dfrac{1}{8} + \dfrac{3}{5}$ **c.** $\sqrt[4]{\dfrac{16}{81}}$

d. $\dfrac{3x^2 - 2y}{y^3 + 5}$ for $x = -2$ and $y = 3$

Solution: **a.** If your graphing utility has a separate negative key, you will need to enter the negative sign of a number differently than the operation of subtraction. Notice that the subtraction and negative symbols look different on the screen display.

$$-6.3 + (-4.5) + 2.13 - (-12.6)$$

Negative   Negative   Negative   Subtraction

```
-6.3+(-4.5)+2.13
-(-12.6)
              3.93
```

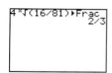

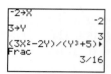

b. When entering fractions into a graphing utility, we use the division key, so $\dfrac{1}{8}$ is entered as $1 \div 8$ and appears as 1/8. Therefore, we can enter the expression as 1/8 + 3/5. Do we need to use parentheses to group the fractions? According to the order of operations, division is performed first, followed by addition. Because the graphing utility uses the standard order of operations, this expression will be evaluated correctly by the graphing utility whether or not parentheses are used, as demonstrated on the screen to the left. Remember that some graphing utilities have a fraction feature that allows answers to be displayed in fractional form.

c. Take care to enter this expression so that the entire fraction is included under the root symbol. On many graphing utilities, this often means enclosing the fraction in parentheses, as shown to the left.

d. This expression can be evaluated with a graphing utility in several ways. One way is to first store values of the variables x and y using the store feature (on some graphing utilities, the $\rightarrow$ symbol on the screen represents the store feature). When the algebraic expression is entered in terms of x and y, the graphing utility evaluates the expression for the given values of the variables, as shown at the left. Notice that the numerator and denominator of the expression are grouped using parentheses, since each contains more than one term. Explore on your graphing utility how exponents are entered. Does it have special square and cube keys/commands, or must the general exponent symbol $\wedge$ be used?

INTRODUCTION TO TABLES

In Chapter 2, we learn to model with tables. Most graphing utilities are capable of making tables based on an expression. To create a table with a graphing utility, we define the expression, the minimum x-value at which the table starts, and the increment in the x-values.

EXAMPLE 2 Use the table feature of a graphing utility to complete the following table giving the weekly allowance for a child aged x years in a certain family.

Age	x	5	8	11	14	17
Weekly allowance	$0.6x - 1.5$					

Solution: Begin by entering the expression $0.6x - 1.5$ into your graphing utility's Y= editor (or its equivalent). Next format the table using a table setup menu. From the table above, we set the table to start at the x-value 5 and set the increment in x-values at 3. Then display the table.

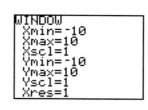
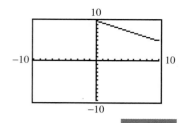

Do you think the window we chose gives a good view of the graph of the equation? We usually try to use a window that shows all distinguishing features of the graph of the equation. A better window might be one that gives a better view of the y-intercept and shows the x-intercept, as shown below:

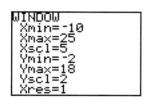
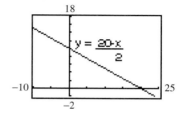

EXAMPLE 6   The equation that models the cost of manufacturing stereo systems for a certain company is $y = 200x + 5000$, where y gives the cost, in dollars, to manufacture a total of x stereos. Use a graphing utility to graph the equation.

Solution: Graph $y_1 = 200x + 5000$. To do so, enter the equation in the Y= editor and choose an appropriate window. The standard window, for example, is not an appropriate window for this equation because this portion of the rectangular coordinate system contains no part of the graph of y.

To view a part of the graph of the equation that gives meaning to the values of x and y, recall that x is the number of stereos and y is manufacturing cost. Since we cannot manufacture a negative number of stereos, let's view the x-axis from 0 to 25. When $x = 0$, notice that $y = 200(0) + 5000$, or $y = 5000$. When $x = 25$, $y = 200(25) + 5000$, or $y = 10,000$. Although other windows are appropriate, we show the graph of y_1 in a [0, 25, 1] by [5,000, 10,000, 1000] window. From the graph or a related table, we can see the costs, y, for manufacturing x stereos. For example, to manufacture 5 stereos, the cost is $6000.

PRACTICE PROBLEM 6
The cost equation for manufacturing bicycles is $y = 30x + 600$, where y gives the cost, in dollars, to manufacture a total of x bicycles. Use a graphing utility to graph the equation.

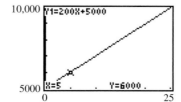
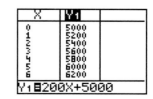

A graphing utility can also be used to solve equations in one variable graphically.

EXAMPLE 7   Solve the equation $6x + 13 = -4(x + 1) - 17$ graphically.

Solution: There are many ways to solve this equation with a graphing utility.

PRACTICE PROBLEM 7
Solve each equation
graphically.

a. $5x - 3(2x - 1)$
$= 3(x - 5)$

b. $2(x + 4)$
$= 8(3 - x) - 30$

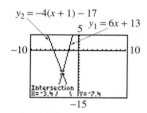

$y_2 = -4(x + 1) - 17$
$y_1 = 6x + 13$

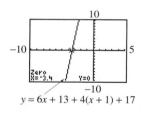

$y = 6x + 13 + 4(x + 1) + 17$

INTERSECTION-OF-GRAPHS METHOD

For the intersection-of-graphs method, graph

$$y_1 = \text{left side of equation and}$$
$$y_2 = \text{right side of equation.}$$

The x-coordinate of a point of intersection of the two graphs is a solution of the equation. The graphing utility's trace feature can be used to approximate a point of intersection (zooming can help refine the approximation) or an intersect feature (often in the Calc, Calculate, or G-Solve menu) can be used to find the coordinates of a point of intersection, to the tolerance level of your graphing utility. The graph at the left shows the solution using this method as $x = -3.4$.

X-INTERCEPT METHOD

Another way to solve this equation is the x-intercept method. First, write the equation so that one side is 0. For this example, we show the equivalent equation in an appropriate window. The x-intercept of the graph is the solution of the equation. The graphing utility's trace feature can be used to approximate the x-intercept, or a zero or root finder feature can be used to find the value of the x-intercept, accurate to the tolerance level of your graphing utility. The graph at the left shows the solution using this method as being $x = -3.4$. Notice that the same solution is obtained with either method.

SYSTEMS OF EQUATIONS

In Chapter 5, we learn to solve systems of linear equations. With a graphing utility, we can solve systems of linear equations, as well as systems of nonlinear equations graphically. The concept is similar to the intersection-of-graphs method for solving an equation as shown in Example 7. With a system of equations, we graph each equation in the system on the same set of axes and find any points of intersection.

EXAMPLE 8   Solve the system of equations graphically.

$$\begin{cases} 3x + y = -5 \\ 2x - 3y = 10 \end{cases}$$

Solution: The first step is to solve each equation for y. Then enter the equations into the Y= editor and graph the equations. The intersect feature or a combination of tracing and zooming can be used to find the point of intersection. Note from the

PRACTICE PROBLEM 8
Solve each system of equations graphically. If necessary, round to two decimal places.

$$\begin{cases} 2x - 8y = 9 \\ 3x + 5y = -5 \end{cases}$$

screen below that the solution to this system is approximately $(-0.45, -3.64)$. Solving this system algebraically gives the exact solution $\left(-\dfrac{5}{11}, -\dfrac{40}{11}\right)$.

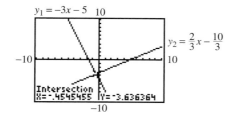

INTRODUCTION TO MATRICES

In Chapter 5, we also learn how to use a matrix as a problem-solving tool when solving a system of linear equations. Many graphing utilities have matrix operation capabilities. For instructions on entering a matrix into a graphing utility, see the next example.

EXAMPLE 9   Enter the augmented matrix of the system into a graphing utility.

$$2x + 3y = 4z + 4$$
$$x + 5 = 2y$$
$$5x + 3z = 1$$

Solution: Begin by rewriting the system so that each equation is in standard form and the variables are aligned in columns. Then form the augmented matrix based on this written system.

PRACTICE PROBLEM 9
Enter the augmented matrix of the system into a graphing utility.

$$\begin{cases} 5x - 8y = 40 \\ 4y = -7 - 3x \end{cases}$$

$$\begin{cases} 2x + 3y - 4z = 4 \\ x - 2y = -5 \\ 5x + 3z = 1 \end{cases} \rightarrow \begin{bmatrix} 2 & 3 & -4 & | & 4 \\ 1 & -2 & 0 & | & -5 \\ 5 & 0 & 3 & | & 1 \end{bmatrix}$$

Note that the coefficients of the variables in each equation correspond to the entries (also called elements) in the first three columns of the matrix. Any variable missing from the equation corresponds to an entry of 0 in the matrix. The constants on the right side of each equation make up the fourth column of the matrix.

To enter a matrix into a graphing utility, select a matrix name to be edited and define the dimensions of the matrix in the form n rows $\times$ m columns. For this example, we have a 3×4 matrix. Now, begin entering elements of the matrix. Some graphing utilities use double subscripts to identify positions of elements within the matrix. For instance, i, j denotes the row number i and the column number j so that the element's position is the jth column of the ith row.

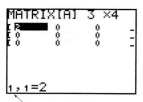 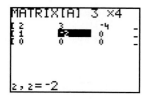 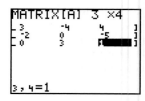

Entry for row 1 column 1	Entries in the matrix continue	The last entry of the matrix
The matrix name A was chosen. A is defined as a 3×4 matrix (3 rows by 4 columns).	across the first row. The 2, 2 entry is entered as -2.	3, 4 is entered as 1.

RECOGNIZING LINEAR AND QUADRATIC MODELS

In Section 9.7, we learn to recognize linear and quadratic models. One feature of a graphing utility that helps us to recognize these models is the capability of plotting data points. Lists, as introduced earlier in this appendix, can be used to plot points.

EXAMPLE 10   The following table lists the amount of United States import trade from Mexico in millions of dollars for the given years.

YEAR	1985	1990	1993	1994
Imports (millions of dollars)	19,132	30,172	39,930	49,493

Plot the points representing these data on a rectangular coordinate system.

Solution: There are four basic steps to plotting points, as shown below:

PRACTICE PROBLEM 10
Use a graphing utility to plot the data below from Christy Mathewson's major league career with the New York Giants.

YEAR	WALKS PER 9 INNINGS
1900	4.2
1901	2.6
1902	2.4
1903	2.5
1904	1.9
1905	1.7
1906	2.6
1907	1.5
1908	1.0

Enter the data. Enter x-coordinates into L_1 and corresponding y-coordinates into L_2.

Access the statistical features of your graphing utility to indicate the type of graph. This procedure can vary a lot depending on the graphing utility.

Set a window for the data so that the Xmin and Xmax include all values in the first list, L_1, and the Ymin and Ymax include all values in the second list, L_2.

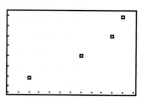

Plot the data.

CONIC SECTIONS

In Chapter 10, we learn about the parabola, circle, ellipse, and hyperbola. These conic sections can be graphed using a graphing utility. Because in most cases the equations of these conic sections are not functions of x, you may need to graph two equations to obtain the graph of a conic section. Let's practice graphing conic sections with a graphing utility.

EXAMPLE 11   Use a graphing utility to graph the conic section.

$$\frac{x^2}{16} - \frac{y^2}{4} = 1$$

Solution: Solve the equation for y.

PRACTICE PROBLEM 11
Use a graphing utility to graph each conic section.

a. $5x^2 + y^2 = 36$

b. $\dfrac{y^2}{9} - \dfrac{x^2}{25} = 1$

$$\frac{x^2}{16} - \frac{y^2}{4} = 1$$

$$x^2 - 4y^2 = 16 \qquad \text{Multiply both sides by 16.}$$

$$x^2 - 16 = 4y^2$$

$$\pm\sqrt{x^2 - 16} = 2y \qquad \text{Take the square root of each side.}$$

$$\pm 0.5\sqrt{x^2 - 16} = y \qquad \text{Multiply both sides by } \tfrac{1}{2}, \text{ or } 0.5.$$

To save time entering these equations, we can enter them as $y_1 = 0.5\sqrt{x^2 - 16}$ and $y_2 = -y_1$. The graph is shown below in a $[-9, 9, 1]$ by $[-6, 6, 1]$ window.

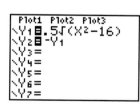
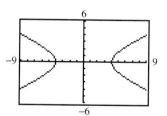

TECHNOLOGY NOTE
Some graphing utilities have a variables menu that allows you to use the names of functions, such as y_1 or y_2, in other expressions. Consult your owner's manual for specific instructions on how to use this feature.

SEQUENCES

In Chapter 12, we learn about different types and uses of sequences and series. The use of sequences and series capabilities differs from graphing utility to graphing utility. Some graphing utilities have a sequence mode that gives special treatment of sequences in graphs, tables, and expressions on the home screen. Different graphing utilities have different ways of designating the variable used in the general term of a sequence. For instance, some graphing utilities allow the use of the special variable n in the general term of a sequence. However, the graphing utility may only allow the use of this special variable in certain functions, for instance, in the Y= editor or table editor, but not in functions on the home screen.

D ANSWERS TO SELECTED EXERCISES

CHAPTER 1

Review of Real Numbers and Algebraic Expressions

Exercise Set 1.1

1. Repeating   **3.** Terminating   **5.** Rational   **7.** Irrational   **9.** Rational   **11.** $\{3, 0, \sqrt{36}\}$ **13.** $\{3, \sqrt{36}\}$ **15.** $\{\sqrt{7}\}$ **17.** True
19. False   **21.** True   **23.** $\sqrt{25}$ is exactly 5, but $\sqrt{10}$ produces an infinite, nonrepeating decimal (of which the screen only shows part).
25. $\{1, 2, 3, 4, 5\}$ **27.** $\{11, 12, 13, 14, 15, 16\}$ **29.** $\{0\}$ **31.** $\{0, 2, 4, 6, 8\}$ **33.**

$$-1 \quad 0 \quad 2 \quad 4 \quad 6$$

35.

$$-1 \quad -\tfrac{1}{2} \quad 0 \quad \tfrac{1}{2}\,\tfrac{2}{3} \quad 1$$

37.

$$-10 \quad -8 \quad -6 \quad -4 \quad -2$$

39. For any set A, there is no element that is

in $\{\ \}$ but not in A.   **41.** $\in$ **43.** $\notin$ **45.** $\notin$ **47.** $\notin$ **49.** True   **51.** True   **53.** False   **55.** False   **57.** True   **59.** False

61. Every natural number n can be written as $\frac{n}{1}$. But $-3 = \frac{-3}{1}$ is not a natural number, for example.   **63.** $2x$ **65.** $2x + 5$ **67.** $x - 10$

69. $x + 2$ **71.** $\frac{x}{11}$ **73.** $3x + 12$ **75.** $x - 17$ **77.** $2(x + 3)$ **79.** $\dfrac{5}{4 - x}$

81. \$355,000,000; \$1,392,000,000; \$1,025,000,000; \$1,526,000,000

Exercise Set 1.2

1. $4c = 7$ **3.** $3(x + 1) = 7$ **5.** $\frac{n}{5} = 4n$ **7.** $z - 2 = 2z$ **9.** $>$ **11.** $=$ **13.** $<$ **15.** $7x \le -21$ **17.** $-2 + x \ne 10$ **19.** $2(x - 6) > \frac{1}{11}$

21.

$$-4\ -3\ -2\ -1\ \ 0\ \ 1\ \ 2\ \ 3\ \ 4$$
$(-\infty, -2)$

23.

$$-4\ -3\ -2\ -1\ \ 0\ \ 1\ \ 2\ \ 3\ \ 4$$
$[1, \infty)$

25.

$$-8\ -6\ -4\ -2\ \ 0\ \ 2\ \ 4\ \ 6\ \ 8$$
$(4.7, \infty)$

27.

$$-4\ -3\ -2\ -1\ \ 0\ \ 1\ \ 2\ \ 3\ \ 4$$
$\left(-\infty, 3\tfrac{1}{2}\right]$

29.

$$-8\ -6\ -4\ -2\ \ 0\ \ 2\ \ 4\ \ 6\ \ 8$$
$(-2, 5]$

31.

$$-4\ -3\ -2\ -1\ \ 0\ \ 1\ \ 2\ \ 3\ \ 4$$
$[1.3, 2.3)$

33.

$$-8\ -6\ -4\ -2\ \ 0\ \ 2\ \ 4\ \ 6\ \ 8$$
$\left[-5\tfrac{1}{4}, 0\right]$ **35.** 6.2   **37.** $-\frac{4}{7}$ **39.** $\frac{2}{3}$ **41.** 0   **43.** $\frac{1}{5}$ **45.** $-\frac{1}{8}$ **47.** -4

49. Undefined   **51.** $\frac{8}{7}$ **53.** Zero. For every real number x, $0 \cdot x \ne 1$, so 0 has no reciprocal. It is the only real number that has no
reciprocal because if $x \ne 0$, then $x \cdot \dfrac{1}{x} = 1$ by definition.   **55.** $y + 7x$ **57.** $w \cdot z$ **59.** $\dfrac{x}{5} \cdot \dfrac{1}{3}$ **61.** No; $2 - 1 \ne 1 - 2$, for example.

63. Associative Property of Addition   **65.** Commutative Property of Multiplication   **67.** $(7 \cdot 5) \cdot x$ **69.** $x + (1.2 + y)$

71. $14(z \cdot y)$ **73.** $12 - (5 - 3) = 10; (12 - 5) - 3 = 4$; Subtraction is not associative.   **75.** $3x + 15$ **77.** $16a + 8b$
79. $12x + 10y + 4z$ **81.** $6 + 3x$ **83.** 0 **85.** 7 **87.** $(10 \cdot 2)y$ **89.** $a(b + c) = ab + ac$
91. 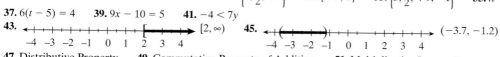   **93.**   **95.**
97. $[5, \infty)$ **99.** $(-\infty, 10)$ **101.** $[2, 8]$

Exercise Set 1.3

1. -2 **3.** 4 **5.** 0 **7.** -3 **9.** $-(-2) = 2; -|-2| = -2$ **11.** 5 **13.** -24 **15.** -11 **17.** -4 **19.** $\frac{4}{3}$ **21.** -2 **23.** -60
25. 80 **27.** 3 **29.** 0 **31.** -8 **33.** $-\frac{3}{7}$ **35.** $\frac{1}{21}$ **37.** -49 **39.** 36 **41.** -8
43. $-3^2 = -(3 \cdot 3) = -9; (-3)^2 = (-3)(-3) = 9$ **45.** 7 **47.** $\frac{1}{3}$ **49.** 4 **51.** 3 **53.** 3 **55.** -12 **57.** 15 **59.** -12.2 **61.** 0
63. -1 **65.** -21 **67.** $-\frac{1}{2}$ **69.** -6 **71.** -17 **73.** -1 **75.** 2 **77.** 10 **79.** $-\frac{5}{27}$ **81.** $\frac{13}{35}$ **83.** 4205 meters   **85.** 3.1623
87. 2.8107 **89.** 15.6% **91.** 17.7% **93.** long term; short term is very volatile.   **95.** b   **97.** d   **99.** Yes. Two players have 6 points each, (third player has 0 points), or two players have 5 points each, (third has 2 points).

Exercise Set 1.4

1. 48 **3.** -1 **5.** -3 **7.** 14.4 **9.** -135 **11.** -2.1 **13.** $-\frac{1}{3}$ **15.** 17 **17.** $-\frac{79}{15}$ **19.** $-\frac{5}{14}$ **21.** -0.588 **23.** 13
25. 65 **27.** -2 **29.** 16.6 **31.** -0.61 **33.** -121.2 **35.** $x = -2.7; y = 0.9; x^2 + 2y = 9.09$ **37.** $a = 4; b = -2.5; a^2 + b^2 = 22.25$
39. a. $(1/2)x = 1; 1/2x = 1$ **b.** The same; same order of operations   **41. a.** $18, 22, 28, 208$ **b.** Increase   **43. a.** $600, 150, 105$
b. Decrease   **45.** $12, 24, 48, 96$ **47.** $\$36,909.60$ **49.** $0.05n$ **51.** $0.05n + 0.1d$ **53.** $6.49x$ **55.** $\$0.0825p$ **57.** $25 - x$
59. $180 - x$ **61.** $x + 2$ **63.** $\$31 + 0.29x$ **65.** $-15x + 18$ **67.** $2k + 10$ **69.** $-3x + 5$ **71.** $4x + 9$ **73.** $4n - 8$ **75.** -24
77. $2x + 10$ **79.** $1.91x + 4.32$ **81.** $15.4z + 31.11$ **83.** 10 million   **85.** 42 million   **87.** Increasing

Chapter 1 Review

1. $3x$ **3.** $2x - 1$ **5.** $\{-2, 0, 2, 4, 6\}$ **7.** $\{\}$ **9.** $\{..., -1, 0, 1, 2\}$ **11.** False   **13.** True   **15.** True   **17.** True   **19.** True
21. True   **23.** True   **25.** True   **27.** True   **29.** $\left\{5, \frac{8}{2}, \sqrt{9}\right\}$ **31.** $\{\sqrt{7}, \pi\}$ **33.** $\left\{5, \frac{8}{2}, \sqrt{9}, -1\right\}$ **35.** $n + 2n = -15$
37. $6(t - 5) = 4$ **39.** $9x - 10 = 5$ **41.** $-4 < 7y$
43. $[2, \infty)$ **45.** $(-3.7, -1.2)$
47. Distributive Property   **49.** Commutative Property of Addition   **51.** Multiplicative Inverse Property
53. Associative Property of Multiplication   **55.** -0.6 **57.** -1 **59.** $\frac{1}{0.6} = 1.\overline{6}$ **61.** 1 **63.** $(3 + x) + (7 + y)$
65. $2 \cdot \frac{1}{2}$, for example   **67.** $7 + 0$ **69.** Distributive Property   **71.** $=$ **73.** $>$ **75.** -35 **77.** 0.31 **79.** 13.3 **81.** 0 **83.** 0
85. -5 **87.** 4 **89.** 9 **91.** 3 **93.** $-\frac{32}{135}$ **95.** $-\frac{5}{4}$ **97.** $\frac{5}{8}$ **99.** -1 **101.** -57 **103.** -4 **105.** $\frac{5}{7}$ **107.** $\frac{1}{5}$ **109.** -5
111. 5 **113.** $-3x + 13$ **115.** $-3x + 44$ **117.** $324,000$ beats

Chapter 1 Test

1. False   **2.** False   **3.** False   **4.** False   **5.** True   **6.** False   **7.** -3 **8.** 43 **9.** -225 **10.** -2 **11.** 1 **12.** 12 **13.** 1
14. a. $17.25, 40.25, 69.00, 115.00$ **b.** Increase   **15.** $2|x + 5| = 30$ **16.** $\frac{(6 - y)^2}{7} < -2$ **17.** $\frac{9z}{|-12|} \neq 10$ **18.** $3\left(\frac{n}{5}\right) = -n$
19. $20 = 2x - 6$ **20.** $-2 = \frac{x}{x + 5}$ **21.** Distributive Property   **22.** Associative Property of Addition   **23.** Additive Inverse Property
24. Multiplication Property of Zero   **25.** $0.05n + 0.1d$ **26.** $-14x + 24$ **27.** $14y$ **28.** 64.3118

29.

Diameter	d	2	3.8	10	14.9
Radius	r	1	1.9	5	7.45
Circumference	πd	6.28	11.94	31.42	46.81
Area	πr^2	3.14	11.34	78.54	174.37

30.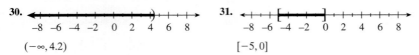

$(-\infty, 4.2)$

31.

$[-5, 0]$

CHAPTER 2
Basic Concepts of Algebra

Section 2.1 Mental Math

1. $8x + 21$ **3.** $6n - 7$ **5.** $-4x - 1$

Exercise Set 2.1

1. $\{-0.9\}$ **3.** $\{6\}$ **5.** $\{-1.1\}$ **7.** $\{-5\}$ **9.** $\{0\}$ **11.** $\{2\}$ **13.** $\{-9\}$ **15.** $\left\{-\dfrac{10}{7}\right\}$ **17. a.** $4x + 5$ **b.** $\{-3\}$
c. Answers may vary.   **19.** $\left\{\dfrac{1}{6}\right\}$ **21.** $\{4\}$ **23.** $\{1\}$ **25.** $\left\{\dfrac{40}{3}\right\}$ **27.** $\{n \mid n$ is a real number$\}$ **29.** $\{x \mid x$ is a real number$\}$
31. Answers may vary.   **33.** $6x - 5 = 4x - 21; \{-8\}$ **35.** $3(n - 3) + 2 = -1 + n; \{3\}$ **37.** $\{4.2\}$ **39.** $\{2\}$ **41.** $\{-2\}$ **43.** $\{0\}$
45. $\{5\}$ **47.** $\{\ \}$ **49.** $\left\{\dfrac{1}{8}\right\}$ **51.** $\{0\}$ **53.** $\{29\}$ **55.** $\{-8\}$ **57.** $\left\{\dfrac{20}{11}\right\}$ **59.** $\{\ \}$ **61.** $\{4\}$ **63.** $\{17\}$ **65.** $\left\{\dfrac{4}{5}\right\}$ **67.** $4y$ **69.** $3z + 3$
71. $0.15x + 0.30$ **73.** $10x + 3$ **75.** $K = -11$ **77.** $K = 24$ **79.** $\{-4.86\}$ **81.** $\{1.53\}$ **83.** 13   **85.** Not a fair game
87. 7   **89.** 0   **91.** -7

Exercise Set 2.2

1. -5 **3.** 225 and 45   **5.** 78   **7.** 1.92   **9.** Approximately 408 million acres   **11.** 51,700   **13.** 20%   **15.** 117
17. B737-200 has 112 seats; B767-300 has 224 seats   **19.** \$430   **21.** Square's side is 18 cm; Triangle's side is 24 cm
23. Width: 6 cm; Length: 14 cm   **25.** 5 years   **27.** Width: 8.4 meters; Height: 47 meters   **29.** 64°, 32°, 84°   **31.** 660 million hours
33. Fallon: 89406; Fernley: 89408; Gardnerville: 89410   **35.** Any three consecutive integers   **37.** 38°, 38°, 104°   **39.** 59,974
41. Greater use of calling cards and voice recognition technology, for example.   **43.** 3.6 million   **45.** 25 skateboards   **47.** 800 books
49. It loses money.   **51.** Answers may vary.   **53.** Answers may vary.   **55.** 6   **57.** 208   **59.** -55 **61.** 3195   **63.** 121.8152

Section 2.3 Mental Math

1. f   **3.** b   **5.** d

Exercise Set 2.3

1. 343, 512, 729, 1000, 1331   **a.** 729 cm^3 **b.** 1331 ft^3 **c.** 7 in.   **3.** 298, 301.50, 305, 308.50   **a.** 44 hours   **b.** 35 hours
c. 37 hours   **5. a.** $y_1 = 5.25x$ **b.** $y_2 = 7x$ **c.** Rough draft: 5.25, 6.56, 7.88, 9.19, 10.50, 11.81, 13.13, 14.44, 15.75;
Manuscript: 7.00, 8.75, 10.50, 12.25, 14.00, 15.75, 17.50, 19.25, 21.00   **d.** \$6.56   **e.** \$75.25   **7.** Circumference: 12.57, 31.42, 131.95, 591.88;
Area: 12.57, 78.54, 1385.44, 27,877.36   **a.** 188.50 yd; 2827.43 yd^2 **b.** 493.23 mm; 19,359.28 mm^2 **9. a.** $y_1 = 1100 + 0.0005x$
b. 1150, 1200, 1250, 1300, 1350, 1400, 1450   **c.** \$16,200   **11. a.** $y_1 = 12.96 + 1.10x$ **b.** 12.96, 14.06, 15.16, 16.26, 17.36, 18.46, 19.56
13. $1\dfrac{1}{8}$ cups, $\dfrac{3}{8}$ cup, $\dfrac{3}{4}$ cup, $4\dfrac{1}{2}$ cups, $\dfrac{3}{16}$ tsp.

15. a.

Year	5	10	15	20
First job	$19,500	$24,500	$29,500	$34,500
Second job	$22,875	$25,750	$28,625	$31,500

b. The second job

17. a. $y_1 = 15 + 0.01924x$ **b.** 15, 24.62, 34.24, 43.86, 53.48   **19. a.** $y_1 = 0.65x$ **b.** $19.47, $23.37, $12.51, $25.97, $11.67, $6.47, $16.74

Preferred prices may vary.

21.

Diameter	Area	Circumference	Cost (Area)	Cost (Circumference)
6	28.274	18.850	$1.41	$4.71
12	113.10	37.699	$5.65	$9.42
18	254.47	56.549	$12.72	$14.14
24	452.39	75.398	$22.62	$18.85
30	706.86	94.248	$35.34	$23.56
36	1017.9	113.10	$50.89	$28.27

23. a. 1569, 1606, 1611, 1584, 1525   **b.** 1613 ft   **25. a.** Anne: $49.00; Michelle: $50.00   **b.** Whenever the distance is less than $333\frac{1}{3}$ miles.
27. a. $C = 90 + 75x$ **b.** 240, 390, 540, 690   **29. a.** 3.5 ft   **b.** 5 ft   **c.** 10 ft   **31. a.** 6.73 or 6:44   **b.** 4; July 1   **c.** 7:14   **d.** 6:17
e. No; a model of all sunrise times would be wavy, not a parabola.   **33.** {9.1}   **35.** {x|x is a real number}   **37.** $\left\{\frac{2}{3}\right\}$

Section 2.4 Mental Math

1. $y = -2x + 5$ **3.** $a = 5b + 8$ **5.** $k = h - 5j + 6$

Exercise Set 2.4

1. $\frac{D}{r} = t$ **3.** $\frac{I}{PT} = R$ **5.** $y = \frac{9x - 16}{4}$ **7.** $\frac{P - 2L}{2} = W$ **9.** $\frac{J + 3}{C} = A$ **11.** $\frac{W}{h - 3t^2} = g$ **13.** $\frac{T - 2C}{AC} = B$ **15.** $\frac{C}{2\pi} = r$
17. $\frac{E}{I} - R = r$ **19.** $\frac{2s}{n} - a = L$ **21.** $w = \frac{uv}{v - u}$ **23.** $\frac{3st^4 - N}{5s} = v$ **25.** $r = \frac{S - a}{S}$ **27.** $\frac{S - 2LW}{2L + 2W} = H$
29. $4703.71, $4713.99, $4719.22, $4722.74, $4724.45   **31.** 12 times a year; you earn more interest   **33. a.** $7313.97   **b.** $7321.14
c. $7325.98   **35.** 40°C   **37. a.** 14 ft × 28 ft   **b.** 392 ft²   **39.** 3.6 hours or 3 hours, 36 minutes   **41.** 171 packages   **43.** 0.42 feet
45. 60; 108; 135; 144; 150   **47.** 0.25 sec.   **49.** To receive from and transmit to fixed locations on the earth, for example.
51. a. 5 in. × 11 in. × 2 in.   **b.** 110 in.³   **53.** $1831.96   **55.** 2 gallons   **57. a.** 1174.86 m³   **b.** 310.34 m³   **c.** 1485.20 m³
59. 2.25 hours or 2 hours, 15 minutes   **61.**   **63.** $\frac{1}{8}$ **65.** $\frac{1}{8}$ **67.** $\frac{1}{2}$ **69.** $\frac{3}{8}$ **71.** 0   **73.** 0
75. {-3, -2, -1}   **77.** {-3, -2, -1, 0, 1}

	AU from sun
Mercury	0.388
Venus	0.723
Earth	1.000
Mars	1.523
Jupiter	5.202
Saturn	9.538
Uranus	19.193
Neptune	30.065
Pluto	39.505

Exercise Set 2.5

1. $519.43   **3.** $458.86   **5.** 10.7%   **7.** 14.1%   **9.** 85   **11.** Growth/Income: $5250; Small Company/Aggressive Growth: $3750;
International: $3000; Bonds: $3000   **13.** Answers may vary.   **15.** 1314 ft   **17.** 1131.5 ft   **19.** 6.8   **21.** 6.9   **23.** 85 5
27. 70 and 71   **29.** 9   **31.** 21, 21, 20   **33.** Police officer   **35.** 940   **37.** Answers may vary.   **39.** The De.
43. Democrat: 50%; Republican: 38%; Independent: 10%   **45.** Answers may vary.   **47.** $210.54   **49.** $58.0ε

53. 17.6 **55.**

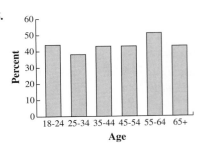

57. Answers may vary. **59.** No
61. The mean is lowered; the median and mode are unchanged
63. 5, 9, 13, 17, 21; 2; 4 **65.** 37, 32, 27, 22, 17; 1; −5

Chapter 2 Review

1. {3} **3.** $\left\{-\dfrac{45}{14}\right\}$ **5.** {0} **7.** {6} **9.** {$h|h$ is a real number} **11.** { } **13.** {−3} **15.** $\left\{\dfrac{96}{5}\right\}$ **17.** {32} **19.** {8} **21.** { } **23.** {2}

25. $x = -7$ **27.** 52 **29.** \$20,000 **31.** No such odd integers exist. **33.** 258 miles **35.** 5 plants, \$200 **37.** 34, 39.36, 49, 55.36

39. Coast charge: \$9.10, \$10.20, \$11.30, \$12.40; Cross Gates charge: \$12.00, \$13.50, \$15.00, \$16.50 **41.** $W = \dfrac{V}{LH}$

43. $y = \dfrac{5x + 12}{4}$ **45.** $m = \dfrac{y - y_1}{x - x_1}$ **47.** $r = \dfrac{E}{I} - R$ **49.** $g = \dfrac{T}{r + vt}$ **51.** $B = \dfrac{2A}{h} - b$ **53.** $b = \dfrac{ca}{a - c}$ **55.** $y = \dfrac{28}{3}x$

57. a. \$3695.27 **b.** \$3700.81 **59. a.** −40°F, 5°F, 50°F, 140°F **b.** 212°F **c.** 32°F **61.** 16 packages **63.** 58 mph
65. 23.56, 46.18, 79.19, 94.25, 128.28 **67.** Mean: 40.6; Median: 40; Mode: None **69.** Mean: 5.97; Median: 6.05; Modes: 4.9 and 6.8
71. Mean: 0.53; Median: 0.6; Modes: 0.6 and 0.8 **73.** Mean: 587.5; Median: 489; Mode: None
75. Federal loans: \$28.67 billion; Institutional: 10.06 billion; Federal grants/work study: \$8.55 billion; State: \$3.02 billion
77. Less than 10; 29% **79.** The larger the company, the more likely employees are to have health benefits.

Chapter 2 Test

1. {10} **2.** {1} **3.** { } **4.** {$n|n$ is a real number} **5.** {12} **6.** $\left\{-\dfrac{80}{29}\right\}$ **7.** $y = \dfrac{3x - 8}{4}$ **8.** $n = \dfrac{9m}{7}$ **9.** $g = \dfrac{S}{t^2 + vt}$

10. $C = \dfrac{5}{9}(F - 32)$ **11.** 9.6 **12.** \$425 million **13.** 8 **14.** 850 sunglasses **15.** \$3542.27

16. a. \$1,900, \$1,950, \$2,000, \$2,050, \$2,100 **b.** \$2,050 **c.** \$14,000 **17.** 800 ft^2 **18. a.** At 2 and 3 seconds **b.** 120 ft
c. At 2.5 seconds **d.** 3.0 seconds **19.** 4.75 **20.** 5 **21.** 5 **22.** 80 lb **23.** 114 lb **24.** 1990
25. Consumption of both fresh fruits and fresh vegetables is increasing.

Chapter 2 Cumulative Review

1. *(Sec. 1.1, Ex. 1)* **a.** {2, 3, 4, 5} **b.** {101, 102, 103, ...}
2. *(Sec. 1.1, Ex. 3)* **a.** True **b.** False **c.** False **d.** False **e.** True
3. *(Sec. 1.2, Ex. 2)* **a.** > **b.** = **c.** < **d.** < **4.** *(Sec. 1.3, Ex. 9)* **a.** 3 **b.** 2 **c.** 2
5. *(Sec. 1.2, Ex. 6)* **a.** −8 **b.** $-\dfrac{1}{5}$ **c.** 9.6 **6.** *(Sec. 1.2, Ex.7)* **a.** $\dfrac{1}{11}$ **b.** $-\dfrac{1}{9}$ **c.** $\dfrac{4}{7}$
7. *(Sec. 1.2, Ex. 10)* $6x + 3y$ **8.** *(Sec. 1.2, Ex. 11)* **a.** Commutative property of addition **b.** Distributive property
c. Additive inverses **d.** Double negative property **e.** Associative property of addition
9. *(Sec. 1.3, Ex. 2)* **a.** −14 **b.** −4 **c.** 5 **d.** −10.2 **e.** $\dfrac{1}{4}$ **f.** $-\dfrac{5}{21}$
10. *(Sec. 1.3, Ex. 5)* **a.** 8 **b.** $-\dfrac{1}{3}$ **c.** −9 **d.** 0 **e.** $-\dfrac{2}{11}$ **f.** 42 **g.** 0
11. *(Sec. 1.4, Ex. 2)* **a.** 2 **b.** 9 **c.** −1 **d.** −12 **12.** *(Sec. 1.4, Ex. 6)* 128, 160, 192
13. *(Sec. 1.4, Ex. 9)* **a.** $-2x + 4$ **b.** $8yz$ **c.** $4z + 6.1$ **14.** *(Sec. 2.1, Ex. 1)* {2}
15. *(Sec. 2.3, Ex. 2)* **a.** $45 + 30x$ **b.** \$75, \$105, \$135, \$165, \$195, \$225
16. *(Sec. 2.3, Ex. 3)* Gift: \$39.95, Tax: \$2.80, Total Cost: \$42.75, Gift: \$45.00, Tax: \$3.15, Total Cost: \$48.15,
Gift: \$62.75, Tax: 4.39, Total Cost: \$67.14

(Sec. 2.2, Ex. 2) 4 **18.** *(Sec. 2.4, Ex. 1)* $h = \dfrac{V}{lw}$ **19.** *(Sec. 2.4, Ex. 4)* −20, −12.22, 0, 21.11, 37.78
(Sec. 2.5, Ex. 2) Median = 83.5, Mode = 89

CHAPTER 3

Introduction to Graphs and Functions

Section 3.1 Mental Math

1. $A(5,2)$ **3.** $C(3,-1)$ **5.** $E(-5,-2)$ **7.** $G(-1,0)$

Exercise Set 3.1

1. $(3,2)$ in quadrant I   **3.** $(-5,3)$ in quadrant II   **5.** $\left(5\frac{1}{2},-4\right)$ in quadrant IV   **7.** $(0,3.5)$ on y-axis

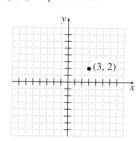 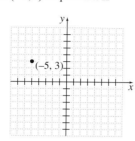 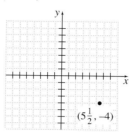 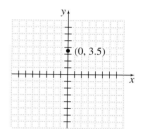

9. $(-2,-4)$ in quadrant III

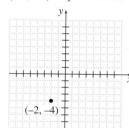

11. Quadrant IV   **13.** x-axis   **15.** Quadrant III   **17.** $(2,8)$ in quadrant I
19. $(1,-4)$ in quadrant IV   **21.** Possible answer: $[-10,10,1]$ by $[0,20,1]$
23. Possible answer: $[-100,100,10]$ by $[-100,100,10]$ **25.** c   **27.** d
29. $4\sqrt{2}\approx 5.66;(3,4)$ **31.** $3\sqrt{5}\approx 6.71;\left(1\frac{1}{2},1\right)$ **33.** $\sqrt{13}\approx 3.61;\left(-3,8\frac{1}{2}\right)$
35. $5;\left(-\frac{1}{2},6\right)$ **37.** $7+2\sqrt{17}+\sqrt{89}\approx 24.68$
39. $4\sqrt{61}\approx 31.24$ **41.** Scalene   **43.** Scalene   **45.** Isosceles

47. $AB=5\sqrt{2}, BC=\sqrt{10}, AC=2\sqrt{10}, BC^2+AC^2=AB^2$ **49.** $y=3x+5$ **51.** $y=x^2+2$
53. $x=-5$ **55.** $x=-\frac{1}{10}$ **57.**

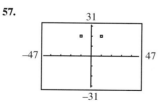

59.

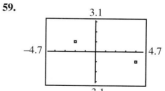

Exercise Set 3.2

1. No; Yes   **3.** Yes; Yes   **5.** Yes; Yes   **7.** Yes; No   **9.** Yes; Yes   **11.** Linear; line   **13.** Linear; line   **15.** Linear; line
17. Nonlinear; V-shaped   **19.** Linear; line   **21.** Nonlinear; parabola   **23.** Nonlinear; parabola   **25.** Linear; line   **27.** Linear; line
29. Nonlinear; V-shaped   **31.** Nonlinear; cubic   **33.** Nonlinear; V-shaped   **35.** Linear; line   **37.** $y=-2x+10$
39. $y=-\frac{(7x+4)}{3}$ **41.** y-intercept: $(0,-5.6)$; x-intercepts: $(-2,0)$ and $(1.4,0)$ **43.** d   **45.** c   **47.** a   **49.** d   **51.** c   **53.** b   **55.** c

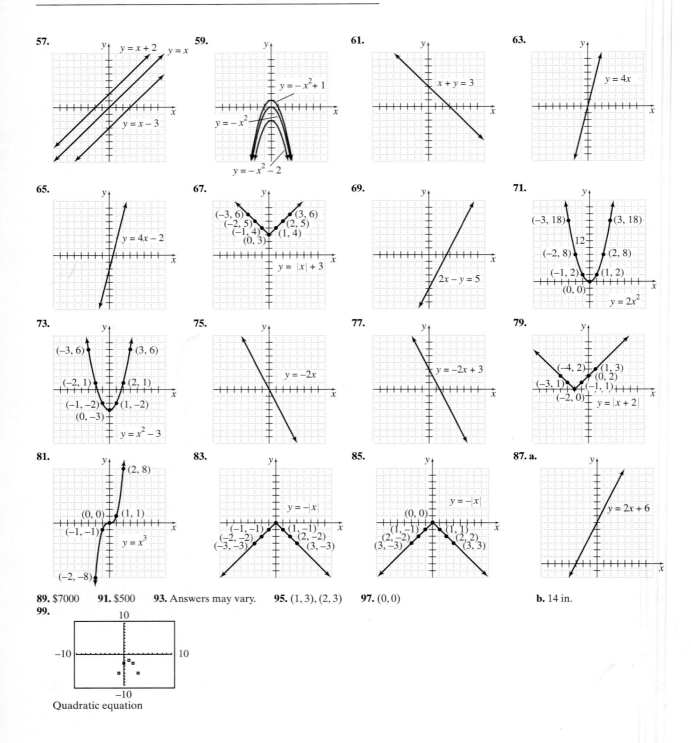

57. **59.** **61.** **63.**

65. **67.** **69.** **71.**

73. **75.** **77.** **79.**

81. **83.** **85.** **87. a.**

89. $7000 **91.** $500 **93.** Answers may vary. **95.** $(1, 3), (2, 3)$ **97.** $(0, 0)$ **b.** 14 in.

99.

Quadratic equation

Exercise Set 3.3

1. Domain $= \{-1, 0, -2, 5\}$; Range $= \{7, 6, 2\}$; Function   **3.** Domain $= \{-2, 6, -7\}$; Range $= \{4, -3, -8\}$; Not a function

5. Domain $= \{1\}$; Range $= \{1, 2, 3, 4\}$; Not a function   **7.** Domain $= \left\{\frac{3}{2}, 0\right\}$; Range $= \left\{\frac{1}{2}, -7, \frac{4}{5}\right\}$; Not a function   **9.** Domain $= \{-3, 0, 3\}$;

Range $= \{-3, 0, 3\}$; Function   **11.** Domain $= \{-1, 1, 2, 3\}$; Range $= \{2, 1\}$; Function   **13.** Domain $= \{$Colorado, Alaska, Delaware,

Illinois, Connecticut, Texas$\}$; Range $= \{6, 1, 20, 30\}$; Function   **15.** Function   **17.** Not a function   **19.** Function   **21.** Function

23. Not a function   **25.** Domain $= [0, \infty)$; $(-\infty, \infty)$; Not a function   **27.** Domain $= [-1, 1]$; Range $= (-\infty, \infty)$; Not a function

29. Domain $= (-\infty, \infty)$; Range $= (-\infty, -3] \cup [3, \infty)$; Not a function   **31.** Domain $= [2, 7]$; Range $= [1, 6]$; Not a function

33. Domain $= \{-2\}$; Range $= (-\infty, \infty)$; Not a function   **35.** Domain $= (-\infty, \infty)$; Range $= (-\infty, \infty)$; Function   **37.** Answers may vary.

39. Yes   **41.** No   **43.** Yes   **45.** Yes   **47.** Yes   **49.** No   **51.** 15   **53.** 38   **55.** 16   **57.** 0   **59. a.** 0   **b.** 1   **c.** -1

61. a. -5 **b.** -5 **c.** -5 **63. a.** 5.1   **b.** 15.5   **c.** 9.533   **65. a.** 23   **b.** 23   **c.** 33   **67.** 7   **69.** $55, $90, $125, $160, $195

71. a. 36   **b.** 64   **c.** 84   **d.** 96   **e.** 100   **f.** 0   **73. a.** $6.5 billion   **b.** $6.701 billion   **75.** $25.18 billion   **77.** $f(x) = x + 7$

79. 2744 in.^{3} **81.** 166.38 cm   **83.** 163.2 mg

85. $(0, 5), (-5, 0), (1, 6)$ **87.** $(0, 2), \left(\frac{8}{7}, 0\right), \left(\frac{12}{7}, -1\right)$ **89.** $(0, 0), (0, 0), (-1, -6)$ **91.** Yes; 170 m

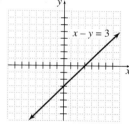

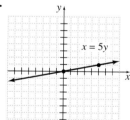

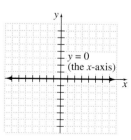

Exercise Set 3.4

1. c   **3.** d

5.

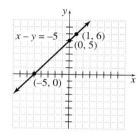

7.

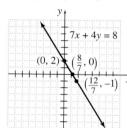

9.

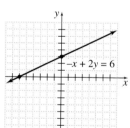

11.

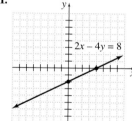

13. Answers may vary.

15.

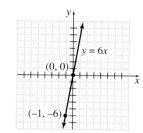

17.

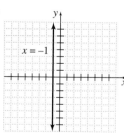

19.

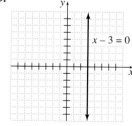

21. For y_2, shift the graph of y_1 3 units up. For y_3, shift the graph of y_1 5 units down.   **23.** $y = -2x + 4$

25. c   **27.** a

29.

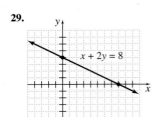

31.

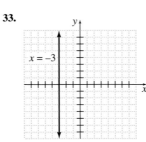

33.

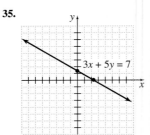

35.

37.

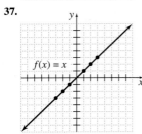

39.

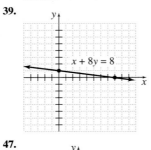

41.

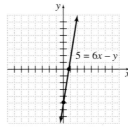

43.

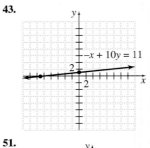

45.

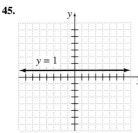

47.

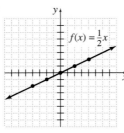

49.

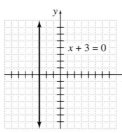

51.

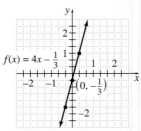

53.

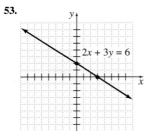

55. b and d

57. a. $(0, 500)$; If no tables are produced, 500 chairs can be produced.
b. $(750, 0)$; If no chairs are produced, 750 tables can be produced.   **c.** 466 chairs

59. a. \$64

b.

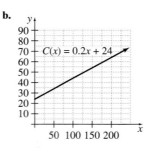

c. The line has a positive slope.

61. a. \$2581.80   **b.** 2004   **c.** Answers may vary.   **63.** $\dfrac{3}{2}$ **65.** 6   **67.** $-\dfrac{6}{5}$

69.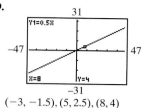

$(-3, -1.5), (5, 2.5), (8, 4)$

71.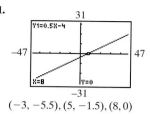

$(-3, -5.5), (5, -1.5), (8, 0)$

Section 3.5 Mental Math

1. Upward   **3.** Horiztonally

Exercise Set 3.5

1. $\frac{9}{5}$; Upward   **3.** $-\frac{7}{2}$; Downward   **5.** $-\frac{5}{6}$; Downward   **7.** $\frac{1}{3}$; Upward   **9.** $-\frac{4}{3}$; Downward   **11.** 0; Neither   **13.** $-\frac{2}{3}$ **15.** 1.5

17. 0.2   **19.** ℓ_2 **21.** ℓ_2 **23.** ℓ_2 **25.** $m = 5, b = -2$ **27.** $m = -2, b = 7$ **29.** $m = \frac{2}{3}, b = -\frac{10}{3}$ **31.** $m = \frac{1}{2}, b = 0$ **33.** a

35. b   **37.** Undefined   **39.** 0   **41.** Undefined   **43.** If the graph moves downward as we go from left to right, then the slope is negative. If the graph moves upward as we go from left to right, then the slope is positive. Horizontal lines have a slope of 0, and vertical lines have undefined slope.   **45.** $m = -1, b = 5$ **47.** $m = \frac{6}{5}, b = 6$ **49.** $m = 3, b = 9$ **51.** $m = 0, b = 4$ **53.** $m = 7, b = 0$

55. $m = 0, b = -6$ **57.** Undefined, no y-intercept   **59.** Neither   **61.** Parallel   **63.** Perpendicular

65. Two lines, both with positive slopes, cannot be perpendicular because the product of two positive numbers is never -1.   **67.** $\frac{3}{2}$

69. $-\frac{1}{2}$ **71.** $8\frac{1}{3}\%$ **73.** $\frac{3}{25}$ **75.** Answers may vary.   **77.** $-\frac{7}{2}$ **79.** $\frac{2}{7}$ **81.** $\frac{5}{2}$ **83.** $-\frac{2}{5}$ **85.** 1.5 yards per second

87.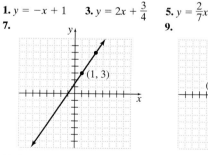

89. $y = 5x + 32$ **91.** $y = -3x - 30$

Section Set 3.6 Mental Math

1. $m = -4, b = 12$ **3.** $m = 5, b = 0$ **5.** $m = \frac{1}{2}, b = 6$ **7.** Parallel   **9.** Neither

Exercise Set 3.6

1. $y = -x + 1$ **3.** $y = 2x + \frac{3}{4}$ **5.** $y = \frac{2}{7}x$

7.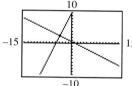

9.

11.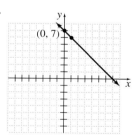

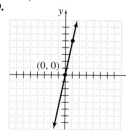

13. $3x - y = 1$ **15.** $2x + y = -1$ **17.** $x - 2y = -10$ **19.** $9x + 10y = -27$ **21.** $2x + y = 3$ **23.** $2x - 3y = -7$

25. $f(x) = 3x - 6$ **27.** $f(x) = -2x + 1$ **29.** $f(x) = -\frac{1}{2}x - 5$ **31.** $f(x) = \frac{1}{3}x - 7$ **33.** Answers may vary.   **35.** -2 **37.** 2

39. -2 **41.** $y = -4$ **43.** $x = 4$ **45.** $y = 5$ **47.** $f(x) = 4x - 4$ **49.** $f(x) = -3x + 1$ **51.** $f(x) = -\frac{3}{2}x - 6$ **53.** $2x - y = -7$
55. $f(x) = -x + 7$ **57.** $x + 2y = 22$ **59.** $2x + 7y = -42$ **61.** $4x + 3y = -20$ **63.** $x = -2$ **65.** $x + 2y = 2$ **67.** $y = 12$
69. $8x - y = 47$ **71.** $x = 5$ **73.** $f(x) = -\frac{3}{8}x - \frac{29}{4}$ **75. a.** $R(x) = 32x$ **b.** 128 ft per second **77. a.** $P(x) = 3075x + 97,500$
b. \$125,175 **79. a.** $P(x) = 12,000x + 18,000$ **b.** \$102,000 **c.** 9 years **81. a.** $y = -1000x + 13,000$ **b.** 9500 Fun Noodles
83. **85.** **87.** $\{14\}$ **89.** $\left\{\frac{7}{2}\right\}$ **91.** $\left\{-\frac{1}{4}\right\}$

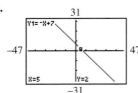

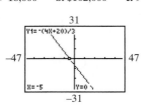

Section 3.7 Mental Math

1. $x = 6$ **3.** $x = 5$ **5.** $x = 0$

Exercise Set 3.7

1. $\{2\}$ **3.** $\{-1\}$ **5.** $\{-2\}$ **7.** $\{6\}$ **9.** $\{x \mid x$ is a real number$\}$ **11.** $\{\ \}$ **13.** $\{x \mid x$ is a real number$\}$ **15.** $\{\ \}$
17. $\{x \mid x$ is a real number$\}$ **19.** $\{25.8\}$ **21.** $\{3.5\}$ **23.** $\{-3.28\}$ **25.** $\{\ \}$ **27.** $\{4\}$ **29.** $\{8\}$ **31.** $\{x \mid x$ is a real number$\}$
33. $\{-0.32\}$ **35.** $\{0.6\}$ **37.** $\{17\}$ **39.** $\{0.8\}$ **41. a.** The second consultant **b.** The first consultant **c.** A 6-hr job
43. a. The first agency's cost is lower. **b.** The second agency's cost is lower. **c.** 60 miles **45.** $(12, 22)$ **47.** Greater than
49. $\{1.43\}$ **51.** $\{1.51\}$ **53.** $\{0.91\}$ **55.** $\{22.87\}$ **57.** The equation is false for all real numbers.
59. $(-\infty, -4)$ **61.** $[2, \infty)$

63. $(-3, 5)$ **65.** $(-\infty, 3.5]$

Chapter 3 Review

1.

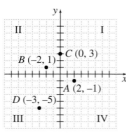

3. $\sqrt{197}$ **5.** $\sqrt{130}$ **7.** 16.60 **9.** $(-5, 5)$ **11.** $\left(-\frac{15}{2}, 1\right)$
13. $\left(\frac{1}{20}, -\frac{3}{16}\right)$ **15.** No; Yes **17.** Yes; Yes **19.** $(7, 44)$ **21.** Linear **23.** Not linear

25.

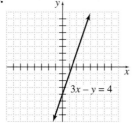

27.

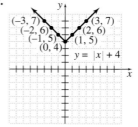

29.

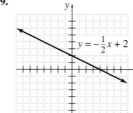

31. d **33.** c
35. Domain: $\left\{-\frac{1}{2}, 6, 0, 25\right\}$;
Range: $\left\{\frac{3}{4}, -12, 25\right\}$; Function
37. Domain: $\{2, 4, 6, 8\}$;
Range: $\{2, 4, 5, 6\}$; Not a function
39. Domain: $(-\infty, \infty)$;
Range: $(-\infty, -1] \cup [1, \infty)$;
Not a function

41. Domain: $(-\infty, \infty)$; Range: $\{4\}$; Function   **43.** -3 **45.** 18   **47.** -3 **49.** 381 pounds
51.

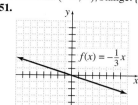
$f(x) = -\frac{1}{3}x$

53. C   **55.** B   **57.**

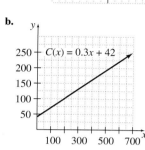
$4x + 5y = 20$

59.

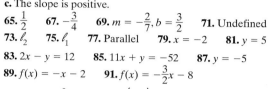
$4x - y = 3$

61.

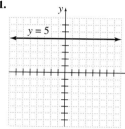
$y = 5$

63. a. \$87   **b.**

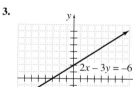
$C(x) = 0.3x + 42$

c. The slope is positive.
65. $\frac{1}{2}$ **67.** $-\frac{3}{4}$ **69.** $m = -\frac{2}{7}, b = \frac{3}{2}$ **71.** Undefined
73. ℓ_2 **75.** ℓ_1 **77.** Parallel   **79.** $x = -2$ **81.** $y = 5$
83. $2x - y = 12$ **85.** $11x + y = -52$ **87.** $y = -5$
89. $f(x) = -x - 2$ **91.** $f(x) = -\frac{3}{2}x - 8$
93. $f(x) = -\frac{3}{2}x - 1$ **95.** $\left\{-\frac{2}{5}\right\}$ **97.** $\{\}$ **99.** $\{-13\}$
101. $\{-2.43\}$

Chapter 3 Test

1.

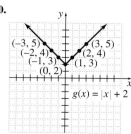
$C(-1, 6)$
Quadrant II
$B(4, 0)$
$A(6, -2)$
Quadrant IV

2. $(-6, -3)$ **3.**

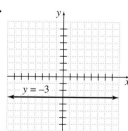
$2x - 3y = -6$

4.

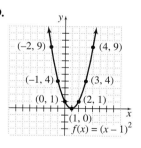
$4x + 6y = 7$

5.

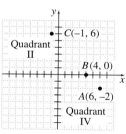
$f(x) = \frac{2}{3}x$

6.

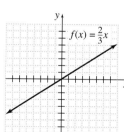
$y = -3$

7. $-\frac{3}{2}$ **8.** $m = -\frac{1}{4}, b = \frac{2}{3}$ **9.**

$(-2, 9)$ $(4, 9)$
$(-1, 4)$ $(3, 4)$
$(0, 1)$ $(2, 1)$
$(1, 0)$
$f(x) = (x - 1)^2$

10.

$(-3, 5)$ $(3, 5)$
$(-2, 4)$ $(2, 4)$
$(-1, 3)$ $(1, 3)$
$(0, 2)$
$g(x) = |x| + 2$

11. $2\sqrt{26}$ units   **12.** $\left(-\frac{1}{2}, \frac{3}{10}\right)$ **13.** $y = -8$ **14.** $x = -4$ **15.** $y = -2$ **16.** $3x + y = 11$
17. $5x - y = 2$ **18.** $f(x) = -\frac{1}{2}x$ **19.** $f(x) = -\frac{1}{3}x + \frac{5}{3}$ **20.** $f(x) = -\frac{1}{2}x - \frac{1}{2}$ **21.** Neither
22. b   **23.** a   **24.** d   **25.** c   **26.** Domain: $(-\infty, \infty)$; Range: $\{5\}$; Function
27. Domain: $\{-2\}$; Range: $(-\infty, \infty)$; Not a function   **28.** Domain: $(-\infty, \infty)$; Range: $[0, \infty)$; Function
29. Domain: $(-\infty, \infty)$; Range: $(-\infty, \infty)$; Function
30. a. \$17,110   **b.** \$24,190   **c.** 2009   **31.** $\left\{-\frac{29}{17}\right\}$ **32.** $\{0.15\}$

Chapter 3 Cumulative Review

1. *(Sec. 1.3, Ex. 4)* **a.** 6   **b.** −7

2. *(Sec. 1.2, Ex. 4)* **a.** $(-\infty, 0]$ **b.** $(2, \infty)$

3. *(Sec. 1.4, Ex. 4)* **a.** 16   **b.** 436   **c.** 49.8643   **4.** *(Sec. 2.7, Ex. 3)* $\{-4\}$ **5.** *(Sec. 2.2, Ex. 1)* 23, 49

6. *(Sec. 2.4, Ex. 2)* $y = \frac{2x}{3} + \frac{7}{3}$ **7.** *(Sec. 3.1, Ex. 1)*

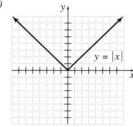

a. Quadrant IV   **b.** No quadrant   **c.** Quadrant II
d. No quadrant   **e.** Quadrant III   **f.** Quadrant I
8. *(Sec. 3.2, Ex. 7)* $\sqrt{61}$ units

9. *(Sec. 3.2, Ex. 7)*

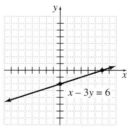

10. *(Sec. 3.3, Ex. 1)* **a.** Domain: $\{2, 0, 3\}$; Range: $\{3, 4, -1\}$
b. Domain: $\{-4, -3, -2, -1, 0, 1, 2, 3\}$; Range: $\{1\}$
c. Domain: {Erie, Escondido, Gary, Miami, Waco}; Range: $\{104, 109, 117, 359\}$
11. *(Sec. 3.3, Ex. 5)* **a.** Yes   **b.** Yes   **c.** No   **d.** Yes   **e.** No   **f.** Yes

12. *(Sec. 3.4, Ex. 3)*   **13.** *(Sec. 3.4, Ex. 6)*

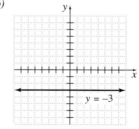

14. *(Sec. 3.5, Ex. 2)* $-\frac{13}{8}$ **15.** *(Sec. 3.5, Ex. 4)* **a.** $m = \frac{3}{4}$, y-intercept $= -1$ **b.** Increasing

16. *(Sec. 3.6, Ex. 3)* $3x + y = -2$ **17.** *(Sec. 3.6, Ex. 5)* $y = 3$ **18.** *(Sec. 3.7, Ex. 1)* $\{3\}$

19. *(Sec. 3.7, Ex. 6)* $\{\}$ **20.** *(Sec. 3.7, Ex. 3)* $\{0.22\}$

CHAPTER 4

Inequalities and Absolute Value

Exercise Set 4.1

1. $(-\infty, -3)$ **3.** $[0.3, \infty)$ **5.** $(5, \infty)$

7. $(-2, 5)$

9. $(-1, 5)$

11. Use a parenthesis when the end-value is not included; use a bracket when the end-value is included.

13. $(-\infty, 4)$ **15.** $[-3, \infty)$ **17.** $\{\ \}$

19. $(-1, \infty)$

21. $(-\infty, 2]$

23. $(-\infty, -4.7)$

25. $\left(-\infty, -\dfrac{1}{6}\right]$

27. $(-\infty, 11]$

29. $\left[\dfrac{8}{3}, \infty\right)$

31. $(-13, \infty)$

33. $(-\infty, 7]$

35. $(-\infty, \infty)$

37. $\{\ \}$

39. Answers may vary.

41. $(6, \infty)$

43. $\left[-\dfrac{1}{10}, \infty\right)$

45. $(-\infty, -1]$

47. $(0, \infty)$

49. $(-2, \infty)$

51. $\left[-\dfrac{3}{5}, \infty\right)$

53. $[-9.6, \infty)$

55. $(38, \infty)$

57. $[0, \infty)$

59. $(-\infty, -5]$

61. $\left(-\infty, \dfrac{1}{4}\right)$

63. $(-\infty, -1]$

65. $\left[-\dfrac{79}{3}, \infty\right)$

67. She must score at least 30 on the final exam. **69.** The plane can carry a maximum of 1040 lbs of luggage and cargo. **71.** At most 17 ounces can be mailed for \$4.00. **73.** Plan 1 is more economical than Plan 2 if more than 200 calls are made. **75.** $F \geq 932°$
77. a. 1993 **b.** Answers may vary. **79.** Increasing **81.** 35.25 lb/person/year **83.** 2012 **85.** The lines will intersect. **87.** 2, 3, 4
89. 2, 3, 4, $\cdots$

91. $[0, 5]$

93. $\left(-\dfrac{1}{2}, \dfrac{3}{2}\right)$

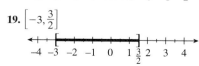

Exercise Set 4.2

1. $\{2, 3, 4, 5, 6, 7\}$ **3.** $\{4, 6\}$ **5.** $\{\cdots, -2, -1, 0, 1, \cdots\}$ **7.** $\{5, 7\}$
9. $(-2, 5)$

11. $[6, \infty)$

13. $(-\infty, -3]$

15. $(11, 17)$

17. $[1, 4]$

19. $\left[-3, \dfrac{3}{2}\right]$

21. $[-21, -9]$

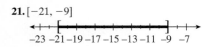

23. $(-\infty, -1) \cup (0, \infty)$

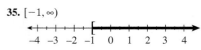

25. $[2, \infty)$

27. $(-\infty, \infty)$

29. Answers may vary.

31. $(-1, 2)$

33. $(-\infty, \infty)$

35. $[-1, \infty)$

37. $[-5, \infty)$

39. $\left[\dfrac{3}{2}, 6\right]$

41. $\left(\dfrac{5}{4}, \dfrac{11}{4}\right)$

43. $\{\ \}$

45. $(-7, \infty)$

47. $(-\infty, -3]$

49. $(-\infty, 1] \cup \left(\dfrac{29}{7}, \infty\right)$

51. $\{\ \}$

53. a. $(-5, 2.5)$ **b.** $(-\infty, -5) \cup (2.5, \infty)$ **55. a.** $[2, 9]$ **b.** $(-\infty, 2] \cup [9, \infty)$

57. $\left[-\dfrac{1}{2}, \dfrac{3}{2}\right)$

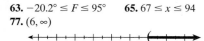

59. $\left(-\dfrac{4}{3}, \dfrac{7}{3}\right)$

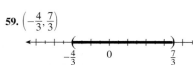

61. $(6, 12)$

63. $-20.2° \le F \le 95°$ **65.** $67 \le x \le 94$ **67.** 1992 and 1993   **69.** -12 **71.** -4 **73.** $x = -7, 7$ **75.** $x = 0$
77. $(6, \infty)$

79. $[3, 7]$

81. $(-\infty, -1)$

Section 4.3 Mental Math

1. 7   **3.** -5 **5.** -6 **7.** 12

Exercise Set 4.3

1. $\{7, -7\}$ **3.** $\{4.2, -4.2\}$ **5.** $\{7, -2\}$ **7.** $\{8, 4\}$ **9.** $\{5, -5\}$ **11.** $\{3, -3\}$ **13.** $\{0\}$ **15.** $\{\ \}$ **17.** $\left\{\dfrac{1}{5}\right\}$ **19.** $|x| = 5$ **21.** $\left\{9, -\dfrac{1}{2}\right\}$
23. $\left\{-\dfrac{5}{2}\right\}$ **25.** Answers may vary.   **27.** $\{-1, 4\}$ **29.** $\{2.5\}$ **31.** $\{\ \}$ **33.** $\left\{0, \dfrac{14}{3}\right\}$ **35.** $\{2, -2\}$ **37.** $\{\ \}$ **39.** $\{7, -1\}$ **41.** $\{\ \}$
43. $\{\ \}$ **45.** $\left\{-\dfrac{1}{8}\right\}$ **47.** $\left\{\dfrac{1}{2}, -\dfrac{5}{6}\right\}$ **49.** $\left\{2, -\dfrac{12}{5}\right\}$ **51.** $\{3, -2\}$ **53.** $\left\{-8, \dfrac{2}{3}\right\}$ **55.** $\{\ \}$ **57.** $\{4\}$ **59.** $\{13, -8\}$ **61.** $\{3, -3\}$
63. $\{8, -7\}$ **65.** $\{2, 3\}$ **67.** $\left\{2, -\dfrac{10}{3}\right\}$ **69.** $\left\{\dfrac{3}{2}\right\}$ **71.** $\{\ \}$ **73.** $\{-1.52, 2.83\}$ **75.** $\{0.08, 1.49\}$ **77.** Answers may vary.   **79.** 35%
81. 4.39 billion   **83.** $-2, -1, 0, 1, 2$, for example   **85.** No solution

Exercise Set 4.4

1. $[-4, 4]$

3. $(1, 5)$

5. $(-5, -1)$

7. $[-10, 3]$

9. $[-5, 5]$

11. $\{\ \}$

13. $[0, 12]$

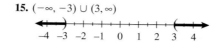

15. $(-\infty, -3) \cup (3, \infty)$

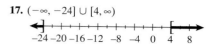

17. $(-\infty, -24] \cup [4, \infty)$

19. $(-\infty, -4) \cup (4, \infty)$

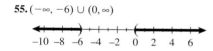

21. $(-\infty, \infty)$

23. $\left(-\infty, \frac{2}{3}\right) \cup (2, \infty)$

25. $\{0\}$

27. $\left(-\infty, -\frac{3}{8}\right) \cup \left(-\frac{3}{8}, \infty\right)$

29. $|x| < 7$

31. $|x| \le 5$

33. $[-2, 2]$

35. $(-\infty, -1) \cup (1, \infty)$

37. $(-5, 11)$

39. $\left(-\infty, \frac{2}{3}\right) \cup (2, \infty)$

41. $\{\,\}$

43. $(-\infty, \infty)$

45. $[-2, 9]$

47. $(-\infty, -11] \cup [1, \infty)$

49. $(-\infty, 0) \cup (0, \infty)$

51. $(-\infty, \infty)$

53. $\left[-\frac{1}{2}, 1\right]$

55. $(-\infty, -6) \cup (0, \infty)$

57. $\left(-\frac{31}{5}, \frac{11}{5}\right)$

59. $[-1, 8]$

61. $\left[-\frac{23}{8}, \frac{17}{8}\right]$

63. a. $\{-5, 11\}$ **b.** $(-5, 11)$ **c.** $(-\infty, -5] \cup [11, \infty)$ **65. a.** $\{-8, 4\}$ **b.** $[-8, 4]$ **c.** $(-\infty, -8) \cup (4, \infty)$
67. $(-2, 5)$ **69.** $\{5, -2\}$ **71.** $(-\infty, -7] \cup [17, \infty)$ **73.** $\left\{-\frac{9}{4}\right\}$ **75.** $(-2, 1)$ **77.** $\left\{\frac{4}{3}, 2\right\}$
79. $\{\,\}$ **81.** $\left\{-\frac{17}{2}, \frac{19}{2}\right\}$ **83.** $\left(-\infty, -\frac{25}{3}\right) \cup \left(\frac{35}{3}, \infty\right)$ **85.** $3.45\text{ m} < x < 3.55\text{ m}$ **87.** Answers may vary.
89. $\frac{1}{6}$ **91.** 0 **93.** 1 **95.** $\frac{8}{3}$ **97.** 0

Exercise Set 4.5

1.

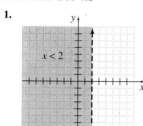

3.

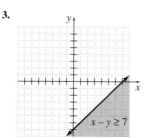

5.

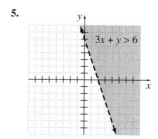

7.

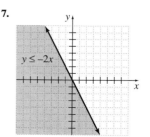

9.

11.

13. Use a dashed boundary line with $<$ or $>$.

15.

17.

19.

21.

23.

25.

27.

29.

31.

33.

35.

37.

39.

41.

43.

45.

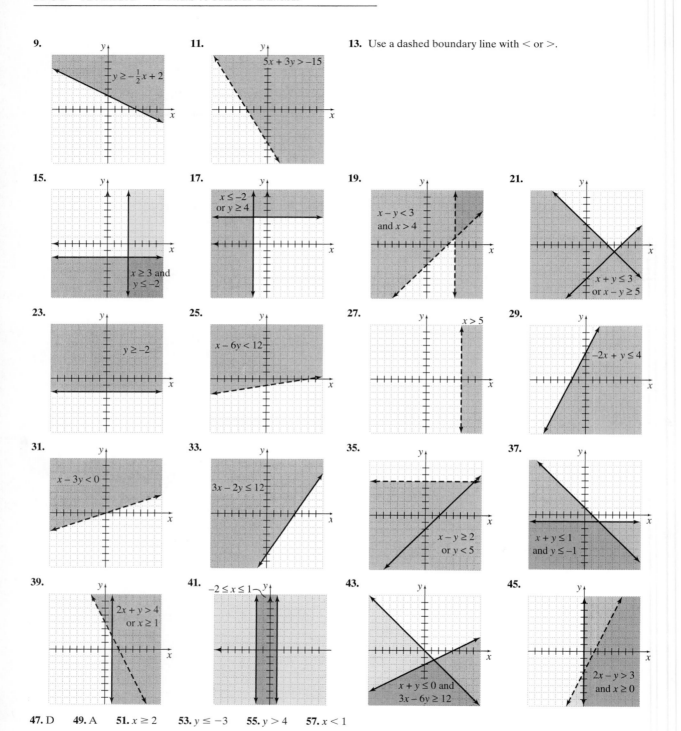

47. D   **49.** A   **51.** $x \geq 2$ **53.** $y \leq -3$ **55.** $y > 4$ **57.** $x < 1$

59. $x \leq 20$ and $y \geq 10$

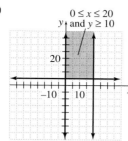

61.

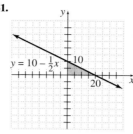

63. 9   **65.** 25   **67.** -16 **69.** $\frac{4}{49}$

71. Domain: $(-\infty, 2] \cup [2, \infty)$; Range: $(-\infty, \infty)$;
This relation is not a function.

Chapter 4 Review

1. $(3, \infty)$ **3.** $(-4, \infty)$ **5.** $\left(-\infty, \frac{34}{14}\right]$ **7.** $\left(\frac{1}{2}, \infty\right)$ **9.** $[-19, \infty)$ **11.** More than 35 lbs   **13.** At least 9.6   **15.** $[2, 2.5]$ **17.** $\left(\frac{1}{8}, 2\right)$

19. $\left(\frac{7}{8}, \frac{27}{20}\right]$ **21.** $(-5, 2]$ **23.** $\left(\frac{11}{3}, \infty\right)$ **25.** $\{16, -2\}$ **27.** $\{0, -9\}$ **29.** $\left\{2, -\frac{2}{3}\right\}$ **31.** $\{\ \}$ **33.** $\{3, -3\}$ **35.** $\left\{5, -\frac{1}{3}\right\}$

37. $\left\{7, -\frac{8}{5}\right\}$ **39.** $\left(-\frac{8}{5}, 2\right)$ **41.** $(-\infty, -3) \cup (3, \infty)$

43. $\{\ \}$

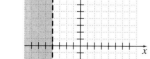

45. $\left(-\infty, -\frac{22}{15}\right] \cup \left[\frac{6}{5}, \infty\right)$

47. $(-\infty, -27) \cup (-9, \infty)$

49.

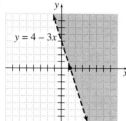

$y = 4 - 3x$

51.

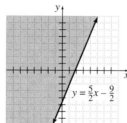

$y = \frac{5}{2}x - \frac{9}{2}$

53.

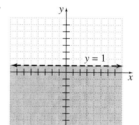

$y = 1$

55.

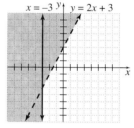

$x = -3$ $y = 2x + 3$

Chapter 4 Test

1. $(-\infty, -3)$ **2.** $(-3, \infty)$ **3.** $(5, \infty)$ **4.** $[2, \infty)$ **5.** $\left(\frac{3}{2}, 5\right]$ **6.** $(-\infty, -2) \cup \left(\frac{4}{3}, \infty\right)$ **7.** $[5, \infty)$ **8.** $[4, \infty)$ **9.** $[-3, -1)$

10. $\left\{\frac{2}{3}, 1\right\}$ **11.** $\{\ \}$ **12.** $\left\{-\frac{13}{3}, -1\right\}$ **13.** $\{-7, 13\}$ **14.** $\left[-\frac{5}{7}, 1\right]$

15.

$x = -4$

16.

$y = -2$

17.

$y = 2x - 5$

18.

$y = -\frac{1}{2}x + \frac{3}{2}$ $y = -4$

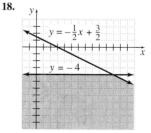

19. 2013 and the years following it

Chapter 4 Cumulative Review

1. *(Sec. 1.1, Ex. 2)* **a.** True   **b.** True   **c.** False

2. *(Sec. 1.2, Ex. 5)* **a.** $(-5, -1)$

-8 -7 -6 -5 -4 -3 -2 -1 0

b. $[-1.5, 4)$

-4 -3 -2 -1 0 1 2 3 4

3. *(Sec. 1.3, Ex. 3)* **a.** -6 **b.** -7 **c.** -16 **d.** 20.5 **e.** $\frac{1}{6}$ **f.** 0.94 **g.** -3

4. *(Sec. 1.3, Ex. 6)* **a.** -5 **b.** 3 **c.** $-\frac{1}{8}$ **d.** -4 **e.** $\frac{1}{4}$ **f.** Undefined

5. *(Sec. 1.3, Ex. 8)* **a.** 3 **b.** 5 **c.** $\frac{1}{2}$ **6.** *(Sec. 1.4, Ex. 1)* **a.** 23 **b.** 18 **c.** -1 **d.** $\frac{12}{5}$

7. *(Sec. 1.4, Ex. 7)* **a.** $0.25x$ dollars **b.** $2x$ grams **c.** $156x$ dollars **d.** $0.09x$ dollars **8.** *(Sec. 2.1, Ex. 8)* $(-\infty, \infty)$

9. *(Sec. 2.2, Ex. 3)* $2350 **10.** *(Sec. 2.4, Ex. 3)* $b = \dfrac{2A - Bh}{h}$

11. *(Sec. 3.2, Ex. 5)*

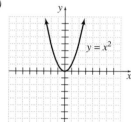

$y = x^2$

12. *(Sec. 3.2, Ex. 8)*

$y = |x| - 3$

13. *(Sec. 3.5, Ex. 1)* 1 **14.** *(Sec. 4.1, Ex. 3)* **a.** $\left(-\infty, \dfrac{3}{2}\right]$ **b.** $(-3, \infty)$ **15.** *(Sec. 4.1, Ex. 6)* $(-\infty, \infty)$

16. *(Sec. 4.1, Ex. 7)* $4500 **17.** *(Sec. 4.2, Ex. 1)* $(-\infty, 4)$ **18.** *(Sec. 4.2, Ex. 3)* $\left[-9, -\dfrac{9}{2}\right]$ **19.** *(Sec. 4.2, Ex. 4)* $\left(-\infty, \dfrac{13}{5}\right] \cup [4, \infty)$

20. *(Sec. 4.3, Ex. 1)* $\{-2, 2\}$ **21.** *(Sec. 4.4, Ex. 2)* $(4, 8)$ **22.** *(Sec. 4.4, Ex. 6)* $(-\infty, \infty)$

CHAPTER 5

Systems of Equations

Exercise Set 5.1

1. $\{(1.5, 1)\}$ **3.** $\{(-2, 0.75)\}$ **5.** $\{(2, -1)\}$ **7.** $\{(1, 2)\}$ **9.** $\{\ \}$ **11.** No **13.** $\{(2, 8)\}$ **15.** $\{(0, -9)\}$ **17.** $\{(-5, 3)\}$

19. $\left\{\left(\dfrac{5}{2}, \dfrac{5}{4}\right)\right\}$ **21.** $\{(1, -2)\}$ **23.** $\{(7, 2)\}$ **25.** $\{\ \}$ **27.** $\{(x, y)|3x + y = 1\}$ **29.** Answers may vary. **31.** $\left\{\left(\dfrac{3}{2}, 1\right)\right\}$ **33.** $\{(2, -1)\}$

35. $\{(-5, 3)\}$ **37.** $\{(x, y)|x + 3y = 4\}$ **39.** $\{\ \}$ **41.** $\left\{\left(\dfrac{1}{2}, \dfrac{1}{5}\right)\right\}$ **43.** $\{(8, 2)\}$ **45.** $\{(x, y)|x = 3y + 2\}$ **47.** $\left\{\left(-\dfrac{1}{4}, \dfrac{1}{2}\right)\right\}$ **49.** $\{(3, 2)\}$

51. $\{(7, -3)\}$ **53.** $\{\ \}$ **55.** $\{(-2, 1)\}$ **57.** $\{(2.11, 0.17)\}$ **59.** $\{(0.57, -1.97)\}$ **61.** 5000 ties; $21 **63.** Supply is greater than demand. **65.** $\{(1875, 4687.5)\}$ **67.** Makes money **69.** For values of $x > 1875$ **71.** 2003 **73.** True **75.** False

77. $6y - 4z = 25$ **79.** $x + 10y = 2$ **81.** $\left\{\left(\dfrac{1}{4}, 8\right)\right\}$ **83.** $\left\{\left(\dfrac{1}{3}, \dfrac{1}{2}\right)\right\}$ **85.** $\left\{\left(\dfrac{1}{4}, -\dfrac{1}{3}\right)\right\}$ **87.** $\{\ \}$

Exercise Set 5.2

1. $\{(-2, 5, 1)\}$ **3.** $\{(-2, 3, -1)\}$ **5.** $\{(x, y, z)|x - 2y + z = -5\}$ **7.** $\{\ \}$ **9.** Answers may vary. One possibility is $\{3x = -3, 2x + 4y = 6, x - 3y + z = -11\}$ **11.** $\{(0, 0, 0)\}$ **13.** $\{(-3, -35, -7)\}$ **15.** $\{(6, 22, -20)\}$ **17.** $\{\ \}$ **19.** $\{(3, 2, 2)\}$

21. $\{(x, y, z)|x + 2y - 3z = 4\}$ **23.** $\{(-3, -4, -5)\}$ **25.** $\{(12, 6, 4)\}$ **27.** $\{(1, 1, -1)\}$ **29.** 15 and 30 **31.** $\{5\}$ **33.** $\left\{-\dfrac{5}{3}\right\}$

35. $\{(1, 1, 0, 2)\}$ **37.** $\{(1, -1, 2, 3)\}$

Exercise Set 5.3

1. 10 and 8 **3.** Plane: 520 mph; Wind: 40 mph **5.** 20 quarts of 4%; 40 quarts of 1% **7.** 9 large frames; 13 small frames **9.** -10 and -8 **11.** Tables: $.80; Pens: $0.20 **13.** Plane: 630 mi/hr; Wind: 90 mi/hr **15.** 5 in., 7 in., 7 in., and 10 in. **17.** 18, 13, and 9

19. $2,000 worth of sales   **21.** $1.90 for template; $0.75 for pencil; $2.25 for paper   **23.** 750 units   **25.** 750
29. a. $R(x) = 31x$ **b.** $C(x) = 15x + 500$ **c.** 32 baskets   **31.** $x = 40; y = 70$ **33. a.** 100 miles   **b.** $64.0(
35. 120 liters of 25%; 60 liters of 40%; 20 liters of 50%   **37.** $a = 1, b = -2, c = 3$ **39.** $x = 95, y = 123, z = 7$
41. $a = 0.28; b = -3.71; c = 12.83; 2.12$ **43.** $3y + 8z = 18$ **45.** $-5x - 5z = -16$ **47.** $\frac{3}{8}$ **49.** $\frac{5}{8}$

Exercise Set 5.4

1. $\{(2, -1)\}$ **3.** $\{(-4, 2)\}$ **5.** $\{(-2, 5, -2)\}$ **7.** $\{(1, -2, 3)\}$ **9.** $\{\}$ **11.** $\{(x, y)|x - y = 3\}$ **13.** $\{(4, -3)\}$ **15.** $\{(2, 1, -1)\}$
17. $\{(9, 9)\}$ **19.** $\{\}$ **21.** $\{\}$ **23.** $\{(1, -4, 3)\}$ **25.** Tablets: $0.80; Pens: $0.20   **27.** 9 large; 13 small
29. $1400 at 6.2%, $1200 at 7.5%, and $2200 at 8%   **31.** Function   **33.** Not a function   **35.** -13 **37.** -36 **39.** 0

Exercise Set 5.5

1. 26   **3.** -19 **5.** 0   **7.** $\{(1, 2)\}$ **9.** $\{(x, y)|3x + y = 1\}$ **11.** $\{(9, 9)\}$ **13.** 8   **15.** 0   **17.** 54   **19.** $\{(-2, 0, 5)\}$ **21.** $\{(6, -2, 4)\}$
23. 16   **25.** 15   **27.** $\frac{13}{6}$ **29.** 0   **31.** 56   **33.** Zero   **35.** 5   **37.** $\{(-3, -2)\}$ **39.** $\{\}$ **41.** $\{(-2, 3, -1)\}$ **43.** $\{(3, 4)\}$
45. $\{(-2, 1)\}$ **47.** $\{(x, y, z)|x - 2y + z = -3\}$ **49.** $\{(0, 2, -1)\}$ **51.** $6x - 18$ **53.** $9x - 15$
55.   **57.**   **59.** -125 **61.** 24

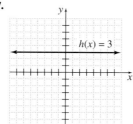

Chapter 5 Review

1. $\{(-3, 1)\}$ **3.** $\{\}$ **5.** $\left\{\left(3, \frac{8}{3}\right)\right\}$ **7.** $\{(2, 0, 2)\}$ **9.** $\left\{\left(-\frac{1}{2}, \frac{3}{4}, 1\right)\right\}$ **11.** $\{\}$ **13.** $\{(1, 1, -2)\}$ **15.** 10, 40, and 48   **17.** 63 and 21
19. 58 mph, 65 mph   **21.** 20 liters of 10%, 30 liters of 60%   **23.** 17 pennies, 20 nickels, and 16 dimes
25. Two sides 22 cm each, third side 29 cm   **27.** $\{(-3, 1)\}$ **29.** $\left\{\left(-\frac{2}{3}, 3\right)\right\}$ **31.** $\left\{\left(\frac{5}{4}, \frac{5}{8}\right)\right\}$ **33.** $\{(1, 3)\}$ **35.** $\{(1, 2, 3)\}$ **37.** $\{(3, -2, 5)\}$
39. $\{(1, 1, -2)\}$ **41.** -17 **43.** 34   **45.** $\left\{\left(-\frac{2}{3}, 3\right)\right\}$ **47.** $\{(-3, 1)\}$ **49.** $\{\}$ **51.** $\{(1, 2, 3)\}$ **53.** $\{(2, 1, 0)\}$ **55.** $\{\}$

Chapter 5 Test

1. 34   **2.** -6 **3.** $\{(1, 3)\}$ **4.** $\{\}$ **5.** $\{(2, -3)\}$ **6.** $\{(x, y)|5x + 2y = 5\}$ **7.** $\{(-1, -2, 4)\}$ **8.** $\{\}$ **9.** $\left\{\left(\frac{7}{2}, -10\right)\right\}$ **10.** $\{(2, -1)\}$
11. $\{(3, 6)\}$ **12.** $\{(3, -1, 2)\}$ **13.** $\{(5, 0, -4)\}$ **14.** $\{(x, y)|x - y = -2\}$ **15.** $\{(5, -3)\}$ **16.** $\{(-1, -1, 0)\}$ **17.** $\{\}$ **18.** 275 frames
19. 53 double; 27 single   **20.** 5 gallons of 10%; 15 gallons of 20%

Chapter 5 Cumulative Review

1. *(Sec. 1.2, Ex. 3)* **a.** $5 + y \geq 7$ **b.** $11 \neq z$ **c.** $20 < 5 - 2x$ **2.** *(Sec. 1.2, Ex. 8)* $5 + 7x$
3. *(Sec. 1.4, Ex. 3)* **a.** 1.6   **b.** -28.7 **c.** 25.23   **4.** *(Sec. 2.1, Ex. 5)* $\{2\}$
5. *(Sec. 3.3, Ex. 6)* **a.** Domain: $[-3, 5]$; Range: $[-2, 4]$; Function   **b.** Domain: $[-4, 4]$; Range: $[-2, 2]$; Not a function
c. Domain: $(-\infty, \infty)$; Range: $[3, \infty)$; Function   **d.** Domain: $(-\infty, \infty)$; Range: $(-\infty, \infty)$; Function
e. Domain: $(-\infty, \infty)$; Range: $[-2, \infty)$; Function   **6.** *(Sec. 3.5, Ex. 6)* Undefined   **7.** *(Sec. 3.6, Ex. 4)* $f(x) = \frac{5}{8}x - \frac{5}{2}$
8. *(Sec. 4.1, Ex. 2)* $\{x|x \geq -10\}, [-10, \infty)$,

.sec. 4.2, Ex. 2) $(-3, 2)$ **10.** *(Sec. 4.3, Ex. 2)* $\left\{-2, \frac{4}{5}\right\}$ **11.** *(Sec. 4.4, Ex. 5)* $(-\infty, -4] \cup [10, \infty)$ **12.** *(Sec. 4.4, Ex. 8)* $\{-1\}$

3. *(Sec. 4.5, Ex. 1)*   **14.** *(Sec. 4.5, Ex. 2)*

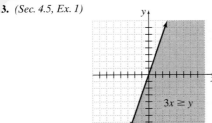

$3x \geq y$

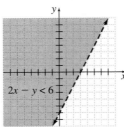

$2x - y < 6$

15. *(Sec. 5.1, Ex. 4)* $\left\{\left(-4, \frac{1}{2}\right)\right\}$ **16.** *(Sec. 5.1, Ex. 6)* $\{\}$ **17.** *(Sec. 5.2, Ex. 1)* $\{(-4, 2, -1)\}$ **18.** *(Sec. 5.3, Ex. 2)* 52 mph, 47 mph

19. *(Sec. 5.4, Ex. 2)* $\{(1, -1, 3)\}$ **20.** *(Sec. 5.5, Ex. 3)* -2

CHAPTER 6

Exponents, Polynomials, and Polynomial Functions

Section 6.1 Mental Math

1. $\frac{5}{xy^2}$ **3.** $\frac{a^2}{bc^5}$ **5.** $\frac{x^4}{y^2}$

Exercise Set 6.1

1. 4^5 **3.** x^8 **5.** $-140x^{12}$ **7.** $-20x^2y$ **9.** $-16x^6y^3p^2$ **11.** -1 **13.** 1 **15.** 6 **17.** Answers may vary.   **19.** a^3 **21.** x

23. $-13z^4$ **25.** $-6a^4b^4c^6$ **27.** $\frac{1}{16}$ **29.** $\frac{1}{x^8}$ **31.** $\frac{5}{a^4}$ **33.** $\frac{1}{x^7}$ **35.** $4r^8$ **37.** 1 **39.** x^{5+7a} **41.** x^{2t-1} **43.** x^{4a+7}

45. z^{6x-7} **47.** x^{6t-1} **49.** $\frac{13}{36}$ **51.** 9 **53.** x^{15} **55.** $10x^{10}$ **57.** $\frac{1}{z^3}$ **59.** y^4 **61.** $\frac{3}{x}$ **63.** -2 **65.** r^8 **67.** $\frac{1}{x^9y^4}$

69. $\frac{b^7}{9a^7}$ **71.** $\frac{6x^{16}}{5}$ **73.** 3.125×10^7 **75.** 1.6×10^{-2} **77.** 6.7413×10^4 **79.** 1.25×10^{-2} **81.** 5.3×10^{-5} **83.** 0.0000000036

85. $93,000,000$ **87.** $1,278,000$ **89.** $7,350,000,000,000$ **91.** 0.000000403 **93.** Rewrite the number as the product of a number between 1 and 10 (including 1 but not 10) and 10 to an appropriate integer power. Answers may vary.   **95.** 9.18×10^8 km

97. 4.8×10^{10} operations   **99.** 1.0×10^{-3} sec   **101.** 100 **103.** $\frac{27}{64}$ **105.** 64 **107.** $\frac{1}{16}$

Section 6.2 Mental Math

1. x^{20} **3.** x^9 **5.** y^{42} **7.** z^{20} **9.** z^{18}

Exercise Set 6.2

1. $\frac{1}{9}$ **3.** $\frac{1}{x^{36}}$ **5.** $\frac{1}{y^5}$ **7.** $9x^4y^6$ **9.** $16x^{20}y^{12}$ **11.** $\frac{c^{18}}{a^{12}b^6}$ **13.** $\frac{y^{15}}{x^{35}z^{20}}$ **15.** $\frac{1}{a^2}$ **17.** $4a^8b^4$ **19.** $\frac{x^4}{4z^2}$ **21.** Yes; $a = \pm 1$

23. $\frac{1}{125}$ **25.** $\frac{1}{x^{63}}$ **27.** $\frac{343}{512}$ **29.** $16x^4$ **31.** $-\frac{y^3}{64}$ **33.** $4^8x^2y^6$ **35.** $\frac{36}{p^{12}}$ **37.** $-\frac{a^6}{512x^3y^9}$ **39.** $\frac{x^{14}y^{14}}{a^{21}}$ **41.** $\frac{x^4}{16}$ **43.** 64

45. $\frac{1}{y^{15}}$ **47.** $\frac{2}{p^2}$ **49.** $\frac{3}{8x^8y^7}$ **51.** $\frac{1}{x^{30}b^6c^6}$ **53.** $\frac{25}{8x^5y^4}$ **55.** $\frac{2}{x^4y^{10}}$ **57.** x^{9a+18} **59.** x^{12a+2} **61.** b^{10x^2-4x} **63.** y^{15a+3}

65. $16x^{4t+4}$ **67.** 1.45×10^9 **69.** 8.0×10^{15} **71.** 4×10^{-7} **73.** 3×10^{-1} **75.** 2×10^1 **77.** 1.0×10^1 **79.** 8×10^{-5}

81. 1.1×10^7 **83.** $8.877840909 \times 10^{20}$ **85.** $0.002 = 2 \times 10^{-3}$ sec   **87.** 1.331928×10^{13} tons   **89.** $\frac{8}{x^6y^3}$ m^3

91. 2.5808×10^{-5} m^2 **93.** Answers may vary.   **95.** $-3m - 15$ **97.** $-3y - 5$ **99.** $-3x + 5$

Exercise Set 6.3

1. 0   **3.** 2   **5.** 3   **7.** Binominal of degree 1   **9.** Trinomial of degree 2   **11.** Monomial of degree 3   **13.** Degree 3; none of these
15. Answers may vary.   **17.** 57   **19.** 499   **21. a.** 2 and 3   **b.** 2 and 3   **23.** 5.7 sec   **25. a.** The second egg   **b.** 1 sec
27. $6y$ **29.** $11x - 3$ **31.** $xy + 2x - 1$ **33.** $18y^2 - 17$ **35.** $3x^2 - 3xy + 6y^2$ **37.** $x^2 - 4x + 8$ **39.** $y^2 + 3$ **41.** $-2x^2 + 5x$
43. $-2x^2 - 4x + 15$ **45.** $4x - 13$ **47.** $x^2 + 2$ **49.** $12x^3 + 8x + 8$ **51.** $7x^3 + 4x^2 + 8x - 10$ **53.** $-18y^2 + 11yx + 14$
55. $-x^3 + 8a - 12$ **57.** $5x^2 - 9x - 3$ **59.** $-3x^2 + 3$ **61.** $2x^3 + 3x^2 + 8xy^2 - 3$ **63.** $7y^2 - 3$ **65.** $5x^2 + 22x + 16$
67. $-q^4 + q^2 - 3q + 5$ **69.** $15x^2 + 8x - 6$ **71.** $x^4 - 7x^2 + 5$ **73.** $4x^{2y} + 2x^y - 11$ **75.** \$80,000   **77. a.** \$1066.14   **b.** \$2650.84
c. \$5097.54   **d.** No; $f(x)$ is not linear.   **79.** \$26,000   **81. a.** 284 ft   **b.** 536 ft   **c.** 756 ft   **d.** 944 ft   **83.** 19 sec
85. $5x^2 + 3x - 4$ **87.** $-4x^2 + 9x$ **89.** $-4x^2 - 9x - 18$ **91. a.** $(0, 4)$ **b.** $(0, -3.9)$ **c.** $(0, 0)$ **93.** A; $(0, -2)$ **95.** D; $(0, -2)$
97. $(6x^2 + 14y)$ units   **99.** $15x - 10$ **101.** $-2x^2 + 10x - 12$ **103. a.** $2a - 3$ **b.** $-2x - 3$ **c.** $2x + 2h - 3$ **105. a.** $4a$ **b.** $-4x$
c. $4x + 4h$ **107. a.** $4a - 1$ **b.** $-4x - 1$ **c.** $4x + 4h - 1$

Exercise Set 6.4

1. $-12x^5$ **3.** $12x^2 + 21x$ **5.** $-24x^2y - 6xy^2$ **7.** $-4xa^3b - 4ya^3b + 12ab$ **9.** $2x^2 - 2x - 12$ **11.** $2x^4 + 3x^3 - 2x^2 + x + 6$
13. $15x^2 - 7x - 2$ **15.** $15m^3 + 16m^2 - m - 2$ **17.** Answers may vary.   **19.** $x^2 + x - 12$ **21.** $10x^2 + 11xy - 8y^2$
23. $3x^2 + 8x - 3$ **25.** $9x^2 - \frac{1}{4}$ **27.** $x^2 + 8x + 16$ **29.** $36y^2 - 1$ **31.** $9x^2 - 6xy + y^2$ **33.** $9b^2 - 36y^2$ **35.** $16b^2 + 32b + 16$
37. $4s^2 - 12s + 8$ **39.** $x^2y^2 - 4xy + 4$ **41.** Answers may vary.   **43.** $9x^2 + 18x + 5$ **45.** $10x^5 + 8x^4 + 2x^3 + 25x^2 + 20x + 5$
47. $49x^2 - 9$ **49.** $9x^3 + 30x^2 + 12x - 24$ **51.** $16x^2 - \frac{2}{3}x - \frac{1}{6}$ **53.** $36x^2 + 12x + 1$ **55.** $x^4 - 4y^2$ **57.** $-30a^4b^4 + 36a^3b^2 + 36a^2b^3$
59. $2a^2 - 12a + 16$ **61.** $49a^2b^2 - 9c^2$ **63.** $m^2 - 8m + 16$ **65.** $9x^2 + 6x + 1$ **67.** $y^2 - 7y + 12$
69. $2x^3 + 2x^2y + x^2 + xy - x - y$ **71.** $9x^4 + 12x^3 - 2x^2 - 4x + 1$ **73.** $12x^3 - 2x^2 + 13x + 5$ **75.** $5x^2 + 25x$ **77.** $x^4 - 4x^2 + 4$
79. $x^3 + 5x^2 - 2x - 10$ **81. a.** $6x + 12$ **b.** $9x^2 + 36x + 35$ **83.** $\pi(7y^3 - 42y^2 + 63y)$ cm^3 **85.** $c^2 - 3c$ **87.** $a^2 + 7a + 10$
89. $a^2 - 2ab + b^2 - 3a + 3b$ **91. a.** $a^2 + 2ah + h^2 + 2a + 2h + 1$ **b.** $a^2 + 2a + 1$ **c.** $2ah + h^2 + 2h$ **93.** $-6y^4z^{3n} + 3yz^n$
95. $x^{2a} - y^{4b}$ **97.** $\frac{3}{2}$ **99.** $-\frac{1}{7}$ **101.** Function

Section 6.5 Mental Math

1. 6   **3.** 5   **5.** x **7.** $7x$

Exercise Set 6.5

1. a^3 **3.** y^2z^2 **5.** $3x^2y$ **7.** $5xz^3$ **9.** $6(3x - 2)$ **11.** $4y^2(1 - 4xy)$ **13.** $2x^3(3x^2 - 4x + 1)$ **15.** $4ab(2a^2b^2 - ab + 1 + 4b)$
17. $(6 + 5a)(x + 3)$ **19.** $(2x + 1)(z + 7)$ **21.** $(3x - 2)(x^2 + 5)$ **23.** Answers may vary.   **25.** $(a + 2)(b + 3)$ **27.** $(a - 2)(c + 4)$
29. $(x - 2)(2y - 3)$ **31.** $(4x - 1)(3y - 2)$ **33.** $2\pi r(r + h)$ **35.** $A = P(1 + RT)$ **37. a.** $h(t) = -16t(t - 4)$ **b.** 48 ft
c. Answers may vary.   **39.** $3(2x^3 + 3)$ **41.** $x^2(x + 3)$ **43.** $4a(2a^2 - 1)$ **45.** $-4xy(5x - 4y^2)$ **47.** $5ab^2(2ab + 1 - 3b)$
49. $3b(3ac^2 + 2a^2c - 2a + c)$ **51.** $(4x - 3)(y - 2)$ **53.** $(2x + 3)(3y + 5)$ **55.** $(y - 5)(x + 3)$ **57.** $(2a - 3)(3b - 1)$
59. $(6x + 1)(2y + 3)$ **61.** $(2m - 1)(n - 8)$ **63.** $3x^2y^2(5x - 6)$ **65.** $(x + 2)(2x + 3y)$ **67.** $(5x - 3)(x + y)$ **69.** $(x^2 + 4)(x + 3)$
71. $(x^2 - 2)(x - 1)$ **73.** d   **75.** $55x^7$ **77.** $125x^6$ **79.** $x^2 - 3x - 10$ **81.** $x^2 + 5x + 6$ **83.** $y^2 - 4y + 3$
85. $x^n(x^{2n} - 2x^n + 5)$ **87.** $2x^{3a}(3x^{5a} - x^{2a} - 2)$

Section 6.6 Mental Math

1. 2 and 5   **3.** 8 and 3

Exercise 6.6

1. $(x + 6)(x + 3)$ **3.** $(x - 4)(x - 8)$ **5.** $(x + 12)(x - 2)$ **7.** $(x - 6)(x + 4)$ **9.** $3(x - 2)(x - 4)$ **11.** $4z(x + 2)(x + 5)$
13. $2(x + 18)(x - 3)$ **15.** ± 5 and ± 7 **17.** $(x + 5), (x - 3)$ **19.** $(x - 2), (x - 6)$ **21.** $(2x - 3)(x - 4)$ **23.** $(2x - 3)^2$
25. $2(3x - 5)(2x + 5)$ **27.** $y^2(3y + 5)(y - 2)$ **29.** $2x(3x^2 + 4x + 12)$ **31.** $(x + z)(x + 7z)$ **33.** $(2x + y)(x - 3y)$
35. $2(7y + 2)(2y + 1)$ **37.** $(x + 4)(x - 5); -4, 5$ **39.** $(x - 4)(x - 7); 4, 7$ **41.** $b = \pm 8$ and ± 16 **43.** $(x^2 + 3)(x^2 - 2)$
45. $(5x + 2)(5x + 8)$ **47.** $(x^3 - 4)(x^3 - 3)$ **49.** $(a - 3)(a + 8)$ **51.** $x(3x + 4)(x - 2)$ **53.** $(x - 27)(x + 3)$ **55.** $(x - 18)(x + 3)$
57. $3(x - 1)^2$ **59.** $(3x + 1)(x - 2)$ **61.** $(4x - 3)(2x - 5)$ **63.** $3x^2(3x + 2)(2x + 1)$ **65.** $3(a + 2b)^2$ **67.** Prime
69. $(2x + 13)(x + 3)$ **71.** $(3x - 2)(2x - 15)$ **73.** $(x^2 - 6)(x^2 + 1)$ **75.** $x(3x + 1)(2x - 1)$ **77.** $(4a - 3b)(3a - 5b)$ **79.** $(3x + 5)^2$

81. $y(3x - 8)(x - 1)$ **83.** $2(x + 3)(x - 2)$ **85.** $(x + 2)(x - 7)$ **87.** $(2x^3 - 3)(x^3 + 3)$ **89.** $2x(6y^2 - z)^2$ **91.** $x^2(x + 5)(x + 1)$
93. $3x(5x - 1)(2x + 1)$ **95.** $x^3 - 8$ **97.** -9 **99.** -8 **101.** $(x^n + 2)(x^n + 8)$ **103.** $(x^n - 6)(x^n + 3)$ **105.** $(2x^n + 1)(x^n + 5)$
107. $(2x^n - 3)^2$

Exercise Set 6.7

1. $(x + 3)^2$ **3.** $(2x - 3)^2$ **5.** $3(x - 4)^2$ **7.** $x^2(3y + 2)^2$ **9.** $(x + 5)(x - 5)$ **11.** $(3 + 2z)(3 - 2z)$ **13.** $(y - 5)(y + 9)$
15. $4(4x + 5)(4x - 5)$ **17.** $(x + 3)(x^2 - 3x + 9)$ **19.** $(z - 1)(z^2 + z + 1)$ **21.** $(m + n)(m^2 - mn + n^2)$
23. $(x - 3)(x^2 + 3x + 9)y^2$ **25.** $(a + 2b)(a^2 - 2ab + 4b^2)b$ **27.** $(5y - 2x)(25y^2 + 10xy + 4x^2)$ **29.** $(x + 3 + y)(x + 3 - y)$
31. $(x - 5 - y)(x - 5 + y)$ **33.** $(2x + 1 - z)(2x + 1 + z)$ **35.** $(3x + 7)(3x - 7)$ **37.** $(x^2 + 9)(x + 3)(x - 3)$
39. $(x + 4 - 2y)(x + 4 + 2y)$ **41.** $(x + 2y - 3)(x + 2y + 3)$ **43.** $(x - 1)(x^2 + x + 1)$ **45.** $(x + 5)(x^2 - 5x + 25)$ **47.** Prime
49. $(2a + 3)^2$ **51.** $2y(3x - 1)(3x + 1)$ **53.** $(x^2 - y)(x^4 + x^2y + y^2)$ **55.** $(x + 8 - x^2)(x + 8 + x^2)$ **57.** $3y^2(x^2 + 3)(x^4 - 3x^2 + 9)$
59. $(x + y + 5)(x^2 + 2xy + y^2 - 5x - 5y + 25)$ **61.** $(2x - 1)(4x^2 + 20x + 37)$ **63.** $\pi R^2 - \pi r^2 = \pi(R - r)(R + r)$
65. $\frac{4}{3}\pi R^3 - \frac{4}{3}\pi(6)^3 = \frac{4}{3}\pi(R - 6)(R^2 + 6R + 36)$ mm^3 **67.** $(x - 4 - y)(x - 4 + y)$ **69.** $x(x - 1)(x^2 + x + 1)$ **71.** $2xy(7x - 1)$
73. $4(x + 2)(x - 2)$ **75.** $(3x - 11)(x + 1)$ **77.** $4(x + 3)(x - 1)$ **79.** $(2x + 9)^2$ **81.** $(2x + 3y)(4x^2 - 6xy + 9y^2)$
83. $8x^2(2y - 1)(4y^2 + 2y + 1)$ **85.** $(x + 5 + y)(x^2 + 10x + 25 - xy - 5y + y^2)$ **87.** $(5a - 6)^2$ **89.** $(x + 3)^2$ **91.** $(m - 7)^2$
93. $(x - 4)^2$ **95. a.** $(x + 1)(x^2 - x + 1)(x - 1)(x^2 + x + 1)$ **b.** $(x + 1)(x - 1)(x^4 + x^2 + 1)$ **c.** Answers may vary.   **97.** $\{-7\}$
99. $\{3\}$ **101.** $\{0\}$ **103.** $\{-4\}$ **105.** $(x^n - 6)(x^n + 6)$ **107.** $(5x^n - 9)(5x^n + 9)$ **109.** $(x^n - 5)(x^n + 5)(x^{2n} + 25)$

Section 6.8 Mental Math

1. $\{3, -5\}$ **3.** $\{3, -7\}$ **5.** $\{0, 9\}$

Exercise Set 6.8

1. $\left\{-3, \frac{4}{3}\right\}$ **3.** $\left\{\frac{5}{2}, -\frac{3}{4}\right\}$ **5.** $\{-3, -8\}$ **7.** $\left\{\frac{1}{4}, -\frac{2}{3}\right\}$ **9.** $\{1, 9\}$ **11.** $\left\{\frac{3}{5}, -1\right\}$ **13.** $\{0\}$ **15.** $\{6, -3\}$ **17.** $\left\{\frac{2}{5}, -\frac{1}{2}\right\}$ **19.** $\left\{\frac{3}{4}, -\frac{1}{2}\right\}$
21. $\left\{-2, 7, \frac{8}{3}\right\}$ **23.** $\{0, -3, 3\}$ **25.** $\{-1, 1, 2\}$ **27.** Answers may vary.   **29.** $\left\{-\frac{7}{2}, 10\right\}$ **31.** $\{0, 5\}$ **33.** $\{5, -3\}$ **35.** $\left\{\frac{1}{3}, -\frac{1}{2}\right\}$
37. $\{9, -4\}$ **39.** $\left\{\frac{4}{5}\right\}$ **41.** $\{0, -5, 2\}$ **43.** $\left\{0, \frac{4}{5}, -3\right\}$ **45.** $\{\ \}$ **47.** $\{-7, 4\}$ **49.** $\{4, 6\}$ **51.** $\left\{-\frac{1}{2}\right\}$ **53.** $\{-3, 3, -4\}$
55. $\{0, -5, 5\}$ **57.** $\{-6, 5\}$ **59.** $\left\{0, -\frac{1}{3}, 1\right\}$ **61.** $\left\{0, -\frac{1}{3}\right\}$ **63.** $\left\{-\frac{7}{8}\right\}$ **65.** $\left\{\frac{31}{4}\right\}$ **67.** $\{1\}$ **69.** Answers may vary.
71. -11 and -6 or 6 and 11 **73.** 75 ft   **75.** 105 units   **77.** 12 cm and 9 cm   **79.** 3 ft   **81.** Ascending 2 sec, descending 3 sec
83. D   **85.** A   **87.** C

Chapter 6 Review

1. 4   **3.** -4 **5.** 1   **7.** $-\frac{1}{16}$ **9.** $-x^2y^7z$ **11.** $\frac{1}{a^9}$ **13.** $\frac{1}{x^{11}}$ **15.** $\frac{1}{y^5}$ **17.** -3.62×10^{-4} **19.** 410,000   **21.** $\frac{a^2}{16}$ **23.** $\frac{1}{16x^2}$
25. $\frac{1}{8^{18}}$ **27.** $-\frac{1}{8x^9}$ **29.** $-\frac{27y^6}{x^6}$ **31.** $\frac{xz}{4}$ **33.** $\frac{2}{27z^3}$ **35.** $2y^{x-7}$ **37.** -2.21×10^{-11} **39.** $\frac{x^3y^{10}}{3z^{12}}$ **41.** 5   **43.** $12x - 6x^2 - 6x^2y$
45. $4x^2 + 8y + 6$ **47.** $8x^2 + 2b - 22$ **49.** $12x^2y - 7xy + 3$ **51.** $x^3 + x - 2xy^2 - y - 7$ **53.** 58   **55.** $x^2 + 4x - 6$
57. $(6x^2y - 12x + 12)$ cm   **59.** $-12a^2b^5 - 28a^2b^3 - 4ab^2$ **61.** $9x^2a^2 - 24xab + 16b^2$ **63.** $15x^2 + 18xy - 81y^2$
65. $x^4 + 18x^3 + 83x^2 + 18x + 1$ **67.** $16x^2 + 72x + 81$ **69.** $-9a^2 + 6ab - b^2 + 16$ **71.** $(9y^2 - 49z^2)$ square units
73. $16x^2y^2z - 8xy^zb + b^2$ **75.** $8x^2(2x - 3)$ **77.** $2ab(3b + 4 - 2ab)$ **79.** $(6a - 5)(a + 3b)$ **81.** $(x - 6)(y + 3)$ **83.** $(p - 5)(q - 3)$
85. $x(2y - x)$ **87.** $(x - 4)(x + 20)$ **89.** $3(x + 2)(x + 9)$ **91.** $(3x + 8)(x - 2)$ **93.** $(15x - 1)(x - 6)$ **95.** $3(x - 2)(3x + 2)$
97. $(x + 7)(x + 9)$ **99.** $(x^2 - 2)(x^2 + 10)$ **101.** $(x - 9)(x + 9)$ **103.** $6(x - 3)(x + 3)$ **105.** $(4 + y^2)(2 - y)(2 + y)$
107. $(x - 7)(x + 1)$ **109.** $(y + 8)(y^2 - 8y + 64)$ **111.** $(1 - 4y)(1 + 4y + 16y^2)$ **113.** $2x^2(x + 2y)(x^2 - 2xy + 4y^2)$
115. $(x - 3 - 2y)(x - 3 + 2y)$ **117.** $(4a - 5b)^2$ **119.** $\left\{\frac{1}{3}, -7\right\}$ **121.** $\left\{0, 4, \frac{9}{2}\right\}$ **123.** $\{0, 6\}$ **125.** $\left\{-\frac{1}{3}, 2\right\}$ **127.** $\{-4, 1\}$
129. $\{0, 6, -3\}$ **131.** $\{0, -2, 1\}$ **133.** $-\frac{15}{2}$ or 7   **135.** 5 sec

Chapter 6 Test

1. $\dfrac{1}{81x^2}$ **2.** $-12x^2z$ **3.** $\dfrac{3a^7}{2b^5}$ **4.** $-\dfrac{y^{40}}{z^5}$ **5.** 6.3×10^8 **6.** 1.2×10^{-2} **7.** 0.000005 **8.** 9.0×10^{-4} **9.** $-5x^3 - 11x - 9$

10. $-12x^2y - 3xy^2$ **11.** $12x^2 - 5x - 28$ **12.** $25a^2 - 4b^2$ **13.** $36m^2 + 12mn + n^2$ **14.** $2x^3 - 13x^2 + 14x - 4$ **15.** $4x^2y(4x - 3y^3)$

16. $(x - 15)(x + 2)$ **17.** $(2y + 5)^2$ **18.** $3(2x + 1)(x - 3)$ **19.** $(2x + 5)(2x - 5)$ **20.** $(x + 4)(x^2 - 4x + 16)$

21. $3y(x + 3y)(x - 3y)$ **22.** $6(x^2 + 4)$ **23.** $(x + 3)(x - 3)(y - 3)$ **24.** $\left\{4, -\dfrac{8}{7}\right\}$ **25.** $\{-3, 8\}$ **26.** $\left\{-\dfrac{5}{2}, -2, 2\right\}$

27. $(x - 2y)(x + 2y)$ **28. a.** 960 ft   **b.** 953.44 ft   **c.** 11 sec

Chapter 6 Cumulative Review

1. *(Sec. 1.1, Ex. 4)* **a.** 8x   **b.** $8x + 3$ **c.** $\dfrac{x}{-7}$ **d.** $2x - (-1.6)$ **2.** *(Sec. 1.3, Ex. 1)* **a.** 3   **b.** 5   **c.** -2 **d.** -8 **e.** 0

3. *(Sec. 2.1, Ex. 4)* {1}   **4.** *(Sec. 3.5, Ex. 3)* $\dfrac{2}{3}$ **5.** *(Sec. 3.5, Ex. 5)* **a.** Parallel   **b.** Neither   **c.** Perpendicular

6. *(Sec. 4.1, Ex. 4)* $\left\{x \mid x > \dfrac{5}{2}\right\}$ **7.** *(Sec. 4.5, Ex. 3)*

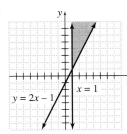

$y = 2x - 1$ $x = 1$

8. *(Sec. 4.5, Ex. 4)*

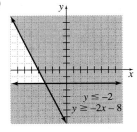

$y \le -2$ $y \ge -2x - 8$

9. *(Sec. 5.1, Ex. 5)* $\{(-2, 2)\}$ **10.** *(Sec. 5.4, Ex. 3)* {}   **11.** *(Sec. 5.5, Ex. 1)* **a.** -2 **b.** -10

12. *(Sec. 6.1, Ex. 1)* **a.** 2^7 **b.** x^{10} **c.** y^7 **13.** *(Sec. 6.2, Ex. 1)* **a.** x^{35} **b.** 2^6, or 64   **c.** $\dfrac{1}{5^2}$ or $\dfrac{1}{25}$ **d.** y^{12}

14. *(Sec. 6.3, Ex. 2)* **a.** 3, trinomial   **b.** 3, monomial   **c.** 2, binomial   **d.** 4, none of these

15. *(Sec. 6.4, Ex. 2)* **a.** $10x^2 - 8x$ **b.** $-12x^4 + 18x^3 - 3x^2$ **c.** $-7x^3y^2 - 3x^2y^2 + 11xy$ **16.** *(Sec. 6.5, Ex. 5)* $(x - 5)(2 + 3a)$

17. *(Sec. 6.6, Ex. 7)* $2xy(2x - 1)(3x - 4)$ **18.** *(Sec. 6.7, Ex. 5)* $(x + 2)(x^2 - 2x + 4)$ **19.** *(Sec. 6.8, Ex. 5)* $\left\{-\dfrac{1}{6}, 3\right\}$

20. *(Sec. 6.8, Ex. 8)* 9 sec

CHAPTER 7

Rational Expressions

Exercise Set 7.1

1. $\dfrac{10}{3}, -8, -\dfrac{7}{3}$ **3.** $-\dfrac{17}{48}, \dfrac{2}{7}, -\dfrac{3}{8}$ **5.** $\{x \mid x \text{ is a real number}\}$ **7.** $\{t \mid t \text{ is a real number and } t \ne 0\}$ **9.** $\{x \mid x \text{ is a real number and } x \ne 7\}$

11. $\{x \mid x \text{ is a real number and } x \ne 0, x \ne -3\}$ **13.** $\{x \mid x \text{ is a real number and } x \ne 0, x \ne -2, x \ne 1\}$ **15.** $\{x \mid x \text{ is a real number and } x \ne 2,$

$x \ne -2\}$ **17.** Answers may vary.   **19.** $\dfrac{5x^2}{9}$ **21.** $\dfrac{x^4}{2y^2}$ **23.** $\dfrac{1}{2(q - 1)}$ **25.** 1   **27.** $\dfrac{-1}{1 + x}$ **29.** $\dfrac{-1}{x - 7}$ **31.** $\dfrac{4}{3}$ **33.** -2

35. $\dfrac{x + 1}{x - 3}$ **37.** $\dfrac{2(x + 3)}{x - 3}$ **39.** $\dfrac{3}{x}$ **41.** $\dfrac{1}{x - 2}$ **43.** $\dfrac{x + 1}{x^2 + 1}$ **45.** $x^2 - 4$ **47.** $x - 4$ **49.** $\dfrac{2x + 1}{x - 1}$ **51.** $-x^2 - 5x - 25$

53. $\dfrac{4x^2 + 6x + 9}{2}$ **55.** $\dfrac{x + 5}{x^2 + 5}$ **57.** $\dfrac{10y^2z}{4y^3z}$ **59.** $\dfrac{3x^2 + 15x}{2x^2 + 9x - 5}$ **61.** $\dfrac{x^2 - 4}{x + 2}$ **63.** $\dfrac{30m^2}{6m^3}$ **65.** $\dfrac{35}{5m - 10}$ **67.** $\dfrac{y^2 + 8y + 16}{y^2 - 16}$

69. $\dfrac{12x^2 + 24x}{x^2 + 4x + 4}$ **71.** $\dfrac{x^2 - 2x + 4}{x^3 + 8}$ **73.** $\dfrac{ab - 3a}{ab - 3a + 2b - 6}$

75. $\{x|x$ is a real number and $x \neq 2, x \neq -2\}$ **77.** $\{x|x$ is a real number and $x \neq \frac{1}{2}, x \neq -4\}$ **79.**

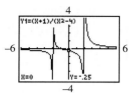

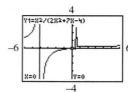

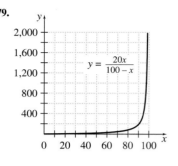

$$y = \frac{20x}{100 - x}$$

81. a. $\{x|0 \leq x < 100\}$ **b.** \$42,857.14   **c.** \$150,000; \$400,000
d. \$900,000; \$1,900,000; \$9,900,000   **83. a.** 4 cans   **b.** 9 cans
c. No; $\frac{3 + x}{4 + x}$ can never equal 1.   **85.** $\frac{1}{18}$ **87.** 3   **89.** $\frac{2}{7}$ **91.** 65 and over

Exercise Set 7.2

1. $\frac{x}{2}$ **3.** $\frac{c}{4}$ **5.** $\frac{x}{3}$ **7.** $-\frac{4}{5}$ **9.** $\frac{-6a}{2a + 1}$ **11.** $\frac{1}{6}$ **13.** $-\frac{8}{3}$ **15.** $\frac{5}{x - 2}$ sq. m   **17.** $\frac{x}{2}$ **19.** $\frac{4}{c}$ **21.** $\frac{x}{3}$ **23.** $\frac{5}{3}$

25. $\frac{4}{(x + 2)(x + 3)}$ **27.** $\frac{1}{2}$ **29.** $\frac{49y^7}{2x^2}$ **31.** $(f + g)(x) = x^2 + 5x + 1$ **33.** $(f - g)(x) = x^2 - 5x + 1$ **35.** $\left(\frac{g}{f}\right)(x) = \frac{5x}{x^2 + 1}$

37. $\frac{-2}{3x^3y^2}$ **39.** $\frac{4}{ab^6}$ **41.** $\frac{1}{4a(a - b)}$ **43.** $\frac{x^2 + 5x + 6}{4}$ **45.** $\frac{3}{2(x - 1)}$ **47.** $\frac{4a^2}{a - b}$ **49.** $\frac{2x^2 - 18}{5(x^2 - 8x - 15)}$ **51.** $\frac{x + 2}{x + 3}$

53. $\frac{3b}{a - b}$ **55.** $\frac{3a}{a - b}$ **57.** $\frac{1}{4}$ **59.** -1 **61.** $\frac{8}{3}$ **63.** $\frac{8(a - 2)}{3(a + 2)}$ **65.** $\frac{8}{x^2 y}$ **67.** $\frac{(y + 5)(2x - 1)}{(y + 2)(5x + 1)}$ **69.** $\frac{15a + 10}{a}$

71. $\frac{5x^2 - 2}{(x - 1)^2}$ **73.** $(f + g)(x) = x^2 - 2x + 2$ **75.** $\left(\frac{f}{g}\right)(x) = -\frac{2x}{x^2 + 2}$ **77.** $(g - h)(x) = x^2 - 4x - 1$ **79.** $(h + f)(x) = 2x + 3$

81. $f(a + b) = -2a - 2b$ **83.** $\left(\frac{f}{h}\right)(x) = \frac{-2x}{4x + 3}$ **85.** Answers may vary.   **87.** $\frac{(x + 2)(x - 1)^2}{x^5}$ ft.   **89.** $P(x) = R(x) - C(x)$ **91.** $\frac{7}{5}$

93. $\frac{1}{12}$ **95.** $\frac{11}{16}$ **97.** $2x^2(x^n + 2)$ **99.** $\frac{1}{10y(y^n + 3)}$ **101.** $\frac{y^n + 1}{2(y^n - 1)}$

Exercise Set 7.3

1. $-\frac{3}{x}$ **3.** $\frac{2 + x}{x - 2}$ **5.** $x - 2$ **7.** $\frac{1}{2 - x}$ **9.** $\frac{x^2}{x^2 + 10x + 25}$ sq ft   **11.** $35x$ **13.** $x(x + 1)$ **15.** $(x + 7)(x - 7)$

17. $6(x + 2)(x - 2)$ **19.** $(a + b)(a - b)^2$ **21.** Answers may vary.   **23.** $\frac{17}{6x}$ **25.** $\frac{35 - 4y}{14y^2}$ **27.** $\frac{-13x + 4}{(x + 4)(x - 4)}$

29. $\frac{2x + 4}{(x - 5)(x + 4)}$ **31.** 0 **33.** $\frac{x^2}{x - 1}$ **35.** $\frac{1 - x}{x - 2}$ **37.** $\frac{y^2 + 2y + 10}{(y - 2)(y - 4)(y + 4)}$ **39.** $\frac{5(x^2 + x - 4)}{(3x + 2)(x + 3)(2x - 5)}$

41. $\frac{x^2 + 5x + 21}{(x - 2)(x + 1)(x + 3)}$ **43.** $\frac{5 - 2x}{2(x + 1)}$ **45.** $\frac{2(x^2 + x - 21)}{(x + 3)^2(x - 3)}$ **47.** $\frac{3}{x^2 y^3}$ **49.** $-\frac{5}{x}$ **51.** $\frac{25}{6(x + 5)}$ **53.** $\frac{-2x - 1}{x^2(x - 3)}$

55. $\frac{2ab - b^2}{(a + b)(a - b)}$ **57.** $\frac{2x + 16}{(x + 2)^2(x - 2)}$ **59.** $\frac{5a + 1}{(a + 1)^2(a - 1)}$ **61.** Answers may vary.   **63.** Answers may vary.

65. $\frac{2x^2 + 9x - 18}{6x^2}$ **67.** $\frac{4}{3}$ **69.** $\frac{4a^2}{9(a - 1)}$ **71.** 4 **73.** $\frac{6x}{(x + 3)(x - 3)^2}$ **75.** $\frac{-4}{x - 1}$ **77.** $\frac{-32}{x(x + 2)(x - 2)}$

79. a. Both rows: -1, -2, $-7, 8, 3$; yes   **b.** $\frac{2x^2 + 3x}{4x^2 - 9} + \frac{6}{2x - 3} = \frac{x(2x + 3)}{(2x - 3)(2x + 3)} + \frac{6}{2x - 3} = \frac{x + 6}{2x - 3}$

81.   **83.**   **85.**

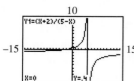

X	Y1	Y2
-5	3	3
-4	2	2
-3	1	1
-2	0	0
-1	-1	-1
0	-2	-2
1	-3	-3

Y2 = -(2+X)

87.

X	Y1	Y2
-5	-.5794	-.5794
-4	-.6952	-.6952
-3	-.869	-.869
-2	-1.159	-1.159
-1	-1.738	-1.738
0	-3.476	-3.476
1	ERROR	ERROR

Y2 = 73/(21(X-1))

89. $(f \cdot g)(x) = \frac{18}{x^3 + 2x^2 - 4x - 8}$ **91.** $(f - g)(x) = \frac{3x^2 - 6x + 12}{x(x - 2)(x + 2)}$ **93.** 5   **95.** $5 - y^2$ **97.** 5   **99.** 3   **101.** 2   **103.** 25 feet

105. $\frac{13}{4y}$ **107.** $\frac{3 - 16x}{48x^2}$ **109.** $\frac{9x + 7}{x^4}$

Exercise Set 7.4

1. $\frac{5}{6}$ **3.** $\frac{8}{5}$ **5.** 4   **7.** $\frac{7}{13}$ **9.** $\frac{4}{x}$ **11.** $\frac{9x-18}{9x^2-4}$ **13.** $\frac{1-x}{x+1}$ **15.** $\frac{xy^2}{y^2+x^2}$ **17.** $\frac{2b^2+3a}{b^2-ab}$ **19.** $\frac{x}{x^2-1}$ **21.** $\frac{x+1}{x+2}$ **23.** $\frac{10}{69}$

25. $\frac{2(x+1)}{2x-1}$ **27.** $\frac{x(x+1)}{6}$ **29.** $\frac{x}{2-3x}$ **31.** $\frac{-y}{x+y}$ **33.** $\frac{-2x^3}{(x-y)y}$ **35.** $\frac{2x+1}{y}$ **37.** $\frac{x-3}{9}$ **39.** $\frac{1}{x+2}$ **41.** $\frac{x}{5x-10}$

43. $\frac{x-2}{2x-1}$ **45.** $-\frac{x^2+4}{4x}$ **47.** $\frac{x-3y}{x+3y}$ **49.** $\frac{1+a}{1-a}$ **51.** $\frac{6xy+x^2}{2y}$ **53.** $\frac{5a}{2a+4}$ **55.** $5xy^2+2x^2y$ **57.** $\frac{xy}{5y+2x}$ **59.** $\frac{xy}{y+x}$

61. x^2+x **63. a.** $\frac{1}{a+h}$ **b.** $\frac{1}{a}$ **c.** $\frac{\frac{1}{a+h}-\frac{1}{a}}{h}$ **d.** $\frac{-1}{a(a+h)}$ **65. a.** $\frac{3}{a+h+1}$ **b.** $\frac{3}{a+1}$ **c.** $\frac{\frac{3}{a+h+1}-\frac{3}{a+1}}{h}$

d. $\frac{-3}{(a+h+1)(a+1)}$ **67.** $\frac{x^2y^2}{4}$ **69.** $-9x^3y^4$ **71.** $\{-4,14\}$ **73.** $(-4,14)$ **75.** $\frac{x-1}{x}$ **77.** $2x$ **79.** $3a^2+4a+4$

Exercise Set 7.5

1. $2a+4$ **3.** $3ab+4$ **5.** $2x^2+3x-2$ **7.** x^2+2x+1 **9.** (x^4+2x-6) meters   **11.** $(x+1)+\frac{1}{x+2}$ **13.** $2x-8$

15. $x-\frac{1}{2}$ **17.** $2x^2-\frac{1}{2}x+5$ **19.** $3x-7$ in.   **21.** No remainder   **23.** $\frac{5b^5}{2a^3}$ **25.** x^3y^3-1 **27.** $a+3$ **29.** $x-2+\frac{2}{x-5}$

31. $5a-3b$ **33.** $x+5-\frac{10}{2x+3}$ **35.** $x+2xy+1$ **37.** $x-5$ **39.** $2x+3$ **41.** $(2x^2-8x+38)-\frac{156}{x+4}$ **43.** $(3x+3)-\frac{1}{x-1}$

45. $-2x^3+3x^2-x+4$ **47.** $(3x^3+5x+4)-\frac{2x}{x^2-2}$ **49.** $x-\frac{5}{3x^2}$ **51.** $P(1)=4$; Remainder $=4$

53. $P(-3)=372$; Remainder $=372$ **55.** Answers may vary.   **57.** $(3x^2+10x+8)+\frac{4}{x-2}$ **59.** $=$ **61.** $=$ **63.** $[-7,9]$

65. $(-\infty,-3)\cup(2,\infty)$ **67.** $2x^2+\frac{1}{2}x-5$ **69.** $\left(2x^3+\frac{9}{2}x^2+10x+21\right)+\frac{42}{x-2}$ **71.** $3x^4-2x$

Exercise Set 7.6

1. $x+8$ **3.** $x-1$ **5.** $x^2-5x-23-\frac{41}{x-2}$ **7.** $4x+8+\frac{7}{x-2}$ **9.** 3   **11.** 73   **13.** -8 **15.** $x^2+\frac{2}{x-3}$

17. $6x+7+\frac{1}{x+1}$ **19.** $2x^3-3x^2+x-4$ **21.** $3x-9+\frac{12}{x+3}$ **23.** $3x^2-\frac{9}{2}x+\frac{7}{4}+\frac{47}{8x-4}$ **25.** $3x^2+3x-3$

27. $3x^2+4x-8+\frac{20}{x+1}$ **29.** x^2+x+1 **31.** $x-6$ **33.** 1   **35.** -133 **37.** 3   **39.** $-\frac{187}{81}$ **41.** $\frac{95}{32}$ **43.** Answers may vary.

45. $(x+3)(x^2+4)=x^3+3x^2+4x+12$ **47.** 0   **49.** $x^3+2x^2+7x+28$ **51.** $(x-1)$ meters   **53.** $\left\{\frac{13}{3}\right\}$ **55.** $\{-3,1\}$ **57.** $\{-1\}$

59. $(2y+1)(4y^2-2y+1)$ **61.** $(a-3)(a^2+3a+9)$ **63.** $(x+y)(x-1)$ **65.** $2x(x+4)(x-4)$

Exercise Set 7.7

1. $\{72\}$ **3.** $\{2\}$ **5.** $\{6\}$ **7.** $\{2\}$ **9.** $\{3\}$ **11.** $\{\}$ **13.** $\{15\}$ **15.** $\{4\}$ **17.** $\{\}$ **19.** $\{1\}$ **21.** $\{1\}$ **23.** $\{-3\}$ **25.** $\left\{\frac{5}{3}\right\}$

27. $\{2,10\}$ **29.** $\{2\}$ **31.** $\{3\}$ **33.** $\{\}$ **35.** $\{\}$ **37.** $\{-1\}$ **39.** $\{9\}$ **41.** $\{1,7\}$ **43.** $\left\{\frac{1}{10}\right\}$ **45.** 800 sharpeners

47. $\left\{-\frac{2}{3}\right\}$ **49.** $\frac{5}{2x}$ **51.** $-\frac{y}{x}$ **53.** $\frac{-a^2+31a+10}{5(a-6)(a+1)}$ **55.** $\left\{-\frac{3}{13}\right\}$ **57.** $\frac{-a-8}{4a(a-2)}$ **59.** $\frac{x^2-3x+10}{2(x+3)(x-3)}$

61. $\{x\mid x \text{ is a real number and } x\neq-1, x\neq2\}$ **63.** $\frac{22z-45}{3z(z-3)}$ **65.** $\left\{-\frac{1}{4},\frac{1}{9}\right\}$ **67.** $\{2,3\}$ **69.** $\{1.39\}$ **71.** $\{-0.08\}$

73. $\frac{2}{x}=\frac{10}{5}$; Verify $x=1$ **75.** $\frac{6x+7}{2x+9}=\frac{5}{3}$; Verify $x=3$ **77.** 73 and 74   **79.** 2 and $\frac{1}{2}$ **81.** -9 **83.** -9 **85.** 1%   **87.** 67%

89. $\{-1,0\}$ **91.** $\{-2\}$

Exericse Set 7.8

1. $C=\frac{5}{9}(F-32)$ **3.** $R=\frac{R_1R_2}{R_1+R_2}$ **5.** $n=\frac{2S}{a+L}$ **7.** $b=\frac{2A-ah}{h}$ **9.** $T_2=\frac{P_2V_2T_1}{P_1V_1}$ **11.** $f_2=\frac{f_1f}{f_1-f}$ **13.** $L=\frac{n\lambda}{2}$

15. $c=\frac{2L\omega}{\theta}$ **17.** 1 or 5   **19.** 5   **21.** 6 ohms   **23.** $\frac{15}{13}$ ohms   **25.** 15.6 hrs   **27.** 10 min.   **29.** 200 mph   **31.** 15 mph

33. -8 and -7 **35.** 36 minutes   **37.** 45 mph and 60 mph   **39.** 10 mph, 8 mph   **41.** $t=2$ hrs   **43.** 135 mph   **45.** 12 miles

47. $2\frac{2}{9}$ hours   **49.** Jet = 3 hrs; Car = 4 hrs   **51.** $1\frac{1}{2}$ minutes   **53.** 63 mph   **55.** $\{-5\}$ **57.** $\{2\}$ **59.** 25%   **61.** 84%
63. Answers may vary.

Exercise Set 7.9

1. $A = kB$ **3.** $X = \dfrac{k}{Z}$ **5.** $N = kP^2$ **7.** $T = \dfrac{k}{R}$ **9.** $P = kR$ **11.** $A = 45$ **13.** $V = 24 \text{ m}^3$ **15.** $H = 10$ **17.** $I = 72$ amps

19. $x = kyz$ **21.** $r = kst^3$ **23.** $Q = 7$ **25.** $M = \dfrac{48}{5}$ **27.** $B = 2$ **29.** $W = 2.7$ **31.** $W = 4.05$ lbs   **33.** \$0.80   **35.** 108 cars

37. 16 lbs   **39.** 7.91 tons   **41.** multiplied by 9   **43.** divided by 4   **45.** multiplied by 2

47. $4, 2, 1, \dfrac{1}{2}, \dfrac{1}{4}$ **49.** $20, 10, 5, \dfrac{5}{2}, \dfrac{5}{4}$ **51.** $C = 8\pi$ in.; $A = 16\pi$ in.^{2} **53.** $C = 18\pi$ cm; $A = 81\pi$ cm^2

55. $\dfrac{9}{5}$ **57.** undefined

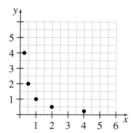

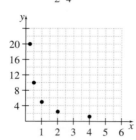

59.

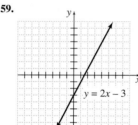

$y = 2x - 3$

61.

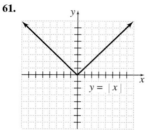

$y = |x|$

63.

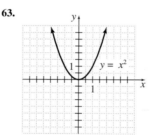

$y = x^2$

Chapter 7 Review

1. $\{x \mid x$ is a real number$\}$ **3.** $\{x \mid x$ is a real number and $x \neq 5\}$ **5.** $\{x \mid x$ is a real number and $x \neq 0, x \neq -8\}$ **7.** $\dfrac{x^2}{3}$ **9.** $\dfrac{9m^2 p}{5}$ **11.** $\dfrac{1}{5}$

13. $\dfrac{1}{x - 1}$ **15.** $\dfrac{2(x - 3)}{x - 4}$ **17. a.** \$119   **b.** \$77   **c.** Decrease   **19.** $\dfrac{2x^3}{z^3}$ **21.** $\dfrac{2}{5}$ **23.** $\dfrac{1}{6}$ **25.** $\dfrac{3x}{16}$ **27.** $\dfrac{3c^2}{14a^2 b}$

29. $\dfrac{(x + 5)(x + 4)}{3}$ **31.** $\dfrac{7(x - 4)}{2(x - 2)}$ **33.** $-\dfrac{1}{x}$ **35.** $\dfrac{8}{9a^2}$ **37.** $\dfrac{6}{a}$ **39.** $-x - 6$ **41.** $\dfrac{2x + 1}{x - 5}$ **43.** $60x^2 y^5$ **45.** $5x(x - 5)$ **47.** $\dfrac{2}{5}$

49. $\dfrac{2}{x^2}$ **51.** $\dfrac{1}{x - 2}$ **53.** $\dfrac{5x^2 - 3y^2}{15x^4 y^3}$ **55.** $\dfrac{-x + 5}{x^2 - 1}$ **57.** $\dfrac{2x^2 - 5x - 4}{x - 3}$ **59.** $\dfrac{3x^2 - 7x - 4}{27x^3 - 64}$ **61.** $\dfrac{-12}{x(x + 1)(x - 3)}$

63. $\dfrac{14x - 40}{(x + 4)^2(x - 4)}$ **65.** $\dfrac{2}{3}$ **67.** $\dfrac{2}{15 - 2x}$ **69.** $\dfrac{y}{2}$ **71.** $\dfrac{20x - 15}{2(5x^2 - 2)}$ **73.** $\dfrac{5xy + x}{3y}$ **75.** $\dfrac{1 + x}{1 - x}$ **77.** $\dfrac{x - 1}{3x - 1}$ **79.** $-\dfrac{x^2 + 9}{6x}$

81. a. $\dfrac{3}{a + h}$ **b.** $\dfrac{3}{a}$ **c.** $\dfrac{\frac{3}{a + h} - \frac{3}{a}}{h}$ **d.** $\dfrac{-3}{a(a + h)}$ **83.** $\dfrac{9b^2 z^3}{4a}$ **85.** $\dfrac{3}{b} + 4b$ **87.** $2x^3 + 6x^2 + 17x + 56 + \dfrac{156}{x - 3}$

89. $x^2 + \dfrac{7}{2}x - \dfrac{1}{4} + \dfrac{15}{8\left(x - \frac{1}{2}\right)}$ **91.** $3x^2 + 6$ **93.** $3x^2 - \dfrac{5}{2}x - \dfrac{1}{4} - \dfrac{5}{8\left(x + \frac{3}{2}\right)}$ **95.** $x^2 + 3x + 9 - \dfrac{54}{x - 3}$

97. $3x^3 - 6x^2 + 10x - 20 + \dfrac{50}{x + 2}$ **99.** -9323 **101.** $\dfrac{365}{32}$ **103.** $\{6\}$ **105.** $\{2\}$ **107.** $\left\{\dfrac{3}{2}\right\}$ **109.** $\left\{\dfrac{5}{3}\right\}$ **111.** $\left\{-\dfrac{1}{3}, 2\right\}$

113. $a = \dfrac{2A}{h} - b$ **115.** $R = \dfrac{E}{I} - r$ **117.** $A = \dfrac{LH}{k(T_1 - T_2)}$ **119.** 7   **121.** -10 and -8 **123.** 12 hours   **125.** 490 mph

127. 8 mph   **129.** 4 mph   **131.** $\{-10\}$ **133.** 9   **135.** 3.125 ft^3

Chapter 7 Test

1. $\{x|x \text{ is a real number and } x \neq 1\}$ **2.** $\{x|x \text{ is a real number and } x \neq -3, x \neq -1\}$ **3.** $\dfrac{5x^3}{3}$ **4.** $-\dfrac{7}{8}$ **5.** $\dfrac{x}{x+9}$ **6.** $\dfrac{x+2}{5}$ **7.** $\dfrac{5}{3x}$

8. $\dfrac{4a^3b^4}{c^6}$ **9.** $\dfrac{x+2}{2(x+3)}$ **10.** $\dfrac{-4(2x+9)}{5}$ **11.** $\dfrac{3}{x^3}$ **12.** -1 **13.** $\dfrac{5x-2}{(x-3)(x+2)(x-2)}$ **14.** $\dfrac{30-x}{6(x-7)}$ **15.** $\dfrac{3}{2}$ **16.** $\dfrac{1}{5}$ **17.** $\dfrac{64}{3}$

18. $\dfrac{(x-3)^2}{x-2}$ **19.** $\dfrac{4xy}{3z} + \dfrac{3}{z} + \dfrac{1}{3x}$ **20.** $(x^5 + 5x^4 + 8x^3 + 16x^2 + 33x + 63) + \dfrac{128}{x-2}$ **21.** $4x^3 - 15x^2 + 47x - 142 + \dfrac{425}{x+3}$ **22.** 91

23. $x^2 - 7x + 12$ **24.** $x^3 - 6x^2 + 5x$ **25.** $\{7\}$ **26.** $-\dfrac{29}{21}; \{2, -2\}$ **27.** $\{8\}$ **28.** $\dfrac{7a^2 + b^2}{4a - b}$ **29.** 5 **30.** $\dfrac{6}{7}$ hrs   **31.** 16 **32.** 9

33. 256 feet

Chapter 7 Cumulative Review

1. *(Sec. 1.2, Ex. 1)* **a.** $x + 5 = 20$ **b.** $2(3 + y) = 4$ **c.** $x - 8 = 2x$ **d.** $\dfrac{z}{9} = 3(z - 5)$

2. *(Sec. 1.2, Ex. 12)* **a.** Multiplicative inverse   **b.** Double negative property   **c.** Distributive property

3. *(Sec. 2.2, Ex. 4)* 62 cm, 62 cm, 25 cm   **4.** *(Sec. 3.1, Ex. 5)* $\left(-1, \dfrac{3}{2}\right)$

5. *(Sec. 3.2, Ex. 4)*

6. *(Sec. 3.6, Ex. 7)* $2x + 3y = 20$ **7.** *(Sec. 4.3, Ex. 3)* $\{-20, 24\}$

8. *(Sec. 4.4, Ex. 3)* $\left\{x \middle| -2 \leq x \leq \dfrac{8}{5}\right\}$ **9.** *(Sec. 5.2, Ex. 2)* $\{\ \}$ **10.** *(Sec. 6.1, Ex. 2)* **a.** $15x^7$ **b.** $-8x^4p^{12}$

11. *(Sec. 6.2, Ex. 4)* **a.** $\dfrac{z^2}{9x^4y^{20}}$ **b.** $\dfrac{27a^4x^6}{2}$ **12.** *(Sec. 6.4, Ex. 6)* $6x^2 - 29x + 28$ **13.** *(Sec. 6.6, Ex. 4)* $2(n^2 - 19n + 40)$

14. *(Sec. 6.5, Ex. 8)* $(n^2 + 1)(m^2 - 2)$ **15.** *(Sec. 6.8, Ex. 6)* $\{-2, 0, 2\}$ **16.** *(Sec. 7.1, Ex. 1)* **a.** $\$102.60$ **b.** $\$12.60$

17. *(Sec. 7.2, Ex. 2)* **a.** $-\dfrac{4x^2}{2x+1}$ **b.** $-(a+b)^2$ **18.** *(Sec. 7.4, Ex. 3)* $\dfrac{xy + 2x^3}{y - 1}$ **19.** *(Sec. 7.5, Ex. 2)* $\dfrac{7a}{2b} - 1$

20. *(Sec. 7.7, Ex. 3)* $\{\ \}$ **21.** *(Sec. 7.8, Ex. 3)* 8 ft   **22.** *(Sec. 7.9, Ex. 1)* $\dfrac{1}{6}; 15$

CHAPTER 8

Rational Exponents, Radicals, and Complex Numbers

Exercise Set 8.1

1. 10 **3.** $\dfrac{1}{2}$ **5.** 0.01 **7.** -6 **9.** x^5 **11.** $4y^3$ **13.** $\dfrac{1}{2}$ **15.** -2 **17.** Not a real number   **19.** $-5x^3$ **21.** $2x$ **23.** $|z|$ **25.** x

27. $|x - 5|$ **29.** $2|z|$ **31.** $10|(2x - y)^3|$ **33.** Answers may vary.   **35.** -11 **37.** $2x$ **39.** y^6 **41.** $5ab^{10}$ **43.** $-3x^3$ **45.** a^4b

47. $-2x^2y$ **49.** $\dfrac{5}{7}$ **51.** $\dfrac{x}{2y}$ **53.** $-\dfrac{z^7}{3x}$ **55.** $\dfrac{x}{2}$ **57.** Rational, 3 **59.** Irrational, 6.083 **61.** Rational, 13 **63.** Rational, 2

65. x^5 **67.** x^6 **69.** $9|x|$ **71.** $-12|y^7|$ **73.** $|x + 2|$ **75.** $\sqrt{3}$ **77.** -1 **79.** -3 **81.** $\sqrt{7}$ **83.** $\{x|x \geq 0\}; C$ **85.** $\{x|x \geq 3\}; D$

87. All real numbers; A   **89.** All real numbers; B   **91.** $-32x^{15}y^{10}$ **93.** $-60x^7y^{10}z^5$ **95.** $\dfrac{1}{2}x^9y^5$

Exercise Set 8.2

1. $\sqrt{49} = 7$ **3.** $\sqrt[3]{27} = 3$ **5.** $\sqrt[4]{\dfrac{1}{16}} = \dfrac{1}{2}$ **7.** $\sqrt{169} = 13$ **9.** $2\sqrt[3]{m}$ **11.** $\sqrt{9x^4} = 3x^2$ **13.** $\sqrt[3]{-27} = -3$ **15.** $-\sqrt[4]{16} = -2$

17. $(\sqrt[4]{16})^3 = 8$ **19.** $(\sqrt[3]{-64})^2 = 16$ **21.** $\sqrt[4]{-16}$, not a real number   **23.** $\sqrt[5]{(2x)^3}$ or $(\sqrt[5]{2x})^3$ **25.** $\sqrt[3]{(7x+2)^2}$ or $(\sqrt[3]{7x+2})^2$

27. $\left(\sqrt{\dfrac{16}{9}}\right)^3 = \dfrac{64}{27}$ **29.** $\dfrac{1}{8^{4/3}} = \dfrac{1}{16}$ **31.** $\dfrac{1}{(-64)^{2/3}} = \dfrac{1}{16}$ **33.** $\dfrac{1}{(-4)^{3/2}}$, not a real number   **35.** $\dfrac{1}{x^{1/4}}$ **37.** $a^{2/3}$ **39.** $\dfrac{5x^{3/4}}{7}$

41. Answers may vary.   **43.** $3^{1/2}$ **45.** $y^{5/3}$ **47.** $4^{1/5}y^{7/5}$ **49.** $(y+1)^{3/2}$ **51.** $2x^{1/2} - 3y^{1/2}$ **53.** $\sqrt{x}$ **55.** $2\sqrt{x}$ **57.** $\sqrt{x+3}$

59. $\sqrt{xy}$ **61.** $a^{7/3}$ **63.** $\dfrac{8u^3}{v^9}$ **65.** $-b$ **67.** $y - y^{7/6}$ **69.** $2x^{5/3} - 2x^{2/3}$ **71.** $4x^{2/3} - 9$ **73.** $x^{8/3}(1 + x^{2/3})$ **75.** $x^{1/5}(x^{1/5} - 3)$

77. $x^{-1/3}(5 + x)$ **79.** $a^{1/3}$ **81.** $x^{1/5}$ **83.** 1.6818   **85.** 5.6645   **87.** $25 \cdot 3$ **89.** $16 \cdot 3$ **91.** $8 \cdot 2$ **93.** $27 \cdot 2$

Exercise Set 8.3

1. $\sqrt{14}$ **3.** $\sqrt[3]{36}$ **5.** $\sqrt{6x}$ **7.** $\sqrt{\dfrac{14}{xy}}$ **9.** $\sqrt[4]{20x^3}$ **11.** $\sqrt{2}$ **13.** $\sqrt[3]{8} = 2$ **15.** x^2y **17.** $24m^2$ **19.** $4\sqrt{2}$ **21.** $4\sqrt[3]{3}$

23. $25\sqrt{3}$ **25.** $\dfrac{\sqrt{6}}{7}$ **27.** $2\sqrt{5}$ **29.** $\dfrac{\sqrt[3]{4}}{3}$ **31.** $\dfrac{\sqrt{2}}{7}$ **33.** $10x^2\sqrt{x}$ **35.** $2y^2\sqrt[3]{2y}$ **37.** $a^2b\sqrt[4]{b^3}$ **39.** $y^2\sqrt{y}$ **41.** $5ab\sqrt{b}$

43. $-2x^2\sqrt[5]{y}$ **45.** $x^4\sqrt[3]{50x^2}$ **47.** $-4a^4b^3\sqrt{2b}$ **49.** $\dfrac{\sqrt{5x}}{2y}$ **51.** $\dfrac{-z^2\sqrt[3]{z}}{3x}$ **53.** $\dfrac{x\sqrt[4]{x^3}}{2}$ **55.** $3x^3y^4\sqrt{xy}$ **57.** $5r^3s^4$ **59.** $\dfrac{x\sqrt{y}}{10}$

61. $\dfrac{\sqrt[4]{8}}{x^2}$ **63.** $\sqrt[6]{72}$ **65.** $\sqrt[15]{343y^5}$ **67.** $\sqrt[6]{x^5}$ **69.** $\sqrt[6]{125r^3s^2}$ **71. a.** 20π sq. cm   **b.** 211.57 sq. feet   **73.** 2,022,426,050 square feet

75. a. 6.63 thousand tires   **b.** 7.07 thousand tires   **c.** Answers may vary.   **77.** $14x$ **79.** $2x^2 - 7x - 15$ **81.** y^2 **83.** $-3x - 15$

Section 8.4 Mental Math

1. $6\sqrt{3}$ **3.** $3\sqrt{x}$ **5.** $12\sqrt[3]{x}$

Exercise Set 8.4

1. $-2\sqrt{2}$ **3.** $10x\sqrt{2x}$ **5.** $17\sqrt{2} - 15\sqrt{5}$ **7.** $-\sqrt[3]{2x}$ **9.** $5b\sqrt{b}$ **11.** $\dfrac{31\sqrt{2}}{15}$ **13.** $\dfrac{\sqrt[3]{11}}{3}$ **15.** $\dfrac{5\sqrt{5x}}{9}$ **17.** $14 + \sqrt{3}$

19. $7 - 3y$ **21.** $6\sqrt{3} - 6\sqrt{2}$ **23.** $-23\sqrt[3]{5}$ **25.** $2b\sqrt{b}$ **27.** $20y\sqrt{2y}$ **29.** $2y\sqrt[3]{2x}$ **31.** $6\sqrt[4]{11} - 4\sqrt[4]{11}$ **33.** $4x\sqrt[4]{x^3}$

35. $\dfrac{2\sqrt{3}}{3}$ **37.** $\dfrac{5x\sqrt[3]{x}}{7}$ **39.** $\dfrac{5\sqrt{7}}{2x}$ **41.** $\dfrac{\sqrt[3]{2}}{6}$ **43.** $\dfrac{14x\sqrt[3]{2x}}{9}$ **45.** $15\sqrt{3}$ in.   **47.** $\sqrt{35} + \sqrt{21}$ **49.** $7 - 2\sqrt{10}$ **51.** $3\sqrt{x} - \sqrt{3x}$

53. $6x - 13\sqrt{x} - 5$ **55.** $\sqrt[3]{a^2} + \sqrt[3]{a} - 20$ **57.** $6\sqrt{2} - 12$ **59.** $2 + 2x\sqrt{3}$ **61.** $-16 - \sqrt{35}$ **63.** $x - y^2$ **65.** $3 + 2\sqrt{3x} + x^2$

67. $5x - 3\sqrt{15x} - 3\sqrt{10x} + 9\sqrt{6}$ **69.** $-\sqrt[3]{4} + 2\sqrt[3]{2}$ **71.** $\sqrt[3]{x^2} - 4\sqrt[6]{x^5} + 8\sqrt[3]{x} - 4\sqrt{x} + 7$ **73. a.** $22\sqrt{5}$ ft   **b.** 150 sq. ft.

75. $x - 7$ **77.** $\dfrac{7}{x + y}$ **79.** $2a - 3$ **81.** $\dfrac{-2 + \sqrt{3}}{3}$

Exercise Set 8.5

1. $\dfrac{\sqrt{3}}{3}$ **3.** $\dfrac{\sqrt{5}}{5}$ **5.** $\dfrac{4\sqrt[3]{9}}{3}$ **7.** $\dfrac{3\sqrt{2x}}{4x}$ **9.** $\dfrac{3\sqrt[3]{2x}}{2x}$ **11.** $-2(2 + \sqrt{7})$ **13.** $\dfrac{-7(\sqrt{x} + 3)}{x - 9}$ **15.** $-5 + 2\sqrt{6}$

17. $\dfrac{2a + \sqrt{ab} + 2\sqrt{a} + \sqrt{b}}{4a - b}$ **19.** $\dfrac{3\sqrt[3]{3a}}{a}$ **21.** $\dfrac{3\sqrt[3]{4}}{2}$ **23.** $\dfrac{2\sqrt{21}}{7}$ **25.** $-\dfrac{8(1 - \sqrt{10})}{9}$ **27.** $\dfrac{x - \sqrt{xy}}{x - y}$ **29.** $\dfrac{5}{\sqrt{15}}$ **31.** $\dfrac{6}{\sqrt{10}}$

33. $\dfrac{2x}{7\sqrt{x}}$ **35.** $\dfrac{5y}{\sqrt[3]{100xy}}$ **37.** $\dfrac{3}{10 + 5\sqrt{7}}$ **39.** $\dfrac{x - 9}{x - 3\sqrt{x}}$ **41.** $\dfrac{2}{\sqrt{10}}$ **43.** $\dfrac{2x}{11\sqrt{2x}}$ **45.** $\dfrac{7}{2\sqrt[3]{7^2}}$ **47.** $\dfrac{-7}{12 + 6\sqrt{11}}$ **49.** $\dfrac{3x^2}{10\sqrt[3]{9x}}$

51. $\dfrac{6x^2y^3}{\sqrt{6z}}$ **53.** $\dfrac{1}{3 + 2\sqrt{2}}$ **55.** $\dfrac{x - 1}{x - 2\sqrt{x} + 1}$ **57.** Answers may vary.   **59.** $\dfrac{\sqrt{A\pi}}{2\pi}$ **61.** Answers may vary.   **63.** $\{5\}$

65. $\left\{-\dfrac{1}{2}, 6\right\}$ **67.** $\{2, 6\}$

Exercise Set 8.6

1. $\{8\}$ **3.** $\{7\}$ **5.** $\{\ \}$ **7.** $\{7\}$ **9.** $\{6\}$ **11.** $\left\{-\dfrac{9}{2}\right\}$ **13.** $\{29\}$ **15.** $\{4\}$ **17.** $\{-4\}$ **19.** $\{\ \}$ **21.** $\{7\}$ **23.** $3\sqrt{5}$ ft   **25.** $2\sqrt{10}$ m

27. $\{9\}$ **29.** $\{50\}$ **31.** $\{\ \}$ **33.** $\left\{\dfrac{15}{4}\right\}$ **35.** $\{7\}$ **37.** $\{5\}$ **39.** $\{-12\}$ **41.** $\{9\}$ **43.** $\{-3\}$ **45.** $\{1\}$ **47.** $\{1\}$ **49.** $\left\{\dfrac{1}{2}\right\}$

51. Answers may vary.   **53.** $2\sqrt{131} \approx 22.9$ m   **55.** $\sqrt{100.84} \approx 10.0$ mm   **57.** $\{3.19\}$ **59.** $\{\ \}$ **61.** $\{3.23\}$ **63.** 17 feet   **65.** 13 feet

67. 1191 feet   **69.** 2743 deliveries   **71.** $50\sqrt{145} \approx 602.08$ lb   **73.** ≈ 0.51 km   **75.** Not a function   **77.** Not a function

79. Not a function   **81.** $-\dfrac{20 + 16y}{3y}$ **83.** $\dfrac{x + y}{x - y}$ **85.** $\{-1, 2\}$ **87.** $\{-8, -6, 0, 2\}$

Section 8.7 Mental Math

1. $9i$ **3.** $i\sqrt{7}$ **5.** -4 **7.** $8i$

Exercise Set 8.7

1. $2i\sqrt{6}$ **3.** $-6i$ **5.** $24i\sqrt{7}$ **7.** $-3\sqrt{6}$ **9.** $6 - 4i$ **11.** $-2 + 6i$ **13.** $-2 - 4i$ **15.** $18 + 12i$ **17.** 7 **19.** $12 - 16i$ **21.** $-4i$

23. $\dfrac{28}{25} - \dfrac{21}{25}i$ **25.** $4 + i$ **27.** Answers may vary.   **29.** $-\sqrt{14}$ **31.** $-5\sqrt{2}$ **33.** $4i$ **35.** $i\sqrt{3}$ **37.** $2\sqrt{2}$ **39.** 1 **41.** i

43. $-i$ **45.** -1 **47.** 63 **49.** $2 - i$ **51.** 20 **53.** $6 - 3i\sqrt{3}$ **55.** 2 **57.** $-5 + \dfrac{16}{3}i$ **59.** $17 + 144i$ **61.** $\dfrac{3}{5} - \dfrac{1}{5}i$ **63.** $5 - 10i$

65. $\dfrac{1}{5} - \dfrac{8}{5}i$ **67.** $8 - i$ **69.** $40°$ **71.** $x^2 - 5x - 2 - \dfrac{6}{x - 1}$ **73.** 5 **75.** 14 **77.** 16.7%

Chapter 8 Review

1. 9 **3.** -2 **5.** $-\dfrac{1}{7}$ **7.** -6 **9.** $-a^2b^3$ **11.** $2ab^2$ **13.** $\dfrac{x^6}{6y}$ **15.** $|x|$ **17.** -27 **19.** $-x$ **21.** $5|(x - y)^5|$ **23.** $-x$

25.

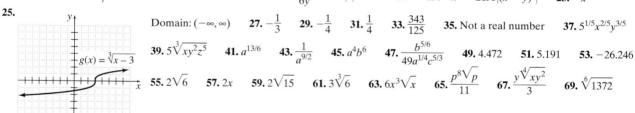

Domain: $(-\infty, \infty)$ **27.** $-\dfrac{1}{3}$ **29.** $-\dfrac{1}{4}$ **31.** $\dfrac{1}{4}$ **33.** $\dfrac{343}{125}$ **35.** Not a real number   **37.** $5^{1/5}x^{2/5}y^{3/5}$

39. $5\sqrt[3]{xy^2z^5}$ **41.** $a^{13/6}$ **43.** $\dfrac{1}{a^{9/2}}$ **45.** a^4b^6 **47.** $\dfrac{b^{5/6}}{49a^{1/4}c^{5/3}}$ **49.** 4.472 **51.** 5.191 **53.** -26.246

55. $2\sqrt{6}$ **57.** $2x$ **59.** $2\sqrt{15}$ **61.** $3\sqrt[3]{6}$ **63.** $6x^3\sqrt{x}$ **65.** $\dfrac{p^8\sqrt{p}}{11}$ **67.** $\dfrac{y\sqrt[4]{xy^2}}{3}$ **69.** $\sqrt[6]{1372}$

71. a. $\dfrac{5}{\sqrt{\pi}}$ or $\dfrac{5\sqrt{\pi}}{\pi}$ meters   **b.** 5.75 inches   **73.** $xy\sqrt{2y}$ **75.** $3a\sqrt[4]{2a}$ **77.** $\dfrac{3\sqrt{2}}{4x}$ **79.** $-4ab\sqrt[4]{2b}$ **81.** $x - 6\sqrt{x} + 9$

83. $4x - 9y$ **85.** $\sqrt[3]{a^2} + 4\sqrt[3]{a} + 4$ **87.** $a + 64$ **89.** $\dfrac{\sqrt{3x}}{6}$ **91.** $\dfrac{2x^2\sqrt{2x}}{y}$ **93.** $-\dfrac{10}{3} - \dfrac{5}{3}\sqrt{7}$ **95.** $-5 + 2\sqrt{6}$ **97.** $\dfrac{6}{\sqrt{2y}}$

99. $\dfrac{4x^3}{y\sqrt{2x}}$ **101.** $\dfrac{x - 25}{-3\sqrt{x} + 15}$ **103.** $\{\ \}$ **105.** $\{\ \}$ **107.** $\{16\}$ **109.** $\sqrt{241}$ **111.** 4.24 feet   **113.** $-i\sqrt{6}$ **115.** $-\sqrt{10}$

117. $-13 - 3i$ **119.** $10 + 4i$ **121.** $1 + 5i$ **123.** 87 **125.** $-\dfrac{1}{3} + \dfrac{1}{3}i$

Chapter 8 Test

1. $6\sqrt{6}$ **2.** $-x^{16}$ **3.** $\dfrac{1}{5}$ **4.** 5 **5.** $\dfrac{4x^2}{9}$ **6.** $-a^6b^3$ **7.** $\dfrac{8a^{1/3}c^{2/3}}{b^{5/12}}$ **8.** $a^{7/12} - a^{7/3}$ **9.** $|4xy|$ **10.** -27 **11.** $\dfrac{3\sqrt{y}}{y}$

12. $\dfrac{8 - 6\sqrt{x} + x}{8 - 2x}$ **13.** $\dfrac{\sqrt[3]{b^2}}{b}$ **14.** $\dfrac{6 - x^2}{8(\sqrt{6} - x)}$ **15.** $-x\sqrt{5x}$ **16.** $4\sqrt{3} - \sqrt{6}$ **17.** $x + 2\sqrt{x} + 1$ **18.** $\sqrt{6} - 4\sqrt{3} + \sqrt{2} - 4$

19. -20 **20.** 23.685 **21.** 0.019 **22.** $\{2, 3\}$ **23.** $\{\ \}$ **24.** $\{6\}$ **25.** $i\sqrt{2}$ **26.** $-2i\sqrt{2}$ **27.** $-3i$ **28.** 40 **29.** $7 + 24i$

30. $-\frac{3}{2} + \frac{5}{2}i$ **31.** $\frac{5\sqrt{2}}{2}$ **32.** Domain: $[-2, \infty)$ **33.** 27 mph **34.** 360 ft

Chapter 8 Cumulative Review

1. *(Sec. 1.2, Ex. 9)* $(4 \cdot 9)y; 36y$ **2.** *(Sec. 2.1, Ex. 9)* **a.** $2x + 1$ **b.** $12x - 3$ **3.** *(Sec. 3.6, Ex. 6)* $x = 2$
4. *(Sec. 4.4, Ex. 7)* $\{x|x \le -3 \text{ or } x \ge 9\}$ **5.** *(Sec. 5.1, Ex. 7)* $\{(0, -5)\}$
6. *(Sec. 5.3, Ex. 3)* 42 liters of 30% solution and 28 liters of 80% solution **7.** *(Sec. 6.3, Ex. 3)* **a.** -4 **b.** 11
8. *(Sec. 6.5, Ex. 7)* $(b - 6)(a + 2)$ **9.** *(Sec. 6.6, Ex. 5)* $(2x + 5)(x + 3)$ **10.** *(Sec. 6.7, Ex. 11)* **a.** $(p^2 + 1)(5 + q)$ **b.** $(3x + 4)^2$
c. Prime **11.** *(Sec. 7.1, Ex. 3)* **a.** $\frac{3y^4}{x}$ **b.** $\frac{1}{5x - 1}$ **12.** *(Sec. 7.2, Ex. 3)* **a.** $\frac{x^6}{15y^2}$ **b.** $-\frac{m^2}{15(m + 2)}$
13. *(Sec. 7.3, Ex. 1)* **a.** $\frac{5 + x}{7}$ **b.** $\frac{3x}{2}$ **c.** $x - 7$ **d.** $-\frac{1}{3y^2}$ **14.** *(Sec. 7.4, Ex. 2)* **a.** $\frac{x(x - 2)}{2(x + 2)}$ **b.** $\frac{x}{y}$
15. *(Sec. 7.5, Ex. 5)* $2x - 3 - \frac{3}{3x - 5}$ **16.** *(Sec. 7.7, Ex. 1)* $\left\{-\frac{3}{2}\right\}$ **17.** *(Sec. 7.8, Ex. 4)* $2\frac{2}{9}$ hr; No
18. *(Sec. 8.1, Ex. 1)* **a.** 6 **b.** 0 **c.** $\frac{2}{7}$ **d.** 0.5 **e.** x^3 **f.** $3x^5$ **g.** -9 **19.** *(Sec. 8.2, Ex. 1)* **a.** $\sqrt{4} = 2$ **b.** $\sqrt[3]{64} = 4$ **c.** $\sqrt[4]{x}$
d. $\sqrt[6]{0} = 0$ **e.** $-\sqrt{9} = -3$ **f.** $\sqrt[3]{81x^8} = 3x^2$ **g.** $\sqrt[3]{5y}$ **20.** *(Sec. 8.3, Ex. 3)* **a.** $5\sqrt{2}$ **b.** $2\sqrt[3]{3}$ **c.** $\sqrt{26}$ **d.** $2\sqrt[4]{2}$
21. *(Sec. 8.4, Ex. 3)* **a.** $5\sqrt{3} + 3\sqrt{10}$ **b.** $\sqrt{35} + \sqrt{5} - \sqrt{42} - \sqrt{6}$ **c.** $21x - 7\sqrt{5x} + 15\sqrt{x} - 5\sqrt{5}$ **d.** $49 - 8\sqrt{3}$ **e.** $2x - 25$
22. *(Sec. 8.5, Ex. 1)* **a.** $\frac{3\sqrt{15}}{5}$ **b.** $\frac{8\sqrt{x}}{3x}$ **c.** $\frac{\sqrt[3]{4}}{2}$ **23.** *(Sec. 8.6, Ex. 4)* $\{3\}$

CHAPTER 9

Quadratic Equations and Functions

Exercise Set 9.1

1. $\{-4, 4\}$ **3.** $\{-\sqrt{7}, \sqrt{7}\}$ **5.** $\{-3\sqrt{2}, 3\sqrt{2}\}$ **7.** $\{-\sqrt{10}, \sqrt{10}\}$ **9.** $\{-8, -2\}$ **11.** $\{6 - 3\sqrt{2}, 6 + 3\sqrt{2}\}$ **13.** $\left\{\frac{3}{2} - \sqrt{2}, \frac{3}{2} + \sqrt{2}\right\}$
15. $\{-3i, 3i\}$ **17.** $\{-\sqrt{6}, \sqrt{6}\}$ **19.** $\{-2i\sqrt{2}, 2i\sqrt{2}\}$ **21.** $\{1 - 4i, 1 + 4i\}$ **23.** $\{-7 - \sqrt{5}, -7 + \sqrt{5}\}$ **25.** $\{-3 - 2i\sqrt{2}, -3 + 2i\sqrt{2}\}$
27. $x^2 + 16x + 64 = (x + 8)^2$ **29.** $z^2 - 12z + 36 = (z - 6)^2$ **31.** $p^2 + 9p + \frac{81}{4} = \left(p + \frac{9}{2}\right)^2$ **33.** $x^2 + x + \frac{1}{4} = \left(x + \frac{1}{2}\right)^2$ **35.** $\pm 8x$
37. $\pm 5z$ **39.** $\{-5, -3\}$ **41.** $\{-3 - \sqrt{7}, -3 + \sqrt{7}\}$ **43.** $\left\{-\frac{1}{2} - \frac{\sqrt{5}}{2}, -\frac{1}{2} + \frac{\sqrt{5}}{2}\right\}$ **45.** $\{-1 - \sqrt{6}, -1 + \sqrt{6}\}$
47. $\left\{2 - \frac{\sqrt{30}}{3}, 2 + \frac{\sqrt{30}}{3}\right\}$ **49.** $\left\{\frac{3}{2} - \frac{\sqrt{11}}{2}, \frac{3}{2} + \frac{\sqrt{11}}{2}\right\}$ **51.** $\left\{-4, \frac{1}{2}\right\}$ **53.** $\{-1, 5\}$ **55.** $\{-4 - \sqrt{15}, -4 + \sqrt{15}\}$
57. $\left\{-1 - \frac{\sqrt{21}}{3}, -1 + \frac{\sqrt{21}}{3}\right\}$ **59.** $\left\{-1, \frac{5}{2}\right\}$ **61.** $\{-1 + i, -1 - i\}$ **63.** $\{3 - \sqrt{6}, 3 + \sqrt{6}\}$ **65.** $\{-2 + i\sqrt{2}, -2 - i\sqrt{2}\}$
67. $\left\{-\frac{3}{2} + \frac{7\sqrt{5}}{10}, -\frac{3}{2} - \frac{7\sqrt{5}}{10}\right\}$ **69.** $\left\{\frac{1}{4} + \frac{\sqrt{47}}{4}i, \frac{1}{4} - \frac{\sqrt{47}}{4}i\right\}$ **71.** $\{-5 - i\sqrt{3}, -5 + i\sqrt{3}\}$ **73.** $\{-1.27, 6.27\}$ **75.** $\{-1.10, 0.90\}$
77. No; the graph shows no solutions. **79.** 20% **81.** 11% **83.** Answers may vary. **85.** Simple **87.** 7.97 seconds
89. 5.96 seconds **91.** 6 inches **93.** $\frac{27\sqrt{2}}{2}$ inches **95.** 2828 lamps **97.** $\frac{7}{5}$ **99.** $\frac{1}{5}$ **101.** $5 - 10\sqrt{3}$ **103.** $\frac{3}{4} - \frac{\sqrt{7}}{2}$ **105.** $2\sqrt{7}$
107. $\sqrt{13}$

Exercise Set 9.2

1. $\{-6, 1\}$ **3.** $\left\{-\frac{3}{5}, 1\right\}$ **5.** $\{3\}$ **7.** $\left\{-\frac{7}{2} - \frac{\sqrt{33}}{2}, -\frac{7}{2} + \frac{\sqrt{33}}{2}\right\}$ **9.** $\left\{\frac{1}{8} - \frac{\sqrt{57}}{8}, \frac{1}{8} + \frac{\sqrt{57}}{8}\right\}$ **11.** $\left\{\frac{7}{6} - \frac{\sqrt{85}}{6}, \frac{7}{6} + \frac{\sqrt{85}}{6}\right\}$
13. $\{1 - \sqrt{3}, 1 + \sqrt{3}\}$ **15.** $\left\{-\frac{3}{2}, 1\right\}$ **17.** $\left\{\frac{3}{2}, -\frac{\sqrt{11}}{2}, \frac{3}{2} + \frac{\sqrt{11}}{2}\right\}$ **19.** $\{-6, 1\}$ **21.** $\left\{\frac{3}{8} - \frac{\sqrt{87}}{8}i, \frac{3}{8} + \frac{\sqrt{87}}{8}i\right\}$
23. $\{-2 - \sqrt{11}, -2 + \sqrt{11}\}$ **25.** $\left\{\frac{-5 - i\sqrt{5}}{10}, \frac{-5 + i\sqrt{5}}{10}\right\}$
27. $\frac{-b + \sqrt{b^2 - 4ac}}{2a} + \frac{-b - \sqrt{b^2 - 4ac}}{2a} = \frac{-b + \sqrt{b^2 - 4ac} - b - \sqrt{b^2 - 4ac}}{2a} = \frac{-2b}{2a} = -\frac{b}{a}$ **29.** Two real solutions

31. One real solution   **33.** Two real solutions   **35.** Two complex solutions   **37.** $\left\{-\frac{5}{2} - \frac{\sqrt{17}}{2}, -\frac{5}{2} + \frac{\sqrt{17}}{2}\right\}$ **39.** $\left\{\frac{5}{2}, 1\right\}$

41. $\left\{\frac{3}{2} - \frac{\sqrt{29}}{2}, \frac{3}{2} + \frac{\sqrt{29}}{2}\right\}$ **43.** $\left\{-\frac{1}{6} - \frac{\sqrt{19}}{6}, -\frac{1}{6} + \frac{\sqrt{19}}{6}\right\}$ **45.** $\{-3 - 2i, -3 + 2i\}$ **47.** $\left\{\frac{-1 - i\sqrt{23}}{4}, \frac{-1 + i\sqrt{23}}{4}\right\}$ **49.** $\{1\}$

51. $\left\{\frac{19}{2} - \frac{\sqrt{345}}{2}, \frac{19}{2} + \frac{\sqrt{345}}{2}\right\}$ **53.** Two real solutions   **55.** Two complex solutions

57. $4\sqrt{10}$ ft by $10\sqrt{10}$ ft; approximately 12.6 ft by 31.6 ft   **59.** Base: $2\sqrt{42} \approx 13.0$ cm; Height: $\sqrt{42} \approx 6.5$ cm

61. Width: $20\sqrt{2} \approx 28.3$ in.; Length: $30\sqrt{2} \approx 42.4$ in.   **63. a.** $5\sqrt{10}$ cm by $15\sqrt{10}$ cm; approximately 15.8 cm by 47.4 cm

b. $40\sqrt{10} \approx 126.5$cm   **65.** $\{0.6, 2.4\}$ **67.** $\{-1.7, 3.6\}$ **69.** 2.8 seconds   **71.** From Sunday to Monday   **73.** Wednesday

75. $f(4) = 33$; yes   **77.** $\left\{\frac{11}{5}\right\}$ **79.** $\{15\}$ **81.** $(x^2 + 5)(x + 2)(x - 2)$ **83.** $(z + 3)(z - 3)(z + 2)(z - 2)$ **85.** $\left\{\frac{\sqrt{3}}{3}\right\}$

87. $\left\{\frac{-\sqrt{2} - i\sqrt{2}}{2}, \frac{-\sqrt{2} + i\sqrt{2}}{2}\right\}$ **89.** $\left\{\frac{\sqrt{3} - \sqrt{11}}{4}, \frac{\sqrt{3} + \sqrt{11}}{4}\right\}$

Exercise Set 9.3

1. $\{3 + \sqrt{7}, 3 - \sqrt{7}\}$ **3.** $\left\{\frac{3}{4} + \frac{\sqrt{57}}{4}, \frac{3}{4} - \frac{\sqrt{57}}{4}\right\}$ **5.** $\left\{\frac{1}{2} + \frac{\sqrt{29}}{2}, \frac{1}{2} - \frac{\sqrt{29}}{2}\right\}$ **7.** $\left\{1, -\frac{1}{2} + \frac{\sqrt{3}}{2}i, -\frac{1}{2} - \frac{\sqrt{3}}{2}i\right\}$

9. $\left\{0, -3, \frac{3}{2} + \frac{3\sqrt{3}}{2}i, \frac{3}{2} - \frac{3\sqrt{3}}{2}i\right\}$ **11.** $\{4, -2 + 2i\sqrt{3}, -2 - 2i\sqrt{3}\}$ **13.** $\{-2i, 2i, -2, 2\}$ **15.** $\left\{-\frac{1}{2}, \frac{1}{2}, -i\sqrt{3}, i\sqrt{3}\right\}$ **17.** $\{-3, 3, -2, 2\}$

19. $\{125, -8\}$ **21.** $\left\{-\frac{4}{5}, 0\right\}$ **23.** $\left\{-\frac{1}{8}, 27\right\}$ **25.** $\left\{-\frac{2}{3}, \frac{4}{3}\right\}$ **27.** $\left\{-\frac{1}{125}, \frac{1}{8}\right\}$ **29.** $x^3 - 39x + 70 = 0$, for example

31. $\{\sqrt{3}, -\sqrt{3}, \sqrt{2}, -\sqrt{2}\}$ **33.** $\left\{-\frac{3}{2} + \frac{\sqrt{201}}{6}, -\frac{3}{2} - \frac{\sqrt{201}}{6}\right\}$ **35.** $\{2, 3\}$ **37.** $\{-4, 2 + 2i\sqrt{3}, 2 - 2i\sqrt{3}\}$ **39.** $\{125, 27\}$

41. $\{-3i, 3i, 1\}$ **43.** $\left\{\frac{1}{8}, -8\right\}$ **45.** $\left\{-\frac{1}{2}, \frac{1}{3}\right\}$ **47.** $\left\{5, -\frac{5}{2} + \frac{5\sqrt{3}}{2}i, -\frac{5}{2} - \frac{5\sqrt{3}}{2}i\right\}$ **49.** $\{-3\}$ **51.** $\{\sqrt{5}, -\sqrt{5}, -2i, 2i\}$

53. $\left\{-3, \frac{3}{2} + \frac{3\sqrt{3}}{2}i, \frac{3}{2} - \frac{3\sqrt{3}}{2}i\right\}$ **55.** $\{6, 12\}$ **57.** $\left\{-\frac{1}{3}, \frac{1}{3}, \frac{\sqrt{6}}{3}i, -\frac{\sqrt{6}}{3}i\right\}$ **59.** 8.53 hours   **61.** 12 or -8

63. Toshiba: 16.5 hours; IBM: 15.5 hours   **65.** $(-\infty, 3]$ **67.** $(-5, \infty)$ **69.** Domain: $\{x | x \text{ is a real number}\}$; Range: $\{y | y \text{ is a real number}\}$; It is a function.   **71.** Domain: $\{x | x \text{ is a real number}\}$; Range: $\{y | y \geq 1\}$; It is a function.

Exercise Set 9.4

1. $(-\infty, -5) \cup (-1, \infty)$ **3.** $[-4, 3]$ **5.** $[2, 5]$ **7.** $\left(-5, -\frac{1}{3}\right)$ **9.** $(2, 4) \cup (6, \infty)$ **11.** $(-\infty, -4] \cup [0, 1]$

13. $(-\infty, -3) \cup (-2, 2) \cup (3, \infty)$ **15.** $(-7, 2)$ **17.** $(-1, \infty)$ **19.** $(-\infty, -1] \cup (4, \infty)$ **21.** Answers may vary.

23. $(-\infty, 2) \cup \left(\frac{11}{4}, \infty\right)$ **25.** $(0, 2] \cup [3, \infty)$ **27.** $(-\infty, -7) \cup (8, \infty)$ **29.** $\left[-\frac{5}{4}, \frac{3}{2}\right]$ **31.** $(-\infty, 0) \cup (1, \infty)$ **33.** $(-\infty, -4] \cup [4, 6]$

35. $\left(-\infty, -\frac{2}{3}\right] \cup \left[\frac{3}{2}, \infty\right)$ **37.** $\left(-4, -\frac{3}{2}\right) \cup \left(\frac{3}{2}, \infty\right)$ **39.** $(-\infty, -5] \cup [-1, 1] \cup [5, \infty)$ **41.** $\left(-\infty, -\frac{5}{3}\right) \cup \left(\frac{7}{2}, \infty\right)$ **43.** $(0, 10)$

45. $(-\infty, -4) \cup [5, \infty)$ **47.** $(-\infty, -6] \cup (-1, 0] \cup (7, \infty)$ **49.** $(-\infty, 1) \cup (2, \infty)$ **51.** $(-\infty, -8] \cup (-4, \infty)$ **53.** $(-\infty, 0] \cup \left(5, \frac{11}{2}\right]$

55. $(0, \infty)$ **57.** $(-\infty, -1) \cup (0, 1)$ **59.** $(2, 11)$

61.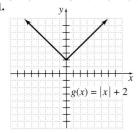

63.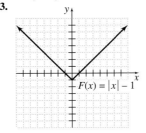

65.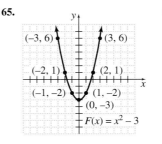

67.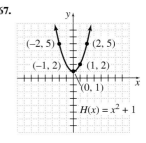

Section 9.5 Mental Math

1. $(0, 0)$ **3.** $(2, 0)$ **5.** $(0, 3)$ **7.** $(-1, 5)$

Exercise Set 9.5

1.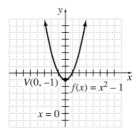

$V(0, -1)$ $f(x) = x^2 - 1$ $x = 0$

3.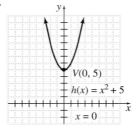

$V(0, 5)$ $h(x) = x^2 + 5$ $x = 0$

5.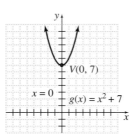

$V(0, 7)$ $x = 0$ $g(x) = x^2 + 7$

7.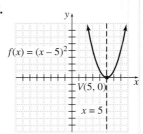

$f(x) = (x - 5)^2$ $V(5, 0)$ $x = 5$

9.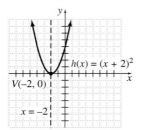

$h(x) = (x + 2)^2$ $V(-2, 0)$ $x = -2$

11.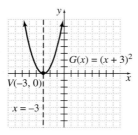

$G(x) = (x + 3)^2$ $V(-3, 0)$ $x = -3$

13.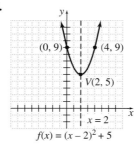

$(0, 9)$ $(4, 9)$ $V(2, 5)$ $x = 2$ $f(x) = (x - 2)^2 + 5$

15.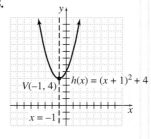

$V(-1, 4)$ $h(x) = (x + 1)^2 + 4$ $x = -1$

17.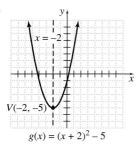

$x = -2$ $V(-2, -5)$ $g(x) = (x + 2)^2 - 5$

19.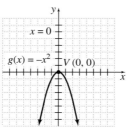

$x = 0$ $g(x) = -x^2$ $V(0, 0)$

21.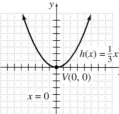

$h(x) = \frac{1}{3}x^2$ $V(0, 0)$ $x = 0$

23.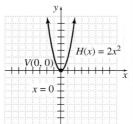

$H(x) = 2x^2$ $V(0, 0)$ $x = 0$

25.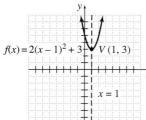

$f(x) = 2(x - 1)^2 + 3$ $V(1, 3)$ $x = 1$

27.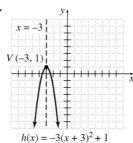

$x = -3$ $V(-3, 1)$ $h(x) = -3(x + 3)^2 + 1$

29.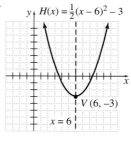

$H(x) = \frac{1}{2}(x - 6)^2 - 3$ $V(6, -3)$ $x = 6$

31.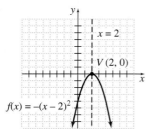

$x = 2$ $V(2, 0)$ $f(x) = -(x - 2)^2$

33.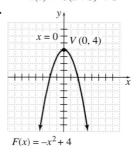

$x = 0$ $V(0, 4)$ $F(x) = -x^2 + 4$

35.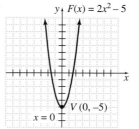

$F(x) = 2x^2 - 5$ $V(0, -5)$ $x = 0$

37.
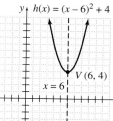
$h(x) = (x - 6)^2 + 4$
$V(6, 4)$
$x = 6$

39.
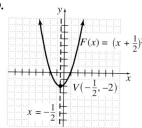
$F(x) = \left(x + \frac{1}{2}\right)^2 - 2$
$V\left(-\frac{1}{2}, -2\right)$
$x = -\frac{1}{2}$

41.
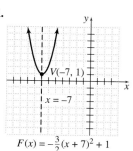
$V(-7, 1)$
$x = -7$
$F(x) = -\frac{3}{2}(x + 7)^2 + 1$

43.
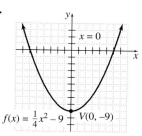
$x = 0$
$f(x) = \frac{1}{4}x^2 - 9$ $V(0, -9)$

45.
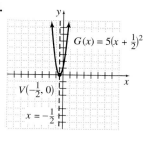
$G(x) = 5\left(x + \frac{1}{2}\right)^2$
$V\left(-\frac{1}{2}, 0\right)$
$x = -\frac{1}{2}$

47.
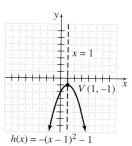
$x = 1$
$V(1, -1)$
$h(x) = -(x - 1)^2 - 1$

49.
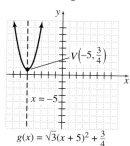
$V\left(-5, \frac{3}{4}\right)$
$x = -5$
$g(x) = \sqrt{3}(x + 5)^2 + \frac{3}{4}$

55. $f(x) = 5(x - 2)^2 + 3$
57. $f(x) = 5(x + 3)^2 + 6$

51.
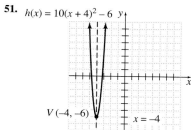
$h(x) = 10(x + 4)^2 - 6$
$V(-4, -6)$
$x = -4$

53.
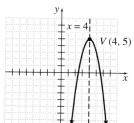
$x = 4$
$V(4, 5)$
$f(x) = -2(x - 4)^2 + 5$

59.
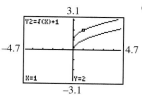

61.
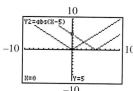

63.
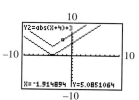

65.
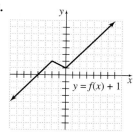
$y = f(x) + 1$

67.
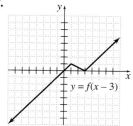
$y = f(x - 3)$

69.
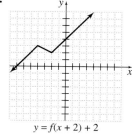
$y = f(x + 2) + 2$

71. $x^2 + 8x + 16$ **73.** $z^2 - 16z + 64$ **75.** $y^2 + y + \frac{1}{4}$
77. $\{-6, 2\}$ **79.** $\{-5 - \sqrt{26}, -5 + \sqrt{26}\}$
81. $\{4 - 3\sqrt{2}, 4 + 3\sqrt{2}\}$

Exercise Set 9.6

1. $(-4, -9)$ **3.** $(5, 30)$ **5.** $(1, -2)$ **7.** $\left(\frac{1}{2}, \frac{5}{4}\right)$

9. Axis of symmetry: $x = -2$

Vertex: $(-2, -9)$

x-intercepts: $(-5, 0), (1, 0)$

y-intercept: $(0, -5)$

11. Axis of symmetry: $x = 1$

Vertex: $(1, 0)$

x-intercept: $(1, 0)$

y-intercept: $(0. -1)$

13. Axis of symmetry: $x = 0$

Vertex: $(0, -4)$

x-intercepts: $(-2, 0), (2, 0)$

y-intercept: $(0, -4)$

15. Axis of symmetry: $x = -\frac{1}{2}$

Vertex: $\left(-\frac{1}{2}, -4\right)$

x-intercepts: $\left(\frac{1}{2}, 0\right), \left(-\frac{3}{2}, 0\right)$

y-intercept: $(0, -3)$

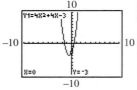

17. Axis of symmetry: $x = -4$
Vertex: $(-4, -1)$
x-intercepts: $(-3, 0), (-5, 0)$
y-intercept: $(0, 15)$

19. Axis of symmetry: $x = 3$
Vertex: $(3, -4)$
x-intercepts: $(1, 0), (5, 0)$
y-intercept: $(0, 5)$

21. Axis of symmetry: $x = 2$
Vertex: $(2, 1)$
No x-intercepts
y-intercept: $(0, 5)$

23. Axis of symmetry: $x = -1$
Vertex: $(-1, 3)$
No x-intercepts
y-intercept: $(0, 5)$

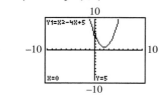

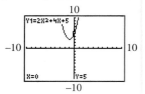

25. Axis of symmetry: $x = 3$

Vertex: $(3, 18)$

x-intercepts: $(0, 0), (6, 0)$

y-intercept: $(0, 0)$

27. Axis of symmetry: $x = 0$

Vertex: $(0, 1)$

No x-intercepts

y-intercepts: $(0, 1)$

29. Axis of symmetry: $x = 1$

Vertex: $(1, -16)$

x-intercepts: $(5, 0), (-3, 0)$

y-intercept: $(0, -15)$

31. Axis of symmetry: $x = \frac{1}{2}$

Vertex: $\left(\frac{1}{2}, \frac{5}{4}\right)$

x-intercepts: $(0, 0), (1, 0)$

y-intercept: $(0, 0)$

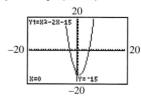

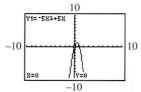

33. Axis of symmetry: $x = 1$

Vertex: $(1, -11)$

No x-intercepts

y-intercept: $(0, -12)$

35. Axis of symmetry: $x = 2$

Vertex: $(2, 3)$

No x-intercepts

y-intercept: $(0, 15)$

37. Axis of symmetry: $x = -\frac{1}{2}$

Vertex: $\left(-\frac{1}{2}, -6\frac{1}{4}\right)$

x-intercepts: $(-3, 0), (2, 0)$

y-intercept: $(0, -6)$

39. Axis of symmetry: $x = -\frac{3}{4}$

Vertex: $\left(-\frac{3}{4}, 36\frac{1}{8}\right)$

x-intercepts: $\left(\frac{7}{2}, 0\right), (-5, 0)$

y-intercept: $(0, 35)$

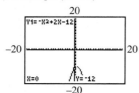

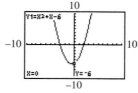

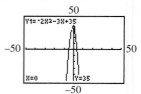

41. 144 ft   **43.** 16 feet   **45.** 30 and 30

47. Upward   **49.** Upward   **51.** -0.84 **53.**   1.43

Vertex: $(-5, -10)$
x-intercepts: $(-8.2, 0), (-1.8, 0)$
y-intercept: $(0, 15)$

Vertex: $(1, 4)$
No x-intercepts
y-intercept: $(0, 7)$

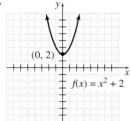

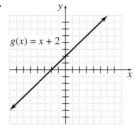

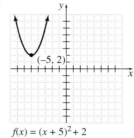

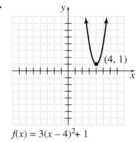

55.

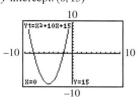

$f(x) = x^2 + 2$

57.

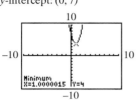

$g(x) = x + 2$

59.

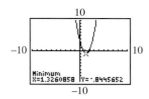

$f(x) = (x + 5)^2 + 2$

61.

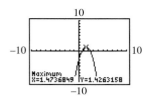

$f(x) = 3(x - 4)^2 + 1$

63.

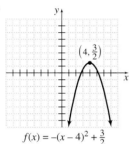

$f(x) = -(x - 4)^2 + \frac{3}{2}$

Exercise Set 9.7

1. Linear   **3.** Neither   **5.** Linear   **7.**

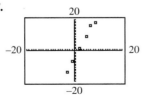

$y = 3x - 5$

9.

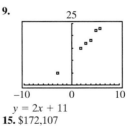

$y = 2x + 11$

11.

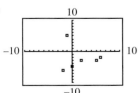

$y = \frac{1}{2}x - 5$

13.

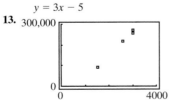

$y = 118.75x - 89,143.25$

15. \$172,107

17.

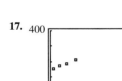

$y = 0.08x + 150$

19. 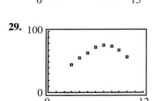 $y = 0.62x + 0.04$ **21.** Quadratic   **23.** Linear   **25.** Quadratic   **27.** Quadratic

29. 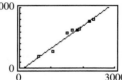 $y = -\frac{33}{16}(x - 7)^2 + 78$ **31.** 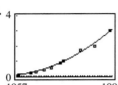 **a.** 0.46   **b.** 3.53

33. 1990: 21, 0; 1991: 21.5, 0.2; 1992: 22.0, −0.3; 1993: 23.6, −2.0   **35.** $791.75   **37.** $663.75   **39.** $y = -(x - 3)^2 + 10$

41. $y = -\frac{26}{9}(x - 4)^2 + 9$ **43.** $y = -\frac{13}{25}(x + 3)^2 + 12$ **45. a.** 1987   **b–c.** 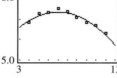   **d.** 1988; Yes

47. a.–b. 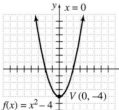   The line passes through or near all the points.   **c.** 542 points

49. a. $28,620, $36,720, $60,480   **b.** $s_{\text{proj.}} = 1.08 s_{\text{pres.}}$ **51.** $\{-3, 5\}$ **53.** $\left\{-\frac{3}{2}, 5\right\}$ **55.** $\{10\}$ **57.** −8   **59.** 5

Chapter 9 Review

1. $\{14, 1\}$ **3.** $\left\{\frac{4}{5}, -\frac{1}{2}\right\}$ **5.** $\{-7, 7\}$ **7.** $\left\{\frac{2}{9}, -\frac{4}{9}\right\}$ **9.** $\left\{-\frac{3}{2} + \frac{\sqrt{5}}{2}, -\frac{3}{2} - \frac{\sqrt{5}}{2}\right\}$ **11.** $\left\{-\frac{3}{8} + \frac{\sqrt{7}}{8}i, -\frac{3}{8} - \frac{\sqrt{7}}{8}i\right\}$

13. $75\sqrt{2}$ or 106.1 miles   **15.** Two real solutions   **17.** One real solution   **19.** Two real solutions   **21.** One real solution

23. Two complex solutions   **25.** $\{0, -5\}$ **27.** $\left\{1, -\frac{5}{2}\right\}$ **29.** $\left\{\frac{1 - i\sqrt{35}}{9}, \frac{1 + i\sqrt{35}}{9}\right\}$ **31.** $\left\{1, \frac{9}{4}\right\}$ **33.** $(6 + 6\sqrt{2})$ cm

35. $\{-4, 2 - 2i\sqrt{3}, 2 + 2i\sqrt{3}\}$ **37.** $\left\{-\frac{8\sqrt{7}}{7}, \frac{8\sqrt{7}}{7}\right\}$ **39.** $\left\{-\frac{16}{5}, 1\right\}$ **41.** $\{8, 64\}$ **43.** $\left\{-\frac{1}{5}, \frac{1}{4}\right\}$ **45.** −5   **47.** $[-5, 5]$

49. $\left(-\infty, -\frac{5}{4}\right] \cup \left[\frac{3}{2}, \infty\right)$ **51.** $(5, 6)$ **53.** $(-\infty, -6) \cup \left(-\frac{3}{4}, 0\right) \cup (5, \infty)$ **55.** $(-5, -3) \cup (5, \infty)$ **57.** $\left(-\frac{6}{5}, 0\right) \cup \left(\frac{5}{6}, 3\right)$

59.   **61.**   **63.**   **65.**

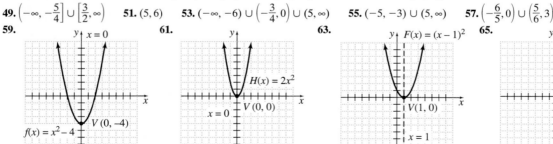

$V(4, -2)$ $f(x) = (x - 4)^2 - 2$

67. 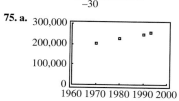 **69.** 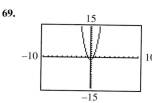 **71.** Vertex: $\left(-\frac{5}{6}, \frac{73}{12}\right)$ **73.** The numbers are both 210.

x-intercepts: $(-2.26, 0), (0.59, 0)$
y-intercept: $(0, 4)$

75. a. 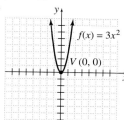 **b.** 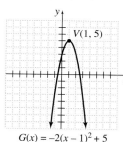 $y = 2412(x - 1980) + 226{,}548$ **c.** 274,788 million

Chapter 9 Test

1. $\left\{\frac{7}{5}, -1\right\}$ **2.** $\{-1 - \sqrt{10}, -1 + \sqrt{10}\}$ **3.** $\left\{\frac{1}{2} + \frac{\sqrt{31}}{2}i, \frac{1}{2} - \frac{\sqrt{31}}{2}i\right\}$ **4.** $\{3 + \sqrt{7}, 3 - \sqrt{7}\}$ **5.** $\left\{-\frac{1}{7}, -1\right\}$ **6.** $\left\{\frac{3}{2} + \frac{\sqrt{29}}{2}, \frac{3}{2} - \frac{\sqrt{29}}{2}\right\}$

7. $\{-2 + \sqrt{11}, -2 - \sqrt{11}\}$ **8.** $\{3, -3, i, -i\}$ **9.** $\{1, -1, i, -i\}$ **10.** $\{6, 7\}$ **11.** $\{3 + \sqrt{7}, 3 - \sqrt{7}\}$ **12.** $\left\{1 + \frac{\sqrt{6}}{2}i, 1 - \frac{\sqrt{6}}{2}i\right\}$

13. $\left(-\infty, -\frac{3}{2}\right) \cup (5, \infty)$ **14.** $(-\infty, -5) \cup (-4, 4) \cup (5, \infty)$ **15.** $(-\infty, -3) \cup (2, \infty)$ **16.** $(-\infty, -3) \cup [2, 3)$

17. 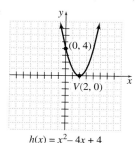 **18.** **19.** **20.**

21. $2 + \sqrt{46}$ or 8.8 feet **22.** $5 + \sqrt{17}$ or 9.12 hours **23. a.** 272 feet **b.** 5.12 seconds $y = 302.5x - 589{,}204$ **c.** 15,796 million

24. a. 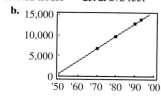 **b.**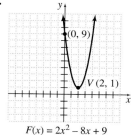

Chapter 9 Cumulative Review

1. *(Sec. 1.4, Ex. 10)* **a.** $2xy - 2$ **b.** $2x^2 + 23$ **c.** $3.1x - 0.3$ **2.** *(Sec. 2.1, Ex. 7)* $\{\}$ **3.** *(Sec. 3.3, Ex. 7)* **a.** 5 **b.** 7 **c.** 4

4. *(Sec. 4.3, Ex. 4)* $\{-1, 1\}$ **5.** *(Sec. 6.1, Ex. 6)* **a.** $\frac{1}{x^{11}}$ **b.** p^7 **c.** $\frac{1}{4}$ **d.** $\frac{y^7}{5x^8}$ **e.** $\frac{3}{x^7}$

6. *(Sec. 6.3, Ex. 5)* **a.** $-5x^2 - 6x$ **b.** $8xy - 3x$ **7.** *(Sec. 6.6, Ex. 8)* $(4x + 3y)^2$ **8.** *(Sec. 6.8, Ex. 7)* $\{-5, -1, 1\}$

9. *(Sec. 7.1, Ex. 6)* **a.** $\frac{15x^2 y^2}{10xy^3}$ **b.** $\frac{6x^2 - x - 1}{2x^2 - 11x + 5}$ **10.** *(Sec. 7.2, Ex. 5)* **a.** $3x - 4$ **b.** $-x + 2$ **c.** $2x^2 - 5x + 3$ **d.** $\frac{x-1}{2x-3}, x \neq \frac{3}{2}$

11. *(Sec. 7.3, Ex. 4)* **a.** $\frac{-4x + 15}{x - 3}$ **b.** $\frac{4}{x - y}$ **12.** *(Sec. 7.5, Ex. 4)* $2x - 5$ **13.** *(Sec. 7.6, Ex. 1)* $2x^2 + 5x + 2 + \frac{7}{x - 3}$

14. *(Sec. 8.1, Ex. 3)* **a.** 3 **b.** -3 **c.** -5 **d.** -81 **e.** $4x$ **15.** *(Sec. 8.2, Ex. 4)* **a.** $x^{1/5}$ **b.** $(17x^2 y^5)^{1/3}$ **c.** $(x - 5a)^{1/2}$

d. $3(2p)^{1/2} - 5p^{2/3}$ **16.** *(Sec. 8.4, Ex. 2)* **a.** $\frac{5\sqrt{5}}{12}$ **b.** $\frac{5\sqrt[3]{7x}}{2}$ **17.** *(Sec. 8.5, Ex. 4)* $\frac{x - 4}{5(\sqrt{x} - 2)}$ **18.** *(Sec. 8.6, Ex. 5)* $\left\{\frac{2}{9}\right\}$

19. *(Sec. 8.6, Ex. 7)* **a.** Approximately 45.8 ft   **b.** Yes   **20.** *(Sec. 9.1, Ex. 2)* $\{-1 + 2\sqrt{3}, -1 - 2\sqrt{3}\}$

21. *(Sec. 9.2, Ex. 2)* $\left\{\dfrac{2 + \sqrt{10}}{2}, \dfrac{2 - \sqrt{10}}{2}\right\}$ **22.** *(Sec. 9.3, Ex. 3)* $\{2, -2, i, -i\}$ **23.** *(Sec. 9.4, Ex. 6)* $[-2, 3)$

24. *(Sec. 9.5, Ex. 3)*

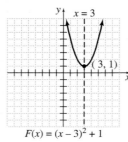

$F(x) = (x - 3)^2 + 1$

25. *(Sec. 9.6, Ex. 4)* The rock's maximum height is $6\frac{1}{4}$ feet which was reached in $\frac{5}{8}$ second.

CHAPTER 10

Conic Sections

Section 10.1 Mental Math

1. Upward   **3.** To the left   **5.** Downward

Exercise Set 10.1

1.

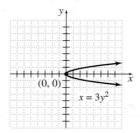

$x = 3y^2$

3.

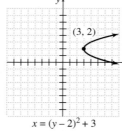

$x = (y - 2)^2 + 3$
x-intercept: $(7, 0)$

5.

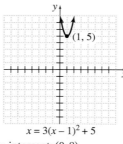

$x = 3(x - 1)^2 + 5$
y-intercept: $(0, 8)$

7.

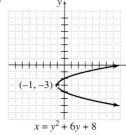

$x = y^2 + 6y + 8$
x-intercept: $(8, 0)$;
y-intercept: $(0, -2), (0, -4)$

9.

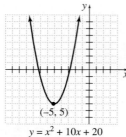

$y = x^2 + 10x + 20$
y-intercept: $(0, 20)$;
x-intercepts: $(-5 - \sqrt{5}, 0)$,
$(-5 + \sqrt{5}, 0)$

11.

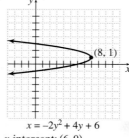

$x = -2y^2 + 4y + 6$
x-intercept: $(6, 0)$
y-intercepts: $(0, -1), (0, 3)$

13.

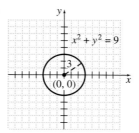

$x^2 + y^2 = 9$

15.

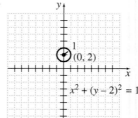

$x^2 + (y - 2)^2 = 1$

17.

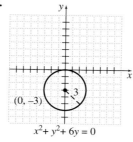

$(x - 5)^2 + (y + 2)^2 = 1$

19.

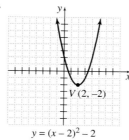

$x^2 + y^2 + 6y = 0$

21.

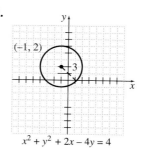

$x^2 + y^2 + 2x - 4y = 4$

23.

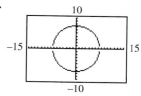

25.

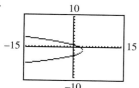

27.

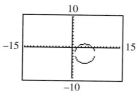

29.

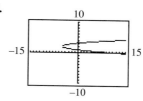

31. $(x - 2)^2 + (y - 3)^2 = 36$
33. $x^2 + y^2 = 3$
35. $(x + 5)^2 + (y - 4)^2 = 45$
37. Answers may vary.

39.

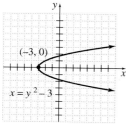

$(-3, 0)$
$x = y^2 - 3$

41.

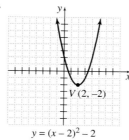

$V(2, -2)$
$y = (x - 2)^2 - 2$

43.

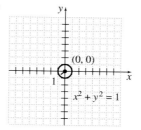

$(0, 0)$
$x^2 + y^2 = 1$

45.

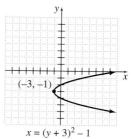

$(-3, -1)$
$x = (y + 3)^2 - 1$

47.

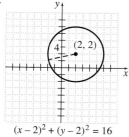

$(2, 2)$
$(x - 2)^2 + (y - 2)^2 = 16$

49.

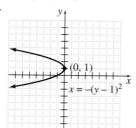

$(0, 1)$
$x = -(y - 1)^2$

51.

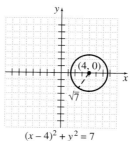

$(4, 0)$
$\sqrt{7}$
$(x - 4)^2 + y^2 = 7$

53.

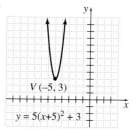

$V(-5, 3)$
$y = 5(x+5)^2 + 3$

55.

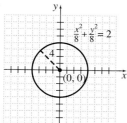

$\dfrac{x^2}{8} + \dfrac{y^2}{8} = 2$
$(0, 0)$

57.

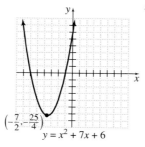

$\left(-\dfrac{7}{2}, -\dfrac{25}{4}\right)$
$y = x^2 + 7x + 6$

59.

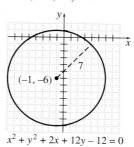

$(-1, -6)$
$x^2 + y^2 + 2x + 12y - 12 = 0$

61.

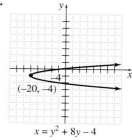

$(-20, -4)$
$x = y^2 + 8y - 4$

63.

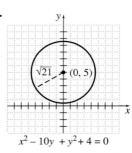

$x^2 - 10y + y^2 + 4 = 0$

65.

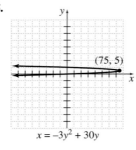

$x = -3y^2 + 30y$

67.

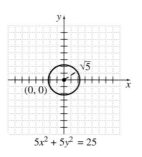

$5x^2 + 5y^2 = 25$

69.

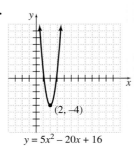

$y = 5x^2 - 20x + 16$

71. Yes, it is.   **73.** $y = -\dfrac{2}{125}x^2 + 40$

75.

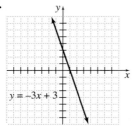

77.

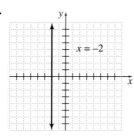

79. $\dfrac{\sqrt{10}}{4}$ **81.** $2\sqrt{5}$

Exercise Set 10.2

1.

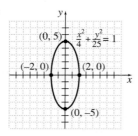

3.

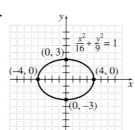

5.

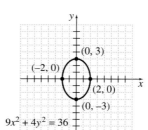

7.

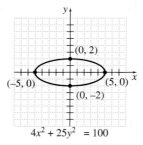

9.

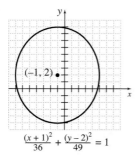

$\dfrac{(x+1)^2}{36} + \dfrac{(y-2)^2}{49} = 1$

11.

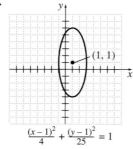

$\dfrac{(x-1)^2}{4} + \dfrac{(y-1)^2}{25} = 1$

13.

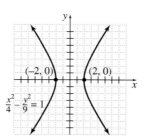

15.

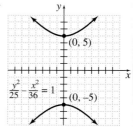

17.

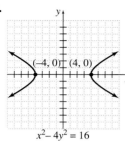

$$x^2 - 4y^2 = 16$$

19.

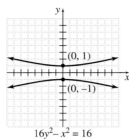

$$16y^2 - x^2 = 16$$

21. Answers may vary.

23. Parabola

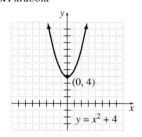

$$y = x^2 + 4$$

25. Ellipse

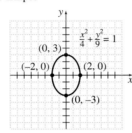

$$\frac{x^2}{4} + \frac{y^2}{9} = 1$$

27. Hyperbola

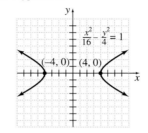

$$\frac{x^2}{16} - \frac{y^2}{4} = 1$$

29. Circle

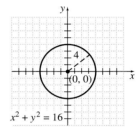

$$x^2 + y^2 = 16$$

31. Parabola

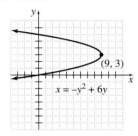

$$x = -y^2 + 6y$$

33. Ellipse

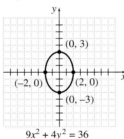

$$9x^2 + 4y^2 = 36$$

35. Hyperbola

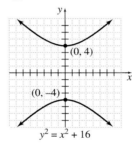

$$y^2 = x^2 + 16$$

37. Parabola

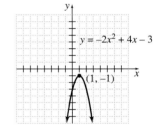

$$y = -2x^2 + 4x - 3$$

39. $(1{,}782{,}000{,}000, 356{,}400{,}000)$

41.

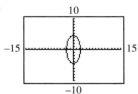

43.

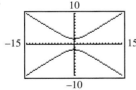

45.

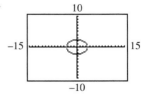

47.

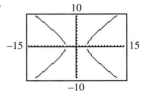

49. Hyperbola; Center $(-2, 1)$

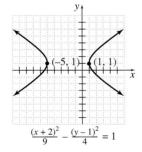

$$\frac{(x + 2)^2}{9} - \frac{(y - 1)^2}{4} = 1$$

51. Hyperbola; Center $(0, -4)$

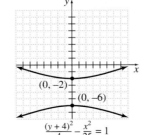

$$\frac{(y + 4)^2}{4} - \frac{x^2}{25} = 1$$

53. Ellipse; Center $(3, 2)$

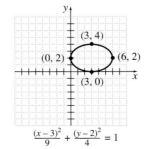

$$\frac{(x - 3)^2}{9} + \frac{(y - 2)^2}{4} = 1$$

55. $(-\infty, 5)$ **57.** $[4, \infty)$

59. $-2x^3$ **61.** $-5x^4$

Exercise Set 10.3

1. $\{(3, -4), (-3, 4)\}$ **3.** $\{(\sqrt{2}, \sqrt{2}), (-\sqrt{2}, -\sqrt{2})\}$ **5.** $\{(4, 0), (0, -2)\}$ **7.** $\{(\sqrt{5}, 2), (\sqrt{5}, -2), (-\sqrt{5}, 2), (-\sqrt{5}, -2)\}$ **9.** $\{\}$
11. $\{(3, 6), (1, -2)\}$ **13.** $\{(-5, 25), (2, 4)\}$ **15.** $\{\}$ **17.** $\{(1, -3)\}$ **19.** $\{(1, -2), (1, 2), (-1, -2), (-1, 2)\}$ **21.** $\{(0, -1)\}$
23. $\{(1, 3), (-1, 3)\}$ **25.** $\{(\sqrt{3}, 0), (-\sqrt{3}, 0)\}$ **27.** $\{\}$ **29.** $\{(6, 0), (-6, 0), (0, -6)\}$ **31.** 0, 1, 2, 3, or 4 **33.** ± 9 and ± 7
35. 15 cm by 19 cm **37.** 15 thousand compact discs; price, $3.75
39.
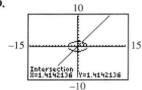
41.
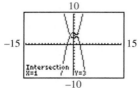
43.
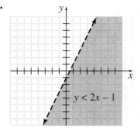
45.
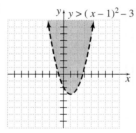

47. $(8x - 25)$ inches **49.** $(4x^2 + 6x + 2)$ meters

Exercise Set 10.4

1.
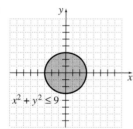
3.
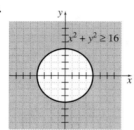
5.
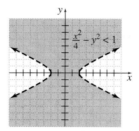
7.
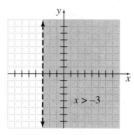

9.
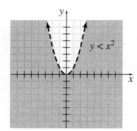
11.
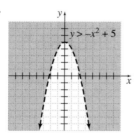
13.
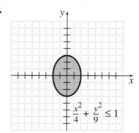
15.
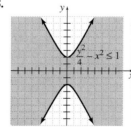

17.
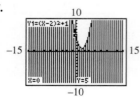
19.
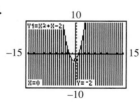
21. Answers may vary.
23.
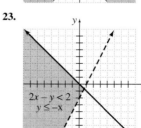

25.

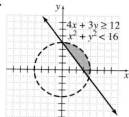

$$4x + 3y \geq 12$$
$$x^2 + y^2 < 16$$

27.

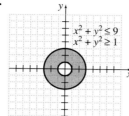

$$x^2 + y^2 \leq 9$$
$$x^2 + y^2 \geq 1$$

29.

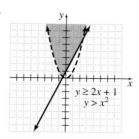

$$y \geq 2x + 1$$
$$y > x^2$$

31.

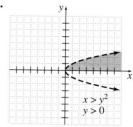

$$x > y^2$$
$$y > 0$$

33.

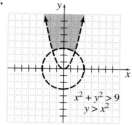

$$x^2 + y^2 > 9$$
$$y > x^2$$

35.

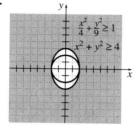

$$\frac{x^2}{4} + \frac{y^2}{9} \geq 1$$

$$x^2 + y^2 \geq 4$$

37.

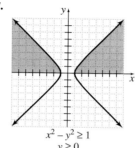

$$x^2 - y^2 \geq 1$$
$$y \geq 0$$

39.

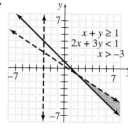

$$x + y \geq 1$$
$$2x + 3y < 1$$
$$x > -3$$

41.

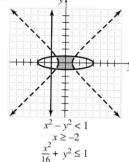

$$x^2 - y^2 < 1$$
$$x \geq -2$$
$$\frac{x^2}{16} + y^2 \leq 1$$

43. Not a function   **45.** Function   **47.** 1   **49.** $3a^2 - 2$

Chapter 10 Review

1. $(x + 4)^2 + (y - 4)^2 = 9$ **3.** $(x + 7)^2 + (y + 9)^2 = 11$

5.

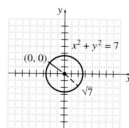

$$x^2 + y^2 = 7$$
$(0, 0)$
$\sqrt{7}$

7.

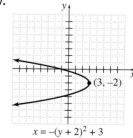

$(3, -2)$
$$x = -(y + 2)^2 + 3$$

9.

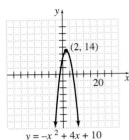

$y = -x^2 + 4x + 10$

11.

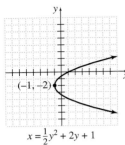

$x = \frac{1}{2}y^2 + 2y + 1$

13.

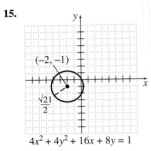

$x^2 + y^2 + 2x + y = \frac{3}{4}$

15.

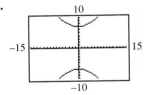

$4x^2 + 4y^2 + 16x + 8y = 1$

17.

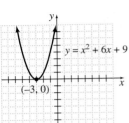

$y = x^2 + 6x + 9$

19. $(x - 5.6)^2 + (y + 2.4)^2 = 9.61$ **21.**

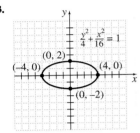

$x^2 - \frac{y^2}{4} = 1$

23.

$\frac{y^2}{4} + \frac{x^2}{16} = 1$

25.

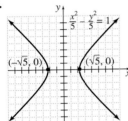

$\frac{x^2}{5} - \frac{y^2}{5} = 1$

27.

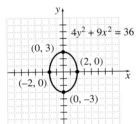

$4y^2 + 9x^2 = 36$

29.

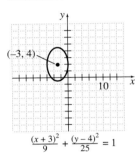

$\frac{(x + 3)^2}{9} + \frac{(y - 4)^2}{25} = 1$

31.

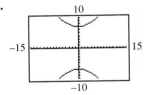

33.

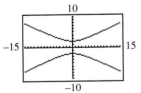

35.

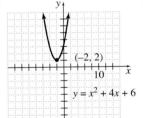

$y = x^2 + 4x + 6$

37.

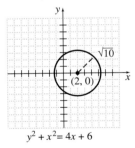

$y^2 + x^2 = 4x + 6$

39.

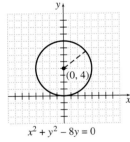

$x^2 + y^2 - 8y = 0$

41.

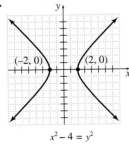

$x^2 - 4 = y^2$

43.

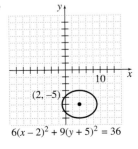

$6(x - 2)^2 + 9(y + 5)^2 = 36$

45.

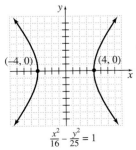

$\frac{x^2}{16} - \frac{y^2}{25} = 1$

47. $\{(1, -2), (4, 4)\}$
49. $\{(-1, 1), (2, 4)\}$
51. $\{(2, 2\sqrt{2}), (2, -2\sqrt{2})\}$
53. $\{(-1, -3), (-1, 3), (1, -3),$
$\quad (1, 3)\}$
55. $\{(1, 4)\}$
57. 15 feet by 10 feet

59.
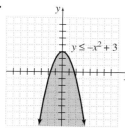
$y \le -x^2 + 3$

61.
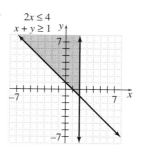
$x^2 - y^2 < 1$

63.
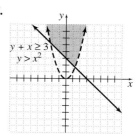
$2x \le 4$
$x + y \ge 1$

65.
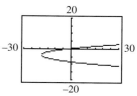
$y + x \ge 3$
$y > x^2$

67.
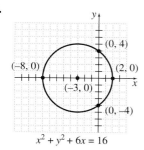
$x^2 - y^2 \le 1$
$x^2 + y^2 < 4$

Chapter 10 Test

1.
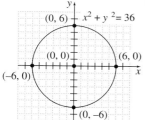
$x^2 + y^2 = 36$
$(0, 6)$
$(0, 0)$
$(6, 0)$
$(-6, 0)$
$(0, -6)$

2.
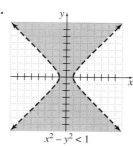
$x^2 - y^2 = 36$
$(-6, 0)$
$(6, 0)$

3.
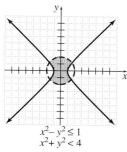
$16x^2 + 9y^2 = 144$
$(0, 4)$
$(-3, 0)$
$(3, 0)$
$(0, -4)$

4.
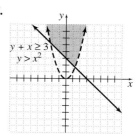
$(0, 16)$
$(4, 0)$
$y = x^2 - 8x + 16$

5.
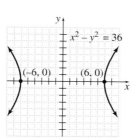
$(0, 4)$
$(-8, 0)$
$(2, 0)$
$(-3, 0)$
$(0, -4)$
$x^2 + y^2 + 6x = 16$

6.
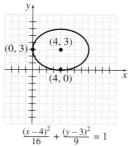
$(0, 3)$
$(4, 3)$
$(4, 0)$
$\dfrac{(x-4)^2}{16} + \dfrac{(y-3)^2}{9} = 1$

7.
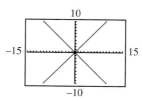

8.
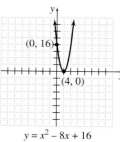

9. $\{(12, -5), (-12, 5)\}$ **10.** $\{(5, 1), (5, -1), (-5, 1), (-5, -1)\}$ **11.** $\{(1, 2), (6, 12)\}$ **12.** $\{(1, 1), (-1, -1)\}$

13.

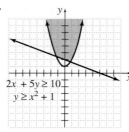

14.

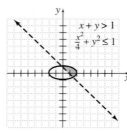

15.

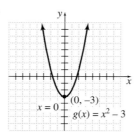

16.

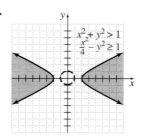

17. B   **18.** Width = 30 feet; Height = 10 feet

Chapter 10 Cumulative Review

1. *(Sec. 1.3, Ex. 7)* **a.** 9   **b.** $\frac{1}{16}$ **c.** -25 **d.** 25   **e.** -125 **f.** -125 **2.** *(Sec. 2.1, Ex. 6)* $\left\{\frac{21}{11}\right\}$ **3.** *(Sec. 2.4, Ex. 5)* \$11,607.55

4. *(Sec. 3.3, Ex. 11)* **a.** Approximately \$11.5 billion   **b.** \$13,600 million   **5.** *(Sec. 3.6, Ex. 1)* $y = \frac{1}{4}x - 3$ **6.** *(Sec. 4.3, Ex. 6)* { }

7. *(Sec. 4.4, Ex. 4)* $\{x|x \le -3 \text{ or } x \ge 3\}$ **8.** *(Sec. 5.1, Ex. 9)* $\left\{\left(-\frac{21}{10}, \frac{3}{10}\right)\right\}$ **9.** *(Sec. 5.3, Ex. 5)* $30°, 40°, 110°$

10. *(Sec. 6.1, Ex. 3)* **a.** 1   **b.** -1 **c.** 1   **d.** 2   **11.** *(Sec. 6.2, Ex. 3)* **a.** $\frac{y^6}{4}$ **b.** x^9 **c.** $\frac{49}{4}$ **d.** $\frac{y^{16}}{25x^5}$

12. *(Sec. 6.3, Ex. 7)* $12x^3 - 12x^2 - 9x + 2$ **13.** *(Sec. 6.6, Ex. 2)* $(x-5)(x-7)$ **14.** *(Sec. 7.2, Ex. 4)* $\frac{2(4x^2 - 10x + 25)}{x^2 + 1}$

15. *(Sec. 7.5, Ex. 1)* $10x^2 - 5x + 20$ **16.** *(Sec. 8.1, Ex. 2)* **a.** 1   **b.** -4 **c.** $\frac{2}{5}$ **d.** x^2 **e.** $-2x^3$

17. *(Sec. 8.5, Ex. 3)* **a.** $\frac{14}{3\sqrt{35}}$ **b.** $\frac{2x}{\sqrt[3]{20xy}}$ **18.** *(Sec. 9.1, Ex. 1)* $\{5\sqrt{2}, -5\sqrt{2}\}$ **19.** *(Sec. 9.4, Ex. 1)* $\{x|-3 < x < 3\}$

20. *(Sec. 9.5, Ex. 1)* **a.**

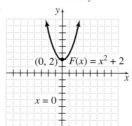

b.

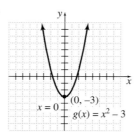

21. *(Sec. 10.3, Ex. 1)* $\{(2, \sqrt{2})\}$

CHAPTER 11

Exponential and Logarithmic Functions

Exercise Set 11.1

1. 42   **3.** -18 **5.** 0   **7.** $(f \circ g)(x) = 25x^2 + 1; (g \circ f)(x) = 5x^2 + 5; (f \circ f)(x) = x^4 + 2x^2 + 2$

9. $(f \circ g)(x) = 2x + 11; (g \circ f)(x) = 2x + 4; (f \circ f)(x) = 4x - 9$ **11.** $(f \circ g)(x) = -8x^3 - 2x - 2; (g \circ f)(x) = -2x^3 - 2x + 4$

13. $(f \circ g)(x) = \sqrt{-5x + 2}; (g \circ f)(x) = -5\sqrt{x} + 2$ **15.** $H(x) = (g \circ h)(x)$ **17.** $F(x) = (h \circ f)(x)$ **19.** $G(x) = (f \circ g)(x)$

21. One-to-one; $f^{-1} = \{(-1, -1), (1, 1), (2, 0), (0, 2)\}$ **23.** One-to-one; $h^{-1} = \{(10, 10)\}$

25. One-to-one; $f^{-1} = \{(12, 11), (3, 4), (4, 3), (6, 6)\}$ **27.** Not one-to-one   **29.** One-to-one;

Rank (input)	1	49	12	2	45
State (output)	CA	VT	VA	TX	SD

31. a. 3   **b.** 1   **33. a.** 1   **b.** -1 **35.** One-to-one

37. Not one-to-one   **39.** One-to-one   **41.** Not one-to-one

43. $f^{-1}(x) = x - 4$

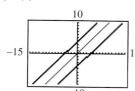

45. $f^{-1}(x) = \dfrac{x+3}{2}$

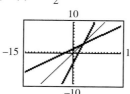

47. $f^{-1}(x) = \dfrac{3x+4}{12}$

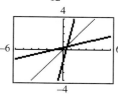

49. $f^{-1}(x) = \sqrt[3]{x}$

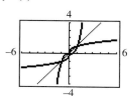

51. $f^{-1}(x) = 5x + 2$

53. $g^{-1}(x) = \sqrt{x-5}$

51.

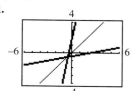

53.

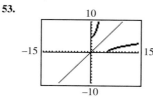

55. $(f \circ f^{-1})(x) = x; (f^{-1} \circ f)(x) = x$

57. $(f \circ f^{-1})(x) = x; (f^{-1} \circ f)(x) = x$

59. a. $\left(-2, \frac{1}{4}\right), \left(-1, \frac{1}{2}\right), (0,1), (1,2), (2,5)$

b. $\left(\frac{1}{4}, -2\right), \left(\frac{1}{2}, -1\right), (1,0), (2,1), (5,2)$

c-d.

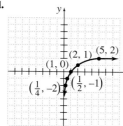

61. $f^{-1}(x) = \dfrac{x-1}{3}$

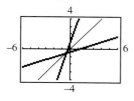

63. $f^{-1}(x) = x^3 - 3$

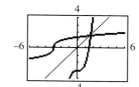

65. 5   **67.** 8

69. $\dfrac{1}{27}$ **71.** 9

73. $\sqrt{3} \approx 1.73$

Exercise Set 11.2

1.

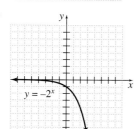

3.

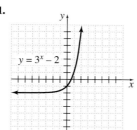

5.

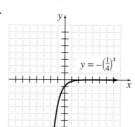

7.

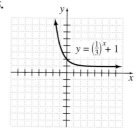

9.

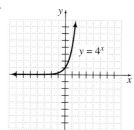

11.

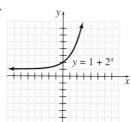

13.

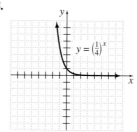

15.

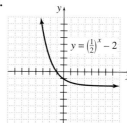

17. $b^1 = b$ **19.** $\{3\}$ **21.** $\left\{\frac{3}{4}\right\}$ **23.** $\left\{\frac{8}{5}\right\}$ **25.** $\left\{-\frac{2}{3}\right\}$ **27.** $\{4\}$ **29.** $\left\{\frac{3}{2}\right\}$ **31.** $\left\{-\frac{1}{3}\right\}$ **33.** $\{-2\}$ **35.** $y = 3^x$ **37.** $y = \left(\frac{1}{2}\right)^x$

39. 24.6 lb   **41.** 519 rats   **43.** 1.1 g   **45.** 16,190,000 residents   **47.** \$7621.42   **49.** \$4065.59   **51. a.** 20.2 lb   **b.** 18.6 lb

c. 16.5 lb   **53.** 1.827   **55.** $\{4\}$ **57.** $\{\ \}$ **59.** $\{2,3\}$ **61.** $\{3\}$ **63.** $\{-1\}$

Exercise Set 11.3

1. 3   **3.** −2   **5.** $\frac{1}{2}$ **7.** −1   **9.** 0   **11.** 4   **13.** 2   **15.** 5   **17.** 4   **19.** −3   **21.** Answers may vary.   **23.** {2}   **25.** {81}

27. {7}   **29.** {−3}   **31.** {−3}   **33.** {2}   **35.** {2}   **37.** $\left\{\frac{27}{64}\right\}$ **39.** {10}   **41.** 3   **43.** 3   **45.** 1

47.
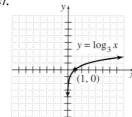

49.
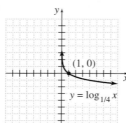

51.
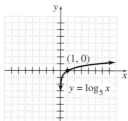

53.
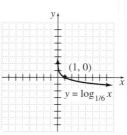

55.
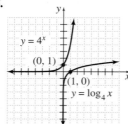

57.
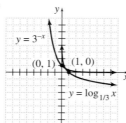

59. 0.0827   **61.** 2 and 3   **63.** −1   **65.** $x - 5$

67. 3   **69.** $\dfrac{y - 9}{y^2 - 1}$

Exercise Set 11.4

1. $\log_5 14$ **3.** $\log_4(9x)$ **5.** $\log_{10}(10x^2 + 20)$ **7.** $\log_5 3$ **9.** $\log_2\left(\frac{x}{y}\right)$ **11.** $\log_4 4 = 1$ **13.** $\log_2 25$ **15.** $\log_5(x^3 z^6)$

17. $\log_{10}\left(\dfrac{x^3 - 2x}{x + 1}\right)$ **19.** $\log_3 4 + \log_3 y - \log_3 5$ **21.** $3 \log_2 x - \log_2 y$ **23.** $\frac{1}{2}\log_b 7 + \frac{1}{2}\log_b x$ **25.** 0.2   **27.** 1.2   **29.** 0.23

31. $\log_4 35$ **33.** $\log_3 4$ **35.** $\log_7 \frac{9}{2}$ **37.** $\log_4 48$ **39.** $\log_2 \dfrac{x^{7/2}}{(x + 1)^2}$ **41.** $\log_8 x^{16/3}$ **43.** $\log_7 5 + \log_7 x - \log_7 4$

45. $3 \log_5 x + \log_5(x + 1)$ **47.** $2 \log_6 x - \log_6(x + 3)$ **49.** 1.29   **51.** −0.68   **53.** −0.125   **55.** True   **57.** False   **59.** False

61.

63. −1   **65.** $\frac{1}{2}$

Exercise Set 11.5

1. 0.9031   **3.** 0.3636   **5.** 0.6931   **7.** −2.6367   **9.** 1.1004   **11.** 1.6094   **13.** 1.6180   **15.** 0 is not in the domain of $\log x$.   **17.** 2

19. −3   **21.** 2   **23.** $\frac{1}{4}$ **25.** 3   **27.** 2   **29.** −4   **31.** $\frac{1}{2}$ **33.** ln 50 is larger.   **35.** $10^{1.3} \approx 19.9526$ **37.** $\frac{1}{2}10^{1.1} \approx 6.2946$

39. $e^{1.4} \approx 4.0552$ **41.** $\dfrac{4 + e^{2.3}}{3} \approx 4.6581$ **43.** $10^{2.3} \approx 199.5262$ **45.** $e^{-2.3} \approx 0.1003$ **47.** $\dfrac{-1 + 10^{-0.5}}{2} \approx -0.3419$

49. $\dfrac{e^{0.18}}{4} \approx 0.2993$ **51.** 1.5850   **53.** −2.3219   **55.** 1.5850   **57.** −1.6309   **59.** 0.8617   **61.** 4.2   **63.** 5.3   **65.** $3656.38

67. $2542.50

69.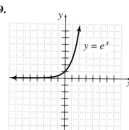

71.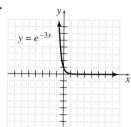

73.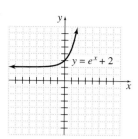

75.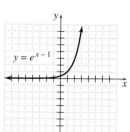

77.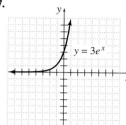

79.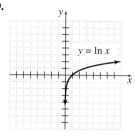

81.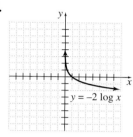

83.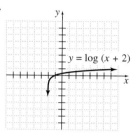

85.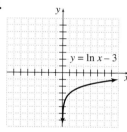

87.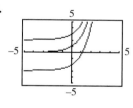

89. $\left\{\frac{4}{7}\right\}$ **91.** $x = \frac{3y}{4}$ **93.** $\{-6, -1\}$ **95.** $\{(2, -3)\}$

Exercise Set 11.6

1. $\frac{\ln 6}{\ln 3} \approx 1.6309$ **3.** $\frac{\ln 3.8}{2 \ln 3} \approx 0.6076$ **5.** $3 + \frac{\ln 5}{\ln 2} \approx 5.3219$ **7.** $\frac{\ln 5}{\ln 9} \approx 0.7325$ **9.** $-7 + \frac{\ln 3}{\ln 4} = -6.2075$ **11.** $\frac{4 + \dfrac{\ln 11}{\ln 7}}{3} \approx 1.7441$

13. $\frac{1}{6} \ln 5 \approx 0.2682$ **15.** $\{11\}$ **17.** $\{9, -9\}$ **19.** $\left\{\frac{1}{2}\right\}$ **21.** $\left\{\frac{3}{4}\right\}$ **23.** $\{2\}$ **25.** $\left\{\frac{1}{8}\right\}$ **27.** $\{11\}$ **29.** $\{4, -1\}$ **31.** $\left\{\frac{1}{5}\right\}$

33. $\{100\}$ **35.** $\left\{-\frac{5}{2} + \frac{\sqrt{33}}{2}\right\}$ **37.** $\left\{\frac{192}{127}\right\}$ **39.** $\left\{\frac{2}{3}\right\}$ **41.** 103 wolves   **43.** 10,250,000 inhabitants

45. a.

t	$600(1 + 0.12/12)^{12t}$
1	643.37
2	689.88
3	739.76
4	793.23
5	850.58
6	912.06
7	978.00

b. 10 years
47. 1.7 years
49. 8.8 years
51. 55.7 inches
53. 11.9 lb/in.²
55. 3.2 miles
57. 12 weeks
59. 18 weeks

61. $-\frac{5}{3}$ **63.** $\frac{17}{4}$

65. $f^{-1}(x) = \frac{x - 2}{5}$

Chapter 11 Review

1. $x^2 + 2x - 1$ **3.** 18 **5.** -2 **7.** One-to-one; $h^{-1} = \{(14, -9), (8, 6), (12, -11), (15, 15)\}$

9. One-to-one;

Rank (input)	2	4	1	3
Region (output)	West	Midwest	South	Northeast

11. a. 3 **b.** 7 **13.** Not one-to-one

15. Not one-to-one **17.** $f^{-1}(x) = \dfrac{x - 11}{6}$

19. $q^{-1}(x) = \dfrac{x - b}{m}$ **21.** $r^{-1}(x) = \dfrac{2}{13}(x + 4)$ **23.**

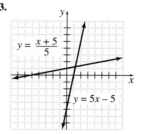

25. $\{3\}$ **27.** $\left\{-\dfrac{4}{3}\right\}$ **29.** $\left\{\dfrac{3}{2}\right\}$

31.

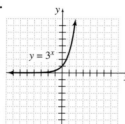

33.

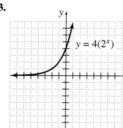

35. \$2963.11 **37.** The results are the same.

39. $\log_2\left(\dfrac{1}{16}\right) = -4$ **41.** $0.4^3 = 0.064$ **43.** $\{9\}$ **45.** $\{3\}$

47. $\{3\}$ **49.** $\{-2\}$ **51.** $\{9\}$ **53.** $\{2\}$ **55.** $\{-8, 1\}$ **57.**

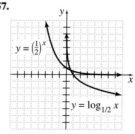

59. $\log_2 18$

61. $\log\left(\dfrac{3}{2}\right)$ **63.** $\log_5 2$ **65.** $\log_3(x^4 + 2x^3)$ **67.** $\log_4(x + 5) - 2\log_4 x$ **69.** $\log_7 y + 3\log_7 z - \log_7 x$ **71.** -0.11 **73.** -0.8239

75. 1.5326 **77.** -1 **79.** 4 **81.** $\left\{\dfrac{1}{3}e^{1.6}\right\}$ **83.** $\left\{\dfrac{-1 + e^2}{3}\right\}$ **85.** 1.22 mm **87.** 1.2619 **89.** \$1307.51 **91.** $\dfrac{1}{3}\dfrac{\ln 5}{\ln 6} \approx 0.2994$

93. $\dfrac{\dfrac{\ln 9}{\ln 4} - 2}{3} \approx -0.1383$ **95.** $\dfrac{\dfrac{\ln 3}{\ln 8} + 2}{4} \approx 0.6321$ **97.** $-5 + \dfrac{\ln\left(\frac{2}{3}\right)}{\ln 4} \approx -5.2925$ **99.** $\{0.9\}$ **101.** $\left\{\dfrac{-3e^2}{3 - e^2}\right\}$ **103.** $\{\ \}$

105. 241,308,968 **107.** 21 years **109.** 157 years **111.** 8.5 years **113.** $\{2.82\}$

Chapter 11 Test

1. 5   **2.** $x - 7$ **3.** $x^2 - 6x - 2$ **4.**

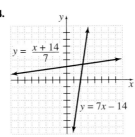

$y = \dfrac{x + 14}{7}$

$y = 7x - 14$

5. One-to-one   **6.** Not a function   **7.** One-to-one;
$f^{-1}(x) = -\dfrac{x - 6}{2}$ **8.** One-to-one; $f^{-1} = \{(0, 0), (3, 2), (5, -1)\}$
9. Not one-to-one   **10.** $\log_3 24$ **11.** $\log_5\left(\dfrac{x^4}{x + 1}\right)$
12. $\log_6 2 + \log_6 x - 3 \log_6 y$ **13.** -1.53 **14.** 1.0686 **15.** $\{-1\}$
16. $\dfrac{-5 + \dfrac{\ln 4}{\ln 3}}{2} \approx -1.8691$ **17.** $\left\{\dfrac{1}{9}\right\}$ **18.** $\left\{\dfrac{1}{2}\right\}$ **19.** $\{22\}$

20. $\left\{\dfrac{25}{3}\right\}$ **21.** $\left\{\dfrac{43}{21}\right\}$ **22.** -1.0979 **23.**

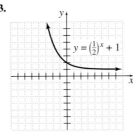

$y = \left(\dfrac{1}{2}\right)^x + 1$

24.

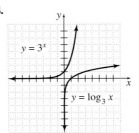

$y = 3^x$

$y = \log_3 x$

25. \$5234.58   **26.** 6 years

27.

t	y
1	58,501
2	60,042
3	61,624
4	63,247
5	64,913

28. 15 years   **29.** 1.2%   **30.** $\{3.95\}$

Chapter 11 Cumulative Review

1. *(Sec. 1.4, Ex. 8)* **a.** $20 - x$ **b.** $8 + x$ **c.** $90 - x$ **d.** $x + 1$ **e.** $x - 2$ **2.** *(Sec. 3.2, Ex. 2)* **a.** \$1580   **b.** More than \$1000
3. *(Sec. 3.4, Ex. 5)*   **4.** *(Sec. 4.1, Ex. 5)* $\left(-\infty, -\dfrac{7}{3}\right]$

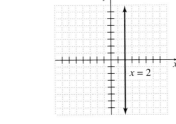

$x = 2$

5. *(Sec. 5.1, Ex. 8)* $\{(x, y) \mid -5x - 3y = 9\}$ or equivalently $\{(x, y) \mid 10x + 6y = -18\}$ **6.** *(Sec. 5.2, Ex. 3)* $\left\{\left(\dfrac{1}{2}, 0, \dfrac{3}{4}\right)\right\}$
7. *(Sec. 6.1, Ex. 4)* **a.** x^3 **b.** 5^6 **c.** $5x$ **d.** $\dfrac{6y^2}{7}$ **8.** *(Sec. 6.3, Ex. 8)* $12x^3 - 12x^2 - 9x + 2$
9. *(Sec. 6.4, Ex. 1)* **a.** $10x^9$ **b.** $-7xy^{15}z^9$ **10.** *(Sec. 6.4, Ex. 4)* $4x^4 + 8x^3 + 39x^2 + 14x + 56$ **11.** *(Sec. 6.7, Ex. 2)* $3x(a - 2b)^2$
12. *(Sec. 6.8, Ex. 2)* $\left\{-5, \dfrac{1}{2}\right\}$ **13.** *(Sec. 7.7, Ex. 2)* $\{-2\}$

14. *(Sec. 8.2, Ex. 2)* **a.** $(\sqrt{4})^3 = 8$ **b.** $-(\sqrt[4]{16})^3 = -8$ **c.** $(\sqrt[3]{-27})^2 = 9$ **d.** $\left(\sqrt{\dfrac{1}{9}}\right)^3 = \dfrac{1}{27}$ **e.** $\sqrt[5]{(4x - 1)^3}$
15. *(Sec. 8.7, Ex. 3)* **a.** $13 - 8i$ **b.** $3 - 4i$ **c.** 58 **16.** *(Sec. 9.4, Ex. 4)* $(-\infty, -2) \cup (1, 5)$

17. *(Sec. 9.5, Ex. 4)*

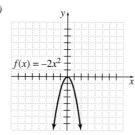

18. *(Sec. 10.1, Ex. 6)*

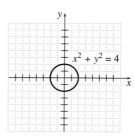

19. *(Sec. 11.1, Ex. 1)* **a.** $25; 7$ **b.** $x^2 + 6x + 9; x^2 + 3$ **20.** *(Sec. 11.2, Ex. 3)* **a.** $\{4\}$ **b.** $\left\{\frac{3}{2}\right\}$ **c.** $\{6\}$ **d.** $\{1.431\}$

21. *(Sec. 11.4, Ex. 5)* **a.** 1.11 **b.** 1.36 **c.** 0.215 **22.** *(Sec. 11.5, Ex. 7)* $\left\{\frac{e^5}{3}\right\}; \{\approx 49.4711\}$ **23.** *(Sec. 11.6, Ex. 3)* $\{2\}$

CHAPTER 12

Sequences, Series, and the Binomial Theorem

Exercise Set 12.1

1. $5, 6, 7, 8, 9$ **3.** $-1, 1, -1, 1, -1$ **5.** $\frac{1}{4}, \frac{1}{5}, \frac{1}{6}, \frac{1}{7}, \frac{1}{8}$ **7.** $2, 4, 6, 8, 10$ **9.** $-1, -4, -9, -16, -25$ **11.** $2, 4, 8, 16, 32$

13. $7, 9, 11, 13, 15$ **15.** $-1, 4, -9, 16, -25$ **17.** 75 **19.** 118 **21.** $\frac{6}{5}$ **23.** 729 **25.** $\frac{4}{7}$ **27.** $\frac{1}{8}$ **29.** -95 **31.** $-\frac{1}{25}$

33. $a_n = 4n - 1$ **35.** $a_n = -2^n$ **37.** $a_n = \frac{1}{3^n}$ **39.** 48 ft, 80 ft, and 112 ft   **41.** $a_n = 0.10(2^{n-1}); \$819.20$

43. 75 cases, 150 cases, 300 cases, 600 cases, 1200 cases, 2400 cases, 4800 cases   **45.** 50 sparrows in 2000; extinct in 2006

47. $1, 0.7071, 0.5774, 0.5, 0.4472$ **49.** $2, 2.25, 2.3704, 2.4414, 2.4883$

51.

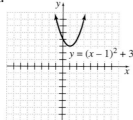

53.

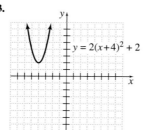

55. $\sqrt{13}$ **57.** $\sqrt{41}$

Exercise Set 12.2

1. $4, 6, 8, 10, 12$ **3.** $6, 4, 2, 0, -2$ **5.** $1, 3, 9, 27, 81$ **7.** $48, 24, 12, 6, 3$ **9.** 33 **11.** -875 **13.** -60 **15.** 96 **17.** -28

19. 1250 **21.** 31 **23.** 20 **25.** $a_1 = \frac{2}{3}; r = -2$ **27.** Answers may vary.   **29.** $a_1 = 2; d = 2$ **31.** $a_1 = 5; r = 2$ **33.** $a_1 = \frac{1}{2}; r = \frac{1}{5}$

35. $a_1 = x; r = 5$ **37.** $a_1 = p; d = 4$ **39.** 19 **41.** $-\frac{8}{9}$ **43.** $\frac{17}{2}$ **45.** $\frac{8}{81}$ **47.** -19 **49.** $a_n = 4n + 50; 130$ seats

51. $a_n = 6(3)^{n-1}; 6, 18, 54, 162, 486$ **53.** $162, 54, 18, 6, 2; a_n = 162\left(\frac{1}{3}\right)^{n-1}$; the 6th bounce   **55.** $a_n = 125n + 1875; \$3375$ **57.** 25 grams

59. $\$11{,}782.40; \$5891.20, \$2945.60, \$1472.80$ **61.** $19.652, 19.618, 19.584, 19.550$ **63.** Answers may vary.   **65.** $\frac{11}{18}$ **67.** 40 **69.** $\frac{907}{495}$

Exercise Set 12.3

1. -2 **3.** 60 **5.** 20 **7.** $\frac{73}{168}$ **9.** $\frac{11}{36}$ **11.** 60 **13.** 74 **15.** 62 **17.** $\frac{241}{35}$ **19.** $\sum_{i=1}^{5}(2i-1)$ **21.** $\sum_{i=1}^{4}4(3)^{i-1}$
23. $\sum_{i=1}^{6}(-3i+15)$ **25.** $\sum_{i=1}^{4}\frac{4}{3^{i-2}}$ **27.** $\sum_{i=1}^{7}i^2$ **29.** -24 **31.** -13 **33.** 82 **35.** -20 **37.** -2 **39.** $1, 2, 3, \ldots, 10$; 55 trees
41. $a_n=6(2)^{n-1}$; 96 units   **43.** 78.93 ft   **45.** 30 opossums; 68 opossums   **47.** 6.25 lb; 93.75 lb   **49.** 16.4 in.; 134.5 in.
51. a. $2+6+12+20+30+42+56$ **b.** $1+2+3+4+5+6+7+1+4+9+16+25+36+49$ **c.** They are equal; 168   **d.** True
53. 10 **55.** $\frac{10}{27}$ **57.** 45 **59.** 90

Exercise Set 12.4

1. 36 **3.** 484 **5.** 63 **7.** 2.496 **9.** 55 **11.** 16 **13.** 24 **15.** $\frac{1}{9}$ **17.** -20 **19.** $\frac{16}{9}$ **21.** $\frac{4}{9}$ **23.** 185 **25.** $\frac{381}{64}$
27. $-\frac{33}{4}$ **29.** $-\frac{75}{2}$ **31.** $\frac{56}{9}$ **33.** 4000, 3950, 3900, 3850, 3800; 3450 cars; 44,700 cars
35. Firm A (Firm A, \$265,000; Firm B, \$254,000)   **37.** \$39,930; \$139,230   **39.** 20 min.; 123 min.   **41.** 180 feet
43. Player A, 45 points; Player B, 75 points   **45.** \$3050   **47.** \$10,737,418.23   **49.** $0.8+0.08+0.008+\cdots;\frac{8}{9}$ **51.** Answers may vary.
53. 720 **55.** 3 **57.** $x^2+10x+25$ **59.** $8x^3-12x^2+6x-1$

Exercise Set 12.5

1. $m^3+3m^2n+3mn^2+n^3$ **3.** $c^5+5c^4d+10c^3d^2+10c^2d^3+5cd^4+d^5$ **5.** $y^5-5y^4x+10y^3x^2-10y^2x^3+5yx^4-x^5$
7. Answers may vary.   **9.** 8 **11.** 42 **13.** 360 **15.** 56 **17.** $a^7+7a^6b+21a^5b^2+35a^4b^3+35a^3b^4+21a^2b^5+7ab^6+b^7$
19. $a^5+10a^4b+40a^3b^2+80a^2b^3+80ab^4+32b^5$ **21.** $q^9+9q^8r+36q^7r^2+84q^6r^3+126q^5r^4+126q^4r^5+84q^3r^6+36q^2r^7+9qr^8+r^9$
23. $1024a^5+1280a^4b+640a^3b^2+160a^2b^3+20ab^4+b^5$ **25.** $625a^4-1000a^3b+600a^2b^2-160ab^3+16b^4$
27. $8a^3+36a^2b+54ab^2+27b^3$ **29.** $x^5+10x^4+40x^3+80x^2+80x+32$ **31.** $5cd^4$ **33.** d^7 **35.** $-40r^2s^3$ **37.** $6x^2y^2$
39. $30a^9b$ **41.**

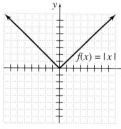
Not one-to-one

43.

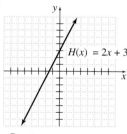
One-to-one

45.

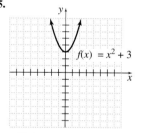
Not one-to-one

Chapter 12 Review

1. $-3, -12, -27, -48, -75$ **3.** $\frac{1}{100}$ **5.** $a_n=\frac{1}{6n}$ **7.** 144 ft, 176 ft, 208 ft   **9.** 450, 1350, 4050, 12,150; 36,450
11. $-2, -\frac{4}{3}, -\frac{8}{9}, -\frac{16}{27}, -\frac{32}{81}$ **13.** 111 **15.** -83 **17.** $a_1=3; d=5$ **19.** $a_n=\frac{3}{10^n}$ **21.** Geometric; $a_1=\frac{8}{3}; r=\frac{3}{2}$
23. Geometric; $a_1=7x; r=-2$ **25.** 8, 6, 4.5, 3.4, 2.5, 1.9; good   **27.** $a_n=2^{n-1}$; \$512; \$536,870,912
29. $a_n=900+(n-1)150$; \$1650/month   **31.** $1+3+5+7+9=25$ **33.** $\frac{1}{4}+\left(-\frac{1}{6}\right)+\frac{1}{8}=\frac{5}{24}$ **35.** -4 **37.** -10 **39.** $\sum_{i=1}^{6}3^{i-1}$
41. $\sum_{i=1}^{4}\frac{1}{4^i}$ **43.** $a_n=20(2)^n$; n represents the number of 8-hr periods; 1280 yeast cells   **45.** Job A, \$23,900; Job B, \$23,800   **47.** 150
49. 900 **51.** -410 **53.** 936 **55.** 10 **57.** -25 **59.** \$30,417.50; \$99,867.50   **61.** \$58; \$553   **63.** 2696 mosquitoes   **65.** $\frac{5}{9}$
67. $x^5+5x^4z+10x^3z^2+10x^2z^3+5xz^4+z^5$ **69.** $16x^4+32x^3y+24x^2y^2+8xy^3+y^4$
71. $b^8+8b^7c+28b^6c^2+56b^5c^3+70b^4c^4+56b^3c^5+28b^2c^6+8bc^7+c^8$ **73.** $256m^4-256m^3n+96m^2n^2-16mn^3+n^4$ **75.** $35a^4b^3$

Chapter 12 Test

1. $-\frac{1}{5}, \frac{1}{6}, -\frac{1}{7}, \frac{1}{8}, -\frac{1}{9}$ **2.** $-3, 3, -3, 3, -3$ **3.** 247 **4.** 39,999 **5.** $a_n = \frac{2}{5^n}$ **6.** $a_n = (-1)^n 9n$ **7.** 155 **8.** -330 **9.** $\frac{144}{5}$

10. 1 **11.** 10 **12.** -60 **13.** $a^6 - 6a^5 b + 15a^4 b^2 - 20a^3 b^3 + 15a^2 b^4 - 6ab^5 + b^6$ **14.** $32x^5 + 80x^4 y + 80x^3 y^2 + 40x^2 y^3 + 10xy^4 + y^5$

15. $y^8 + 8y^7 z + 28y^6 z^2 + 56y^5 z^3 + 70y^4 z^4 + 56y^3 z^5 + 28y^2 z^6 + 8yz^7 + z^8$

16. $128p^7 + 448p^6 r + 672p^5 r^2 + 560p^4 r^3 + 280p^3 r^4 + 84p^2 r^5 + 14pr^6 + r^7$ **17.** 925 people; 250 people

18. $1 + 3 + 5 + 7 + 9 + 11 + 13 + 15$; 64 shrubs **19.** 33.75 cm; 218.75 cm **20.** 320 cm **21.** 304 feet; 1600 ft **22.** $\frac{14}{33}$

Chapter 12 Cumulative Review

1. *(Sec. 2.1, Ex. 2)* $\left\{\frac{38}{7}\right\}$ **2.** *(Sec. 3.3, Ex. 2)* **a.** Yes **b.** No **c.** Yes **3.** *(Sec. 3.6, Ex. 8)* $f(x) = \frac{3}{2}x - 2$

4. *(Sec. 4.4, Ex. 1)* $(-7, 7)$ **5.** *(Sec. 5.1, Ex. 10)* 500 **6.** *(Sec. 5.3, Ex. 1)* 7, 11

7. *(Sec. 6.1, Ex. 5)* **a.** $\frac{1}{25}$ **b.** $\frac{2}{x^3}$ **c.** $\frac{1}{3x}$ **d.** $\frac{1}{m^{10}}$ **e.** $\frac{1}{27}$ **f.** $\frac{11}{18}$ **g.** t^5 **8.** *(Sec. 6.3, Ex. 9)* $12z^5 + 3z^4 - 13z^3 - 11z$

9. *(Sec. 6.7, Ex. 7)* $(y - 4)(y^2 + 4y + 16)$ **10.** *(Sec. 7.3, Ex. 5)* $\dfrac{7x^2 - 9x - 13}{(2x + 1)(x - 5)(3x - 2)}$

11. *(Sec. 7.5, Ex. 6)* $3x^2 + 2x + 3 + \dfrac{-6x + 9}{x^2 - 1}$ **12.** *(Sec. 8.3, Ex. 2)* **a.** $\frac{5}{7}$ **b.** $\frac{2}{3}$ **c.** $\frac{\sqrt{x}}{3}$ **d.** $\dfrac{\sqrt[4]{3}}{2y}$ **13.** *(Sec. 8.6, Ex. 2)* $\left\{-1, -\frac{1}{9}\right\}$

14. *(Sec. 9.2, Ex. 1)* $\left\{-\frac{1}{3}, -5\right\}$ **15.** *(Sec. 9.6, Ex. 2)*

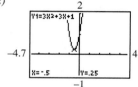

Vertex: $(-0.5, 0.25)$;
No x-intercepts;
y-intercept: $(0, 1)$

16. *(Sec. 10.2, Ex. 1)*

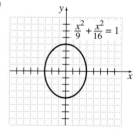

17. *(Sec. 10.3, Ex. 3)* $\{(2, \sqrt{3}), (-2, \sqrt{3}), (2, -\sqrt{3}), (-2, -\sqrt{3})\}$

18. *(Sec. 11.1, Ex. 3)* **a.** Yes **b.** No **c.** Yes **d.** No **e.** No **19.** *(Sec. 11.2, Ex. 5)* $2505.09

20. *(Sec. 11.3, Ex. 2)* **a.** $\{-1\}$ **b.** $\{125\}$ **c.** $\{5\}$ **d.** $\{0\}$ **e.** $\{0\}$ **21.** *(Sec. 11.6, Ex. 4)* $\left\{\frac{2}{99}\right\}$

22. *(Sec. 12.1, Ex. 2)* **a.** $-\frac{1}{3}$ **b.** $\frac{1}{24}$ **c.** $\frac{1}{300}$ **d.** $-\frac{1}{45}$ **23.** *(Sec. 12.2, Ex. 5)* $19{,}200 + 800n$; $22,400

24. *(Sec. 12.5, Ex. 4)* $x^5 + 10x^4 + 40x^3 + 80x^2 + 80x + 32$

APPENDIX A

Review of Angles, Lines, and Special Triangles

1. 71°   **3.** 19.2°   **5.** $78\frac{3}{4}°$ **7.** 30°   **9.** 149.8°   **11.** $100\frac{1}{2}°$ **13.** $m\angle 1 = 110°, m\angle 2 = 70°, m\angle 3 = 70°, m\angle 7 = 110°, m\angle 6 = 70°,$

$m\angle 5 = 110°, m\angle 4 = 70°$ **15.** 90°   **17.** 90°   **19.** 90°   **21.** 90° and 45°   **23.** 90° and 73°   **25.** 90° and $50\frac{1}{4}°$ **27.** $x = 6$

29. $x = 4.5$ **31.** $c = 10$ **33.** $b = 12$

APPENDIX C

Using a Graphing Utility

Practice Problem 1

a. 75.1   **b.** $-\dfrac{23}{36}$ **c.** $\dfrac{11}{6}$ **d.** $\dfrac{9}{4}$

Practice Problem 2

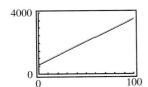

Practice Problem 3

Median, 18; Mean, 38.5

Practice Problem 4

a. Many windows are possible. One is $X_{min} = 0, X_{max} = 30, X_{scl} = 5, Y_{min} = -20, Y_{max} = 25, Y_{scl} = 5.$

b. Many windows are possible. One is $X_{min} = -1.5, X_{max} = 1, X_{scl} = 0.5, Y_{min} = -1, Y_{max} = 1.5, Y_{scl} = 0.5$

Practice Problem 5

a.

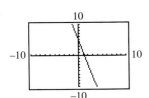

b.

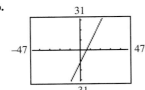

Practice Problem 6

Practice Problem 7

a. $x = 4.5$ **b.** $x = -1.4$

Practice Problem 8

$(0.15, -1.09)$

Practice Problem 9

Practice Problem 10

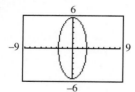   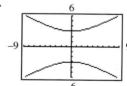

Practice Problem 11

a.

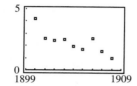

b.

Photo Credits

CHAPTER 1	**CO** Larry Mulvehill / Rainbow
CHAPTER 2	**CO** Bachmann / The Image Works **p. 64** Howard Bluestein / Photo Researchers, Inc. **p. 85** Photo Researchers
CHAPTER 3	**CO** Gamma-Liaison, Inc.
CHAPTER 4	**CO** J. Sohm / The Image Works
CHAPTER 5	**CO** Tom Ives / The Stock Market **p. 295** Tom Ives / The Stock Market
CHAPTER 6	**CO** Myrleen Ferguson / PhotoEdit **p. 314** Chris Butler / Science Photo Library / Photo Researchers, Inc.
CHAPTER 7	**CO** Jim Corwin / Photo Researchers, Inc.
CHAPTER 8	**CO** Richard Megna / Fundamental Photographs **p. 499** Steve Gottlieb / FPG International **p. 507** Richard Megna / Fundamental Photographs
CHAPTER 9	**CO** Ken Fisher / Tony Stone Images
CHAPTER 10	**CO** Bob Daemmrich Photos / The Image Works
CHAPTER 11	**CO** Nino Mascardi / The Image Bank
CHAPTER 12	**CO** Chuck Savage / The Stock Market

INDEX

GEOMETRIC FORMULAS

RECTANGLE

Perimeter: $P = 2l + 2w$
Area: $A = lw$

SQUARE

Perimeter: $P = 4s$
Area: $A = s^2$

TRIANGLE

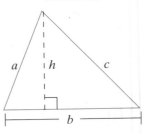

Perimeter: $P = a + b + c$

Area: $A = \dfrac{1}{2}bh$

SUM OF ANGLES OF TRIANGLE

$A + B + C = 180°$
The sum of the measures of the three angles is 180°.

RIGHT TRIANGLE

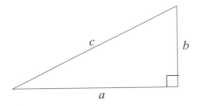

Perimeter: $P = a + b + c$

Area: $A = \dfrac{1}{2}ab$

One 90° (right) angle

PYTHAGOREAN THEOREM (FOR RIGHT TRIANGLES)

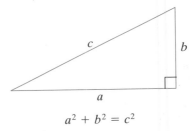

$a^2 + b^2 = c^2$

ISOSCELES TRIANGLE

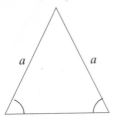

Triangle has:
two equal sides and
two equal angles.

EQUILATERAL TRIANGLE

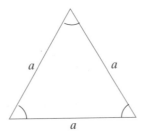

Triangle has:
three equal sides and
three equal angles.
Measure of each angle is 60°.

TRAPEZOID

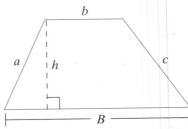

Perimeter: $P = a + b + c + B$

Area: $A = \dfrac{1}{2}h(B + b)$